T0142888

Lecture Notes in Computer Science 1296

Edited by G. Goos, J. Hartmanis and J. van Leeuwen

Advisory Board: W. Brauer   D. Gries   J. Stoer

Springer
Berlin
Heidelberg
New York
Barcelona
Budapest
Hong Kong
London
Milan
Paris
Santa Clara
Singapore
Tokyo

Gerald Sommer   Kostas Daniilidis
Josef Pauli (Eds.)

Computer Analysis of Images and Patterns

7th International Conference, CAIP '97
Kiel, Germany, September 10-12, 1997
Proceedings

 Springer

Series Editors

Gerhard Goos, Karlsruhe University, Germany

Juris Hartmanis, Cornell University, NY, USA

Jan van Leeuwen, Utrecht University, The Netherlands

Volume Editors

Gerald Sommer
Kostas Daniilidis
Josef Pauli
Christian-Albrechts-Universität zu Kiel, Institut für Informatik
Preußerstr. 1-9, D-24105 Kiel, Germany
E-mail: (gs/kd/jp)@informatik.uni-kiel.de

Cataloging-in-Publication data applied for

Die Deutsche Bibliothek - CIP-Einheitsaufnahme

Computer analysis of images and patterns : 7th international
conference ; proceedings / CAIP '97, Kiel, Germany, September 10 -
12, 1997. Gerald Sommer ... (ed.). - Berlin ; Heidelberg ; New York ;
Barcelona ; Budapest ; Hong Kong ; London ; Milan ; Paris ; Santa
Clara ; Singapore ; Tokyo : Springer, 1997
   (Lecture notes in computer science ; Vol. 1296)
   ISBN 3-540-63460-6

CR Subject Classification (1991): I.4, I.5, I.3.3, I.3.7, J.2

ISSN 0302-9743
ISBN 3-540-63460-6 Springer-Verlag Berlin Heidelberg New York

© Springer-Verlag Berlin Heidelberg 1997
Printed in Germany

Typesetting: Camera-ready by author
SPIN 10546359   06/3142 – 5 4 3 2 1 0   Printed on acid-free paper

Preface

This volume presents the papers accepted for the 7th International Conference on Computer Analysis of Images and Patterns, CAIP'97 held in Kiel, Germany, September 10-12, 1997. The CAIP conference series was founded twelve years ago in Berlin. The goal of the conference was the establishment of a forum for the exchange of ideas between Western and Eastern-bloc scientists in the areas of image processing and pattern recognition. Fortunately, the political circumstances hindering such contacts were eliminated. However, economic circumstances do still not allow scientists from Central and Eastern Europe to cover the participation expenses of conferences around the world. Although the scientific quality of CAIP is comparable to other well known conferences, CAIP differs from them in the insistence on supporting and keeping high the participation of C & E European scientists.

We received 150 submissions which were reviewed double blind in responsibility of three members of the program committee. Final decisions were based on the reviewers' ratings and were taken in electronic communication with the members of the steering committee. The final program consisted of 61 oral and 31 poster presentations.

The geographic spectrum of the submissions spanned over 30 countries with 67 submissions from Central and Eastern European countries. The proportion of the latter among the accepted papers is 25 out of 92. We succeeded in entirely covering the expenses of 16 scientists from C & E Europe thanks to the Deutsche Forschungsgemeinschaft and partially supporting another 13 from the same area. The high quality of submissions and the active participation of C & E Europe researchers proves the need to continue the CAIP conference series in the same spirit as until now.

Our thanks go to all the authors for the high quality of their work and for their cooperation in meeting the time and layout schedules. In the name of the steering committee we would like to thank the program committee members and the additional reviewers for their time and their conscientious work. We thank the International Association for Pattern Recognition for taking CAIP'97 under its auspices, the Christian-Albrechts-Universität Kiel for hosting the conference, and the industrial and institutional sponsors for financial support. Springer-Verlag Heidelberg and especially Alfred Hofmann are gratefully acknowledged for publishing the CAIP proceedings in the LNCS series. Last but not least, the conference could not take place without the extraordinary commitment of the local organizing committee.

Kiel, June 1997   Gerald Sommer

Conference Chair: Gerald Sommer, Germany

Program Committee

R. Bajcsy, USA	R. Klette, New Zealand
I. Bajla, Slovakia	J.J. Koenderink, Netherlands
V. Chernov, Russia	W. Kropatsch, Austria
D. Chetverikov, Hungary	I. Pitas, Greece
V. Di Gesù, Italy	R. Mohr, France
J.O. Eklundh, Sweden	G. Sagerer, Germany
A. Gagalowicz, France	W. Skarbek, Poland
G.L. Gimel'farb, Ukraine	F. Solina, Slovenia
V. Hlaváč, Czech Republic	S. Stiehl, Germany
J. Kittler, United Kingdom	S.W. Zucker, USA

Steering Committee

V. Hlaváč, Czech Republic (Chair)
D. Chetverikov, Hungary
R. Klette, New Zealand
F. Solina, Slovenia
G. Sommer, Germany

Organizing Committee

G. Sommer (Chair)
J. Bruske (Accom., Publicity)
K. Daniilidis (Program)
M. Hansen (Local Arrangements)
U. Mahlmeister (Techn. Support)
F. Maillard (Secretary)
J. Pauli (Finance)

Additional Reviewers

F. Ackermann	P. Gros	V. Myasnikov	L. Spreeuwers
M. Berger	M. Haindl	N. Nikopoulos	M. Sramek
H. Bischof	M. Hanajik	T. Pajdla	P. Sturm
J. Blanc	R. Horaud	M. Pappas	T. Svoboda
A. Bors	A. Jaklic	F. Pernus	I. Tastl
U. Cahn v. Seelen	I. Kakadiaris	M. Petrou	W. Triggs
S. Carlson	G. Kamberova	I. Pitas	S. Tsekeridou
G. Celeux	E. Kolomiyets	L. Quan	M. Urban
V. Chatzis	C. Kotropoulos	P. Raudin	C. Vavoulidis
H. Christensen	S. Kovacic	K. Rohr	P. Venetianer
S. Christy	B. Lamiroy	R. Sablatnig	J. Verestoy
L. Chudy	Z. Lan	R. Sara	T. Werner
C. Cotsaces	A. Leonardis	C. Schnörr	V. Witkovsky
G. Csurka	J. Müller	V. Sergeyev	V. Zyka
G. Fink	J. Maver	W. Skarbek	
J. Fojtik	C. Menard	V. Smatny	

Sponsors: Sun Microsystems, COSYCO GmbH/Datacube, Siemens AG, Isatec GmbH, Sparkasse Kiel, TELEROB GmbH, Deutsche Forschungsgemeinschaft, Förderverein der TF Kiel, Wissenschaftsministerium SH, Institut für Informatik und Praktische Mathematik der Christian-Albrechts-Universität Kiel.

Table of Contents

Pattern Analysis

Object Recognition and Tracking

Invariants

Applications

Shape

Texture Analysis

Motion and Calibration

Low Level Processing I

Structure from Motion

Low Level Processing II

Stereo and Correspondence

Segmentation and Grouping

Object Recognition

Mathematical Morphology

Pose Estimation

Face Analysis

Poster Session I

Poster Session II

Computational Complexity Reduction in Eigenspace Approaches*

Aleš Leonardis[1,2] and Horst Bischof[1]

[1] Vienna University of Technology, Institute for Automation
Pattern Recognition and Image Processing Group
Treitlstraße 3/1832, A-1040 Vienna, Austria
{ales,bis}@prip.tuwien.ac.at
[2] University of Ljubljana, Faculty of Computer and Information Science
Tržaška 25, SI-1001 Ljubljana, Slovenia
ales.leonardis@fri.uni-lj.si

Abstract. Matching of appearance-based object representations using eigenimages is computationally very demanding. Most commonly, to recognize an object in an image, parts of the input image are projected onto the eigenspace and the recovered coefficients indicate the presence or absence of a particular object. In general, the process is sequentially applied to the entire image. In this paper we discuss how to alleviate the problems related to complexity. First, we propose to use a *focus-of-attention (FOA) detector* which is intended to select candidate areas of interest with minimal computational effort. Only at these areas we then recover the coefficients of eigenimages. Secondly, we propose to employ a *multiresolution* approach. However, this requires that we depart from the standard way of calculating the coefficients of the eigenimages which relies on the orthogonality property of eigenimages. Instead we calculate them by solving a system of linear equations. We show the results of our approach on real data.

1 Introduction and Motivation

The appearance-based approaches to vision problems have recently received a renewed attention in the vision community due to their ability to deal with combined effects of shape, reflectance properties, pose in the scene, and the illumination conditions [7]. Besides, the appearance-based representations can be acquired through an automatic learning phase which is not the case with traditional shape representations. The approach has led to a variety of successful applications, e.g., illumination planning [8], visual positioning and tracking of robot manipulators [9], visual inspection [11], "image spotting" [6], and human face recognition [10, 2]. As stressed by its proponents, the major advantage of

* This work was supported by a grant from the Austrian National Fonds zur Förderung der wissenschaftlichen Forschung (No. S7002MAT). A. L. also acknowledges the support from the Ministry of Science and Technology of Republic of Slovenia (Projects J2-6187, J2-8829).

the approach is that both learning as well as recognition are performed using just two-dimensional brightness images without any low- or mid-level processing.

The appearance-based methods consist of two stages. In the first stage a set of images (templates, i.e., *training samples*) is obtained. These images usually encompass the appearance of a single object under different orientations [11], different illumination directions [8], or multiple instances of a class of objects, e.g., faces [10]. The sets of images are normally highly correlated. Thus, they can be efficiently compressed using Karhunen-Loève transform (i.e., PCA) [1], resulting in a low-dimensional eigenspace.

In the second stage, given an input image, the recognition system projects parts of the input image, (i.e., subimages of the same size as training images), to the eigenspace. In the absence of specific cues, e.g., when motion can be used to pre-segment the image, the process is sequentially applied to the entire image. The recovered coefficients indicate the particular instance of an object and/or its position, illumination, etc.

Several drawbacks have been identified with this classical approach which rests on direct appearance-based matching [7]: complexity is very high (subimages are projected to the eigenimages at each image point) and besides, the approach is non-robust with respect to outliers, occlusion, and segmentation.

Recently, we proposed an approach to solve the problems related to non-robustness [5]. The main idea of the approach lies in the way how the coefficients of the eigenimages are determined. Instead of computing the coefficients by a projection of the data onto the eigenimages, we extract them by a hypothesize-and-test paradigm using subsets of image points. Competing hypotheses are then subject to a selection procedure based on the Minimum Description Length (MDL) principle. The approach enables us not only to reject outliers and to cope with occlusions but also to simultaneously deal with multiple classes of eigenimages. However the problems with the computational complexity still remain.

In this paper we discuss how to alleviate the problems related to complexity. First, we propose to use a *focus-of-attention (FOA) detector* which is specifically tuned to the members of a class of templates and intended to select candidate areas of interest with minimal computational effort. Only at these areas we then recover the coefficients of eigenimages. Secondly, we propose to employ a *multiresolution* approach. However, this requires that we depart from the standard way of calculating the coefficients of the eigenimages which relies on the orthogonality property of eigenimages and requires to use all points in the eigenimage. Instead we calculate them by solving a system of linear equations embedded in the hypothesize-and-test paradigm.

The paper is organized as follows: In the next section we introduce the notation and give a brief overview of the previous work. In section 3 we describe the *focus-of-attention (FOA) detector* which is tuned to a class of templates to output in a single pass the areas-of-interest. The extension of the robust recovery of eigenimage coefficients to multiple resolutions is explained in Section 4. Experimental results are presented in Section 5. We conclude with a summary.

2 Previous Work

Here we briefly review the standard method and our new method to the recovery of coefficients of eigenimages which will serve as the basis to explain our approach. We follow the derivation from [5].

Let $\mathbf{y} = [y_1, \ldots, y_m]^T \in \mathbb{R}^m$ be an individual template, and $\mathcal{Y} = \{\mathbf{y}_1, \ldots \mathbf{y}_n\}$ be a set of templates[3]. To simplify the notation we assume $\mathcal{Y}$ to be normalized, having zero mean. Let $\mathbf{Q}$ be the covariance matrix of the vectors in $\mathcal{Y}$; we denote the eigenvectors of $\mathbf{Q}$ by $\mathbf{e}_i$, and the corresponding eigenvalues by λ_i, $1 \leq i \leq n$. It is well known that $< \mathbf{e}_i, \mathbf{e}_j >= 0, i \neq j$ and $< \mathbf{e}_i, \mathbf{e}_i >= 1$, where $<>$ stands for a scalar product. We assume that the eigenvectors are ordered in the descending order with respect to the corresponding eigenvalues λ_i. Then, depending on the correlation among the templates in $\mathcal{Y}$, only p, $p < n$, eigenvectors are needed to represent the $\mathbf{y}_i$ to a sufficient degree of accuracy as a linear combination of eigenvectors $\mathbf{e}_i$

$$\tilde{\mathbf{y}} = \sum_{i=1}^{p} a_i(\mathbf{y})\mathbf{e}_i \ . \tag{1}$$

The space spanned by the first p eigenvectors is called the *eigenspace*.

According to the standard approach, to recover the parameters a_i during the matching stage, a data vector $\mathbf{x}$ is projected onto the eigenspace

$$a_i(\mathbf{x}) =< \mathbf{x}, \mathbf{e}_i >= \sum_{j=1}^{m} x_j e_{ij} \quad 1 \leq i \leq p \ . \tag{2}$$

$\mathbf{a}(\mathbf{x}) = [a_1(\mathbf{x}), \ldots, a_p(\mathbf{x})]^T$ is the point in the eigenspace obtained by projecting $\mathbf{x}$ onto the eigenspace. Let us call the $a_i(\mathbf{x})$ coefficients of $\mathbf{x}$. The reconstructed data vector $\tilde{\mathbf{x}}$ can be written as

$$\tilde{\mathbf{x}} = \sum_{i=1}^{p} a_i(\mathbf{x})\mathbf{e}_i \ . \tag{3}$$

To make the problem of calculating the coefficients of the eigenimages more robust (and not depending on a complete set of data), we proposed a different approach [5]. In order to calculate the coefficients a_i (Eq. 2) only p points $\mathbf{r} = (r_1, \ldots r_p)$ are needed. The coefficients a_i are obtained by simply solving the following system of linear equations:

$$x_{r_i} = \sum_{j=1}^{p} a_j(\mathbf{x}) e_{jr_i} \quad 1 \leq i \leq p \ . \tag{4}$$

The considerations regarding Eq. (4) hold if all eigenimages are taken into account, i.e., $p = n$, and if there is no noise in the data x_{r_i}. However, if each template is approximated only by a linear combination of a subset of eigenimages

---

[3] Throughout the paper a simple vector notation is used since the extension to 2-D is straightforward. We also assume that $m > n$.

and there is also noise present in the data, then Eq. (4) can no longer be used, but rather one has to solve an overconstrained system of equations in the least squares sense using k data points ($p < k << m$). Thus the solution vector $\mathbf{a}$ is sought which minimizes

$$E(\mathbf{r}) = \sum_{i=1}^{k}(x_{r_i} - \sum_{j=1}^{p} a_j(\mathbf{x})e_{jr_i})^2 \ . \tag{5}$$

Of course, the minimization of Eq. (5) can only produce correct values for coefficient vector $\mathbf{a}$ if the set of points r_i does not contain outliers, i.e., not only extreme noisy points but also points belonging to different backgrounds or some other templates due to occlusion. Therefore, the solution has to be sought in a robust manner. Thus, multiple sets of points are randomly selected and the minimum least squares problem, Eq. (5) is solved. The obtained coefficient vectors are treated only as hypotheses which are then evaluated, both from the point of view of the error and from the number of compatible points. This leads to a possibly redundant set of hypotheses, which is then resolved by the selection procedure [5].

In this paper we primarily focus on mechanisms that preserve the robustness and at the same time reduce the computational complexity. In particular, we want to prevent that the coefficients need to be calculated at each point in the image and to extend the approach to multiple resolutions which is a well-known general approach to reduce the computational complexity.

3 Focus-of-Attention Detector

The first step to reduce the overall computational complexity of matching representations based on eigenimages is to design a *focus-of-attention (FOA) detector* which is intended to select candidate areas of interest with minimal computational effort, thus avoiding the need to recover the coefficients of eigenimages at each image point. The detector should operate on raw image data and output a set of locations in the image containing an object of interest with some probability. We require the detector to be efficient and to be tuned in such a way to detect possible objects with probability close to 1. While it should also try to minimize the number of false alarms, the major goal is not to miss any real targets. The constraints of computational efficiency and low miss-rate suggest that a linear filtering operation should be used. The problem of finding the best linear filter to detect deterministic or random signals in white noise is well-known and the solution is a *matched filter*. For example, Fayyad *et al.* [4] used a matched filter to detect volcanos in astronomy data.

A matched filter $\mathbf{f}$ is constructed simply by averaging the normalized examples of a class of templates. The response of the matched filter is then obtained by the normalized cross-correlation between $\mathbf{f}$ and each normalized image patch.

4 Multiresolution Approach

Another well-known technique to reduce the computational complexity is the multiresolution approach [3]. The idea is to systematically reduce the resolution of the image and then seek the solution of the problem on the coarsest resolution. This solution is then refined through finer resolutions.

The FOA-detector described in the previous section can easily be implemented on multiple resolutions using subsampled detector on subsampled images. However, the computation of coefficients of eigenimages on multiple resolutions is not that straightforward. In fact, the standard eigenspace method cannot be applied in an ordinary multiresolution way because it relies on the orthogonality of the eigenvectors. Namely, reducing the resolution of the eigenvectors, i.e., subsampling, destroys the orthogonality. That is the reason why Yoshimura and Kanade [11] who proposed fast template matching based on normalized correlation by using multiresolution eigenimages had to build a multiresolution image structure where eigenimages in each resolution layer were computed from a set of templates in that layer. This is both computationally costly and requires additional storage space.

We propose a different approach which requires only a single set of eigenimages obtained on the finest resolution which are then simply subsampled. This is possible since recovering the coefficients by solving a linear system of equations does not require orthogonality—therefore we can also subsample the eigenvectors. To be more specific, suppose that we use in Eq. (5) a set of points r_i which are also present in the subsampled image—then the recovered coefficients are the same.

5 Experimental Results

Here we show how we tested the proposed procedures on a set of "biscuit animals" (see Fig. 1). In addition we have a set of rotated ducks and another set of rotated camels under 32 orientations (the figures are omitted due to lack of space).

5.1 Focus-of-Attention Detector

Fig. 2 shows the filters which represent the FOA-detectors of the animal class and the duck class, respectively. We applied these detectors to an image with considerable cluttered background. After thresholding the result, we get the responses in the areas outlined in Fig. 3. We also get a few false (positive) responses but the majority of the detected pixels lies on the objects we are looking for. Compared to searching the whole image, only 3.7% (animals) and 0.4% (duck) of the pixels need to be investigated. One should also note the discriminatory power of the "duck-detector", which yields very low responses on the other animals despite some similarities between the patterns.

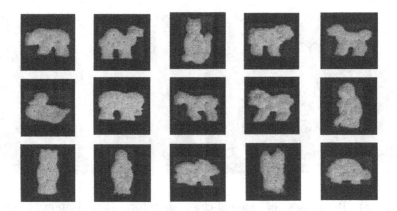

Fig. 1. Set of training templates (a class of biscuit-animals).

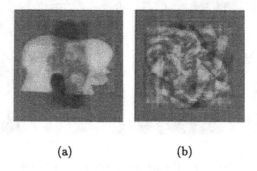

(a) (b)

Fig. 2. (a) FOA-detector for the class of animals; (b) FOA-detector for the class of rotated ducks.

Fig. 4 (a) shows the result obtained with subsampled detectors, which speeds up the processing by about a factor of 4. The quality of the results is very similar to the one obtained by the original-size detectors (compare with Fig. 3 (a)).

Fig. 4 (b) shows the result of applying the animal-class detector on an image with occluded objects. Even in this case we get a response on all the animals (even on the 70% occluded camel).

5.2 Multiresolution

As stated in section 4, when recovering the coefficients of the eigenvectors by solving a systems of linear equations, we do not require the orthogonality of the eigenvectors and can therefore apply the method on multiple resolutions. Fig. 5 shows a turtle image and an occluded turtle image. Both images as well as the eigenimages of the animals have been subsampled by a factor of 4 and the

Fig. 3. (a) Responses of animal-class detector. (b) Responses of rotated-duck-class detector.

Fig. 4. (a) Responses of animal-class detector on a subsampled image. (b) Responses of animal-class detector on partially occluded patterns.

coefficients have been recovered. Figs. 6 (a)-(b) show the back projected results. One can see that in both cases the turtle has been recovered. For comparison, in Figs. 6 (c)-(d) we show the results of applying the standard method (which requires the orthogonality) in the resampled case.

6 Conclusions

We discussed different means how to reduce the computational complexity when dealing with detecting objects represented with eigenimages. First we proposed to use a *focus-of-attention (FOA) detector* which is intended to select candidate areas of interest with minimal computational effort. Such a detector (matched filter), which is specifically tuned to the members of a class of templates drastically reduces the search area where the coefficients of eigenimages have to be recovered. Secondly, by adopting a different approach to the recovery of coefficients we have shown that an ordinary *multiresolution schema* can be utilized—achieving significant computational savings without compromising the robustness.

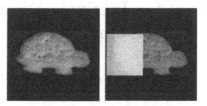

Fig. 5. Turtle-image: complete and occluded.

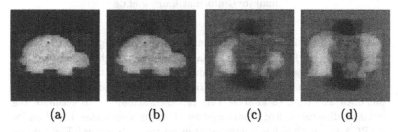

| (a) | (b) | (c) | (d) |

Fig. 6. Reconstructed turtle-image from the *resampled*: (a) complete image, (b) occluded image, (c) complete image using *standard* eigenimage-reconstruction procedure, (d) occluded image using *standard* eigenimage-reconstruction procedure.

References

1. T. W. Anderson. *An Introduction to Multivariate Statistical Analysis.* Wiley, 1958.
2. D. Beymer and T. Poggio. Face recognition from one example view. In *Proceedings of 5th ICCV'95.* IEEE Computer Society Press, 1995.
3. H. Bischof. *Pyramidal Neural Networks.* Lawrence Erlbaum Associates, 1995.
4. U. M. Fayyad, P. J. Smyth, M. C. Burl, and P. Perona. Learning to catalog science images. In S. K. Nayar and T.Poggio, editors, *Early Visual Learning*, pages 237–268. Oxford University Press, 1996.
5. A. Leonardis and H. Bischof. Dealing with occlusions in the eigenspace approach. In *Proc. of CVPR96*, pages 453–458. IEEE Computer Society Press, 1996.
6. H. Murase and S. K. Nayar. Image spotting of 3D objects using parametric eigenspace representation. In G. Borgefors, editor, *The 9th SCIA*, volume 1, pages 323–332, Uppsala, Sweden, June 1995.
7. H. Murase and S. K. Nayar. Visual learning and recognition of 3-D objects from appearance. *International Journal of Computer Vision*, 14:5–24, 1995.
8. H. Murase and S.K. Nayar. Illumination planning for object recognition using parametric eigenspaces. *IEEE Trans. on PAMI*, 16(12):1219–1227, 1994.
9. S. K. Nayar, H. Murase, and S. A. Nene. Learning, positioning, and tracking visual appearance. In *IEEE Int. Conf. on Robotics and Automation*, San Diego, 1994.
10. M. Turk and A. Pentland. Eigenfaces for recognition. *Journal of Cognitive Neuroscience*, 3(1):71–86, 1991.
11. S. Yoshimura and T. Kanade. Fast template matching based on the normalized correlation by using multiresolution eigenimages. In *Proceedings of IROS'94*, pages 2086–2093, 1994.

An Algorithm for Intrinsic Dimensionality Estimation

J. Bruske, G. Sommer

Computer Science Institute
Christian-Albrechts University zu Kiel, Germany
email:jbr@informatik.uni-kiel.de

Abstract. In this paper a new method for analyzing the *intrinsic dimensionality* (ID) of low dimensional manifolds in high dimensional feature spaces is presented. The basic idea is to first extract a low-dimensional representation that captures the *intrinsic topological structure* of the input data and then to analyze this representation, i.e. to estimate the intrinsic dimensionality. Compared to previous approaches based on local PCA the method has a number of important advantages: First, it can be shown to have only *linear time complexity* w.r.t. the dimensionality of the input space (in contrast to the cubic complexity of the conventional approach) and hence becomes applicable even for very high dimensional input spaces. Second, it is *less sensitive to noise* than former approaches, and, finally, the extracted representation can be directly used for further data processing tasks including auto-association and classification.

The presented method for ID estimation is illustrated on a synthetic data set. It has also been successfully applied to ID estimation of full scale image sequences, see [BS97].

1 Introduction

Adopting the classification in [JD88], there are two primary approaches for estimating the intrinsic dimensionality[1]. The first one is the *global approach* in which the swarm of patterns is unfolded or flattened in the d-dimensional space. Benett's algorithm [Ben69] and its successors as well as variants of MDSCAL [Kru64] for intrinsic dimensionality estimation belong to this category. The second approach is a *local* one and tries to estimate the intrinsic dimensionality directly from information in the neighborhood of patterns without generating configurations of points or projecting the pattterns to a lower dimensional space. Pettis' [PBJD79], Fukunaga and Olsen's [FO71] as well as Trunk's [Tru76] and Verveer and Duin's method [VD95] belong to this category.

Our approach belongs to the second category as well and is based on optimally topology preserving maps (OTPMs) and local principal component analysis (PCA) using a number of evenly distributed pointers in the manifold. It is

---

[1] The intrinsic, or topological, dimensionality of N patterns in an n-dimensional space refers to the minimum number of "free" parameters needed to generate the patterns [JD88]. It essentially determines whether the n-dimensional patterns can be described adequately in a subspace (submanifold) of dimensionality $m < n$.

conceptually similar to that of Fukunaga and Olsen, [FO71], using local PCA as well, but by utilizing OTPMs can be shown to better scale with high dimensional input spaces (linear instead of cubic) and to be more robust against noise. The local subspaces as revealed by local our PCA can be directly used for data modeling, as e.g. in [KL94].

2 Foundations

In this section we want to make the reader familiar with the basic ingredients of our algorithm for ID estimation to be presented in the next section. We fist introduce $OTPMs$, the underlying representation, and then turn to efficient PCA for $m < n$ points, the underlying method used for analyzing the $OTPM$, and finally comment on the problem of estimating the ID by local PCA, the general approach of our algorithm.

2.1 Constructing Optimally Topollogy Preserving Maps

Optimally Topology Preserving Maps ($OTPMs$) are closely related to Martinetz' Perfectly Topology Preserving Maps (PTPMs) [MS94] and are constructed in just the same way. The only reason to introduce them separatly is that in order to form a PTFM the pointers must be "dense" in the manifold M. Without prior knowledge this assumption cannot be checked, and in practice it will rarely be valid. $OTPMs$ emerge if just the construction method for PTFMs is applied without checking for the density condition. Only in favourable cases one will obtain a PTFM (probably without noticing). $OTPMs$ are nevertheless optimal in the sense of the topographic function introduced by Villmann in [VDM94]: In order to measure the degree of topology preservation of a graph G with an associated set of pointers S. Villmann effectively constructs the $OTPM$ of S and compares G with the $OTPM$. By construction, the topographic function just indicates the highest (optimal) degree of topology preservation if G is an $OTPM$.

Definition 1 OTPM. Let $p(x)$ be a probability distribution on the input space R^n, $M = \{x \in R^n | p(x) \neq 0\}$ a manifold of feature vectors, $T \subseteq M$ a training set of feature vectors and $S = \{c_i \in M | i = 1, \ldots, N\}$ a set of pointers in M.

We call the undirected graph $G = (V, E)$, $|V| = N$, an *optimally topology preserving map of S given the training set T*, $OTPM_T(S)$, if

$$(i, j) \in E \Leftrightarrow \exists x \in T \, \forall k \in V \setminus \{i, j\} : \max\{\| c_i - x \|, \| c_j - x \|\} \leq \| c_k - x \|$$

Corolary 1 *If $T = M$ and if S is dense in M then $OTPM_T(S)$ is a PTFM.*

Note that the definition of the $OTPM$ is constructive: For calculating the $OTPM_T(S)$ simply pick $x \in T$ according $p_T(x)$, calculate the best and second best matching pointers, c_{bmu} and c_{smu}, and connect bmu with smu. If repeated infinitely often, G will converge to $OTPM_T(S)$ with probability one. This procedure is just the essence of Martinetz' Hebbian learning rule.

For use in intrinsic dimensionality estimation and elsewhere, $OTPM_T(S)$ has two important properties. First, it does indeed only depend on the intrinsic dimensionality of T, i.e. it is independent of the dimensionality of the input space. Embedding T into some higher dimensional space does not alter the graph. Second, it is invariant against scaling and rigid transformations (translations and rotations). Just by definition it is the representation that optimally reflects the intrinsic (topological) structure of the data.

2.2 Efficient PCA for fewer points than dimensions

With $S = \{c_i \in R^n | i = 1, \ldots, N\}$ and $A^T = [c_1 - \bar{c}, \ldots, c_N - \bar{c}]$ the basic trick from linear algebra for $N < n$ is to calculate the PCA of $\hat{\Sigma} = AA^T$ instead of a PCA of the original covariance matrix $\Sigma = A^T A$. The eigenvalues of Σ, $\mu_1, \ldots, \mu_N$, are then identical to the eigenvalues $\nu_1, \ldots, \nu_N$ of $\hat{\Sigma}$ and the eigenvectors of Σ, $u_1, \ldots, u_N$, can be calculated from the eigenvectors $v_1, \ldots, v_N$ of $\hat{\Sigma}$ by setting $u_i = A^T v_i$. This can be simply checked by

$$\hat{\Sigma} v_i = \nu_i v_i \Leftrightarrow AA^T v_i = \nu_i v_i \Leftrightarrow A^T AA^T v_i = \nu_i A^T v_i \Leftrightarrow \Sigma(A^T v_i) = \nu_i A^T v_i$$

Since the PCA of the $N \times N$ matrix $\hat{\Sigma}$ can be calculated in $O(N^3)$, [PTVF88], and $\hat{\Sigma} = AA^T$ clearly can be computed in time $O(N^2 n)$, it takes only time $O(N^2 n + N^3)$ instead of $O(n^3)$ to calculate the PCA of the covariance matrix of S.

2.3 On the problem of ID estimation with local PCA

We assume the data points $x \in T$ to be noisy samples of a vector valued function $f : R^r \to R^n$

$$x = f(k) + \eta \tag{1}$$

where $k = [k_1, \ldots, k_r]$ is an r-dimensional parameter vector and η denotes the noise. The function f can be imagined to describe an r-dimensional hypersurface S in n-dimensional space. The effect of noise is to render the surface not infinitely thin (see [VD95]). Within a small region a linear approximation of the data set (such as provided by the eigenvectors of local PCAs) is only valid if the largest variance in directions n_j perpendicular to S is much smaller than the smallest variance in directions s_i of S, i.e.

$$\frac{\min_i Var(s_i)}{\max_j Var(n_j)} \gg 1. \tag{2}$$

Here, $Var(s_i)$, the intra-surface variance, depends on the size of the local region and $Var(n_j)$ depends on the variance caused by the noise *and* the fact that S cannot be exactly represented as a linear surface. This leads to a basic dilemma for any ID estimation algorithm based on local PCA: If the region is too large, $Var(n_j)$ might be high due to the non-linear nature of S. If, on the other hand, the region is too small, the noise is still there and will eventually dominate $Var(s_i)$. The solution is to search for the region size that gives the best compromise.

Closely related to the problem of noise is the problem of having available only a limited set of data. In order to make local PCA approaches work, the data set has to be large enough to represent the non-linearities and to allow for filtering out the noise.

3 Dimensionality Analysis with $OTPMs$

The basic procedure for intrinsic dimensionality analysis with $OTPMs$ works as follows: To find a set S of N pointers which reflects the distribution of T we first employs a clustering algorithm for T whose output are N cluster centers. Then we calculate the graph G as the optimal topology preserving map of S given T. The final step is to perform for each node v_i a principal component analysis of the correlation matrix of the difference vectors $c_j - c_i$ of the pointers c_j associated with the nodes v_j adjacent to v_i in G. The result of this analysis, i.e. eigenvalues and vectors for each node, is the output of the procedure and subjected to further analysis. Provided the complexity of the clustering algorithm is independent of the intrinsic dimensionality d, the serial time complexity is $O(n + m(d, T, S)^3)$, where $m(d, T, S)$ is the maximum number of direct neighbors of a node in the $OTPM$ as depending on the intrinsic dimensionality, the training set T and the set of pointers S. Bounds on $m(d, T, S)$ or even a functional form are hard to derive, yet m stays constant for constant ID, is independent of the input dimension n, and experiments confirm that it is indeed small for small IDs.

In the rest of this section we will first comment on the use of clustering algorithms and then extend the procedure to derive our actual ID estimation method.

3.1 Clustering in TPCA

The reason for clustering the data prior to construction of the $OTPM$ and not just drawing N pointers randomly from T is twofold: First the distribution of the pointers should reflect the underlying distribution $p_T(x)$ as accurately as possible and second we would like to eliminate noise on the data. Any vector quantization algorithm which aims at minimizing the (normalized) quantization error

$$J = \frac{1}{n} \sum_{i=1}^{N} \int_{V_i} \| x - c_i \|^2 p(x)dx, \tag{3}$$

where V_i denotes the Voronoi cell of c_i, is a good choice since by minimizing the total variance it will preferably place the pointers within the manifold M and filter out orthogonal noise. This holds because as long as criterion (2) is fulfilled placing pointers within the surface and hence reducing the intra-surface variance causes the largest decrease in J. From information theory it is known that it also produces a distribution of pointers which reflects the probability density, e.g. [Zad82].

3.2   The ID estimation procedure

Eventuallywe must decide how many dominant eigenvalues exist in each local region, i.e. what size an eigenvalue as obtained by each local PCA must exceed to indicate an associated intra-surface eigenvector. This amounts to determining a threshold. We adopted the $D\alpha$ criterion from Fukunaga et. al. [FO71] which regards an eigenvalue μ_i as significant if

$$\frac{\mu_i}{\max_j \mu_j} > \alpha\%. \tag{4}$$

If no prior knowledge is available, different values of α have to be tested. Otherwise, knowledge of the largest noise component can be used to calculate α.

A second problem is that due to the noise/non-linearity dilemma mentioned in section 2.3 we do not know the optimal local region sizes in advance and, in particular, do not know the optimal number of pointers N. Monitoring the development of the local eigenvalues for a growing number of pointers ($N = 1,...$) and searching for characteristic transitions is the most natural way to proceed. In this case, one does not want to cluster all the $N + 1$ pointers from scratch but rather would like to incrementally build on the existing N clusters, i.e. just add one new cluster and modify the existing ones if necessary. Using the LBG vector quantization algorithm. [LBG80], we start with $N = 1$ and add a new pointer by first searching the cluster with highest intra cluster variance. In this cluster we then search for the training sample x with the highest quantization error, add a new pointer at x, take this configuration of $N + 1$ pointers as the new starting configuration for the LBG algorithm and run $tpca$ for the $N + 1th$ round. This procedure of first searching for the worst quantized cluster helps to alleviate problems with outliers which could lead to multiple insertions at the same point if only the worst quantized example was considered.

Finally, if we have reason to believe that the data set has constant intrinsic dimensionality (i.e. has been generated by one function and not by a mixture of functions) our estimate of the intrinsic dimensionality will be the average of all local ID estimates together with its standard deviation. The ID estimate and its standard deviation is then plotted versus the number of pointers N, with different plots resulting from different choices of α. In the next section we will demonstrate that these plots actually do give very fine and characteristic hints on the ID of the data set. Our estimation procedure is interactive because the user has to choose a set of thresholds α and the final decision on the ID depends on his inspection of the ID plots. Yet for reasons already indicated and further illustrated in the next section. without prior knowledge a fully automated procedure based on local PCA which outputs the ID estimate given the data set does not make sense.

4 Experimental Results

In order to provide an impression of the characteristics of our ID estimation procedure we here apply it to a mixture of noisy data sets of different intrinsic structure and dimensionality. The data set is descibed and illustrated in figure 1.

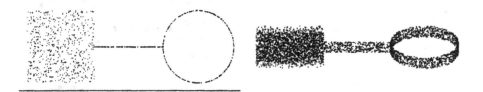

Fig. 1. Two views of the Square-Line-Circle data set. The 3d data set consists of 5000 random dots within a circle, a line and a square in the xy-plane with uniform noise in the z-direction. The noise has a variance of 1/12. The data density is approximately uniform over the data set. Left: View on the xy-plane, Right: Rotation of 60° around x-axis

Figure 2 shows the ID estimation procedure in progress for a growing number of pointers on the D10 level. From top to bottom, left to right with 5, 10, 20, 35, 45, and 70 nodes in the $OTPM$. Dark circles indicate a local ID estimate of one, medium dark circles an estimate of two and light circles of three (D10 criterion). For only five nodes the $OTPM$ indicates a one dimensional connection structure for the circle and the line and a two dimensional one for the square, identical to the ID estimates (by local PCA of the $OTPM$). For 10 nodes the $OTPM$ has already grasped the intrinsic structure of the data set. For 20 nodes we also get the correct local ID estimates for the line-data and the square but the ID estimate of the circle data is still two instead of one. This is due to the curvature (non-linearity) of the circle. From 35 to 45 nodes even the true ID of the circle is revealed because the number of pointers has now become large enough for a linear approximation of the circle on the D10 level. For even higher numbers of pointers the distribution of pointers as obtained by the LBG algorithm will eventually approximate the noise, i.e. leave the surface. From now on (see figure 2 for 70 nodes) the ID will be overestimated.

The mean squared quantization error for the Square-Line-Circle data set

$$mse = \frac{1}{|T|} \sum_{i=1}^{N} \sum_{x \in V_i} \| x - c_i \|^2 \tag{5}$$

for e.g. $N = 45$ nodes is 0.29 which is only about three times the variance of the noise. Subtracting the noise variance, only two times the noise variance remains for the average local intra-surface variance. Clearly, a simple local PCA approach as e.g. that of Fukunaga et al. (taking the unfiltered data as input to the local PCA) would not yield the correct local ID estimates on a D10 level for that local region size but would detect the noise variance as a second or third most significant eigenvalue on any level. This is what we refer to as the increased robustness against noise and the increased discrimination ability of our procedure.

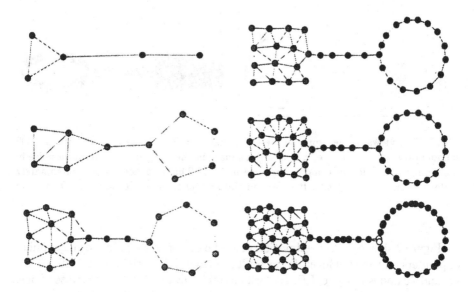

Fig. 2. Local ID estimation for the Square-Line-Circle data set for a growing number of pointers (nodes in the $OTPM$) on the D10 level. From top to bottom, left to right: 5, 10, 20, 35, 45, 70 nodes. Dark circles indicate a local ID estimate of one, medium dark circles an estimate of two and light circles of three dimensions.

Further applications of our ID estimation technique, including ID estimation of a sequence of full scale images, can be found in [BS97]. Due to limited space they had to be omitted here.

5 Conclusion

We have presented an algorithm for estimating the intrinsic dimensionality of low dimensional submanifolds embedded in high dimensional feature spaces. The algorithm belongs to the category of local ID-estimation procedures, is based on local PCA and directly extends and improves its predecessor, the algorithm of Fukunaga and Olsen, [FO71], in terms of computational complexity and noise sensitivity. The main ideas are first to cluster the data, second to construct an $OTPM$ and third to use the $OTPM$ and not the data itself for local PCA.

Besides tests on an illustrative artificial data set (this article) the procedure has been successfully applied to ID-estimation of image sequences with image resolutions of up to 256×256 pixels. [BS97]. Such application is out of reach for conventional ID-estimation procedures based on local PCA and to the best of our knowledge has not been tackled before.

OTPMs together with eigenvectors and eigenvalues returned by local PCA are not only useful for ID estimation but can be used for linear approximation of the data and construction of auto-associators in quite an obvious way. Such associators will work by projecting new data to the local subspaces spanned by the eigenvectors, i.e. by projecting to a linear approximation of the manifold.

References

[Ben69]   R. S. Bennett. The intrinsic dimensionality of signal collections. *IEEE Transactions on Information Theory*, 15:517–525, 1969.

[BS97]   J. Bruske and G. Sommer. Intrinsic dimensionality estimation with optimally topology preserving maps. Technical Report 9703, Inst. f. Inf. u. Prakt. Math. Christian-Albrechts-Universitaet zu Kiel, 1997. (submitted to IEEE PAMI).

[FO71]   K. Fukunaga and D. R. Olsen. An algorithm for finding intrinsic dimensionality of data. *IEEE Transactions on Computers*, 20(2):176–183, 1971.

[JD88]   A. K. Jain and R. C. Dubes. *Algorithms for Clustering Data*. Prentince Hall, 1988.

[KL94]   N. Kambhatla and T.K. Leen. Fast non-linear dimension reduction. In *Advances in Neural Information Processing Systems, NIPS 6*, pages 152–159, 1994.

[Kru64]   J. B. Kruskal. Multidimensional scaling by optimizing goodness of fit to a nonmetric hypothesis. *Psychometrika*, 29:1–27, 1964.

[LBG80]   Y. Linde, A. Buzo. and R.m. Gray. An algorithm for vector quantizer design. *IEEE Transaction on Communications*, 28(1):84–95, 1980.

[MS94]   T. Martinetz and K. Schulten. Topology representing networks. In *Neural Networks*, volume 7. pages 505–522, 1994.

[PBJD79] K. Pettis, T. Bailey. T. Jain, and R. Dubes. An intrinsic dimensionality estimator from near-neighbor information. *IEEE Transactions on Pattern Analysis and Machine Intelligence PAMI*, 1:25–37, 1979.

[PTVF88] W.H. Press, S.A. Teukolsky, W.T. Vetterling, and B.P. Flannery. *Numerical Recipes in C - The Art of Scientific Computing*. Cambridge University Press, 1988.

[Tru76]   G. V. Trunk. Statistical estimation of the intrinsic dimensionality of a noisy signal collection. *IEEE Transactions on Computers*, 25:165–171, 1976.

[VD95]   P. J. Verveer and R. P.W. Duin. An evaluation of intrinsic dimensionality estimators. *IEEE Transactions on Pattern Analysis and Machine Intelligence PAMI*, 17(1):81–86, 1995.

[VDM94]   T. Villmann, R. Der. and T. Martinetz. A novel aproach to measure the topology preservation of feature maps. *ICANN*, pages 289–301, 1994.

[Zad82]   P. L. Zador. Asymptotic quantization error of continuous signals and the quantization dimension. *IEEE Transactions on Information Theory*, 28(2):139–149. 1982.

Fully Unsupervised Clustering Using Centre-Surround Receptive Fields with Applications to Colour-Segmentation

Eric Pauwels*, Peter Fiddelaers and Florica Mindru

ESAT-VISICS, K.U.Leuven
K. Mercierlaan 94, B-3001 Leuven, BELGIUM
e-mail: eric.pauwels@esat.kuleuven.ac.be

Abstract. In this paper we argue that the emphasis on *similarity-matching* within the context of Content-based Image Retrieval (CBIR) highlights the need for improved and reliable clustering-algorithms. We propose a fully unsupervised clustering algorithm that is obtained by changing the non-parametric density estimation problem in two ways. Firstly, we use cross-validation to select the appropriate width of the convolution-kernel. Secondly, using kernels with a positive centre and a negative surround (DOGs) allows for a better discrimination between clusters and frees us from having to choose an arbitrary cut-off threshold. No assumption about the underlying data-distribution is necessary and the algorithm can be applied in spaces of arbitrary dimension. As an illustration we have applied the algorithm to colour-segmentation problems.

1 Introduction

Due to the rapidly growing ability to store and access large image-databases, we witness a surge in the interest in *content-based image retrieval* (CBIR), see e.g. [3, 4] and the references therein. The aim is to devise reliable algorithms that will automate, or at least assist in, *classifying, annotating and retrieving* images on the basis of visually salient and relevant information. This will then allow the user to efficiently search large image- (or multi-media) databases for images that are an adequate response to an image-oriented query or that are *visually similar* to a given input-image. Apart from their obvious commercial value, such search-engines would contribute significantly to the realisation of genuine *visual intelligence*, since they would allow an autonomous agent to compare new visual input with visually similar images in its database and make inferences about the new image by drawing on the annotations that are tagged to the retrieved images.

In contrast to more conventional database-searches (for text or records), querying image-databases is much more challenging for essentially two reasons.

---

* Postdoc Research Fellow, FWO, Belgium.

First, unless we are dealing with a highly constrained set of images, it is virtually impossible (not to mention, mindbogglingly inefficient) to try and capture the visual experience into words. (The old Chinese adage "An image is worth a thousand words" says it all.) *Secondly,* in most image-databases searches we are looking for results that are visually *similar but not necessarily identical* to the query-image. From an abstract point of view, this emphasis-shift means that quite a large class of CBIR-related problems can be reformulated in terms of *clustering-problems in an appropriate feature-space.* Clustering is therefore pushed into the lime-light since it is obvious that efficient clustering-algorithms can provide us with a powerful tool that can be wielded at different levels: *segmentation within a single image:* given image features (such as colour, texture, etc...) such algorithms can be used to segment the image and highlight regions of interest; *summarizing and classifying collections of images in a database:* searching for clusters and checking cluster-validity can assist in the selection of meaningful features that allow grouping images in classes that reflect common characteristics.

As the preceding considerations have underlined the *importance of clustering within the CBIR-context,* we now turn briefly to this topic for a bird's-eye view that will identify some shortcomings when it comes to CBIR-related applications. Clustering (and its close relative, unsupervised learning) has a long and distinguished history during which an extensive body of techniques have been developed (see e.g. [1, 2]). Broadly speaking, most of the available algorithms can be classified as being of either *statistical* or *discrete mathematics* (i.e. combinatorial/graph-theoretical) flavour. The latter tend to rely on different sorts of distance- or similarity-matrices which are used to construct linkages between points that in some sense are optimal (e.g. minimal spanning trees). As for the former, more often than not, the assumption is that the data-distribution can be modelled as a mixture of a (small) number of different densities. From this viewpoint, clustering boils down to *density-estimation followed by mode-searching.*

Unfortunately, in many applications for computer vision or CBIR, the datasets that need to be clustered exhibit two characteristics that make straightforward applications of classical clustering techniques problematic. *First,* there is the fact that many clusters are highly non-Gaussian; e.g. data-clouds often have a long, slender and curved appearance. This, invalidates many of the statistical techniques that assume underlying Gaussian distributions. (For a lucid account of some of the problems encountered in high-dimensional density-estimation related to image- and texture-processing, we refer to Popat and Picard [6].) *Secondly,* given the sheer number of pixels in an image, or images in a database, mapping the data into the appropriate feature-spaces usually yields a huge number of data-points (often in excess of 10^4) that need to be grouped. This over-abundance of points tends to render most combinatorial and graph-theoretical approaches intractable.

The aim of this paper is to outline a simple clustering-strategy that is essentially parameter-free and takes advantage of the abundance of data points typically encountered in CBIR-applications. The clustering-algorithm we propose has a statistical flavour in that we try to locate clusters by constructing a *density-function* that reflects the spatial distribution of the data-points. Since in many applications for computer vision or CBIR the datasets that need to be clustered are non-Gaussian, it stands to reason to opt for *non-parametric density estimation*. More specifically, this means that the density will be generated by convolving the dataset with a predefined kernel G of width σ. (Notice that there is no objection to this kernel being Gaussian, as it reflects the local "influence" of a single data-point, rather than the spatial structure of an entire cluster!) Rephrased in this context, the density-estimation problem is reduced to selecting an appropriate kernel G and an acceptable width σ (and it is here that most density-based clustering-algorithms introduce some adhoc selection-criterion).

Our contribution in this paper is twofold:

1. In section 2.1 we will show that we can use the data-abundance to our advantage in that it will allow us to generate a number of independent samples, all reflecting the same underlying structure. Hence, an hypothesis generated by one sample can be *cross-validated* using an independent sample. This will allow us to fix an appropriate value for σ.

2. Secondly, in section 2.2 we will discuss how to improve the discrimination power by using *kernels* that have a *centre-surround* or difference-of-Gaussians (DOG) *structure* (see Fig.3). Not only does this judicious choice of kernels (inspired by neurophysiological data) allows us to disentangle poorly separated clusters, but it will also take care of the threshold-selection problem that vexes more classical approaches of density estimation (for a quick preview, have a look at Fig. 2).

Given these two improvements, the clustering-process no longer depends on a single operator-specified parameter and can therefore proceed *completely unsupervised*. Moreover, the bulk of the computational burden is concentrated in calculations of convolution-type and can therefore be easily carried out by many processors working in *parallel*. Hence, neither the dimension of the clusterspace, nor the number of datapoints need be a limiting factor. To illustrate the efficacy of our approach we will conclude by applying it to colour-segmentation problems.

2 Description of clustering algorithm

2.1 Scale-selection through cross-validation

For notational convenience we will expound the theory for data in the 1-dimensional space $\mathbb{R}$; however, the extension to higher dimensional data is straightforward. We assume that we are provided with a dataset of N points in $\mathbb{R}$ which we need to cluster. For later reference we will assume that the N datapoints are distributed according to a density ϕ. Since in most practical vision-applications, N is very large (often $N > 10^4$), we will base the clustering on a randomly generated subsample η of n real numbers, with $n \ll N$. Restricting our attention to the subsample η allows us to draw a second independent random sample from the same dataset for cross-validation later on (see ξ-sample below).

As explained in the introduction, a first step towards clustering these points is the construction of an estimate f of the (unknown) density ϕ from which they were drawn. One obvious candidate for f is a (Gaussian) *kernel estimator* based on the sample η. This is obtained by centering a Gaussian kernel G_σ with width σ at each of the data-points η_i and adding all kernel-densities together. More formally, for 1-dimensional data $\eta = \{\eta_1, \eta_2, \ldots, \eta_n\}$ and a fixed standard-variation $\sigma > 0$ this would give rise to a density f_σ on $x \in \mathbb{R}$ defined by

$$f_\sigma(x \mid \eta) := \frac{1}{n} \sum_{i=1}^{n} G_\sigma(x; \eta_i) \quad \text{where} \quad G_\sigma(x; \eta_i) = \frac{1}{\sqrt{2\pi\sigma^2}} \, e^{-(x-\eta_i)^2/2\sigma^2} \quad (1)$$

is a Gaussian density of width σ. For fixed σ, it is easy to show that this estimator is in fact a *maximum likelihood density estimator* (MLE) within the class $\mathcal{L}_\sigma$ of densities that are finite linear combination of Gaussians with variance σ^2 (see e.g. [7]). This means that within this class, f_σ maximizes the likelihood $L(f \mid \eta)$ of a density f with respect to the given sample $\eta = (\eta_1, \ldots, \eta_n)$:

$$f_\sigma = \text{Arg} \max_{f \in \mathcal{L}_\sigma} L(f \mid \eta) \quad \text{with} \quad L(f \mid \eta) = \prod_{i=1}^{n} f(\eta_i). \quad (2)$$

Intuitively, this is obvious since by the very construction of f_σ, the local maxima of f_σ will occur at the data-points, so that the value of L in (2) is obtained by multiplying all the (locally) maximal values $f_\sigma(\eta_i)$.

The problem, of course, is that the relevant value for σ is not known in advance. To solve this problem we take recourse to a scheme of *resampling*. The motivation for this approach is that in computer-vision applications, unlike in classical statistics, there often is an over-abundance of data-points. This makes it possible to draw several *independent* samples from the *same underlying distribution*. These independent samples can then be used to test hypotheses about the underlying distribution generated by the first sample.

In order to elaborate this idea further we will assume that apart from the "mother-sample" $\eta = \{\eta_1, \ldots, \eta_n\}$ (that was used to generate the density f_σ via (1)) we have a second independent sample $\xi = \{\xi_1, \ldots, \xi_n\}$, drawn from the same population of size N. The so-called test-sample ξ is now used to *determine the appropriate value for σ via cross-validation*.

Loosely speaking, the basic idea of the method we propose is that we fix σ at the *smallest* value for which there no longer is a "significant" difference between $L(f_\sigma \mid \xi)$ and $L(f_\sigma \mid \eta)$. Indeed, this would be indicative of the fact that f_σ reflects the genuine structure of the underlying density ϕ rather than chasing the peculiarities of the η-sample.

To be sure, as it stands this is an over-simplification. For one thing, since f_σ is constructed adding Gaussians that are centered at the datapoints η_i, the local maxima of f_σ will tend to coincide with these data-points, which results in an over-inflation of the likelihood $L(f_\sigma \mid \eta)$. It makes more sense to compare

$L(f_\sigma \,|\, \xi)$ with the likelihood one would expect for a "typical" sample drawn from f_σ. Since it is easier (though completely equivalent, of course) we reformulate this question in terms of the *log-likelihood*

$$\ell_n \equiv \ell(f_\sigma \,|\, \xi) := \log L(f_\sigma \,|\, \xi) = \sum_{i=1}^{n} \log f_\sigma(\xi_i).$$

One reason for introducing logarithms is that it converts the term of interest in a sum of independent, identically distributed random variables for which the distribution is approximately normal (courtesy of the Central Limit Theorem). This means that the distribution of ℓ_n is (up to first approximation) completely characterized by its mean and variance, the values of which are specified in the following lemma (the proof amounts to little more than rewriting the definitions).

Lemma 1. *Let* $\xi = (\xi_1, \ldots, \xi_n)$ *be a sample of size* n *generated from a density* ϕ. *Denote the log-likelihood of the sample by*

$$\ell_n \equiv \ell_n(\phi \,|\, \xi) = \sum_{i=1}^{n} \log \phi(\xi_i)$$

Then the mean and variance of ℓ_n *can be readily expressed in terms of the density* ϕ:

$$E(\ell_n) = n \int_{\mathbb{R}} \phi(x) \log \phi(x)\, dx = -nH(\phi), \tag{3}$$

where $H(\phi)$ *is the entropy of* ϕ, *and*

$$Var(\ell_n) = n \int_{\mathbb{R}} \phi(x) \left[\log \phi(x) + nH(\phi)\right]^2 dx. \tag{4}$$

□

So if f_σ captures the essential structure of the underlying ϕ from which both η and ξ are drawn, we expect that the log-likelihood $\ell_n(f_\sigma \,|\, \xi)$ of the test-sample will be approximately equal to theoretically predicted value $E(\ell_n) = -nH(f_\sigma)$, and the difference between the two can be measured in terms of $\sqrt{Var(\ell_n)}$, i.e. the expected variability on $E(\ell_n)$. These values can be computed using numerical integration of the relevant formulae for f_σ.

Put differently, the optimal kernel-width is determined by selecting the smallest σ for which $\ell_n(f_\sigma \,|\, \xi)$ enters the appropriate error-band about the predicted value $E(\ell_n)$ (see Fig.1 (right)); notice how the green $\ell_n(f_\sigma \,|\, \eta)$ and purple $\ell_n(f_\sigma \,|\, \xi)$ line converge for $\sigma \to \infty$. Notice also how for large σ the log-likelihood of the test-sample exceeds $E(\ell_n)$. The reason for this becomes clear if one realizes that large values for σ produce a heavy-tailed density f_σ with an extensive support, so that most points in the ξ-sample are now situated near the centre of the density. However, $E(\ell_n)$ reflects the log-likelihood of a *typical* sample of f_σ, for which the sample-points are much more dispersed over the whole support of f_σ.

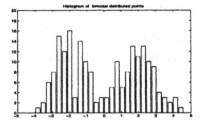
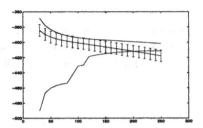

Fig. 1. *Left:* Histogram of the $n = 200$ points which constitute a (bimodal) "mother-sample" η. *Right:* The horizontal axis specifies the width σ of the Gaussian kernel G_σ (cfr. eq.(1)). The green curve represents $\ell_n(f_\sigma|\eta)$, the log-likelihood of f_σ w.r.t. η-sample. Notice how the green curve lies consistently above the red curve which represents $E(\ell_n)$, i.e. the expected log-likelihood of a typical sample from f_σ; the error-bars have a length $\sqrt{Var(\ell_n)}$ (cfr. Lemma 1). Finally, the purple curve represents the log-likelihood $\ell_n(f_\sigma|\xi)$ of an independent test-sample ξ with respect to f_σ. The relevant value for σ is obtained when there no longer is a significant difference between the observed $\ell_n(f_\sigma|\xi)$ and the expected value $E(\ell_n)$.

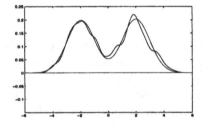
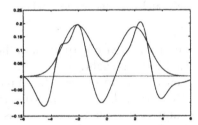

Fig. 2. *Left:* The "real" underlying distribution ϕ (blue) together with density $f_\sigma(x|\eta)$ (red) estimated using Gaussian kernel G_σ, with σ determined by cross-validation. The number of clusters depends on cutting-threshold for f_σ. Cutting this density above or below the 0.05-threshold will yield two, respectively a single cluster. *Right:* Blue line as above; The "density" f_σ (red) recovered using receptive field K_σ. We get the cluster-cores by cutting f_σ at 0-level.

2.2 Centre-surround convolution kernels

A cursory glance at the top panel of Fig.2 reveals two things. First of all, the above-expounded strategy to determine an appropriate value for σ beautifully recovers the bimodal structure of the underlying density ϕ. So from the point of view of *automatic density-estimation*, this methodology delivers the goods. However, from the point of view of *automatic clustering*, the vexed problem of thresholding remains to be addressed. In this particular case the critical threshold-value seems to be approximately 0.05: cutting the density above or below that value will produce two, respectively one cluster. To eliminate this unsatisfactory state of affairs, we change the structure of convolution-kernel G_σ. *Once an optimal value σ for the width is obtained via cross-validation*, instead of using a Gaussian we apply a kernel K_σ that is actually a *difference of Gaussians*

(DOG): $K_\sigma = G_\sigma - G_{2\sigma}$. Since the positive contribution to K_σ has a smaller variance than the negative one, K_σ has the appearance of a mexican hat (see Fig.3), thus mimicking the characteristic structure of the *centre-surround recep-tive fields* found e.g. in retinal ganglion cells. As can be seen, the ratio of the

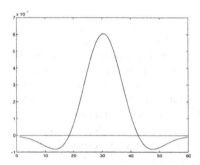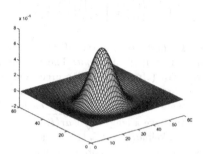

Fig. 3. The generic structure of a simple centre-surround or DOG receptive field in one (left) and two dimensions (right).

negative to positive contribution to the receptive field is taken to be 2. Although this choice seems slightly arbitrary, it is inspired by neurophysiological data; e.g. we know that for receptive fields of retinal ganglion cells of rats this ratio equals 1.6 . Because of their intrinsic structure (i.e. an excitatory centre combined with an inhibitory surround) these kernels will peak in regions where points cluster, but simultaneously tend to suppress weaker information in their neighbourhood (due to the inhibitory surround). This means that clusters in the data will au-tomatically stand out and clear the regions around them so that they become more salient. In particular, even in small gaps between neighbouring clusters the density obtained by using K_σ as a convolution filter, will dip below zero (see right panel of Fig.2). This relieves us from having to choose a threshold: the cluster-cores (i.e. the central region of the cluster) are recovered by thresholding the resulting density f_σ at zero. (In practice we take the threshold equal to the small positive value generated as the response of a single point: this way we avoid that single isolated points are selected as cluster-centers). The points outside the cluster-cores are assigned using one of the standard classification techniques such as Mahalanobis-distance or region-growing.

3 Results

To illustrate the feasibility of this approach we have applied it to the important problem of colour-segmentation and background-foreground separation. Fig. 4 gives the reader some feeling for the sort of clusters that are encountered. The left column shows two RGB colour images of rather challenging scenes. The difficulty is further underscored by the elongated form of the pixel-clouds in (RGB) colour-space (the projection on the RB-plane is shown in the right column). Using the center-surround shaped receptive fields allows the unsupervised clustering algorithm to discriminate between the adjacent clusters (whereupon each pixel is assigned an average colour characteristic for the cluster to which it belongs: see middle row).

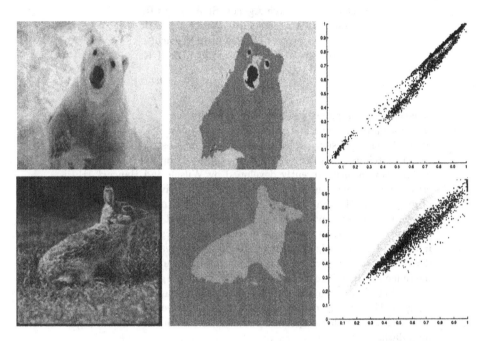

Fig. 4. Application of clustering algorithm to segmentation of colour-images. For more details, see main text.

References

1. A.K. Jain and R.C. Dubes: *Algorithms for Clustering Data.* Prentice Hall, 1988.
2. Leonard Kaufman and Peter J. Rousseeuw: *Finding Groups in Data: An Introduction to Cluster Analysis.* J. Wiley and Sons, 1990.
3. W. Niblack, R. Barber, W. Equitz, M. Flickner, E. Glasman, D. Petkovic, P. Yanker, C. Faloutsos and G. Taubin:
4. A. Pentland, R.W. Picard and S. Sclaroff: *Photobook: Content-Based Manipulation of Image Databases.* SPIE Storage and Retrieval Image and Video Databases II, No. 2185, Feb 6-10, 1994, San Jose.
5. R.W. Picard: *A Society of Models for Video and Image Libraries.* M.I.T. Media Lab Technical Report No.
6. K. Popat and R.W. Picard: *Cluster-Based Probability Model and Its Application to Image and Texture Processing.* IEEE Trans. on Image Processing, Vol.6, No.2, Feb. 1997, pp. 268-284.
7. J.R. Thompson and R.A. Tapia: *Nonparametric Function Estimation, Modeling and Simulation.* SIAM, 1990.

Multi-sensor Fusion with Bayesian Inference

Mark L. Williams[1], Richard C. Wilson[2], and Edwin R. Hancock[2]

[1] Defence Research Agency, St Andrews Road,
Malvern, Worcestershire, WR14 3PS, UK.
[2] Department of Computer Science,
University of York, York, Y01 5DD, UK.

Abstract. This paper describes the development of a Bayesian framework for multiple graph matching. The study is motivated by the plethora of multi-sensor fusion problems which can be abstracted as multiple graph matching tasks. The study uses as its starting point the Bayesian consistency measure recently developed by Wilson and Hancock. Hitherto, the consistency measure has been used exclusively in the matching of graph-pairs. In the multiple graph matching study reported in this paper, we use the Bayesian framework to construct an inference matrix which can be used to gauge the mutual consistency of multiple graph-matches. The multiple graph-matching process is realised as an iterative discrete relaxation process which aims to maximise the elements of the inference matrix. We experiment with our multiple graph matching process using an application vehicle furnished by the matching of aerial imagery. Here we are concerned with the simultaneous fusion of optical, infra-red and synthetic aperture radar images in the presence of digital map data.

1 Introduction

Graph matching is a critical image interpretation tool for high and intermediate level vision [14,13,12,15]. In essence the technique allows symbolic image abstractions to be matched against one-another. It has been widely exploited in applications such as stereo matching and relational model matching. In fact, it was Barrow and Popplestone who first demonstrated the practicality of relational graphs as a flexible image representation [1]. More recently, the methodological basis has attracted renewed interest in two distinct areas. The first of these addresses the issue of how to compute a reliable distance measure between corrupted relational descriptions. Contributions here include the use of binary attribute relations by Christmas, Kittler and Petrou [9], the efficient structural hashing idea of Messmer and Bunke [10], and, Wilson and Hancock's use of both constraint filtering [19] and graph editing [20] operations to control structural corruption. The second issue concerns the efficient search for optimal matches. Recent efforts under this heading include the use of mean-field annealing by Suganathan et al [16], the soft-assign technique of Gold and Rangarajan [6,7] and Cross et al's [3,4] use of genetic search.

Despite this recent activity, the advances in methodology have been confined almost exclusively to the matching of graph pairs. This is an important omission. Multi-sensor fusion tasks pose an important practical challenge in applications such as remote-sensing and robotics, and are naturally couched as multiple graph matching problems [17,18]. However, although principled multi-sensor fusion algorithms are to hand, these invariably pose the matching process as one of feature registration or clustering [8] rather than relational graph-matching. This means that the bulk of available algorithms are effectively low-level and non-contextual. Relational matching algorithms offer the advantage of drawing on structural constraints in the matching process. This means that they do not for instance need to draw on an explicit transformational model between the different sensors being deployed in the scene recovery task. In other words, they are calibration-free.

The aim in this paper is to extend our existing methodology to the case of matching multiple-graphs. We have already addressed to dual issues of modelling relational distance [19–21] and efficiently locating optimal matches [2–5]. In particular, our framework delivers the probability distribution for pairwise matches using a Bayesian model of matching errors. The key to extending this framework to the case of multiple graphs is a means of assessing relational similarity beyond the level of graph-pairs. We meet this goal by measuring the compositional consistency of the pairwise matches. This process draws on a triangle of inference. If node A in graph the graph indexed l matches onto the node B in the graph indexed m and this node in turn matches onto the node C in the graph indexed n, then for compositional consistency we make the inference that the node A in graph l must also match onto the node C in graph n. Rather than imposing a series of hard-tests we assess the degree of compositional consistency using a Bayesian inference matrix. Multiple graph matches are sought so as to maximise the elements of this matrix.

2 Theory

The basic task confronting us is to find the pairwise matches between a set of N relational graphs. Suppose that the graph indexed l is denoted by $G_l = (V_l, E_l)$ where V_l is the set of nodes and E_l is the set of edges. The complete set of graphs is denoted by $\Gamma = \{G_l; l = 1, ..., N\}$. The pairwise match between the graph indexed l and the graph indexed m is represented by the function $f_{l,m} : V_l \rightarrow V_m$ from the nodes of the graph G_l to the nodes of the graph G_m.

In our previous work we showed how the consistency of match between graph pairs could be gauged using the probability distribution for matching errors [19–21]. The distribution is derived from first principles and assumes that matching errors occur with a uniform probability P_e. According to our model, the probability that the node a in graph G_l matches onto the nodes b in graph G_m is given by the following sum of exponential terms

$$P(f_{l,m}(a) = b) = \frac{K_a}{|\Theta_b|} \sum_{S \in \Theta_b} \exp\left[-k_e H(S, R_a)\right] \tag{1}$$

The physical quantity underpinning this distribution is the Hamming distance $H(S, R_a)$. This quantity counts the number of errors between the match $R_a = \{f(b); (a, b) \in E_l\}$ residing on the contextual neighbourhood on the node a in graph G_l and each of the set of feasible configurations $S \in \Theta_b$ that can be formed by permuting the nodes in the contextual neighbourhood of the node b from graph G_m. The constants $K_a = (1 - P_e)^{|R_a|}$ and $k_e = \ln \frac{1-P_e}{P_e}$ are regulated by the uniform probability of matching errors P_e.

To extend our methodology from matching graph pairs to the case of multiple graph matching, we require a means of augmenting our consistency measure. In particular, we require a way ensuring that the system of pairwise matching functions are mutually consistent over the multiple graphs. In developing our new framework we appeal to the idea inducing inference triangles by forming compositions of the pairwise matching functions. In other words, we aim to ensure that if $f_{l,m}(a) = b$ and $f_{m,n}(b) = c$ then the composition results the consistent match $f_{l,n}(a) = c$. However, rather than imposing this transitive closure condition as a hard constraint, we aim to capture it in a softened Bayesian manner. We measure the consistency of the match $f_{l,n}(a) = c$ by computing the fuzzy-inference matrix

$$I_{l,n}(a, c) = \sum_{\substack{G_m \in \Gamma \\ m \neq l, m \neq n}} \sum_{b \in V_m} P(f_{l,m}(a) = b) P(f_{m,n}(b) = c) \tag{2}$$

The matrix elements effectively accumulate the consistency of match over the set of transitive matching paths which lead from the node a of the graph G_l to the node c of graph G_n. These paths are specified by the set of intermediate matches $f_{l,m}(a) = b$ residing on the nodes of the graph G_m. The matrix averages over both the possible intermediate graphs and the set of possible matching configurations residing on the intermediate graph.

To update the matches residing on the multiple graphs, we select the configuration which maximises the relevant inference matrix element. In other words

$$f_{l,n}(a) = \arg\max_{c \in V_n} I_{l,n}(a, c) \tag{3}$$

In our implemenatation of the multi-graph matching process is realised by parallel deterministic updates along the lines adopted in our previous work on discrete relaxation [19]. However, we acknowledge that this simplistic optimisation process is prone to converegence to local optima. Suffice to say that the work reported in this paper is aimed at proof-of-concept of the multiple graph-matching technique. In other words, our aim is to demonstrate the utility of the fuzzy inference matrix $I_{l,n}(a, c)$. Studies aimed at improving the optmisation process are in hand and will be reported in due course.

3 Experiments

There are two aspects of our evaluation of the multiple graph-matching method. The first of these is a study focussed on the matching of three different modalities of aerial imagery against a cartographic map. The sensing modalities are optical, infra-red line-scan and synthetic aperture radar. The second aspect of our study is concerned with assessing the sensitivity of the matching process on simulation data. The aim here is to illustrate the effectiveness of the matching process on data with known ground-truth.

3.1 Real-world Data

We illustrate the effectiveness of our multiple graph matching process on an application furnished by remotely sensed data. We use four overlapping data sets. The data under study consists of partially overlapped images and a cartographic map of the imaged area. Figure 1a shows an optical image of a rural region surrounding an airstrip. Figure 1b shows an infra- red line scan image of the airstrip, containing less of the surrounding region. Figure 1c is a synthetic aperture radar (SAR) image of the airstrip and some of its surroundings. Figure 1d is the cartographic map of area. The matching problem uses only the rural region to the left of the main runway in each image. It is important to stress that the runway is used only as a fiducial structure for registering the ground-truth. It does not feature in any of our matching experiments.

The matching process is posed as one of finding correspondences between linear hedge-structures in the four data-sets. Relational graphs are constructed by computing the constrained Delaunay graphs for the relevant sets of line-segments. The graphs corresponding to the images in Figures 1a-1d are shown in Figures 2a-2d. It is worth stressing the difficulties in registering the graph by-eye. In other words, the multiple graph-matching probelm used to evaluate our inference-process is a demanding one. The optical image gave rise to a 133 node graph, there were 68 nodes in the graph derived from the SAR data and 63 nodes in the graph derived from the infra-red image. The region of map selected gave rise to 103 separate hedge sections each represented by a node in the map graph. To give some idea of the difficulty of the multi-graph registration process, it is worth making some comments on the qualitative features of the images under match. The clarity of the optical image made it easy to relate its salient structure to those in the map, and to a lesser extent the SAR and infra-red images. The weakest link in the inference chain is provided by the matching of the SAR and infra-red images. These two images contain quite different structure. More importantly, their potential matches in the optical image are relatively disjoint. In-fact, ground-truthing the matches between the SAR image and the infra-red image proved to be a difficult. The label-error probabilities used in the experimental evaluation are computed using the overlap statistics for the four data-sets.

Because it represents the weakest link, we commence our discussion of the experimental results by considering the match of the SAR and infra-red images.

Out of the 68 nodes in the SAR graph 28 had valid matches to nodes in the infra-red graph. Initially none of the SAR graph nodes matched correctly to the nodes in the graph derived from the infra-red image. The matching process using no inference was unable to recover any correct matches. When the inference process was included in the matching process the final match recovered all of the possible correct matches.

Out of the 103 nodes in the map graph 83 had valid matches to nodes in the graph derived from the optical image. Initially none of the map graph nodes were correctly matched and conventional graph to graph matching was unable to improve on this. By including the inference terms into the matching process a final match including 58 correct matches was achieved.

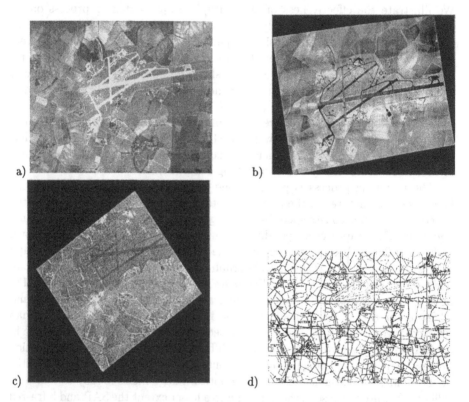

Fig. 1. a) Optical image, b) Infra-red line-scan image, c) Synthetic aperture radar image, d)Cartographic data.

3.2 Sensitivity Study

The inference process described in Section 2 is applicable to the simultaneous matching of an arbitrary number of graphs. However, the simplest case involves

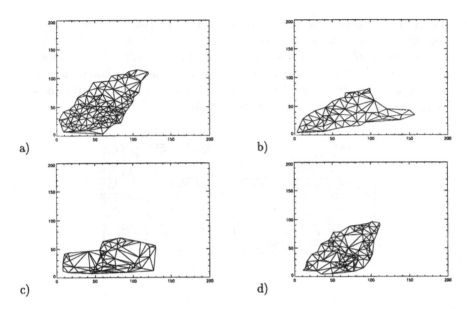

Fig. 2. a) Graph for the optical image, b) Graph for the infra-red line-scan image, c) Graph for the synthetic aperture radar image, d) Graph for the cartographic data.

the matching of three graphs. This allows us to illustrate the performance advantages offered by the inference process under varying noise conditions.

The data used in this study has been generated from random point-sets. The Voronoi tesselations of the random point-sets are used to construct Delaunay graphs. By adding a predetermined fraction of extraneous nodes to the original point-sets we simulate the effects of graph-corruption due to segmentation errors. We base our study on 50 node graphs. We attempt matches under corruption levels of 10%, 20%, 30%, 40% and 50%. We have also included initialisation errors in our simulation experiments. Our study is aimed at demonstrating the improvements that can be realised if a third inference graph is used to supplement the matching of two poorly initialised graphs. In the case of the poorly initialised graphs only 10& of the matches are correct. In both cases the number of correct matches to the inference graph is initially 50%. In other words, the aggregate initialisation error associated with the inference route is 25%.

The results of our simulation study are summarised in Table 1. Here we show a comparison of different multiple graph matching schemes under varying corruption and fixed initialisation error. Each entry in the table represents the average over ten experiments. In the first three rows of the table we list, the fractional corruption, the best achievable fraction of correct matches, together with the fraction of initialisation errors. In the fourth row we show the results when pairwise graph matching is attempted without inference. The fifth row shows the result of including the third graph as an intermediate state in the triangle of inference. In this case the matching to the inference graph is kept fixed and only the matches between the original graph are updated. The results are consistently better than the pairwise match provided that the level of corruption does not exceed 40%. Finally, the sixth row shows the result obtained if we also update the matches to the inference graph. Although there are fluctuations from one corruption level to another, the average performance obtained with updated inference is better than its static counterpart shown in the fifth row.

Corruption	0	10	20	30	40	50
Best possible	100	90	80	70	60	50
Fraction initially correct	25	25	25	25	25	25
No inference	96	17.4	10.4	8.6	9.6	8.8
Fixed inference	100	58.6	39.6	27.6	15.4	7.8
Updated inference path	100	77.8	45.2	26.0	10.8	9.6

Table 1. Results of Simulation Study.

4   Conclusions

We have detailed a fuzzy inference process for multiple graph matching. The basic idea underpinning the method is to use the recently reported Bayesian consistency measure of Wilson and Hancock to construct an inference matrix for multiple graph matching. This matrix allows strong pairs of graph matches to enhance the accuracy of weaker matches through a process of inducing inference triangles. The utility of our new multi-graph inference technique is demonstrated on the matching of aerial images obtained in three different sensor modalities against a digital map. Here the inference process is demonstrated to render the matching of otherwise unmatchable infra-red and SAR images feasible.

There are a number of ways in which this study is being extended. In the first instance, the optimisation process adopted in our present study is a deterministic discrete relaxation procedure [19]. One of the possibilities for enhancing the optimisation process is to adopt a stoachastic search procedure. For instance in a recent paper we have shown how genetic search can be used to improve the performance of matching graph pairs.

References

1. Barrow H.G. and R.J. Popplestone, "Relational Descriptions in Picture Processing", *Machine Intelligence*, **6**, 1971.
2. Cross A.D.J., R.C. Wilson and E.R. Hancock, "Discrete Relaxation on a Boltzmann Machine", *Proceedings ICANN 95*, pp. 491–496,1995.
3. Cross A.D.J., R.C.Wilson and E.R. Hancock, "Genetic Search for structural matching", *Proceedings ECCV96*, **LNCS 1064**, pp. 514–525, 1996.

4. Cross A.D.J. and E.R.Hancock, "Relational Matching with stochastic optimisation" *IEEE Computer Society International Symposium on Computer Vision*, pp. 365–370, 1995.

5. Finch A.M., Wilson R.C. and Hancock E.R., "Softening Discrete Relaxation", *to appear in Neural Information Processing Systems 9*, MIT Press 1997.

6. Gold S., A. Rangarajan and E. Mjolsness, "Learning with pre-knowledge: Clustering with point and graph-matching distance measures", *Neural Computation*, **8**, pp. 787–804, 1996.

7. Gold S. and A. Rangarajan, "A Graduated Assignment Algorithm for Graph Matching", *IEEE PAMI*, **18**, pp. 377–388, 1996.

8. Hathaway R.J., J.C. Bezdek, W. Pedrycz, "A parametric model for fusing heterogeneous fuzzy data", *IEEE Transactions on Fuzzy Systems*, **4**, pp. 270–281, 1996.

9. Kittler J., W.J. Christmas and M.Petrou, "Probabilistic Relaxation for Matching Problems in Machine Vision", *Proceedings of the Fourth International Conference on Computer Vision*, pp. 666-674, 1993.

10. Messmer B.T. and H. Bunke, " Efficient error-tolerant subgraph isomorphism detection", *Shape, Structure and Pattern Recognition, edited by D. Dori and A. Bruckstein*, pp. 231-240, 1994.

11. Mjolsness E., G.Gindi and P. Anandan, "Optimisation in model matching and perceptual organisation", *Neural Computation*, **1**, pp. 218–219, 1989.

12. Sanfeliu A. and Fu K.S., "A Distance Measure Between Attributed Relational Graphs for Pattern Recognition", *IEEE SMC*, **13**, pp 353–362, 1983.

13. Shapiro L. and R.M.Haralick, "Structural Description and Inexact Matching", *IEEE PAMI*, **3**, pp 504–519, 1981.

14. Shapiro L. and R.M.Haralick, "A Metric for Comparing Relational Descriptions", *IEEE PAMI*, **7**, pp 90-94, 1985.

15. Simic P., "Constrained nets for graph matching and other quadratic assignment problems", *Neural Computation*, **3** , pp. 268–281, 1991.

16. Suganathan P.N., E.K. Teoh and D.P. Mital, "Pattern Recognition by Graph Matching using Potts MFT Networks", *Pattern Recognition*, **28**, pp. 997–1009, 1995.

17. Tang Y.C. and C.S.G. Lee, "A Geometric Feature Relation Graph Formalism for Consistent Sensor Fusion", *IEEE SMC*, **22**, pp 115–129, 1992.

18. Tang Y.C. and C.S.G. Lee, "Optimal strategic recognition of objects based on candidate discriminating graph with coordinated sensors", *IEEE SMC*, **22**, pp. 647–661, 1992.

19. Wilson R.C., Evans A.N. and Hancock E.R., "Relational Matching by Discrete Relaxation", *Image and Vision Computing*, **13**, pp. 411–421, 1995.

20. Wilson R.C and Hancock E.R., "Relational Matching with Dynamic Graph Structures", *Proceedings of the Fifth International Conference on Computer Vision*, pp. 450-456, 1995.

21. Wilson R.C. and E.R. Hancock, "Gauging relational consistency and correcting structural errors", *IEEE Computer Society Computer Vision and Pattern Recognition Conference*, pp. 47-54, 1996.

MORAL – A Vision-Based Object Recognition System for Autonomous Mobile Systems *

Stefan Lanser, Christoph Zierl, Olaf Munkelt, Bernd Radig

Technische Universität München
Forschungsgruppe Bildverstehen (FG BV), Informatik IX
Orleansstr. 34, 81667 München, Germany
email: {lanser,zierl,munkelt,radig}@informatik.tu-muenchen.de

Abstract. One of the fundamental requirements for an autonomous mobile system (*AMS*) is the ability to navigate within an à priori known environment and to recognize task-specific objects, i.e., to identify these objects and to compute their 3D pose relative to the AMS. For the accomplishment of these tasks the AMS has to survey its environment by using appropriate sensors. This contribution presents the vision-based 3D object recognition system MORAL[1], which performs a model-based interpretation of single video images of a CCD camera. Using appropriate parameters, the system can be adapted dynamically to different tasks. The communication with the AMS is realized transparently using *remote procedure calls*. As a whole this architecture enables a high level of flexibility with regard to the used hardware (computer, camera) as well as to the objects to be recognized.

1 Introduction

In the context of autonomous mobile systems (*AMS*) the following tasks can be solved using a vision-based object recognition system: Recognition of task-specific objects (e.g., doors, trashcans, or workpieces), localization of objects (estimation of the three-dimensional pose relative to the AMS) in order to support a manipulation task, and navigation in an à priori known environment (estimation of the three-dimensional pose of the AMS in the world). These tasks can be formalized by an interpretation

$$\mathcal{I} = \langle\ obj, \{(I_{j_1}, M_{i_1}), \dots, (I_{j_k}, M_{i_k})\}, (\mathcal{R}, \mathcal{T})\ \rangle \tag{1}$$

with *obj* the object hypothesis, (I_{j_l}, M_{i_l}) the correspondence between image feature I_{j_l} and model feature M_{i_l}, and $(\mathcal{R}, \mathcal{T})$ the estimated 3D pose of the object.

---

* This work was supported by *Deutsche Forschungsgemeinschaft* within the *Sonderforschungsbereich 331*, "*Informationsverarbeitung in autonomen, mobilen Handhabungssystemen*", project L9.

[1] **M**unich **O**bject **R**ecognition **A**nd **L**ocalization

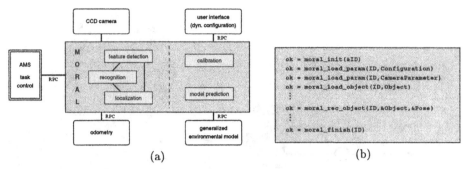

Fig. 1. (a) Overview of MORAL and (b) a typical RPC sequence.

The object recognition system MORAL presented in this contribution accomplishes these tasks by comparing the predicted model features with the features extracted from a video image [LMZ95]. The underlying 3D models are polyhedral approximations of the environment provided by a hierarchical world model. This contribution mainly deals with rigid objects. The extension of this work to the recognition of articulated objects, i.e., objects consisting of multiple rigid components connected by joints, is shown briefly in Sec. 3.

Related work to the recognition task can be found in [IK88], [Gri90], [DPR92], and [Pop94]. The problem of the self-localization of an AMS is considered in [FHR+90], [CK94], and [KP95]. Common to both tasks are the use of a geometric model and the basic localization procedure, i.e., the determination of the 6 DOF pose of the AMS relative to the world or to an object, respectively.

2 System Architecture

The presented object recognition system (see Fig. 1 (a)) is implemented as a RPC server (*remote procedure call*), which is called by any client process, especially by the AMS task control system. The standardized RPCs allow a hardware independent use of the MORAL system. Therefore MORAL can easily be applied on different platforms. By using the same mechanism, MORAL communicates (optionally) with other components, e.g., the *generalized environmental model*. With the help of specific RPCs MORAL can be dynamically configured, and can thus be flexibly adapted to modified tasks. The internal structure of MORAL essentially consists of five modules, which are implemented in ANSI-C and C++, respectively.

2.1 Calibration

In order to obtain the 3D object pose from the grabbed video image, the internal camera parameters (mapping the 3D world into pixels) as well as the external camera parameters (pose of the CCD camera relative to the manipulator or the vehicle) have to be determined with sufficient accuracy.

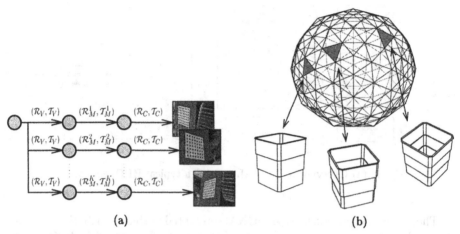

(a)   (b)

Fig. 2. (a) Estimation of the camera pose based on known relative movements of the robot manipulator; (b) each triangle of the tesselated Gaussian sphere defines a 2D view of an object.

Internal Camera Parameters. The proposed approach uses the model of a pinhole camera with radial distortions to map 3D point in the scene into 2D pixels of the video image [LT88]. It includes the internal parameters as well as the external parameters $\mathcal{R}$, a matrix describing the orientation, and $\mathcal{T}$, a vector describing the position of the camera in the world.

In the first stage of the calibration process the internal camera parameters are computed by simultanously evaluating images showing a 2D calibration table with N circular marks P_i taken from K different viewpoints. This *multiview calibration* [LZ95] minimizes the distances between the projected 3D midpoints of the marks and the corresponding 2D points in the video images. The 3D pose $\mathcal{R}, \mathcal{T}$ of the camera is estimated during the minimization process. Thus, only the model of the calibration table itself has to be known à priori.

Hand-Eye Calibration. Once the internal camera parameters have been determined, the 3D pose of the camera relative to the *tool center point* is estimated in the second stage of the calibration process (*hand-eye calibration*).

In the case of a camera mounted on the manipulator of a mobile robot the 3D pose of the camera $(\mathcal{R}, \mathcal{T})$ is the composition of the pose of the robot $(\mathcal{R}_V, \mathcal{T}_V)$, the relative pose of the manipulator $(\mathcal{R}_M, \mathcal{T}_M)$, and the relative pose of the camera $(\mathcal{R}_C, \mathcal{T}_C)$, see Fig. 2 (a). The unknown pose $(\mathcal{R}_C, \mathcal{T}_C)$ is determined by performing controlled movements $(\mathcal{R}_M^k, \mathcal{T}_M^k)$ of the manipulator similar to [Wan92], for details see [LZ95].

Since the used 2D calibration table is mounted on the mobile robot itself, the manipulator can move to the different viewpoints for the *multiview calibration* automatically. Thus, the calibration can be accomplished in only a few minutes.

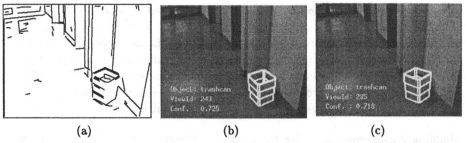

(a)   (b)   (c)

Fig. 3. (a) Detected image features; (b-c) the two highest ranked hypotheses according to the image features in (a) and the object model in Fig. 2 (b).

2.2 Model Prediction

The recognition of rigid objects is based on a set of characteristic 2D views (*multiview representation*). Here, a set of 320 perspective 2D views is determined using the tesselated Gaussian sphere. These views contain the model features used for establishing the correspondences (I_{j_i}, M_{i_i}) part of $\mathcal{I}$ in Eq. (1). This model prediction is provided by the *Generalized Environmental Model (GEM)* [HS96] based on the boundary representation of a polyhedral 3D object. The tesselated Gaussian sphere and three 2D views of the object TRASHCAN are shown in Fig. 2 (b).

Referring to Eq. (1), *obj* is not limited to a specific object. In the case of vision-based navigation it means the environment of the AMS. Instead of using a set of characteristic 2D views, the expected 2D model features are provided by GEM according to the *expected position*.

2.3 Feature Detection

The detection of image features is based on the image analysis system HORUS [ES97]. HORUS provides a large number of different image operators which are controlled by MORAL according to the performed task. Object-specific knowledge is used to control the image segmentation process. This knowledge comprises different types, e.g., lines, faces, arcs, and features like color of the foreground or background (Fig. 3 (a)). The feature detection module can be parameterized either by the modules of MORAL itself or by the clients which call services of MORAL via the RPC mechanism, e.g., the AMS task control or the user interface.

2.4 Object Recognition

The aim of the object recognition module is to identify objects and to determine their rough 3D pose by searching for the appropriate 2D model view matching the image. This is done by establishing correspondences between image features extracted from the CCD image and 2D model features of an object. Obviously, if a rough 3D pose is à priori known, the recognition module is "bypassed" and the localization module is called directly (e.g., continous self-localization).

Building Associations. The first step in the object recognition process is to build a set of *associations*. An association is defined as a quadruple (I_j, M_i, v, c_a), where I_j is an image feature, M_i is a model feature, v is one of the characteristic 2D model views of an object, and c_a is a confidence value of the correspondence between I_j and M_i. This value is obtained by traversing aspect-trees [Mun95] or by a geometrical comparison of the features incorporating constraints based on the topological relations **Vertex** , **OvlPar** , and **Cont** [LZ96].

Building Hypotheses. In order to select the "correct" view of an object the associations are used to build *hypotheses* $\{(obj, \mathcal{A}_i, v_i, c_i)\}$. For each 2D view v_i all corresponding associations with sufficient confidence are considered. From this set of associations the subset of associations $\mathcal{A}_i$ with the highest rating forming a *consistent labeling* of image features is selected. That means that the geometric transformations in the image plane between model and image features in $\mathcal{A}_i$ are similar, i.e., they form a cluster in the transformation space. Furthermore, the image features have to fulfil the same topological constraints as their corresponding model features [LZ96]. The confidence value c_i depends on the confidence values of the included associations and the percentage of mapped model features. The result of the described recognition process is a ranked list of possible hypotheses (see Fig. 3 (b-c)) which are verified and refined by the localization module.

A view in a hypothesis determines one translational and two rotational degrees of freedom of the object pose. Additionally, while building the hypotheses a scale factor and a translational vector in the image plane is computed. Using this weak perspective, a rough 3D position of the object is derived. Note, that some of the hypotheses from the object recognition module may be "incorrect". These hypotheses are rejected by the subsequent verification and refinement procedure performed by the localization module.

2.5 Localization

The aim of the localization module is the determination of all 6 DOF of the CCD camera relative to an object or to the environment. The input for this module is a rough 3D pose and a set of 3D model lines. This input can be either a model of the environment (provided by GEM) or of an specific object (result of the previous object recognition).

Based on known uncertainties of the pose and/or the position of the model lines, specific search spaces for the extraction of image lines are computed [LL95]. Final correspondences are established using an interpretation-tree approach [Gri90]. The traversion of the tree is dynamically adapted due to topological constraints of the actually involved correspondences. During this process model and image lines are aligned and therefore the input pose is refined. Similar to [Low91], this is performed by a weighted least squares technique minimizing

$$\sum_{k=1}^{m} e\left(M_{i_k}, I_{j_k}, (\mathcal{R}, \mathcal{T})\right)^2 \tag{2}$$

Fig. 4. AMS used for the experiments.

using an appropriate error function e. For example, if e measures the 3D distance between a 3D model line $M_{i_k} = \langle M_{i_k}^1, M_{i_k}^2 \rangle$ and the plane which is spanned by the corresponding image line $I_{j_k} = \langle I_{j_k}^1, I_{j_k}^2 \rangle$ and the focus of the camera (similarly to [KH94]), Eq. (2) transforms to

$$\sum_{k=1}^{m} W_{i_k} \sum_{l=1}^{2} [\hat{n}_{j_k}^T \cdot \mathcal{R} \cdot (M_{i_k}^l - \mathcal{T})]^2$$

$$\hat{n}_{j_k} = \frac{n_{j_k}}{\|n_{j_k}\|} , \quad n_{j_k} = \begin{pmatrix} I_{j_k}^1 \\ f \end{pmatrix} \times \begin{pmatrix} I_{j_k}^2 \\ f \end{pmatrix}$$

with f the focal length of the used CCD camera.

If only coplanar features are visible, which are seen from a large distance compared to the size of the object, the 6 DOF estimation is quite unstable because some of the pose parameters are highly correlated. In this case à priori knowledge of the orientation of the camera with respect to the ground plane of the object might be used to determine two angular degrees of freedom. Naturally, this approach decreases the flexibility of the system. Tilted objects cannot be handled any longer. A more flexible solution is the use of a second image of the scene taken from a different viewpoint with known relative movement of the camera (*motion stereo*). By simultanously aligning the model to both images the flat minimum of the 6 DOF estimation can be avoided. Note, that for well structured objects with some non-coplanar model features a 6 DOF estimation based on a single video image yields good results as well.

3 Applications

In the following several applications of MORAL in the context of autonomous mobile systems within the joint research project *SFB 331* are described. The easy integration of MORAL into these different platforms (see Fig. 4) with different cameras[2] and tasks shows the flexibility of our system. A typical sequence of RPC calls to MORAL is outlined in Fig. 1 (b).

An example for the vision-based navigation of an AMS in a known environment with MORAL is shown in Fig. 5. Using the input pose (given by the

---

[2] Note, that in the AMS shown on the right the camera is mounted at the robot hand.

  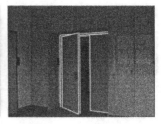

Fig. 5. Vision-based navigation, model specific search spaces, and determination of the rotational joint state of the object DOOR.

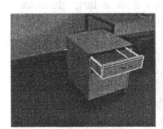  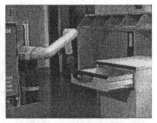

Fig. 6. Examples for the vision-based 3D object recognition with MORAL.

Fig. 7. Recognition of an articulated object: Guided by the hierarchical structure of the object model the unknown joint configuration of the drawer cabinet is determined recursively to enable an robot to open the upper drawer.

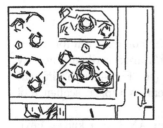  

Fig. 8. The pose estimation of workpieces enables the grasping by a mobile robot.

odometry of the AMS), the predicted 3D model features, and the image grabbed from the CCD camera, MORAL determines the exact pose of the AMS relative to the world. Thereby, the uncertainty of the AMS pose leads to model specific search spaces. The second image in Fig. 5 shows an example for three selected model lines with a pose uncertainty of 10° for the roll angle. On a standard SPARC 10 workstation the whole navigation task (image preprocessing and pose estimation) takes less than 2 seconds.

Some results for the object recognition are presented in Fig. 6: On the left side the refinement of the rough 3D pose in Fig. 3 (b-c) is shown. This refinement is the result of the 3D localization module described in 2.5. The other images in Fig. 6 show the 3D object recognition of the objects TOOLCARRIAGE and DRAWERCABINET. All of these tasks are performed in approx. 3 to 10 seconds on a SPARC 10, depending on the number of image and model features.

Articulated objects are handled using a hierarchical representation and recognition scheme [HLZ97]. The object model consists of rigid 3D subcomponents linked by *joints*. Each joint state within the *joint configuration* of an object indicates the pose of a subcomponent relative to its parent component. On the right side of Fig. 5 the result of the identification of a rotational joint state is shown: After recognizing the door-frame the aim of this task is to determine the opening angle (*rotational joint state*) of the door-wing. Using this result, the AMS can determine whether it can pass the door or has to open it. A further example of the recognition of an articulated object is shown in Fig. 7: After recognizing the static component of the object DRAWERCABINET, the translational joint states corresponding to the two drawers are computed following the object hierarchy.

A further application of MORAL – the grasping of a workpiece by a mobile robot – is presented in Fig. 8. Since the exact pose of the object is à priori unknown, the robot guiding system calls MORAL to determine the object pose. Using the CCD camera mounted in the gripper exchange system, MORAL computes the requested pose. In this case the *motion stereo* approach described in Sec. 2.5 is used to simultanously align the model to two images.

4 Conclusion

This paper presented the object recognition system MORAL, which is suitable for handling vision-based tasks in the context of autonomous mobile systems. Based on à priori known 3D models, MORAL recognizes objects in single video images and determines their 3D pose. Alternatively, the 3D pose of the camera relative to the environment can be computed, too. The flexibility of the approach has been demonstrated by several online experiments on different platforms.

Future work will focus on the following topics: Using an indexing scheme, the object recognition module should be enabled to handle a larger model database; in addition to the use of CAD models, the 3D structure of objects should be reconstructed by images taken from different viewpoints; finally curved lines have to be integrated as new features, both during the model prediction and the feature detection.

References

[CK94]   H. Christensen and N. Kirkeby. Model-driven vision for in-door navigation. *Robotics and Autonomous Systems*, 12:199–207, 1994.

[DPR92] S. Dickinson, A. Pentland, and A. Rosenfeld. From Volumes to Views: An Approach to 3-D Object Recognition. *CVGIP: Image Understanding*, 55(2):130–154, March 1992.

[ES97]   Wolfgang Eckstein and Carsten Steger. Architecture for Computer Vision Application Development within the HORUS System. *Journal of Electronic Imaging*, 6(2):244–261, April 1997.

[FHR+90] C. Fennema, A. Hanson, E. Riseman, J. R. Beveridge, and R. Kumar. Model-Directed Mobile Robot Navigation. *IEEE Trans. on Systems, Man, and Cybernetics*, 20(6):1352–1369, November 1990.

[Gri90] W. Eric L. Grimson. *Machine Vision for Three Dimensional Scenes*, chapter Object Recognition by Constrained Search, pages 73–108, 1990.

[HLZ97] A. Hauck, S. Lanser, and C. Zierl. Hierarchical Recognition of Articulated Objects from Single Perspective Views. In *Computer Vision and Pattern Recognition*. IEEE Computer Society Press, 1997. To appear.

[HS96]   A. Hauck and N. O. Stöffler. A Hierarchical World Model with Sensor- and Task-Specific Features. In *International Conference on Intelligent Robots and Systems*, pages 1614–1621, 1996.

[IK88]   K. Ikeuchi and T. Kanade. Automatic Generation of Object Recognition Programs. *IEEE Trans. on Computers*, 76(8):1016–1035, August 1988.

[KH94]   R. Kumar and A. R. Hanson. Robust Methods for Pose Determination. In *NSF/ARPA Workshop on Performance versus Methodology in Computer Vision*, pages 41–57. University of Washington, Seattle, 1994.

[KP95]   A. Kosaka and J. Pan. Purdue Experiments in Model-Based Vision for Hallway Navigation. In *Workshop on Vision for Robots in IROS'95*, pages 87–96. IEEE Press, 1995.

[LL95]   S. Lanser and T. Lengauer. On the Selection of Candidates for Point and Line Correspondences. In *International Symposium on Computer Vision*, pages 157–162. IEEE Computer Society Press, 1995.

[LMZ95] S. Lanser, O. Munkelt, and C. Zierl. Robust Video-based Object Recognition using CAD Models. In IAS-4, pages 529–536. IOS Press, 1995.

[Low91] D. G. Lowe. Fitting Parameterized Three-Dimensional Models to Images. *IEEE Trans. on PAMI*, 13(5):441–450, 1991.

[LT88]   R. Lenz and R. Y. Tsai. Techniques for Calibration of the Scale Factor and Image Center for High Accuracy 3D Machines Metrology. *IEEE Trans. on Pattern Analysis and Machine Intelligence*, 10(5):713–720, 1988.

[LZ95]   S. Lanser and Ch. Zierl. Robuste Kalibrierung von CCD-Sensoren für autonome, mobile Systeme. In *Autonome Mobile Systeme*, Informatik aktuell, pages 172–181. Springer-Verlag, 1995.

[LZ96]   S. Lanser and C. Zierl. On the Use of Topological Constraints within Object Recognition Tasks. In *13th ICPR*, volume 1, pages 580–584, 1996.

[Mun95] O. Munkelt. Aspect-Trees: Generation and Interpretation. *Computer Vision and Image Understanding*, 61(3):365–386, May 1995.

[Pop94] A. R. Pope. Model-Based Object Recognition. Technical Report TR-94-04, University of British Columbia, January 1994.

[Wan92] C. C. Wang. Extrinsic Calibration of a Vision Sensor Mounted on a Robot. *Transactions on Robotics and Automation*, 8(2):161–175, April 1992.

Real–Time Pedestrian Tracking in Natural Scenes

J. Denzler, H. Niemann

Universität Erlangen–Nürnberg
Lehrstuhl für Mustererkennung (Informatik 5)
Martensstr. 3, D–91058 Erlangen

Abstract In computer vision real–time tracking of moving objects in natural scenes has become more and more important. In this paper we describe a complete system for data driven tracking of moving objects. We apply the system to tracking pedestrians in natural scenes. No specialized hardware is used. To achieve the necessary efficiency several principles of active vision, namely selection in space, time, and resolution are implemented. For object tracking, a contour based approach is used which allows contour extraction and tracking within the image frame rate on general purpose architectures. A pan/tilt camera is steered by a camera control module to pursue the moving object. A dedicated attention module is responsible for the robustness of the complete system. The experiments over several hours prove the robustness and accuracy of the whole system. Tracking of pedestrians in a natural scene has been successful in 79% of the time.

1 Introduction

The development of machines which interact with their environment, has become more and more important in the past years. One important aspect of those machines, especially if they are able to move autonomously, is the capability to detect and to track moving objects, even if the system is moving itself, too.

There exist some sophisticated systems for tracking moving objects in real–time. [11] present a stereo camera system, which pursues moving objects in indoor natural scenes. The approach of [2] realizes a contour tracker, which pursues a moving toy car in a laboratory environment with cooperative background. Tracking cars on highways in real–time has been the topic of [10, 9]. An optical flow based realization of real–time tracking can be found in [3].

All these system have in common, that they use specialized hardware like pipelined image processors or transputer networks. Only a few systems are known for real–time tracking on general purpose architectures, for example [1, 6]. In this paper we describe a complete system for data driven detection and tracking of moving objects in real–time on a general purpose processor. Principles of active vision are included in different modules of the system, in order to cope with real–time constraints. No specialized hardware is used. We demonstrate the quality of the system for tracking moving pedestrians in outdoor scenes. Due to the nature of outdoor scenes the system has to cope with changing illuminations, reflections and other moving objects.

The paper is structured as follows. In Sect. 2 we give an overview over the system and describe the two stages in which the system operates. Sect. 3 describes in more detail the tracking algorithm. Sect. 4 demonstrates the quality of the system for natural scenes, even during sunshine, rain and snow fall. The papers concludes in Sect. 5 with a discussion of the approach and an outlook to future work.

2 System Overview

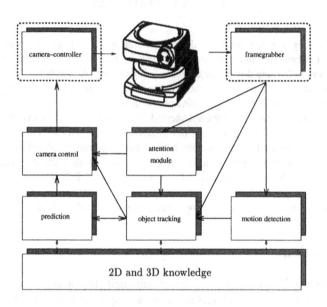

Figure1. Overview of the complete system

In Figure 1 an overview of the complete system for data driven tracking is shown. Two hardware dependent modules can be seen, the frame-grabber and the physically control of the camera. The other modules, the motion detection module, the object tracking module, the prediction module, the camera control module, and the attention module are implemented in C++ and use general purpose processors. In addition, a 2D- and 3D knowledge base is available, where problem dependent knowledge can be introduced into the system. This kind of knowledge is not used for the application presented in this paper.

The system runs in two stages: an *initialization stage*, where the motion detection modules detects motion in the scene, and a *tracking stage*. During the initialization stage, the camera is static to allow for a simple difference image operation for motion detection in the scene. For the 128 × 128 low resolution difference image a binary image is computed. This is smoothed to get a few significant regions in the image, where changes have occurred. The largest of these regions in size is selected and used as initialization for the tracking stage.

During the tracking stage all modules work together and communicate with each other. This is indicated by the connections between the modules. The object tracking module (a detailed description follows in Sect. 3) tracks the contour of the moving object, and uses predictions computed by the prediction module. Only a small window of the camera image containing the moving object is processed. The center of gravity of the object's contour is sent to the camera control module to steer the camera. The motion detection module computes with a low frame rate independently moving objects, using knowledge about the camera motion [8]. The robustness of the whole system is realized by the attention module. This module watches over the system to detect errors during tracking. For this, in certain intervals features are computed for the contour of the moving object (x– and y–moments of the contour). Rapid changes in the moments give hints for errors in the contour extraction.

As soon as an error is detected, the attention module stops tracking, sends a signal to the camera control module to stop camera motion, and then switches back to the initialization stage.

3 Object Tracking

For object tracking we have developed a new algorithm, called *active rays*. This algorithm is motivated by active contours and is based on the principle of contour extraction and tracking. In contrast to [9], who uses also a radial representation of the contour, we define an internal energy similar to the approach of active contours, which couples neighboring contour points.

For active rays, we first have to define a reference point $m = (x_m, y_m)^T$, which has to lie inside the image contour. An active ray $\varrho_m(\phi, \lambda)$ is defined on the image plane (x, y) as a 1D function depending on those gray values $f(x, y)$ of the image, which are on a straight line from the image point m in direction ϕ

$$\varrho_m(\phi, \lambda) = f(x_m + \lambda \cos(\phi), y_m + \lambda \sin(\phi)), \tag{1}$$

with $0 \le \lambda \le n_\phi$, where n_ϕ is given by the diagonal of the image. Now, by assuming,

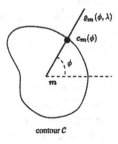

Figure 2. Representation of a contour point by active rays

that each ray only hits the object contour once, we can identify a point of the contour by the parameter $\lambda^*(\phi) \geq 0$

$$\lambda^*(\phi) = \underset{\lambda}{\mathrm{argmin}}\left(-|\nabla f(x_m + \lambda\cos(\phi), y_m + \lambda\sin(\phi))|^2\right)$$

$$= \underset{\lambda}{\mathrm{argmin}}\left(-\left|\frac{\partial}{\partial\lambda}\varrho_m(\phi,\lambda)\right|^2\right), \tag{2}$$

with $0 \leq \phi < 2\pi$. If the ray hits the contour many times (for example for concave contours) we have to introduce a set of values $\lambda^*(\phi)$. This is not important for the application described here. The representation by active rays is illustrated in Figure 2. The step, which lead to (2), is motivated by the assumption, that an edge in 2D can also be found by a gradient search in the corresponding 1D signal. Of course, edges which are in the direction ϕ from the reference point cannot be found on the ray $\varrho_m(\phi, \lambda)$. The experiments will show, that this case is not relevant in practice. Having the optimal value for $\lambda^*(\phi)$ the contour point $c_m(\phi)$ in the image plane can easily be computed by

$$c_m(\phi) = (x_m + \lambda^*(\phi)\cos(\phi), y_m + \lambda^*(\phi)\sin(\phi)), \tag{3}$$

with $0 \leq \phi < 2\pi$. The most important aspect of this approach, especially for real–time applications, is the reduction of the contour point search from the 2D image plane to a 1D signal. This reduces the computation time which will be shown in the experimental part of this paper.

In Figure 3, left the extracted contour of an object and in Figure 3, right the function $\lambda^*(\phi)$ are shown. One can observe, that the function $\lambda^*(\phi)$ is smooth for the angles which corresponds to the correctly extracted contour $(0 - 4/3\pi)$. Then, an error can be seen, both in the extracted contour and in the function $\lambda^*(\phi)$. For $\phi \in [4/3\pi, 3/2\pi[$ the function is not smooth, because a wrong contour has been extracted. This is no surprise. Looking at equation (2) one can see, that up to now, the contour points are calculated without taking into account neighboring contour elements. Thus, we need to introduce some linkage between neighboring contour points to take into consideration that normally contours are coherent in space, i.e. that contours are smooth. A usual approach to connect neighboring contour points together is to introduce an internal energy similar to the active contour approach.

An internal energy which handles the above mentioned demands is

$$E_i(c_m(\phi)) = \frac{\alpha(\phi)|\frac{d}{d\phi}\lambda(\phi)|^2 + \beta(\phi)|\frac{d^2}{d\phi^2}\lambda(\phi)|^2}{2}. \tag{4}$$

This energy also depends only on a 1D function, compared to active contours [7], where the corresponding energy depends on a 2D one. This energy can be formally derived from the internal energy definition of active contours. This is beyond the scope of this paper and is published elsewhere [5]. Now we have an energy, which describes contour point candidates for each ray and an energy, which connects the rays to get a smooth contour. Similar to active contours we define a total energy E

$$E = \int_0^{2\pi}\left[\frac{\alpha(\phi)|\frac{d}{d\phi}\lambda(\phi)|^2 + \beta(\phi)|\frac{d^2}{d\phi^2}\lambda(\phi)|^2}{2} - \left|\frac{d}{d\lambda}\varrho_m(\phi,\lambda)\right|^2\right]d\phi. \tag{5}$$

 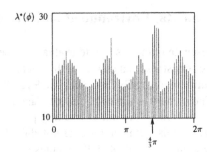

Figure3. Left: 2D contour extracted by active rays. Right: 1D function λ^* of the corresponding 2D contour shown on the left.

The contour extraction can then be described as an energy minimization problem. Using the variational approach the Euler–Lagrange differential equation

$$\alpha(\phi)\frac{d^2}{d\phi^2}\lambda(\phi) - \beta(\phi)\frac{d^4}{d\phi^4}\lambda(\phi) + \frac{d}{d\lambda}\left|\frac{d}{d\lambda}\varrho_m(\phi,\lambda)\right|^2 = 0$$

must be solved. Again, this differential equation depends on a 1D function, in contrast to the same differential equation for active contours.

Some remarks must be done regarding the reference point m. As already mentioned, this point must lie inside the object contour, but the position may be arbitrary. For a prediction step, a unique position would be of great advantage. Thus, we always choose the center of gravity of the contour

$$m = 1/2\pi \int_0^{2\pi} c_m(\phi)\,d\phi \qquad (6)$$

for the reference point. If this equation does not hold for an actual reference point and an extracted contour, we update the reference point using equation (6), and restart the contour extraction with the updated reference point's position.

But how can we compute the reference point for the first image, if we do not know the contour of the object? This is no problem, because we apply this algorithm to object tracking. Assuming a static camera during motion detection, we always have a coarse idea, where a moving object is located in the image. The approach we have chosen is a difference image algorithm (see the motion detection module in the previous Sect.), which computes areas in the image, where changes occur. The center of such an area is taken a initial reference point m for the first image.

Up to now we have shortly summarized the theory of active rays for the continuous case. By applying active rays for contour extraction in images we have to go to the discrete case. Two approaches for the discretization are possible: fixed sampling rate $\Delta\phi$ for the angle ϕ or the so called any–time behavior. Due to lack of space the reader is referred to [4]

4   Experimental Environment and Results

We have chosen an outdoor scene to evaluate the performance of the complete system as well as the robustness and quality of the tracking algorithm by active rays.

The system runs on a SGI Onyx with two R10000 processors without any specialized hardware. The camera is a pan/tilt device Canon VCC1 which is connected with the workstation via RS232. A similar tracking performance can be achieved with a HP 735/99 MHz, but this architectures does not allow for full frame rate image grabbing.

We have manually evaluated 40 minutes of tracking. Every 20th frame has been written to disk after the experiments, i.e. 3000 images have been evaluated. We judged visually, whether or not the active ray has correctly extracted the moving object. There have been 73 automatic initializations in which 34 initializations fail. The reason is, that the center of gravity of the binary motion region not always lie inside the moving object. For example, for two objects moving close together, the binary region covers both object and the center of gravity lies between the objects. This has been expected, because we apply a very fast and simple algorithm for motion detection.

After correct initialization (39 times), we could correctly track 20 moving persons for 8680 images as long as they have been visible (average number of images: 434, i.e. 17 seconds). Caused by an error in the tracking algorithm, in 19 cases the object was lost after an average number of 265 images, i.e. after 10 seconds which results in another 5040 correctly tracked images. Most of the errors occur near the end of the place, because there are a lot of strong background edges (the trees).

After loosing the object, the attention module needs an average number of 91 images (i.e. 3.6 sec), for detecting the loss of the object; in 26 of the 39 cases the detection time was below one second. Hence, in 3560 images the system has not tracked a moving person, although it has been in the tracking stage. Thus, we get a total success rate for tracking of 79 %.

In Figure 4 some typical result can be seen. In the last row, the moving person moves outside the field of view which the camera can cover by pan/tilt movements. In Figure 5 some additional results can be seen. In the first row the effect of the data driven tracking is illustrated. The system tracks the moving contour of the small snow-plow, because it cannot distinguish between contours of moving pedestrians and some other contours. For this, knowledge is necessary which can be added into the 2D and 3D knowledge base (see Sect. 2). Tracking one person out of the crowd also works, if the initialization is correct (Figure 5, second row).

5   Conclusion and Future Work

In this paper we have presented a complete system for data driven object tracking. Several principles of active vision are implemented, which allows — in contrast to [9] — for real–time tracking without specialized hardware. The important aspect of this system are the tracking module, which uses a new algorithm for contour extraction and tracking working within the image frame rate, and the attention module, which watches over the complete system to detect errors. Thus, a very robust tracking over a long time can be performed. The system satisfies two key components, stated by [11], namely

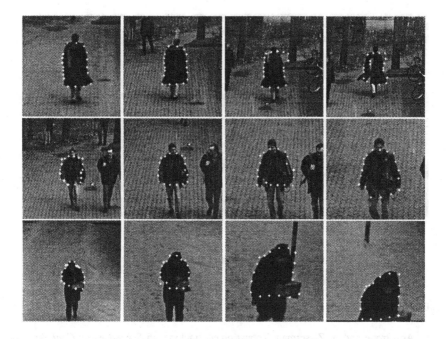

Figure4. Results for tracking moving pedestrians

- the continuous operation over time and real–time response to different types of events and
- an open and expandable design due the object–oriented implementation.

The emphasis on real–time performance on general purpose hardware has led us to consider simple algorithms, especially for the motion detection and motion tracking module. However, we argue, that the combination of simple methods and an attention module, which detects errors, can lead to a robust performance. Our experiments on natural scenes support this claim.

The system can be applied to gate control problems. The tracking algorithm is robust against changes in the camera parameters, especially against zooming during tracking. This means, that during the tracking a synchronous zooming toward the moving object can be done, for example to extract features of the face to identify the person. This will be our short time goal. In addition, we are working on integrating task specific knowledge into the knowledge base of the system, in order to increase the contour extraction and tracking result. One idea is to limitate the normalized shape of the contour, i.e. the function $\lambda^*(\phi)$, to certain values, which corresponds to typical contours of moving pedestrians.

Figure5. Results for tracking moving pedestrians

References

1. M. Armstrong and A. Zisserman. Robust object tracking. In *Second Asian Conference on Computer Vision*, pages I/58–I/62, Singapore, 1995.
2. R. Curwen and A. Blake. Dynamic contours: Real–time active splines. In A. Blake and A. Yuille, editors, *Active Vision*, pages 39–58. MIT Press, Cambridge, Massachusetts, London, England, 1992.
3. K. Daniilidis, M. Hansen, C. Krauss, and G. Sommer. Auf dem Weg zum künstlichen aktiven Sehen: Modellfreie Bewegungsverfolgung durch Kameranachführung. In *DAGM 1995, Bielefeld*, pages 277–284, 1995.
4. J. Denzler. *Active Vision for Real–Time Object Tracking.* Dissertation, Technische Fakultät, Universität Erlangen–Nürnberg, Erlangen, in preparation, 1997.
5. J. Denzler and H. Niemann. Echtzeitobjektverfolgung mit aktiven Strahlen. In *Mustererkennung, 1996*, pages 84–91, Heidelberg, 1996.
6. G.D. Hager and K. Toyama. X vision: Combining image warping and geometric constraints for fast visual tracking. In A. Blake, editor, *Computer Vision - ECCV 96*, pages 507–517, Berlin, Heidelberg, New York, London, 1996. Lecture Notes in Computer Science.
7. M. Kass, A. Wittkin, and D. Terzopoulos. Snakes: Active contour models. *International Journal of Computer Vision*, 2(3):321–331, 1988.
8. D. Murray and A. Basu. Motion tracking with an active camera. *IEEE Transactions on Pattern Analysis and Machine Intelligence*, 16(5):449–459, 1994.
9. S.M. Smith and J.M. Brady. Asset-2: Real–time motion segmentation and shape tracking. *IEEE Transactions on Pattern Analysis and Machine Intelligence*, 17(8):814–820, 1995.
10. F. Thomanek and E.D. Dickmanns. Autonomous road vehicle guidance in normal traffic. In *Second Asian Conference on Computer Vision*, pages III/11–III/15, Singapore, 1995.
11. T. Uhlin, P. Nordlund, A. Maki, and J.O. Eklundh. Towards an active visual observer. In *International Conference on Computer Vision*, pages 679–686, Cambridge, Massachusetts, 1995.

Non-rigid Object Recognition Using Principal Component Analysis and Geometric Hashing

Kridanto Surendro and Yuichiro Anzai

Department of Computer Science, Keio University
3-14-1 Hiyoshi, Kohoku-ku, Yokohama 223, Japan.

Abstract. A novel approach is proposed to recognize non-rigid 3D objects from 2D images using principal component analysis and geometric hashing. For all of the models that we want to be able to recognize, we calculate the statistic of point features using principal component analysis and then, calculate the invariants of them. In recognition stage, we calculate the needed invariants from an unknown image and used as indexing keys to retrieve from the model base the possible matches with the model features. We hypothesize the existence of an instance of the model if a model's features scores enough hits on the vote count.

1 Introduction

An object recognition system finds objects in the real world from an image of the world, using object models which are known a priori. This task is surprisingly difficult. Humans perform object recognition effortlessly and instantaneously, but algorithmic description of this task for implementation on machines has been very difficult.

This paper describes an algorithm of non-rigid 3D objects recognition from 2D images using the combination of principal component analysis and geometric hashing. Principal component analysis is used to model shape of the non-rigid objects while geometric hashing is used to derive a hypothesis that an instance of object models occur in the scene.

The organization of this paper is as follows. Section 2 and section 3 discusses the principal component analysis and geometric hashing methods respectively. In section 4 we describe our proposed algorithm for non-rigid objects recognition. Section 5 shows experimental results that prove the applicability of our method and in section 6 we draw conclusion, discuss the problem, and discuss our future research interests.

2 Principal Component Analysis

Principal component analysis has been used as a method for building flexible model of the shape of non-rigid objects [1, 7]. The method offers $O[n]$ computational complexity and has been applied [8] to a wide variety of examples in medical images and industrial components.

The method works by examining statistics of the co-ordinates of the label-led point features over the training shapes. In order to be able to compare equivalents points from different shapes, the set of training shapes are normalized using the algorithm [7] as follows.

1. Scale, rotate, and translate each of the shapes in the set to align to the first shape.
2. Calculate the mean of the transformed shapes.
3. Scale, rotate, and translate the mean to align to the first shape.
4. Scale, rotate, and translate each of the shapes again to match to the adjusted mean.
5. Repeat no.2-4, until convergence.

Once a set of aligned shapes is available the mean shape, $\bar{x}$, is calculated using

$$\bar{x} = \frac{1}{N} \sum_{i=1}^{N} x_i, \tag{1}$$

where

$$x_i = (\ x_{i1},\ y_{i1},\ x_{i2},\ y_{i2},\ ...,\ x_{in},\ y_{in}\ )^T$$

is a vector describing the n points of the ith shape in the set, and N is number of shape in the training set.

The modes of variation can be found by applying principal component analysis to the deviations from the mean. For each shape the vector dx_i is calculated, where

$$dx_i = x_i - \bar{x}. \tag{2}$$

The 2n x 2n covariance matrix S is then calculated using

$$S = \frac{1}{N} \sum_{i=1}^{N} dx_i dx_i^T. \tag{3}$$

The eigenvectors of the covariance matrix correspond to modes of variation of the training data, such that

$$S\ p_k = \lambda_k\ p_k \tag{4}$$

where p_k is the eigenvectors k = 1 ... 2n, and λ_k is the eigenvalue of S, $\lambda_k \geq \lambda_{k+1}$. And,

$$p_k^T\ p_k = 1. \tag{5}$$

Moreover, the eigenvector of the covariance matrix corresponding to the largest eigenvalue describe the most significant mode of variation.

Any shape in the training set can be approximated using the mean shape and a weighted sum of these deviations obtained from the first t modes

$$x = \bar{x} + P\ b \tag{6}$$

where $\mathbf{P} = (\mathbf{p}_1 \ \mathbf{p}_2 \ ... \ \mathbf{p}_t)$ is the matrix of the first t eigenvectors, and $\mathbf{b} = (b_1 \ b_2 \ ... \ b_t)^T$ is a vector of weights.

The eigenvectors are orthogonal, $\mathbf{P}^T\mathbf{P} = I$, therefore

$$\mathbf{b} = \mathbf{P}^T (\mathbf{x} - \bar{\mathbf{x}}). \tag{7}$$

3 Geometric Hashing

Geometric hashing [11, 12] is intended for 2-D and 3-D object recognition in cluttered scenes under various viewing assumptions, and for different sensor types. Objects are modeled as sets of local features, some of which might be occluded in the scene image. Recognition is defined as a mapping between a subset of scene features and a subset of model features for some model in the database. These mappings are constrained to belong to a predefined set of geometric transformations T. The precise class of possible transformations depends on knowledge and assumptions about the sensor and object types, the camera and object configuration, and possible geometric transformations between the two [10].

Geometric hashing performs recognition with a set of point features. Geometric constraints between these local features in the models are encoded in a lookup table that done in a preprocessing stage. During this stage, that usually performed off-line, the locations of the table are filled with entries containing references to object modes and some additional parameters. This stage can thus afford to be computationally expensive.

During recognition, the models listed in the indexed entries are collected into a list of candidate models. The most likely of the candidate models are selected on the basis of the number of votes they have received. Since the indexing result only established a weak link between the unknown image and a model, verification is needed to determine the quantitative relationship between the two. Verification can be calculated by solving the matching problem that is establishing a correspondence between the image and model features.

4 The Algorithm

In this section we will present the algorithm of our non-rigid object recognition system. The algorithm can be separated in preprocessing stage and recognition stage. During the preprocessing stage, that can be considered "off-line", the model base is constructed using the invariant of point features of the mean shapes resulted from principal component analysis method. In the recognition stage, the algorithm uses the hash table data structure that was prepared during the preprocessing stage and try to match with the invariants of an unknown image.

Globally, the preprocessing stage can be described, as follow.

1. **Shape representation.** We do some image preprocessing to extract the region of the object models from the background, and then traced the regions to produce a chain of boundary points. The list of boundary points

is interpolated using uniform cubic B-spline [3] to get the unoccluded representation of image shapes. Point features are then located uniformly over the region of shape using discrete mapping method [1].

2. **Component analysis.** Point features of each shape are aligned and normalized in the same way so that they correspond as closely as possible. In our approach, shapes are aligned and normalized again the first shape of training set. The normalized point features are then analyzed using principal component analysis. The results are vector of the mean shape and matrix of the eigenvectors that can be used to generate the new shapes of the model.

3. **Setting up model base.** Invariants of the point features are computed using the mean shape of object models, this mean that only one shape mode is involved. Given a set of n point features belonging to one of the mean shape object models, M_i, and let p_1 and p_2 be an ordered pair of points from that set. The orthonormal basis of the system O_{xy} defined by the basis pair will be $p_x = p_2 - p_1$ and $p_y = Rot(p_x)$ respectively. Any point p in the plane can be represented in this basis, namely, there is a unique pair of scalars (u,v) such that

$$p - p_0 = u p_x + v p_y, \qquad (8)$$

where $p_0 = \frac{p_1 + p_2}{2}$ is the midpoint between p_1 and p_2. In other words, the center of the coordinate system O_{xy} coincides with p_0.

The parameters u and v are the coordinates of the point p in the coordinate system defined by the basis pair. If transformations T applied to all of the n points, then with respect to the new basis (Tp_x, Tp_y), the transformed coordinates of Tp will again be (u,v). That is, (u,v) is invariant with respect to transformation T. Details of these calculation described in [4].

Once these invariants calculated we label them with model, M_i, from which they originated, and stored this model-basis combination in hash table.

If rigid object models are considered to be involved and will be added to the model base, then for each of them we compute shape representation, calculate invariants, and set up model base.

During the recognition, given a scene represented by n points do the following:

1. Choose an arbitrary ordered pair, (p_1, p_2), of point features in the scene, and compute the coordinates of the remaining point features in the coordinate system O_{xy} that the selected basis defines.

2. For each such coordinates and for some neighborhood of them check the appropriate hash table bin, and cast a vote for the model and the basis.

3. Rank the pairs of the model and the basis based on the number of times they had been retrieved from the hash-table.

4. For all those passing a threshold verify the match between the model points mapped onto scene coordinates and the scene points, using all available point.

5. If not all scene point have been identified jump to step (1) and repeat the procedure using a different image basis pair.

5 Experimental Results

In this section we present the results of experiments we conducted on our method. We built a model base consists of 15 object models where 12 of them are assumed to be rigid objects and three of them are non-rigid objects. Fig. 1 shows some of the rigid objects used in our experiments and Fig. 2 shows the examples of non-rigid objects.

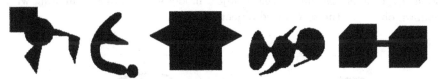

Fig. 1. Images of some rigid object models used in experiments.

Images of the kangaroos and elephants on Fig.2 are generated manually while the robot toys are image photographs taken by a camera. We do some image preprocessing to extract objects from the background.

Fig. 2. Subset of non-rigid object models used in experiments.

For each object models we choose 20 point features and calculate the invariants of them to build the model base. In our method the point features are invariants under similarity transformations.

5.1 The Recognition

In experiments using elephant shape as an object to be recognized, the images used for the experiments are obtained from drawings similar with images on the training shapes. The images are traced, linked, and interpolated to get the boundary region of shapes. The point features are located uniformly as described in the previous section.

Fig. 3 left shows image of the 'unknown' object that will be recognized, and Fig. 3 right shows the result overlap with the mean shape. Number of point features used to represent the unknown object are 30 points, and we try to match them with invariants of the point features in the hash table. During the recognition stage we access the hash table at the appropriate bin and some neighbor bins surrounding it. We select three object models that received higher vote as hypothesis, and finally select object model with the best matching with unknown object as the recognized object.

Fig. 3. Recognizing the elephant.

Fig. 4 shows the example of rigid object recognition. Here, unknown object is represent using 30 point features.

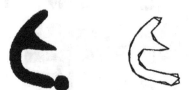

Fig. 4. Recognizing rigid object.

It is known that our model matching strategy uses only local features, therefore it should be able to recognize objects that are partially occluded. Experiment with partially occluded object is shown in Fig. 5, where two objects, the kangaroo and elephant, are appeared on the same image. In this experiment number of point features used is 60 points because when we used point features less than 60, we got a wrong result.

After the voting process, we found that both of the kangaroo and elephant received higher votes. We could only recognize one of the objects because we always select the best matching object as the recognized object.

Fig. 5. Recognizing object that partially occluded.

6 Discussion and future research

This paper represents our preliminary ideas and experiments in non-rigid object recognition. We based our method on the combination of principal component analysis and geometric hashing. By using this method we can model shape of the object from a set of learning shapes, store rigid and non-rigid objects model on the same model base, hypothesize the occurrence of object in the scene, and recognize the unknown object by matching it with the objects model.

Experiments with rigid and non-rigid objects shown that our proposed method can be used to recognize them. While experiment with objects that partially occluded, our method could recognize one of the objects because we only select the best matching between the hypothesis and the object models. Our method could recognized the partially occluded objects because we increased number of the point features of the unknown objects, therefore there are possibilities to match some points from partially occluded objects with invariants points of the models.

In our method, point features are uniformly located over the boundary of the object shape, and invariants of them are derived using the mean shape of non-rigid objects. In the recognition stage, point features of the unknown object also located uniformly therefore we must search over the neighbors of the point features to match with invariants of the object models in model base. In order to reduce the search, we are of a certain opinion that improving the index method could overcome the problem.

Based on those problem, our future research will be focused on the efforts of how to improve the indexing method and how to exploit the existence of matrix eigenvector.

References

1. A.M. Baumberg and D.C. Hogg, *Learning Flexible Models from Image Sequences*, Report 93.36, Division of A.I., School of Computer Studies, University of Leeds.
2. B. Lamiroy and P. Gros, "Rapid Object Indexing and Recognition using Enhanced Geometric Hashing," *Lectures Notes in Computer Science 1064*, (B. Buxton and R. Cipolla, Ed.), pp. 59-70, Springer-Verlag, Berlin, 1996.

3. D.F. Rogers, and J.A. Adams, *Mathematical Elements for Computer Graphics,* 2nd ed. McGraw-Hill International Editions,1990.

4. I. Rigoutsos, *Massively Parallel Bayesian Object Recognition,* Ph.D thesis, New York University, 1992.

5. P. Havaldar, G. Medioni, and F. Stein, "Extraction of Groups for Recognition," in *Lectures Notes in Computer Science 800,* (Jan-Olof Eklundh, Ed.), pp. 251-261, Springer-Verlag, Berlin, 1994.

6. R. Chin and C.R. Dryer, "Model-Based Recognition in Robot Vision," *Computing Surveys,* Vol. 18, no.1.

7. T.F Cootes, C.J. Taylor, D.H. Cooper, and J. Graham, "Training Models of Shape from Sets of Examples," In *Proc. British Machine Vision Conference,* pp. 9-18, Springer-Verlag, 199 2.

8. T.F Cootes, A. Hill, C.J. Taylor, and J. Haslam, "The Use of Active Shape Models for Locating Structures in Medical Images," *Image and Vision Computing,* Vol. 12, No. 6, pp. 355-366, 1994.

9. W.E.L. Grimson, *Object Recognition by Computer - The Role of Geometric Constraints,* The MIT Press, Cambridge, MA, USA, 1990.

10. Y.C. Hecker, and R.M. Bolle, "On Geometric Hashing and the Generalized Hough Transform," *IEEE Transactions on System, Man, and Cybernetics,* Vol.24, No.9, pp.1328-1338, 1994.

11. Y. Lamdan, J.T. Schwartz, and H.J. Wolfson, "Affine Invariant Model-based Object Recognition," *IEEE Transactions on Robotics and Automation,* Vol. 6, No.5, pp. 578-589, 1990.

12. Y. Lamdan, and H.J. Wolfson, Geometric hashing: A general and efficient model-based recognition scheme. In *Proceedings of the Second International Conference on Computer Vision,* pages 238-249, 1988. Tarpon Springs (Fl.)

Object Identification with Surface Signatures

Adnan A. Y. Mustafa
Dept. of Mechanical and Industrial Engineering
Kuwait University
PO Box 5969 - Safat -13060 - State of KUWAIT
symymus@kuc01.kuniv.edu.kw

Linda G. Shapiro
Dept. of Computer Science and Engineering
University of Washington
Seattle, WA 98195, USA
Shapiro@cs.washington.edu

Mark A. Ganter
Dept. of Mechanical Engineering
University of Washington
Seattle, WA 98195, USA
ganter@cs.washington.edu

Abstract. In this paper we describe a model-based object identification system. Given a set of 3D objects and a scene containing one or more of these objects, the system identifies which objects appear in the scene by matching surface signatures. Surface signatures are statistical features which are uniform for a uniform surface. Two types of surfaces are employed; curvature signatures and spectral signatures. Furthermore, the system employs an inexpensive acquisition setup consisting of a single CCD camera and two light sources. The system has been tested on 95 observed-surfaces and 77 objects of varying degrees of curvature and color with good results.

1. Introduction

In this paper we describe an inexpensive system that employs a single CCD camera and two light sources for identifying 3D objects appearing in cluttered scenes. The system input consists of two color images of the scene illuminated by two different white light sources. Color photometric stereo is employed to produce a surface normal map of the scene which is then integrated with the spectral data (i.e. color data) to produce a curvature-spectral description of object surfaces. Finally, the system matches the extracted surface description to model data accessed from the model database (constructed through training) to produce hypotheses about the objects appearing in the scene. The main features employed for matching are surface characteristics that are invariant to change in pose, are unaffected by shadows from adjacent or occluding objects, and are not sensitive to changes in lighting; we call these features *surface signatures*.

2. Related Literature

Many matching techniques have been introduced in the literature, such as template matching (e.g. [1]) and feature point matching (e.g. [2]). Template matching has many limitations as it is sensitive to rotation and scaling. Feature point matching techniques which match *interesting* object points to their model counterparts are more robust. These features may either be global or local. A major problem with employing global features is that matching fails if the matching data is incomplete as may occur as a result of occlusion or shadowing effects. However, by employing local features the effects of missing data can be reduced.

Surface curvature is an intrinsic feature that can play an important role in matching. Many strategies for classifying surfaces have been proposed; Besl and Jain [3] employed the mean and Gaussian curvature to classify surfaces, Haralick et al. [4] employed topographic labels based on gradients, Hessians and directional derivatives, and Newman et al. [5] identified quadric surfaces using surface position and surface normals. Surface color is another feature that can be employed for matching. However, most of the research dealing with color has concentrated on segmentation (e.g. [6] [7]) and very little has been applied to matching and recognition. More recently the advantages of exploiting color data for recognition has become apparent [8] [9] [10]. Finlayson et. al [11] presented a color angular indexing approach which uses distributions of three color angles and three color-edge angles for recognition. Higuchi et. al [12] modified the extended Gaussian image (*EGI*) to represent not only object curvature but also object color. Grewe and Kak [13] employed shape and color for object recognition.

3. Surface Signatures

Surface signatures are employed to characterize surfaces. A surface signature (S) is a feature vector that reflects the probability of occurrence of a feature for a given surface. The motivation for using signatures is that they are invariant to change in pose and are not affected by partial occlusion or shadows. This invariance stems from the fact that signatures are statistical features that are uniform for a uniform surface.

In our work we employ two types of surface signatures: curvature signatures that characterize curvature characteristics and spectral signatures that characterize spectral characteristics (i.e. color). The curvature signatures consist of three signatures: the *angle curvature signature* (S_{AC}), the *Gaussian curvature signature* (S_{GC}) and the *mean curvature signature* (S_{MC}), which characterize the surface distributions of the crease angle, the Gaussian curvature and the average mean curvature, respectively. The crease angle between two adjacent surface normals is defined as the angle between the two normals. Similarly, the spectral signatures consist of three signatures: the *red spectral signature* (S_{RS}), the *green spectral signature* (S_{GS}) and the *blue spectral signature* (S_{BS}), which characterize the surface distributions of the red color band, the green color band, and the blue color band, respectively. Figure 1 shows surface signatures for two different surfaces.

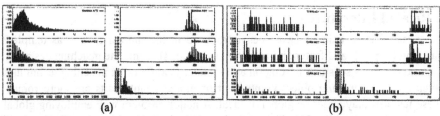

(a) (b)

Fig. 1. Surface signatures (a) *Banana* signatures: Curvature signatures (right): from top to bottom; S_{AC}, S_{MC} and S_{GC}. Spectral signatures (left): from top to bottom; S_{RS}, S_{GS} and S_{BS}. (b) *Corn* signatures. Curvature signatures (right): from top to bottom; S_{AC}, S_{MC} and S_{GC}. Spectral signatures (left): from top to bottom; S_{RS}, S_{GS} and S_{BS}.

4. Signature Construction

A description of the surface signature construction procedure was presented in [14]. A brief description is presented here.

4.1 Constructing Curvature Signatures

The construction of curvature signatures is based on generating surface normals for the objects appearing in the scene. This is accomplished by employing color photometric stereo to produce a surface normal map of the scene. The normal map is then segmented into individual surfaces to produce a segmented image which identifies the entities to be taken as surfaces. The segmented image is produced by identifying regions where large crease angles exist as this indicates large change in curvature and possible surface boundaries. Curvature signatures are then constructed for each surface by calculating the crease angle, the Gaussian curvature and the mean curvature (at each surface normal) and constructing their respective curvature signatures; S_{AC}, S_{GC} and S_{MC}.

4.2 Constructing Spectral Signatures

The spectral signatures are constructed by processing the image scenes to generate a refined spectral image (specularities removed). The spectral image is masked with the segmented image to identify independent spectral surfaces. Spectral histograms for the red spectral band, the green spectral band and the blue spectral bands are then constructed to generate their respective surface spectral signature; S_{RS}, S_{GS} and S_{BS}.

5. Signature Matching

Matching surfaces between surfaces extracted from image scenes to model surfaces is accomplished by comparing their signatures using four error metrics. These metrics compare the surface signature profiles based on distance, variance, spread and correlation [15]. These four error metrics are the combined to form the *signature match error* (*E*), that gives an indication of the measure of similarity between two signatures. The surface matching process constitutes of two processes; matching surface curvature and matching surface spectral attributes. Both processes employ the signature match error at their core. Finally, the results of the curvature match and spectral match are combined to produce the surface error.

5.1 Surface Curvature Matching

Differences between two surfaces based on curvature attributes is measured by matching their three curvature signatures (S_{AC}, S_{GC} and S_{MC}) which measures the differences (i.e. errors) based on the crease angle, the Gaussian curvature and the mean curvature. These errors are combined to produce the *surface curvature match error* (*E_C*),

$$E_C(s_1, s_2) = \mathbf{E}_C \mathbf{W}_C^T \tag{1}$$

where $\mathbf{E}_C = (E_{AC} \quad E_{GC} \quad E_{MC})$, $\mathbf{W}_C = (w_{AC} \quad w_{GC} \quad w_{MC})$ and s_i denotes surface *i*. E_{AC}, E_{GC} and E_{MC} are the angle, the Gaussian and the mean signature match errors, respectively. w_{AC}, w_{GC} and w_{MC} are error weights for the angle, the Gaussian and the mean curvature signature errors, respectively. Small error values of E_C indicate strong similarity between surfaces based on curvature attributes.

5.2 Surface Spectral Matching

Differences between two surfaces based on spectral attributes is measured by matching their three spectral signatures (S_{RS}, S_{GS}, and S_{BS}) which measures the errors based on the surface spectral content. These errors are combined to produce the *surface spectral match error* (E_S),

$$E_S(s_1, s_2) = \mathbf{E}_S \mathbf{W}_S^T \tag{2}$$

where $\mathbf{E}_S = (E_{RS} \quad E_{GS} \quad E_{BS} \quad E_{rS})$ and $\mathbf{W}_S = (w_{RS} \quad w_{GS} \quad w_{BS} \quad w_{rS})$. E_{RS}, E_{GS} and E_{BS} are the red, green and blue spectral signature match errors, respectively. E_{rS} is the *spectral ratio error* of the spectral signatures (measures the hue of the object). w_{RS}, w_{GS} and w_{BS} are the red, green and blue spectral signature weights, respectively, which sum to unity. w_{rS} is the weight of the spectral ratio error. Small error values of E_S indicate strong similarity between surfaces based on spectral attributes.

5.3 Surface Matching

To measure differences between surfaces based on both curvature and spectral attributes, both the curvature signature match error and the spectral signature match error are combined to produce the *surface match error* ($\overline{E}$),

$$\overline{E}(s_1, s_2) = \mathbf{E}\mathbf{W}^T \tag{3}$$

where $\mathbf{E} = (E_C \quad E_S)$ and $\mathbf{W} = (w_C \quad w_S)$. w_C and w_S are the surface curvature and surface spectral weight factors, respectively. Small error values of $\overline{E}$ indicate strong similarity between surfaces based on their surface curvature and spectral attributes.

5.4 Error Weight Selection

Specifying values for the weights w_{AC}, w_{GC} and w_{MC} depend on the object set being analyzed. For example, if the object set consists of flat surfaces and objects with cylindrical surfaces of different sizes, then the Gaussian curvature signature weight (w_{GC}) should be assigned a value much smaller than the other curvature weights since all objects will have Gaussian curvature signature distributions close to zero. Similarly, if the object set consists of surfaces with colors that have, for example, red and green components only, then w_{BS} can be assigned a value of zero without any loss of accuracy. Specifying values for the surface weights w_C and w_S depends, once again, on the object set being analyzed. If the object set consists of objects that are more easily distinguishable based on object color than object curvature and shape then $w_C < w_S$. In [15] we present a method to determine the optimum weight values for a given object set.

6. Performance Metrics

Several performance metrics are employed to evaluate the matching performance.

6.1 The Relative Performance Index (*RPI*)

The *relative performance index* (*RPI*) of a model-surface to an observed-surface is the relative surface match error of the model-surface to the observed-surface with respect to all model-surface errors in the model set,

$$RPI(s_1, s_2) = \frac{\overline{E}(s_1, s_2) - \overline{E}_{\min}(s_1)}{\overline{E}_{\max}(s_1) - \overline{E}_{\min}(s_1)} \tag{4}$$

where $\overline{E}(s_1, s_2)$ is the surface match error between s_1 and s_2, $\overline{E}_{\min}(s_1)$ and $\overline{E}_{\max}(s_1)$ are the minimum and the maximum surface match errors for s_1 in the model set, respectively. Hence, the *RPI* of a model-surface indicates how well a model-surface matched to the observed-surface with respect to other model-surfaces in the set.

6.2 The *Model Hypothesis Error* (Θ)

The *model hypothesis error* (Θ) evaluates incorrect object hypotheses by measuring how well the correct model matches to an object appearing in a scene,

$$\Theta(m) = \frac{1}{n} \sum_{i=1}^{n} RPI(\Psi^{-1}(s_i), s_i) \tag{5}$$

where s_i is the *i*th surface of model m in the scene, $\Psi^{-1}(s)$ denotes the correct observed-surface match to the model-surface s and n is the number of visible model-surfaces of model m in the scene.

6.3 The *Surface Error Performance Index* (*SEPI*)

The *surface error performance index* (*SEPI*) of a scene is a measure of how well all observed-surfaces match to their correct model-surfaces for a given scene,

$$SEPI = \frac{1}{n} \sum_{i=1}^{n} \overline{E}(s_i, \Psi(s_i)) \tag{6}$$

where $\Psi(s)$ denotes the correct surface match to observed-surface s and n is the number of observed-surfaces in the scene.

6.4 The *Surface Matching Performance Index* (*SMPI*)

The *surface matching performance index* (*SMPI*) of a scene is a measure of how well all observed-surfaces match to their correct model-surfaces, with respect to all model-surfaces in the set,

$$SMPI = (1 - \frac{1}{n} \sum_{i=1}^{n} RPI(s_i, \Psi(s_i))) \cdot 100\% \tag{7}$$

where $\Psi(s)$ is the correct model-surface match to observed-surface s, and n is the number of observed-surfaces in the scene. A high *SMPI* value indicates that most of the observed-surfaces match very well to their correct model-surfaces, with respect to all model surfaces in the set.

6.5 The *Model Hypothesis Performance Index* (*MHPI*)

The *model hypothesis performance index* (*MHPI*) of a scene is a measure of how well correct models are identified for a given scene,

$$MHPI = (1 - \frac{1}{N} \sum_{m=1}^{N} \Theta(\Omega(m))) \cdot 100\% \tag{8}$$

where $\Omega(m)$ is the correct *model-match* to object m and N is the number of objects in the scene. A high *MHPI* value indicates that most objects are correctly identified for the scene.

7. Results and Discussion

We have constructed an object recognition system based on matching surface signatures [16] [17]. The system takes two color images of the scene (taken from two different viewpoints), processes the scene data to produce surface curvature signatures and surface color signatures, matches the signatures to model data and finally produces a hypothesis about the objects appearing in the scene. There is no restriction on the orientation, location or pose of objects in the scenes and occlusion and shadows are permitted. We present below our results using two different sets.

7.1 The Toy set

The first set, the *Toy* set, consists of seven real 3D geometric objects: a *blue cylinder*, a *cyan cylinder*, two *white elliptical cylinders* (slightly different in size), a *red elliptical cylinder*, a *red triangular prism* and a *blue hexagonal prism*. Three types of surfaces are present in this set; flat surfaces, cylindrical surfaces and elliptic surfaces. Four colors are also present in the set. Simple observation of the object set reveals that surface curvature will be the only way that the blue cylinder can be distinguished from the *blue hexagonal prism*, or the *red elliptical cylinder* from the *red triangular prism*, or the two *white elliptical cylinders* from each other. Similarly, spectral characteristics will determine which flat surface can be identified as belonging to which object. Hence, it is obvious that employing either curvature data or color data is insufficient to identify objects correctly, and integrating both types of data is necessary for correct hypotheses.

Tests consisted of 39 surfaces and 26 objects appearing in seven cluttered scenes. Of the 26 objects appearing in these scenes, 9 objects where both partially occluded and in partial shadow and 4 other objects where in partial shadow. A typical scene is shown in Figure 2. Surface signature matching reveals that the surfaces matched fairly well to their model surfaces as values of *SEPI* were in the range 0.09 – 0.25 with an average *SEPI* value of 0.16. This implies that surfaces matched to their correct model-surfaces with an average surface error of 0.16. Scenes that had large *SEPI* values (*SEPI* > 0.2), had these errors due to the fact that the observed-surfaces in these scenes were mostly occluded or in shadow. Values of *SMPI* for the scenes were in the range 83% – 100% with an average value of 93.4% for the whole set, indicating that 93.4% of the surfaces appearing in the scenes were correctly matched to their model surfaces.

The overall model identification performance for the whole set, as measured by *MHPI*, was an impressive 99.8% (all scenes except one had a *MHPI* value of 100%). In other words, out of the 26 objects appearing in the scenes, objects were correctly hypothesized 25 out of 26 times. For the single case when the correct model was not correctly hypothesized due to partial visibility of the object, the correct model produced the next best match with a hypothesis error of only 0.06.

Fig. 2. A *Toy* image scene.   **Fig. 3.** A *Fruit* image scene

7.2 The *Fruit* set

The second set, the *Fruit* set, consists of seven plastic fruit models: an *apple*, a *banana, corn, grape,* a *lemon, pepper, strawberry*, and a *tomato*. A sample test scene from the *Fruit* set was shown in Figure 3. These objects differ from the objects of the *Toy* set by having complex shape. Object colors consisted of red, yellow, purple and green. For simplicity, we will assume each object to consist of a single surface. Surface signatures for the banana and corn surfaces were shown in Figure 1.

Tests for this set consisted of 30 observed-surfaces and 30 objects. Once again several objects were partially occluded and/or in shadow. Although objects of this set had complex shapes, the surface matching errors were lower than those obtained for the toy model set (overall *SEPI* for the whole set was 0.07). This is because surfaces in the set are of the same general type, curved surfaces, whereas in the other set two types of surfaces, curved and flat surfaces, were present. Hence, the contribution of curvature errors to the overall surface error for this set were less than those for the other set. Values of *SMPI* for this set were in the range of 96% – 100%, with an average *SMPI* value of 98.4%, indicating good surface matching results. With good matching results, good identification was also possible; the system was able to correctly hypothesize 98.4% of the objects appearing in the scenes.

8. Conclusion and Future Work

We have described a surface matching approach that integrates both curvature and color information for object identification. Surfaces were matched by matching their surface signatures. Surface signatures are statistical features that are uniform for a uniform surface. Two types of signatures were employed; color signatures and curvature signatures. Tests conducted on objects of various curvature and colors with partial occlusion and shadows permitted have produced excellent results; over 90% of surfaces extracted from cluttered image scenes where correctly matched. This led to identification success rates of over 90%.

Although, only two types of surface signatures were employed in this work, the concept of surface signatures can be easily extended to other types of features, such as shape and texture, which is the focus of current research. In addition, the current surface signature types defined can be expanded. For instance, the number of spectral signatures can be expanded to include other parts of the electromagnetic spectrum (e.g. infra-red).

References

1. Anuta, Paul 1970. "Spatial Registration of Multispectral and Multitemporal Digital Imagery Using FFT Techniques", *IEEE Transaction on Geoscience Electronics*, 8 (4), Oct., pp. 353-368.
2. Ranade, S. and Rosenfeld A. 1980. "Point Pattern Matching in Relaxation", Pattern Recognition, 12, pp. 269-275.
3. Besl, P. and Jain, R. 1985. "Intrinsic and Extrinsic Surface Characterization". In the *Conference on Computer Vision and Pattern Recognition*, New York, NY, June.
4. Haralick, R. and Watson, L. and Laffey, T. 1983. "The Topographic Primal Sketch", *Int. J. Robot Res.*, 2 (1), 50-72.
5. Newman, T., Flynn, P. and Jain, A. 1993. "Model-Based Classification of Quadric surfaces", *CVGIP: Image Understanding*, 58 (2), September, pp. 235-249.
6. Haralick, R. and Kelly, G. 1969. "Pattern Recognition With Measurement Space And Spatial Clustering For Multiple Images", *Proc. IEEE* 57, 654-445, April.
7. Ohta, Y., Kanade, K. and Saki, T. 1980. "Color information for region Segmentation". In *Computer Vision, Graphics and Image Processing*, 13, 224-241.
8. Healey, G., Shafer, S. and Wolff, L., eds. 1992. *Physics-Based Vision: Color*, Jones and Bartlett Publishers, Boston.
9. Swain, M. and Ballard, D. 1991. "Color Indexing", In *the International Journal of Computer Vision*, 7 (11), pp. 12-32.
10. Stricker, M. and Orengo, M. 1995. "Similarity of Color Images". In *Storage and Retrieval for Image and Video Databeases III*, Volume 2420, pp. 381-392.
11. Finlayson, G., Chatterjee, S. and Funt, B. "Color Angular Indexing". In the Fourth European Conference Computer Vision, Cambridge, UK, April 1996, pp. 16-27.
12. Higuchi, K., Delingette, H. and Ikeuchi, K. 1994. "Merging Multiple Views Using a Spherical Representation", *Second CAD-Based Vision Workshop*, Champion, Penn, Feb. 8-11, pp. 17-26.
13. Grewe, L. and Kak, A. 1994. "Interactive Learning of Multiple Attribute Hash Table for Fast 3D Object Recognition", *Second CAD-Based Vision Workshop*, Champion, Penn, Feb. 8-11, pp. 17-26.
14. Mustafa, A. A., Shapiro, L. G. and Ganter, M. A. 1996. "3D Object Recognition from Color Intensity Images". In *the 13th International Conference On Pattern Recognition*, Vienna, Austria, August 25-30.
15. Mustafa, A. A., Shapiro, L. G. and Ganter, M. A. 1997. "Matching Surface Signatures for Object Recognition". To appear in *the eighth Scandinavian Conference On Image Analysis*, Lappeenranta, Finland, June 8-11.
16. Mustafa, Adnan. A. 1995. "Object Identification with Surface Signatures using Color Photometric Stereo". Ph.D. dissertation, Mechanical Engineering Department, *University of Washington*, March.
17. Mustafa, A. A., Shapiro, L. G. and Ganter, M. A. 1997. "3D Object Identification with Color and Curvature Signatures". To appear in *Pattern Recognition*.

Computing Projective and Permutation Invariants of Points and Lines

Gabriella CSURKA[1] and Olivier FAUGERAS[2]

[1] INRIA Rhône-Alpes, 655 Av. de l'Europe, 38330 Montbonnot Saint Martin, France
[2] INRIA Sophia-Antipolis, 2004 Route des Lucioles, 06902 Sophia Antipolis, France

Abstract. Until very recently it was believed that visual tasks require camera calibration. More recently it has been shown that various visual or visually-guided robotics tasks may be carried out using only a projective representation characterized by the projective invariants. This paper studies different algebraic and geometric methods of computation of projective invariants of points and/or lines using only informations obtained by a pair of uncalibrated cameras. We develop combinations of those projective invariants which are insensitive to permutations of the geometric primitives of each of the basic configurations and test our methods on real data in the case of the six points configuration.

1 Introduction

Researchers in computer vision and robotics are more and more interested on developing reliable algorithms for uncalibrated cameras. The basic reason is that a number of visual or visually-guided robotics tasks concerning for example obstacle detection, tracking or visual servoing may be carried out using only a projective representation [13,12,7].

Still the study of every geometry is based on the study of properties which are invariant under the corresponding group of transformations, the projective geometry is characterized by the projective invariants. Consequently, this paper is consecrated to the study of projective invariants of various configurations of points and/or lines in 3D space. We compare different algebraic and geometric methods where the computation is done either in the space or using image measurements. In the 3D case, we suppose that we have an arbitrary three-dimensional projective representation of the object obtained by explicit projective reconstruction. In the image case, we suppose that the only information we have is the image measurements and the epipolar geometry of the views and we compute three-dimensional invariants without any explicit reconstruction.

As non minimal configurations of points and/or lines can be decomposed into minimal sub-configurations with invariants, and these invariants characterize the invariants of the original configuration it is sufficient to study only minimal configurations. In 3D projective space, a point is defined by three parameters, a line by four and a projective transformation by 15. Hence, m points and n lines generally have $3m + 4n - 15$ invariants, and the minimal configurations of points and lines are six points (3 invariants), four points and a line (1 invariant), three points and two lines (2 invariants), two points and three lines (3 invariants). Four lines actually have 2 rather then 1 invariant because they have an isotropy subgroup of dimension 1 [5,2]. We do not need to study the case of two points and three lines, as they define six planes which correspond to six points in the dual space with the same number (3) of invariants.

Consequently, to compute the invariants of an arbitrary configuration of points and/or lines it is sufficient to be able to compute the invariants of four basic configurations: six points, four points and a line, three points and two lines, and four lines. For each configuration we show how to compute invariants in 3D projective space and in the images, using both algebraic and geometric approaches.

As these invariants generally depend on the order of the points and lines, we will also look for characteristics which are both projective and permutation invariants and finally experiments with real data are presented.

2 Algebraic Approach

Consider eight points in space represented by their homogeneous coordinates $M_{i,i=1..8} = (x_i, y_i, z_i, t_i)^\top$, and compute the following ratio of determinants:

$$I = \frac{[M_1\,M_2\,M_3\,M_4][M_5\,M_6\,M_7\,M_8]}{[M_{\dot 1}\,M_{\dot 2}\,M_{\dot 3}\,M_{\dot 4}][M_{\dot 5}\,M_{\dot 6}\,M_{\dot 7}\,M_{\dot 8}]}, \tag{1}$$

where $\dot k$ denotes the value $\sigma(k)$ for an arbitrary permutation σ of $\{1, 2, \ldots, 8\}$. It is easy to show that I is invariant under projective transformations and does not depend on the homogeneous representatives chosen for the projective points. Using the Grassmann-Cayley algebra [4] described below we can compute I only from a pair of images obtained with weakly calibrated cameras without explicitly reconstructing the scene [1,3] :

$$\frac{(\beta_{12-34}^\top \mathbf{F}\alpha_{12-34})(\beta_{56-78}^\top \mathbf{F}\alpha_{56-78})\overline{(\beta_{\dot 1}^\top \mathbf{F}\alpha_2)}\,\overline{(\beta_{\dot 3}^\top \mathbf{F}\alpha_4)}\,\overline{(\beta_{\dot 5}^\top \mathbf{F}\alpha_6)}\,\overline{(\beta_{\dot 7}^\top \mathbf{F}\alpha_8)}}{(\beta_{\dot 1 \dot 2 - \dot 3 \dot 4}^\top \mathbf{F}\alpha_{\dot 1 \dot 2 - \dot 3 \dot 4})(\beta_{\dot 5 \dot 6 - \dot 7 \dot 8}^\top \mathbf{F}\alpha_{\dot 5 \dot 6 - \dot 7 \dot 8})(\beta_{\dot 1}^\top \mathbf{F}\alpha_2)(\beta_{\dot 3}^\top \mathbf{F}\alpha_4)(\beta_{\dot 5}^\top \mathbf{F}\alpha_6)(\beta_{\dot 7}^\top \mathbf{F}\alpha_8)} \tag{2}$$

where, $\alpha_{i_1 i_2 - i_3 i_4} = \alpha_{i_1}\alpha_{i_2} \wedge \alpha_{i_3}\alpha_{i_4}$, $\beta_{i_1 i_2 - i_3 i_4} = \beta_{i_1}\beta_{i_2} \wedge \beta_{i_3}\beta_{i_4}$ and $\overline{(\beta_i^\top \mathbf{F}\alpha_j)}$ stands for the expression $sign(\beta_j^\top \mathbf{F}\alpha_i)\sqrt{-(\beta_j^\top \mathbf{F}\alpha_i)(\beta_i^\top \mathbf{F}\alpha_j)}$.

Using the ratio (1), we can compute invariants for each of the basic configurations, and from (2) result a direct formula to compute them from image measurements :

Six point configurations Using (1) and (2), we can deduce the following three invariants for the configuration of six points $A_{i,i=1..6}$ in 3D projective space :

$$I_1 = \frac{[A_1\,A_3\,A_2\,A_5][A_3\,A_4\,A_2\,A_6]}{[A_1\,A_3\,A_2\,A_6][A_3\,A_4\,A_2\,A_5]} = \frac{(\beta_{13-25}^\top \mathbf{F}\alpha_{13-25})(\beta_{34-26}^\top \mathbf{F}\alpha_{34-26})}{(\beta_{13-26}^\top \mathbf{F}\alpha_{13-26})(\beta_{34-25}^\top \mathbf{F}\alpha_{34-25})}$$

$$I_2 = \frac{[A_1\,A_2\,A_3\,A_5][A_1\,A_4\,A_3\,A_6]}{[A_1\,A_2\,A_3\,A_6][A_1\,A_4\,A_3\,A_5]} = \frac{(\beta_{12-35}^\top \mathbf{F}\alpha_{12-35})(\beta_{14-36}^\top \mathbf{F}\alpha_{14-36})}{(\beta_{12-36}^\top \mathbf{F}\alpha_{12-36})(\beta_{14-35}^\top \mathbf{F}\alpha_{14-35})} \tag{3}$$

$$I_3 = \frac{[A_2\,A_3\,A_1\,A_5][A_2\,A_4\,A_1\,A_6]}{[A_2\,A_3\,A_1\,A_6][A_2\,A_4\,A_1\,A_5]} = \frac{(\beta_{23-15}^\top \mathbf{F}\alpha_{23-15})(\beta_{24-16}^\top \mathbf{F}\alpha_{24-16})}{(\beta_{23-16}^\top \mathbf{F}\alpha_{23-16})(\beta_{24-15}^\top \mathbf{F}\alpha_{24-15})}$$

One line and four points Denoting the four points by $A_{i,i=1..4}$ and the line by L we obtain (cf. (1) and (2)) that the following ratio :

$$I = \frac{[L\,A_1\,A_3][L\,A_2\,A_4]}{[L\,A_1\,A_4][L\,A_2\,A_3]} = \frac{(\beta_{0-13}^\top F\alpha_{0-13})(\beta_{0-24}^\top F\alpha_{0-24})\overline{(\beta_1^\top F\alpha_4)}\,\overline{(\beta_2^\top F\alpha_3)}}{(\beta_{0-14}^\top F\alpha_{0-14})(\beta_{0-23}^\top F\alpha_{0-23})(\beta_1^\top F\alpha_3)(\beta_2^\top F\alpha_4)} \quad (4)$$

where $[L\,A_i\,A_j] = [P\,Q\,A_i\,A_j]$ for P and Q arbitrary two distinct points on the line, α_i, β_i, l, l' are the projections of $A_{i,i=1..4}$ and L in the two images, $\alpha_{0-ij} = l \times (\alpha_j \times \alpha_i)$ and $\beta_{0-ij} = l' \times (\beta_j \times \beta_i)$.

Three Points and Two Lines The following two ratios are independent invariants for configurations of three points $A_{i,i=1..3}$ and two lines $L_{k,k=1,2}$:

$$I_1 = \frac{[L_1\,A_1\,A_2][L_2\,A_1\,A_3]}{[L_1\,A_1\,A_3][L_2\,A_1\,A_2]} = \frac{(\beta_{1-12}^\top F\alpha_{1-12})(\beta_{2-13}^\top F\alpha_{2-13})}{(\beta_{1-13}^\top F\alpha_{1-13})(\beta_{2-12}^\top F\alpha_{2-12})}$$

$$I_2 = \frac{[L_1\,A_1\,A_2][L_2\,A_2\,A_3]}{[L_1\,A_2\,A_3][L_2\,A_1\,A_2]} = \frac{(\beta_{1-12}^\top F\alpha_{1-12})(\beta_{2-23}^\top F\alpha_{2-23})}{(\beta_{1-23}^\top F\alpha_{1-23})(\beta_{2-12}^\top F\alpha_{2-12})}$$

Configurations of Four Lines Consider four lines $L_{i,i=1..4}$ in 3D projective space. This configuration ha 16 parameters, so naively we might expect to have $16 - 15 = 1$ independent invariant. However, the configuration has a 1D isotropy subgroup, so there are actually two independent invariants [5,2] :

$$I_1 = \frac{[L_1\,L_2][L_3\,L_4]}{[L_1\,L_4][L_2\,L_3]} = \frac{(\beta_{1-2}^\top F\alpha_{1-2})(\beta_{3-4}^\top F\alpha_{3-4})}{(\beta_{1-4}^\top F\alpha_{1-4})(\beta_{2-3}^\top F\alpha_{2-3})}$$

$$I_2 = \frac{[L_1\,L_3][L_2\,L_4]}{[L_1\,L_4][L_2\,L_3]} = \frac{(\beta_{1-3}^\top F\alpha_{1-3})(\beta_{2-4}^\top F\alpha_{2-4})}{(\beta_{1-4}^\top F\alpha_{1-4})(\beta_{2-3}^\top F\alpha_{2-3})} \quad (5)$$

3 Geometric Approach

Alternatively, one can also take a geometric approach to compute invariants for each configurations. All of these invariants can be expressed in terms of cross ratios, which are the fundamental projective invariants. Cross ratios are also invariant under projection from n-dimensional space to m-dimensional space too, where $m < n$. This is important, because it corresponds to the case of the projection of the world on the camera's retinal plane.

There are several ways to compute projective invariants for configurations of points and lines. Here we give just one example, for each configuration:

Six point configurations The basic idea is to construct a pencil of planes using the six points. Taking two of the points, for example A_1 and A_2, the four planes defined by A_1, A_2 and $A_k, k = 3..6$ belong to the pencil of planes through the line $A_1 A_2$. Their cross ratio is an invariant of the configuration. Taking other pairs of points for the axis of the pencil gives further cross ratios [2] :

$$\{A_2 A_3 A_1, A_2 A_3 A_4; A_2 A_3 A_5, A_2 A_3 A_6\} = I_1$$

$$\{A_1 A_3 A_2, A_1 A_3 A_4; A_1 A_3 A_5, A_1 A_3 A_6\} = I_2 \quad (6)$$

$$\{A_1 A_2 A_3, A_1 A_2 A_4; A_1 A_2 A_5, A_1 A_2 A_6\} = I_3$$

One Line and Four Points It is easy to see that the four planes $LA_{i,i=1..4}$ belong to a pencil so they define an invariant cross ratio λ, such that $\lambda = I$.

Three Points and Two Lines The invariants obtained by the algebraic approach can also be obtained as follows [2]. Cut the two lines with the plane $A_1A_2A_3$ defined by the three points to give R_1 and R_2. The five coplanar points define a pair of independent invariants, for example $\{A_1A_2, A_1A_3; A_1R_1, A_1R_2\} = I_1$ and $\{A_2A_1, A_2A_3; A_2R_1, A_2R_2\} = I_2$.

Four Lines Assume that the four lines $L_{i,i=1..4}$ are in general position, in the sense that they are skew none of them can be expressed as a linear combination of three others (there is a list of particular cases in [2]). Then there exist exactly two transversals T_1 and T_2 intersecting them [11,5]. Note the intersection points by $A_{i,i=1..4}$ and $B_{i,i=1..4}$. The cross ratios $\lambda_1 = \{A_1, A_2; A_3, A_4\}$ and $\lambda_2 = \{B_1, B_2; B_3, B_4\}$ are invariant for this configuration and the relation between $I_{i,i=1,2}$ and $\lambda_{j,j=1,2}$ are: $I_1 = 1 - \lambda_1 - \lambda_2 + \lambda_1\lambda_2$ and $I_2 = \lambda_1\lambda_2$.

To compute these invariants in 3D space we need an arbitrary projective reconstruction of the points. For the lines we use their Plücker representation [11]. However, the invariants can be also computed directly from the images by using the fact that the cross ratio is preserved under perspective projections. In the first three cases, the basic idea is to find the intersection of a line and a plane directly in the images. This can be done by using the homography induced by the 3D plane between the images or by solving some epipolar constraints [2]. The case of the four lines is more complicated (see [2]).

4 Projective and Permutation Invariants

This section generalizes the work of Meer et al. [8] and Morin [9] on the invariants of five coplanar points to 3D configurations of points and lines.

The invariants of the previous sections depend on the ordering of the underlying point and line primitives However, if we want to compare two configurations, we often do not know the correct relative ordering of the primitives. If we work with order dependent invariants, we must compare all the possible permutations of the elements. For example, if $I_{k,k=1..3}$ are the three invariants of a set of points $A_{i,i=1..6}$ and we would like to know if another set $B_{i,i=1..6}$ is projectively equivalent, we must compute the triplet of invariants for each of the 6!=720 possible permutations of the B_i This is a very costly operation, which can be avoided if we look for permutation-invariant invariants.

The projective invariants we computed were all expressed as cross ratios. To obtain permutation invariant cross ratios of four elements, it is sufficient to take an arbitrary symmetric function of the six possible values ($\lambda, 1 - \lambda, \frac{1}{\lambda}, \frac{1}{1-\lambda}, \frac{\lambda}{\lambda-1}$ and $\frac{\lambda-1}{\lambda}$) we can obtain for the cross ratio of four points taking them in arbitrary order. For example, Meer et al. [8] propose the bounded function:

$$J(\lambda) = \frac{J_1 J_0}{J_1 - J_2} = \frac{2\lambda^6 - 6\lambda^5 + 9\lambda^4 - 8\lambda^3 + 9\lambda^2 - 6\lambda + 2}{\lambda^6 - 3\lambda^5 + 3\lambda^4 - \lambda^3 + 3\lambda^2 - 3\lambda + 1}, \tag{7}$$

In the case of six points we have the three invariants $I_{k,k=1..3}$ given in (3). Consider the values:

$$
\begin{array}{cccccc}
J(I_1) & J(\frac{I_2}{I_3}) & J(\frac{I_2-1}{I_3-1}) & J(\frac{I_3(I_2-1)}{I_2(I_3-1)}) & J(\frac{I_3-I_1}{I_3-I_2}) & \\
J(I_2) & J(\frac{I_1}{I_3}) & J(\frac{I_1-1}{I_3-1}) & J(\frac{I_3(I_1-1)}{I_1(I_3-1)}) & J(\frac{I_2(I_3-I_1)}{I_1(I_3-I_2)}) & (8)\\
J(I_3) & J(\frac{I_1}{I_2}) & J(\frac{I_1-1}{I_2-1}) & J(\frac{I_2(I_1-1)}{I_1(I_2-1)}) & J(\frac{(I_2-1)(I_3-I_1)}{(I_1-1)(I_3-I_2)}) &
\end{array}
$$

Note that the values of I_1, I_2, I_3 depend on the order of the points, but any other ordering gives the same J_{ij} in different ordering. For this reason, we sort the 15 values after computing them to give a vector that characterizes the configuration independently of the order of underlying points (see the proof in [2]).

Now, consider the configuration of a line and four unordered points. This case is very simple case because we have a cross ratio, so it is enough to take the value of the function $J(I)$ where I is the invariant given by (4). Similarly to compute permutation invariants of four lines in space, we apply J to the pair of cross ratios λ_1 and λ_2 given Sect. 3. It turn out that the order of the two sets of intersection points of the four lines with the two transversals change in the same way under permutations of the lines. When λ_1 and λ_2 are complex conjugates, $J(\lambda_1)$ and $J(\lambda_2)$ are also complex conjugates, but if we want to work with real invariants it is sufficient to take $J_1 = J(\lambda_1) + J(\lambda_2)$ and $J_2 = J(\lambda_1)J(\lambda_2)$.

In the case of three points and two lines as we have no trouble distinguishing lines from points, mixed permutations of points and lines are not necessary to be considered. Considering permutations on points and the possibility to interchange the lines we obtain that the sorted vector of elements $J(I_1)$, $J(I_2)$ and $J(\frac{I_1}{I_2})$ is a projective and permutation invariant for this configuration.

5 Experimental results

We present here some experimental results that test the different methods of computing invariants for six points configurations. The computational methods for the other minimal configurations are similar, so we will not show the results for them in detail here. Further results and comparisons using synthetic data can be found in [2].

We have evaluated the methods on two series of real images (see Fig. 1). Among the extracted points, we select those with correspondences in every image and form some sets of six points to test (noted by $A1 \ldots A7$).

We will use the following names for methods:
- A3D – 3D algebraic approach: compute $I_{i,i=1..3}$ using (3).
- G3D – 3D geometric approach: compute cross ratios of planes using (6).
- A2D – 2D algebraic approach: compute $I_{i,i=1..3}$ in the images using (3).
- G2D – 2D geometric approach: compute cross ratios of planes directly in the images.

In the case of A3D and G3D we used the three-dimensional coordinates of projective points reconstructed by the method of triangulation presented in [6]. First note that A3D and G3D give essentially the same results (see Fig. 3(a)). This means that they handle image noise in the same way. This is because there is an important error spread during reconstruction, but negligible error when computing the invariants from the reconstructed points. So, we will only compare methods A3D, A2D and G2D.

The results of Fig. 2 and Fig. 3(a) show that A3D/G3D is the most stable method tested, and G2D the least. This confirms the conclusions of the papers [10] concerning projective reconstruction. Indeed, the invariants of six points correspond to the projective coordinates of the sixth point in the basis of the other fives. It was shown in [10,2] that explicit reconstruction (A3D/G2D) produces more reliable results than implicit methods (A2D, G2D), and that the algebraic implicit method (A2D) is more stable than the geometric one (G2D).

However, if we use the *epipolar constraint correction algorithm*[1] on image correspondences before computing the invariants we obtain essentially identical results for all the methods (Fig. 3(b)). These results (see Fig. 3(c)) are more stable than the results obtained without epipolar constraint correction (Fig. 2). However, some configurations seem to be more stable than others and the results are not always sufficiently accurate to be used in further computation.

Consider now the invariants shown in Fig. 3(c) and compute the sorted list of permutation invariants J_{ij} given by (8) for the seven sets of six points. We compare the sorted vectors by computing the means of the Euclidean distances between them across all the images (†). We also compute the cluster means for the seven configurations Ai and for each element corresponding to a configurations we look which is the nearest cluster mean (‡) :

†	A1	A2	A3	A4	A5	A6	A7
A1	.0047						
A2	.0496	.0305					
A3	.0394	.0193	.0054				
A4	.0260	.0683	.0505	.0176			
A5	.0247	.0677	.0506	.0155	.0051		
A6	.0343	.0401	.0283	.0465	.0472	.0032	
A7	.0625	.0995	.0805	.0346	.0276	.0799	.002

‡	A1	A2	A3	A4	A5	A6	A7
A1	42	1	0	1	0	1	0
A2	5	32	1	2	0	3	2
A3	2	3	39	0	0	1	0
A4	10	2	0	31	2	0	0
A5	1	0	0	0	43	0	1
A6	2	10	2	9	0	22	0
A7	0	0	0	1	0	0	44

One can see that generally the best score is for the same set of invariants so we can conclude that if we have good estimates of point correspondences and the epipolar geometry, they can be used to discriminate between configurations.

Conclusion

The invariants of complex configurations can be computed from minimal configurations into which they can be decomposed. So it was sufficient to treat only the case of four minimal configurations: six points, four points and a line, three points and two lines and four lines, for which we gave different methods to compute the invariants.

There are basically two approaches - algebraic and geometric - and in each case we showed the relations between the resulting invariants. We analyzed the

---

[1] This algorithm presented in [6], modify each pair of correspondence to verify exactly the epipolar constraints.

computation of these invariants in 3D space assuming that the points and lines had already been reconstructed in an arbitrary projective basis, and we also gave methods to compute them directly from correspondences in a pair of images. In the second case the only information needed is the matched projections of the points and lines and the fundamental matrix. For each basic configuration we also gave permutation and projective invariant

Finally, we tested our methods on real data. The experiments showed that although the invariants are sensitive to noise, if we have good estimates of point correspondences and the epipolar geometry they can be used to discriminate between configurations.

Acknowledgment We would like to thank Bill Triggs for his carefully reading of the draft of this manuscript.

References

1. S. Carlsson. Multiple image invariants using the double algebra. In J. Mundy and A. Zissermann, editors, *Proceeding of the DARPA–ESPRIT workshop on Applications of Invariants in Computer Vision, Azores, Portugal*, pages 335–350, October 1993.
2. G. Csurka. *Modelisation projective des objets tridimensionnels en vision par ordinateur*. Thèse de doctorat, Université de Nice – Sophia Antipolis, April 1996.
3. G. Csurka and O. Faugeras. Computing three-dimensional projective invariants from a pair of images using the grassmann-cayley algebra. *Image and Vision Computing*, 1997. to appear.
4. P. Doubilet, G.C. Rota, and J. Stein. On the foundations of combinatorial theory: Ix combinatorial methods in invariant theory. *Studies in Applied Mathematics*, 53:185–216, 1974.
5. R. Hartley. Invariants of lines in space. In *Proceedings of DARPA Image Understanding Workshop*, pages 737–744, 1993.
6. R. Hartley and P. Sturm. Triangulation. In *Proceedings of ARPA Image Understanding Workshop, Monterey, California*, pages 957–966, November 1994.
7. R. Horaud, F. Dornaika, and B. Espiau. Visually guided object grasping. IEEE *Transactions on Robotics and Automation*, 1997. submitted.
8. P. Meer, S. Ramakrishna, and R. Lenz. Correspondence of coplanar features through p^2-invariant representations. In *Proceedings of the 12th International Conference on Pattern Recognition, Jerusalem, Israel*, pages A-196–202, 1994.
9. L. Morin. *Quelques contributions des invariants projectifs à la vision par ordinateur*. Thèse de doctorat, Institut National Polytechnique de Grenoble, January 1993.
10. C. Rothwell, O. Faugeras, and G. Csurka. A comparison of projective reconstruction methods for pairs of views. *Computer Vision and Image Understanding*, 1997. to appear.
11. J.G. Semple and G.T. Kneebone. *Algebraic Projective Geometry*. Oxford Science Publication, 1952.
12. U. Uenohara and T. Kanade. Geometric invariants for verification in 3-d object tracking. In *Proceedings of the IEEE/RSJ International Conference on Intelligent Robots and Systems, Osaka, Japan*, volume II, pages 785–790, November 1996.
13. C. Zeller and O. Faugeras. Applications of non-metric vision to some visual guided tasks. In *Proceedings of the 12th International Conference on Pattern Recognition, Jerusalem, Israel*, 1994.

Fig. 1. A set of ten images of the Arcades' Square of Valbonne.

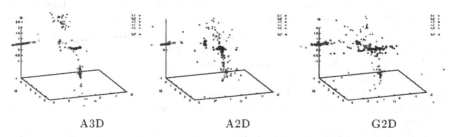

A3D　　　　　　　　　A2D　　　　　　　　　G2D

Fig. 2. The invariants $(I1, I2, I3)$ obtained for the seven chosen sets of six points $(A1, A2, \ldots A7)$ by methods A3D, A2D and G2D using all possible pairs of view (45).

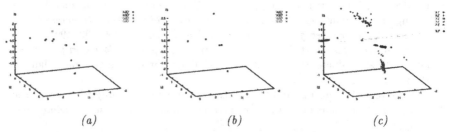

(a)　　　　　　　　　*(b)*　　　　　　　　　*(c)*

Fig. 3. The invariants $(I1, I2, I3)$ obtained for the seven chosen set of six points $(A1, A2, \ldots A7)$ by methods A3D, G3D, A2D and G2D using the first pair of view of the "square" scene. The invariants were computed *(a)* without and *(b)* with the epipolar constraint correction. *(c)* $(I1, I2, I3)$ from all pairs of images of the scene using method A2D with epipolar constraint correction.

Point Projective and Permutation Invariants

Tomáš Suk *and* Jan Flusser *

Institute of Information Theory and Automation,
Academy of Sciences of the Czech Republic,
Pod vodárenskou věží 4, 182 08 Praha 8, The Czech Republic,
E-mail: suk@utia.cas.cz, flusser@utia.cas.cz

Abstract. The paper deals with features of a point set which are invariant with respect to projective transform. First, projective invariants for five-point sets, which are simultaneously invariant to the permutation of the points, are derived. They are expressed as functions of five-point cross-ratios. Possibilities of the choice of their roots are referred and their normalization is discussed.

Then, the invariants for more than five points are derived. Stability and discriminability of the features is demonstrated by numerical experiments.

1 Introduction

One of the important tasks in image processing and computer vision is a recognition of objects on images captured under different viewing angles. If the scene is planar, then the distortion between two frames can be described by *projective transform*

$$x' = (a_0 + a_1 x + a_2 y)/(1 + c_1 x + c_2 y)$$
$$y' = (b_0 + b_1 x + b_2 y)/(1 + c_1 x + c_2 y) , \tag{1}$$

where x, y are the coordinates in the first frame and x', y' are the coordinates in the second one.

This contribution deals with point-based features for recognition of projectively deformed point sets. The points must be labeled by some way, but the correspondence between the points can be unknown and then the correct labeling of the points is also unknown. Therefore the features are designed as invariant to the choice of labeling of the points, i.e. to the permutation of the points.

The plane transformed by the projective transform contains a straight line, which is not mapped into the second plane (more precisely it is mapped into the infinity) and which divides the plane into two parts. If the transform has form (1), then it is the line

$$1 + c_1 x + c_2 y = 0 \tag{2}$$

If all points (the whole point set) lay in one part of the plane, then some additional theorems dealing with topology of the set holds for the transform, e.g.

---

* This work has been supported by the grant No. 102/95/1378 of the Grant Agency of the Czech Republic.

the convex hull is preserved during the projective transform in that case. It can be used for construction of invariants with less computational complexity [1].

This paper deals with the general case of the projective transform, when the points can lay in both parts of the plane, the convex hull is not preserved and all possibilities of the positions of the points must be taken care of. The only constraint is that the points do not lay directly at the straight line (2).

A projective invariant can be defined for five points minimally. The simplest invariant is a five-point cross-ratio

$$\varrho(1,2,3,4,5) = \frac{P(1,2,3)P(1,4,5)}{P(1,2,4)P(1,3,5)}, \qquad (3)$$

where $P(1,2,3)$ is the area of the triangle with vertices 1,2 and 3.

The work [2] deals with five-point projective and permutation invariants. It proposes to use median of all possible values. The more precise description of relations between various values of the cross-ratio during permutation of the five points can be found in [3]. Our first attempt to solve this problem was published in [4].

2 Five-point permutation invariants

The five-point cross-ratio (3) can be used for construction of the invariants to the choice of labeling.

2.1 The derivation of the invariants

Reiss in [2] uses the function $\varrho + \varrho^{-1}$, which is unstable near zero. If some triplet of five points in (3) is collinear, then the function $\varrho + \varrho^{-1}$ is infinite. Thus the more suitable function is $\psi = \frac{2}{\varrho+\varrho^{-1}} = \frac{2\varrho}{\varrho^2+1}$. If ϱ or ϱ^{-1} is zero, then the function ψ is zero.

If invariance to the choice of labeling of the points is required, some symmetric function must be used for considering all possibilities of labeling. We can use additions and multiplications

$$
\begin{aligned}
F'_+(1,2,3,4,5) &= \psi(1,2,3,4,5) + \psi(1,2,3,5,4) + \psi(1,2,4,5,3), \\
F'_\cdot(1,2,3,4,5) &= \psi(1,2,3,4,5) \cdot \psi(1,2,3,5,4) \cdot \psi(1,2,4,5,3).
\end{aligned}
\qquad (4)
$$

It is suitable to use e.g. the function $F_+ = 16(\frac{1}{5} - \frac{1}{6-F'_+})$ instead of F'_+ because of higher discriminability of the invariants and normalize $F'_\cdot$ by the constant $F_\cdot = -\frac{3}{8}F'_\cdot$ to the interval from 0 to 1. Functions F_+ and $F_\cdot$ are invariant to the choice of labeling of last four points, but the point 1 must be common at all cross-ratios. To obtain full invariance to the choice of labeling, we must alternate all five points as common ones

$$
\begin{aligned}
I_{s_1,s_2}(1,2,3,4,5) = F_{s_1}(1,2,3,4,5)s_2 F_{s_1}(2,3,4,5,1)s_2 F_{s_1}(3,4,5,1,2)s_2 \\
s_2 F_{s_1}(4,5,1,2,3)s_2 F_{s_1}(5,1,2,3,4),
\end{aligned}
\qquad (5)
$$

where s_1 and s_2 are either sign $+$ or $\cdot$.

The set of n points has $2n$ degrees of freedom and projective transform has 8 parameters. Therefore the set can have only

$$m = 2n - 8 \tag{6}$$

independent invariants to projective transform. That is why only two from four invariants I_{++}, $I_{.+}$, $I_{+.}$ and $I_{..}$ can be independent.

2.2 The roots of the invariants

Lenz and Meer [3] deal with the five-point projective and permutation invariants in detail. They discovered that if the common point stays the same and the other points are permuted, then the values of the cross-ratios are

$$\varrho_1 = \varrho, \ \varrho_2 = \frac{1}{\varrho}, \ \varrho_3 = 1 - \varrho, \ \varrho_4 = \frac{1}{1 - \varrho}, \varrho_5 = \frac{\varrho}{\varrho - 1}, \ \varrho_6 = \frac{\varrho - 1}{\varrho}. \tag{7}$$

If we construct a function $F(\varrho)$, which has the same value for all these values of ϱ ($\varrho_1, \varrho_2, \cdots, \varrho_6$), it is invariant to the permutation of four points. If we change the common point, we receive another value of the cross-ratio, let us say σ, and a function

$$K(\varrho, \sigma) = F(\varrho) + F(\sigma) + F(\frac{\varrho}{\sigma}) + F(\frac{\varrho - 1}{\sigma - 1}) + F(\frac{\varrho(\sigma - 1)}{\sigma(\varrho - 1)}) \tag{8}$$

is the five-point projective and permutation invariant.

As our research implies, if the function $F(\varrho)$ is constructed as the quotient of two polynomials and ϱ is its root, then each of the values $\varrho_1, \varrho_2, \cdots, \varrho_6$ must be its root. We can consider it in form

$$F(\varrho) = \frac{P_1(\varrho)}{P_2(\varrho)}, \tag{9}$$

$$P_1(\varrho) = (\varrho - \varrho_1)(\varrho - \varrho_2)(\varrho - \varrho_3)(\varrho - \varrho_4)(\varrho - \varrho_5)(\varrho - \varrho_6) \tag{10}$$

and similarly $P_2(\varrho)$ (when all roots differ from zero).

It is useful in practice, when the invariants are defined everywhere. It means the polynomial $P_2(\varrho)$ in the denominator has only imaginary roots, because $F(\varrho)$ is undefined at points of real roots of $P_2(\varrho)$. There are proposed two such invariants in [3]

$$F_{14} = \frac{2\varrho^6 - 6\varrho^5 + 9\varrho^4 - 8\varrho^3 + 9\varrho^2 - 6\varrho + 2}{\varrho^6 - 3\varrho^5 + 3\varrho^4 - \varrho^3 + 3\varrho^2 - 3\varrho + 1}, \tag{11}$$

$$F_{15} = \frac{(\varrho^2 - \varrho + 1)^3}{\varrho^6 - 3\varrho^5 + 5\varrho^3 - 3\varrho + 1}. \tag{12}$$

We can express $P_2(\varrho)$ in form

$$\begin{aligned} P_2(\varrho) = &(\varrho - a_1 - b_1 i)(\varrho - a_1 + b_1 i)(\varrho - a_2 - b_2 i) \\ &(\varrho - a_2 + b_2 i)(\varrho - a_3 - b_3 i)(\varrho - a_3 + b_3 i) \end{aligned} \tag{13}$$

and we can derive from (7) that the following conditions must be satisfied for roots of $P_2(\varrho)$

$$a_1^2 + b_1^2 = 1,$$
$$(a_2 - 1)^2 + b_2^2 = 1, \quad a_2 = 1 - a_1, \quad b_2 = b_1, \tag{14}$$
$$a_3 = \tfrac{1}{2}, \quad b_3 = \frac{b_1}{(a_1-1)^2+b_1^2} = \frac{b_2}{a_2^2+b_2^2},$$

see Fig. 1 left. The order of the roots can be changed. Our invariants has form in this context

$$
F'_+ = \frac{2\varrho}{\varrho^2+1} + \frac{2(1-\varrho)}{(1-\varrho)^2+1} + \frac{2\frac{\varrho}{\varrho-1}}{\frac{\varrho^2}{(\varrho-1)^2}+1} =
$$
$$
= 2\frac{\varrho^6-3\varrho^5+3\varrho^4-\varrho^3+3\varrho^2-3\varrho+1}{2\varrho^6-6\varrho^5+11\varrho^4-12\varrho^3+11\varrho^2-6\varrho+2},
$$
$$
F. = -\frac{3}{8}\frac{(-8)\varrho^2(1-\varrho)^2}{(1+\varrho^2)(2-2\varrho+\varrho^2)(1-2\varrho+2\varrho^2)} = \tag{15}
$$
$$
= \frac{3\varrho^2(1-\varrho)^2}{2\varrho^6-6\varrho^5+11\varrho^4-12\varrho^3+11\varrho^2-6\varrho+2},
$$
$$
F_+ = 16(\tfrac{1}{5} - \frac{1}{6-F'_+}) = \frac{8}{5}\frac{\varrho^2(1-\varrho)^2}{(\varrho^2-\varrho+1)^3}.
$$

The roots of the numerators of the invariants are 0 and 1, that of the denominator of $F.$ are $1\pm i$, $\pm i$ and $0.5\pm0.5i$ and the roots of the denominator of F_+ are $0.5\pm i\frac{\sqrt{3}}{2}$.

The function F_+ was chosen instead of F'_+ because of its different denominator than $F.$, what promises better discriminability of the invariants. The invariants F_+ and $F.$ were chosen because their roots are farther from the imaginary axis than that of F_{14} and F_{15} and therefore they promise to be more stable and also generalization for more than 5 points is easier.

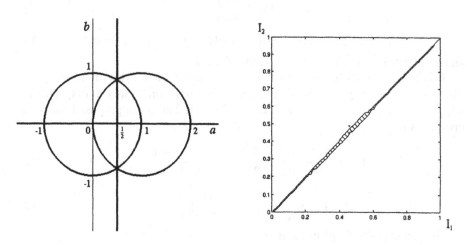

Fig. 1. Possible values of roots (left) and the invariants I_1, I_2 (right)

2.3 The normalization of the invariants

The invariants $I_1 = I_{++}$ and $I_2 = I_{.+}$ corresponding the functions F_+ and $F_.$ utilize the space ineffectively (see Fig. 1 right).

A better result can be reached as the sum and the difference (see Fig. 2 left)

$$I_1' = (I_1 + I_2)/2, \qquad I_2' = (I_1 - I_2 + 0.006) \cdot 53, \qquad (16)$$

but good utilization of the given range is reached by subtraction and division

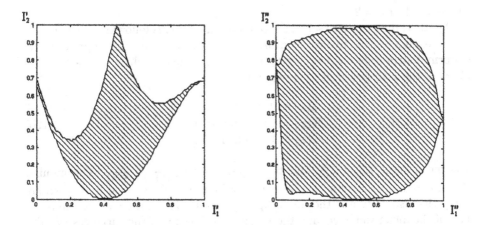

Fig. 2. Possible values of the invariants I_1', I_2' (left) and I_1'', I_2'' (right)

by the polynomials (see Fig. 2 right)

$$I_1'' = I_1', \qquad I_2'' = \frac{1 - I_2' + p(I_1')}{d(I_1')}, \qquad (17)$$

where $p(I_1') = 10.110488 \cdot I_1'^6 - 27.936483 \cdot I_1'^5 + 31.596612 \cdot I_1'^4 - 16.504259 \cdot I_1'^3 - 0.32251158 \cdot I_1'^2 + 3.0473587 \cdot I_1' - 0.66901966$. If $I_1' < 0.475$ then $d(I_1') = 17.575974 \cdot I_1'^4 - 16.423212 \cdot I_1'^3 + 9.111527 \cdot I_1'^2 - 0.43942294 \cdot I_1' + 0.016542258$ else $d(I_1') = 3.9630392 \cdot I_1'^4 - 13.941518 \cdot I_1'^3 + 21.672754 \cdot I_1'^2 - 17.304971 \cdot I_1' + 5.6198814$.

3 Invariants for more than five points

There is a number of solutions of the problem of the generalization of the invariants from the previous section to n points ($n > 5$). One of them with good experimental results can be addition of powers of the invariants I_1'', I_2'' over all possible combinations of 5 from n points C_5^n

$$I_{1,k} = \sum_{Q \in C_5^n} I_1''^k(Q), \qquad I_{2,k} = \sum_{Q \in C_5^n} I_2''^k(Q), \qquad k = 1, 2, \ldots, n - 4. \quad (18)$$

The number of these invariants is chosen as $2n - 8$ according to (6). The computing complexity is approximately $\binom{n}{5}T$, i.e. $O(n^5)$, where T is the computing complexity of one five-point invariant. This result is $5! = 120$ times faster than the solution in [4]. The number of added values is very high and that is why a normalization of the results is suitable. To preserve values inside acceptable intervals, we can use

$$I'_{s,k} = \sqrt[k]{\frac{1}{\binom{n}{5}} \sum_{Q \in C_5^n} I''^k_s(Q)}, \tag{19}$$

where $s = 1$ or $s = 2$.

Another, may be more sophisticated, normalization is following. We can suppose the values of the five-point invariants I''_1, I''_2 as random with approximately uniform distribution on the interval from 0 to 1. Then the distribution function $F_k(x)$ of the k-th power of the invariant is

$$\begin{aligned}
F_k(x) &= 0 \quad \text{from } -\infty \text{ to } 0 \\
F_k(x) &= x^{\frac{1}{k}} \text{ from } 0 \text{ to } 1 \\
F_k(x) &= 1 \quad \text{from } 1 \text{ to } \infty.
\end{aligned} \tag{20}$$

The mean value $\mu_k = \frac{1}{1+k}$ and the variance $\sigma_k^2 = \frac{k^2}{(1+k)^2(1+2k)}$. The number of summands in the sum (18) is relatively high and the central limit theorem imply the distribution of the sum is approximately Gauss, the mean value is the sum of the mean values μ_k and the variance is the sum of the variances σ_k^2. The given range is the best utilized in case of uniform distribution of the resulting invariants and therefore we can normalize the invariants with Gauss distribution function

$$G(x; \mu, \sigma) = \frac{1}{\sqrt{2\pi}\sigma} \int_{-\infty}^{x} e^{-\frac{(\xi-\mu)^2}{2\sigma^2}} d\xi \tag{21}$$

with the mean value $\mu = \binom{n}{5}\frac{1}{1+k}$, the variance $\sigma^2 = \binom{n}{5}\frac{k^2}{(1+k)^2(1+2k)}$ and the standard deviation $\sigma = \frac{k}{1+k}\sqrt{\binom{n}{5}\frac{1}{1+2k}}$

$$I''_{s,k} = G(I_{s,k}; \mu, \sigma). \tag{22}$$

An approximate substitution of the function G is used in practice.

4   Numerical experiment

The following numerical experiment was carried out to observe the stability and discriminability of the invariants (19) and (22).

A set of 11 points was created by a pseudo-random generator. They are uniformly distributed from 0 to 511. Independent noise with Gauss distribution with zero mean value and gradually increasing standard deviation was added to the coordinates both of all points and of one point. In the first experiment the standard deviation σ_2 increased from 0 to 9.5 with the step 0.5, in the second one σ_1 increased from 0 to 190 with the step 10. The distances d between original and noisy set in the space of the invariants were computed. To obtain a curve of dependency of the distance on the noise standard deviation the averages of many experiments were used. The average of gradually 10, 20, 100 and 1000 experiments was used and only the curve from 1000 values was little enough dependent on the concrete realization of the noise. The result of the invariants (19) normalized by the average and by the root is in Fig. 3 left.

The scale of the horizontal axis is different for both cases so the curves were as similar as possible. The ratio of the scales is 7.36; it means if two sets differ by only one point, then the distance of the points must be at least approximately 7.36 times greater than the noise standard deviation to be possible to distinguish both sets. If the noise standard deviation is greater than approximately 9, i.e. about 2 % of the coordinate range, then the sets cannot be distinguished at all.

The results of the invariants (22) normalized by the Gauss distribution function is in Fig. 3 right.

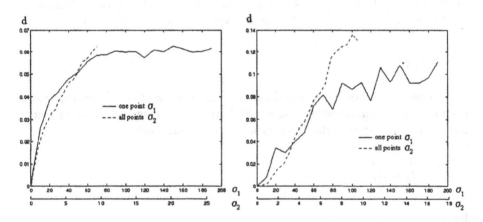

Fig. 3. The normalization by means of the average and the root (left) and by means of the Gauss distribution function (right)

The result is similar to previous case, the ratio of the scales is 11.18, that means little bit worse discriminability. The main difference is the scale on the vertical axis, which is about twice greater. It means these invariants utilize the given range better.

5 Summary

The features for recognition of the projectively deformed point sets were derived. They are invariant simultaneously with respect to projective transform and to the permutation of the points. They are expressed as functions of five-point cross-ratios. Possibilities of the choice of the roots of the polynomials in the denominators of the invariants and the normalization of these invariants are shown. The generalization for more than five points is made and their stability is studied. The performance of the features is demonstrated by numerical experiments.

6 References

1. N. S. V. Rao, W. Wu, C. W. Glover, "Algorithms for Recognizing Planar Polygonal Configurations Using Perspective Images", *IEEE Trans. on Robotics and Automation*, Vol. 8, No. 4, pp. 480-486, 1992.
2. T. H. Reiss, "Recognition Planar Objects Using Invariant Image Features", Lecture Notes in Computer Science 676, Springer, 1993.
3. R. Lenz, P. Meer, "Point configuration invariants under simultaneous projective and permutation transformations", *Pattern Recognition*, Vol. 27, No. 11, pp. 1523-1532, 1994.
4. T. Suk, J. Flusser, "The Features for Recognition of Projectively Deformed Images", *Proc. ICIP'95*, Washington, Vol. III, pp. 348-351, 1995.

7 Appendix

Sometimes a task how to save and load information about combinations to and from a memory can be need to solve. We have got the combinations k elements from n and we can save this information by following way:

$a = 0$
for $i_1 := 0; i_1 < n$
for $i_2 := i_1 + 1; i_2 < n$

$\vdots$

for $i_k := i_{k-1} + 1; i_k < n$
$\{ m[a] := $ information $(i_1, i_2, \ldots, i_k)$
$a := a + 1 \}$

and then we need to load the information about k-tuple $(i_1, i_2, \ldots, i_k)$. That is why we need to compute the address a from the k-tuple. If we sort the indices by size so it holds

$$i_1 < i_2 < \ldots < i_k$$

then this address can be computed as

$$a = \binom{n}{k} - 1 + \sum_{j=1}^{k} \sum_{m=0}^{k-j+1} (-1)^{m+1} \binom{n}{k-j-m+1} \binom{i_j+m}{m}. \qquad (23)$$

Computing 3D Projective Invariants from Points and Lines

J. Lasenby[1] and E. Bayro-Corrochano[2]

[1] Cambridge University Engineering Department, Trumpington St, Cambridge, U.K.,
jl@eng.cam.ac.uk
[2] Computer Science Institute. Christian Albrechts University, Kiel. Germany,
edb@informatik.uni-kiel.de

Abstract. In this paper we will look at some 3D projective invariants for both point and line matches over several views and, in the case of points, give explicit expressions for forming these invariants in terms of the image coordinates. We discuss whether such invariants are useful by looking at their formation on simulated data.

1 Introduction

The following sections will derive the forms for some 3D projective invariants using the system of *Geometric Algebra* (GA). Geometric algebra is a coordinate-free approach to geometry based on the algebras of Grassmann [5] and Clifford [3] – the approach we adopt here was pioneered by David Hestenes [7,6]. We will use the GA framework for projective geometry and geometric invariance outlined in [9,10]. A more extensive discussion of the formation of invariants in GA is given in [11].

2 Geometric Algebra – a Brief Outline

An n-dimensional geometric algebra is a graded linear space. As well as vector addition and scalar multiplication we have a non-commutative product which is associative and distributive over addition – this is the **geometric** or **Clifford** product. A further distinguishing feature of the algebra is that any vector squares to give a scalar. The geometric product of two vectors a and b is written ab and can be expressed as a sum of its symmetric ($a{\cdot}b$) and antisymmetric ($a{\wedge}b$) parts

$$ab = a{\cdot}b + a{\wedge}b. \qquad (1)$$

The inner product of two vectors is the standard *scalar* or *dot* product. The outer or wedge product of two vectors is a new quantity we call a **bivector**. We think of a bivector as a directed area in the plane containing a and b, formed by sweeping a along b. Thus, $b \wedge a$ will have the opposite orientation making the wedge product anticommutative. The outer product is immediately generalizable to higher dimensions – for example, $(a \wedge b) \wedge c$, a **trivector**, is interpreted as the oriented volume formed by sweeping the area $a \wedge b$ along vector c. The outer product of k vectors is a k-vector, which has *grade* k. Sums of objects of different grades are called *multivectors* and GA provides a system in which we can efficiently manipulate multivectors. In a space of 3 dimensions we can construct a trivector $a \wedge b \wedge c$, but no 4-vectors exist. The highest grade element in a space is called the **pseudoscalar**. The unit pseudoscalar is denoted by I.

3 Projective Space and the Projective Split

Points in real 3D space will be represented by vectors in $\mathcal{E}^3$, a 3D space with a Euclidean metric. Since any point on a line through some origin O will be mapped to a single point in the image plane, we associate a point in $\mathcal{E}^3$ with a line in a 4D space, R^4. We then define basis vectors: $(\gamma_1, \gamma_2, \gamma_3, \gamma_4)$ in R^4 and $(\sigma_1, \sigma_2, \sigma_3)$ in $\mathcal{E}^3$ and identify R^4 and $\mathcal{E}^3$ with the geometric algebras of 4 and 3 dimensions. We require that vectors, bivectors and trivectors in R^4 will represent points, lines and planes in $\mathcal{E}^3$. Choosing γ_4 as a selected direction in R^4, we can then define a mapping which associates the bivectors $\gamma_i \gamma_4$, $i = 1, 2, 3$, in R^4 with the vectors σ_i, $i = 1, 2, 3$, in $\mathcal{E}^3$. This process of association is called the **projective split**. To ensure $\sigma_i^2 = +1$ we are forced to assume a non-Euclidean metric for the basis vectors in R^4. We choose to use $\gamma_4^2 = +1$, $\gamma_i = -1$, $i = 1, 2, 3$.

For a vector $\mathbf{X} = X_1 \gamma_1 + X_2 \gamma_2 + X_3 \gamma_3 + X_4 \gamma_4$ in R^4 the projective split is obtained by taking the geometric product of $\mathbf{X}$ and γ_4. This leads to the association of the vector x in $\mathcal{E}^3$ with the bivector $\mathbf{X} \wedge \gamma_4 / X_4$ in R^4 so that

$$x = \frac{X_1}{X_4}\gamma_1\gamma_4 + \frac{X_2}{X_4}\gamma_2\gamma_4 + \frac{X_3}{X_4}\gamma_3\gamma_4 = \frac{X_1}{X_4}\sigma_1 + \frac{X_2}{X_4}\sigma_2 + \frac{X_3}{X_4}\sigma_3, \qquad (2)$$

which $\Rightarrow x_i = \frac{X_i}{X_4}$, for $i = 1, 2, 3$. The process of representing x in a higher dimensional space can therefore be seen to be equivalent to using **homogeneous coordinates**, $\mathbf{X}$, for x.

3.1 Projective Geometry and Algebra in Projective Space

We now look at the basic projective geometry operations of *meet* and *join*, and briefly discuss algebra in projective space. For more detail the reader is referred to [8,11,9]

Any pseudoscalar P can be written as $P = \alpha I$ where α is a scalar, so that

$$PI^{-1} = \alpha II^{-1} = \alpha \equiv [P]. \qquad (3)$$

This *bracket* is precisely the bracket of the Grassmann-Cayley algebra. We then define the dual, A^*, of an r-vector A as

$$A^* = AI^{-1}. \qquad (4)$$

The **join** $J = A \bigwedge B$ of an r-vector A and an s-vector B by

$$J = A \wedge B \qquad \text{if } A \text{ and } B \text{ are linearly independent,} \qquad (5)$$

while it can be shown that the *meet* of A and B can be written as

$$A \vee B = (A^* \wedge B^*)I = (A^* \wedge B^*)(I^{-1}I)I = (A^* \cdot B). \qquad (6)$$

The join and meet can be used to describe lines and planes and to intersect these quantities. Consider three non-collinear points, P_1, P_2, P_3, represented by vectors x_1, x_2, x_3 in $\mathcal{E}^3$ and by vectors $\mathbf{X}_1$, $\mathbf{X}_2$, $\mathbf{X}_3$ in R^4. The line L_{12} joining points P_1 and P_2, and the plane Φ_{123} passing through points P_1, P_2, P_3, can be expressed in R^4 by the following bivector and trivector respectively:

$$L_{12} = \mathbf{X}_1 \wedge \mathbf{X}_2 \qquad \Phi_{123} = \mathbf{X}_1 \wedge \mathbf{X}_2 \wedge \mathbf{X}_3. \qquad (7)$$

In $\mathcal{E}^3$ the intersection of a line and a plane, two planes and two lines can be dealt with entirely using the meet operation. Details and derivations are given in [11].

4 3-D Projective Invariants from Multiple Views

Given six general 3D points P_i, $i = 1...,6$, represented by vectors $\{x_i, X_i\}$ in $\mathcal{E}^3$ and R^4, we can form 3D projective invariants. One such invariant is

$$Inv1 = \frac{[X_1X_2X_3X_4][X_4X_5X_2X_6]}{[X_1X_2X_4X_5][X_3X_4X_2X_6]}. \tag{8}$$

If one can express the bracket $[X_iX_jX_kX_l]$ in terms of the image coordinates of the points, then this invariant will be readily computable. Some recent work which has addressed this problem has utilized the Grassmann-Cayley (CG) algebra [2,4]. In [2] invariants were computed from a pair of images in terms of the image coordinates and the fundamental matrix, F, using the CG-algebra. Despite the clarity of the derivations in [2], some degree of confusion has arisen when subsequent workers have tried to implement these invariants with real data [4]. In the following sections we will look at how we would derive, using the GA formalism, explicit expressions for the invariants in terms of the experimental data and discuss why this confusion has arisen.

Consider the scalar S_{1234} formed from the bracket of 4 points

$$S_{1234} = [X_1X_2X_3X_4] = (X_1{\wedge}X_2{\wedge}X_3{\wedge}X_4)I_4^{-1} = (X_1{\wedge}X_2){\wedge}(X_3{\wedge}X_4)I_4^{-1}. \tag{9}$$

The quantities $(X_1 \wedge X_2)$ and $(X_3 \wedge X_4)$ represent the lines joining points P_1 & P_2, and P_3 & P_4. a_0 and b_0 are the centres of projection of the two cameras and the two camera image planes are defined by the two sets of vectors $\{a_1, a_2, a_3\}$ and $\{b_1, b_2, b_3\}$. The projection of points $\{P_i\}$ through the centres of projection onto the image planes are given by the vectors $\{a_i'\}$ and $\{b_i'\}$. Note that the vectors, a_i, b_i, etc., are vectors in $\mathcal{E}^3$; we let the representations of these vectors in R^4 be $A_i, B_i, A_i', B_i'...$, etc.

It can be shown [11] that we are able to reproduce the result given in [2], namely that it is possible to write the bracket of the 4 points (in R^4) as

$$S_{1234} = [X_1X_2X_3X_4] \equiv [A_0B_0A_{1234}'B_{1234}'], \tag{10}$$

where A_{1234}' is the 4D representation of a_{1234}', the intersection of the lines joining $\{a_1'$ & $a_2'\}$ and $\{a_3'$ & $a_4'\}$. In [11] equation (10) is obtained by splitting up the bracket into $X_1 \wedge X_2$ and $X_3 \wedge X_4$, and then expressing each of these lines (bivectors) as the meet of two planes (trivectors). When we take ratios of brackets to form invariants, the same decomposition of $X_i \wedge X_j$ must occur in the numerator and denominator so the factors, due to the choices of the γ_4 components, cancel. In the case of $Inv1$ given in equation (8), we have

$$Inv1 = \frac{\{(X_1{\wedge}X_2){\wedge}(X_3{\wedge}X_4)\}I_4^{-1}\{(X_4{\wedge}X_5){\wedge}(X_2{\wedge}X_6)\}I_4^{-1}}{\{(X_1{\wedge}X_2){\wedge}(X_4{\wedge}X_5)\}I_4^{-1}\{(X_3{\wedge}X_4){\wedge}(X_2{\wedge}X_6)\}I_4^{-1}} \tag{11}$$

so we see that this decomposition rule has been obeyed. Consider now the invariant which can be thought of as arising from 4 points and a line (since the line $X_1{\wedge}X_2$ appears in each bracket), namely

$$Inv2 = \frac{[X_1X_2X_3X_4][X_1X_2X_5X_6]}{[X_1X_2X_3X_5][X_1X_2X_4X_6]}. \tag{12}$$

We note that we can simply rearrange equation (12) into the form of equation (8) and decompose into bivectors to obtain the following

$$Inv2 \equiv \frac{[A_0B_0A'_{1423}B'_{1423}][A_0B_0A'_{1526}B'_{1526}]}{[A_0B_0A'_{1523}B'_{1523}][A_0B_0A'_{1426}B'_{1426}]}.$$ (13)

So far the invariants have been derived in 4D using the 4D definition of the fundamental matrix; we therefore need to correctly transfer the expression to 3D. Expanding the bracket in equation (10) by expressing the intersection points in terms of the As and Bs ($A'_i = \alpha_{ij}A_j$ and $B'_i = \beta_{ij}B_j$) and defining a matrix $\tilde{F}$ such that

$$\tilde{F}_{ij} = [A_0B_0A_iB_j]$$ (14)

and vectors $\alpha_{1234} = (\alpha_{1234,1}, \alpha_{1234,2}, \alpha_{1234,3})$ and $\beta_{1234} = (\beta_{1234,1}, \beta_{1234,2}, \beta_{1234,3})$, it is easy to see that we can write $S_{1234} = \alpha^T{}_{1234}\tilde{F}\beta_{1234}$ [2]. The ratio

$$Inv1 = \frac{(\alpha^T{}_{1234}\tilde{F}\beta_{1234})(\alpha^T{}_{4526}\tilde{F}\beta_{4526})}{(\alpha^T{}_{1245}\tilde{F}\beta_{1245})(\alpha^T{}_{3426}\tilde{F}\beta_{3426})}$$ (15)

is therefore an invariant. We now wish to express $Inv1$ in terms of the observed image coordinates and the fundamental matrix calculated from these coordinates. A point P_i will be projected onto points a'_i and b'_i in image planes 1 and 2, which can be written as

$$a'_i = a_1 + \lambda_i(a_2 - a_1) + \mu_i(a_3 - a_1) = \delta_{i1}a_1 + \delta_{i2}a_2 + \delta_{i3}a_3$$ (16)

so that $\sum_{j=1}^{3}\delta_{ij} = 1$. Similarly, we have $b'_i = \epsilon_{i1}b_1 + \epsilon_{i2}b_2 + \epsilon_{i3}b_3$ (so that $\sum_{j=1}^{3}\epsilon_{ij} = 1$). Using the projective split we can now write the α_{ij}'s and β_{ij}'s in terms of the δ_{ij}'s and ϵ_{ij}'s:

$$\alpha_{ij} = \frac{A'_i \cdot \gamma_4}{A_j \cdot \gamma_4}\delta_{ij} \qquad \beta_{ij} = \frac{B'_i \cdot \gamma_4}{B_j \cdot \gamma_4}\epsilon_{ij}$$ (17)

The 'fundamental' matrix $\tilde{F}$ is such that $\alpha^T{}_i\tilde{F}\beta_i = 0$, if α_i and β_i correspond to the same world point P_i. Given more than eight pairs of corresponding observed points in the two planes, (δ_i, ϵ_i), $i = 1, .., 8$, we can form an 'observed' fundamental matrix F such that

$$\delta^T{}_iF\epsilon_i = 0.$$ (18)

This F can be found by some method such as the Longuet-Higgins 8-point algorithm [12] or, more correctly, by some method which gives an F which has the true structure [13]. Therefore, if we define $\tilde{F}$ by

$$\tilde{F}_{kl} = (A_k \cdot \gamma_4)(B_l \cdot \gamma_4)F_{kl}$$ (19)

then it follows from equation (17) that

$$\alpha_{ik}\tilde{F}_{kl}\beta_{il} = (A'_i \cdot \gamma_4)(B'_i \cdot \gamma_4)\delta_{ik}F_{kl}\epsilon_{il}.$$ (20)

Therefore an $\tilde{F}$ defined as in equation (19) will also act as a fundamental matrix in R^4.

According to the above, we can write the invariant as

$$Invl = \frac{(\delta^T{}_{1234}F\epsilon_{1234})(\delta^T{}_{4526}F\epsilon_{4526})\phi_{1234}\phi_{4526}}{(\delta^T{}_{1245}F\epsilon_{1245})(\delta^T{}_{3426}F\epsilon_{3426})\phi_{1245}\phi_{3426}} \tag{21}$$

where $\phi_{pqrs} = (\mathbf{A}'_{pqrs}\cdot\gamma_4)(\mathbf{B}'_{pqrs}\cdot\gamma_4)$. The ratio of the $\delta^T F\epsilon$ terms using only the observed coordinates and the estimated fundamental matrix, will therefore not be an invariant – one must include the factors ϕ_{1234} etc. It is easy to show [11] that these factors can be formed as follows:

Since a'_3, a'_4 and a'_{1234} are collinear we can write $a'_{1234} = \mu_{1234}a'_4 + (1 - \mu_{1234})a'_3$. Then, by expressing $\mathbf{A}'_{1234}$ as the intersection of the line joining $\mathbf{A}'_1$ and $\mathbf{A}'_2$ with the plane through $\mathbf{A}_0, \mathbf{A}'_3, \mathbf{A}'_4$ we can projective split and equate terms to give

$$\frac{(\mathbf{A}'_{1234}\cdot\gamma_4)(\mathbf{A}'_{4526}\cdot\gamma_4)}{(\mathbf{A}'_{3426}\cdot\gamma_4)(\mathbf{A}'_{1245}\cdot\gamma_4)} = \frac{\mu_{1245}(\mu_{3426} - 1)}{\mu_{4526}(\mu_{1234} - 1)}. \tag{22}$$

We obtain the values of μ from the images. The factors $\mathbf{B}'_{pqrs}\cdot\gamma_4$ are found in a similar way so that if $b'_{1234} = \lambda_{1234}b'_4 + (1 - \lambda_{1234})b'_3$ etc., the overall expression for the invariant becomes

$$Invl = \frac{(\delta^T{}_{1234}F\epsilon_{1234})(\delta^T{}_{4526}F\epsilon_{4526})}{(\delta^T{}_{1245}F\epsilon_{1245})(\delta^T{}_{3426}F\epsilon_{3426})} \frac{\mu_{1245}(\mu_{3426} - 1)}{\mu_{4526}(\mu_{1234} - 1)} \cdot \frac{\lambda_{1245}(\lambda_{3426} - 1)}{\lambda_{4526}(\lambda_{1234} - 1)}. \tag{23}$$

While the above has adopted the approach of forming all invariants in 4D and then finding the equivalent expression in 3D, the approach outlined in [2] gave the invariant in the form of equation (15), but did indeed *define* α_{1234} as follows:

$$\alpha_1\alpha_2\wedge\alpha_3\alpha_4 \tag{24}$$

where the '$\wedge$' in this equation is the *meet* of the Cayley-Grassmann algebra. Thus, α_{1234} is **not** the homogeneous coordinate vector of the intersection point of the two lines in the image plane joining $\mathbf{A}'_1$ & $\mathbf{A}'_2$ and $\mathbf{A}'_3$ & $\mathbf{A}'_4$, but rather some multiple of that vector, given by equation (24). It can be easily shown that computing the invariant using equation (24) and the corresponding expressions for the other intersection points, produces exactly those correction factors arrived at by us in equation (23). It is therefore likely that the past confusion over the formation of the invariants has been soley due to the misinterpretation of the nature of the quantities α_{ijkl} and β_{ijkl}; however, the derivation we have presented here is totally unambiguous and, by clearly distinguishing between 3- and 4D quantities, cannot be misinterpreted.

5 3D Projective Invariants for Lines

Consider again the projective invariant $Inv2$. Splitting equation (12) into bivectors gives

$$Inv2 = \frac{[L_1\wedge L_2][L_3\wedge L_4]}{[L_1\wedge L_4][L_3\wedge L_2]} \tag{25}$$

where, $L_1 = \mathbf{X}_1\wedge\mathbf{X}_3$, $L_2 = \mathbf{X}_2\wedge\mathbf{X}_4$, $L_3 = \mathbf{X}_1\wedge\mathbf{X}_6$ and $L_4 = \mathbf{X}_2\wedge\mathbf{X}_5$. We thus have an invariant of four lines (provided the lines are not coplanar). Following the notation used in [1] we can express each of these lines as an intersection of planes:

$$L_1 = l_i^{13} l_j^{13'}(\Phi_i^A \vee \Phi_j^B) \qquad L_2 = l_k^{24} l_m^{24'}(\Phi_k^A \vee \Phi_m^B) \tag{26}$$

$$L_3 = l_n^{16} l_p^{16'}(\Phi_n^A \vee \Phi_p^B) \qquad L_4 = l_q^{25} l_r^{25'}(\Phi_q^A \vee \Phi_r^B). \tag{27}$$

In this expression $\Phi_1^A = A_0 \wedge A_2 \wedge A_3$, $\Phi_2^A = A_0 \wedge A_3 \wedge A_1$ etc., and the ls and l's are the line coordinates (equivalent to the homogeneous line coordinates) defined by,

$$A_1' \wedge A_2' = l_i^{12} L_i^A \qquad B_1' \wedge B_2' = l_i^{12'} L_i^B \tag{28}$$

where, $L_1^A = A_2 \wedge A_3$, $L_2^A = A_3 \wedge A_1$ and $L_3^A = A_1 \wedge A_2$ etc. We can now write $L_1 \wedge L_2$ as

$$L_1 \wedge L_2 = l_i^{13} l_j^{13'} l_k^{24} l_m^{24'} \{(\Phi_i^A \vee \Phi_j^B) \wedge (\Phi_k^A \vee \Phi_m^B)\} = S_{ijkm} l_i^{13} l_j^{13'} l_k^{24} l_m^{24'} \tag{29}$$

where we define the 4th rank tensor S_{ijkm} by

$$S_{ijkm} = \{(\Phi_i^A \vee \Phi_j^B) \wedge (\Phi_k^A \vee \Phi_m^B)\}. \tag{30}$$

It can be shown that S has only 9 independent elements which are, of course, the elements of F. S relates pairs of intersecting lines in two images via the following equation;

$$S_{ijkm} l_i^{ab} l_j^{ab'} l_k^{ac} l_m^{ac'} = 0, \tag{31}$$

where l_i^{ab} are the line coordinates of the line joining points a and b in the first image etc. Thus, according to equations (25) and (29), given two pairs of intersecting lines ((13)&(16) and (24)&(25)), we can form the following 3D projective invariant:

$$Inv2 = \frac{(S_{ijkm} l_i^{13} l_j^{13'} l_k^{24} l_m^{24'})(S_{npqr} l_i^{16} l_j^{16'} l_k^{25} l_m^{25'})}{(S_{ijqr} l_i^{13} l_j^{13'} l_k^{25} l_m^{25'})(S_{npkm} l_i^{16} l_j^{16'} l_k^{24} l_m^{24'})}. \tag{32}$$

We note that the above is equivalent to the determination of the invariant of 4 lines given in [2]:

$$Inv2 = \frac{(l^T{}_{12} \tilde{F} l_{12}')(l^T{}_{34} \tilde{F} l_{34}')}{(l^T{}_{14} \tilde{F} l_{14}')(l^T{}_{32} \tilde{F} l_{32}')} \tag{33}$$

where $l_{ij} = l_j \times l_i$ etc, with l_i the homogeneous line coordinates. A fuller discussion of the subject of 3D projective invariants from lines will be given elsewhere.

6 Experiments

Here we investigate the formation of the 3D projective invariants from sets of 6 matching image points – in particular we look at their stability in noisy environments.

The simulated data was a set of 38 points taken from the vertices of a wireframe house and viewed from three different camera positions. From three sets of 6 points (non coplanar) we form $Inv1$ for each set over views 1 & 2, 1 & 3 and 2 & 3. During the simulations the world points are projected onto the image planes and then gaussian noise is added. Figure 1 shows results for the three sets of points chosen. In figure 1, a). c), e) we plot the value of the invariant with increasing noise. In a), c), and e) the invariant was formed using an F calculated via a linear least-squares method from a set of 30 matching points. Figure 1 b), d) and f) show the same invariants formed this time by taking the noisy point matches but the true value of F (i.e. that formed in the noiseless case). The true values of the invariants for the three sets of lines were 0.655, 0.402 and 8.99.

For small values of the noise the invariants can be calculated accurately. In greater noise large variations are possible for some invariants whereas other invariants are relatively robust. Figure 1 indicates that uncertainties in the calculation of F will significantly affect the invariant in some cases. It is also apparent that the formation of this invariant is more accurate between some pairs of views than between others. We should expect this since altering the view may mean that the 6 points move closer to some unstable or degenerate configuration. In summary it appears that the type of invariant described here may be useful for data which is not noisy but that the degradation in the presence of significant noise may render it ineffective for real images.

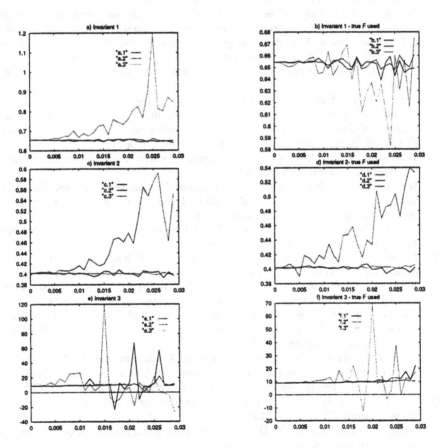

Fig. 1. Plots showing the behaviour of the 3D invariant between three different pairs of views with increasing noise. The solid, dashed and dotted lines show the invariant formed between views 1 & 2, 1 & 3 and 2 & 3 respectively (denoted by $a.1$, $a.2$, $a.3$ etc. in the key). The x-axis shows the standard deviation of the gaussian noise used.

7 Conclusions

This paper outlines a framework for projective geometry and the algebra of incidence which is then used to discuss the formation of projective invariants. Explicit expressions are given for one sort of 3D invariant using image points and the behaviour of this invariant is investigated for a variety of simulated scenarios. Such invariants may be useful in low noise, but in cases of greater uncertainty there may be too many problems for the invariants to be useful over a wide range of possible circumstances.

References

1. Bayro-Corrochano, E., Lasenby, J. and Sommer, G. 1996. Geometric Algebra: a framework for computing point and line correspondences and projective structure using n-uncalibrated cameras. *Proceedings of the International Conference on Pattern Recognition (ICPR'96), Vienna, August 1996.*

2. Carlsson, S. 1994. The Double Algebra: and effective tool for computing invariants in computer vision. *Applications of Invariance in Computer Vision.* Lecture Notes in Computer Science 825; Proceedings of 2nd-joint Europe-US workshop, Azores, October 1993. Eds. Mundy, Zisserman and Forsyth. Springer-Verlag.

3. Clifford, W.K. 1878. Applications of Grassmann's extensive algebra. *Am. J. Math.* 1: 350–358.

4. Csurka, G. and Faugeras, O. 1995. Computing three-dimensional projective invariants from a pair of images using the Grassmann-Cayley algebra. Proceedings of Europe-China Workshop on *Geometric Modeling and Invariants for Computer Vision,* Ed. Roger Mohr and Wu Chengke, Xi'an, China, April 1995.

5. Grassmann, H. 1877. Der ort der Hamilton'schen quaternionen in der ausdehnungslehre. *Math. Ann.,* 12: 375.

6. Hestenes, D. 1986. New Foundations for Classical Mechanics *D. Reidel,* Dordrecht.

7. Hestenes, D. and Sobczyk, G. 1984. Clifford Algebra to Geometric Calculus: A unified language for mathematics and physics. *D. Reidel,* Dordrecht.

8. Hestenes, D. and Ziegler, R. 1991. Projective Geometry with Clifford Algebra. *Acta Applicandae Mathematicae,* 23: 25–63.

9. Lasenby, J., Bayro-Corrochano, E., Lasenby, A. and Sommer, G. 1996. A new methodology for computing invariants in computer vision. *Proceedings of the International Conference on Pattern Recognition (ICPR'96), Vienna. August 1996.*

10. Lasenby, J., Bayro-Corrochano. E., Lasenby, A. and Sommer. G. 1996. A New Framework for the Computation of Invariants and Multiple-View Constraints in Computer Vision. *Proceedings of the International Conference on Image Processing (ICIP).* Lausanne, September 1996.

11. Lasenby, J., Bayro-Corrochano, E., Lasenby, A. and Sommer, G. 1996. A New Methodology for the Computation of Invariants in Computer Vision. Cambridge University Engineering Department Technical Report, CUED/F-INENG/TR.244.

12. Longuet-Higgins, H.C. 1981. A computer algorithm for reconstructing a scene from two projections. *Nature,* 293: 133–138.

13. Luong, Q-T. and Faugeras, O.D. 1996. The fundamental matrix: theory, algorithms and stability analysis. *IJCV,* 17: 43–75.

$2D \rightarrow 2D$ Geometric Transformation Invariant to Arbitrary Translations, Rotations and Scales

Lucian Muresan *

JINR, Dubna, Moscow Region, 141980 Russia

Abstract. This paper presents a geometric transformation invariant to arbitrary translations, rotations and scales changes. The output contains only intrinsic information on the internal structure of pattern consisting of pairs of angles. This approach does not use specific feature points of the curve as curvature maxima, torsion extrema or curvature discontinuity. The quasi-linearity feature of this transformation allows recognition of two or more images without segmentation. It also allows the reconstruction of image when a part of it is missing or the recognition when extra noise curves are on the same scene.

1 Introduction

In visual pattern recognition and segmentation the objects to be recognized are subjected to various forms of transformation involving shift, scale, rotation and deformation. During the last couple of years it has become increasingly popular to use invariance in computer vision. Many invariant $2D$ algorithms have been suggested in the literature on visual pattern recognition. These algorithms differ in the degrees of capabilities. Invariant signatures was used for planar shape recognition under viewing distortions and partial occlusion [5],[6]. The article [3] describes an algorithm for $3D$ object hypothesis generation from a single image. The combinatorial properties of the triplets of line segments are used to define an index to a model library. Order and incidence relations have the invariance properties that make them especially interesting for general recognition problems. Relation between $3D$ and $2D$ invariants is described in [8]. It established polynomial relations between the invariants of certain mixed sets of $3D$ points and lines and the invariants of their $2D$ projected images. Direct ways for projective reconstruction in the case of constrained scenes are discussed in [4]. In [7] there are many invariant examples for a geometric configuration containing points, lines or conics. The article [2] discusses some feature relations invariant to $T - R - S$ as angle between lines, logarithm ratio of line lengths, using specific local feature points to establish the similarity-invariant coordinate system. Article [1] presents a system for matching and pose $3D$ curves making use of constraints not only on certain feature points but also on the entire curve.

---

* Home Institute: Institute of Physical Chemistry Spl. Independentei No. 202, Bucharest, Romania

The goal of this paper is to define and investigate a geometrical transformation $\Omega : 2D \to 2D$ invariant to arbitrary translations (T), rotations (R) and scaling (S). The input space noted *Retina* is a $R \times R$ plane and the input image I_R is a geometrical configuration of any type of curves. The output space noted as *Cortex* is a $R \times R$ plane and $I_C = \Omega(I_R)$ is a mixed points and curves which is insensitive if I_R was translated, rotated or scaled. All possible $(T \circ R \circ S)(I_R)$ "converge" to the same image $\Omega(I_R)$. For one image we study the situations when a part of the image is missing or when the other noise-curves are presented at the same time . For two images presented simultaneously as a scene $I_R = I_{R1} \cup I_{R2}$ we study the behavior of Ω when I_{R1} and I_{R2} are independently shifted, rotated or scaled. The dependence of transformation to small deformations is presented. The transformation Ω defines equivalence classes of objects in input space, where each member of the equivalence class (I_R) has the same $\Omega(I_R)$ representation in angle space.

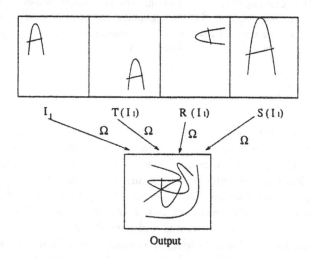

I_1 $T(I_1)$ $R(I_1)$ $S(I_1)$

Ω Ω Ω Ω

Output

Fig. 1. $T - R - C$ invariance. I_R -input pattern, $T(I_R)$ translated $R(I_R)$ rotated and $S(I_R)$ scaled

2 $T - R - S$ invariance

In the context of this article, we consider a geometric transformation as an unconstrained geometric operation. It is not essential if the connectivity is preserved within the image. An explicit definition of Ω will be given for each point $P_R(x, y) \in I_R$ the corresponding points $\{P_{C1}, P_{C2}\} \in I_C$. Ω is one to many type, one point can have many representation points. The invariance with respect to translations, rotations and scaling of a geometrical operation Ω is defined by relations:

$$\Omega(T(I_R)) = \Omega(I_R) \qquad \Omega(R(I_R)) = \Omega(I_R) \qquad \Omega(S(I_R)) = \Omega(I_R) \quad (1)$$

for all T, R, C and for all I. Eqs.1 are illustrated in Fig.1. All images converge modulo $T \circ R \circ S$ to the same image I_C.

3 Useful properties of $T - R - S$ invariant representation

In this section we define other important properties of a $T - R - S$ invariant representation useful for visual pattern recognition, image segmentation and selective attention.

3.1 The quasi-linearity

When two patterns I_{R1} and I_{R2} are on the scene, the input of Ω is that scene denoted by $I_{R1} \cup I_{R2}$. In this case each image represents noise for the other image. The problem is to recognize I_{R1} and I_{R2} in such a scene (Fig.2).

The quasi-linearity property of Ω is defined by:

$$\Omega(I_{R1} \cup I_{R2}) = \Omega(I_{R1}) \cup \Omega(I_{R2}) \cup Interference(I_{R1}, I_{R2}) \quad (2)$$

where $\Omega(I_{R1} \cup I_{R2})$ is the invariant representation of the scene, $\Omega(I_{R1})$ and $\Omega(I_{R2})$ are the invariant representations of I_{R1}, I_{R2} respectively.

$Interference(I_{R1}, I_{R2})$ is the part of the output and is generated only when both I_{R1} and I_{R2} are simultaneously present on $Retina$. The $Interference(I_{R1}, I_{R2})$ term depends on the relative position of I_{R1} and I_{R2}.

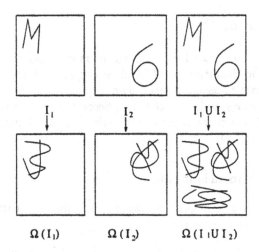

Fig. 2. Quasi-linearity property. Separated images I_1, I_2 and the scene with both images $I_1 \cup I_2$.

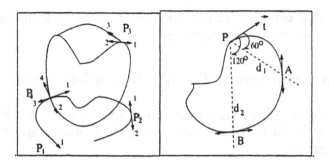

Fig. 3. Different semi-tangents (left) and the invariant rule (right)

If we translate, rotate or scale independently I_{R1} and I_{R2} on the scene, this term only will change (Eq.3).

$$\Omega(I_{R1}^{trs} \cup I_{R2}^{trs}) = \Omega(I_{R1}) \cup \Omega(I_{R2}) \cup Interference(I_{R1}^{trs}, I_{R2}^{trs}) \qquad (3)$$

with:

$$I_{R1}^{trs} = (T_1 \circ R_1 \circ S_1)I_{R1} \qquad I_{R2}^{trs} = (T_2 \circ R_2 \circ S_2)I_{R2}$$

The quasi-linearity feature of a geometrical transformation allows the recognition of two or more images without the segmentation. It also allows the reconstruction of an image when part of it is missing or the recognition when extra noise curves are present.

4 The definition of Ω

In this section we give the procedure to assign pairs of angles to each point of the input pattern. The definition of Ω involves two steps. First we defined a rule $T - R - S$ invariant to assign to each point $P(x, y) \in I_R$ one or more pairs of points $(A(x_A, y_A), B(x_B, y_B))$, $A, B \in I_R$. Second, we use each pair (A, B) and the point P to define pairs of invariant angles (α_k, β_k). Finally, we assign to $P(x, y)$ the pairs (α_k, β_k).

4.1 The $T - R - S$ invariant rule

A rule $P \overset{rule}{\Longrightarrow} (A, B)$ which assigns other two points of the same pattern A, B to P is $T - R - S$ invariant if, we have :

$$T \circ R \circ S(P) \overset{rule}{\Longrightarrow} (T \circ R \circ S(A), T \circ R \circ S(B)) \qquad for \ all \ T \circ R \circ S \qquad (4)$$

There are many possible rules with this property. In this section we define one of them.

In each point P of a curve, we can define one semi-tangent $\vec{t}_P$. Usually there are two, with 180° difference between them. At the intersection of two curves the number of semi-tangents can be three or more (Fig.3 left) . The angle between two semi-tangents is invariant. Let's denote the semi-straight line at 60° clockwise from $\vec{t_P}$ by d_1 and the same for 120° by d_2 (Fig.3 right). The intersections of d_1 and curves of the input pattern are the points $\{A_1, A_2, ...\}$ and of d_1 and the curves are the points $\{B_1, B_2, ...\}$. Now we can define the invariant rule: for each point P and all semi-tangents $\{t_q\}_{q=1,2..}$ of P we assign the pairs of points $\{(A_k^q, B_j^q)\}_{k,j=1,2..}$ to P (Fig.3 right). If d_1 or d_2 give no intersections for that $\vec{t_P}$, there are no pairs defined in this case.

4.2 The pairs of angles and Ω definition

Having three points of pattern, we can define a pair of angles corresponding to these three points. If A, B are two points that are given by $T - R - S$ invariant rule for P and $\vec{t_P}, \vec{t_A}, \vec{t_B}$ are semi-tangents at P, A, B respectively, then a pair of angles is :

$$(\alpha, \beta)_P \quad with \quad \alpha = (\vec{t_P}, \vec{t_A}) \quad \beta = (\vec{t_P}, \vec{t_B}) \tag{5}$$

For the point P there are one or more pairs (α_k, β_k). These angles are invariants. Now we can define the transformation Ω. For each point P, Ω assigns all corresponding pairs of angles $(\alpha_k, \beta_k)_P$ defined above :

$$(x_P, y_P) \xrightarrow{\Omega} \{(\alpha_1, \beta_1), (\alpha_2, \beta_2), (\alpha_3, \beta_3)...\} \tag{6}$$

For the curves which can be parameterized it is simple to obtain an explicit mathematical expression of Ω. There are no constrains related to the type, dimensions and number of curves of input pattern. There are no theoretical constraint of the type of translations rotations and scales.

5 Properties of Ω. Computer simulation

With C implementation of this algorithm under XWidows platform using the OSF/Motif interface we obtained the output of Ω for input curves that can be parameterized.

The main propertiy is the $T-R-S$ invariance. In Fig.4 one pattern I_R (a) was translated $T(I_R)$ (b), rotated $R(I_R)$ (c) and scaled $S(I_R)$ (d) . The corresponding invariant representations of patterns (a), (b), (c) and (d) are displayed in the bottom pictures.

A local deformation of the input pattern I_R gives a small deformation of the output representation $\Omega(I_R)$ (Fig.5). The quasi-continuity property is important, if there is a deformation-invariant algorithm. Giving the output to a deformation-invariant algorithm this $T - R - S$ invariant algorithm will perform $T - R - S$ and deformation invariant transformation.

Two images I_{R1} and I_{R2} where presented on *Retina* as a scene $I_{R1} \cup I_{R2}$. Figure 6 shows the *Cortex* corresponding to the image I_{R1}(a), to the image I_{R2}

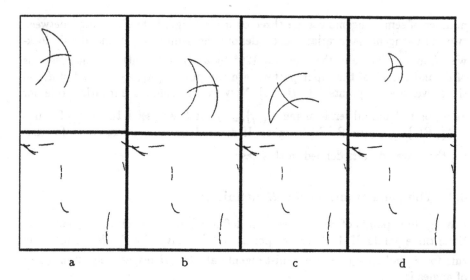

Fig. 4. Results of $T - R - C$ invariance for Ω. The input pattern is I_R (a), after translation $T(I_R)$ (b), rotation $R(I_R)$ (c) and scale $S(I_R)$ (d). Bottom output of the Ω

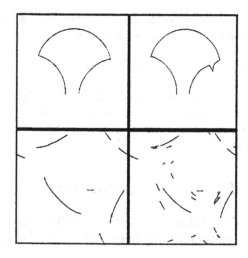

Fig. 5. Results of quasi-continuity of Ω. Small changes of input I_R give small changes of $\Omega(I_R)$.

(b) and the scene $I_{R1} \cup I_{R2}$ (c). We can see the presence of $\Omega(I_{R1})$ and $\Omega(I_{R2})$ on $\Omega(I_{R1} \cup I_{R2})$ (Fig.6 c). The interference term changes. if I_{R1} and I_{R2} are independently translated, rotated or scaled but the representations of I_{R1} and I_{R2} are not changed (Fig.6 d).

This property can be used in other algorithms as image segmentation, selective attention, analyzing sequences of images of a rigid object and reconstruction of an image when a part of it is missing . When a part of I_R is missing we obtain

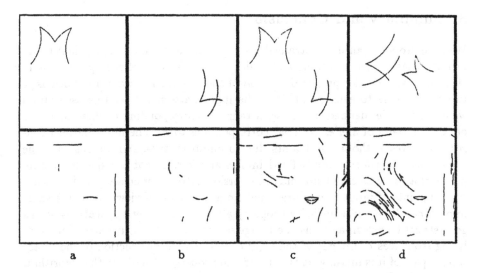

Fig. 6. Results of quasi-linearity of Ω. Images I_{R1} (a), I_{R2} (b), scene $I_{R1} \cup I_{R2}$ (c) and after I_{R1} and I_{R2} are independently translated, rotated and scaled (d)

Fig. 7. Behavior of Ω when a part of input image I_R (a) is missing (b)

Fig.7. The entire pattern I_R (Fig.7 left) with $\Omega(I_R)$ is on the right and the case when a part of it is missing is in (Fig.7 right). If I_R^{part} is I_R without a part of it we can write:

$$if \quad I_R^{part} \subset I_R \quad then \quad \Omega(I_R^{part}) \subset \Omega(I_{R1}) \tag{7}$$

An associative memory can be designed using this feature.

6 Summary and Conclusion

We have proposed an unconstrained geometrical transformation invariant to arbitrary translations, rotations and scales. The invariance of the approach is in effect achieved by the proper definition of the rule for invariant associations of two points A, B to each point P of the input pattern . After this association we use the angles defined by the semi-tangents corresponding to these points to obtain the pairs of angles (α_k, β_k) corresponding to P. The input pattern consists of curves without constraints due to numbers, type and dimensions. The geometrical transformation defined in this article does not use specific feature points of a curve as curvature maxima, torsion extrema or curvature discontinuity. All points of input curves are equal from this point of view having the same importance. The quasi-linearity property allows many other applications of this geometrical transformation as the image segmentation, selective attention, analyzing sequences of images of the rigid object, and reconstruction of the image when a part of it is missing. It is not a trivial property. Usually, in the algorithms presented in the literature on invariant representation of curves, when the input pattern contains several modeled curves, different curves have to be separated a priori.

Acknowledgement

I am extremely grateful to Ioana Bugariu and Ivan Bajla for helpful comments.

References

1. Li,S.Z.: Matching: Invariant to Translations, Rotations and Scale Changes. Pattern Recognition bf 25 (1992) 583-594
2. Li,S.Z.: Invariant Representation, Matching and Pose Estimation of $3D$ Space Curves under Similarity Transformations. Preprint Pattern Recognition accepted for publication (1996)
3. Carlsson,S.: Combinatorial Geometry for Shape Representation and Indexing. Proc. International Workshop on Object Representation for Computer Vision (1996) April 13-14 Cambrige England.
4. Carlsson,S: View Variation and Linear Invariants in $2D$ and $3D$. Tech. rep. Royal Institute of Technology (1995)
5. Bruckstein, A.M., Rivlin, E. and Weiss,I.: Scale Space Local Invariants CIS Report 9503, Technion-Israel Institute Of Technology (1995)
6. Bruckstein, A.M., Holt, R.J., Netravali, A.N. and Richardson, T.J.: Invariant Signatures for Planar Shape Recognition under Partial Occlusion CVCIP: Image Understanding 58 (1993) 49-65
7. Mundy, J.L., Zisserman, A. and Forsyth, D. (eds.): Applications of Invariance in Computer Vision Lecture Notes in Computer Science Springer-Verlang 825 (1994)
8. Maybank, S.J.: Relation Between $3D$ Invariants and $2D$ Invariants Proc. IEEE Workshop on Representation of Visual Scenes MIT, Cambrige, MA. June 24 (1995)

Extraction of Filled-in Data from Colour Forms

I.Aksak[1], Ch. Feist[2] ,V.Kiiko[1], R.Knoefel[2],
V.Matsello[1], V.Oganovskij[3], M.Schlesinger[1], D.Schlesinger[3], G.Stanke[2]

[1]Institute of Cybernetics, Ukrainian Academy of Sciences, 40, Prospect Akademika Glushkova, Kiev, Ukraine; e-mail: schles@image.kiev.ua
[2]Gesellschaft zur Foerderung angewandter Informatik, Rudower Chaussee 5, Gebaeude 13.7, D-12484 Berlin; e-mail: stanke@gfai.de
[3]National Technical University, (Kiev Polytechnic Institute), 252056 Kiev, Ukraine, Prospekt Peremohy, 37

Abstract. The paper describes an intelligent system for extraction filled-in data from colour conventional forms. Filled-in characters in these forms can touch or cross the form frames. The empty form description is acquired in this system by interactive learning followed by automatic processing and extraction filled-in data from input forms. Some steps of input form processing (form coding and form dropout) are carried out for any input form and other (form registration and broken characters reconstruction) are needed if the colour of the user filled-in data coincides with the colour of the empty form only. Experimental results showing the performance of the system are described.

1 Introduction

Intelligent form analysis is one of the most actual problems in document image processing. A typical filled-in form consists of pre-printed data (form frames, symbols and machine pre-printed characters) and user filled-in data (machine-typed or hand-written characters and check marks). Distinguishing of the filled-in data from the pre-printed ones and extraction these data from the input form is the major task in form processing.

Existing form processing systems can be categorised into two groups. The systems from the first group either use interactive tools for manually extracting filled-in data [1] or process specially designed forms [2]. The systems from the second group [3-7] are intelligent systems for processing of conventional forms, which have not been constructed and/or filled-in specially to facilitate reading by machine. As a result these forms usually contain filled-in data that touch or cross form frames or pre-printed entities. The problem of extraction filled-in data in these systems involves the following issues: empty form template, input form registration and pre-printed data dropout, reconstruction of broken strokes introduced during separation of form frames.

Form template means the capturing of empty form structure or description. The form structure knowledge is acquired by interactive learning in [4,5] and by automatic recognition in [6] using block adjacency graph (BAG) representation of this form.

Form registration is the process of aligning or calibrating an incoming form with the previously defined empty form template. During form registration the translational and rotational differences between the form template and the input form are minimized. Therewith, translation and rotation of input form image are carried out in [4,5] and the same operations but at the rectangles, describing pre-printed strokes in empty form, are carried out in [6].

Form dropout is the process of matching some objects or parts of input form as corresponding to pre-printed entities in empty form and deleting these objects.

Because some character strokes touch or cross form frames, these strokes will be broken after separating the form frame. The method of reconstruction of these strokes, based on skeleton, is introduced in [3]. In [4,5] the same process is performed by copying the pixels that lie just above and below the deleted line through the region where the line is being erased. In [6] form frame separation and character reconstruction are implemented by means of BAG.

The systems [3-7] have some distinctive features but are similar in that they deal with photocopied (in black and white) bilevel images of the forms only. At the same time beyond a doubt that the colour filled-in form has essentially more useful information when the colour of user pen differs from the colour of the empty form. In this case it is hopeful to extract filled-in data without need to reconstruct broken strokes and to handle the most difficult problem for [3-7], when filled-in characters touch or cross not only form frames but pre-printed text and objects also.

We propose a system for automatic extraction filled-in data from colour forms. In Section 2 the main parts of the system are described. Some of these parts (form colour coding and dropout) are used for every form and others (form registration and broken strokes reconstruction) are needed only if the colour of filled-in data coincides with the colour of the empty form. The performance of the system is shown in Section 3.

2 Filled-in data extraction

The applied content of filled-in data extraction problem is as following. There is a picture (in common case colour) of the empty (not filled) form. There is also a picture of this form with some inscriptions, added by a man, which are in general case coloured too. On the base of the certain information about the empty form the machine technology for extraction these inscriptions (filled-in data) is to be developed. In other words, it is necessary to transform a filled document into the picture, which would be produced, if the inscriptions would be written by a black pen on a white background. Processing steps of the system that solve this task are described in the following sections.

2.1 Form template

To extract filled-in data the following information about the empty form in general case must be acquired:

a). List C of colours, which exist on the document, and the coding table, that for every three-tupple (r,g,b) of bytes, defining the intensities of red, green and blue lights, defines the set $C' \subset C$ of colours, which this (r,g,b) can be represented by. It is supposed, that the variables r, g and b take their values from the interval [0, 31], so the coding table consists of $32 \times 32 \times 32$ lists and each list has a length not more than 16 elements. On the base of this coding table the initial rgb-presentation of the empty form must be transformed to another one, in which for every pixel t the certain list $e(t)$ of the names of colours is defined.

b). List of horizontal and vertical lines of the empty form frame. These lines are represented by coordinates and the colour of the corresponding "bounding" rectangles in this form.

c). List of text strokes in the empty form. This list is represented in the same format like the list of lines.

d). List of rectangles covering the characteristic parts of the form. These rectangles are used while aligning the incoming form with the previously defined form template.

This information is generated interactively on the base of filled or empty form, that was scanned in *rgb*-codes. Form template is constructed only once, and then it is used for automatic processing of input filled-in forms. Separate steps of this processing are described in Sections 2.2-2.4. The principles of construction of the coding table from *rgb*-codes to colour codes are as follows.

Let r,g,b be variables that define the quantities of red, green and blue lights in pixels of the scanned picture. Let *rgb* be three-tuple of these variables and *RGB* be the set of all such three-tuples, i.e. the set of points with integer coordinates in the 3-D cube. For subsequent processing it is necessary to transform the *rgb*-representation of the scanned incoming form into the following one. Let C be the set of the names of colours, observed in the picture of the empty form, for example $C=\{white, black, brown, yellow\}$. For analysis of the filled-in form it is necessary to indicate for every pixel of this form not the *rgb* values but some list of the names of colours (for some pixels this list can be empty), i.e. some subset $C' \subset C$. For such transformation of *rgb* presentation to C' one it is necessary to define in the 3-D *RGB* cube for every colour $c \in C$ the region *RGB(c)*, containing all rgb-tuples, which this colour can be represented by. The region *RGB(c)* is defined for every colour c, as the region of the certain type on the base of the learning information from user. Every region *RGB(c)* is the polyhedron, every side of which is parallel to the one of the following thirteen planes:

$$r = 0, \; g = 0, \; b = 0, \; r+g = 0, \; g+b = 0, \; b+r = 0, \; r-g = 0, \; g-b = 0, \; b-r = 0,$$
$$r+g+b = 0, \; r-g+b = 0, \; r+g-b = 0, \; -r+g+b = 0;$$

The learning information is received from the user who observes *rgb*-presentation of the picture and points out some pixels having by his opinion the certain colour c. On the base of this information the program defines the set $U(c)$ of indicated by user *rgb*-tuples and then constructs the region *RGB(c)*, which represents the minimal polyhedron of defined above type and includes the set $U(c)$.

2.2 Colour coding
Colour coding operation has the colour picture as input data, in which for each pixel the *rgb*-code is given. This representation of the picture is transformed to another one, in which for each pixel the list of names of colours, which are observed in this pixel, is given. On the next steps of the software the picture is presented only in such representation. The colour coding operation is based on previously constructed during form template coding table.

2.3 Form registration
The form registration consists of two parts: coarse and fine registration. The first part results in such displacement and rotation of the whole input form picture, that it will in the best way correspond to the previously defined set of lines in the empty form. The second part results in correction of the position and parameters of separate lines in empty form template for every incoming filled-in form, because the positions of lines and their thicknesses in the input form can be slightly different from the lines in the empty form.

The procedure of rotation of the picture is rather non-trivial part of aligning the input form with the empty one. It is known, that pixels coordinates after rotation are not integer and if to perform the rounding of these coordinates not correctly, the situation can occur, when either more than one pixels or neither can be mapped into the certain pixel of the picture after rotation. In this case the rotation turns to be irreversible that, firstly, contradicts to a reasonable understanding of the rotation and, secondly, destroys the initial information about the picture. To avoid these lacks we represent the rotation as the following three successive operations:

1) displacement of the rows of the picture: $x_{fir} = x + k1*y$; $y_{fir} = y$;

2) displacement of the columns of the picture: $x_{sec} = x_{fir}$; $y_{sec} = y_{fir} + k2*x_{fir}$;

3) displacement of the rows of the picture: $x_{fin} = x_{sec} + k3*y_{sec}$; $y_{fin} = y_{sec}$.

Values k1, k2 and k3 are parameters of these operations. If the following relations

a) $1 + k2*k3 = \cos\varphi$; b) $k1 + k3 + k1*k2*k3 = \sin\varphi$;

c) $-k2 = \sin\varphi$; d) $1 + k1*k2 = \cos\varphi$

are hold, the equalities $x_{fin} = x_\varphi$ and $y_{fin} = y_\varphi$ are valid for every x and y. A solution of these relations is $k1 = tg(\varphi/2)$; $k2 = -\sin\varphi$; $k3 = tg(\varphi/2)$ and implementation of image rotation as a superposition of specified three operations turns to be reversible for digitized images.

The procedure of fine registration of the filled-in form is as follows. Let C be a list of colours and ρ be a coloured picture. It means, that for every pixel t the list of colours $\rho(t) \subset C$ is defined. Let L be the following set of horizontal lines. Every line $l \in L$ is determined by the three-tuple (h_1, h_2, h_3), such that $h_1 + h_2 + h_3 = h$, $h_1 > 0$, $h_2 > 0$, $h_3 > 0$. This three-tuple determines a picture of the line l that contains five horizontal stripes B_1, L_1, S, L_2, B_2, as it is shown in Fig. 1a.

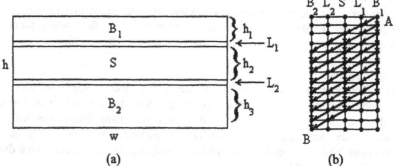

(a)   (b)

Fig. 1. Five stripes (a) of the horizontal line and the graph (b) representing the set of these lines.

The stripe B_1 (B_2) consists of h_1 (h_3) rows of pixels; for every pixel t from B_1 or B_2 holds $l(t) = B$, where B is the predefined list of background colours. Both L_1 and L_2 consist of one row of pixels and for every of these pixels t any colour is allowable, i.e. $l(t) = C$. The stripe S consists of h_2 rows of pixels and for every pixel $t \in S$ holds $l(t) = \{a\}$, where a is the predefined colour for the given line. The set of all possible pairs of positions of the stripes L_1 and L_2 defines the set L of pictures, i.e. the set of lines. For every line l and every picture ρ their dissimilarity $R(\rho, l)$ is defined by the following way. A pixel t will be referred as a strange one, if it satisfies the condition $\left(\rho\left(t\right) = \varnothing\right) \vee \left(\rho\left(t\right) \not\subset l(t)\right)$. The dissimilarity $R(\rho,l)$ is a total amount of the strange pixels in the picture.

The problem under solution is as following: for the given picture ρ the line $l^* \in L$ is to be found, which minimizes the dissimilarity $R(\rho,l)$, i.e. $l^* = \arg\min\limits_{l} R(\rho,l)$. Time for direct calculation of this expression is of order $h^3 \times w$, because an amount of lines in L is of order h^2 and calculation of R for every line requires $h \times w$ operations. Because the problem under consideration is reduced, as it will be shown below, to searching shortest path on the certain oriented weighted graph, the time for solution of this problem can be of order $h \times w$.

The set of lines L may be represented by the following oriented graph G, shown in Fig. 1b.

The nodes of this graph are represented by the nodes of two-dimensional grid. Every node has coordinates $(i,k), k \in \{B_1, L_1, S, L_2, B_2\}, 0 \le i < h$. It means that every column of graph G corresponds to the certain stripe and every row corresponds to the certain row in the picture. The edges of the graph are displayed by the arrows in Fig. 1. Every path from the node A to the node B corresponds to the certain line from the set L. This isomorphism is defined by the following sentence: if the path on this graph passes through the node (i,k), the i-th row of pixels belongs to the stripe k.

The dissimilarity function $R(\rho,l)$ must be represented by the weights of nodes, so that for any line $l \in L$ the value $R(\rho,l)$ is equal to the total weight of the path, which the line l is represented by. This requirement is satisfied if the weight $v(i,k)$ of the node (i,k) is defined by the following way: $v(i,k) = 0$, if $k=L_1$ or $k=L_2$, and $v(i,k)$ is the amount of the strange pixels in the i-th row for other k.

The calculation weights of all nodes takes a time of order the number $w \times h$ of the pixels in the picture. For the above described case of graph the search of the best path on the base of dynamic programming takes a time proportional to the number of graphs nodes, i.e. to the number h of rows in the picture. The program solves the above considered task for every line in the empty form and has as a result the new position for the line in this form.

2.4 Form dropout

This operation is carried out on the base of the following two pictures:

   a) the registered input form picture, in which for each pixel t the list $\rho\left(t\right)$ of colours is given, which are observed in this pixel;

   b) the empty form picture, in which for each pixel t the certain list $e(t)$ of names of colours is also given. This list consists of the names of the colours in the pixel before writing the inscriptions in the document.

Form dropout consists in the following transformation of the input form picture. The colour of every pixel t in this picture is changed to black colour, if the following condition

$$\left(\rho\left(t\right) = \varnothing\right) \vee \left(\rho\left(t\right) \not\subset e\left(t\right)\right)$$

is satisfied, and to white colour otherwise. The first part in this condition means, that rgb-code in the pixel t of the picture under analysis does not correspond to any colour, permitted in the empty form. The second condition means that in the pixel *t* of the input form some colour is observed, which is not allowable for the given pixel, although this colour is allowable for the form at all.

2.5 Broken stroke reconstruction

In the areas, where the inscriptions touch or cross form frames, they share pixels. During the form frame separation these pixels will be erased, if the colours of inscriptions and of form frames coincide, and as a result some of touching or crossing characters will be broken. In this case after the form frame separation we perform reconstruction of the broken strokes.

For every horizontal line of form frame this reconstruction consists of the following stages. The first stage results in noise removing on the upper and lower boundaries of erased line. During the second stage thicknesses and directions of line segments touching to upper and lower boundaries of this line are defined. These line segments consist of the runs of horizontal black pixels to the right and to the left of which white pixels are located. Two runs are called neighbouring if one of the pixels of the first run is neighbouring to some pixel of the second one. The direction of the line segment is defined over the sequence (from 2 to 5) neighbouring runs located either above of the upper or below of the lower boundaries of the line under processing.

The final and the most important stage consists in joining (filling the gap) of some of these line segments. At the beginning different pairs of line segments correspondingly above and below of the erased line are considered and two line segments of the pair are joined if they are crossed (taking in consideration their thicknesses) while extending one or both these line segments into erased region. Filling the gap between two line segments results in replacing of white pixels in black ones inside quadrangle between these line segments if their thickness are approximately the same. Otherwise joining is carried out by extending of the line segment with the lesser thickness. After joining line segments at upper and lower boundaries the pairs of neighbouring line segments both locating at upper or lower boundary are considered. These line segments are joined if certain conditions are satisfied, which depend on the directions, thicknesses and the distance between line segments.

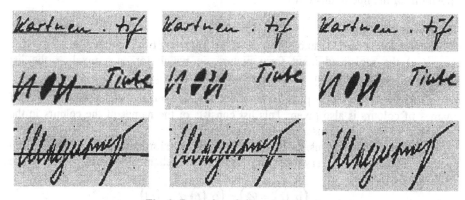

Fig. 2. Examples of character reconstruction.

A few examples of horizontal strokes reconstruction are given in Fig. 2. First column in this figure shows segments of the input image, second column shows the result after the form frame separation, and the last column shows the reconstructed characters. The reconstruction of characters having broken parts after separation of vertical lines of form frame is similar to considered above.

The resulted picture of extracted inscriptions contains inevitably the noise in the form of thin lines, whose thickness is one pixel, arising on the boundary of lines and inscriptions of the document. This noise is eliminated by rather obvious procedure.

3 Experimental results

The system has been developed for the OS MS Windows and tested on more than 30 forms of 6 different types filled-in by hand. These forms were filled in by different people, using ball-point pen, pencil or ink pen, and were scanned with a small amount of skew ($< 10°$) at 240 dpi. The example of filled-in blue ink-printed form is shown in Fig. 3a. This form was filled in by blue pen in the upper part of the form, by red pen in the middle and by green pen in the lower parts of this form. The picture of the form is of size 1380 x 1900 (15 x 20 cm) and as it is shown in Fig. 3a the hand-written characters touch the form frames in all the four directions (top, bottom, left and right). Processing of this picture was carried out under the assumption that the colour of user's pen does not differ essentially from the colour of the empty form and because of that the whole number of processing operations were implemented. As a result it takes tens of seconds on a Pentium IBM PC to process this picture. Otherwise if the colour of the user's pen does not coincide with the colours in the empty form, the operations of input form registration and broken characters reconstruction have not to be used, and because of that the running time is drastically reduced.

The form dropout results for Fig. 3a are shown in Fig. 3b. These results are occurred to be better in comparison with presented in [6], because extracted filled-in data in Fig. 3b contain no pre-printed entities, but the system [6] takes lesser time for processing the input form. While the subsequent modification of the system we would like to continue developing of the methods of automatic coding and registering of coloured forms, and as a result to reduce the time of their processing.

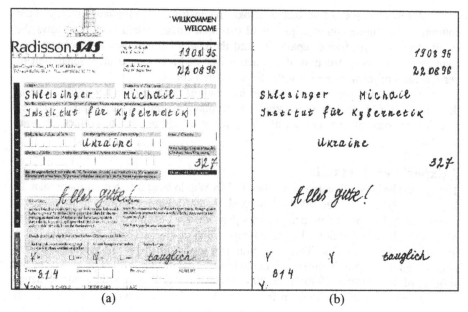

Fig. 3. Input form (a) and result (b) of extraction filled-in data.

Acknowledgement

This research was fullfiled in a multinational project and was sponsored by the Ministry of Education, Science, Research and Technology of Germany which is gratefully acknowledged.

References

[1] Leedham and D. Monger, "Evaluation of an Interactive Tool for Handwritten Form Description", Proc. Third Internationall Conf. Document Analysis and Recognition, pp. 1,185-1,188, Montreal, 1995.

[2] S.L. Taylor, R. Fritzson and J.A. Pastor, "Extraction of Data from Preprinted Forms", Machine Vision and Applications, vol. 5, pp. 211-222, 1992.

[3] G. Maderlechner, "Symbolic Subtraction from Fixed Formatted Graphics and Text from Filled In Forms", Machine Vision and Applications, vol. 3, pp.457-459, 1990.

[4] R. Casey and D.Ferguson, "Intelligent Forms Processing", IBM Systems Journal 29(3), pp. 435-450, 1990.

[5] R. Casey, D. Ferguson, K. Mohiuddin and E. Walach, "Intelligent Forms Processing System", Machine Vision and Applications, vol. 5, pp. 143-155, 1992.

[6] Bin Yu and Anil K. Jain, "A Generic System for Form Dropout", IEEE Trans. Pattern Analysis and Machine Intelligence, vol. 18, pp. 1127-1134.

[7] K. Franke, "Unterschriftenverifikation mit SIC Natura", Bildanswertung für Handel, Banken und Behörden, Berlin, 19./20.02.97, Fraunhofer IPK - Workshop.

Improvement of Vessel Segmentation by Elastically Compensated Patient Motion in Digital Subtraction Angiography Images

Thorsten M. Buzug, Cristian Lorenz and Jürgen Weese

Philips Research Hamburg, Röntgenstraße 24-26, 22335 Hamburg, Germany
buzug@pfh.research.philips.com

Abstract. Digital subtraction angiography is a standard diagnosis tool for the examination of vessels. For this method X-ray images (*contrast images*) are taken from the patient while a radio-opaque contrast agent is injected through a catheter. The first image of such a scene is usually taken before injection and is called *mask image*. In clinical routine the mask image is manually shifted to perform a rough motion compensation. Then the corrected mask is subtracted from the contrast image to erase all disturbing permanent structures like e.g. bones or organs. In the vessel diagnosis chain a vessel segmentation is often applied to the subtraction results. Real DSA-images only corrected via an interactive-shift routine still suffer from motion artifacts which may lead to false results from the segmentation step. Especially, for the abdomen where the patient motion is very complex it is illustrated how residual artifacts result in mis-segmentations. In the present paper we demonstrate that an affine transformation, and particularly, an elastic transformation yield an excellent patient motion compensation which is a sufficient basis for the segmentation algorithm. We describe a registration procedure based on the estimation of a motion vector field. Additionally, we outline a new vessel segmentation algorithm.

1 Introduction

Digital subtraction angiography is a well-known technique for vessel diagnosis and has been applied for many years [1]. A sequence of X-ray images of the patient is taken during injection of opaque dye, which is used as X-ray contrast agent. Usually, the first image of such a sequence is taken before dye injection and is called *mask* image, and all other images of the sequence are consequently called *contrast* images. The main idea of DSA is the subtraction of the mask image from one of the contrast images to get rid of the disturbing projection pattern of organs, bones, catheters, surgical implants etc. Ideally, the subtraction result should show the contrasted vessels exclusively. However, real digital subtraction angiography images still suffer from motion artifacts which are a problem for vessel segmentation algorithms. Though the patient is asked not to move during image acquisition, there are residual movements as e.g. heart beat (on the short time scale) and peristaltic motion of the intestine (on the large time scale) which cannot be controlled by the patient. In some cases the resulting motion artifacts are so strong that smaller vessels-of-interest cannot be recognized at all. For that reason enormous efforts have been put into the image processing problems of digital subtraction angiography and especially into the registration task of DSA, i.e. registration of mask and contrast image prior to subtraction to compensate for patient movements. An important requirement of the DSA registration task is that the region-of-interest considered within registration should contain the interesting vascular structures. This, however, is also one of the main problems in DSA, because the contrasted vascular structures can be regarded as a gray-value distortion that makes the images to be compared dissimilar. Recently, we have shown that a template matching procedure leads to a very accurate motion vector field, if histogram-based similarity measures are employed [2]. Especially, the energy similarity measure derived from one-dimensional histograms of the difference image is a very smooth and significant objective function [3].

In section 2 the DSA enhancement algorithm with its underlying mathematics will be briefly outlined. To demonstrate the effect of motion correction prior to subtraction, the method is applied to an image pair from clinical practice. It is shown that the quality of the affinely corrected DSA image can be further improved using an elastic approach based on thin-plate splines [4]. In section 3 the results of a vessel segmentation approach applied to the subtraction images are presented. The method is based on the analysis of the Hessian matrix, i.e. the second derivatives, as proposed for detection of line structures by Koller et al. [5]. Since the Hessian matrix, its eigenvalues and vectors depend strongly on the chosen scale we additionally use the so called γ-parameterized normalized derivatives, introduced by Lindeberg [6] to detect the optimal scale corresponding to the line-structure of interest. The new combination of these techniques leads to the proposed vessel segmentation algorithm that is applied to the DSA images presented in section 2.

2 Motion Compensation

DSA image enhancement is generally a registration task. Mask and contrast image must be properly aligned prior to subtraction to reduce degradation (cloud-like artifacts) in the image caused by patient motion. In [2] an algorithm for that purpose has been proposed. The algorithm is semi-automatic in the sense that the clinician interactively chooses an arbitrarily sized rectangle that includes the interesting vascular structures as region-of-interest, and consists of the detection of homologous landmarks and the approximation of an appropriate motion model as outlined below.

2.1 Detection of Homologous Landmarks

Firstly, the interactively defined, rectangular region-of-interest in the contrast image is divided into a number of non-overlapping templates with predefined size that cover the entire region. It has been mentioned in [7] that angiographic images are usually of low contrast with no sharp edges. For that reason the matching criterion utilized by template matching techniques can fail. This problem can be solved by a template-exclusion technique that takes into account the amount of contrast variations inside a template. Therefore, as an additional step, those templates are excluded that are located in regions with insufficient contrast variation [2]. Secondly, for the templates defined in the previous step, the corresponding templates in the mask image must be identified. This is done by optimization of an appropriate similarity measure. The resulting shifts are used to establish a motion vector field. Fig. 1 shows the motion vector field for an interactively selected rectangular region-of-interest in an image pair of the abdomen. The success of a kidney transplantation is verified by injection of contrast agent into the kidney vessel tree. The images were acquired using a Philips Integris V3000 system. The main problem for the estimation of the motion vector field is the choice of an appropriate

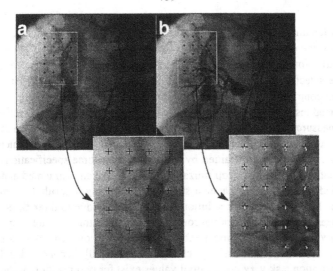

Fig. 1. Mask (a) and contrast image (b) of abdomen fluoroscopies showing the success of a kidney transplantation. In the rectangular region-of-interest (blown-up for better visualization in both images) the distribution of landmarks can be seen as black-colored crosses. Additionally, the motion-vector field - visualized in the blown-up only - is given by white lines attached to the crosses in the contrast image.

similarity measure, because mask and contrast image are dissimilar by injection of contrast agent. Recently, similarity measures have been introduced, obtained from weighted one-dimensional histograms that are optimally adapted to the dissimilarities of DSA [3]. Such measures are defined in a 3-step procedure that consists in: (i) subtraction of images inside the template, (ii) calculation of gray-value histogram of the difference image and normalization of the histogram according to $\Sigma p_k = 1$, where p_k is the fraction of pixels with gray-value g_k. The fraction of pixels depends on the shift parameters r and s: $p_k = p_k(r,s)$, and (iii) evaluation of similarity measure $M(r,s) = \Sigma h(p_k)$, using an appropriately defined weighting function $h(p)$. The histogram-based similarity measures are motivated by the observation that for optimal registration of mask and contrast image the difference image shows low contrast variation in the area of the template, whereas in the case of misregistration the contrast variation is larger. The situation corresponding to a perfect registration, i.e. a very peaky histogram, must have the maximum distance from the worst case, i.e. an equally distributed histogram. That distance is evaluated by the similarity measure. It has been mathematically proven that any strictly convex weighting function is appropriate to measure the degree of misregistration [3]. And it has further been shown that this class of similarity measures is better adapted to the gray-value distortions in DSA images than other well-known similarity measures such as e.g. cross correlation [8], sum of absolute differences [9], and deterministic sign change [10]. In ref. [3] the quality of the histogram-based measures is compared to other frequently used similarity measures. As a result the energy similarity measure - the sum of the squared histogram values, i.e. $h(p_k) = p_k^2$ - turned out to be the most suitable measure for template matching and is consequently used in this paper. Optimization of the energy measure for each template leads to the motion vector field shown in fig. 1b. It yields a set of homologous landmarks - or control points - for mask and contrast image. These points are essentially the centers of the respective templates, and serve as basis for fitting an appropriate transformation as outlined in the next subsection. The transformation is only valid for the selected region-of-interest and is hence applied only in that region. In a final step the contrast and corrected mask image are subtracted yielding a DSA image that is enhanced inside the region-of-interest. The optimization of the objective functions (similarity measures) is an important step in the DSA registration task. On the one hand such a procedure must be robust and reliable to obtain accurate shift parameters. On the other hand we are strongly restrained by the computation-time specifications and therefore, computationally expensive optimization methods like e.g. simulated annealing cannot be applied. In the present paper a simple hill-climbing algorithm is employed for optimization. This is an adequate solution for DSA mainly due to three facts. Firstly, the similarity measures based on strictly convex weighted histograms are very smooth, so the probability for trapping into local extrema is relatively low. That is in contrast to the situation e.g. for the deterministic sign change measure [10]. Secondly, for this particular DSA registration task very good initial values exist for optimization of the shift parameters, because starting at zero shift is almost always a good choice. Finally, we have thrown away 20 percent of the templates that show lowest contrast variation of the background pattern as can be seen in fig. 1. The contrast variation is measured by the entropy values of the mask image at the locations of the templates [2]. Therefore, for the remaining templates a significant optimum of the objective function can be expected.

2.2 Estimation of Transformation

The set of homologous landmarks (i.e. the template centers, sometimes also called *control points*) resulting from the template matching is straightforwardly used to estimate the parameters of an appropriate transformation f relating the points of the mask image to the corresponding points in the contrast image inside the region-of-interest:

$$(x',y') \rightarrow (x,y) = \left(f_x(x',y'), f_y(x',y') \right),$$ (1)

where (x',y') and (x,y) are the coordinates of the contrasted image and the mask image, respectively. As a first result the image pair, introduced in fig. 1, is subtracted after motion correction with an affine transformation. The affine correction can be written as

$$x = a_{11} + a_{12}x' + a_{13}y'$$
$$y = a_{21} + a_{22}x' + a_{23}y'$$ (2)

As there are far more than three homologous landmarks, this is an overdetermined problem and a singular value decomposition is used to produce a solution that is the best result in the least-squares sense [11]. The affine transformation can cope with distortions like translation, rotation, scaling and skewing, and as shown in fig. 2, the patient-induced distortions can be significantly reduced. However, patient motion is often more complex. Especially for the presented abdominal images, it can be shown that residual artifacts are visible in the subtraction images even after an affine correction. Therefore, higher-order registration procedures must be applied. In this paper, we present the results of an elastic registration based on thin-plate splines. For the elastic matching approach Bookstein´s thin-plate spline method [4] is applied which can be considered as an interpolation. The set of homologous landmarks is the same as for the affine transformation, indicated in the mask and in the contrast image (see fig. 1). $Q_k=(x_k,y_k)$ is a set of points in the mask image and $P_k=(x'_k,y'_k)$ the corresponding set in the contrast image (k=1,...,N, where N is the number of landmarks or control points). For the thin-plate spline approach the function f in eq. (1) can be written as

$$f_.(x'_k,y'_k)=a_{.1}+a_{.2}x'_k+a_{.3}y'_k+\sum_{i=1}^{N}w_{.i}U\left(\|(x'_k,y'_k)-P_i\|\right)\cdot$$ (3)

The point denotes the index for the x- or y-coordinate and the radial basis function used in the thin-plate approach is

$$U(r)=r^2\log(r^2) \text{ , where } r=\sqrt{x^2+y^2}$$ (4/5)

is the Euclidean distance. In addition to these equations the method also requires boundary conditions which ensure that terms are removed that grow faster than linear terms as one moves far away from the data. This system of equations is solved by singular value decomposition [11]. Fig. 2 demonstrates the success of the motion correction prior to subtraction inside the region-of-interest. In the contrast image the region-of-interest is marked with a rectangle. The fig. 2a to 2c present the subtraction results using manual shift-correction, affine correction and elastic correction, respectively. It can be seen that without an appropriate correction prior to subtraction the vessel tree is heavily distorted by the cloud-like artifacts that seem from motion of air bubbles in the bowels (two artifacts are indicated with arrows). Especially the smaller vessels are entirely covered by these distortions. The affine correction leads to a significant improvement of the subtraction quality (see fig. 2b). Here, the smaller vessels are clearly visible. However, some residual distortions remain (also indicated with an arrow). As mentioned above the affine transformation can cope with distortions like translation, rotation, scaling and skewing. Therefore, it is expected that an elastic approach based on thin-plate splines leads to further improvements of the DSA-image quality, especially for more complex distortions as foreseen for abdominal images. Fig. 2c shows that also the residual artifacts can be removed using the elastic correction approach. A general problem is, however, that distortions are inherently 3-dimensional. There are 3-dimensional patient motions leading to distortions which cannot be described by a position dependent shift in the 2-dimensional projection images. Therefore, it is not expected that DSA-registration algorithms can ever be developed to the extent that they remove all motion artifacts.

3 Vessel Segmentation

For the vessel segmentation of the DSA images presented in the previous section a multi-scale line-filter based on the second derivatives has been used. A significant second derivative across a line-structure can be expected, since the gray-value increases rapidly from the line border to the centerline and decreases again to the opposite border. Longitudinally to a line-structure only insignificant second derivatives should occur. An analysis of the Hessian matrix can thus be used to detect line structures [5]. In the 2D case, a dark line structure on a bright background is characterized by a large positive second derivative across the line. This is reflected by a Hessian matrix having a large positive eigenvalue and a small eigenvalue of either sign. In addition, the eigenvector

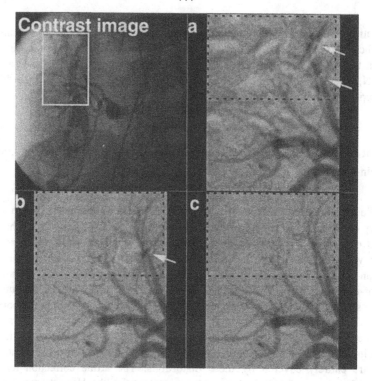

Fig. 2. Upper left image: Contrast image with indication of region-of-interest. The marked rectangle in the contrast image reflects the area where the correction is applied. (a) DSA image with manual shift-correction, (b) affine correction, and (c) elastic correction prior to subtraction. The arrows indicate residual artifacts. The black-dashed rectangles indicate the regions where the vessel segmentation algorithm - discussed in section 3 - is applied.

corresponding to the smaller eigenvalue in magnitude can be used as an estimate for the longitudinal direction of the line-structure. Prior to the calculation of the second derivatives, the image is blurred by convolution with a Gaussian kernel of width σ. The Hessian matrix, its eigenvalues and vectors depend strongly on the chosen scale defined by σ. To detect the optimal scale corresponding to the line-structure of interest we use the so called *γ-parameterized normalized derivatives* $\partial / \partial_{x,\gamma-norm} = \sigma^\gamma \partial / \partial_x$ introduced by Lindeberg [6]. The factor σ^γ accounts for the decrease of derivatives due to the blurring. To detect the optimal scale, the normalized derivatives are evaluated for several σ. The optimal scale corresponds to the largest response of the normalized derivative. In our case we are interested in the second derivative. For a one-dimensional line-profile we define the line-response function $r(x,\sigma)$:

$$r(x,\sigma) = \sigma^\gamma \cdot \frac{d^2 f(x,\sigma)}{dx^2} \tag{6}$$

with $f(x,\sigma) = p(x) * g(x,\sigma)$ being the convolution of the line profile p(x) and a Gaussian kernel of width σ. In the case of a line-structure of a two-dimensional image the eigenvalue λ_1 of the Hessian matrix larger in magnitude is used instead of the second derivative:

$$r_{2D}(x,\sigma) = \sigma^\gamma \cdot \lambda_1(x.\sigma) \tag{7}$$

For the parameter γ a value of $\gamma = 3/2$ is chosen, because it leads to an estimation of the correct width of a Gaussian profile. The response function $r(x,\sigma)$ is well suited for the scale estimation of a line-structure. However, $r(x,\sigma)$ should not be used directly as line filter function. This is due to the fact that edge-like structures produces a response comparable to line-structures which is in general an unwanted effect. To distinguish lines from edges we define therefore an *edge-indicator* as the ration between gradient and eigenvalue larger in magnitude of the Hessian:

$$E_{2D}(x,\sigma_{opt}) = \frac{\left|\nabla f(x.\sigma_{opt})\right|}{\sigma \cdot \lambda_1(x.\sigma_{opt})} \tag{8}$$

E_{2D} is large in case of an edge-like structure and small in case of a line-structure. This behavior can be used to construct a line-measure function μ_{2D} that allows to suppress the edge-response:

$$\mu_{2D}(x,\sigma_{opt}) = \left| r(x,\sigma_{opt}) \right| \cdot e^{-\left|E_{2D}(x.\sigma_{opt})\right|} \tag{9}$$

Fig. 3 shows the application of the line-measure μ_{2D} to the subtraction images (fig. 2a-c) generated with the methods described above. The first column of fig. 3 shows the application of the edge-indicator, column two of the response function and column three of the line-measure to the DSA images. The results displayed in row a-c correspond to (manually) shift, affine and elastically corrected DSA images from fig. 2a-c. Due to the suppression of edge-like structures, the line-measure enhances the center-line of the vessel. The arrows indicate segmentation errors that are due to motion artifacts of the subtraction images. The artifacts of the manually corrected DSA image caused a rather strong

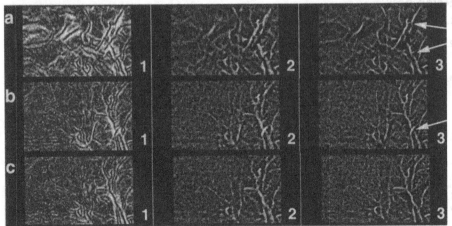

Fig. 3. Rows a-c show the results of the line-filter applied to the subtraction images displayed in fig. 2a-c. The first column shows the edge-indicator $E_{2D}(x,\sigma_{opt})$, the second column the response-function $r_{2D}(x,\sigma_{opt})$ and the third column the line-measure $\mu_{2D}(x,\sigma_{opt})$ which is a combination of the former two. The arrows indicate residual artifacts.

apparent vessel (upper arrow in fig. 3 a-3). This error disappeared in the segmentation result of the affine corrected image in (fig. 3 b-3). The arrow in fig. 3b-3 indicates a smaller error caused by a residual artifact of the affine correction. The segmentation result of the elastically corrected image (fig. 3c-3) is free from both errors.

4 Conclusion

The quality of abdominal subtraction images is enhanced by the application of a motion vector field based registration that suppresses patient motion. We employed the energy similarity measure obtained from one-dimensional histograms to estimate the motion vectors. Homologous landmarks are identified in the mask and contrast image which are straightforwardly used for the estimation of the parameters of an appropriate motion model. We compared a manual shift-correction with an affine and an elastic motion correction approach. While the affine correction leads to an improvement of the image quality it can be demonstrated that a further enhancement is obtained if an elastic approach is used. The use of the thin-plate spline interpolation method is enabled by the fact that the motion vector field is of excellent quality. Outliers in the motion vector field would have drastically decreased the image quality. In the present paper, the DSA-diagnosis chain is completed by the application of a new vessel segmentation or filter algorithm, respectively. The method is applied to the three correction approaches mentioned above.

Acknowledgments

The authors would like to thank Dr. L. J. Schultze Kool, University Hospital Leiden, for providing us with the data sets. The algorithm was implemented on an experimental version of the EasyVision workstation from Philips Medical Systems and we would like to thank Integrated Clinical Solution (EasyVision/EasyGuide™) Advanced Development, Philips Medical Systems, Best, for helpful discussions.

References

1. W. A. Chilcote, M. T. Modic, W. A. Pavlicek et. al., *Digital subtraction angiography of the carotid arteries: A comparative study in 100 patients*, Radiology **139** (1981) 287.
2. T. M. Buzug and J. Weese, *Improving DSA images with an automatic algorithm based on template matching and an entropy measure*, CAR'96, H. U. Lemke, M. W. Vannier, K. Inamura and A. G. Farman (Eds.), (Elsevier, Amsterdam, 1996) p. 145.
3. T. M. Buzug, J. Weese, C. Fassnacht and C. Lorenz, *Image registration: Convex-weighted functions for histogram-based similarity measures*, CVRMed/MRCAS'97, J. Troccaz, E. Grimson and R. Mösgen (Eds.), Lecture Notes in Computer Science **1205** (Springer, Berlin, 1997) p. 203.
4. F. L. Bookstein, *Principal warps: Thin-plate splines and the decomposition of deformations*, IEEE Trans. PAMI **11** (1989) 567.
5. T. M. Koller, G. Gerig, G. Székely and D. Dettwiller, *Multiscale Detection of Curvilinear Structures in 2D and 3D Image Data*, 5th International Conference on Computer Vision (Cambridge, 1995) p. 864.
6. T. Lindeberg, *On scale selection for differential operators*, 8th SCIA (1993) p. 857.
7. B. C. S. Tom, S. N. Efstratiadis, A. K. Katsaggelos, *Motion estimation of skeletonized angiographic images using elastic registration*, IEEE Trans. Med. Imaging **13** (1994) 450.
8. W. K. Pratt, *Correlation techniques of image registration*, IEEE Trans. on AES, AES-10 (1974) 353.
9. J. M. Fitzpatrick, D. R. Pickens, H. Chang, Y. Ge and M. Özkan, *Geometrical transformations of density images*, SPIE **1137** (1989) 12.
10. A. Venot and V. Leclerc, *Automated correction of patient motion and gray values prior to subtraction in digitized angiography*, IEEE Trans. on Med. Im. **4** (1984) 179.
11. W. H. Press, B. P. Flannery, S. A. Teukolsky and W. T. Vetterling, *Numerical Recipes in C* (Cambridge University Press, Cambridge, 1990) p. 534.

Three-Dimensional Quasi-binary Image Restoration for Confocal Microscopy and Its Application to Dendritic Trees

Andreas Herzog[1], Gerald Krell[1], Bernd Michaelis[1], Jizhong Wang[2],
Werner Zuschratter[2] and Katharina Braun[2]

1) INSTITUTE FOR MEASUREMENT TECHNOLOGY AND ELECTRONICS
OTTO-VON-GUERICKE UNIVERSITY OF MAGDEBURG
PO Box 4120, D-39016 Magdeburg, Germany
2) FEDERAL INSTITUTE FOR NEUROBIOLOGY MAGDEBURG
Brenneckestrasse 6, D-39118 Magdeburg, Germany

Abstract. For the analysis of learning processes and the underlying changes of the shape of excitatory synapses (spines), 3-D volume samples of selected dendritic segments are scanned by a confocal laser scanning microscope. The images are unsharp because of the (direction dependent) resolution limit. A simple deconvolution is not sufficient for the needed resolution.

Therefore parametric model for the dendrite and the spines is created to reconstruct structures and edge positions with a resolution smaller than one voxel. The tree-like structure of the nerve cell serves as a-priori information. Simple geometrical elements are connected to a model that is adapted for size and position in sub-pixel domain. To estimate the deviation between the microscope image and the model, the model is sampled with the same resolution as the microscope image and convolved by the microscope point spread function (PSF). During an iterative process the model parameters are optimised. The result is a binary image of higher resolution without strong distortions by PSF.

1 Introduction

The synaptic changes in certain brain areas are important for various kinds of learning processes. It has been shown that during different forms of learning (filial imprinting), a special type of excitatory synapses, the spine synapses, do not only change in number but also in size and shape. Theoretical considerations suggest that the size and the three dimensional shape of spine synapses may influence their electrical, biochemical and molecular properties. Thus, changes of these morphological parameters may reflect learning-related subtle changes of synaptic efficiency, but so far, quantitative morphological data for these hypotheses are lacking [1,2].

At the state of the art the required resolution can be achieved only by electron microscopy or confocal laser scanning microscopy. However, the number of required tissue sections and images and the time consumption for preparation of specimens are in favour of the confocal method. For this approach, selected neurons are injected with a fluorochrome (e.g. Lucifer yellow). The cells are completely filled including all dendritic branches with spines and also the axon. Serial confocal laserscans with maximum spatial resolution reveal fine morphological details of dendrites and the various spine types.

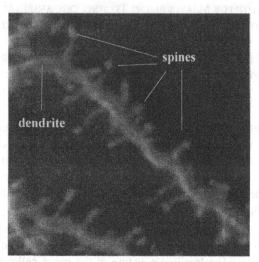

Fig. 1 Part of a nerve cell with a dendrite and spines

Fig. 1 shows a section of an example image with a dendritic branch and spines. It is scanned by a Leica TCS 4D microscope with a sampling distance of 0.1 μm in each direction. The image is produced by the shadow projection tool of IMARIS [3]. The interesting structures (spine heads and necks) range from 0.5 μm to 1.3 μm. A typical image consists of 512×512×150 samples in a volume of about 50×50×15 μm³.

Since the structures of interest are of a size close to the resolution limit of microscope, a correction of scanning errors is necessary. In the three-dimensional confocal laser scanning microscope the direction-dependent pulse response leads to a lower bandwidth in z direction compared to x and y directions [4]. Further, processing of the image without prior restoration would cause shape errors. When measuring or comparing structures whose orientation vary in the 3-D space a significant error must be expected without correction.

2 Quasi-binary image reconstruction

2.1 General method

The aim of the presented work is the generation of a binary image $b(x,y,z)$ for the analysis of the interesting objects. The fluoro-chrome concentration within the cell can be considered as homogeneous in a local neighbourhood, the inside of the structure is clearly demarcated from the outside and the thickness of transition is small compared to the degradation caused by the imaging acquisition. Therefore the original intensity distribution (object function) can be assumed as quasi-binary. But the degradation by the image acquisition results in the real microscope image.

The main idea of the proposed method is the comparison of the microscope image F with an estimated image $\hat{F}$ that is obtained convoluting the actual estimate of the binary image b by the microscope point spread function (PSF) h. The PSF can be calculated by microscope parameters or measured by test bodies [5]. The approximation error in a considered region (ROI) depends on the local parameters $p_1, p_2 \ldots p_k$ which describe the binary image b and the estimated image $\hat{F}$:

$$Q = \sum_{ROI} \left(F(x,y,z) - \hat{F}(x,y,z,p_1,p_2 \ldots p_k) \right)^2 \Rightarrow \min. \qquad (1)$$

The kind and number of parameters are variable. The algorithm starts with an initial estimation of parameters and is refined iteratively (Fig. 2).

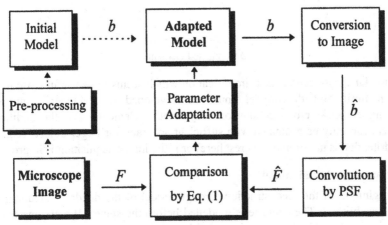

Fig. 2 Model adaptation overview

2.2 Continuous parametric model

The type of the model is variable and depends on the shape and size of the analysed objects. The simplest way is to consider every voxel as an independent parameter with the binary value of 1 or 0. The problems are the large number and the only possible binary variation of parameters. By using neighbourhood relations of the voxels as constraints (a-priori knowledge) the results become better. Additional the resolution is limited by the voxel size.

In our application we work at the resolution limit of the microscope. For a description of small structures like spine necks, the switching of one voxel causes a large change of the local error. To include subpixel information we leave the voxel based domain and describe the objects vector based. This description is independent from the voxel raster. We have to add as much a-priori knowledge as possible. If we assume a circular cross-section we can use a model with simple geometrical bodies. The tree-like structure of the nerve cell serves as a-priori information for the connection of the elements.

The basic element is a cylinder with hemispheres of the same radius at the ends. The centres of the cylinder top and bottom shall be called "element points". A basic model element is completely described by the coordinates of the two element points and the radius. The model is created by successively connecting elements at their element points (see Fig. 3). The overlapping of the connected elements reduces artefacts between them. The parametric model includes subpixel information and variability of edge positions.

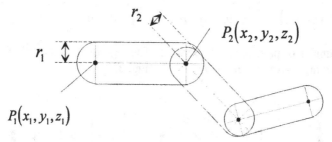

Fig. 3 Connected elements

Using circular cross-sections is a simplification which seems to be useful close to the resolution limit. But this model could be developed to a higher complexity introducing elliptical cross-sections or unsymmetric elements. But the additional parameters can only be estimated with sufficient accuracy for larger objects (e.g. for the dendrite, that is not of main interest here) or if the image acquisition is improved.

Estimation of the initial model

At the beginning of the reconstruction a rough model of the dendrite including the spines is established. The dendrite is modelled before the spines. We assume that in the image is only a single interesting dendrite without additional branches. At the beginning all elements are of the same size. Starting with an interactively defined element a tracking algorithm follow the dendrite.

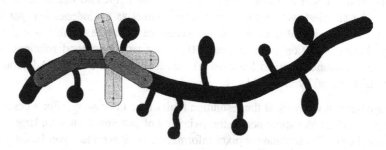

Fig. 4 Growth of the initial dendrite model

The algorithm search for the next element with the best direction (Fig. 4) to optimise a quality factor which depends on:
- the average gray level in the volume of the microscope image corresponding to the actual position of the element and
- the direction change of the new element regarding the preceding and the initial element.

The direction with the best quality will be the direction of the next element. The algorithm is repeated until one of the stopping conditions are reached:

- the average gray level is below a threshold
- the element touches the image border.

The creation of the spine model starts with a virtual hull in a distance from the dendrite (The dendrite is adapted to the microscope image without considering the spines first.). All spines with significant length will pierce this hull. For the initial spine model we define the first element of each spine between the dendrite skin and the hull. The element points on the hull (break through points) are indicated by local gray level maximums in the corresponding position of microscope image (Fig. 5). The distance between the hull and the dendrite depends on the smoothness of the dendrite. If the distance is too low, artefacts of the dendrite skin could be interpreted as spines.

Fig. 5 Local gray level maximums on the hull indicating spine element points

The spine begins at the dendrite skin. If we assume a maximal length for the elements, the search region for the connection to the dendrite is limited. The points on the dendrite skin in this area are tested for being a reasonable starting point for the spine. Quality criteria are:

- the average gray level and

- the direction of the element regarding the dendrite

Following elements will be created by a tracking algorithm like the dendrite elements.

If necessary the initial model can be edited interactively to correct detection errors. In particular, broken spine necks and closely situated spines are hard to detect automatically. This way, all significant spines are represented by the model.

Gray value representation of the model

To estimate the quadratic error between the microscope image and the model, the model is sampled with the same resolution as the microscope image and convolved by the microscope PSF.

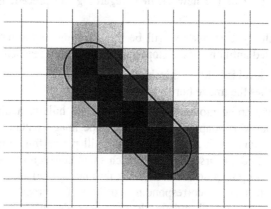

Fig. 6 Model sampling

The microscope has an orthogonal scanning raster. Similarly the spatially continuous binary model b can be sampled:

$$\hat{b}(x_0, y_0, z_0) = \frac{1}{\Delta x \Delta y \Delta z} \int_{-\frac{1}{2}}^{\frac{1}{2}} \int_{-\frac{1}{2}}^{\frac{1}{2}} \int_{-\frac{1}{2}}^{\frac{1}{2}} b(x_0 + v\Delta x, y_0 + \mu\Delta y, z_0 + \eta\Delta z) \, dv \, d\mu \, d\eta \qquad (2)$$

The voxels at the border of the elements are assigned gray levels equal to the approximate solution of the integral (Eq. (2)). This corresponds with the volume part of the voxel that is occupied by the model. The inner voxels keep the binary value (maximal gray level).

As a result we obtain a sampled model image ($\hat{b}$) with the same scanning raster as the microscope image. The variability of the model is provided by the graylevel values. To consider the influence of the PSF the model image is discretely convolved by the estimated PSF of the scanning system (Fig. 2).

Adaptation of the model parameters

To adapt the model parameters we can compare the convolved model image $\hat{F}$ with the microscope image F (Fig. 7). Strictly speaking a global objective criterion for the whole image (Eq. (1)) should be analysed for optimisation. But this would lead to an immensely large number of parameters. As an approximation the whole image is divided into regions of interest around the elements. The parameter change of an element effects the gray level distribution in the convoluted image in a region around it (ROI). The parameters of the elements can be changed continuously. Therefore usual optimisation algorithms like the gradient descent can be applied for adaptation.

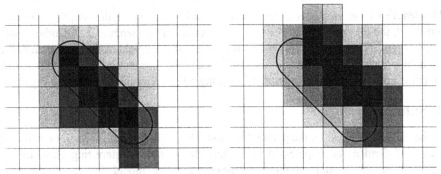

Fig. 7 Comparing gray levels in a region of interests (ROI) around an element;

convolved model image of the actual model (left) and

microscope image (right); line represents the element border

But in the current application we assume certain constraints (convex elements, simple PSF model etc.) to simplify and accelerate the solution. For example relationships between radius change and the average gray levels or between position change and the centres of gravity in the both images can be used to get an estimate for parameter change (Fig. 7).

Fig. 8 Adapted model. left: high resolution rendering; right: sampled with same scanning

resolution as the microscope image (Fig. 1)

Because the ROIs of the elements overlap, the relationships between them must be considered. The parameter changes in one iteration have to be sufficiently small. The gray-level conversion and convolution of the whole model after each iteration leads to a merger of the otherwise independent objective criteria. Therefore in some cases additional the adaptation of the whole system is needed after the approximate estimation.

The length of the elements is measured in each iteration. If the length exceeds a threshold, the element is split into two elements of half the length. On the other hand, if the length of an element falls below the radius of the neighbourhood elements, the element is removed. The iteration is terminated if the reconstruction error falls below of a threshold. Fig. 8 shows a result of adapting a model for the nerve cell in Fig. 1.

3 Summary

In the chosen application, the parametric model $b(x,y,z)$ represents a highly abstracted image of dendrite and spines. A-priori knowledge of the interesting structures is included in the model. Higher complexity for the model (e. g. elliptical or unsymmetric cross-sections) can be easily added by this method. But this does not seem appropriate because of the uncertainties of image acquisition (resolution limit, space-variant PSF in the biological specimen, noise etc.). The described method was applied successfully to the morphological analysis of some thousands of spines [1,7].

This work was supported by LSA grant (665A/2384) and DFG/BMBF grant (INK 15 A1 / TP A 4).

References

1. H. Scheich, E. Wallhäußer-Franke, K. Braun, "Does synaptic selection explain auditory imprinting?" *Memory: Organization and Locus of Change*, Oxford University Press, pp. 114-159, 1991.
2. D. A. Rusakov, M. G. Stewart, M. Sojka,G. Richter-Levin, T.V.P Bliss: Dendritic spines form 'collars' in hippocampal granule cells. in Neuroreport Vol. 6 No 11 31 July, pages 1557-1667, 1995.
3. Imaris 2.2.4 Reference Manual, Bitplan AG, Technopark Zürich, Technopark-strasse 30, 8005 Zürich, Switzerland 1994.
4. J. B. Pawley (editor), *Handbook of Biological Confocal Microscopy*. Plenum Press, New York, revised edition 1989.
5. H. T. M. van der Voort, K. C. Strasters: Restoration of Confocal Images for Quantitative Image Analysis. Journal of Microscopy vol. 178, pp. 165-181, 1995.
6. A. Herzog; G. Sommerkorn; U. Seiffert; B. Michaelis; K. Braun; W. Zuschratter, "Rekonstruktion und Klassifikation dendritischer Spines aus konfokalen Bilddaten" *Proceedings des Aachener Workshops „ Bildverarbeitung in der Medizin 08.11.-09.11.1996"*, Springer, pp. 65 - 70, 1996.
7. G. Sommerkorn, U. Seiffert, D. Surmeli, A. Herzog, B. Michaelis, K. Braun.: Classification of 3D dendritic Spines using SOM. International Conference of Artificial Neural Networks and Genetic Algorithms (ICANNGA97), Norwich, England 2.4. - 4.4. 1997.

Mosaicing of Flattened Images from Straight Homogeneous Generalized Cylinders

Adrian G. Borş[1], William Puech[2],

Ioannis Pitas[1], and Jean-Marc Chassery[2]

[1] Department of Informatics, University of Thessaloniki, Box 451,
54006 Thessaloniki, Greece - {adrian,pitas}@zeus.csd.auth.gr
[2] TIMC-IMAG Laboratory, Institut Albert Bonniot, Domaine de la Merci,
38706 La Tronche Cedex, France - {William.Puech,Jean-Marc.Chassery}@imag.fr

Abstract. This paper presents a new method for reconstructing paintings from component images. A set of monocular images of a painting from a straight homogeneous generalized cylinder is taken from various viewpoints. After deriving the surface localization in the camera coordinate system, the images are backprojected on the curved surface and flattened. We derive the perspective distortion of the scene in the case when it is mapped on a cylindrical surface. Based on the result of this study we derive the necessary number of views in order to represent the entire scene depicted on a cylindrical surface. We propose a matching-based mosaicing algorithm for reconstructing the scene from the curved surface. The proposed algorithm is applyed on paintings.

1 Introduction

In this study we consider images taken by monocular vision. Let us consider a painting on a straight homogeneous generalized cylinder [1]. We identify the localization parameters of the painted surface by considering the projections of two parallels in the given image. We calculate the common normal of the parallels projections and we derive the localization parameters. Based on the localization parameters we backproject the image on the curved surface and afterwards we flatten it [2].

The difference in perspective distortion has been used for computing the shape of the curved surface from texture information [3]. In this paper we analyze the geometrical distortions caused by the perspective view in the case of images painted on cylindrical surfaces.

Mosaicing is a well known technique used for representing images of paintings [4, 5]. Distortions caused by the painting surface shape must be corrected before the mosaicing. In [6] the perspective projection distortions caused by rotating a camera around its focal point are corrected by projecting the images onto a Gaussian sphere which is flattened on a plane tangent to it. Mosaicing assemble a set of images, each representing details of a certain region of the painting, in order to reconstruct the entire scene. The methods used in [4, 6] employ the matching of manually selected points. In this study we propose an automatic mosaicing method [7] based on region matching [5]. In the case when the images are obtained by flattening the representations of a curved surface, the image regions with large distortions, caused by perspective projection, are excluded from matching. We evaluate the bounds of the necessary number of views in order to represent the entire painting from a cylinder. The proposed method is applied in painting visualization and the result can be further used for painting restoration.

2   Curved Surface Localization and Flattening

In order to perform the localization from a single perspective view, we limit our study to the case when the surface has a curvature different than zero in only one direction. The localization is described by three rotation angles : θ_x, θ_y and θ_z. These rotation angles provide the relationship between the camera coordinate system and the coordinate system of the curved surface. Two projection axes must be localized [2]. First we find the projection of the revolution axis, and afterwards we derive the position of the second axis corresponding to the projection of one particular parallel. In order to find the projection of the revolution axis, we identify the common normal $P_1 P_2$ of two parallel curves which are projected in the image, as shown in Figure 1 and described in [2]. The slope of the straight line $P_1 P_2$ gives us the axis' direction. In the image coordinate system (u, v), the equation of this axis is :

$$v = A_1.u + B_1, \tag{1}$$

where A_1 and B_1 are the coefficients of the straight line $P_1 P_2$. From the equation (1) we derive the rotation angle θ_z corresponding to the angle between u and the revolution axis, as shown in Figure 1:

$$\theta_z = \arctan(A_1). \tag{2}$$

Among all the parallels on the curved surface, only one is projected on the image as a straight line. This straight line belongs to the plane passing through that parallel and through viewpoint, and defines the second axis. In order to obtain the two other rotation angles, θ_x and θ_y, let us consider P, a point located on the curve passing through either P_1 or P_2. We define the curvature at a point P_i as :

$$K_i = \lim_{P \to P_i} \frac{\alpha(P) - \alpha(P_i)}{|\widehat{PP_i}|}, \tag{3}$$

where $i \in \{1, 2\}$, $\alpha(P_i)$ is the angle of the tangent to P_i, and $|\widehat{PP_i}|$ is the length of the arc between P and P_i. We denote by P_0, the point belonging to the revolution axis, where the curvature K_0 equals zero. Considering (u'_0, v'_0) the coordinates of P_0 we obtain the equation of the second axis :

$$v = -\frac{1}{A_1}.u + \left(v'_0 + \frac{u'_0}{A_1}\right). \tag{4}$$

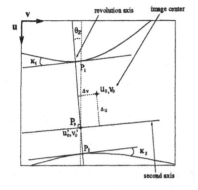

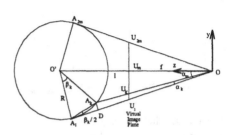

Figure 1. The two axes derived from parallel curves.

Figure 2. The cross-section representation through cylinder and image.

We denote by (Δ_u, Δ_v) the vector distance between (u_0, v_0) and P_0, the intersection of the two axes, as shown in Figure 2. The two rotation angles are then given by :

$$\begin{cases} \theta_x = \arctan(\frac{\Delta_v}{f.k}) \\ \theta_y = \arctan(\frac{\Delta_u}{f.k}), \end{cases} \tag{5}$$

where f is the focal distance and k is the resolution factor. Based on the localization parameters we backproject the image on the 3D surface, we match the point P_0 with the image center (u_0, v_0) and we obtain the virtual image. In the virtual image, P_0 is projected to the image center and the projection of the revolution axis is vertical. After backprojecting the image onto the curved surface, we flatten it in order to obtain a new image without geometrical distortions caused by the surface curvature [2, 7].

3 The Perspective Distortion Analysis when the Image is Backprojected on a Cylindrical Surface

The flattening method described in the previous Section recovers the distortions caused by the geometry of the painted surface, but not the distortions caused by perspective projection. Let us consider the image of a cylindrical surface constructed such that the axes of the camera coordinate system coincide with the object axes. The focal axis z is perpendicular to the revolution axis of the cylinder. The radius of the cylinder is denoted by R, and the viewpoint O is situated at a distance l from the revolution axis of the cylinder, as shown in the cross-section representation from Figure 2. The projection of the arc $|\widehat{A_1 A_{2m}}|$ to the virtual image plane is the line segment $|U_1 U_{2m}|$. The horizontal cross-section through image is discretized in $2m$ equal-sized intervals :

$$|U_k U_{k-1}| = |U_{k-1} U_{k-2}| = 1 \text{ pixel for } k = 3, \ldots, 2m. \tag{6}$$

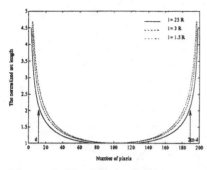
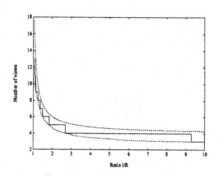

Figure 3. The length of the arcs each of them corresponding to an equal-sized image segment.

Figure 4. The necessary number of views.

Let us denote the angles $\widehat{A_1OA_k}$, $\widehat{A_1O'A_k}$ by α_k and β_k. Based on geometrical considerations, we express the length of the line segment $|A_kD|$ in Figure 2, in two different ways. From the triangles $O'A_1A_k$ and A_1A_kD we derive :

$$|A_kD| = 2R\sin^2\left(\frac{\beta_k}{2}\right). \tag{7}$$

$|A_kD|$ is calculated from the triangle A_kDO :

$$|A_kD| = (\sqrt{l^2 - R^2} - R\sin(\beta_k))\tan(\alpha_k). \tag{8}$$

From the triangle OU_kU_m we have :

$$\tan(\alpha_m - \alpha_k) = \frac{|U_mU_k|}{|U_mO|} = \frac{|U_mU_k|}{|U_1U_m|}\frac{|U_1U_m|}{|U_mO|} \tag{9}$$

Based on classical geometry properties in triangles OU_1U_m, $OO'A_1$ and by using (6) we find :

$$\tan(\alpha_m - \alpha_k) = \frac{m-k}{m}\frac{|O'A_1|}{|A_1O|} = \frac{m-k}{m\sqrt{\mu^2 - 1}} \tag{10}$$

where we denote $\mu = \frac{l}{R}$ and the angle $\widehat{O'OA_1}$ by α_m. Afterwards, we derive $\tan(\alpha_k)$ with respect to the number of pixels k :

$$\tan(\alpha_k) = \frac{k\sqrt{\mu^2 - 1}}{m\mu^2 - k}. \tag{11}$$

From (7), (8), and (11), we obtain :

$$2\sin^2\left(\frac{\beta_k}{2}\right) = (\sqrt{\mu^2 - 1} - \sin(\beta_k))\frac{k\sqrt{\mu^2 - 1}}{m\mu^2 - k} \tag{12}$$

for $k = 1, \ldots, 2m$. After deriving the angles β_k from (12), we compute the arc of the cylinder corresponding to an image segment of constant size :

$$|\widehat{A_kA_{k-1}}| = (\beta_k - \beta_{k-1})R. \tag{13}$$

The normalized arc length $|\widehat{A_kA_{k-1}}| / |\widehat{A_mA_{m-1}}|$, calculated from (13) is represented in Figure 3 for $m = 100$, when $\mu \in \{1.5; 3; 25\}$. From this plot we observe that arcs of different length from the cylindrical surface are projected to segments with the same length in the image plane.

4   The Estimation of the Necessary Number of Views

Let us consider a set of images all taken at the same distance l from the cylinder's axis, each two neighboring images having an overlapping region. The region from the cylindrical surface which projects in the image without significant distortion contains neighboring arcs having small size variation from each other :

$$\left| \frac{|\widehat{A_k A_{k-1}}| - |\widehat{A_{k-1} A_{k-2}}|}{|\widehat{A_m A_{m-1}}| - |\widehat{A_{m-1} A_{m-2}}|} \right| = \left| \frac{\beta_k + \beta_{k-2} - 2\beta_{k-1}}{\beta_m + \beta_{m-2} - 2\beta_{m-1}} \right| \le \delta \qquad (14)$$

where $|\widehat{A_k A_{k-1}}|$ and $|\widehat{A_{k-1} A_{k-2}}|$ are evaluated in (13), and δ is a small constant, measuring the difference in the arc length variation, representing a distortion measure.

As it can be observed from Figure 3, the neighboring images are likely to be differently distorted in the overlapping regions. The regions located towards the margins of the cylindrical surface representation are likely to contain larger distortions than the regions situated near the cylinder's axis projection. Let us consider that the minimal distortion condition from (14) is verified for $k = d, \ldots, 2m - d$, where d is the pixel index for which we obtain the equality in (14).

Each two images must overlap on a region corresponding to an angle larger than $2\beta_d$, i.e. where the distortions according to (14) are small enough. If we consider that each pixel in the scene should be projected in two neighboring images at most, we obtain the maximum number of images. The minimal and the maximal numbers of images required to represent the entire scene are :

$$\frac{\pi}{\arccos\left(\frac{R}{l}\right) - \beta_d} < n < \frac{2\pi}{\arccos\left(\frac{R}{l}\right)} \qquad (15)$$

where the angle β_d is derived from (14) and corresponds to the arc $|\widehat{A_1 A_d}|$. The bounds on the number of images to be taken around a cylinder are represented in Figure 4. In the same figure, the ceiling integer value of the minimum necessary number of views is marked by a continuous line. As we observe from this figure, the necessary number of images is large when the distance l from the viewpoint to the cylinders axis is small and decreases at three when is large.

5   The Mosaicing Algorithm

Image mosaicing is employed for assembling a set of images in order to reconstruct an entire scene [5]. The mosaicing approach proposed in this paper is based on matching [5, 7]. Only the part of the overlapping region which contains a small level of distortion, as provided by (14) is considered by the matching algorithm. Let us denote by (du, dv), the displacement between two neighboring images.

(a) Set of images to be mosaiced. (b) The resulting image.

Figure 5. Mosaicing a set of infrared images in order to reconstruct a painting.

As in the case of the block matching algorithms [5] we define a search area $S_u \times S_v$ in the plane uOv. The overlapping part is found based on a search for the best matching between various regions from the two images in the given area :

$$(du, dv) = \arg \min_{k=1,l=1}^{S_u,S_v} (\sum_{i=2n-k}^{2n} \sum_{j=l}^{2m-2d} |pel_p(i - 2n + k, j - l) - pel_{p-1}(i,j)|,$$

$$\sum_{i=2n-k}^{2n} \sum_{j=l}^{2m-2d} |pel_p(i - 2n + k, j) - pel_{p-1}(i, j - l)|), (16)$$

where $pel_p(i,j)$, $pel_{p-1}(i,j)$ are two pixel elements from two successive images $p-1, p$, $(2m - 2d) \times 2n$ is the image size after eliminating the part containing distortions, and d is calculated according to (14).

To calculate the pixel values in the overlapping region we evaluate the minimum distance from a given pixel site (i,j) to the border of the common reliable image region :

$$g_{p-1} = \min\{i, j\} , \ g_p = \min\{dx - i, dy - j\} \tag{17}$$

for $i = 1, \ldots, du$ and $j = 1, \ldots, dv$. In order to ensure a smooth transition, the overlapping area pixels are taken as a weighting of the component images pixels with respect to the distance from the closest nonoverlapping region :

$$f(i,j) = \frac{g_p}{g_p + g_{p-1}} f_p(i,j) + \frac{g_{p-1}}{g_p + g_{p-1}} f_{p-1}(i,j), \tag{18}$$

where $f(i,j)$ and $f_p(i,j)$ denote a pixel element from the mosaiced image and from the pth image, respectively. The proposed procedure can be easily extended to mosaic many images having horizontal and vertical overlapping areas as exemplified on a set of infrared images of an old painting in Figure 5.

(a), (b) Original images of a painted arch ;

(c), (d) The flattened surface representations ;

(e) Result of the mosaicing algorithm ;

Figure 6. Reconstruction of an arch painting by mosaicing.

(a), (b), (c) Original images of a cup ;

(d), (e), (f) The flattened surface representations ;

(g) Result of the mosaicing algorithm ;

Figure 7. Reconstruction of the ceramic decorative pattern by mosaicing.

6 Simulation Results

The proposed algorithm was applied for reconstructing several paintings on convex, concave, or flat surfaces. In Figures 6 (a) and (b) two images representing parts of a Byzantine painting on an arch are shown. The algorithm was applied on the images of a cup represented in Figures 7 (a), (b), (c), as well. These images correspond to a parameter $\mu = l/R = 5.5$. From Figure 4 we observe that we need four images to represent the cup's surface. Due to the cup's handle and because the scene does not cover all around the cup, we use three images. These images present distortions, caused by the painted surface, that depend on the view-angle. We localize the painted surfaces by considering the projections of two parallels representing the edges of the arch or those of the decorative pattern from the cup. After finding the localization of the painted surface, we flatten the projected images of the paintings as shown in Figures 6 (c), (d) and 7 (d), (e), (f), respectively. The mosaicing of flattened representations are displayed in Figures 6 (e) and 7 (g). In color images, the localization parameters and the relative displacements of the images are calculated from the luminance component images. Afterwards, the result is applied on all the color components.

7 Conclusions

In this study we propose a new approach for representing the scenes painted on the surface of straight homogeneous generalized cylinders. We provide a theoretical analysis of the geometrical distortions due to the perspective projection and we derive the necessary number of views to represent entirely the scene painted on a cylindrical surface. We propose a new approach for image mosaicing based on matching. This algorithm is applied for painting visualization.

References

1. J. Ponce, D. Chelberg, W. B. Mann, "Invariant Properties of Straight Homogeneous Generalized Cylinders and their Contours," *IEEE Trans. on PAMI*, vol. 11, no. 9, pp. 951-966, 1989.
2. W. Puech, J.-M. Chassery, I. Pitas, "Curved surface localization in monocular vision," *Pattern Recognition Letters*, 1997 (to appear).
3. B. J. Super, A. C. Bovik, "Shape from texture using local spectral moments," *IEEE Trans. on PAMI*, vol. 17, no. 4, pp. 333-343, Apr. 1995.
4. R. J. Billinge, J. Cupitt, J, Dessipiris, D. Saunders, "A note on an Improved Procedure for the rapid assembly of infrared reflectogram mosaics," *Studies in Conservation*, vol. 38, pp. 92-97, 1993.
5. R. J. Schalkoff, *Digital Image Processing and Computer Vision*. John Wiley, 1989.
6. Ş. Gümüştekin, R. W. Hall, "Mosaic Image Generation on a Flattened Gaussian Sphere," *IEEE Workshop on Applications of Computer Vision*, Sarasota, USA, pp. 50-55, 1996.
7. W. Puech, A. G. Borş, J.-M. Chassery, I. Pitas, "Mosaicing of Paintings on Curved Surfaces," *IEEE Workshop on Applications of Computer Vision*, Sarasota, USA, pp. 44-49, 1996.

Well-Posedness
of Linear Shape-from-Shading Problem

Ryszard Kozera[1] and Reinhard Klette[2]

[1] The University of Western Australia, Department of Computer Science,
Nedlands 6907 WA, Australia
[2] The Auckland University, Tamaki Campus, Computer Science Department,
Private Bag 92019, Auckland, New Zealand

Abstract. We continue to study here a global shape recovery of a smooth surface for which the reflectance map is linear. It was recently proved that under special conditions the corresponding finite difference based algorithms are stable and thus convergent to the ideal solution. The whole analysis was based on the assumption that the problem related to the linear image irradiance equation is well-posed. Indeed, we show in this paper that under certain conditions there exists a unique global C^2 solution (depending continuously on the initial data) to the corresponding Cauchy problem defined over the entire image domain (with non-smooth boundary).

1 Introduction

In this paper we shall analyze *a well-posedness* of a linear image irradiance equation

$$a_1 \frac{\partial u}{\partial x_1}(x_1, x_2) + a_2 \frac{\partial u}{\partial x_2}(x_1, x_2) = E(x_1, x_2) \tag{1}$$

which, in computer vision, models reflectance properties of e.g. maria of the moon or can be used for a local approximation of non-linear reflectance maps. Recently, Kozera [3] and Kozera and Klette [4] have established converegnce and stability results for different sequential algorithms based on combination of central difference, forward and backward difference derivative approximations. Essential to these results, is the assumption that the problem (1) is *well-posed* (see *e.g.* Lax [5]). In this paper we establish sufficient conditions for the existence of a unique global solution u of class C^2 satisfying the corresponding Cauchy problem defined over a rectangular image domain $\Omega \subset \mathbb{R}^2$:

$$(i) \quad E(x_1, x_2) = a_1 \frac{\partial u}{\partial x_1}(x_1, x_2) + a_2 \frac{\partial u}{\partial x_2}(x_1, x_2),$$

$$(ii) \quad u(x_1, 0) = f(x_1) \quad 0 \le x_1 \le a, \text{ for } \operatorname{sgn}(a_1 a_2) \ge 0,$$

$$u(x_1, b) = f(x_1) \quad 0 \le x_1 \le a, \text{ for } \operatorname{sgn}(a_1 a_2) \le 0, \tag{2}$$

$$(iii) \quad u(0, x_2) = g(x_2) \quad 0 \le x_2 \le b,$$

where functions $f \in C^2(0,a) \cap C^0[0,a]$ and $g \in C^2(0,b) \cap C^0[0,b]$ satisfy $f(0) = g(0)$, $E \in C^2(\bar{\Omega})$, and a_1 and a_2 are constants such that $(a_1, a_2) \neq (0,0)$. To simplify consideration (without losing generality) we will assume that $\mathrm{sgn}(a_1 a_2) \geq 0$ and $a_2 \neq 0$. The cases $\mathrm{sgn}(a_1 a_2) \geq 0$ and $a_1 \neq 0$, $\mathrm{sgn}(a_1 a_2) \leq 0$ and $a_2 \neq 0$, and $\mathrm{sgn}(a_1 a_2) \leq 0$ and $a_1 \neq 0$ can be treated analogously.

2 Local existence and uniqueness

In this section we shall remind a result about a local existence of a unique C^2 solution for the linear image irradiance equation defined over a domain with a smooth boundary. A corresponding proof (for a smooth boundary) can be found in literature (see e.g. John [2]). Note that the uniqueness and existence analysis can be performed within the natural class of C^1 functions. As the convergence of the numerical schemes presented in [3] and [4] requires u to be continuously twice-differentiable, the class of C^2 functions shall be considered. We begin the well-posedness analysis by introducing some preliminary notions. For a linear first-order partial differential operator P in two variables

$$Pu(x) = a_1(x)\frac{\partial u}{\partial x_1}(x) + a_2(x)\frac{\partial u}{\partial x_2}(x),$$

where $x = (x_1, x_2) \in \mathbb{R}^2$, let $A(x) = (a_1(x), a_2(x))$. By a *a characteristic form* of P we understand $\sigma_{\widehat{P}}(x_1, x_2)(\xi_1, \xi_2) = a_1(x_1, x_2)\xi_1 + a_2(x_1, x_2)\xi_2$, where $(\xi = (\xi_1, \xi_2) \in \mathbb{R}^2)$. *The normal cone* is a subset of $\mathbb{R}^2$ defined as

$$\mathrm{char}(P,x) = \left\{ (\xi_1, \xi_2) \in \mathbb{R}^2 : \sigma_{\widehat{P}}(x_1, x_2)(\xi_1, \xi_2) = 0 \right\}.$$

Recall now the following:

Definition 1. A smooth curve $\gamma \subset \mathbb{R}^2$ is called a *characteristic* for P at $x \in \gamma$, if a normal vector $\nu(x)$ to γ is not in the normal cone. Moreover, γ is called *non-characteristic* for P if it is not characteristic for P at any point $x \in \gamma$.

It is easy to observe that $\mathrm{char}(P,x) = \{\xi \in \mathbb{R}^2 : \langle A(x), \xi \rangle = 0\}$. The last condition has a simple geometric interpretation. Namely, γ is characteristic for P at $x \in \gamma$ if and only if the vector $A(x)$ belongs to $T_x\gamma$, where $T_x\gamma$ is a tangent space to γ at point x. Consider now the following Cauchy problem

$$a_1(x)u_{x_1}(x) + a_2(x)u_{x_2}(x) = f(x),$$

$$u(x) = \phi(x) \quad \text{over } \gamma \subset \Omega. \tag{3}$$

defined over $\Omega \subset \mathbb{R}^2$. We formulate now the following theorem (for a full proof see [2])

Theorem 2. *Let γ be a C^2 curve, non-characteristic for P defined in (3). Assume that functions a_1, a_2, f, and ϕ are real C^2 functions. Then, in some neighbourhood of γ, there exists exactly one C^2 solution u to Cauchy's problem (3).*

3 Global existence and uniqueness

In this section we discuss the global existence and uniqueness of C^2 solutions to Cauchy's problem defined by (2). We establish sufficient conditions under which this problem is well-posed over the rectangular image domain.

We formulate first the following theorem (for a full proof see [4]):

Theorem 3. *Let Ω be an open domain of $\mathbb{R}^2$ and let $\gamma \subset \partial\Omega$ be a C^2 class curve with the parametrization $\gamma : [s_-, s_+] \rightarrow \gamma = (\gamma_1, \gamma_2) \in \mathbb{R}^2$. Assume, moreover, that γ is non-characteristic for the linear first-order partial differential operator (with constant coefficients) defined by (1). Let, moreover, E be a C^2 class function over $\Omega \cup \partial\Omega$ and let ϕ be a C^2 class function defined over γ. Then the Cauchy problem*

$$Pu(x_1, x_2) = E(x_1, x_2),$$

$$u(x_1, x_2)|_\gamma = \phi(x_1, x_2), \tag{4}$$

has exactly one C^2 solution over domain of influence $\bar{\Omega} = (\Omega \cup \partial\Omega) \cap D_1$, where D_1 is a strip in $\mathbb{R}^2$ (see Figure 1) defined as follows:
1. If $a_1 \neq 0$, then for $x = (x_1, x_2)$

$$D_1 = \left\{ x \in \mathbb{R}^2 : \frac{a_2}{a_1}(x_1 - \gamma_1(s_-)) + \gamma_2(s_-) < x_2 < \frac{a_2}{a_1}(x_1 - \gamma_1(s_+)) + \gamma_2(s_+) \right\}. \tag{5}$$

2. If $a_1 = 0$, then for $x = (x_1, x_2)$

$$D_1 = \left\{ x \in \mathbb{R}^2 : \gamma_1(s_-) < x_1 < \gamma_1(s_+) \right\}. \tag{6}$$

We establish now an existence and uniqueness result for Cauchy's problem in which γ is not *a smooth curve* at one point (as the initial problem (2) is considered over a rectangular image domain). In this case we need to establish sufficient conditions for which solutions to the Cauchy's problem with smooth initial curves *bifurcate* along the line coinciding with characteristic direction that passes through a "non-smoothness point" of γ (see Figure 2). Note that for Ω defined in (2) there is no necessity for the bifurcation in the case when $(a_1, a_2) = (0, 0)$. Thus, in the next theorem we shall assume that also $a_1 \neq 0$.

Theorem 4. *Let $P, E, \Omega,$ and $\bar{\Omega}$ be defined as in Theorem 3. Assume moreover that $cl(\Omega)$ (closure of Ω) is compact and convex, and $\partial\Omega \supset \gamma = \gamma^1 \cup \gamma^2 \cup \{x_0\}$, where $\gamma^1 : (s_-, s_0) \rightarrow \gamma^1 \in \mathbb{R}^2$ and $\gamma^2 : (s_0, s_+) \rightarrow \gamma^2 \in \mathbb{R}^2$ are C^2 curves satisfying*

$$\lim_{s \to s_0^-} \gamma^1(s) = \lim_{s \to s_0^+} \gamma^2(s) = x_0 = (x_{10}, x_{20}).$$

Let ϕ be a continuous function over γ and of class C^2 over $\gamma \setminus \{x_0\}$. Suppose moreover that $\bar{\Omega} = \bar{\Omega}_1 \cup \bar{\Omega}_2 \cup \beta$, where β is the straight line passing through x_0 in direction (a_1, a_2), $\bar{\Omega}_1$ and $\bar{\Omega}_2$ are disjoint subsets defined as in the previous theorem. Assume, moreover, that there exists a unique global C^2 solution u_i over $\bar{\Omega}_i$, for $i = (1, 2)$, to the following Cauchy problems, for $a_1 \neq 0$,

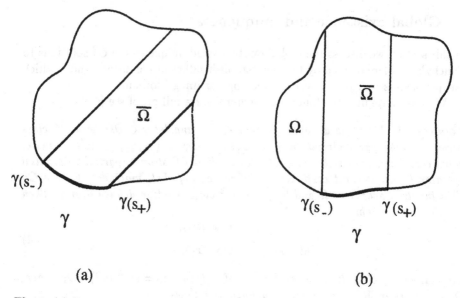

Fig. 1. (a) If $a_1 \neq 0$, then u is globally defined over "tilted" $\bar{\Omega}$ (b) If $a_1 = 0$, then u is globally defined over "vertical" $\bar{\Omega}$

$$Pu(x_1, x_2) = a_1 \frac{\partial u}{\partial x_1}(x_1, x_2) + a_2 \frac{\partial u}{\partial x_2}(x_1, x_2),$$
$$u(x_1, x_2)|_{\gamma_i} = \phi(x_1, x_2), \tag{7}$$

where both γ_1 and γ_2 are also non-characteristic for P. Suppose, moreover, that for each point $(x_1', x_2') \in \beta$ one of the pair of limits

$$\lim_{(x_1, x_2) \in \Omega_1 \to (x_1', x_2')} u_{1x_1}(x_1, x_2) \quad \text{and} \quad \lim_{(x_1, x_2) \in \Omega_2 \to (x_1', x_2')} u_{2x_1}(x_1, x_2) \tag{8}$$

or

$$\lim_{(x_1, x_2) \in \Omega_1 \to (x_1', x_2')} u_{1x_2}(x_1, x_2) \quad \text{and} \quad \lim_{(x_1, x_2) \in \Omega_2 \to (x_1', x_2')} u_{2x_2}(x_1, x_2) \tag{9}$$

exists and the corresponding limits are equal. Assume, moreover, that

$$\lim_{(x_1, x_2) \in \Omega_1 \to (x_1', x_2')} u_{1x_1 x_2}(x_1, x_2) \quad \text{and} \quad \lim_{(x_1, x_2) \in \Omega_2 \to (x_1', x_2')} u_{2x_1 x_2}(x_1, x_2) \tag{10}$$

exist and are equal. Then there exists exactly one C^2 solution u to the Cauchy problem

$$Pu(x_1, x_2) = E(x_1, x_2),$$
$$u(x_1, x_2)|_{\gamma} = \phi(x_1, x_2) \tag{11}$$

over the domain of influence $\bar{\Omega}$.

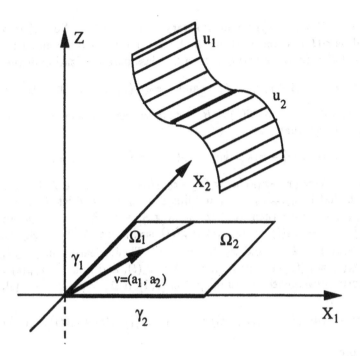

Fig. 2. A bifurcation of two partial solutions u_1 and u_2 along the base characteristic line passing through the point $(0,0)$

Proof. Theorem 3 applied twice to (7) (for $i = 1, 2$) yields immediately uniqueness for (11). Passing to the proof of existence, let (x_1', x_2') be an arbitrary point contained in $\beta \cap \Omega$ and let u_1, u_2 be defined as in (7). The continuity of E combined with (7), (8), (or with (7), (9)), and $a_1 \neq 0$ ensures that both conditions (8) and (9) are met. We can thus safely put

$$\lim_{(x_1, x_2) \in \Omega_1 \to (x_1', x_2')} u_{1x_1}(x_1, x_2) = \lim_{(x_1, x_2) \in \Omega_2 \to (x_1', x_2')} u_{2x_1}(x_1, x_2) = u_{x_1}(x_1', x_2'),$$

$$\lim_{(x_1, x_2) \in \Omega_1 \to (x_1', x_2')} u_{1x_2}(x_1, x_2) = \lim_{(x_1, x_2) \in \Omega_2 \to (x_1', x_2')} u_{2x_2}(x_1, x_2) = u_{x_2}(x_1', x_2').$$

To complete the proof, we first show that both $\lim_{(x_1, x_2) \in \Omega_1 \to (x_1', x_2')} u_1(x_1, x_2)$ and $\lim_{(x_1, x_2) \in \Omega_2 \to (x_1', x_2')} u_2(x_1, x_2)$ exist and are equal. Take $x_0 = (x_{10}, x_{20}) \in$

$\gamma \cup \beta$ and let $B = (x_1', x_2')$ be an arbitrary point of $\beta \cap \Omega$. Note that $\beta \cap \Omega$ is connected as $cl(\Omega)$ is convex (thus the bifurcation takes place along the entire $\beta \subset \Omega$). Let $\bar{\beta} = (\hat{x}_1(t), \hat{x}_2(t)) \subset \beta$ be a C^1 class parametrization defined as

$$\hat{x}_1(t; (x_1', x_2')) = (1-t)x_{10} + tx_1' \quad \text{and} \quad \hat{x}_2(t; (x_1', x_2')) = (1-t)x_{20} + tx_2'.$$

Then clearly $\bar{\beta}(0) = x_0$ and $\bar{\beta}(1) = B$. Note moreover that there exists a constant $k(x_1', x_2') \neq 0$ such that

$$\dot{\hat{x}}_1(t) = x_1' - x_{10} = ka_1 \quad \text{and} \quad \dot{\hat{x}}_2(t) = x_2' - x_{20} = ka_2. \tag{12}$$

Clearly $k(x_1', x_2') = (x_1' - x_{10})/a_1 = (x_2' - x_{20})/a_2$. For each $n \in \mathcal{N}$ let $(A_n, B_n) \in \bar{\Omega}_1$ be such that $\lim_{n \to \infty} A_n = x_0$ and $\lim_{n \to \infty} B_n = B$ and let β_n be a straight line in $\bar{\Omega}_1$ (as Ω convex then $\beta_n \subset \bar{\Omega}_1$) joining A_n with B_n with C^1 parametrization $[0,1] \ni t \to \beta_n(t) = (x_{1n}(t), x_{2n}(t))$ such that $\beta_n(0) = A_n$, $\beta_n(1) = B_n$, and $\dot{\hat{x}}_{1n}(t) = k_n a_1$ and $\dot{\hat{x}}_{2n}(t) = k_n a_2$. Then, clearly, $\lim_{n \to \infty} k_n = k$, $\lim_{n \to \infty} \beta_n(t) = \bar{\beta}(t)$, and $\lim_{n \to \infty} \dot{x}_{1n}(t) = \dot{\hat{x}}_1(t)$, and $\lim_{n \to \infty} \dot{x}_{2n}(t) = \dot{\hat{x}}_2(t)$ with uniform convergence in t running over $[0,1]$. If $\Delta_n^1 = u_1(B_n) - u_1(A_n)$ then

$$\Delta_n^1 = \int_0^1 \left(u_{1x_1}(x_{1n}(t), x_{2n}(t))\dot{x}_{1n}(t) + u_{1x_2}(x_{1n}(t), x_{2n}(t))\dot{x}_{2n}(t) \right) dt.$$

Furthermore,

$$u_1(B_n) = u_1(A_n) + k_n \int_0^1 \left(a_1 u_{1x_1}(x_{1n}(t), x_{2n}(t)) + a_2 u_{1x_2}(x_{1n}(t), x_{2n}(t)) \right) dt,$$

which by (7) and (11) yields

$$u_1(B_n) = u_1(A_n) + k_n \int_0^1 E(x_{1n}(t), x_{2n}(t)) \, dt.$$

The compactness and convexity of $cl(\Omega)$ combined with a global C^2 and thus C^1 class of E defined over $cl(\Omega)$ implies that mapping E is uniformly continuous over $cl(\Omega)$. Thus the sequence $E(x_{1n}(t), x_{2n}(t))$ is uniformly convergent to $E(\hat{x}_1(t), \hat{x}_2(t))$ in t tunning over $[0,1]$. Extend u_1 by continuity to (x_{10}, x_{20}) by putting $u_1(x_{10}, x_{20}) = \phi(x_{10}, x_{20})$. Then, by (7), (8), (9), and Lebesgue's dominated convergence theorem,

$$\lim_{n \to \infty} u_1(B_n) = \phi(x_{10}, x_{20}) + k(x_1', x_2') \int_0^1 E(\hat{x}_1(t; (x_1', x_2')), \hat{x}; (x_1', x_2')_2(t)) \, dt.$$

The latter assures the existence of a continuous extension of u_1 on β given by

$$\bar{u}_1(x_1', x_2') = \phi(x_{10}, x_{20}) + k(x_1', x_2') \int_0^1 E(\hat{x}_1(t; (x_1', x_2')), \hat{x}_2(t; (x_1', x_2'))) \, dt.$$

Similarly

$$\bar{u}_2(x_1', x_2') = \phi(x_{10}, x_{20}) + k(x_1', x_2') \int_0^1 E(\hat{x}_1(t; (x_1', x_2')), \hat{x}_2(t; (x_1', x_2'))) \, dt.$$

Now, by letting

$$u(x_1, x_2) = \begin{cases} u_1(x_1, x_2) & \text{if } (x_1, x_2) \in \bar{\Omega}_1, \\ u_2(x_1, x_2) & \text{if } (x_1, x_2) \in \bar{\Omega}_2, \\ \bar{u}_1(x_1, x_2) = \bar{u}_2(x_1, x_2) & \text{if } (x_1, x_2) \in \beta, \end{cases} \tag{13}$$

we clearly construct a C^1 global solution to (11) over $\bar{\Omega}$, which is C^2 over $\bar{\Omega}_1$ and $\bar{\Omega}_2$ and C^1 along β. To complete the proof, it suffices to show that formula (13) defines C^2 function along β (i.e. u_1 and u_2 bifurcate along β within C^2 class). By differentiating (11) with respect to x_1, we obtain

$$a_1 u_{1x_1x_1}(x_1, x_2) + a_2 u_{1x_1x_2}(x_1, x_2) = E_{x_1}(x_1, x_2), \qquad (x_1, x_2) \in \bar{\Omega}_1$$

$$a_1 u_{2x_1x_1}(x_1, x_2) + a_2 u_{2x_1x_2}(x_1, x_2) = E_{x_1}(x_1, x_2). \qquad (x_1, x_2) \in \bar{\Omega}_2.$$

The continuity of E_{x_1} combined with (10) and $a_1 \neq 0$ implies that

$$\lim_{(x_1,x_2)\in\Omega_1\to(x_1',x_2')} u_{1x_1x_1}(x_1, x_2) = \lim_{(x_1,x_2)\in\Omega_2\to(x_1',x_2')} u_{2x_1x_1}(x_1, x_2) = u_{x_1x_1}(x'),$$

where $x' = (x_1', x_2')$. Similarly we find that

$$\lim_{(x_1,x_2)\in\Omega_1\to(x_1',x_2')} u_{1x_2x_2}(x_1, x_2) = \lim_{(x_1,x_2)\in\Omega_2\to(x_1',x_2')} u_{2x_2x_2}(x_1, x_2) = u_{x_2x_2}(x'),$$

where $x' = (x_1', x_2')$. The latter two equations assure that formula (13) defines a C^2 function along β. The proof is complete. $\square$

We are now ready to construct a global solution to the Cauchy problem (2).

Theorem 5. *Consider the initial problem (2) over $\Omega = \{(x_1, x_2) \in \mathbb{R}^2 : 0 < x_1 < a, 0 < x_2 < b\}$. Assume that $\partial\Omega \supset \gamma = \gamma^1 \cup \gamma^2 \cup \{x_0\}$, where $\gamma_1(s) = (0, s)$ with $-b < s < 0$, $\gamma_2(s) = (s, 0)$ with $0 < s < a$, and $x_0 = (0, 0)$.*

1. *Suppose that $\text{sgn}(a_1 a_2) > 0$. If (8), (9), and (10) are satisfied, then there exists exactly one C^2 class global solution u satisfying (2) defined over the entire rectangular domain of influence $\bar{\Omega}$.*
2. *Suppose $a_1 = 0$. Then there exists exactly one C^2 class global solution u to (2) defined over the entire rectangular domain of influence $\bar{\Omega}$.*

Proof. Suppose first that $\text{sgn}(a_1 a_2) > 0$. Then, by applying Theorem 3 twice to (2), with γ_1 and γ_2 as initial non-characteristic curves, we construct exactly one C^2 solution u_1 (u_2) to (2), over $\bar{\Omega}_1$ ($\bar{\Omega}_2$). Note that the domain of influence $\bar{\Omega}$ can be here represented as $\bar{\Omega}_1 \cup \bar{\Omega}_2 \cup \beta$, where β is the straight line with slope a_2/a_1 passing through point $(0,0)$ (see Figure 2). Note also that Theorem 4 assures that there exists exactly one C^2 global solution u to the problem (2) defined by (13) over $\bar{\Omega}$. Suppose now that $a_1 = 0$. Then by applying Theorem 3 with γ_2 as an initial non-characteristic curve, we can construct a global solution to (2) of class C^2 defined over $\bar{\Omega}$ introduced in Theorem 3 (there is no need for the use of bifuraction Theorem 4). The proof is complete. $\square$

We finally present sufficient conditions for the well-posedness of (2).

Theorem 6. *Assume that the initial assumptions from Theorem 5 are satisfied. Then the Cauchy problem* (2) *is well-posed.*

Proof. The global existence and uniqueness of the solution to (2) is assured by Theorem 5. Note also that for the Cauchy problem (2), the corresponding characteristic system (see [2]) reduces to

$$(i) \quad \frac{\partial x_1}{\partial t}(s,t) = a_1,$$

$$(ii) \quad \frac{\partial x_2}{\partial t}(s,t) = a_2,$$

$$(iii) \quad \frac{\partial v}{\partial t}(s,t) = E\big(x_1(s,t), x_2(s,t)\big),$$

$$(iv) \quad \big(x_1(s,0), x_2(s,0)\big) = g(s),$$

$$(v) \quad v(s,0) = \phi\big(g(s)\big),$$

with $g(s) = (0,-s)$ (for $s \in [-b,0]$) and $g(s) = (s,0)$ (for $s \in [0,a]$), and $\phi\big(g(s)\big) = u(0,-s)$ (for $s \in [-b,0]$) and $\phi\big(g(s)\big) = u(s,0)$ (for $s \in [0,a]$). Combining (i) and (ii) gives $x_1(t,s) = a_1 t + g_1(s)$ and $x_2(t,s) = a_2 t + g_2(s)$, where $g(s) = (g_1(s), g_2(s))$. The latter, together with (iii) and (iv), yields

$$v(t_0, s) = v(s) + \int_0^{t_0} E(a_1 t + g_1(s), a_2 t + g_2(s))dt.$$

The property of integral assures the dependence of the solution v on initial and boundary condition as well as on function E. The proof is complete.□

4   Acknowledgments

The support within the Alexander von Humboldt Grant and Deutsche Forschungsgemeinschaft Project is also acknowledged by both authors.

References

1. Arnold V. I.: Ordinary Differential Equation. MIT Press Cambridge MA (1973)
2. John F.: Partial Differential Equations. Vol. 1 Springer-Verlag New York (1971)
3. Kozera R.: An algorithm for linear shape-from-shading problem. In Proc. 6th International Conference on Computer Analysis of Images and Patterns. Springer-Verlag Berlin-Heidelberg, Prague, Czech Republic (September 1995) 408–415
4. Kozera R. and Klette R.: Finite difference based algorithms for linear shape from shading. Machine Graphics and Vision (to appear)
5. Lax P. D. and Richtmyer R. D.: Survey of the stability of linear finite difference equations. Comm. Pure Appl. Math. 9 (1956) 267–293
6. Maurin K.: Analysis. Vol. 1 PWN-Polish Scientific Publishers (1973)

Comparing Convex Shapes
Using Minkowski Addition

Alexander Tuzikov[1]* and Henk J.A.M. Heijmans[2]

[1] Institute of Engineering Cybernetics, Surganova 6, 220012 Minsk, Belarus,
[2] CWI, P.O. Box 94079, 1090 GB Amsterdam, The Netherlands,
tuzikov@mpen.bas-net.by, Henk.Heijmans@cwi.nl

Abstract. This report deals with similarity measures for convex shapes whose definition is based on Minkowski addition and the Brunn-Minkowski inequality. All measures considered here are invariant under translations. In addition, they may be invariant under rotations, multiplications, reflections, or affine transformations. Restricting oneselves to the class of convex polygons, it is possible to develop efficient algorithms for the computation of such similarity measures. These algorithms use a special representation of convex polygons known as the perimetric measure. Such representations are unique for convex sets and linear with respect to Minkowski addition. Although the paper deals exclusively with the 2-dimensional case, many of the results carry over almost immediately to higher-dimensional spaces.

1 Introduction

Usually two main concepts are used for comparing shapes: distance functions measuring dissimilarity and similarity measures expressing how similar two shapes are. In this report we deal with similarity measures.

Similarity approaches have to be invariant under certain classes of transformations, e.g. similitudes (i.e., translations, rotations, change of scale). Affine transformations are also of great practical value as these can model shape distortions arising when an object is observed by a camera under arbitrary orientations with respect to the image plane. Usually shape normalization technique is first performed to develop a similarity approach which is invariant under a given class of transformations [8, 11].

Different methods for comparing shapes have been proposed in the literature. Among the best known ones are matching techniques [11], in particular contour matching [2], structural matching [3] (which is based on specific structural features), and point set matching [14]. Fourier descriptors derived from contour representations have been used by various authors to describe shape similarity (see [13] for a comprehensive discussion).

In this report we will discuss similarity measures based on Minkowski addition, the Brunn-Minkowski inequality, and the theory of mixed volumes. Most

---

* A. Tuzikov was supported by the Netherlands Organization for Scientific Research (NWO) through a visitor's grant.

of our results apply to arbitrary compact convex sets (in $\mathbf{R}^2$), but efficient algorithms are developed for convex polygons only. In the report results are given without proof. Proofs as well as some additional results are contained in our report [7].

We present first some basic notation. By $\mathcal{K}(\mathbf{R}^2)$, or briefly $\mathcal{K}$, we denote the family of all nonempty compact subsets of $\mathbf{R}^2$. The compact, convex subsets of $\mathbf{R}^2$ are denoted by $\mathcal{C} = \mathcal{C}(\mathbf{R}^2)$, and the convex polygons by $\mathcal{P}(\mathbf{R}^2)$ or just $\mathcal{P}$. We are not interested in the location of a shape $A \subseteq \mathbf{R}^2$ therefore two shapes A and B are said to be *equivalent* if they differ only by translation. Denote this as $A \equiv B$. The *Minkowski addition* of two sets $A, B \subseteq \mathbf{R}^2$ is defined by $A \oplus B = \{a + b \mid a \in A, \ b \in B\}$.

It is well-known [12] that every element A of $\mathcal{C}$ is uniquely determined by its *support function* given by $h(A, u) = \sup\{\langle a, u \rangle \mid a \in A\}, \ u \in S^1$. Here $\langle a, u \rangle$ is the inner product of vectors a and u, and S^1 denotes the unit circle.

Every polygon $P \subseteq \mathbf{R}^2$ can be represented uniquely by specifying the position of one of its vertices and the lengths and directions of all of its edges. Below, p_i will denote the length of edge i and u_i is the vector orthogonal to this edge. By $\angle u_i$ we denote the angle between the positive x-axis and u_i. Since we are not interested in the location of P, it is sufficient to give the sequence $(u_1, p_1), (u_2, p_2), \ldots, (u_n, p_n)$, where $n = n_P$ is the number of vertices of P. We will call this sequence the *perimetric representation* of P and denote it by $\mathbf{M}(P)$. If the polygon is convex, then the order of (u_i, p_i) does not have to be specified in advance, since in this case the normal vectors are ordered counter-clockwise. In this case we can consider $\mathbf{M}(P)$ as a set. We also use the so-called *perimetric measure* $M(P, \cdot)$ representation [10]:

$$M(P, u) = \begin{cases} p_i, & \text{if } u = u_i, \\ 0, & \text{otherwise.} \end{cases}$$

Minkowski addition of two convex polygons can be computed by merging both perimetric representations (see [4, 6]), i.e., the following relation holds:

$$M(P \oplus Q, u) = M(P, u) + M(Q, u), \quad \text{for } P, Q \in \mathcal{P} \text{ and } u \in S^1.$$

We will denote the group of all *affine transformations* on $\mathbf{R}^2$ by G'. If $g \in G'$ and $A \in \mathcal{K}$, then $g(A) = \{g(a) \mid a \in A\}$. We denote also by G the subgroup of G' containing all linear transformations, i.e., transformations g with $g(0) = 0$.

Let us introduce the following notations for subsets of G: by I we denote the isometries, by R the rotations about the origin, by M the multiplications with respect to the origin with a positive factor, by L the (line) reflections (lines passing through the origin), and by S the similitudes (rotations, reflections, multiplications).

If H is a subgroup of G, then H_+ denotes the subgroup of H containing all transformations with positive determinant. For example, $I_+ = R$ and S_+ comprises all multiplications and rotations. If H is a subgroup of G, then the set $\{mh \mid h \in H, \ m \in M\}$ is also a subgroup, which will be denoted by MH.

Denote by r_θ the rotation around the origin over an angle θ, and by l_α the reflection with respect to the line through the origin which makes an angle α with the positive x-axis. It is not difficult to see that the following relations hold: $l_\alpha r_\theta = l_{\alpha - \theta/2}$, $r_\theta l_\alpha = l_{\alpha + \theta/2}$, $l_\beta l_\alpha = r_{2\beta - 2\alpha}$.

2 Mixed volumes and the Brunn-Minkowski inequality

Denote by $V(A)$ the *volume* (or *area*) of a compact set A. It is well-known that for every affine transformation g we have $V(g(A)) = |\det g| \cdot V(A)$, where $\det g$ denotes the determinant of g.

The *mixed volume* $V(A, B)$ of two compact, convex sets $A, B \subseteq \mathbf{R}^2$ is implicitly defined by the following formula for the volume of $A \oplus B$:

$$V(A \oplus B) = V(A) + 2V(A, B) + V(B). \tag{1}$$

Given arbitrary compact, convex sets A and B the mixed volume has the following properties:

$V(A, B) = V(B, A) \geq 0$; if $V(A) > 0$ and $V(B) > 0$ then $V(A, B) > 0$;

$V(A, A) = V(A)$; $V(\lambda A, B) = \lambda V(A, B)$ for every $\lambda > 0$;

$V(g(A), g(B)) = |\det g| \cdot V(A, B)$, for every affine transformation g;

$V(A_1 \oplus A_2, B) = V(A_1, B) + V(A_2, B)$;

$V(A, B)$ is continuous in A and B with respect to the Hausdorff metric.

We use also the following inequality (see [12] for a comprehensive discussion).

Theorem 1 Brunn-Minkowski inequality. *For two arbitrary compact sets $A, B \subseteq \mathbf{R}^2$ the following inequality holds:*

$$V(A \oplus B)^{\frac{1}{2}} \geq V(A)^{\frac{1}{2}} + V(B)^{\frac{1}{2}}, \tag{2}$$

with equality if and only if A and B are convex and homothetic modulo translation, i.e., $B \equiv \lambda A$ for some $\lambda > 0$.

Minkowski's inequality $V(A, B) \geq V(A)^{\frac{1}{2}} V(B)^{\frac{1}{2}}$ follows directly from (1) and (2). Equality holds iff A and B are convex and $B \equiv \lambda A$ for some $\lambda > 0$.

One derives from (2) that $V(A \oplus B) \geq 4 V(A)^{\frac{1}{2}} V(B)^{\frac{1}{2}}$, with equality iff $A \equiv B$ and both sets are convex.

The mixed volume of two convex polygons P, Q can be computed using support functions and perimetric representations. Let the perimetric representation of Q be given by the sequence (v_j, q_j), $j = 1, 2, \ldots, n_Q$. If $h(P, \cdot)$ is the support function of P, then

$$V(P, Q) = \frac{1}{2} \sum_{j=1}^{n_Q} h(P, v_j) q_j. \tag{3}$$

Several formulas for calculating volumes of polyhedra are known in the literature [1]. We use the following formula based on perimetric representation (see [7]).

Let P be a polygon with perimetric representation $(u_1, p_1), (u_2, p_2), \ldots, (u_n, p_n)$ and $\alpha_j = \angle u_j$, then

$$V(P) = \sum_{j=1}^{n} p_i \sin \alpha_i \sum_{j=1}^{i} p_j \cos \alpha_j - \frac{1}{2} \sum_{i=1}^{n} p_i^2 \sin \alpha_i \cos \alpha_i.$$

3 Similarity measures

The goal of this report is to propose a measure for comparing different shapes that can be computed efficiently and that is invariant under a given group H of transformations. For example, if we take for H all rotations, then the result for A and B should be the same as for A and $r(B)$, where $r \in H$. This leads us to the following notion.

Definition 2. Let H be a subgroup of G and $\mathcal{J} \subseteq \mathcal{K}$. A function $\sigma : \mathcal{J} \times \mathcal{J} \to [0, 1]$ is called an H-invariant similarity measure on $\mathcal{J}$ if
(1) $\sigma(A, B) = \sigma(B, A)$;
(2) $\sigma(A, B) = \sigma(A', B')$ if $A \equiv A'$ and $B \equiv B'$;
(3) $\sigma(A, B) = \sigma(h(A), B)$, $h \in H$;
(4) $\sigma(A, B) = 1 \iff B \equiv h(A)$ for some $h \in H$;
(5) σ is continuous in both arguments with respect to the Hausdorff metric.
When H contains only the identity mapping, then σ will be called a *similarity measure*.

Although not stated explicitly in the definition above, it is also required that $\mathcal{J}$ is invariant under H, that is, $h(A) \in \mathcal{J}$ if $A \in \mathcal{J}$ and $h \in H$.

If σ is a similarity measure on $\mathcal{J}$ and H is a subgroup of G, then

$$\sigma'(A, B) = \sup_{h \in H} \sigma(h(A), B)$$

defines an H-invariant similarity measure on $\mathcal{J}$. Here we consider only those subgroups H for which the supremum is attained. Unfortunately, σ' is difficult to compute in many practical situations. Below, however, we present the cases for which this can be done efficiently for convex polygons.

Let H be a given subgroup of G, and define

$$\sigma_1(A, B) = \sup_{h \in H} \frac{4|\det h|^{\frac{1}{2}} V(A)^{\frac{1}{2}} V(B)^{\frac{1}{2}}}{V(A \oplus h(B))}, \tag{4}$$

and

$$\sigma_2(A, B) = \sup_{h \in H} \frac{|\det h|^{\frac{1}{2}} V(A)^{\frac{1}{2}} V(B)^{\frac{1}{2}}}{V(A, h(B))}. \tag{5}$$

Proposition 3. *If H is subgroup of G, then*
(a) σ_1 is an H-invariant similarity measure on $\mathcal{C}$;
(b) σ_2 is an MH-invariant similarity measure on $\mathcal{C}$.

Recall that l_0 is the line reflection with respect to the x-axis.

Proposition 4. *Let σ be a similarity measure on $\mathcal{J}$, and define*
$\sigma'(A, B) = \max\{\sigma(A, B), \sigma(l_0(A), B)\}$.
(a) If σ is R-invariant, then σ' is an I-invariant similarity measure.
(b) If σ is G_+-invariant, then σ' is a G-invariant similarity measure.

4 Similitude and affine invariant similarity measures

We consider similarity measures on $\mathcal{P}$ which are S_+-invariant, i.e., invariant under rotations and multiplications. Towards this goal, we will use the similarity measures defined in (4)–(5) with $H = S_+$ and $H = R$, respectively. In these expressions, the terms $V(P \oplus r_\theta(Q))$ and $V(P, r_\theta(Q))$ play an important role. Let the perimetric representations of the convex polygons P and Q be given by (u_i, p_i), $i = 1, 2, \ldots, n_P$, and (v_j, q_j), $j = 1, 2, \ldots, n_Q$, respectively.

To compute $V(P, r_\theta(Q))$, we use formula (3). Angles θ for which $r_\theta(v_j) = u_i$ are called *critical angles*. The set of all critical angles for P and Q is given by $\{(\angle u_i - \angle v_j) \bmod 2\pi \mid i = 1, 2, \ldots, n_P \text{ and } j = 1, 2, \ldots, n_Q\}$, where $\angle u$ denotes the angle of vector u with the positive x-axis. We denote the critical angles by $0 \le \theta_1^* < \theta_2^* < \cdots < \theta_N^* < 2\pi$. It is evident that $N \le n_P n_Q$.

Proposition 5. *The volume $V(P \oplus r_\theta(Q))$ and the mixed volume $V(P, r_\theta(Q))$ are functions of θ which are concave on $(\theta_k^*, \theta_{k+1}^*)$, for $k = 1, 2, \ldots, N$ and $\theta_{N+1}^* = \theta_1^*$. Here $0 \le \theta_1^* < \theta_2^* < \cdots < \theta_N^* < 2\pi$ are the critical angles of the two convex polygons P and Q.*

Let $H = S_+$ and consider the S_+-invariant similarity measure obtained from (4). We have

$$
\sigma_1(P, Q) = \sup_{\lambda > 0, \ \theta \in [0, 2\pi)} \frac{4\lambda V(P)^{\frac{1}{2}} V(Q)^{\frac{1}{2}}}{V(P \oplus \lambda r_\theta(Q))}
$$

$$
= 4 \left[\inf_{\lambda > 0, \ \theta \in [0, 2\pi)} \frac{V(P) + \lambda^2 V(Q) + 2\lambda V(P, r_\theta(Q))}{\lambda V(P)^{\frac{1}{2}} V(Q)^{\frac{1}{2}}} \right]^{-1}
$$

$$
= 4 \left[\inf_{\lambda > 0} \left\{ \frac{1}{\lambda} \cdot \left(\frac{V(P)}{V(Q)} \right)^{\frac{1}{2}} + \lambda \cdot \left(\frac{V(Q)}{V(P)} \right)^{\frac{1}{2}} \right\} + \inf_{\theta \in [0, 2\pi)} \frac{2V(P, r_\theta(Q))}{V(P)^{\frac{1}{2}} V(Q)^{\frac{1}{2}}} \right]^{-1}.
$$

In order to compute $\sigma_1(P, Q)$ we have to minimize two expressions, one in λ and one in θ. The first expression achieves its minimum at $\lambda = (V(P)/V(Q))^{\frac{1}{2}}$, the second at one of the critical angles associated with P, Q.

The similarity measure σ_2 given by (5) results in

$$
\sigma_2(P, Q) = \sup_{\theta \in [0, 2\pi)} \frac{V(P)^{\frac{1}{2}} V(Q)^{\frac{1}{2}}}{V(P, r_\theta(Q))}. \tag{6}
$$

The measure σ_2 is S_+-invariant (see Proposition 3). As above, the maximum for σ_2 is attained at one of the critical angles associated with P, Q.

Proposition 6. *Given the perimetric representation of the convex polygons P and Q, the time complexity of computing the similarity measures σ_1 and σ_2 is $\mathcal{O}(n_P n_Q(n_P + n_Q))$; here n_P, n_Q are the number of vertices of P and Q, respectively.*

If $H = G$, then the similarity measures σ_1, σ_2 defined in (4) and (5), respectively, are invariant under arbitrary affine transformations. Unfortunately, we do not have efficient algorithms to compute them. However, using the approach of Hong and Tan in [8], we are able to define similarity measures which can be computed efficiently, and which are invariant under a large group of affine transformations, namely G_+, the collection of all linear transformations which have a positive determinant. In combination with Proposition 4, this leads to G-invariant similarity measures.

The basic idea is to transform a set A to its so-called canonical form $A^\bullet$ in such a way that two sets A and B are equivalent modulo a transformation in G_+ if and only if $A^\bullet$ and $B^\bullet$ are equivalent modulo rotation. The definition of the canonical form, as discussed by Hong and Tan [8], is based on the concept of the *ellipse of inertia* known from classical mechanics [5]. The axes of this ellipse of inertia can be found via second order moments of the set [9]. Let a and b be the lengths of the long and short semi-axes of the ellipse of inertia.

Definition 7 [8]. A shape is said to be in *canonical form* if its centroid is positioned at the origin and its ellipse of inertia is a unit circle.

Proposition 8. *Every compact set can be transformed into its canonical form by means of a transformation in G_+, namely by a stretching along the long axis of the ellipse of inertia by a factor $b/(ab)^{1/4}$ and along the short axis by a factor $a/(ab)^{1/4}$.*

Proposition 9. *Let $\sigma : \mathcal{K}_+ \times \mathcal{K}_+ \to [0,1]$ be an R-invariant similarity measure, and define $\sigma^\bullet : \mathcal{K}_+ \times \mathcal{K}_+ \to [0,1]$ by $\sigma^\bullet(A,B) = \sigma(A^\bullet, B^\bullet)$, then $\sigma^\bullet$ is a G_+-invariant similarity measure.*

In [7] we present an algorithm for computation of the perimetric measure of the canonical polygon from the perimetric measure of the original polygon.

Let us conclude with some examples. We consider four, more or less regular, shapes, namely: a triangle, a square, a tetragon with one reflection axis, and a regular octagon. These shapes, along with their canonical forms, are depicted in Figure 1. In this figure, we depict four other convex polygons (and their canonical forms), namely: P, a reflection of P denoted by P_{refl}, a distortion of P denoted by Q (the lower three points have been shifted in the x-direction), and an affine transformation of Q denoted by Q_{aff}.

In Table 1 we compute the similarity measure σ_2 given by (6) which is S_+-invariant. In the first row we compute $\sigma_2(Q, R)$, where R is one of the other polygons depicted in Figure 1. In the third row we compute the values $\sigma_2(Q_{\text{aff}}, R)$. The second row contains the values $\sigma_2^\bullet(Q, R)$, where $\sigma_2^\bullet$ is the G_+-invariant similarity measure obtained from Proposition 9. Observe that we do not compute

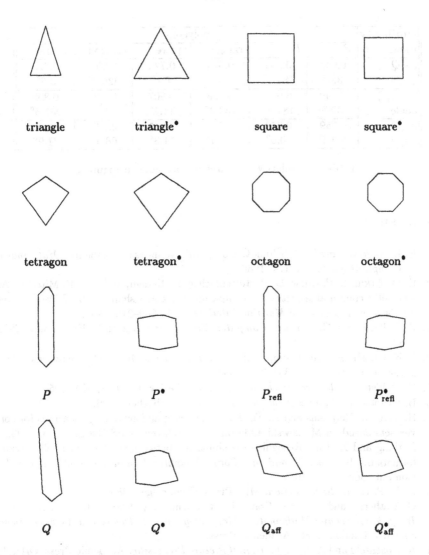

triangle   triangle*   square   square*

tetragon   tetragon*   octagon   octagon*

P $P^\bullet$ P_{refl} $P^\bullet_{\text{refl}}$

Q $Q^\bullet$ Q_{aff} $Q^\bullet_{\text{aff}}$

Fig. 1. Polygons used in this section. Note that $Q^\bullet_{\text{aff}}$ is a rotation of $Q^\bullet$.

$\sigma_2^\bullet(Q_{\text{aff}}, R)$, since these values are identical to $\sigma_2^\bullet(Q, R)$. In Table 1 we also give the angle at which the maximum in expression (6) is achieved.

similarity measures	polygons					
	P	P_{refl}	triangle	square	tetragon	octagon
$\sigma_2(Q,\cdot)$	0.941	0.933	0.674	0.724	0.692	0.725
angle	3.6°	185.2°	348.7°	7.1°	326.7°	3.6°
$\sigma_2^s(Q,\cdot)$	0.949	0.933	0.768	0.907	0.920	0.898
angle	3.8°	184.9°	233.4°	11.5°	322.1°	281.5°
$\sigma_2(Q_{\text{aff}},\cdot)$	0.759	0.749	0.753	0.845	0.847	0.862
angle	260.6°	80.6°	242.1°	11.3°	68.9°	31.9°

Table 1. Similarity measures for polygons in Figure 1.

References

1. E. L. Allgower and P. H. Schmidt. Computing volumes of polyhedra. *Mathematics of Computation*, 46:171–174, 1986.
2. E. M. Arkin, L. P. Chew, D. P. Huttenlocher, K. Kedem, and J. S. B. Mitchell. An efficiently computable metric for comparing polygonal shapes. *IEEE Transactions on Pattern Analysis and Machine Intelligence*, 13:209–216, 1991.
3. D. Ballard and C. Brown. *Computer Vision*. Prentice-Hall, Englewood Cliffs, 1982.
4. P. K. Ghosh. A unified computational framework for Minkowski operations. *Computers and Graphics*, 17:357–378, 1993.
5. H. Goldstein. *Classical Mechanics*. Addison-Wesley, Reading, MA, 1950.
6. B. Grünbaum. *Convex Polytopes*. Interscience Publishers, 1967.
7. H. J. A. M. Heijmans and A. Tuzikov. Similarity and symmetry measures for convex sets based on Minkowski addition. Research report BS-R9610, CWI, 1996.
8. J. Hong and X. Tan. Recognize the similarity between shapes under affine transformation. In *Proc. Second Int. Conf. Computer Vision*, pages 489–493. IEEE Comput. Soc. Press, 1988.
9. B. K. P. Horn. *Robot Vision*. MIT Press, Cambridge, 1986.
10. G. Matheron and J. Serra. Convexity and symmetry: Part 2. In J. Serra, editor, *Image Analysis and Mathematical Morphology, vol. 2: Theoretical Advances*, pages 359–375, London, 1988. Academic Press.
11. A. Rosenfeld and A. C. Kak. *Digital Picture Processing*. Academic Press, Orlando, 2nd edition, 1982.
12. R. Schneider. *Convex Bodies. The Brunn-Minkowski Theory*. Cambridge University Press, Cambridge, 1993.
13. P. J. van Otterloo. *A Contour-Oriented Approach to Digital Shape Analysis*. PhD thesis, Delft University of Technology, Delft, The Netherlands, 1988.
14. M. Werman and D. Weinshall. Similarity and affine invariant distances between 2D point sets. *IEEE Transactions on Pattern Analysis and Machine Intelligence*, 17:810–814, 1995.

Deformation of Discrete Object Surfaces

Yukiko Kenmochi and Atsushi Imiya

Dept. of Information and Computer Sciences, Chiba University
1-33 Yayoi-cho, Inage-ku, Chiba 263, JAPAN

Abstract. We define objects surface in a 3-dimensional (3D) lattice space using the theory of combinatorial topology, and present all possible deformation operations of the object surfaces, such that these operations do not violate the topology of the surfaces.

1 Introduction

In this paper, we define the term "deformation" as the transformation of n-dimensional (nD) objects to nD objects which preserves the topology of the objects. In particular, we focus on deformation of 2D closed surfaces of polyhedral objects. Deformation of object surfaces has been studied for volume segmentation of 3D objects as well as the snake which is a deformable curve around a 2D polygonal figure in a 2D plane [1]. From the viewpoint of other applications, deformation of object surfaces is also useful for describing motion between a sequence of images of 3D objects, for instance myocardial motion.

By comparison with deformation, skeletonization, thinning and localization of the medial axis are described as the transformation of nD objects to mD objects where $m \leq n$ [2, 3, 4]. Because these transformations only preserve the connectivity of points and sometimes do not consider topological structures caused by dimensionalities of objects, m can be less than n. According to our definition, deformation must be a transformation preserving not only connectivity but dimensionality of objects.

In order to describe topological structures of 3D polyhedral objects and their surfaces in a 3D lattice space, we use the theory of combinatorial topology. A similar approach was taken by Kovalevsky [5], but his approach is different from ours, since his approach is voxel based while our approach is lattice-point based. In combinatorial topology, 2D surfaces and 3D objects are represented by sets of triangles and tetrahedra which are regarded as 2D and 3D elements, respectively. If a set of 3D lattice points is given, we can uniquely obtain the representation of polyhedra from the set using the method presented in reference [6]. This combinatorial topology representation enables the classification of all points in 3D objects based on their topological structures [7]. In detail, topological property of every point can be obtained from the local configuration of neighboring points of objects. Using these topological properties of points, we can deform object surfaces without violating their 2D topological structures. In addition, a finite number of all possible operations for deformation are presented as the displacement of a part of an object surface.

2 Discrete Object Surfaces

Let $\mathbf{Z}$ be the set of all integers; $\mathbf{Z}^3$ is defined as a 3D lattice space which consists of points whose three coordinates are all integers. In $\mathbf{Z}^3$, the following 6-, 18- and 26-neighborhoods are defined.

Definition 1. Let $x = (i, j, k)$ be a point in $\mathbf{Z}^3$. Three neighborhoods of x are defined as

$$N_m(x) = \{(p, q, r) \mid (i - p)^2 + (j - q)^2 + (k - r)^2 \leq t, \; p, q, r \in \mathbf{Z}\}, \quad (1)$$

where t is 1, 2, or 3 for m of 6, 18, or 26, respectively.

Depending on each neighborhood, we define elements of 1D curves, 2D surfaces, and 3D objects in $\mathbf{Z}^3$. These elements are called, 1D, 2D and 3D discrete simplexes, respectively [7]. Suppose we define 0D discrete simplexes as isolated points in $\mathbf{Z}^3$, then we can define discrete simplexes whose dimensions are from 1 to 3 as follows.

Definition 2. An nD discrete simplex is defined as a set of points in $\mathbf{Z}^3$, such that the embedded simplex becomes one of nD minimum nonzero regions in $\mathbf{R}^3$ which are bounded by $(n - 1)$D embedded simplexes.

If an n-simplex consists of k points, the embedded simplex is defined as the convex hull of these k points in 3D Euclidean space, $\mathbf{R}^3$. An n-simplex of k points, x_0, x_1, ..., x_k is denoted by $[x_0, x_1, \ldots, x_k]$. Hereafter, we abbreviate nD discrete simplexes and simply call them n-simplexes. According to Definition 2, a finite set of n-simplexes where n can be from 0 to 3 is determined for each neighborhood, as shown in Fig. 1. Note that the congruent n-simplexes that differ from those in Fig. 1 by translation or rotation are omitted from Fig. 1. Since 1-simplexes are regarded as elements of 1D curves in $\mathbf{Z}^3$, a line segment between two neighboring points can be a 1-simplex. The configuration of these two neighboring points depends on the neighborhood, and Fig. 1 shows that there are one, two and three different 1-simplexes for the 6-, 18- and 26-neighborhoods, respectively. According to Definition 2, a 2-simplex must be bounded by a set of 1-simplexes, so that these 1-simplexes hold a 2D minimum nonzero area inside of them. Similarly, 3-simplexes are created, so that every 3-simplex is bounded by a set of 2-simplexes and these 2-simplexes bound a 3D minimum nonzero region.

A function $face$ of an n-simplex, $[x_1, \ldots, x_k]$, denoted by $face([x_1, \ldots, x_k])$ is defined as a set of all discrete simplexes included in the n-simplex, whose dimensions are less than n. For instance, the faces of a 3-simplex are 0-, 1- and 2-simplexes in the 3-simplex. Faces of 0, 1 and 2 dimensions are called 0-, 1- and 2-faces, respectively. If $\mathbf{C}_n$ is a set of n-simplexes, we can define a set of faces of all n-simplexes in $\mathbf{C}_n$ as

$$Face(\mathbf{C}_n) = \bigcup_{[a] \in \mathbf{C}_n} face([a]). \quad (2)$$

Using the function $Face$, we can define 3D objects in $\mathbf{Z}^3$, which are called discrete objects.

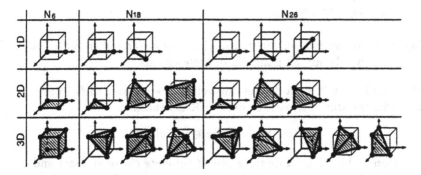

Fig. 1. All possible 1D, 2D and 3D simplexes for the 6-, 18- and 26-neighborhoods.

Definition 3. Let C_3 be a set of connected 3-simplexes. A discrete object is defined as

$$P = C_3 \cup Face(C_3). \tag{3}$$

There are three types of connection between 3-simplexes; two connected 3-simplexes share a 0-, 1- or 2-face in common. The surface of a discrete object is defined as follows.

Definition 4. Let P be a discrete object; the surface of P is defined as

$$\partial P = C_2 \cup Face(C_2), \tag{4}$$

where C_2 is a set of 2-simplexes in P such that every 2-simplex in C_2 belongs to exactly one 3-simplex in P.

The next theorem on discrete object surfaces is derived from Definition 4.

Theorem 5. *Discrete object surfaces have no boundary, that is, they are always closed.*

Proof. According to Definition 3, a discrete object is generated by a process of connecting 3-simplexes with each other. Since every 3-simplex is completely bounded by a set of 2-simplexes with no tear, a set of connected 3-simplexes does not include any tear in the boundary. □

Because of Theorem 5, discrete object surfaces are also called discrete closed surfaces. General discrete surfaces which may have the boundary are defined as well as discrete objects.

Definition 6. Let C_2 be a set of connected 2-simplexes. A discrete surface is defined as

$$S = C_2 \cup Face(C_2). \tag{5}$$

If two 2-simplexes are connected, they share a 0- or 1-face. If a discrete surface includes a 2-simplex whose 1-face is not shared with other 2-simplexes, the discrete surface has the boundary because that 1-face constitutes the boundary.

3 Topology of Points

Using the following two set functions, which are a star and the outer boundary of a star [7], we classify all points in discrete objects.

Definition 7. Let K be a discrete object (discrete closed surface) and x be a point in K; the star of x with respect to K is defined as

$$\sigma(x : K) = \{[a] \mid x \in [a], [a] \in K\}. \tag{6}$$

Definition 8. Let K be a discrete object (discrete closed surface) and x be a point in K; the outer boundary of $\sigma(x)$ with respect to K is defined as

$$\lfloor \sigma(x : K) \rfloor = Face(\sigma(x : K)) \setminus \sigma(x : K). \tag{7}$$

We define two topological types for points of a discrete object as follows.

Definition 9. Let P be a discrete object and x be a point in P. If $\lfloor \sigma(x : P) \rfloor$ constitutes a discrete closed surface, then x is spherical.

Definition 10. Let P be a discrete object and x be a point in P. If $\lfloor \sigma(x : P) \rfloor$ constitutes a discrete surface and it is not closed, then x is semi-spherical.

These two topological types of points are depicted in Fig. 2. Spherical points indicate the interior points of discrete objects and semi-spherical points indicate points on the surfaces of discrete objects, except for intersections of the surfaces. Because points which are intersections of the surfaces are regarded as neither spherical nor semi-spherical, we call these points singular points. Several examples of singular points are also illustrated in Fig. 2. It is obvious that every point on the surface of a discrete object is either semi-spherical or singular, but not spherical. Using these topological properties of points, it is possible to distinguish points on the surface of a discrete object from all points in a discrete object. In addition, it is also possible to extract intersections from points on the surface.

If the surface of a discrete object is given instead of the discrete object, however, we cannot distinguish between intersections and other points on discrete object surfaces. This is because we cannot calculate the above topological types of points from the surface of a discrete object. In order to solve this problem, we define the next topological type of points which we can determine from the surface of a discrete object.

Definition 11. Let ∂P be a discrete object surface and x be a point in ∂P. If $\lfloor \sigma(x : \partial P) \rfloor$ makes a circle, such as a path of adjacent 1-simplexes which does not intersect and whose beginning and end points are the same, then x is cyclic.

Cyclic points are equivalent to semi-spherical points, as shown in Fig. 2. All points on a discrete object surface, except for cyclic points, are called singular points of the surface as well as singular points of the discrete object.

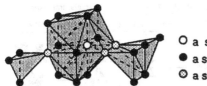

O a spherical point
● a semi-spherical (cyclic) point
⊗ a singular point

Fig. 2. Three topological types of points of discrete objects: spherical, semi-spherical (cyclic) and singular points. The example shown is for the case of the 26-neighborhood.

4 Deformation and Topology

Deformation of discrete object surfaces is defined as the process of expanding or shrinking discrete object surfaces while preserving their topology. If a discrete object surface, ∂P, does not include any singular points, the topology of ∂P is given by an Euler characteristic,

$$E(\partial P) = \mathbf{V} - \mathbf{E} + \mathbf{F}, \tag{8}$$

where $\mathbf{V}$, $\mathbf{E}$ and $\mathbf{F}$ are the numbers of 0-, 1- and 2-simplexes in ∂P. Let $\partial P'$ be a deformation of ∂P. If the relation

$$E(\partial P) - E(\partial P') = 0 \tag{9}$$

holds, it is said that the deformation process preserves the topology of ∂P. If ∂P includes singular points, we deform ∂P so as not to change the number of singular points, their locations and their topological structures, such as the configuration of connected 2-simplexes around the point.

4.1 Deformation Operations

Deformation operations are mainly classified into two classes: one class consists of operations for expanding discrete object surfaces and the other class consists of operations for shrinking discrete object surfaces. We call these two classes of deformations expanding and shrinking deformations, respectively. The process of expanding (shrinking) deformation is equivalent to the process of increasing (decreasing) the volume of a discrete object while preserving its topology. In order to increase (decrease) the volume of a discrete object, we attach (remove) a 3-simplex to (from) the discrete closed surface. This process means that 2-faces in the 3-simplex which are not (are) at the joint become a new part of the discrete closed surface instead of the original part of the discrete closed surface which was (was not) at the joint in the attached (removed) 3-simplex.

Shrinking deformation is the inverse of expanding deformation. Therefore, it is possible to obtain a set of all operations for shrinking deformation if a set of all operations for expanding deformation is defined. Thus, it is sufficient that operations for either expanding or shrinking deformation are provided. In this paper, we show all possible operations for expanding deformation. The following algorithm describes the operation of expanding a discrete object surface using a 3-simplex.

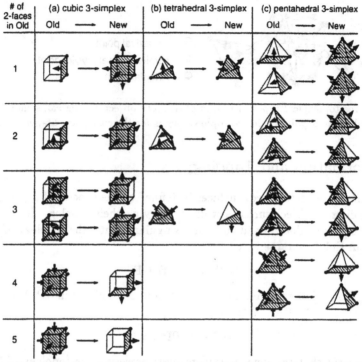

Fig. 3. All elemental operations for expanding deformation of a discrete object surface, ∂P (a) using a cubic 3-simplex for the 6-neighborhood. (b) using a tetrahedral 3-simplex for the 18- and 26-neighborhoods, and (c) using a pentahedral 3-simplex for the 18-neighborhood. Note that ∂P is expanded to $(\partial P \setminus \mathbf{Old}) \cup \mathbf{New}$.

Algorithm 1

input: *A discrete object* $\mathbf{P}$ *and its surface* ∂P.
output: *An expanded surface* $\partial P'$.
begin

1. *Find a configuration of 2-simplexes,* **Old**, *which is illustrated in* Fig. 3, *where* **Old** $\subset \partial P$. *Note that all points in* **Old** *must be cyclic with respect to* ∂P.
2. *A new configuration of 2-simplexes,* **New**, *is obtain from* Fig. 3, *corresponding to* **Old**.
3. *If* $(\mathbf{New} \setminus \mathbf{Old}) \cap \mathbf{P} \neq \emptyset$, *set* **New** = **Old**.
4. *Set* $\partial P' = (\partial P \setminus \mathbf{Old}) \cup \mathbf{New}$.

end

According to Algorithm 1, a nonempty discrete surface **Old** in ∂P is displaced by a nonempty discrete surface **New** which is not in ∂P. These two discrete surfaces, **Old** and **New**, satisfy the following three conditions:

1. **Old** and **New** are included in a 3-simplex;
2. each 2-simplex in the 3-simplex is included in either **Old** or **New**;
3. each of **Old** and **New** must be a discrete surface whose 2-simplexes are connected to each other via their 1-faces.

All possible pairs of **Old** and **New** are illustrated in Fig. 3 for each 3-simplex shape of each neighborhood. For the 6-neighborhood, every 3-simplex forms a cube, and has six square 2-faces, as shown in Fig. 1. Thus, from one to five 2-faces can be included in **Old**. In each case of the numbers of 2-faces in **Old**, there is only one configuration of 2-faces in **Old** and **New**, except for the case of three 2-faces included in **Old**. In this case, there exist two different 2-face configurations in **Old**. For the 18-neighborhood, 3-simplexes are either tetrahedral or pentahedral. In the case of expanding deformation using a tetrahedral 3-simplex, there are three configurations of 2-faces in **Old** and **New**, while there are eight configurations in the case of the expanding deformation using a pentahedral 3-simplex. Similarly, for the 26-neighborhood, there are three configurations of 2-faces in **Old** and **New** because all 3-simplexes are tetrahedral.

Because Algorithm 1 describes only the process of the operation of expanding ∂P using a 3-simplex, we must employ Algorithm 1 n times if n operations are necessary to sufficiently expand ∂P. Obviously, the result of a series of expanding operations depends on the choice of **Old** in step 1 of Algorithm 1 for each operation. Furthermore, according to step 3 of Algorithm 1, ∂P is not expanded if **New**, which is automatically obtained from **Old** in step 2, intersects with ∂P. Evidently, the choice of **Old** in step 1 also governs the possibility of expansion of ∂P for each operation.

4.2 Elemental Operations

Figure 3 indicates that an operation for expanding deformation using a cubic or pentahedral 3-simplex for the 6- or 18-neighborhood can be decomposed into a set of several operations using tetrahedral 3-simplexes for the 26-neighborhood. We show an example of the decomposition of an operation in Fig. 4. Consequently, if a set of all possible operations for expanding deformation using tetrahedral 3-simplexes for the 26-neighborhood is given, we can expand discrete object surfaces for any neighborhood by combining those given operations. In other words, a set of operations using tetrahedral 3-simplexes for the 26-neighborhood can be regarded as a set of elemental operations for expanding deformation.

Fig. 4. An example of deformation operation decomposition; a deformation operation using a cubic 3-simplex is decomposed into six operations using tetrahedral 3-simplexes.

4.3 Topology Preservation by Deformation

Because every deformation operation for the 6- and 18-neighborhoods is decomposed into several deformation operations for the 26-neighborhood, we consider topology preservation by the deformation only in the case of the 26-neighborhood. Let **V**, **E**, **F**, **V'**, **E'**, and **F'** be the numbers of 0-, 1-, and 2-simplexes of **Old** and **New** in Fig. 3, respectively. Table 1 shows these exact numbers for each operation for expanding deformation using a tetrahedral 3-simplex. Then, Eq. (9) is derived for every operation.

# of 2-faces in Old	V E F	V' E' F'
1	3 3 1	4 6 3
2	4 5 2	4 5 2
3	4 6 3	3 3 1

Table 1. The numbers of discrete simplexes of 0, 1 and 2 dimensions in **Old** and **New** of deformation operations by tetrahedral 3-simplexes.

5 Conclusions

In this paper, we introduced combinatorial topology representations of 3D poly-hedral objects and their surfaces, which contains not only connectivity between neighboring points but the 2D topological structures of the points. Using these topological structures, we can deform discrete object surfaces while preserving their topology. We also presented a finite set of all elemental operations for ex-panding deformation and demonstrated that three operations using tetrahedral 3-simplexes for the 26-neighborhood in Fig. 3 are sufficient to compose all other operations for the 6- and 18-neighborhoods.

For application of deformation, the amount of deformation of an object of interset must be measured. In other words, there are constraints on deformation, for instance, positions, geometrical shapes and lengths of objects. Since we only discussed the possible operations for deformation without violating topology in this paper, it is desirable to prepare some spatial metrics to measure the differences between original objects and deformed objects as our future work.

References

1. J. Lachaud and A. Montanvert, Volumic Segmentation Using Hierarchical Representation and Triangulated Surface, Research Report No. 95-37, LIP, ENS de Lyon, October 1995.
2. P. P. Jonker and O. Vermeij, On Skeletonization in 4D Images, *LNCS 1121; P. Perner, P. Wang and A. Rosenfeld (Eds.), Advances in Structural and Syntactical Pattern Recognition*, pp. 79-89, Sprinker-Verlag, 1996.
3. K. Abe, F. Mizutani and C. Wang, Thinning of Gray-Scale Images with Com-bined Sequential and Parallel Conditions for Pixel Removal, *Visual Form Analysis and Recognition*, Editied by C. Arcelli, L. P. Cordella and G. S. Baja, Plenum Press, New York, pp. 1-10, 1991.
4. M. P. P. Schlicher, E. Bouts and P. W. Verbeek, Fast Analytical Medial-Axis Localization in Convex Polyhedra, *IEEE Proceedings of 13th International Conference on Pattern Recognition*, pp. 55-61, 1996.
5. V. A. Kovalevsky, Finite topology as applied to image analysis, Computer Vision, Graphics, and Image Processing, Vol. 46, pp. 141-161, 1989.
6. Y. Kenmochi, A. Imiya, and N. F. Ezquerra, Polyhedra Generation from Lat-tice Points, *LNCS 1176: S. Miguet, A. Montanvert and S. Ubéda (Eds.), Dis-crete Geometry for Computer Imagery*, pp. 127-138, Springer-Verlag, Berlin, Heidelberg, 1996.
7. Y. Kenmochi, A. Imiya, and A. Ichikawa, Discrete Combinatorial Geometry, to appear in Pattern Recognition.

Non-Archimedean Normalized Fields in Texture Analysis Tasks

Vladimir M. Chernov and Andrew V. Shabashev

Image Processing Systems Institute of RAS
151 Molodogvardejskaja st., IPSI RAS, Samara, Russia

e-mail: chernov@sgau.volgacom.samara.su

Abstract. A method of forming features of texture images is proposed. It is based on a new interpretation of digital image as a function defined on integer elements of algebraic numbers quadratic fields. The features are formed on the base of analysis of the image spectral characteristics associated with its metric properties in non-Archimedean metrics.

1 Introduction

The efficiency of spectral methods of digital image processing decreases significantly in geometric properties analysis tasks. The spectral methods support "power" invariance, but they have little sensitivity to local geometric properties of the image. We shall note that these geometric (metric ones included) characteristics of an image do not have to be described adequately in terms of the conventional Euclidean geometry. Moreover, while solving pattern recognition tasks, it may turn out that the most informative features can be obtained from analysis of those metric image properties which can not be interpreted clearly in terms of "ordinary" concepts about distance, closeness etc.

The authors do not intend to discuss alternative approaches to solving particular tasks of texture analysis.

First of all, this paper is purposed to illustrate the possibilities of relatively new (for PR&IP) mathematical techniques - the theory of non-Archimedean normalized fields of algebraic numbers - in tasks of forming the features space. Textures are a convenient testing ground for demonstration of such possibilities.

In concept, principial ideas of the method being described are not absolutely new. Features forming on the base of different algebraic image invariants is one of the main techniques in pattern recognition [1]. Extraction of algebraic image features was considered, for example, in [2], taking into account specific characters of their representation. Different metric structures associated with textures of particular classes are described in [3].

Novelty of approach of this paper is applying non-Archimedean (*p*-adic) metrics - and their extensions onto the discrete plane - to extacting metric invariants of textures of the particular class (*regular textures*). As shown below, these invariants are some binary images (*primitives*) associated with some set of primes and also metrics connected with them. These primitives (*p-adic components of a texture*) can be used directly for recognition by experts or for forming numerical characteristics and, from this on, the features space.

2 Main ideas

Definition 1. Let a real-valued function $x(n_1, n_2)$ be defined on the set

$$\Omega_N = \{\mathbf{n} = (n_1, n_2) : 0 \leq n_1, n_2 \leq N - 1\} \subseteq \mathbb{Z}^2$$

and extended onto $\mathbb{Z}^2$ some way (for example, N-periodically). We shall name this image a *primitive regular texture* or a *primitive*, if there exists such a group $\mathbb{F}$ of discrete plane $\mathbb{Z}^2$ transforms, that the following conditions are valid:

(a) the image $x(\mathbf{n})$ is invariant over the action of the group $\mathbb{F}$:

$$\forall \varphi \in \mathbb{F} \qquad x(\mathbf{n}) = x(\varphi(\mathbf{n}));$$

(b) the group $\mathbb{F}$ is isomorphic to the direct sum of groups:

$$\mathbb{F} = \mathbb{S} \oplus \mathbb{Z} \oplus \ldots \oplus \mathbb{Z}$$

where $\mathbb{S}$ is a group of low order (*the symmetry subgroup*), and the non-finite additive cyclic groups $\mathbb{Z}$ are naturally associated with shifts of the image at vectors from a finitely generated set (*the shifts subgroup*).

Definition 2. Let us assume *real regular texture* to be an image $y(\mathbf{n})$ obtained from $x(\mathbf{n})$ by using a *distorting operator* $\mathbf{P} : y(\mathbf{n}) = \mathbf{P}x(\mathbf{n})$.

1. It is clear that this distortion is not to be "too great" in some metric ρ accepted by the researcher, for example:

$$\rho(y(\mathbf{n}), x(\mathbf{n})) = \sum_{n \in \Omega_N} |y(\mathbf{n}) - x(\mathbf{n})|^2 \leq \varepsilon.$$

In contrast with a real texture, primitive textures have a formalized description and are easier to analyze.

The first idea. Let us figure out a procedure of *texture primitivization*, i.e. development of such a primitive texture for the given one, that different primitives will correspond to different classes of equal textures.

2. It is well-known that prime numbers form the multiplicative basis of the natural numbers semigroup: each natural number can be presented as a product of degrees of prime numbers. On the other hand, the tasks concerned with analysis of additive properties of primes (e.g. their distribution in the positive integers) are quite difficult [4]. It is clear: *additive* properties of the *multiplicative* objects are hard to analyze. "Irregular behavior" of primes leads to the idea

about some "independence" of the features associated with different primes. "The prime numbers play games of chance"[5]. Some approaches to formalization of the naive understanding of prime numbers independence are described in [5,6] with argumentation convincing enough.

First of all, in texture analysis tasks the researcher is interested in geometric properties of an image. Therefore we shall consider image properties in specific geometries associated with prime numbers.

Let p be a prime. Let us assume $\nu_p(a)$ to be the p-adic exponent of an integer number a, that is the largest nonnegative number m for which

$$a \equiv 0 \pmod{p^m}.$$

For a rational number $x = \frac{a}{b}$ the exponent $\nu_p(x)$ is assumed to be

$$\nu_p(x) = \nu_p(a) - \nu_p(b).$$

Let us define a p-adic (*non-Archimedean*) norm on the set of rational numbers by means of relation

$$\|x\|_p = \begin{cases} p^{-\nu_p(x)}, & \text{if } x \neq 0; \\ 0, & \text{if } x \neq 0. \end{cases} \tag{1}$$

Similarly to the fact that the supplement of the field $\mathbb{Q}$ over the usual norm $\|x\|_\infty =| x |$ leads to the forming of real numbers field, its supplement over the norm $\|x\|_p$ leads to the forming of p-adic numbers field $\mathbb{Q}_p$. In a standard way, a metric $\rho_p(x, y)$ generated by the norm is defined:

$$\rho_p(x, y) = \|x - y\|_p, \ x, y \in \mathbb{Q}_p.$$

In particular, two integers are "close" if their difference is divided by a high degree of the given prime p. Similarly to the fact that the norm is supplemented from the field $\mathbb{R}$ onto its quadratic extension - the complex numbers field $\mathbb{C}$:

$$\|z\|_\infty = \sqrt{\text{Im}^2(z) + \text{Re}^2(z)}, \quad z \in \mathbb{C}$$

the p-adic norm is also supplemented onto algebraic extensions of the field $\mathbb{Q}$. For quadratic fields:

$$\mathbb{Q}(\sqrt{-d}) = \{z = x + y\sqrt{-d}; \qquad x, y \in \mathbb{Q}\}$$

the full description of extensions of p-adic valuations is given by the following proposition.

Proposition 1 . *Let d be an integer free of squares, p be a prime, $z = x + y\sqrt{-d}$. Then:*

(a) if $(-d)$ is a quadratic non-residue (mod p), then the function

$$\Psi_{p,d}(z) = \sqrt{p^{-\nu_p(x^2+y^2d)}} = \sqrt{\|x^2 + dy^2\|_p} \tag{2}$$

extends valuation (1) from $\mathbb{Q}$ *onto* $\mathbb{Q}(\sqrt{-d})$;

(b) if $(-d)$ *is a quadratic residue* (mod p) *and* γ *is a solution of the equation* $w^2 = -d$ *in the field* $\mathbb{Q}_p$, *then the functions*

$$\Psi_{p,\gamma}^{\pm}(z) = p^{-\nu_p(x+y\gamma)} = \|x + \gamma y\|_p \qquad (3)$$

extend valuation (1) from $\mathbb{Q}$ *onto* $\mathbb{Q}(\sqrt{-d})$ (see, for example [7]).

Let us identify a pair of arguments (n_1, n_2) of an image $x(n_1, n_2)$ with the element $z = n_1 + n_2\sqrt{-d}$. Pairs (n_1, n_2) with conditions $\Psi_p(n_1 + n_2\sqrt{-d}) \leq p^{-1}$ ("p-adic full spheres of the radius p^{-1}") for different primes p and normalizing functions $\Psi_p = \Psi_{p,d}$, $\Psi_{p,\gamma}^{\pm}$ are shown on Fig. 1.

Fig 1. Examples of overlapping p-adic full spheres

Fig. 2 shows synthesized textures (after acting of a distorting, two images left) and Brodatz textures (on the right).

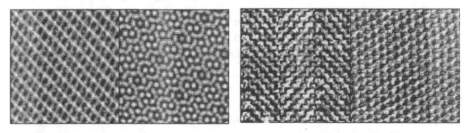

Fig 2. Comparison of synthesized and Brodatz textures

The "texture-like" view of the images leads to the idea about their using to form features while analyzing real (Brodatz, Ref.[8]) textures.

The second idea. Let us figure out a procedure of extraction from texture images components which are invariant over transforms preserving values of the functions Ψ_p for some finite set $\mathfrak{N}_p$ of valuations of fields $\mathbb{Q}(\sqrt{-d})$.

3. Let an image (a texture) $x(\mathbf{n})$ be presented in the form:

$$x(\mathbf{n}) = x_{inv}(\mathbf{n}) + x_{non}(\mathbf{n}),$$

where $x_{inv}(\mathbf{n})$ is a function ("signal") which is invariant over Ψ_p-isometrics for $\forall \Psi_p \in \mathfrak{N}_p$. We shall interpret the function $x_{non}(\mathbf{n})$ as a "noise".

Let $\mathfrak{Is}(\Psi_p)$ be a set of Ψ_p-isometrics, $\mathrm{Card}(\mathfrak{Is})$ be its cardinality. We shall consider an action of the averaging operator $\mathbf{M}_p$ on an image $x(\mathbf{n})$:

$$\mathbf{M}_p x(\mathbf{n}) = \mathrm{Card}(\mathfrak{Is})^{-1} \sum_{J \in \mathfrak{Is}(\Psi_p)} x(J(\mathbf{n})).$$

Taking into account the quite general assumptions about statistical independence of values of the function $x_{non}(\mathbf{n})$, the signal/noise ratio for the averaged image is larger than for the image $x_{non}(\mathbf{n})$. Certainly, accurate numerical estimations are possible to be obtained only for the artificially synthesized images. For real textures the correctness of this assumption can be confirmed only experimentally.

The third idea. Let us choose a class $\mathfrak{N} = \cup \mathfrak{N}_p$, $p \in \mathfrak{P}$ of valuations Ψ that is representative enough. For each of these valuations we shall define a finite set of isometrics, and consider averaged images $\mathbf{M}_p x(\mathbf{n})$ for $\Psi_p \in \mathfrak{N}$. After the thresholding and (if wished) the binarization, we shall obtain a set of textures already primitive. These primitive textures and/or their numerical characteristics can be considered as *features* of real textures, and the convenient pattern recognition methods can be applied.

4. How to choose "good" valuations? It is plausible enough that under the fixed prime p "good" numbers are those for which the main power of the signal $x(\mathbf{n}) = x(n_1, n_2)$ is localized in the points (n_1, n_2) with the condition $\Psi_p(n_1 + n_2\sqrt{-d}) \leq p^{-1}$.

Checking such localization conditions is an ordinary task of the discrete spectral analysis. Indeed, let E be the full power of the signal:

$$E = \sum_{\mathbf{n} \in \Omega} |x(\mathbf{n})|^2, \quad \Omega = \{\mathbf{n} = (n_1, n_2) : 0 \leq n_1, n_2 \leq p^r - 1 \leq N\}.$$

Let, for example, valuation Ψ be like one used in (3) with some γ. Then, let the image be a primitive texture:

$$x(\mathbf{n}) = \begin{cases} \lambda = const, & \text{if } n_1 + \gamma n_2 \equiv 0 \ (\mathrm{mod}\ p); \\ 0, & \text{if } n_1 + \gamma n_2 \neq 0 \ (\mathrm{mod}\ p). \end{cases}$$

In this case the relation for two-dimensional discrete Fourier transform is sequentially converted to the following:

$$\hat{x}(\mathbf{m}) = \sum_{\mathbf{n} \in \Omega} x(\mathbf{n}) \exp\{\frac{2\pi}{p^r}(n_1 m_1 + n_2 m_2)\} =$$

$$p^{-1} \sum_{0 \leq t, n_2 \leq p^r - 1} \lambda \exp\{\frac{2\pi}{p^r}(ptm_1 + n_2(m_2 - \gamma m_1))\}. \tag{4}$$

The latter sum is non-zero only under the conditions

$$m_1 \equiv 0 \ (\mathrm{mod}\ p^{r-1}), \ m_2 \equiv 0 \ (\mathrm{mod}\ p^{r-1}), \ m_2 - \gamma m_1 \equiv 0 \ (\mathrm{mod}\ p^r). \tag{5}$$

Congruences (5) define a Ψ_p-neighborhood of the zero with the diameter equal to p^{-r}. Therefore, taking into account the Parseval equality, we can conclude that the spectrum power is localized inside the neighborhood of the zero with a little Ψ_p-diameter.

In the case of a real texture image, this localization is obviously not so noticeable. But under choosing a "good" valuation, the *main* part of the spectrum power is concentrated around the zero. On this approach, the calculation efforts are caused by definition of the power characteristics of the spectrum parts associated with different valuations for the same prime number p.

3   Algorithm of primitive textures extraction

Step 1. Let $(N \times N)$ be size of the image $x(\mathbf{n}) = x(n_1, n_2)$ being processed. We shall choose a set $\mathfrak{P}$ of prime numbers p, and consider a set of valuations

$$\mathfrak{N} = \bigcup_{p \in \mathfrak{P}} \mathfrak{N}_p, \quad \mathfrak{N}_p = \{\Psi_{p,\gamma}; \quad \gamma = 1, ..., p-1\}, \quad p^2 \leq N.$$

Step 2. For each $p \in \mathfrak{P}$ we shall calculate DFT of the image of size $N_p \times N_p$, where $N_p = p^t \leq N$.

Step 3. It is easy to prove that linear isometries operators preserving the norm (2) have matrices like the following:

$$\begin{pmatrix} a & pb \\ pc & a+pd \end{pmatrix}, \quad \nu_p(a) = 0, \quad a, b, c, d \in \mathbb{Z}; \tag{6}$$

and isometries preserving the norm (3) have matrices like the following:

$$\begin{pmatrix} pa & b \\ pc & d \end{pmatrix}, \quad \nu_p(b), \nu_p(d) = 0; \quad \begin{pmatrix} pa & b \\ c & pd \end{pmatrix} \quad \nu_p(b), \nu_p(c) = 0;$$

$$\begin{pmatrix} a & pb \\ pc & d \end{pmatrix}, \quad \nu_p(a), \nu_p(d) = 0; \quad \begin{pmatrix} a & pb \\ c & pd \end{pmatrix} \quad \nu_p(a), \nu_p(c) = 0. \tag{7}$$

Let $\mathbf{J}$ be some operator of isometry (6) or (7). Then $\mathbf{J}$ induces the dual operator $\mathbf{J}^*$ acting on pairs of indexes of the spectral components (3):

$$\sum_{\mathbf{n} \in \Omega} x(\mathbf{Jn}) \exp\{\frac{2\pi}{p^r}(n_1 m_1 + n_2 m_2)\} = \hat{x}(\mathbf{J}^*(\mathbf{m})).$$

So a linear Ψ_p-isometry of an image induces the linear spectra transform. Therefore, under each $p \in \mathfrak{P}$ and $\Psi \in \mathfrak{N}_p$ we shall find averaged spectra

$$\mathbf{M}_p \hat{x}(\mathbf{m}) = \mathrm{Card}(\mathfrak{I}_s)^{-1} \sum_{\mathbf{J} \in \mathfrak{I}_s(\Psi_p)} \hat{x}(\mathbf{J}^*(\mathbf{n})). \tag{8}$$

For the chosen ε we shall find a set $\mathfrak{P}_0 \subset \mathfrak{P}$ of "good" prime numbers with the condition:

$$p \in \mathfrak{P}_0 \Longleftrightarrow \exists \Psi \in \mathfrak{N}_p : \sum_{\Psi(\mathbf{m}) \geq p^{-2}} |\, M_p \hat{x}(\mathbf{m})\,|^2 \leq \varepsilon. \tag{9}$$

Step 4. For all the primes $p \in \mathfrak{P}_0$ and for $\Psi \in \mathfrak{N}_p$ we shall define functions

$$\hat{y}_p(\mathbf{m}) = \begin{cases} M_p \hat{x}(\mathbf{m}), & \text{if } \Psi(\mathbf{m}) \leq p^{-1}; \\ 0, & \text{if } \Psi(\mathbf{m}) \leq p^{-1} \end{cases}$$

and calculate inverse DFT: $\hat{y}_p(\mathbf{m}) \longmapsto y_p(\mathbf{n})$ and binarization: $y_p(\mathbf{n}) \longmapsto Y_p(\mathbf{n})$.

We consider the set of images $Y_p(\mathbf{n})$ as distorted primitive textures ("Ψ_p-components of a real texture") which are associated with "good" valuations under the condition (8). The obtained images can be used for visual qualitative analysis of the significant components of the given texture. Fig. 3 shows the original test texture (on the left) as well as primitives extracted for "good" primes $p = 13$ and $p = 29$. For "bad" prime $p = 17$ primitives can not be extracted (on the right).

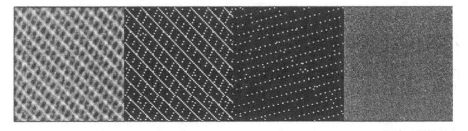

Fig 3. Primitives extraction from the synthesized texture

4 Experimental results

Fig. 4 shows primitives extracted with the described method from some Brodatz textures (left column). The average in (8) for primes $p = 2, ..., 29$ was calculated under $\mathrm{Card}(\mathfrak{J}s) = 15$. In every example the threshold value was chosen taking into account the bimodal character of the histogram obtained.

We shall note peculiarities of the realization of the proposed method. In the paper [9] an algorithm of 2-D DFT ("DFT with a multicovering") is described. Its structure is obviously associated with a covering of summation area with neighborhoods of decreasing Ψ-diameters for all γ at once. This allows to find the set while calculating the DFT.

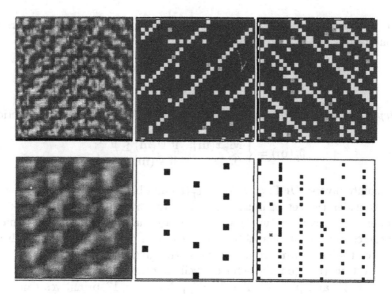

Fig. 4. Extraction of p-adic components from Brodatz textures

5 Acknowledgment

This work was performed with financial support from the Russian Foundation of Fundamental Investigations (Grants 95-01-00367, 97-01-009000).

References

1. Lenz, R.: Group Theoretical Methods in Image Processing. Lecture Note Comp. Sci. Vol. 413. Springer, 1990
2. Liu, K., Huang, Y, S., Ching, Y. S.: Optimal matrix transform for the extraction of algebraic features from images. Int. J. Patt. Recogn. and Artifical Intell. (to appear)
3. Smith, G., Lovell, B.: Metrics for texture classification. Algorithms, Proc. Digital Image Comp.: Techniques and Appl. Brisbane (1995) 223-227
4. Ireland, K., Rosen, M.: A Classical Introduction to Modern Number Theory. Springer, 1982
5. Kac, M.: Statistical Indenpendece in Probability, Analysis and Number Theory. The Math. Ass. of America, 1959
6. Kac, M.: Probability and Related Topics in Physical Sciences. Intersci. Publ., London - New-York, 1958
7. Brodatz, P.: Textures: A Photographic Album for Artists and Designers. Reinhold. NY. 1986
8. Van der Waerden, B. L.: Algebra. 7th ed., Springer, 1966
9. Chernov, V. M.: A Metric Unified Treatment of Two-Dimensional FFT. Proceedings of the 13th International Conference on Pattern recognition. Vienna, (1996), Track B, 662-669

The Radon Transform-Based Analysis
of Bidirectional Structural Textures

Ivan G. Kazantsev

Computing Center, Novosibirsk 630090, Russia

Abstract. In this work directional structural textures that consist of mostly straight lines are considered and the Radon transform–based approach to their directions detection is suggested. Problems of the detection of bidirectional structures are set forth in terms of image reconstruction from projections. In the proposed algorithm, using tomographic criteria of optimal decomposition of the texture image into a sum of two ridge functions, two directions determine the texture structure, are extracted. The selected two projections are jointly most informative and contain main features of regularly arranged bidirectional structures.

1 Introduction

In this work periodic structural textures that consist of mostly straight lines in two directions are of particular interest; we shall refer to them as bidirectional textures and suppose that they can be structurally described by texture elements and their interrelation. The technique we suggest belongs to the Radon (Hough) transform - like methods [1], [2], [3] using large part integrals. There exist approaches [4] which are also based on projection information; they select two main directions which produce a parallelogram grid that isolates each texture element. We start our discussion by considering the image investigated as the sum of structures extended along the linear paths and then, by defining the projection informativity in the Radon space which, for feature detection problems serve us an indicator of image homogeneity direction.

We suggest the Radon transform-based approach to detecting two principal directions of lines in the texture. This method is most succesively applicable to structural textures which can be decomposed into a sum of two ridge functions. The projections taken along with these functions directions are jointly the most informative and they make the best contribution into the texture reconstruction. We derive a functional of informativity which has the maximal response for these directions. The functional is based on the analitical inversion formula of the Radon transform in the case of circular domain and finite number of projections.

The functional has a most simple form when two directions shoud be detected. The question arises: are there exist textures which are sufficiently good expanded in a sum of two ridge functions and the detector maximum is valueable? Preliminary results of computer experiments are presented.

2   Informativity of projection data

Let D be a unit circle in the plane (x, y), $L^2(D)$ is the Lebesgue space of functions with the support D, $f \in L^2(D)$. Denote $J = [0, \pi)$; a tuple $\omega = (\omega_1, ..., \omega_n)$, $\omega_j \in J$ being a set of projection directions. We introduce P_{ω_j}, for some $\omega_j \in J$:

$$P_{\omega_j}[f(x,y)](s) = \int_{-(1-s^2)^{1/2}}^{(1-s^2)^{1/2}} f(s \cdot \cos \omega_j - t \cdot \sin \omega_j, s \cdot \sin \omega_j + t \cdot \cos \omega_j) dt.$$

In the computerized tomography [6] the data, as a rule, are presented in the form of a set $P_\omega f = (P_{\omega_1} f, \ldots, P_{\omega_n} f) \equiv (p_{\omega_1}(s), \ldots, p_{\omega_n}(s))$, for some n–tuple $\omega \in J^n$. We make an assumption that we are dealing with complete projections, i.e., that each $p_{\omega_j}(s)$ is known for all s. The following theorem takes place [5]:

Theorem 1. *Let* $\omega = (\omega_1, \ldots, \omega_n)$ *be a tuple of distinct angles. Let* $f \in L^2(D)$ *and let* H *be the unique function in* $L^2(D)$ *of the smallest norm which satisfies*

$$P_{\omega_j}[f] = P_{\omega_j}[H], j = 1, \ldots, n. \tag{1}$$

Then there exist functions $h_1, \ldots, h_n \in L^2([-1, 1], (1 - t^2)^{-1/2})$ *such that*

$$H(x, y) = \sum_{j=1}^{n} h_j(x \cdot \cos \omega_j + y \cdot \sin \omega_j).$$

The minimum norm solution is the projection of f on $\mathcal{F}^\perp$, an orthogonal complement of the set $\mathcal{F} = \bigcap_{j=1}^{n} N_j$, N_j is the null space of P_{ω_j}, and $L^2(D) = \mathcal{F} \bigoplus \mathcal{F}^\perp$. Hence, the inner product $\langle H, f - H \rangle = 0$ and

$$\|f - H(x, y)\|^2 = \|f\|^2 - \|H(x, y)\|^2$$

holds. The inner product is given by $\langle f, g \rangle = \int_D fg dx dy$. Hence, the quality of reconstruction can be evaluated by the norm $\|H\|$. We shall refer to the quantity $\|H\|^2 \equiv Q(f, \omega)$ as informativity of projection set $P_\omega f$. The method suggested is based on the following Assertions, which could be proved by the direct integration:

Assertion 1

$$Q(f, \omega) = \sum_{j=1}^{n} \int_{-1}^{1} p_j(s) h_j(s) ds.$$

Assertion 2 *Let* $H_\beta(x, y) = h_\beta(x \cos \beta + y \sin \beta)$ *is a backprojected ridge function induced by* $h_\beta(s)$. *Then the projection of* H_β *in the direction* β *has a form*

$$P_\beta[H_\beta(x, y)](s) = 2(1 - s^2)^{1/2} h_\beta(s) ds$$

and the projection of H_β *in the direction* α *is as follows:*

$$P_\alpha[H_\beta(x, y)](s) = \frac{1}{\sin(\alpha - \beta)} \int_{s_1}^{s_1} h_\beta(t) dt,$$

where

$$s_1 = s\cos(\alpha - \beta) - (1 - s^2)^{1/2}\sin(\alpha - \beta),$$

$$s_2 = s\cos(\alpha - \beta) + (1 - s^2)^{1/2}\sin(\alpha - \beta).$$

Assertion 3 *When a function $H(x, y)$ is presented in the form*

$$H(x, y) = \sum_{j=1}^{n} h_j(x\cos\omega_j + y\sin\omega_j),$$

the projection data constraints (1)

$$P_{\omega_j}H = p_j, j = 1, ..., n,$$

are identical to a system of integral equations of the third kind:

$$a(s)h_i(s) + \sum_{j=1, j\neq i}^{n} \frac{1}{\sin(\omega_i - \omega_j)}\int_{s_1}^{s_2} h_j(t)dt = p_i(s), \ i = 1, ..., n \qquad (2)$$

and where

$$a(s) = 2(1 - s^2)^{1/2}, s_1 = s\cos(\omega_i - \omega_j) - (1 - s^2)^{1/2}\sin(\omega_i - \omega_j),$$

$$s_2 = s\cos(\omega_i - \omega_j) + (1 - s^2)^{1/2}\sin(\omega_i - \omega_j).$$

Assertion 4 *The Chebyshev polynomials U_k of the second kind suffices the relation*

$$\int_{s\cos\alpha - \sqrt{(1-s^2)}\sin\alpha}^{s\cos\alpha + \sqrt{(1-s^2)}\sin\alpha} U_k(t)dt = a(s)\frac{\sin((k + 1)\alpha)}{k + 1}U_k(s).$$

Assertion 5 *The solution of system (2) has the following form:*

$$h_i(s) = 1/\pi \sum_{k=0}^{\infty}\sum_{j=1}^{n}\xi_k^{ij}U_k(s)\int_{-1}^{1} p_j(t)U_k(t)dt, \qquad (3)$$

where ξ_k^{ij} are entries of the matrix $\Xi_k = \Lambda_k^{-1}$,

$$\Lambda_k = \begin{bmatrix} 1 & \lambda_k^{12} & \dots & \lambda_k^{1n} \\ \lambda_k^{21} & 1 & \dots & \lambda_k^{2n} \\ \vdots & \vdots & \ddots & \vdots \\ \lambda_k^{n1} & \lambda_k^{n2} & \dots & 1 \end{bmatrix},$$

$$\lambda_k^{ij} = \frac{\sin((k + 1)(\omega_i - \omega_j))}{(k + 1)\sin(\omega_i - \omega_j)}.$$

We have obtained a new formula (3) for the Radon transform inversion.

Assertion 6 *The backprojection technique accuracy can be evaluated provided the reconstructed image norm* $\|f\|$ *is known:*

$$\|f - H\|^2 \equiv \|f\|^2 - Q(f,\omega).$$

Hence, the informativity of the projection set $P_\omega f = (p_{\omega_1}, \ldots, p_{\omega_n})$ *is defined by the value of the functional:*

$$Q(f,\omega) = 1/\pi \sum_{k=0}^{\infty} \sum_{i,j=1}^{n} \xi_k^{ij} \int_{-1}^{1} U_k(s) p_i(s) ds \int_{-1}^{1} U_k(t) p_j(t) dt. \qquad (4)$$

3 Expansion of bidirectional textures into a sum of two ridge functions

In the case of two directions $n = 2$ and an angle $\delta = \omega_1 - \omega_2$, matrices Λ_k and Ξ_k have the following form:

$$\Lambda_k = \begin{bmatrix} 1 & \frac{\sin((k+1)\delta)}{(k+1)\sin\delta} \\ \frac{\sin((k+1)\delta)}{(k+1)\sin\delta} & 1 \end{bmatrix},$$

$$\Xi_k = \left(1 - \frac{\sin^2((k+1)\delta)}{(k+1)^2 \sin^2\delta}\right)^{-1} \begin{bmatrix} 1 & -\frac{\sin((k+1)\delta)}{(k+1)\sin\delta} \\ -\frac{\sin((k+1)\delta)}{(k+1)\sin\delta} & 1 \end{bmatrix}, k > 0,$$

$$\Xi_0 = \Lambda_0^+ = \begin{bmatrix} \frac{1}{4} & \frac{1}{4} \\ \frac{1}{4} & \frac{1}{4} \end{bmatrix}.$$

Since $\int_{-1}^{1} U_0(s) p_1(s) ds = \int_{-1}^{1} U_0(s) p_2(s) ds = \int \int_D f(x,y) dx dy \equiv f_0$, formula (4) can now be written as

$$Q(f,\omega_1,\omega_2) = 1/\pi f_0^2 +$$

$$1/\pi \sum_{k=1}^{\infty} \sum_{i,j=1}^{2} \xi_k^{ij} \int_{-1}^{1} \frac{\sin((k+1)\arccos s) p_i(s)}{\sqrt{1-s^2}} ds \int_{-1}^{1} \frac{\sin((k+1)\arccos t) p_j(t)}{\sqrt{1-t^2}} dt.$$

Integrals in this formula can be calculated using the Gauss-Chebyshev formula:

$$\int_{-1}^{1} \frac{u(x) dx}{\sqrt{1-x^2}} \approx \frac{\pi}{m} \sum_{l=1}^{m} u\left(\cos\left(\frac{2l-1}{2m}\pi\right)\right).$$

If it is known that the angle between structures is δ, an algorithm consists in searching such α_0, that

$$\alpha_0 = \arg\max_\alpha Q(f,\alpha,\alpha+\delta).$$

For our numerical experiments we used two simple test textures – 'parquet'- and 'chess–board'–like images, both with structures in directions 45° and 135°, so that the angle between ridge structures is $\delta = \pi/2$. The images are modelled with the help of standard procedures of CorelDraw 4.0. The informativity $Q(\alpha, \alpha + \pi/2)$ values are computed for the test images (Tables 1 and 2) at equispaced twelve angles α-s. We can observe strong peaks in Q values for the angle $\alpha_0 = \pi/4$ in both cases. We conclude from interrelationship between values $\|f\|^2$ and Q ($\|f\|^2 < Q_{max}$) that the test images investigated are not precisely a sum of two ridge functions, however, the directions of structures are clearly detected.

Table 1. Q-values summary

$\|f\|^2 = 236$												
α:	0	$\frac{\pi}{12}$	$\frac{2\pi}{12}$	$\frac{3\pi}{12}$	$\frac{4\pi}{12}$	$\frac{5\pi}{12}$	$\frac{6\pi}{12}$	$\frac{7\pi}{12}$	$\frac{8\pi}{12}$	$\frac{9\pi}{12}$	$\frac{10\pi}{12}$	$\frac{11\pi}{12}$
$Q(f, \alpha, \alpha + \pi/2)$:	153	150	150	183	150	150	153	150	150	183	150	150
				α_0								

Table 2. Q-values summary

$\|f\|^2 = 107$												
α:	0	$\frac{\pi}{12}$	$\frac{2\pi}{12}$	$\frac{3\pi}{12}$	$\frac{4\pi}{12}$	$\frac{5\pi}{12}$	$\frac{6\pi}{12}$	$\frac{7\pi}{12}$	$\frac{8\pi}{12}$	$\frac{9\pi}{12}$	$\frac{10\pi}{12}$	$\frac{11\pi}{12}$
$Q(f, \alpha, \alpha + \pi/2)$:	58	56	56	89	56	56	58	56	56	89	56	56
				α_0								

References

1. Eichmann G and Kasparis T 1988 Topologically Invariant Texture Descriptors *Computer Vision, Graphics and Image Processing* **41** 267-281
2. Gindi G R and Gmitro A F 1984 Optical feature extraction via the Radon transform *Optical Engineering* **23** No. 5 499 -506
3. Illingworth I and Kittler J 1988 A survey of the Hough transform *Computer Vision , Graphics and Image Processing* **44** 87-116
4. Kim H.-B, Park R.-H 1992 Extracting spatial arrangement of structural textures using projection information *Pattern Recognition* **25** No.3 237-245
5. Logan B F and Shepp L A 1975 Optimal reconstruction of a function from its projections *Duke Math. Journal* **42** 645-59
6. Natterer F 1986 *The Mathematics of Computerized Tomography* (New York: Teubner and Wiley)

Textures and Structural Defects

Dmitry Chetverikov and Krisztián Gede

Computer and Automation Research Institute
1111 Budapest, Kende u.13-17, Hungary
email: mitya@leader.ipan.sztaki.hu

Abstract. Detection of structural defects in textures is addressed as a specific problem of visual texture analysis. A new approach to texture defect detection is proposed. A pilot experimental study shows that the method presented can detect structural imperfections in texture patterns of diverse origin.

1 Introduction

A broad class of automated visual inspection tasks involves *texture inspection* with the purpose of detecting, localizing and classifying defects on surfaces of materials. In their recent survey on visual inspection [13], Newman and Jain treat texture inspection as a distinct class of tasks that includes, among others, inspections of textiles, fabrics and carpets, ceramics and ceramic tiles, banknotes, hardwood, metal and painted surfaces, leather and paper. At the same time, few attempts have been made to understand the common nature of diverse texture inspection problems and propose a unifying approach to defect detection in textures. The capability of the existing texture analysis techniques to detect different texture imperfections has not been investigated. Designing a machine vision system for texture inspection often needs time-consuming and subjective selection of the suitable texture features. Many of the existing systems are highly specialized, with a large number of parameters to be set in a particular application environment.

The difference between texture defect detection and texture segmentation was pointed out in [5] where an initial taxonomy of texture imperfections was introduced. Texture imperfections, or defects, are regions that locally break the homogeneity of texture. The defects themselves do not have textural properties. Texture features computed for defect patches do not have stable values, while feature values of the background texture form a cluster in the feature space. Defects are then detected as outliers. In traditional texture segmentation, feature values of each segment form a cluster and the segments are searched as distinct clusters.

Chetverikov [6] designed a rotationally flexible filter aimed at adaptive defect detection in textures with gradual spatial variation of directionality. Brzakovic et. al. [4] described TEXIS, a system for defect detection in materials characterized by complex textures. TEXIS is introduced as a general texture defect detection system, but it contains many application-specific features and parameters that make it a customized tool for hardwood inspection. Amelung and Vogel [1] addressed the problem of automatic window size selection for texture defect detection operators.

Recently, Sinclair [14] proposed a cluster-based approach to texture analysis and demonstrated its capability to indicate a structural imperfection in a single texture pattern selected from the album [3]. Branca et.al. [2] considered textures composed of oriented structural elements on textural background and define a measure of local directional coherence of the texture flow field in order to detect defects as directional incoherencies. The algorithm was tested on leather patterns. Song et.al. [16] presented a Wigner based approach to detection of synthetic cracks in random and regular textures. The same authors published a survey on defect detection in textures [15] where more references to previous work, mostly applications, can be found.

In section 2, we briefly describe the feature based interaction map (FBIM) approach to texture analysis. Section 3 presents the methodology of defect detection in textures using the structural FBIM filter. Experimental results demonstrating the capability of the FBIM approach to detect defects in different textures are shown in section 4. Finally, the conclusions are drawn.

2 The FBIM approach

The feature based interaction map approach was proposed in [7,10]. This approach uses the EGLDH, the extended graylevel difference histogram that shows the frequencies of the absolute graylevel differences between pixels separated by a spacing vector. In the EGLDH, the magnitude and the angle of the spacing vector are independent, arbitrary parameters while in the conventional graylevel difference histogram (GLDH, see [12]) these parameters are interrelated because of the image raster. Principles, functions and previous applications of the FBIM method are described elsewhere [7,10,9,8,11]. A sketch of the approach is given below.

The polar interaction map $M_{pl}(i,j)$ is the basic entity of the FBIM method. $M_{pl}(i,j)$ is an intensity-coded polar representation of an EGLDH feature, with the rows enumerating the angle α_i, the columns the magnitude (displacement) d_j of the varying spacing vector. $M_{pl}(i,j)$ is computed for a range of angles and magnitudes. The XY interaction map $M_{xy}(k,l)$ is the Cartesian version of $M_{pl}(i,j)$.

The computation of an EGLDH feature and the layout of $M_{pl}(i,j)$ are illustrated in figure 1. In figure 1a, (m,n) scans the pixels of the image, while (x,y) points at a non-integer location specified by the spacing vector. The intensity $I(x,y)$ is obtained by interpolation of the four neighboring pixels. The absolute graylevel difference $|I(m,n) - I(x,y)|$ is used to address and increment a bin of the EGLDH. The standard GLDH features [12] can be used with the EGLDH as well. In this study, the EGLDH feature F is the mean of $|I(m,n) - I(x,y)|$.

Examples of interaction maps are shown in figure 2. $M_{xy}(k,l)$ reflects the structure of a texture pattern in the spatial domain preserving the geometry of the structure. When a pattern is rotated, $M_{xy}(k,l)$ rotates around its center. In $M_{pl}(i,j)$, this means the cyclical shift of the rows. Similarly, scaling the pattern amounts to stretching $M_{pl}(i,j)$ in X direction. Both rotation and scaling are easily implemented with the polar map making $M_{pl}(i,j)$ suitable for orientation- and size-adaptive *structural filtering* [8].

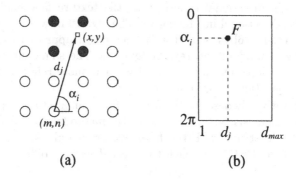

(a)　　　　　　　　　　(b)

Fig. 1. (a) Computing the EGLDH. (b) The layout of the polar interaction map. F is an intensity-coded EGLDH feature.

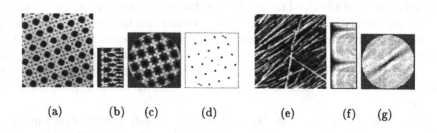

(a)　　　(b)　(c)　　　(d)　　　　(e)　　　(f)　(g)

Fig. 2. Examples of interaction maps. (a) A regular texture. (b) The polar map of (a). (c) The XY map of (a). (d) The structure (blobs) of (c). (e) A linear texture. (f) The polar map of (e). (g) The XY map of (e).

3　Using the FBIM filter to detect texture defects

Structural imperfections are detected in a texture image by matching the polar interaction map of an 'ideal' (reference) texture patch against the map computed in a sliding window. In each position of the window, a measure of dissimilarity between the reference and the current map is obtained and written into the output dissimilarity image. The normalized sum of the pointwise absolute differences between the corresponding pixels of the two maps (the city-block distance) was selected as the dissimilarity measure.

Many natural textures contain gradual variations of orientation and size that may be treated as tolerable or intolerable depending on the application. The FBIM filter can cope with both types of variations by specifying the degrees of variations that should be tolerated. This means that the filter becomes adaptive in orientation or/and size. $M_{pl}(i, j)$ is locally tuned to find the best matching orientation and scale. Technically, a bank of reference maps scaled within the size tolerance range is computed from the reference patch prior to filtering. During the filtering, each of the reference maps is matched against the current map. The best matching scale and orientation are found as those that give the least dissimilarity value, and this value is output.

The FBIM filtered image is negated so that the locations of high dissimilarity are dark blobs. (See figure 3a.) These blobs are enhanced and detected using a simple blob detector that indicates dark blob in a window if the mean intensity in this window is less than that in any of the surrounding windows. The size of the window is the expected blob size. The contrast of the blob is the average difference between the means in the surrounding windows and the mean in the current window. An example of a blob contrast image is demonstrated in figure 3b where darker points indicate stronger response of the blob detector.

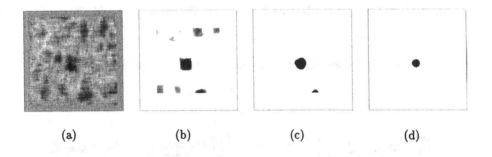

(a)　　　　　(b)　　　　　(c)　　　　　(d)

Fig. 3. Operation of the defect detection algorithm. (a) Result of FBIM filtering of texture 4 of figure 4. (b) Blob contrast image of (a). (c) Result of thresholding of (b) at 55%. (d) Result of thresholding of (b) at 75%.

In this study, we only consider compact defects. Elongated defects will also stand out in the FBIM filtered image, but they should be post-processed in a different way.

Texture defects are detected in the blob contrast image by thresholding this image at a certain percent of the maximum contrast. Figure 3c shows a result of thresholding at the level of 55%. At this level, some pixels are erroneously labeled as defects. At a higher level of 75% (figure 3d), the false detections are discarded as the most contrast blob is the defect. Only the correctly detected pixels remain. The lower the level that separates the two classes of pixels the better the detectability of the signal.

The rest of this paper is devoted to an experimental study aimed at detection of structural texture defects using the FBIM filter.

4 Experimental study

4.1 Experimental protocol

Figure 4 shows the four 256×256 pixel size grayscale test images used in the defect detection experiments, with sample detection results overlaid. The polar and the Cartesian interaction maps of the test textures are shown in figure 5.

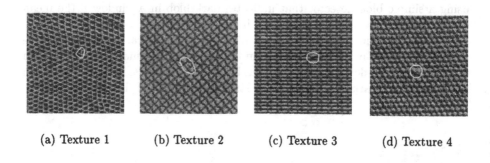

| (a) Texture 1 | (b) Texture 2 | (c) Texture 3 | (d) Texture 4 |

Fig. 4. The test textures with sample detection results overlaid.

Fig. 5. The interaction maps of the test textures (scaled). Each pair of maps describes the respective texture in figure 4.

The test images were selected selected from the album [3]. Each of them contains a single structural imperfection. Texture 1 is a hexagonal cellular structure with a gradual spatial variation in both size and orientation. The defect is perceived as a dark blob on textured background. The size of the defect is comparable to the average size of the cell. Texture 2 is a rectangular cellular structure with limited spatial variations. The hardly visible defect is a local structural distortion. The size of the defect is twice the size of the cell. Texture 3 is a regular rectangular structure. The defect is a local structural distortion whose size is less than the cell size. Finally, texture 4 is a hexagonal structure with very regular arrangement and some variation in the intensity of the elements. The defect is perceived as two elements merged into a bigger one.

The goal of the experimental study is to evaluate the detectability of different texture defects with the FBIM filter. A rectangular mask was manually drawn for each of the four defects. Each mask is set slightly larger than the defect to account for the finite sizes of defects and windows. The *merit of detection Q* is measured by the ratio, in percent, of the pixels detected within the mask to the overall number of the pixels detected in the image. This measure reflects both the signal-to-noise ratio and the localization of the defect. 100% detection means that defect pixels are only indicated. $Q < 100\%$ shows that false positive classifications occurred. False negative classifications are not accounted for since the contours of defects cannot be precisely specified. In most applications, the precise shape of defect is of no particular interest. A threshold can be set in an inspection task to specify the acceptable level of false detections.

The merit of detection depends on the *blob contrast threshold T*. This threshold is relative to the maximum contrast value and is also measured in percents. Other variable parameters of the tests are the *angular resolution $\Delta\alpha$* and the *maximum displacement d_{max}* of the interaction map. The window size W and the size of the blob detector were set to 30. An additional test for $W = 40$ was also run yielding similar results.

To allow for a limited directional variation of the structures, the ±15 degree angular adaptivity was set in all cases. The scaling adaptivity was only used in the control experiment with test pattern 1 where the size variation is considerable.

4.2 Results

The main experimental results are condensed in figure 6. The boxes of figure 6 refer to different textures and angular resolutions $\Delta\alpha$. Each of the boxes shows the intensity-coded merit of detection with white being 100%. In each box, the rows enumerate the growing maximum displacement $d_{max} = 10, 12, ..., 20$, the columns the growing threshold T. The interaction maps were computed for $d = 1, 2, ..., d_{max}$ and $\alpha = 0, \Delta\alpha, ..., 360$.

The 100% detection merit can be achieved for all the test images. A reasonably wide 100% detectability range of thresholds T was obtained for all test textures except the reptile skin where this range is less satisfactory. The reason is the large cell size variation in this texture. When in the control experiment a ±30% scaling adaptivity is used for this texture the results improve significantly, as shown in figure 7 where the white area grows when the adaptivity is switched on.

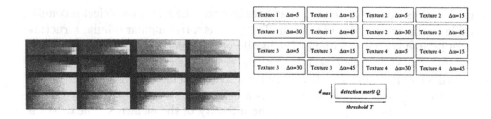

Texture 1 Δα=5	Texture 1 Δα=15	Texture 2 Δα=5	Texture 2 Δα=15
Texture 1 Δα=30	Texture 1 Δα=45	Texture 2 Δα=30	Texture 2 Δα=45
Texture 3 Δα=5	Texture 3 Δα=15	Texture 4 Δα=5	Texture 4 Δα=15
Texture 3 Δα=30	Texture 3 Δα=45	Texture 4 Δα=30	Texture 4 Δα=45

d_{max} | *detection merit Q*

threshold T

Fig. 6. The detection merits and their layout.

Fig. 7. Selected detection merits for texture 1 with the scaling adaptivity 0 (left) and ±30% (right), enlarged.

5 Conclusion

It has been demonstrated that structural texture imperfections of diverse origin can be detected in the framework of a novel approach based on the FBIM structural filtering. For systematic performance evaluation of the approach a large ground truthed image database of texture defects is needed. The initial experiments of this study indicate that adaptivity to variations in directionality and size is strongly desirable. This adaptivity is incorporated in a controllable way in the FBIM filter.

The FBIM approach is especially suitable for detection of structural defects in more or less regular textures. It is less efficient when applied to irregular patterns. Intensity defects like the one in the test pattern 1 are more easily defected as blobs using an appropriate blob detector. The power of the FBIM filter is in its applicability to different tasks by adjustment of the few natural parameters it utilizes.

In setting the filter parameters, the following considerations should be taken into account. The window size W should account for both defect size S and minimum patch size D, i.e., the smallest patch that still represents the background texture. When $S \simeq D$, it is safe to set $W \approx S$. (See [1] for a related discussion.) In the case of multiple defects of different sizes, the usage of resolution pyramids can be a solution. For the interaction map to reflect the structure of the background texture, the maximum spacing d_{max} should extend beyond the texture period. The experimental results indicate that the best angular resolution $\Delta\alpha$ also depends on the background texture. In addition, this parameter can be affected by the trade-off between the detection merit and the speed.

6   Acknowledgment

This work was supported in part by the grants OTKA T14520 and EU INFO COPERNICUS IC15 CT94 0742.

References

1. J. Amelung and K. Vogel. Automated window size determination for texture defect detection. In *Proc. British Machine Vision Conference*, pages 105–114, 1994.
2. A. Branca, M. Tafuri, G. Attolico, and A. Distante. Directionality detection in compositional textures. In *Proc. International Conf. on Pattern Recognition*, pages 830–834. Vol.II, 1996.
3. P. Brodatz. *Textures: a photographic album for artists and designers*. Dover, New York, 1966.
4. D. Brzakovic, H. Beck, and N. Sufi. An approach to defect detection in materials characterized by complex textures. *Pattern Recognition*, 23:99–107, 1990.
5. D. Chetverikov. Texture imperfections. *Pattern Recognition*, 6:45–50, 1987.
6. D. Chetverikov. Detecting defects in texture. In *Proc. International Conf. on Pattern Recognition*, pages 61–63, 1988.
7. D. Chetverikov. Pattern orientation and texture symmetry. In *Computer Analysis of Images and Patterns*, pages 222–229. Springer Lecture Notes in Computer Science vol.970, 1995.
8. D. Chetverikov. Structural filtering with texture feature based interaction maps: Fast algorithm and applications. In *Proc. International Conf. on Pattern Recognition*, pages 795–799. Vol.II, 1996.
9. D. Chetverikov. Texture feature based interaction maps and structural filtering. In *Proc. 20th Workshop of the Austrian Pattern Recognition Group*, pages 143–157. Oldenbourg Verlag, 1996.
10. D. Chetverikov and R.M. Haralick. Texture anisotropy, symmetry, regularity: Recovering structure from interaction maps. In *Proc. British Machine Vision Conference*, pages 57–66, 1995.
11. D. Chetverikov, J. Liang, J. Kőmüves, and R.M. Haralick. Zone classification using texture features. In *Proc. International Conf. on Pattern Recognition*, pages 676–680. Vol.III, 1996.
12. R. M. Haralick and L. G. Shapiro. *Computer and Robot Vision*, volume I. Addison-Wesley, 1992.
13. T.S. Newman and A.K. Jain. A survey of automated visual inspection. *Computer Vision and Image Understanding*, 67:231–262, 1995.
14. D. Sinclair. Cluster-based texture analysis. In *Proc. International Conf. on Pattern Recognition*, pages 825–829. Vol.II, 1996.
15. K. Y. Song, M. Petrou, and J. Kittler. Texture defect detection: A review. In *SPIE Applications of Artificial Intelligence X: Machine Vision and Robotics*, volume 1708, pages 99–106, 1992.
16. K. Y. Song, M. Petrou, and J. Kittler. Wigner based crack detection in texture images. In *Fourth IEE International Conference on Image Processing and its Applications*, pages 315–318, 1992.

Self-Calibration
from the Absolute Conic on the Plane at Infinity

Marc Pollefeys and Luc Van Gool

ESAT-VISICS – K.U.Leuven
Kard.Mercierlaan 94, B3001 Heverlee, Belgium
Marc.Pollefeys@esat.kuleuven.ac.be

Abstract. To obtain a metric reconstruction from images the cameras have to be calibrated. In recent years different approaches have been proposed to avoid explicit calibration. In this paper a new method is proposed which is closely related to some of the existing methods. Some interesting relations between the methods are uncovered. The method proposed in this paper shows some clear advantages. Besides some synthetic experiments a metric model is extracted from a video sequence to illustrate the feasibility of the approach.

1 Introduction

Since a few years it has been shown that it is possible to recover constant intrinsic camera parameters from an uncalibrated image sequence. Translating this theoretical possibility into a working implementation proved to be difficult and several methods emerged. Most of them try to recover geometric entities whose projection stays fixed throughout the sequence. These projections are directly related to the camera intrinsic parameters.

For example Faugeras *et al* [3] and later on Zeller and Faugeras [13] used the absolute conic. In their approach the supporting plane of this conic (i.e. the plane at infinity) is eliminated from the equations. Heyden and Åström [6] and Triggs [12] proposed methods based on the absolute quadric. Other methods were proposed by Hartley [5] and Pollefeys and Van Gool [8].

The approach followed in this article is based on the absolute conic and the plane at infinity. Some nice relationships between this method and the methods based on the absolute quadric will be uncovered. It seems that all these methods are very similar. Our approach naturally results in a different way of dealing with the scale factors which appear in the equations. This is one of its main advantages.

Heyden and Åström [6] consider these scale factors as additional unknowns resulting in convergence problems with longer image sequences (i.e. more unknown scale factors). Triggs [12] eliminates them by cross-multiplying components, thereby introducing other disadvantages.

2 Basic principles

In this paper a pinhole camera model will be used. The following equation expresses the relation between image points and world points.

$$\lambda \mathbf{m} = \mathbf{P}M \tag{1}$$

Here $\mathbf{P}$ is a 3×4 camera matrix, $\mathbf{m}$ and M are column vectors containing the homogeneous coordinates of the image points resp. world points, λ expresses the equality up to a scale factor, in the remainder of the text this is replaced by $\simeq$.

The **projective calibration** of a camera setup can be retrieved from correspondences between the images (see for example [2, 4, 9, 11]). These projective cameras can be chosen as follow:

$$\mathbf{P}_1 = [\mathbf{I}\,|\,0] \text{ and } \mathbf{P}_k = [\mathbf{H}_{1k}\,|\,\mathbf{e}_{1k}] \tag{2}$$

with $\mathbf{H}_{1k}$ the homography for some reference plane (the same for all views) and the epipole $\mathbf{e}_{1k}$ the projection in image k of the first camera position[1]. The **affine representation** corresponds to a particular choice of the possible projective representations. In this case the reference plane has to be the plane at infinity. Its homography can be written as follows:

$$\mathbf{H}_{\infty 1k} \simeq \mathbf{H}_{1k} + \mathbf{e}_{1k}.\mathbf{a} \tag{3}$$

for some $\mathbf{a}$ (which is in general unknown), since all homographies are related through this equation.

In the **metric case** the internal camera parameters have to be taken into account. This yields camera projection matrices of the following form:

$$\mathbf{P}_{E1} = \mathbf{K}\,[\mathbf{I}\,|\,0] \text{ and } \mathbf{P}_{Ek} = \mathbf{K}\,[\mathbf{R}_k\,|\,-\mathbf{R}_k\mathbf{t}_k] \tag{4}$$

with $\mathbf{K}$ the calibration matrix, with $\mathbf{R}_k$ and $\mathbf{t}_k$ representing the orientation and position with respect to the first camera. $\mathbf{K}$ is an upper triangular matrix of the following form:

$$\mathbf{K} = \begin{bmatrix} f_x & s & u_x \\ & f_y & u_y \\ & & 1 \end{bmatrix} \tag{5}$$

with f_x and f_y the relative focal lengths, $\mathbf{u} = (u_x, u_y)$ the principal point and s a skew factor (see for example [5]).

Note that it follows from equation (4) that

$$\mathbf{H}_{\infty 1k} \simeq \mathbf{H}_{\infty k}\mathbf{H}_{\infty 1}^{-1} \simeq \mathbf{K}\mathbf{R}_k\mathbf{K}^{-1} \tag{6}$$

with $\mathbf{H}_{\infty k}$ the homography from the plane at infinity to the kth image plane.

---

[1] Note that the relative scale of $\mathbf{e}_{1k}$ compared to $\mathbf{H}_{1k}$ is not free.

3 Self-calibration

Self-calibration methods try to solve for the intrinsic camera parameters which are contained in $\mathbf{K}$ under the assumption that these are constant throughout the sequence. Self-calibration is often based on fixed entities in the image or in the scene. Affine transformations always keep the plane at infinity Π_∞ fixed. This does not mean that every point in Π_∞ is mapped on the same point, but that they are all mapped in the same plane Π_∞. Euclidean transformations keep –in addition to Π_∞– also some special conic in Π_∞ fixed. This conic is called the absolute conic ω. Closely related to this conic, there also exists a degenerate quadric –called the absolute quadric Ω [10]– which is fixed under all Euclidean transformations.

Most self-calibration techniques try to retrieve one of these fixed entities (or its image) in the projective representation. From that point on metric measurements can be carried out. Often everything is transformed from these non-metric frames to the usual metric frame.

3.1 A fixed image for the absolute conic ω

Because the absolute conic ω is fixed under Euclidean transformations also its image ω_k will be fixed if the same camera is used. This is also valid for the dual of that conic ω_k^{-1}, which will be used here for convenience. Starting from $\omega = \mathbf{I}$, this can be proven as follows:

$$
\begin{aligned}
\omega_1^{-1} &\simeq \mathbf{H}_{\infty 1}\omega^{-1}\mathbf{H}_{\infty 1}^{\mathsf{T}} \simeq \mathbf{K}\mathbf{K}^{\mathsf{T}} \\
\omega_k^{-1} &\simeq \mathbf{H}_{\infty k}\omega^{-1}\mathbf{H}_{\infty k}^{\mathsf{T}} \simeq \mathbf{K}\mathbf{R}_k\mathbf{R}_k^{\mathsf{T}}\mathbf{K}^{\mathsf{T}} \simeq \mathbf{K}\mathbf{K}^{\mathsf{T}}
\end{aligned}
\tag{7}
$$

This fact can be used to calculate the image of the dual of the absolute conic. Because ω lies in Π_∞ the following equations must hold:

$$
\omega_k^{-1} \simeq \mathbf{H}_{\infty 1k}\omega_k^{-1}\mathbf{H}_{\infty 1k}^{\mathsf{T}}
\tag{8}
$$

Exact equality (not up to scale) can be obtained by scaling $\mathbf{H}_{\infty 1k}$ to obtain $\det \mathbf{H}_{\infty 1k} = 1$. In the case where the affine calibration was already obtained equation (8) results in a set of linear equations for the coefficients of ω_k^{-1}. Once ω_k^{-1} is retrieved $\mathbf{K}$ can be obtained from it by Cholesky factorization.

3.2 From projective to metric

If the affine calibration is not known then vector $\mathbf{a}$ (see Eq. 3) has to be retrieved in addition. In the general case the following equations are thus obtained (using an explicit scale factor and omitting the indices):

$$
\lambda\mathbf{K}\mathbf{K}^{\mathsf{T}} = [\mathbf{H} + \mathbf{e}\mathbf{a}]\mathbf{K}\mathbf{K}^{\mathsf{T}}[\mathbf{H} + \mathbf{e}\mathbf{a}]^{\mathsf{T}}
\tag{9}
$$

The problem with these equations is that the scale factors are unknown. It is possible to consider these scale-factors as additional unknowns [6], but this poses additional problems and will make the scheme unworkable for longer image sequences since one has to deal with one additional scale factor per supplementary image.

It is possible to find an easy way of calculating these scale factors as a function of the 3 affine parameters. This is achieved by taking the determinant of the left- and right-hand side of Equation (9):

$$\lambda \det \mathbf{K}\mathbf{K}^\top = \det(\mathbf{H} + \mathbf{ea}) \det \mathbf{K}\mathbf{K}^\top \det(\mathbf{H} + \mathbf{ea})^\top . \tag{10}$$

Which gives us λ:

$$\lambda = \det(\mathbf{H} + \mathbf{ea})^2 \tag{11}$$

which can be factorized as follow:

$$\lambda = (|\mathbf{e}\,\mathbf{h}_2\,\mathbf{h}_3|a_1 + |\mathbf{h}_1\,\mathbf{e}\,\mathbf{h}_3|a_2 + |\mathbf{h}_1\,\mathbf{h}_2\,\mathbf{e}|a_3 + |\mathbf{h}_1\,\mathbf{h}_2\,\mathbf{h}_3|)^2 \tag{12}$$

with $\det \mathbf{H} = |\mathbf{h}_1\,\mathbf{h}_2\,\mathbf{h}_3|$.

By filling in $\mathbf{K}$ as in equation (5) and λ as in equation (12) in equation (9) one obtain $5(n-1)$ equations in 8 unknowns (n being the number of images). Therefore at least 3 images are needed to obtain the calibration from correspondences alone.

These equations can be solved through a nonlinear minimization of the following criterion:

$$\sum_{k=2}^{n} \left\| \frac{1}{\lambda} \mathbf{H}_{\infty 1k} \mathbf{K}\mathbf{K}^\top \mathbf{H}_{\infty 1k}^\top - \mathbf{K}\mathbf{K}^\top \right\|_F \tag{13}$$

The implementation presented in this paper uses a Levenberg-Marquard algorithm. Results seem to be better when $\mathbf{K}$ is normalized to $\|\mathbf{K}\|_F = 1$ where $\|.\|_F$ denotes the Frobenius norm.

The advantage of Equation (12) is that it yields a simple closed form equation for λ. In practice it is more stable to use $\lambda = \frac{\|\mathbf{K}\mathbf{K}^\top\|_F}{\|\mathbf{H}_{\infty 1k}\mathbf{K}\mathbf{K}^\top \mathbf{H}_{\infty 1k}^\top\|_F}$ during minimization. This avoids problems when $\mathbf{H}$ is badly conditioned.

4 Relation with other methods

The method presented in this paper is part of a family of methods which all try to retrieve the absolute entities in projective space (i.e. absolute conic ω or absolute quadric Ω). Once one of these entities is retrieved one can do metric measurements or transform to a metric frame.

Different methods will be discussed here. First the method based on the **Kruppa equations** proposed by Faugeras *et al* [3] and refined by Zeller and Faugeras [13]. In this method the affine parameters are eliminated from the equations. Only the fundamental matrices are needed, not a consistent projective frame for all cameras.

The second method was recently proposed by Heyden and Åström [6]. It is based on the **Kruppa constraints** which relate the dual of the image of the absolute conic to the absolute quadric. It will be shown that these constraints are equivalent with the constraints presented in this paper. Heyden introduced additional unknowns to cope with the scale factors. This strategy only works with a small number of images and is even then suboptimal.

Finally Triggs' method [12] to retrieve the **absolute quadric** is also reviewed. The principles are similar to Heyden's method, but the implementation is different. Scale factors are eliminated by doing cross-multiplication yielding 15 equations (from which only 5 are independent) per camera and doubling the order of the camera intrinsic parameters (i.e. 4th order terms instead of 2nd).

4.1 Kruppa equations

The Kruppa equation can be derived starting from equation (8):

$$\mathbf{KK}^\top \simeq \mathbf{H}_\infty \mathbf{KK}^\top \mathbf{H}_\infty^\top \tag{14}$$

There is an easy way of eliminating the affine parameters a_1, a_2, a_3 from these equations. They can be multiplied with $[\mathbf{e}]_\times$ to the left and $[\mathbf{e}]_\times^\top$ to the right:

$$[\mathbf{e}]_\times \mathbf{KK}^\top [\mathbf{e}]_\times^\top \simeq \mathbf{FKK}^\top \mathbf{F}^\top \tag{15}$$

since the fundamental matrix $\mathbf{F} = [\mathbf{e}]_\times \mathbf{H}_\infty$. From the 5 equations obtained here only 2 are independent. Scale factors are eliminated by cross-multiplication. The disadvantage of this method is that a consistent supporting plane Π_∞ for ω is only indirectly enforced.

4.2 Kruppa constraints

It can be shown that the Kruppa constraints [6] are equivalent with the constraints used in this paper. Starting from equation (9),

$$\lambda \mathbf{KK}^\top = [\mathbf{H} + \mathbf{ea}] \mathbf{KK}^\top [\mathbf{H} + \mathbf{ea}]^\top \ , \tag{16}$$

and rewriting this equation the Kruppa constraints can easily be obtained (using $\tilde{\mathbf{a}} = \mathbf{aK}$):

$$\lambda \mathbf{KK}^\top = [\mathbf{He}] \begin{bmatrix} \mathbf{KK}^\top & \mathbf{KK}^\top \mathbf{a}^\top \\ \mathbf{aKK}^\top & \mathbf{aKK}^\top \mathbf{a}^\top \end{bmatrix} \begin{bmatrix} \mathbf{H}^\top \\ \mathbf{e}^\top \end{bmatrix}$$

$$= \mathbf{P} \begin{bmatrix} \mathbf{KK}^\top & \mathbf{K}\tilde{\mathbf{a}}^\top \\ \tilde{\mathbf{a}}\mathbf{K}^\top & ||\tilde{\mathbf{a}}||^2 \end{bmatrix} \mathbf{P}^\top \ . \tag{17}$$

Equation (17) represents the Kruppa constraints like Heyden presented them in [6].

The problem is that in this form it does not seem possible to calculate the scale factors λ (the trick with the determinants does not work when non-square matrices are involved). Therefore Heyden and Åström [6] deal with them as additional unknowns. This renders this scheme unworkable for more than a few images because of the many additional unknowns.

4.3 The absolute quadric

Triggs' [12] equations are very similar to the Kruppa constraints (Eq. (17)) except that he does not assume $\mathbf{P}_1 = [\mathbf{I}|\mathbf{0}]$

$$\omega_{im}^{-1} \simeq \mathbf{P}\Omega\mathbf{P}^\top \ . \tag{18}$$

The consequence is that the absolute quadric Ω is not directly related to ω_{im}^{-1} through the parameterization anymore. Therefore one has to cope with more unknowns. The constraint rank$(\Omega)=3$ which also followed from the parametrisation in the previous methods now has to be enforced explicitly. The advantage is that all views are treated equally where previous methods implicitly favored the first view. Remark that this can also be achieved with the absolute conic using $\omega_{im}^{-1} \simeq \mathbf{H}_{\infty k}\omega^{-1}\mathbf{H}_{\infty k}^\top$ (with ω the absolute conic *in the scene*). This is used in our method to refine the results and to obtain unbiased results. Experiments have shown that this is especially important for longer image sequences.

5 Experiments

Experiments have been done on both real and synthetic data. First the synthetic data give some insights in the behavior of the method depending on the number of views and the presence of noise. Then the feasibility of the method will be illustrated with some calibration/reconstruction work done on a real video sequence,

5.1 Simulations

The simulations were carried out on sequences of 3, 6 and 10 views. The scene consisted of 50 points uniformly distributed in a unit sphere with its center at the origin. For the calibration matrix the canonical form $\mathbf{K} = \mathbf{I}$ was chosen. The views were taken from all around the sphere and were all more or less pointing towards the origin. The scene points were projected into the images. Gaussian noise with an equivalent standard deviations of 0, 0.1, 0.2, 0.5, 1 and 2 pixels for 500×500 images was added to these projections. For every sequence length and noise level ten sequences were generated. The self-calibration method proposed in this paper was carried out on all these sequences. The results for the camera intrinsic parameters are shown in table 1 for 6 view sequences.

When 6 or 10 views were used the accuracy was very good, even for high amounts of noise (around 1% error for 2 pixels noise). The method almost always converges without problems. For sequences of only 3 views the method gives good results for small amounts of noise, but the error grows when more noise is added. This is due to the fact that in the 3 view case no redundancy is present anymore. In this case the method regularly ends up in an alternative solution.

	f_x	$\frac{f_x}{f_y}$	u_x	u_y	s
0.0	1.0000 ± 0.0000	1.0000 ± 0.0000	0.0000 ± 0.0000	0.0000 ± 0.0000	0.0000 ± 0.0000
0.1	0.9998 ± 0.0005	1.0000 ± 0.0003	0.0001 ± 0.0002	0.0002 ± 0.0002	-0.0001 ± 0.0002
0.2	0.9992 ± 0.0017	1.0001 ± 0.0009	0.0004 ± 0.0009	0.0004 ± 0.0009	0.0000 ± 0.0003
0.5	0.9997 ± 0.0019	0.9998 ± 0.0007	0.0019 ± 0.0017	0.0019 ± 0.0017	-0.0000 ± 0.0006
1.0	1.0014 ± 0.0030	1.0010 ± 0.0022	0.0005 ± 0.0029	-0.0005 ± 0.0029	-0.0004 ± 0.0012
2.0	0.9979 ± 0.0170	0.9986 ± 0.0052	0.0067 ± 0.0108	0.0067 ± 0.0108	0.0023 ± 0.0072

Table 1. Results of synthetic experiments for 6 view sequences (computed internal camera parameters with standard deviation)

5.2 A real video sequence

In this paragraph results obtained from a real sequence are presented. These were recorded with a video camera. Some of the images used for self-calibration are shown in figure 1. The projective camera matrices were obtained following the method described in [1]. These camera matrices were upgraded to metric using the self-calibration method described in this paper and then a 3D model was generated using these cameras and a dense correspondence map obtained as in [7]. Figure 2 contains some perspective views of the reconstruction. To evaluate the metric quality of the reconstruction lines were manually indicated in the reconstruction. The average measured angle was 1.6674 ± 0.9571 degrees for parallel lines and 92.0198 ± 2.2201 degrees for orthogonal lines.

Fig. 1. *Images 3, 9 and 15 of the sequence of the Arenberg castle which was used for the reconstruction*

6 Conclusion

In this article a new method was proposed for self-calibration. It is based on the explicit retrieval of the absolute conic and its supporting plane, the plane at infinity. It was shown that this is theoretically equivalent to solving the Kruppa constraints for the absolute quadric. The advantage of our formulation is that it gives a closed formula for the scale factors. Experiments on real and synthetic data illustrated the feasibility and the accuracy of the method.

Fig. 2. *Some perspective views of the reconstruction*

Acknowledgments

Marc Pollefeys acknowledges a specialization grant from the Flemish Institute for Scientific Research in Industry (IWT). This work was also supported by the EU ACTS project AC074 'VANGUARD'.

References

1. P. Beardsley, P. Torr and A. Zisserman 3D Model Acquisition from Extended Image Sequences, *Proc. ECCV'96*, vol.2, pp.683-695
2. O. Faugeras, What can be seen in three dimensions with an uncalibrated stereo rig, *Proc. ECCV'92*, pp.563-578.
3. O. Faugeras, Q.-T. Luong and S. Maybank. Camera self-calibration: Theory and experiments, *Proc. ECCV'92*, pp.321-334.
4. R. Hartley, Estimation of relative camera positions for uncalibrated cameras, *Proc. ECCV'92*, pp.579-587.
5. R. Hartley, Euclidean reconstruction from uncalibrated views, Applications of invariance in Computer Vision, LNCS 825, Springer-Verlag, 1994.
6. A. Heyden, K. Åström, Euclidean Reconstruction from Constant Intrinsic Parameters *Proc. ICPR'96*.
7. R. Koch, Automatische Oberflächenmodellierung starrer dreidimensionaler Objekte aus stereoskopischen Rundum-Ansichten, PhD thesis, Univ. Hannover, 1996.
8. M. Pollefeys and L. Van Gool, A Stratified Approach to Metric Self-Calibration, *Proc. CVPR'97*.
9. C. Rothwell, G. Csurka and O.D. Faugeras, A comparison of projective reconstruction methods for pairs of views, *Proc. ICCV'95*, pp.932-937.
10. J. G. Semple and G. T. Kneebone, Algebraic Projective Geometry, University Press, Oxford, 1952.
11. M. Spetsakis and Y. Aloimonos, A Multi-frame Approach to Visual Motion Perception *International Journal of Computer Vision*, 6:3, 245-255, 1991.
12. B. Triggs , Autocalibration and the Absolute Quadric, *CVPR'97*.
13. C. Zeller and O. Faugeras, Camera self-calibration from video sequences: the Kruppa equations revisited. Research Report 2793, INRIA, 1996.

A Badly Calibrated Camera in Ego-motion Estimation – Propagation of Uncertainty[*]

Tomáš Svoboda[1] and Peter Sturm[2]

[1] Center for Machine Perception, Czech Technical University, Karlovo nám. 13,
CZ 121-35 Praha 2, Czech Republic, svoboda@cmp.felk.cvut.cz
[2] GRAVIR–IMAG, project MOVI, INRIA Rhône–Alpes, 655, avenue de l'Europe,
38330 Monbonnot, Grenoble, France, peter.sturm@inrialpes.fr

Abstract. This paper deals with the ego-motion estimation (motion of the camera) from two views. To estimate an ego-motion we have to find correspondences and we need a calibrated camera. In this paper we solve the problem how to propagate known camera calibration errors into the uncertainty of the motion parameters. We present a linear estimate of the uncertainty in ego-motion based on the uncertainty in the calibration parameters. We show that the linear estimate is very stable.

1 Introduction

Let us assume to have two images of a static scene captured by the same camera from two different viewpoints. Having at least eight corresponding points and knowing the calibration parameters of the camera, the camera motion can be estimated up to a similarity using a linear algorithm. The similarity reconstruction of the scene can also be done. This problem has been solved over years [6, 11]. To calibrate the camera, some of many methods developed for the off-line camera calibration as Tsai's method [10], can be used.

In this paper we deal with two views and an innacurately calibrated camera. We use a linear method intended for a calibrated camera to estimate the ego-motion. Because of noise in the estimation process, an error analysis is needed. There are two sources of errors: (a) noise in correspondences and (b) noise in the calibration parameters. Weng [11] studied the influence of the noise on the motion parameters. Florou and Mohr [1] used the statistic approach to study reconstruction errors with respect to the calibration parameters. In our paper, we present a linear algorithm to estimate a credibility of the motion parameters, based on the uncertainty in the calibration parameters.

---

[*] This research was supported by Région Rhône Alpes with the program TEMPRA – East Europe between l'INPG Grenoble and the Czech Technical University, the Czech Ministry of Education grant No. VS 96049, grants 102/97/0480, 102/97/0855 and 102/95/1378 of the Grant Agency of the Czech Republic and European Union grant Copernicus CP941068, and the grant of the Czech Technical University No 3097472.

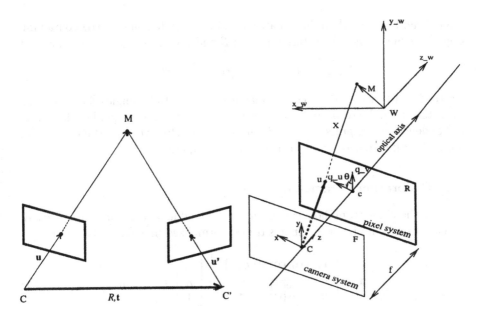

Fig. 1. *Two views, the geometry and the coplanarity constraint. C denotes the center of projection*

Fig. 2. *The pinhole camera, retinal (R) and focal (F) plane and three coordinate systems.*

2 Fundamentals

2.1 Ego-Motion from Point Correspondences

Let the motion between two positions of the camera be desribed by a rotation matrix R and a translation vector $\mathbf{t}$, and let $\mathbf{u}$ and $\mathbf{u}'$ be *normalized image coordinates* of corresponding points, see Figure 1, then the coplanarity constraint can be written as:

$$\mathbf{u}'^T(\mathbf{t} \times R\mathbf{u}) = 0. \tag{1}$$

Introducing the antisymmetric matrix S containing elements of the translation vector $\mathbf{t}$, we can rewrite the coplanarity constraint above as

$$\mathbf{u}'^T SR\mathbf{u} \overset{\triangleright}{=} \mathbf{u}'^T E\mathbf{u} = 0, \tag{2}$$

where E is the *essential matrix*. The equations above was proposed by Longuett Higgins in [6]. The essential matrix E can be reversely decomposed into the motion parameters using method e.g. by Hartley [2]. If E is estimated correctly, it has rank 2 and its nonzero singular values are equal [6]. Using the singular value decomposition we can factorize E as

$$E = UDV^T, \tag{3}$$

where U and V are orthonormal matrices and D is the diagonal matrix containing singular values of E. The rotation matrix R and the matrix S can be computed as

$$S = VZV^T, \ R = UYV^T \text{ or } UY^TV^T, \tag{4}$$

where Z and Y are known constant matrices [2]. The Euler angles characterizing the rotation can be recovered from R using a relationship from [5]. Localized correspondences are expressed in *pixel coordinates*, q. However to estimate motion, normalized image coordinates u is needed, see equation (2).

2.2 Calibration Parameters of a Camera

Assume the classical pinhole camera model. Introducing a *calibration matrix* K we can transform pixel coordinates q into normalized image coordinates u:

$$\mathbf{u} = \begin{bmatrix} \alpha_u & -\alpha_u \cot(\theta) & q_{u_0} \\ 0 & \frac{\alpha_v}{\sin(\theta)} & q_{v_0} \\ 0 & 0 & 1 \end{bmatrix}^{-1} \begin{bmatrix} q_u \\ q_v \\ 1 \end{bmatrix} = K^{-1}\mathbf{q}, \tag{5}$$

where α_u resp. α_v is the "horizontal", resp. the "vertical" scale factor, θ denotes the skew angle, q_{u_0} and q_{v_0} are pixel coordinates of the *principal point*. The principal point is the intersection of the optical axis with the retinal plane, in Figure 2 it is denoted by "c". The angle θ is the angle in the retinal coordinate system. The skew angle θ is often supposed to be very close to $\pi/2$, which is a valid approximation in practice. This approximation is used in the paper. Using K, equation (2) can be rewritten as

$$\mathbf{q}'^T K^{-T} E K^{-1}\mathbf{q} = \mathbf{q}'^T F\mathbf{q} = 0, \tag{6}$$

where F is the *fundamental matrix* [7].

3 Propagation of Uncertainty

Let us suppose that we know the uncertainty of the calibration parameters, i.e. the calibration matrix, the mean, $\overline{K}$, and their covariance matrix, C_K. Now we want to propagate the information in C_K through the whole estimation process. The desired results are covariance matrices of the motion parameters, i.e. C_t for the translation vector, resp. C_R for the rotation matrix, consequently $C_{\mathcal{E}}$ for the Euler angles.

3.1 Covariance of the Essential Matrix

The relationship between F, K and E is determined by the equation (6). Let us suppose that fundamental matrix F is correctly estimated. The calibration matrix has the mean $\overline{K}$ and the covariance matrix C_K. A well known approach is used for the estimation $\overline{E}$ and C_E.

$$\overline{E} = \overline{K}^T F \overline{K}, \tag{7}$$

$$C_E = J_{KE} C_K J_{KE}^T, \text{ where } J_{KE} = \frac{\partial(K^T F K)}{\partial K_{ii}}, \text{ at } K = \overline{K}. \tag{8}$$

3.2 Eigenvalues and Eigenvectors of a Noisy Matrix

The way how to estimate R and t from E has been described by equations (3) and (4). It follows from these equations that the eigenspace analysis of a noisy matrix is needed. Using equation (3) it can be easily verified that

$$EE^T = UD^2U^T, \text{ and } E^TE = VD^2V^T. \tag{9}$$

Both EE^T and E^TE are 3×3 symmetric matrices. The diagonal matrix D contains singular values of E, or $D^2 = \text{diag}(\lambda_1, \lambda_2, \lambda_3)$, where λ_i are eigenvalues of EE^T. Since the eigenvalue problem of a symmetric matrix is always well conditioned [12], it is reasonable to assume that slight variations in E cause slight variations of singular values. However, the same can not be said about the eigenvectors, being the columns of $U = [u_1, u_2, u_3]$, resp. V, [4]. The perturbed matrix $EE^T(\epsilon)$ can be written as

$$EE^T(\epsilon) = (E + \Delta_E)(E^T + \Delta_E^T) = EE^T + \Delta_{EE^T} + \mathcal{O}(\Delta_E^2), \tag{10}$$

where Δ_E denotes error matrix and Δ_{EE^T} is the linear part of the error,

$$\Delta_{EE^T} \cong \Delta_E E^T + E \Delta_E^T. \tag{11}$$

Let $u_1(\epsilon) = u_1 + \delta_{u_1}$, with $\delta_{u_1} \subset \text{span}\{u_2, u_3\}$, be the eigenvector of the perturbed matrix $EE^T(\epsilon)$ associated with the perturbed eigenvalue $\lambda_1(\epsilon)$. Skipping extensive derivation [9, 12], error in egienvectors can be estimated as

$$\delta_{u_1} \cong U \Delta_1 U^T \Delta_{EE^T} u_1, \tag{12}$$

where

$$\Delta_1 = \text{diag}\{0, (\lambda_1 - \lambda_2)^{-1}, (\lambda_1 - \lambda_3)^{-1}\}. \tag{13}$$

The proof of equation (12) can be found in the appendix of [11], or better, together with more theoretical background, in [12]. However, it follows from the equation (13), that the estimate of the errors in the eigenvector will be unstable when λ_1 is very close to λ_2 what is indeed true. The high estimate of the error is caused by an ambiguity of the choice of the eigenvectors. The matrix EE^T is symmetric and thus it has linear divisors, then normalized perturbed eigenvector of $EE^T(\epsilon)$ belonging to the first of the equal eigenvalues has one of the following equivalent form [12]

$$u_1(\epsilon) = [1, k(\epsilon), u_{13}] \tag{14}$$

$$[k(\epsilon), 1, u_{13}]. \tag{15}$$

The eigenvector $\mathbf{u}_2(\epsilon)$ can be expressed in the similar way. The proof is omitted here, the reader is referred to [12]. So, for practical purposes, we rewrite equation (12) as

$$\delta_{\mathbf{u}_1} \cong U_{23}\Delta'_1 U_{23}^T \Delta_{EE^T} \mathbf{u}_1, \tag{16}$$

where

$$\Delta'_1 = \begin{bmatrix} \xi & 0 \\ 0 & \frac{1}{\lambda_1-\lambda_3} \end{bmatrix}, \text{ and } U_{23} = [\mathbf{u}_2, \mathbf{u}_3]. \tag{17}$$

The variable ξ is defined as

$$\xi = \begin{cases} \frac{1}{\lambda_1-\lambda_2} & \text{if } \frac{\lambda_1-\lambda_2}{\text{mean}[\lambda_1,\lambda_2]} > \text{Tol} \\ 0 & \text{otherwise} \end{cases} \tag{18}$$

where Tol is a user defined tolerance. In experiments we use Tol $= 10^{-6}$. The matrices Δ'_2, resp. Δ'_3, are defined similarly. Although this assistance looks quite tricksy, it works very well as we will show in experiments.

Now, return to the question how we can estimate C_U, resp. C_V, having C_E. The equation (16) can be rewritten in matrix form as

$$\delta_{\mathbf{u}_1} = U_{23}\Delta'_1 U_{23}^T [u_{11}I_3, u_{13}I_3, u_{13}I_3]\delta_{EE^T}, \tag{19}$$

where δ_{EE^T} is a vector rearranged from elements of the matrix Δ_{EE^T}, see equation (11), and I_3 is the 3×3 identity matrix. The error matrix Δ_E can be rearranged to the error vector δ_E. From equation (11) we can derive matrix G_{EE^T} such that

$$\delta_{EE^T} = G_{EE^T}\delta_E . \tag{20}$$

Introducing new matrix

$$D_{\mathbf{u}_1} = U_{23}\Delta'_1 U_{23}^T [u_{11}I_3, u_{13}I_3, u_{13}I_3]G_{EE^T}, \tag{21}$$

equation (19) can be rewritten as $\delta_{\mathbf{u}_1} = D_{\mathbf{u}_1}\delta_E$. Under the assumption of the zero mean noise, we have for the covariance matrix of $\mathbf{u}_1$,

$$C_{\mathbf{u}_1} = \mathsf{E}\{\delta_{\mathbf{u}_1}\delta_{\mathbf{u}_1}^T\} = D_{\mathbf{u}_1}\mathsf{E}\{\delta_E\delta_E^T\}D_{\mathbf{u}_1}^T = D_{\mathbf{u}_1}C_E D_{\mathbf{u}_1}^T. \tag{22}$$

The 9×9 covariance matrix C_U is

$$C_U = D_U C_E D_U^T, \tag{23}$$

where D_U is the 9×9 matrix composed from the rows of $D_{\mathbf{u}_i}$, equation (21), matrices. A similar derivation can be done for C_V. The complete derivation can be found in [9].

Having $\overline{U}, C_U$ and $\overline{V}, C_V$, we can compute the mean values of the Euler angles, $\overline{\mathcal{E}}$, resp. translation vector, $\overline{t}$ using similar approach as for $\overline{E}$ and C_E, see equations (7) and (8). A data flow of the algorithm is decribed in Figure 3.

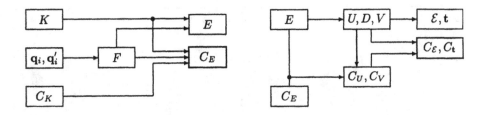

Fig. 3. Algorithm as a data flow.

4 Experiments

In the previous section we have described how to estimate the uncertainty of the essential matrix, and consequently of the motion parameters. Several questions have arisen:

- How accurate is the analytical approach? Remember that we use the first order approximation of the error.

- What happens when the F matrix is not exact, i.e. if there are some errors in the correspondences?

Suppose that the correspondences are correctly localized. The relation between errors in calibration and errors in the motion estimation, resp. estimation of the essential matrix, see eq. (6), depends on a particular camera motion and the type of a scene.

To find answers to the proposed questions we did experiments with real images captured by a camera mounted on a mobile robot. The robot undergoes planar motion and rotates around its vertical axis. We use the program by Z. Zhang [13] to find correspondences and the program by Svoboda and Pajdla [8] to detect outliers (false matches). For the estimation of the motion parameters, we use linear method with point normalization [3]. Two consequent images with correct matches, outliers and motion of correspondences are in Figure 4.

We add the artificial Gaussian noise to camera calibration parameters, i.e. elements of the matrix K, see equation (5). To simulate errors in camera calibration, we have issued from statistical observation presented in [1]. We have concentrated on the reliability of the estimated variance in motion parameters w.r.t statistical observation. There are six lines on each graph in Figure 5. Straight lines denote the linear estimates of the variances of translation vectors resp. Euler angles obtained by our algorithm. The fractional lines with little crosses denote statistical data obtained from 1000 testing cycles. There are also three vectors in the left upper corners: (a) whatCalPar indicates which calibration parameters are corrupted with noise. (b) corrcoef indicates the correlation coefficients between noise in the calibration parameters, significant is the first one which denotes the correlation between α_u and α_v. (c) t resp. Euls indicates the normalized translation vector resp. Euler angles.

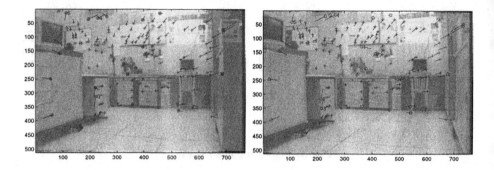

Fig. 4. Image pair. Little crosses denote the correct correspondences, circles denote detected outliers and lines show the correspondence motion.

The results belonging to the image pair from Figure 4 are shown in Figure 5. The good reliability of the linear estimate w.r.t. statistical data can be observed up to 20% Gaussian noise in the camera calibration parameters.

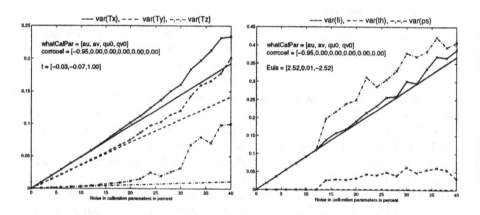

Fig. 5. Variance in translation (left) and Euler angles (right). Lines with little crosses denote the statistical observation.

5 Conclusions

We have presented a linear algorithm for the estimation of the variances of the camera motion parameters. We can characterize the uncertainty in the calibration parameters by their covariance matrix. We used the first order approximation to linearize the nonlinear relationship between the calibration parameters and the motion parameters. The linear estimate of errors in motion was observed to be stable and credible up to 20% noise in the camera calibration parameters and even with some noise in correspondences. The main contribution of the approach proposed is that we can estimate the uncertainty in motion parameters *before* the incident motion is computed.

References

1. Giannoula Florou and Roger Mohr. What accuracy for 3D measurement with cameras? In *International Conference on Pattern Recognition 1996*, pages 354–358, Los Alamitos, California, August 1996. IEEE Computer Society Press.
2. Richard I. Hartley. Estimation of relative camera positions for uncalibrated cameras. In *2nd European Conference on Computer Vision*, pages 579–587. Springer - Verlag, LNCS 588, May 1992.
3. Richard I. Hartley. In defence of the 8-point algorithm. In *Fifth International Conference on Computer Vision*, pages 1064 – 1070. IEEE Copmuter Society Press, 1995.
4. Roger A. Horn and Charles R. Johnson. *Matrix Analysis*. Cambridge University Press, 1985. 1987 reprinted with corrections, 1988,1990,1991,1992,1993.
5. Kenichi Kanatani. *Group-Theoretic Methods in Image Understanding*. Springer-Verlag, 1990.
6. H.C. Longuett-Higgins. A computer algorithm for reconstruction a scene from two projections. *Nature*, 293:133–135, 1981.
7. Quang-Tuan Luong, Rachid Deriche, Olivier Faugeras, and Theo Papadopoulo. On determining the fundamental matrix: Analysis of different methods and experimental results. Research report 1894, INRIA, April 1993.
8. Tomáš Svoboda and Tomáš Pajdla. Eficient motion analysis. Research report, K335/95/95, Czech Technical University, Faculty of Electrical Engineering, October 1995. 29 pages. Available at
    ftp://cmp.felk.cvut.cz/pub/cvl/articles/svoboda/egomot.ps.Z.
9. Tomáš Svoboda and Peter Sturm. What can be done with a badly calibrated camera in ego-motion analysis? Research report CTU-CMP-1996-01, Czech Technical University, Faculty of Electrical Engineering, Center for Machine Perception, September 1996. Available at
    ftp://cmp.felk.cvut.cz/pub/cvl/articles/svoboda/weakcal.ps.Z.
10. Roger Y. Tsai. A versatile camera calibration technique for high-accurancy 3D machine vision metrology using off-the-shelf cameras and lenses. *IEEE Journal of Robotics and Automation*, RA-3(4):323 – 344, August 1987.
11. Juyang Weng, Thomas S. Huang, and Ahuja Narendra. Motion and structure from two perspective views: Algorithms, error analysis, and error estimation. *IEEE Transactions on Pattern Analysis and Machine Inteligence*, 11(5):451–476, May 1989.
12. J.H. Wilkinson. *The Algebraic Eigenvalue Problem*. Oxford University Press, 1965.
13. Zhengyou Zhang, Rachid Deriche, Olivier Faugeras, and Quang-Tuang Luog. A robust technique for matching two uncalibrated images through the recovery of the unknown epipolar geometry. Research report RR-2273, INRIA, May 1994.

6DOF Calibration of a Camera with Respect to the Wrist of a 5-Axis Machine Tool

Stephen Blake[1]

Machine Tool Vision Ltd., 53 Lawrence St., Blackburn, Lancashire, BB2 1QF, England

Abstract. Computer vision systems are being used to guide machine tools fitted with wrist-mounted cameras. The vision system must measure the coordinates of objects with respect to the machine tool world reference frame. Therefore, the system must be calibrated by measuring the coordinates of the camera with respect to the wrist of the machine tool.

The paper shows that the geometry of the camera views—that one is forced to use in practice—cause the calibration problem to be near rank-deficient. The writer thinks that this is the cause of inaccuracy in existing practical work in this area. The paper gives a closed-form solution of the calibration problem which handles the rank-deficiency. The results of the first real tests are reported.

1 Introduction

Computer vision systems are being used to guide machine tools and robots in mechanized assembly and inspection tasks. Applications typically require the vision system to measure the position of target objects relative to the machine tool world reference frame to an accuracy of about $\pm25\mu$m. The field of view of a 512×512 CCD camera is constrained by the accuracy requirement to be no greater than 15×15mm^2 on the target object. The camera must therefore be rigidly fixed to the wrist of the machine tool in order to look at any object in the working volume.

One can obtain the target-to-world coordinates from the sequence target-to-camera, camera-to-wrist, wrist-to-world. The vision system measures target-to-camera coordinates and the machine tool controller outputs wrist-to-world coordinates. This paper considers the calibration task of measuring the 6DOF camera-to-wrist coordinates.

2 Review of Existing Calibration Techniques

In figure 1 TCP$_k$ is the wrist frame of the machine tool and O$_k$ is the camera frame on view k. There are 3 views of the target object T. On view k, the origin of the wrist frame is at point a_k, and the origin of the camera frame is at point b_k. The origin of the reference frame in the target T is at point p.

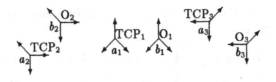

Fig. 1. 3 views of a target

Take any point x_1 in the wrist on view 1. The maneuver from view 1 to view 2 takes $x_1 \to x_2$. Let R_2 be the spinor [6] of the rotation that takes $TCP_1 \to TCP_2$ and let S_2 be the spinor of $O_1 \to O_2$. The maneuvers are

$$x_2 - a_2 = R_2^{-1}(x_1 - a_1)R_2$$
$$x_2 - b_2 = S_2^{-1}(x_1 - b_1)S_2 \ .$$

Eliminating x_2 gives

$$b_2 - a_2 = R_2^{-1}(x_1 - a_1)R_2 - S_2^{-1}(x_1 - b_1)S_2$$

which holds for arbitrary x_1, so that the spinors must be equal,

$$R_2 = S_2 \tag{1}$$

and

$$b_2 - a_2 = R_2^{-1}(b_1 - a_1)R_2 \ . \tag{2}$$

Equations (1) and (2) are the same as the basic equations of the sensor-to-wrist calibration problem described in [1, 2, 3, 4] except that they are expressed in the language of geometric algebra [6].

The references [1, 2, 3, 4] all show that equations (1) and (2) on their own are insufficient to determine a unique solution for the position and orientation of the sensor with respect to the wrist. In order to obtain a unique solution, the maneuver from view 1 to view 3 must be used to give the extra pair of equations

$$R_3 = S_3 \tag{3}$$
$$b_3 - a_3 = R_3^{-1}(b_1 - a_1)R_3 \ , \tag{4}$$

where R_3 and S_3 are the spinors for the rotational parts of the maneuvers $TCP_1 \to TCP_3$ and $O_1 \to O_3$ respectively.

The references [2, 3, 4] give various ways in which equations (1) and (3) can be solved for the rotational part of the sensor-to-wrist transformation. Once this solution has been obtained, one subtracts $b_1 - a_1$ from both sides of equations (2) and (4) to obtain

$$R_2^{-1}(b_1 - a_1)R_2 - (b_1 - a_1) = (b_2 - b_1) - (a_2 - a_1) \tag{5}$$

$$R_3^{-1}(b_1 - a_1)R_3 - (b_1 - a_1) = (b_3 - b_1) - (a_3 - a_1) . \tag{6}$$

Equations (5) and (6) together form an overdetermined linear system for the vector $b_1 - a_1$, which is the shift part of the sensor-to-wrist transformation. The inhomogeneous terms $b_k - b_1$ and $a_k - a_1$ for $k = 2, 3$ on the right hand sides of (5) and (6) are supplied by the sensor and the machine tool controller respectively. References [2, 4] describe practical measurements using wrist-mounted cameras in which the inhomogeneous sensor terms are obtained by measuring the shift of the camera relative to the target object at p (see figure 1) using

$$b_k - b_1 = (b_k - p) - (b_1 - p) . \tag{7}$$

In the writer's experience, this technique does not give a self-consistent calibration. Once the calibration has yielded the camera-to-wrist transformation, one can obtain the target-to-world transformation. One can then predict the target-to-camera transformation for any other view from the sequence target-to-world, world-to-wrist, wrist-to-camera, and compare the prediction with the vision system's measurement of target-to-camera. The writer's vision system uses geometric models of target objects. When the predicted view of the model is displayed over the actual view a shift mismatch of several pixels is obtained.

Furthermore, it is possible to infer that the experiments reported in [2, 4] also suffered from a similar amount of mismatch, although this is not explicitly stated in these papers. If one estimates the size of a pixel projected onto the target object, then the position errors reported in [2, 4] are several pixels. This suggests that these experiments also found self-inconsistency in the calibration. Clearly, one wants a calibration technique that is sufficiently accurate that mismatches between real and predicted views do not occur.

3   The New Calibration Technique

The considerations of the previous section suggest that the raw vision measurements for multiple views are mutually inconsistent. Numerical experiments showed that the orientation and line of sight parts of the target-to-camera measurements are accurate, but the measured ranges of the target from the camera are unreliable. This is because it is not possible to accurately measure the intrinsic camera parameters[1] because of the small field of view (see the Introduction). Consequently, an attempt was made to solve equations (5) and (6) using only line of sight and orientation parts of the raw vision measurements target-to-camera. This is the topic of the next subsection.

---

[1] The intrinsic camera parameters—focal length and image centre—were measured using techniques similar to those used in [4].

3.1 An Equation for the Shift which is Near Rank-Deficient

Using figure 1, introduce ray vectors along the line of sight to the target

$$v_k \propto p - b_k \text{ for } k = 1, 2, 3.$$

as the raw vision measurements. Also introduce the unknown real scale factors c^k for $k = 1, 2, 3$ and note that k is a superscript and not a power. The vectors $c^k v_k$ (the Einstein summation convention is not used in this paper) are the accurate—but currently unknown—shift parts of the target-to-camera transformations. Using figure 1, equations (5) and (6) become

$$R_2^{-1}(b_1 - a_1)R_2 - (b_1 - a_1) + c^2 v_2 - c^1 v_1 = -(a_2 - a_1) \tag{8}$$

$$R_3^{-1}(b_1 - a_1)R_3 - (b_1 - a_1) + c^3 v_3 - c^1 v_1 = -(a_3 - a_1) . \tag{9}$$

This pair of equations constitute a linear system of 6 equations with 6 unknowns. The unknowns are the 3 components of the shift $b_1 - a_1$ and the 3 scale factors c^k for $k = 1, 2, 3$.

Theoretically, it ought to be possible to solve (8) and (9) for the shift $b_1 - a_1$, and the solution ought to be accurate because it does not rely on the lengths of the raw vision measurements being accurate. However, attempts to solve (8) and (9) using the SVD failed because of near rank-deficiency. The linear system is formally of full rank because the writer has obtained an analytic solution to (8) and (9). However, in the numerical solutions, one of the singular values was always very small, with the result that small measurement errors caused the solutions to be non-physical.

3.2 The Cause of the Near Rank-Deficiency

The analytic solution to (8) and (9) proves that the equations are formally of full rank and so the geometry of the views must be the cause of the near rank-deficiency. In practice, the depth of field of the camera is small, so that the distance from the target to the camera has to be approximately constant in all the views. In addition, the field of view is small, so the optical axis of the camera is approximately pointed at the target in all the views. These practical constraints mean that the maneuvers are approximately pure rotations of the wrist/camera combination about the target. In other words $v_k \approx R_k^{-1} v_1 R_k$ for $k = 2, 3$. It shown in the following work that (8) and (9) are rank-deficient in the case of pure rotations about the target. Therefore, one expects that the equations are near rank-deficient in the practical geometry in which the maneuvers are only approximately pure rotations about the target.

In order to prove rank-deficiency, introduce the natural set of basis vectors shown in figure 2 that span the two planes of rotation for the maneuvers. Let θ_k be the rotation vector for the maneuver $TCP_1 \rightarrow TCP_k$ for $k = 2, 3$. This means that vector θ_k points along the axis of rotation of the maneuver and has magnitude equal to the rotation angle in radians. Let i be the unit pseudoscalar of 3D Euclidean space. Geometric algebra [6] writes $R_k = e^{i\theta_k/2}$.

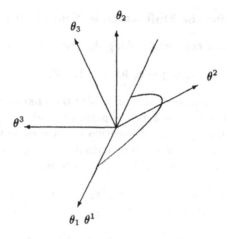

Fig. 2. Definition of basis vectors

Introduce a vector θ_1 along the line of intersection of the planes of rotation in figure 2. The definition is $\theta_1 = (\theta_2 \wedge \theta_3)i^{-1}$. The notation θ_1 is used because $\theta_1, \theta_2, \theta_3$ form a natural basis for the 3D Euclidean space. The dual basis of vectors $\theta^1, \theta^2, \theta^3$ is also needed with the property that $\theta^j \bullet \theta_k = \delta^j_k$ where δ^j_k is the Kronecker delta. Notice that θ_1 and θ^1 are the same vectors up to scale. The vector θ^2 is orthogonal to θ_1 and θ_3. Therefore θ^1 and θ^2 are mutually orthogonal vectors spanning the plane of rotation orthogonal to rotation vector θ_3. Similarly θ^3 and θ^1 are mutually orthogonal vectors spanning the plane of rotation orthogonal to rotation vector θ_2.

Write an arbitrary vector $x = x_\perp + x_\parallel$ as a sum of a rejection vector and projection vector with respect to one of the planes. In this notation $x_\perp$ is a vector along one of the rotation vectors, say θ_2 to be definite. $x_\parallel$ is then a vector in the plane of rotation orthogonal to θ_2 and can be expanded in terms of vectors θ^3 and θ^1.

Apply this splitting to (8) to get a rejection equation and a projection equation.

$$(c^2 v_2 - c^1 v_1)_\perp = -(a_2 - a_1)_\perp \qquad (10)$$

$$R_2^{-1}(b_1 - a_1)_\parallel R_2 - (b_1 - a_1)_\parallel + (c^2 v_2 - c^1 v_1)_\parallel = -(a_2 - a_1)_\parallel \qquad (11)$$

Taking the inner product of the rejection equation (10) with θ_2, and noting that in the ideal case of a pure rotation about the target $v_2 \bullet \theta_2 = v_1 \bullet \theta_2$, one finds that

$$c^2 = c^1 - \frac{(a_2 - a_1) \bullet \theta_2}{v_1 \bullet \theta_2} . \qquad (12)$$

Now substitute the pure rotation relation $v_2 = R_2^{-1} v_1 R_2$ into the projection equation (11) and also replace the scale factor c^2 using (12). The result is

$$R_2^{-1}(b_1 - a_1 + c^1 v_1)_{\|} R_2 - (b_1 - a_1 + c^1 v_1)_{\|} = \frac{(a_2 - a_1) \bullet \theta_2 R_2^{-1} v_{1\|} R_2}{v_1 \bullet \theta_2} - (a_2 - a_1)_{\|} .$$
(13)

A similar equation can be written for the maneuver from view 1 to view 3. It is

$$R_3^{-1}(b_1 - a_1 + c^1 v_1)_{\|} R_3 - (b_1 - a_1 + c^1 v_1)_{\|} = \frac{(a_3 - a_1) \bullet \theta_3 R_3^{-1} v_{1\|} R_3}{v_1 \bullet \theta_3} - (a_3 - a_1)_{\|} .$$
(14)

Equation (13) is a vector equation in the plane spanned by vectors θ^3 and θ^1 and (14) is in the plane spanned by vectors θ^1 and θ^2. Therefore, (13) and (14) are 4 scalar equations for the 3 components of the vector $(b_1 - a_1 + c^1 v_1)$. Suppose one has solved this overdetermined system and that the solution vector is q. Then

$$b_1 = a_1 + q - c^1 v_1$$
(15)

gives the position of the camera and c^1 is an unknown scale factor. The meaning of this equation is shown in figure 3. It shows that when the target T is at the centre of the maneuvers, then camera O_1 is undetermined along a line in the viewing direction from the target. This proves that (8) and (9) are rank-deficient for the ideal geometry in which the maneuvers are pure rotations about the target. The viewing direction line $c^1 v_1$ is the null-space of (8) and (9).

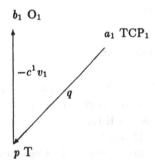

Fig. 3. Camera position

Now consider the practical geometry in which the maneuvers are only approximately pure rotations about the target so that (8) and (9) are near rank-deficient. The significance of the null space is that it does not matter very much where the camera sits in the null space because the equations will still be satisfied to a high accuracy. Although the estimated position of the camera is fairly arbitrary, figure 3 and (15) show that the vector q is in fact the shift part of the target-to-wrist transformation. In other words, the rank-deficient equations accurately fix the position of the target relative to the wrist. One can then accurately obtain the target-to-world transformation from the sequence target-to-wrist, wrist-to-world.

3.3 Handling the Rank-Deficiency

The previous subsection showed that the camera position in the null space is fairly arbitrary. Therefore, set the scale factor $c^1 = 1$ so that v_1 is taken exactly as the measurement of camera-to-target on view 1. One can then obtain a closed form solution of (8) and (9) for the shift part of the camera-to-wrist transformation using the techniques of geometric algebra [6] and by following reasoning similar to that employed in the previous subsection. The result is that the shift is written

$$b_1 - a_1 = x|\theta^1|\theta_1 + y|\theta^2|\theta_2 + +z|\theta^3|\theta_3$$

where x, y, z are scalars given by

$$z + ix = \frac{(b_2 - b_1 - a_2 + a_1) \bullet \theta^3/|\theta^3| + i(b_2 - b_1 - a_2 + a_1) \bullet \theta^1/|\theta^1|}{\exp(i|\theta_2|) - 1}$$

$$x + iy = \frac{(b_3 - b_1 - a_3 + a_1) \bullet \theta^1/|\theta^1| + i(b_3 - b_1 - a_3 + a_1) \bullet \theta^2/|\theta^2|}{\exp(i|\theta_3|) - 1}$$

and $i = \sqrt{-1}$. The solutions for the scale factors (not shown) contain $v_k \bullet \theta_k$ in the denominator for $k = 2, 3$, so one must choose the views to avoid these numbers being close to zero.

4 Results of Experiments

At the time of writing it has not been possible to conduct repeated calibrations in a systematic way in order to report statistical results of the quality of [2, 4].

The first experiments used a Hitachi KPM-1 camera with a 25mm Cosmicar lens stopped down to $F = 11$ fixed to the wrist of a JoMach 16 5-axis machine tool. The $145 \times 190mm^2$ target was grid of squares made by a photographic process on a glass plate.

A touch probe was used to measure the normal to the plate. The vision system was tricked into regarding the target as a block with a depth of 80mm by taking an image of the plate and then moving the camera 80mm along the outward normal to the plate before taking another image. The first image was processed by the vision system as the top face of the block and the second image was processed as the bottom face of the block. Two images were taken in this way for each of the three views to achieve the effect of taking 3 views of a target block of size $145 \times 190 \times 80mm^3$. The same effect could have been achieved by mounting the target plate on a z-translation stage.

After calibration, the camera-to-wrist transformation was used to predict the target-to-world transformation from the sequence target-to-camera, camera-to-wrist, wrist-to-world on view 1. The target-to-camera transformation on other views was then predicted using the sequence target-to-world, world-to-wrist, wrist-to-camera. The prediction was compared with the vision system's measurement of the target-to-camera transformation on the new view. In general, the target-to-camera transformation was not re-measured because the predicted view of the target model displayed over the actual view of the target with a mismatch of no more than one pixel. The table gives the predicted and measured transformations for view 2 and view 3 of a calibration run and the predicted and measured transformations for one other view in which the camera was pointed approximately normally to the calibration plate.

Table 1. Predicted and measured target-to-camera transformations

	roll/deg	pitch/deg	yaw/deg	x/mm	y/mm	z/mm
predicted	32.753	-28.587	-233.230	-0.253	0.754	-749.587
measured	32.752	-28.586	-233.220	-0.256	0.770	-749.439
predicted	-40.165	28.207	123.212	1.984	3.058	-813.909
measured	-40.071	28.085	123.207	1.983	3.057	-813.918
predicted	n/a	n/a	n/a	0.000	0.000	-750.000
measured	2.948	-0.038	-45.492	-0.018	-0.040	-750.037

References

1. Shiu, Y., Ahmad, S.: Calibration of Wrist-Mounted Robotic Sensors by Solving Homogeneous Transform Equations of the Form $AX = XB$. IEEE Trans. Robot. Automat. **5** (1989) 16–29
2. Tsai, R., Lenz, R.: A New Technique for Fully Autonomous and Efficient 3D Robotics Hand/Eye Calibration. IEEE Trans. Robot. Automat. **5** (1989) 345–358
3. Chou, J., Kamel, M.: Finding the Position and Orientation of a Sensor on a Robot Manipulator Using Quaternions. Int. J. Robot. Res. **10** (1991) 240-254
4. Wang, C.: Extrinsic Calibration of a Vision Sensor Mounted on a Robot. IEEE Trans. Robot. Automat. **8** (1992) 161–175
5. Li, M., Betsis D.: Hand-eye calibration. Proc. Int. Conf. on Computer Vision, Boston, MA (1995) 40–46
6. Hestenes, D.: New Foundations for Classical Mechanics (Kluwer 1986)

Automated Camera Calibration and 3D Egomotion Estimation for Augmented Reality Applications

Dieter Koller[1,2,3], Gudrun Klinker[1], Eric Rose[1], David Breen[4], Ross Whitaker[5], and Mihran Tuceryan[6]

[1] Fraunhofer Project Group for AR at ZGDV, Arabellastr.17 (at ECRC), 81925 Munich, Germany
[2] EE Dept., California Inst. of Technology, MC 136-93, Pasadena, CA 91125
[3] Autodesk, Inc., 2465 Latham St., Suite 101, Mountain View, CA 94040
[4] Computer Graphics Lab., California Inst. of Technology, MC 348-74, Pasadena, CA 91125
[5] EE Dept., 330 Ferris Hall, U. of Tennessee, Knoxville, TN 37996-2100
[6] Dept of Comp & Info Science, IUPUI, 723 W. Michigan St, Indianapolis, IN 46202-5132
Email: dieter.koller@autodesk.com

Abstract. This paper addresses the problem of accurately tracking the 3D motion of a monocular camera in a known 3D environment and dynamically estimating the 3D camera location. For that purpose we propose a fully automated landmark-based camera calibration method and initialize a motion estimator, which employes extended Kalman filter techniques to track landmarks and to estimate the camera location at any given time. The implementation of our approach has been proven to be efficient and robust and our system successfully tracks in real-time at approximately 10 Hz. We show tracking results of various augmented reality scenarios.

1 Introduction

Augmented reality (AR) is a technology in which a user's view of the *real* world is enhanced or augmented with additional information generated by a computer. The enhancement may consist of rendered virtual geometric objects placed into the environment, or a display of non-geometric information about existing real objects. Using AR technology, users can interact with a mixed virtual and real world in a natural way.

This paradigm for user interaction and information visualization provides a promising new technology for many applications. AR is being explored within a variety of scenarios. The most active application area is medicine, where AR is used to assist surgical procedures by aligning and merging medical images into video [1; 2; 3; 4]. For manufacturing AR is being used to direct workers wiring an airplane [5]. In telerobotics AR provides additional spatial information to the robot operator [6]. AR may also be used to enhance the lighting of an architectural scene [7], as well as, provide part information to a mechanic repairing an engine [8]. For interior design AR may be used to arrange virtual furniture in a real room [9].

1.1 Technical Contribution

A video-based AR system can be regarded as having two cameras: a real one generating live video of the real environment, and a virtual one producing 3D graphics to be merged with the live video stream. Both cameras must have the same internal and external parameters in order for the real and virtual objects to be properly aligned. This is accomplished by an an initial calibration of the real camera and a dynamic update of its external parameters. From the vision point of view, this is one of the most challenging technical problems that needs to be addressed in order to produce a useful and convincing video-based augmented reality system ([10]).

More traditional augmented reality approaches employ magnetic tracking devices for sensing position and orientation of a moving camera (e.g. [8; 11]). They suffer, however, from (a) limited range (3–5m), (b) interference with ferromagnetic objects of the environment, and (c) lack of portability.

We therefore focused primarily on vision-based algorithms and decided on using landmark-based calibration and tracking to make it more tractable. We are using corners of rectangular patterns attached to a wall as landmarks and track them using extended Kalman filter techniques based on an acceleration-free constant angular velocity and constant linear acceleration motion model. We demonstrate the robustness and accuracy of our tracker within various augmented reality application.

1.2 Related Work

A number of groups have explored the topic of camera tracking for augmented reality. Some researcher [12; 13] have argued that a simple view based, calibration free approach for real-time visual object overlay is sufficient. This is definitely true for certain applications, where no direct metric informations is necessary. *Interactive* AR, however, requires the more complex pose calculation based approach, which allows the decomposition of the image transformation into camera/object pose and the full perspective projection in order to calculate 3D colllision detection and interaction between real and virtual objects, as in [14].

A similar argument is being used in [15] in the context of enhanced reality in medicine [4], where near real-time calibration is performed for each frame based on a few fiducial marks. In [11; 16] a hybrid vision and magnetic system is used to improve the accuracy of tracking a camera over a wide range of motions and conditions.

Tracking known objects in 3D space and ego-motion estimation (camera tracking) have a long history in computer vision (e.g. [17; 18; 19; 20]). Constrained 3D motion estimation is being applied in various robotics and navigation tasks. Much research has been devoted to estimating 3D motion from optical flow fields (e.g. [21]) as well as from discrete moving image features like corners or line segments (e.g. [22; 23; 24]), often coupled with structure-from-motion estimation, or using more than two frames (e.g. [25]). The theoretical problems seem to be well understood, but robust implementation is difficult. The development of our tracking approach and the motion model has mainly been influenced by the one described in [20].

1.3 Outline of the Paper

We start by describing our automated camera calibration procedure in Section 2. In Section 3 we explain the motion model employed by our Kalman filter. The Kalman filter based tracking procedure is then outlined in Section 4. We finally present some results in Section 5 and close with a conclusion in Section 6.

2 Automated Camera Calibration

We propose an automated calibration procedure which determines internal parameters (focal length and focal center according to the standard pinhole camera model) and external camera parameters (camera pose). In the current implementation the internal parameters are fixed during a session and only the external parameters are estimated.

A highly precise camera calibration is required for a good initialization of the tracker. For that purpose we propose a two step calibration procedure, in which we attempt to find the image locations of landmarks placed in the 3D environment at known 3D locations (cf. Figure 3). This addresses the trade-off between high precision calibration and minimal or no user interaction. We use dark rectangular card board as landmarks. In the first step we perform an initial camera calibration based on the image location of the centroids of dark blobs (landmarks) which we extract in the image. This bootstraps the second step consisting of a constraint search for additional image features (corners); thus improving the calibration. We are using the camera calibration algorithm described in [26] and implemented in [27].

The next subsection describes our algorithm for finding dark image blobs. The constrained search for projected model squares is addressed in the context of acquiring measurements for the Kalman filter in Subsection 4.2.

2.1 Finding Dark Image Blobs

The algorithm for finding dark blobs in the image is based on a *watershed* transformation, a morphological operation which decomposes the whole image into connected regions (*puddles*) divided by watersheds (cf. [28]).

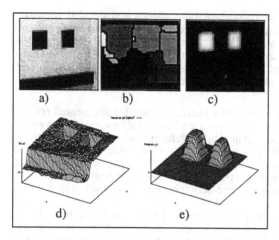

Fig. 1: (a) Subimage with dark squares, (b) watershed transformation with greycoded regions (watershed are drawn in black), (c) greyscale inside operation, measuring the depth of *puddles*, (d) and (e) show 3D plots of images (a) and (c), respectively.

Using this transformation a dark blob surrounded by a bright area provides a strong filter response related to the depth of the *puddle* (cf. Fig. 1). The *deepest* and most compact blobs (puddles) are then matched against the known 3D squares. For this purpose, the squares contain one or more small red squares at known positions, representing binary encodings of the identification numbers of the model squares. The red squares are barely visible in the green and blue channels of the video camera. Thus we can apply a simple variant of a region growing algorithm to the green color channel to determine the borders of each black square. After fitting straight lines to the border, we sample each black square in the red color channel at the supposed locations of the internal red squares to obtain the bit pattern representing the model id. Blobs with invalid identification numbers or with multiple assignments of the same number are discarded. Using this scheme, the tracker can calibrate itself even when some of the model squares are occluded or outside the current field of view (see Figure 6 a)).

3 Motion Model for Rigid Body Motion

Tracking can be stated as estimating the motion parameters according to a known motion model. Our application scenario suggests a fairly irregular camera and object motion within all 6 degrees of freedom[1]. We assume an acceleration free motion of the camera due to the lack of *a priori* knowledge about the forces changing the motion. It is well known that in this case a general motion can be decomposed into a constant translational velocity v_c of the objects centroid c, and a rotation with constant angular velocity ω around an axis through the centroid (of the camera). (cf. Figure 2 and [29]).

The motion equation of a camera point p wrt. world coordinates is then given by: $\dot{p} = v_c + \omega \times (p - c)$. In this equation the constant rotation is wrt. the moving controid and hence no motion invariant wrt. to the world coordinate frame. Instead, the center of rotation itself is moving with $c(t) = c(t_0) + v_c(t - t_0)$. Substituting this into the motion equation produces:

$$\dot{p}(t) = v + \omega \times p + a\,t \qquad (1)$$

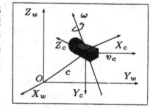

Fig. 2: The world (X_w, Y_w, Z_w) and camera (X_c, Y_c, Z_c) coordinate frames, and translational (v_c) and rotational (ω) velocities.

with $v(t_0) = v_c - \omega \times c(t_0)$ and $a = -\omega \times v_c = $ const. The rotation is now with respect to world coordinates. However, an additional acceleration term a is added. The integration yields (cf. [20; 30]):

$$p(t + \Delta t) = R(\theta)\,p + S(\theta)\,v\,\Delta t + T(\theta)\,a\,\left(\tfrac{\Delta t}{2}\right)^2,$$

---

[1] In an AR application the camera can be hand held or even head mounted so the user is free to move the camera in any direction.

with

$$R(\theta) = I_3 + \frac{\sin\theta}{\theta}\,\Theta + \frac{1-\cos\theta}{\theta^2}\,\Theta^2 = e^\Theta, \quad S(\theta) = I_3 + \frac{1-\cos\theta}{\theta^2}\,\Theta + \frac{\theta - \sin\theta}{\theta^3}\,\Theta^2$$

$$T(\theta) = I_3 + 2\frac{\theta - \sin\theta}{\theta^3}\,\Theta + \frac{\theta^2 - 2(1-\cos\theta)}{\theta^4}\,\Theta^2$$

and $\theta = \omega\,\Delta t = (\theta_x, \theta_y, \theta_z)$, $\theta = ||\theta||$. Θ is the skew-symmetric matrix to vector θ:

4 Tracking and Egomotion Estimation

With tracking our system is able to cope with dynamic scene changes and camera motions. Our tracking approach currently uses the corners of squares attached to moving objects or walls (cf. Figure 3), which have already been used for camera calibration.

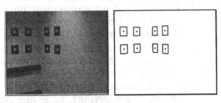

Fig. 3: Our vision-based tracking approach currently tracks the corners of squares. Left: image with eight squares. Right: detected squares only.

Once a complete camera calibration has been performed as described in Section 2, we can switch to the tracking phase, i.e., update the pose and motion parameters of the camera by keeping the internal camera parameter constant. We employ extended Kalman filter (EKF) techniques for optimal pose and motion estimation using the motion described in Eqn. 1.

4.1 Extended Kalman Filter

Our state vector s of the Kalman filter comprises the following 15 components: the position vector t, the rotation vector ϕ, the translational and angular velocity v and ω, respectively, and the translational acceleration a: $s = \{t, \phi, v, \omega, a\}$.

The extended Kalman filter (EKF) equations can be found in most related textbooks, e.g., [31]. We just want to add an implementation note: the standard Kalman filter calculates the gain in conjunction with a recursive computation of the state covariance. This requires a matrix inversion of the dimension of the measurement vector, which can be large as in our application. However, the matrix inversion can be reduced to one of the state dimensions using the *information matrix* formalism. The *information filter* recursively calculates the inverse of the covariance matrix (= information matrix) (cf. [32]):

$$P_k^{+\,-1} = P_k^{-\,-1} + H_k^T R_k^{-1} H_k, \tag{2}$$

where P_k^+ denotes the updated covariance matrix, P_k^- the prediction, H_k the jacobian of the measurement function, and R_k the measurement noise matrix, each at time k. The update equation for the state $\hat{s}_k^+$ then becomes:

$$\hat{s}_k^+ = \hat{s}_k^- + K_k(z_k - h_k(\hat{s}_k^-)) \quad \text{with} \quad K_k = \left(P_k^{-\,-1} + H_k^T R_k^{-1} H_k\right)^{-1} H_k^T R_k^{-1}, \tag{3}$$

which requires the inverse of the updated covariance matrix P_k^+ of Eqn 2. Inverting R_k is straightforward since we assume independent measurements producing a diagonal measurement noise matrix R_k. The transition equation (prediction) can be found in [30].

4.2 Kalman Filter Measurements

Currently we use the image positions of corners of squares as measurements, i.e., our $8 \cdot n$ dimensional measurement vector z comprises the x and y image positions of all of the vertices (corners) of the n squares we find in the image. A measurement z is mapped to the state s by means of the measurement function h: $z = h(s)$.

The image corners are extracted in a multi-step procedure outlined below and in Figure 4. Assume that we are looking for the projection $p_i = l_j \cap l_k$ of the model vertex $v_i = m_j \cap m_k$ which is given by the intersection of the model lines m_j, and m_k (l_j and l_k are the image projections of the model lines m_j and m_k).

- Predict image locations for model lines m_j and m_k.
- Subsample these predicted lines (e.g., into 5 to 10 sample points).
- Find the maximum gradient normal to the line at each of those sample points using a search distance given by the state covariance estimate. We use only 8 possible directions and extract the maximum gradient with sub-pixel accuracy.
- Fit a new line l_j to the extracted maximum gradient points corresponding to the predicted model line m_j.
- Find the final vertex $p_i = l_j \cap l_k$ by intersecting the correspondent lines l_j and l_k.

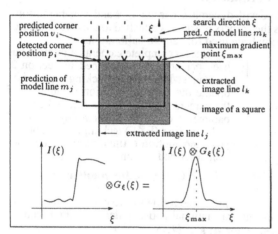

Fig. 4: We use corners as intersections of matched line segments as measurements These line segments are fitted from maximum gradient points which are produced from a one dimensional convolution with a derivative of a gaussian kernel G_ξ normal to the projection of the image line (ξ is a parameterization normal to the line and I is the image intensity).

This procedure allows us to obtain precise image locations without going through lengthy two-dimensional convolutions. Furthermore we use a measurement noise calculated from the covariance of the line segment fitting process. The failure to find certain vertices is detected and indicates either an occlusion not covered by the occlusion reasoning step (described in the next subsection), or a motion not covered by our motion model.

4.3 Occlusion Reasoning and Re-Initialization

Tracking is performed on certain artificial landmarks in the scene. Their visibility can only be corrupted if their image projection falls outside the field of view of the camera or if they are occluded by other *real* objects. Full 3D occlusion reasoning requires monitoring all moving objects and knowledge about the entire 3D geometry, a feature not yet implemented in our system. Figure 6 (a) illustrates an occlusion example. We are currently investigating the use of additional features, such as arbitrary corners or edges which will be added once the tracker has been initialized from the known landmarks.

Failure to find certain landmarks is indicated by a very large measurement noise. Such *unreliable* landmark points are discounted by the Kalman filter. If too many landmark points are labelled as *unreliable*, the tracker re-initializes itsself by re-calibration.

5 Results

The system is currently implemented on Silicon Graphics workstations using SGI's ImageVision and VideoLibrary as well as Performer and OpenGL. It successfully tracks landmarks and estimates camera parameters at approximately 10 Hz with a live PAL-size video stream on a Silicon Graphics Indy.

Our landmarks are black cardboard squares placed on a wall, as seen in Figures 5–6. In a first set of experiments with an initial version of our tracker we recorded an image sequence from a moving camera pointing at the wall. Virtual furniture is then overlayed according to the estimated camera parameter (cf. Figure 5). Since we have a 3D representation of the room and the camera, we are able to perform collision detection between the furniture and the room [14]. The user places the virtual furniture in the augmented scene by interactively pushing it to the wall until a collision is detected. The AR system then automatically lowers the furniture until it rests on the floor.

Fig. 5: Images of a sequence with overlayed virtual furniture. The estimated position of the world coordinate axes is also overlayed on the room corner.

Figure 6 shows various screen-shots from the video screen of the system running in real-time. The figures also exhibit some possible AR applications: 6 b) shows an additional virtual room divider and a reference floor grid; 6 c) visualizes the usual invisible electrical wires inside the wall; 6 d) shows the fire escape routes; 6 e) a (red) arrow (right) shows where to find the fire alarm button, and 6 f) explicitly shows the fire hose as a texture mapped photo of the inside of a cabinet.

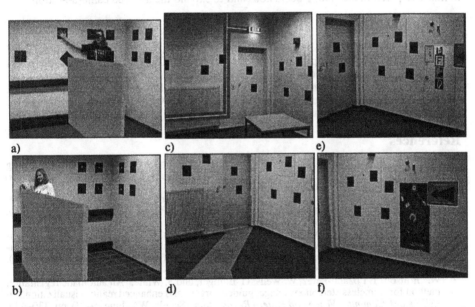

Fig. 6: (a) We successfully track with partial occlusion as long as at least two landmarks (squares) are visible. Models of the occluded landmark as well as a virtual divider have been overlayed to the video. The next images exhibit various AR applications: (b) a virtual room divider and floor grid, (c) electric wires inside the wall, (d) a fire escape route is being shown, (e) a (red) arrow (right) shows where to find the fire alarm button, (f) like (e), but a texture mapped photo of the inside of a cabinet has been superimposed on the cabinet door.

6 Conclusion

In this paper we addressed two major problems of AR applications: (a) the precise alignment of real and virtual coordinate frames for overlay, and (b) capturing the 3D motion of a camera including camera position estimates for each video frame. The latter is especially important for interactive AR applications, where users can manipulate virtual objects in an augmented real 3D environment. This problem has not been tackled successfully before using only video-input measurements.

Intrinsic and extrinsic camera parameters of a real camera are estimated using an automated camera calibration procedure based on landmark detection. These parameter sets are used to align and overlay computer generated graphics of virtual objects onto live video. Since extrinsic camera parameters are estimated separately the virtual objects can be manipulated and placed in the real 3D environment including collision detection with the room boundary or other objects in the scene. We furthermore apply extended Kalman filter techniques for estimating the motion of the camera and the extrinsic camera parameters. Due to the lack of knowledge about the camera movements produced by the user, we simply impose an acceleration-free constant angular velocity and constant linear acceleration-motion to the camera. Angular accelerations and linear jerk caused by the user moving the camera are successfully modeled as process noise.

Robustness has been achieved by using model-driven landmark detection and landmark tracking instead of pure data-driven motion estimation. Real-time performance on an entry level Silicon Graphics workstation (SGI Indy) has been achieved by carefully evaluating each processing step and using lightweight landmark models as tracking features, as well as, well designed image measurement methods in the Kalman filter. The system successfully tracks landmarks and estimates camera parameters at approximately 10 Hz with a live PAL-size video stream on a Silicon Graphics Indy.

Future work will include a fusion of model- and data-driven feature tracking in order to improve performance along occlusions and to expand the allowed camera motion.

7 Acknowledgments

We would like to thank K. Ahlers, C. Crampton, and D.S. Greer of the former UI&V group of ECRC for their help in building our AR system. One of us (D.K.) acknowledges P. Perona (Caltech) for financial support during his stay at Caltech.

This research has been financially supported in part by Bull SA, ICL Plc, Siemens AG, and by the European Community under ACTS Project # AC017 (Collaborative Integrated Communications for Construction).

References

1. M. Bajura, H. Fuchs, and R. Ohbuchi, "Merging virtual objects with the real world: Seeing ultrasound imagery within the patient," in *Computer Graphics (SIGGRAPH '92 Proceedings)* (E. E. Catmull, ed.), vol. 26(2), pp. 203–210, July 1992.
2. W. Lorensen, H. Cline, C. Nafis, R. Kikinis, D. Altobelli, and L. Gleason, "Enhancing reality in the operating room," in *Visualization '93 Conference Proceedings*, (Los Alamitos, CA), pp. 410–415, IEEE Computer Society Press, October 1993.
3. A. State, M. Livingston, W. Garrett, G. Hirota, M. Whitton, E. Pisano, and H. Fuchs, "Technologies for augmented reality systems: Realizing ultrasound-guided needle biopsies," in *Computer Graphics Proceedings, Annual Conference Series:* SIGGRAPH '96 (New Orleans, LA), pp. 439–446, ACM SIGGRAPH, New York, August 1996.
4. W. Grimson, T. Lozano-Perez, W. Wells, G. Ettinger, and S. White, "An automatic registration method for frameless stereotaxy, image, guided surgery and enhanced reality visualization," in *IEEE Conf. Computer Vision and Pattern Recognition*, (Seattle, WA, June 19-23), pp. 430–436, 1994.
5. T. Caudell and D. Mizell, "Augmented reality: An application of heads-up display technology to manual manufacturing processes," in *Proceedings of Hawaii International Conference on System Sciences*, pp. 659–669, January 1992.
6. P. Milgram, S. Zhai, D. Drascic, and J. Grodski, "Applications of augmented reality for human-robot communication," in *Proceedings of IROS '93: International Conference on Intelligent Robots and Systems*, (Yokohama, Japan), pp. 1467–1472, July 1993.
7. C. Chevrier, S. Belblidia, and J. Paul, "Composing computer-generated images and video films: An application for visual assessment in urban environments," in *Computer Graphics: Developments in Virtual Environments (Proceedings of CG International '95 Conference)*, (Leeds, UK), pp. 115–125, June 1995.
8. E. Rose, D. Breen, K. Ahlers, C. Crampton, M. Tuceryan, R. Whitaker, and D. Greer, "Annotating real-world objects using augmented reality," in *Computer Graphics: Developments in Virtual Environments (Proceedings of CG International '95 Conference)*, (Leeds, UK), pp. 357–370, June 1995.

9. K. Ahlers, A. Kramer, D. Breen, P. Chevalier, C. Crampton, E. Rose, M. Tuceryan, R. Whitaker, and D. Greer, "Distributed augmented reality for collaborative design applications," in *Eurographics '95 Proceedings*, (Maastricht, NL), pp. 3–14, Blackwell Publishers, August 1995.

10. G. Klinker, K. Ahlers, D. Breen, P.-Y. Chevalier, C. Crampton, D. Greer, D. Koller, A. Kraemer, E. Rose, M. Tuceryan, and R. Whitaker, "Confluence of computer vision and interactive graphics for augmented reality," *Presence: Teleoperators and Virtual Environments (Special issue on Augmented Reality)*, January 1997.

11. A. State, M. Livingston, W. Garrett, G. Hirota, M. Whitton, E. Pisano, and H. Fuchs, "Superior augmented reality registration by integrating landmark tracking and magnetic tracking," in *Computer Graphics Proceedings, Annual Conference Series:* SIGGRAPH '96 (New Orleans, LA), pp. 429–438, ACM SIGGRAPH, New York, August 1996.

12. M. Uenohara and T. Kanade, "Vision-based object registration for real-time image overlay," *Computers in Biology and Medicine*, vol. 25, pp. 249–260, March 1995.

13. K. Kutulakos and J. Vallino, "Affine object representations for calibration-free augmented reality," in *Virtual Reality Ann. Int'l Symposium (VRAIS '96)*, pp. 25–36, 1996.

14. D. Breen, R. Whitaker, E. Rose, and M. Tuceryan, "Interactive occlusion and automatic object placement for augmented reality," in *Eurographics '96 Proceedings*, (Poitiers, France), pp. 11–22, Elsevier Science Publishers B.V, August 1996.

15. J. Mellor, "Realtime camera calibration for enhanced reality visualization," in *First Int'l Conf. on Computer Vision, Virtual Reality and Robotics in Medicine (CVRMed)*, (Nice, France, April 3–6, 1995, N. Ayache (ed.), Lecture Notes in Computer Science **905**, Springer-Verlag, Berlin, Heidelberg, New York), 1995.

16. M. Bajura and U. Neumann, "Dynamic registration correction in video-Based augmented reality systems," *IEEE Computer Graphics and Applications*, vol. 15, pp. 52–61, September 1995.

17. D. Gennery, "Tracking known three-dimensional objects," in *Proc. Conf. American Association of Artificial Intelligence*, (Pittsburgh, PA, Aug. 18-20), pp. 13–17, 1982.

18. D. Lowe, "Robust model-based motion tracking through the integration of search and estimation," *International Journal of Computer Vision*, vol. 8, no. 2, pp. 113–122, 1992.

19. D. Gennery, "Visual tracking of known three-dimensional objects," *International Journal of Computer Vision*, vol. 7, pp. 243–270, 1992.

20. Z. Zhang and O. Faugeras, *3D Dynamic Scene Analysis*. No. 27 in Springer Series in Information Science, Springer-Verlag, Berlin, Heidelberg, New York, London, Paris, Tokyo, 1992.

21. G. Adiv, "Determining 3-d motion and structure from optical flow generated by several moving objects," *IEEE Transactions on Pattern Analysis and Machine Intelligence*, vol. PAMI-7, pp. 384–401, 1985.

22. T. Huang, "Determining three-dimensional motion and structure from perspective views," in *Handbook of Pattern Recognition and Image Processing*, pp. 333–354, 1986.

23. T. Broida, S. Chandrashekhar, and R. Chellappa, "Recursive 3-d motion estimation from a monocular image sequence," *IEEE Trans. Aerospace and Electronic Systems*, vol. 26, pp. 639–656, 1990.

24. Z. Zhang, "Estimating motion and structure from correspondences of line segments between two perspective images," *IEEE Transactions on Pattern Analysis and Machine Intelligence*, vol. 17, pp. 1129–1139, December 1995.

25. H. Shariat and K. Price, "Motion estimation with more then two frames," *IEEE Transactions on Pattern Analysis and Machine Intelligence*, vol. PAMI-12, pp. 417–434, 1990.

26. J. Weng, P. Cohen, and M. Herniou, "Calibration of stereo cameras using a non-linear distortion model," in *Proc. Int. Conf. on Pattern Recognition*, (Atlantic City, NJ, June 17-21), pp. 246–253, 1990.

27. M. Tuceryan, D. Greer, R. Whitaker, D. Breen, C. Crampton, E. Rose, and K. Ahlers, "Calibration requirements and procedures for a monitor-based augmented reality system," *IEEE Transactions on Visualization and Computer Graphics*, vol. 1, no. 3, pp. 255–273, 1995.

28. J. Barrera, J. Banon, and R. Lotufo, "Mathematical morphology toolbox for the khoros system," in *Conf. on Image Algebra and Morphological Image Processing V, International Symposium on Optics, Imaging and Instrumentation, SPIE's Annual Meeting*, (24-29 July, 1994, San Diego, USA), 1994.

29. H. Goldstein, *Classical Mechanics*. Reading, MA: Addison-Wesley Press, 1980.

30. D. Koller, "A robust vision-based tracking technique for augmented reality applications." in preparation, 1997.

31. A. Gelb, ed., *Applied Optimal Estimation*. Cambridge, MA: MIT Press, 1974.

32. Y. Bar-Shalom and X.-R. Li, *Estimation and Tracking: Principles, Technuques, and Software*. Boston, London: Artech House, 1993.

Optimally Rotation-Equivariant Directional Derivative Kernels*

Hany Farid[1] and Eero P. Simoncelli[2]

[1] University of Pennsylvania, Philadelphia PA 19104-6228, USA
[2] New York University, New York, NY 10012, USA

Abstract. We describe a framework for the design of directional derivative kernels for two-dimensional discrete signals in which we optimize a measure of rotation-equivariance in the Fourier domain. The formulation is applicable to first-order and higher-order derivatives. We design a set of compact, separable, linear-phase derivative kernels of different orders and demonstrate their accuracy.

1 Introduction

A wide variety of algorithms in multi-dimensional signal processing are based on the computation of directional derivatives. For example, gradient measurements are used in computer vision as a first stage of many edge detection, depth-from-stereo, and optical flow algorithms. The motivation for such decompositions usually stems from a desire to locally characterize signals using Taylor series expansions (e.g., [7]).

Derivatives of discretely sampled signals are often computed as differences between neighboring sample values. This type of differentiation arises naturally from the definition of continuous derivatives, and is reasonable when the spacing between samples is well below the Nyquist limit. For example, it is used throughout the numerical analysis literature, where one typically has control over the sample spacing. But such differences are poor approximations to derivatives when the distance between samples is large and cannot be adjusted.

In the digital signal processing community, there has been a fair amount of work on the design of discrete differentiators (see e.g., [9]). This work is usually based on approximating the derivative of a continuous sinc function. The difficulty with this approach is that the resulting kernels typically need to be quite large in order to be accurate.

In addition to the difficulties described above, these two primary methods of differentiation are not well suited for multi-dimensional differentiation. In particular, one often relies on the linear-algebraic properties of multi-dimensional derivatives (gradients) that allow differentiation in arbitrary directions via linear combinations of separable axis derivatives[3]. In the computer vision literature,

---

* This work was supported by ARO Grant DAAH04-96-1-0007, DARPA Grant N00014-92-J-1647, NSF Grant SBR89-20230, and NSF CAREER Grant MIP-9796040 to EPS.

[3] For example, the derivative operator in the direction of unit vector $\hat{u}$ is $u_x \frac{\partial}{\partial x} + u_y \frac{\partial}{\partial y}$.

many authors have used sampled Gaussian derivatives which exhibit better approximations to these algebraic properties than simple differences, but less computationally expensive than sinc functions. Danielsson [3] compared a number of derivative kernels and concluded that the Sobel operators exhibited the most rotation-equivariant behavior. Freeman and Adelson [8] characterized the complete class of rotation-equivariant kernels and termed these "steerable" filters.

We are interested in the optimal design of small separable kernels for efficient discrete differentiation. In previous work [10, 1], we described design techniques for matched pairs of one-dimensional kernels (a lowpass kernel and a differentiator) suitable for multi-dimensional differentiation. Axis derivatives were computed by applying the differentiator along the axis of choice and the lowpass kernel along all remaining axes. The error functional was a weighted least-squares error in the Fourier domain between the differentiator and the derivative of the lowpass kernel. In this paper, we generalize these notions to form a two-dimensional error functional that expresses the desired property of derivative kernels discussed above. This error functional is then minimized to produce a set of optimally rotation-equivariant derivative kernels.

2 Differentiation of Discrete Signals

Differentiation is an operation defined on continuous functions. The computation of derivatives on a discretely sampled function thus requires (at least implicitly) an intermediate interpolation step. The derivative of this interpolated continuous function is then re-sampled at the points of the original sampling lattice.

2.1 Example: Ideal Interpolation

To make this more precise, consider the classical situation in which the sampled function is assumed to have been formed by uniformly sampling a continuous function at the Nyquist rate. In this case, the correct interpolation of the discrete function $f[\cdot]$ is:

$$\overline{f}(x,y) = \sum_{k,l} f[k,l] \cdot c(x - kT, y - lT), \tag{1}$$

where T is the sample spacing (assumed to be identical along both the x and y axes), $\overline{f}(x,y)$ is the interpolated continuous function, and the interpolation function $c(x,y)$ is a separable product of ideal lowpass ("sinc") functions, $c(x,y) = s_T(x)\, s_T(y) = \frac{\sin(\pi x/T)}{\pi x/T}\, \frac{\sin(\pi y/T)}{\pi y/T}$. Assuming that the sum in Equation (1) converges uniformly, we can differentiate both sides of the equation. Without loss of generality, consider the partial derivative with respect to x:

$$D_x\{\overline{f}\}(x,y) = \sum_{k,l} f[k,l] \cdot D_x\{c\}(x - kT, y - lT), \tag{2}$$

where $D_x\{\cdot\}$ indicates a functional that computes the partial derivative of its argument in the x direction. Note that the derivative operator is only being applied to continuous functions, $\overline{f}$ and c.

One arrives at a definition of the derivative of the discrete signal by sampling both sides of the above equation on the original sampling lattice:

$$D_x\{\overline{f}\}(x,y)|_{x=nT,y=mT} = \sum_{k,l} f[k,l] \cdot D_x\{c\}((n-k)T,(m-l)T)$$

$$= \sum_{k,l} f[k,l] \cdot D\{s_T\}((n-k)T) \cdot s_T((m-l)T)$$

$$= \sum_{k,l} f[k,l] \cdot d_T[n-k] \cdot \delta[m-l], \qquad (3)$$

where $d_T[\cdot]$ is the T-sampled sinc derivative, and $\delta[\cdot]$ is the T-sampled sinc (i.e., a Kroenecker delta function). Note that the right side of this expression is a convolution of the discretely sampled function, $f[\cdot]$, with the separable kernel $d_T[n-k]\delta[m-l]$. The continuous interpolation need never be performed.

If the original function was sampled at the Nyquist rate, then convolution with the sampled derivative of the sinc function will return an exact sampled derivative. In practice, however, the coefficients of this kernel decay very slowly and accurate implementation requires very large kernels. In addition, the sinc derivative operator has a large response at high frequencies, making it fragile in the presence of noise.

2.2 Alternative Interpolation Functions

The limitations of the sinc function lead us to consider alternative interpolation functions. Of course, if an interpolator other than the sinc function is used, the resulting derivative may not be that of the original continuous function. However, for many applications this is not a fundamental concern. Consider, for example, the problem of determining the local orientation of an edge. This can be achieved by measuring the gradient vector, which is perpendicular to the edge. If we use an interpolation kernel which preserves the structure of the edge, the gradient direction will still provide the desired information.

Since the separability of the sinc is desirable for computational efficiency (e.g., [5, 4]), we will consider an interpolator that retains this property, and will also assume that the two axes should be treated identically. Thus, the two-dimensional interpolator is written as a separable product, $c(x,y) = d_0(x) \cdot d_0(y)$. The partial derivative (with respect to x) of this interpolator is:

$$D_x\{c\}(x,y) = d_1(x) \cdot d_0(y), \qquad (4)$$

where $d_1(x)$ is the derivative of $d_0(x)$. With this interpolator, the sampled derivative (as in Equation (3)) becomes:

$$D_x\{\overline{f}\}(x,y)|_{x=nT,y=mT} = \sum_{k,l} f[k,l] \cdot d_1((n-k)T) \cdot d_0((m-l)T)$$

$$= \sum_{k,l} f[k,l] \cdot d_1[n-k] \cdot d_0[m-l]. \qquad (5)$$

The discrete derivatives are computed using *two* discrete one-dimensional kernels, $d_0[\cdot]$ and $d_1[\cdot]$, which are the T-sampled versions of $d_0(\cdot)$ and $d_1(\cdot)$, respectively. Note that the separability of the interpolator is retained in the derivative operator. As with the sinc function, we need never make explicit the underlying continuous function $d_0(\cdot)$.

At this point, we could simply choose a continuous function $d_0(\cdot)$, compute its derivative $d_1(\cdot)$, and T-sample the two functions. For example, it is common in computer vision to use a sampled Gaussian and its derivative. However, because the Gaussian is not strictly bandlimited, sampling introduces artifacts, thus destroying the derivative relationship between the resulting kernels. So, instead we choose to simultaneously design a pair of *discrete* kernels that optimally preserve the required derivative relationship.

To design such a pair of discrete kernels, we must state the differential relationship between them. Previously [1], pairs of kernels, $d_0[n]$ and $d_1[n]$, were designed such that their Fourier transforms approximate the correct derivative relationship:

$$j\omega D_0(\omega) = D_1(\omega), \qquad -\pi < \omega < \pi, \tag{6}$$

where capitalized functions correspond to the (continuous but periodic) discrete-space Fourier transform, $D_0(\omega) = \sum_n d_0[n]e^{-j\omega n}$. This constraint states that the derivative (in the Fourier domain) of the kernel $d_0[n]$ is equal to the kernel $d_1[n]$. For the current paper, we wish to impose a similar constraint in the two-dimensional Fourier domain. In particular, the derivative in an arbitrary direction (specified by a unit-vector $\hat{u}$) is:

$$\mathcal{D}_{\hat{u}}\{f\}(x,y) = u_x \mathcal{D}_x\{f\}(x,y) + u_y \mathcal{D}_y\{f\}(x,y), \tag{7}$$

where (u_x, u_y) are the components of $\hat{u}$. Thus, the two-dimensional version of the Fourier domain constraint of Equation (6) is:

$$j(\hat{u} \cdot \hat{\omega})D_0(\omega_x)D_0(\omega_y) = [u_x D_1(\omega_x)D_0(\omega_y) + u_y D_0(\omega_x)D_1(\omega_y)], \tag{8}$$

and should hold for $-\pi < \{\omega_x, \omega_y\} < \pi$ and for all unit vectors $\hat{u}$ (i.e., for all directions). We can now define a weighted least-squares error functional by integrating over these variables:

$$E\{D_0, D_1\} = \int_{\hat{\omega}} W(\hat{\omega}) \cdot \int_{\hat{u}} [j(\omega_x u_x + \omega_y u_y)D_0(\omega_x)D_0(\omega_y) -$$
$$(u_x D_1(\omega_x)D_0(\omega_y) + u_y D_0(\omega_x)D_1(\omega_y))]^2, \tag{9}$$

where $W(\hat{\omega})$ is a weighting function. In order to avoid the trivial (zero) solution, we impose a constraint that the interpolator have unit response at D.C. (i.e., the kernel $d_0[n]$ has unit sum): $D_0(0) = 1$.

2.3 Higher-Order Derivatives

Higher-order derivative kernels may be designed using a similar strategy to that introduced in the previous section. In particular, the Nth-order directional derivative in direction $\hat{u}$ is:

$$\mathcal{D}_{\hat{u}}^{N}\{\overline{f}\}(x,y) = [u_x\mathcal{D}_x + u_y\mathcal{D}_y]^N\{\overline{f}\}(x,y)$$

$$= \sum_{p=0}^{N} b[p,N]u_x^p\, u_y^{(N-p)}\, \mathcal{D}_x^p\mathcal{D}_y^{(N-p)}\{\overline{f}\}(x,y). \quad (10)$$

where $b[p,N] = N!/p!(N-p)!$ is the binomial coefficient. Combining this definition with the interpolation of Equation (2), and sampling both sides gives an expression for the discrete Nth-order directional derivative:

$$\mathcal{D}_{\hat{u}}^{N}\{\overline{f}\}(x,y)|_{x=nT,y=mT} = \sum_{p=0}^{N} b[p,N]u_x^p\, u_y^{(N-p)}.$$

$$\sum_{k,l} f[k,l]d_p[n-k]d_{N-p}[m-l]. \quad (11)$$

This expression is a sum of convolutions with separable kernels composed of a set of discrete one-dimensional derivative (and interpolation) kernels $\{d_p[n] \mid p = 0,1,\ldots,N\}$. As before, we place a constraint on these kernels in the Fourier domain:

$$j^N(\hat{u}\cdot\hat{\omega})^N D_0(\omega_x)D_0(\omega_y) = \sum_{p=0}^{N} b[p,N]u_x^p\, u_y^{(N-p)}\, D_p(\omega_x)D_{(N-p)}(\omega_y). \quad (12)$$

A least-squares error functional is formed by integrating over orientation and the two frequency axes:

$$E\{D_0,D_1,\ldots,D_N\} = \int_{\hat{\omega}} W(\hat{\omega})\cdot$$

$$\int_{\hat{u}} \left(\sum_{p=0}^{N} b[p,N]u_x^p\, u_y^{(N-p)}\left[j^N\omega_x^p\omega_y^{(N-p)}D_0(\omega_x)D_0(\omega_y) - D_p(\omega_x)D_{N-p}(\omega_y)\right]\right)^2 \quad (13)$$

In order to avoid the trivial solution, we again impose a constraint that the interpolator ($d_0[n]$) have unit response at D.C.: $D_0(0) = 1$.

3   Results

The error functionals in Equations (9) and (13) are both fourth-order in the optimization variables and cannot be optimized analytically. In order to obtain solutions, we fix the size of the kernels, and use conjugate gradient descent. As a starting point for the first-order kernels, we use the solution that minimizes the linear one-dimensional constraint of [1]:

$$E\{D_0,D_1\} = \int_{\omega} W'(\omega)[j\omega D_0(\omega) - D_1(\omega)]^2, \quad (14)$$

subject to $D_0(0) = 1$, with a weighting function of $W'(\omega) = 1/(|\omega| + \frac{\pi}{4})$. For the Nth-order kernels, we start from the solution that minimizes the linear one-dimensional set of constraints:

$$(j\omega)^{i-j} D_j(\omega) = D_i(\omega), \qquad (15)$$

for $0 \leq j < i \leq N$, again subject to the constraint that $D_0(0) = 1$. For the weighting functions $W(\hat{\omega})$, we choose a "fractal" weighting of $W(\hat{\omega}) = 1/(\omega_x^2 + \omega_y^2 + \pi^2/16)$.

Based on this design, Table 1 gives a set of first-order derivative kernels of different sizes. Figure 1 shows a comparison of the differentiator $d_1[n]$ with the derivative of the interpolator $d_0[n]$, computed in the Fourier domain. If these kernels were perfectly matched (i.e., $d_1[n]$ is the derivative of $d_0[n]$), then the two curves should coincide. Also shown in this figure is a comparison of these kernels to a variety of other derivative kernels. For a fair comparison, the variance of the Gaussian in this figure was chosen so as to optimize the 1-D constraint of Equation (6). Note that even under these conditions, our kernels better preserve the required derivative relationship between the differentiator and interpolator kernel. Note also that the resulting differentiation kernels are bandpass in nature, and thus less susceptible to noise than typical sinc approximations.

Since the derivative kernels are designed for rotation-equivariance, we consider the application of estimating the orientation of a two-dimensional sinusoidal grating from the horizontal and vertical partial derivatives. The grating had a fixed orientation of 22.5 degrees and spatial frequency in the range $[1, 21]/64$ cycles/pixel. Figure 2 shows the estimation error as a function of spatial frequency for a variety of derivative kernels. In this example, orientation was determined using a total least squares estimator over a 16×16 patch of pixels in the center of the image. Note that the errors for our optimal kernels are substantially smaller than the other filters. The reasonably good performance of the Gaussian is due, in part, to our optimization of its variance. And finally, shown in Table 2 are a set of higher-order derivative kernels of different sizes.

d_0	0.223755	0.552490	0.223755			
d_1	-0.453014	0	0.453014			
d_0	0.092645	0.407355	0.407355	0.092645		
d_1	-0.236506	-0.267576	0.267576	0.236506		
d_0	0.036420	0.248972	0.429217	0.248972	0.036420	
d_1	-0.108415	-0.280353	0	0.280353	0.108415	
d_0	0.013846	0.135816	0.350337	0.350337	0.135816	0.013846
d_1	-0.046266	-0.203121	-0.158152	0.158152	0.203121	0.046266

Table 1. First-order derivative kernels. Shown are pairs of derivative ($d_1[n]$) and interpolator ($d_0[n]$) kernels of various sizes.

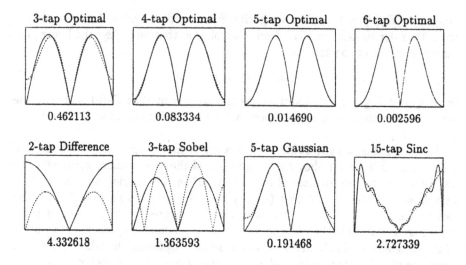

Fig. 1. First-order derivative kernels. Illustrated in each panel are the magnitude of the Fourier transform of the derivative kernel (solid line) and the frequency-domain derivative of the interpolator (dashed line) for our optimally designed kernels (see Table 1). Also illustrated, for comparison, are several other derivative kernels. Beneath the plots are the weighted RMS errors.

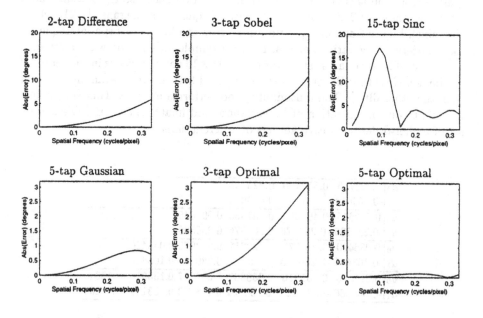

Fig. 2. Differential estimation of orientation. Illustrated is the absolute value of the error in an orientation estimation task for a sinusoidal grating as a function of spatial frequency. The grating was oriented at 22.5°, the angle of maximal error. Note that top and bottom rows have different Y-axis scaling.

d_0	0.026455	0.248070	0.450951	0.248070	0.026455
d_1	-0.097537	-0.309308	0	0.309308	0.097537
d_2	0.236427	0.020404	-0.517610	0.020404	0.236427

d_0	0.004423	0.121224	0.374352	0.374352	0.121224	0.004423
d_1	-0.029091	-0.223003	-0.196061	0.196061	0.223003	0.029091
d_2	0.105303	0.198956	-0.301356	-0.301356	0.198956	0.105303
d_3	-0.230922	0.227030	0.503368	-0.503368	-0.227030	0.230922

d_0	0.002013	0.051225	0.247548	0.398427	0.247548	0.051225	0.002013
d_1	-0.008593	-0.115977	-0.240265	0	0.240265	0.115977	0.008593
d_2	0.033589	0.180366	-0.028225	-0.370850	-0.028225	0.180366	0.033589
d_3	-0.107517	-0.074893	0.469550	0	-0.469550	0.074893	0.107517
d_4	0.201624	-0.424658	-0.252747	0.940351	-0.252747	-0.424658	0.201624

Table 2. Higher-order derivative kernels. Shown are sets of derivative $(d_p[n], p > 0)$ and interpolator $(d_0[n])$ kernels for differentiation orders $N = 2, 3, 4$.

4 Conclusions

Differentiation of discretized signals is a very basic operation, widely used in numerical analysis, image processing, and computer vision. We have described a framework for the design of operators that are efficient (i.e., compact and separable) and optimally equivariant to rotations. This formulation can easily be extended to higher dimensions (e.g., three-dimensional derivatives for motion).

References

1. E P Simoncelli. Design of multi-dimensional derivative filters. In *First Int'l Conf on Image Processing*, Austin, Texas, November 1994.
2. B. Carlsson, A Ahlen, and M. Sternad. Optimal differentiators based on stochastic signal models. 39(2), February 1991.
3. Per-Erik Danielsson. Rotation-invariant linear operators with directional response. In *5th Int'l Conf. Patt. Rec.*, Miami, December 1980.
4. J De Vriendt. Fast computation of unbiased intensity derivatives in images using separable filters. *Int'l Journal of Computer Vision*, 13(3):259–269, 1994.
5. T Vieville and O Faugeras. Robust and fast computation of unbiased intensity derivatives in images. In *ECCV*, pages 203–211. Springer-Verlag, 1992.
6. R Deriche. Fast algorithms for low-level vision. *IEEE Transactions on Pattern Analysis and Machine Intelligence*, 12:78–87, 1990.
7. J J Koenderink and A J van Doorn. Representation of local geometry in the visual system. *Biological Cybernetics*, 55:367–375, 1987.
8. W T Freeman and E H Adelson. The design and use of steerable filters. *IEEE Transactions on Pattern Analysis and Machine Intelligence*, 13(9):891–906, 1991.
9. A V Oppenheim and R W Schafer. *Discrete-Time Signal Processing*. Prentice Hall, 1989.
10. E P Simoncelli. *Distributed Analysis and Representation of Visual Motion*. PhD thesis, EECS Dept., MIT, January 1993.

A Hierarchical Filter Scheme for Efficient Corner Detection

Tobias Stammberger[1,2*], Markus Michaelis[1], Maximilian Reiser[2],
Karl-Hans Englmeier[1]

[1] GSF - National Research Center for Environment and Health,
Ingolstädter Landstr. 1, D-85764 Oberschleißheim, Germany
[2] Institut für Radiologische Diagnostik, Klinikum Großhadern, Marchioninistr. 15,
D-81337 München, Germany

Abstract. There are a number of differential geometric based corner detectors in the literature. These operators require 5 or 9 convolutions with derivative kernels in 2D or 3D respectively, what is expensive in terms of time and memory requirements. In this paper we propose an efficient approach to calculate the response of these operators.

1 Introduction

Corners are important geometric features for a number of matching tasks in computer vision (see e.g. [12], [15]). Our special interest using corners is in matching medical images. In medical image processing there is a growing demand in processing 2D as well as 3D tomographic images. Typical applications are the matching of images of different modalities like CT and MRI or matching images with and without contrast medium.

An important group of corner detectors are so called direct or gray-level based detectors. Most of the direct 2D and 3D corner detectors in the literature compute the differential characteristics of the image by estimating the partial derivatives of the first and second order. For this, Gaussian derivative filters can be used. In practice therefore, one has to convolve the image with 5 or 9 filters (first and second derivatives in 2D or 3D). This is very time consuming, especially in 3D. If all 5 or 9 response images have to be stored like for fast FFT based convolution techniques there are also large memory requirements. In this paper we propose an efficient scheme for corner detection that applies to the 2D as well as to the 3D case.

The idea is the following. Usually major parts of the image contain no strong structures at all and therefore no relevant corner points. However, the usual procedure is to convolve the image with all 5 or 9 filters, store the responses and calculate the curvature image. In our approach we transform the second derivative kernels to a set of orthogonal kernels. One of these kernels turns out to be the Laplace of Gaussian (LoG). This kernel is applicable as an isotropic local activity detector. Only in the neighborhood of positions where the LoG

---

* Corresponding author. E-mail: T.Stammberger@ikra.med.uni-muenchen.de

has a strong response, indicating strong local contrasts, there is a possibility for corner points. We investigate the localization properties of the LoG and the relevant corner detectors. This leads to an extended thresholding for the LoG response. The application of the remaining filters and the calculation of the corner detector response is restricted to the thresholded regions giving a significant speed up without loosing relevant corner points.

2 Corner Detection in Gray-Level Images

There are essentially two different direct corner detection approaches in the literature, both based on differential geometric concepts. The first group of detectors measures the isophote curvature weighted with the gradient magnitude. May be the earliest approach of this type is the one of Kitchen and Rosenfeld [7]:

$$\text{KIT} = \frac{f_{xx}f_y{}^2 + f_{yy}f_x{}^2 - 2f_{xy}f_xf_y}{f_x{}^2 + f_y{}^2} \; . \tag{1}$$

The second group of detectors measures the Gaussian curvature of the intensity surface. The Gaussian curvature K expressed in terms of partial derivatives (denoted by f_x etc.) of the image intensity function $f(x, y)$ calculates to (see e.g. [10])

$$K = \kappa_{min}\kappa_{max} = \frac{f_{xx}f_{yy} - f_{xy}^2}{(1 + f_x^2 + f_y^2)^2} = \frac{\text{DET}}{(1 + f_x^2 + f_y^2)^2} \; . \tag{2}$$

The nominator DET is the determinante of the Hessian matrix of $f(x, y)$ that has been proposed for corner detection by Beaudet [1]. It turns out that DET is the nominator of the Gaussian curvature. In practice one can usually neglect the '1' in the denominator. Then DET is identical to the Gaussian curvature weighted by the gradient to the fourth.

In 3D only one concept of curvature is used in the literature which is the Gaussian curvature of isointensity surfaces. Thirion and Gourdon ([14]) derived the following formula:

$$K = \frac{f_x^2(f_{yy}f_{zz} - f_{yz}^2) + 2f_xf_{xz}(f_yf_{yz} - f_zf_{yy}) + cycl.(x, y, z)}{(f_x^2 + f_y^2 + f_z^2)^2} \; . \tag{3}$$

where $cycl.(x, y, z)$ stands for cyclic permutation of the coordinates. There are many other approaches to corner detection in 2D and 3D that are based on these two concepts ([3], [4], [5], [8], [11]).

3 Decomposition of the Gaussian Derivative Kernels

All corner detection approaches discussed above have in common that they are based on the partial derivatives of the image intensity function. The common way to estimate derivatives in digital images without explicit representation of surface patches, is by convolutions with Gaussian derivative kernels. The normalized, zero mean 1D Gaussian kernel and its two first derivatives are

$$g_0(x) = \frac{1}{\sqrt{2\pi}\sigma}e^{-\frac{x^2}{2\sigma^2}} \quad g_1(x) = -\frac{1}{\sqrt{2\pi}\sigma^3}xe^{-\frac{x^2}{2\sigma^2}} \quad g_2(x) = \frac{1}{\sqrt{2\pi}\sigma^3}(\frac{x^2}{\sigma^2}-1)e^{-\frac{x^2}{2\sigma^2}} \quad (4)$$

The 2D and 3D kernels are built as cartesian separable combinations of the 1D kernels, e.g. $G_{xyz} = g_1(x)g_1(y)g_1(z)$ in 3D, or $G_{xx} = g_2(x)g_0(y)$ in 2D. Evaluating the curvature detectors DET and KIT requires in total 5 convolutions with the Gaussian kernels in 2D and even 9 in 3D.

The idea of our approach is to change from the Gaussian derivative kernels to another set of orthogonal kernels. One set of orthogonal kernels in 2D is given by the following transformation rules:

$$\begin{aligned}
A_0 &= \tfrac{1}{\sqrt{2}}(G_{xx} + G_{yy}) & G_{xx} &= \tfrac{1}{\sqrt{2}}(A_0 + A_1) \\
A_1 &= \tfrac{1}{\sqrt{2}}(G_{xx} - G_{yy}) & G_{yy} &= \tfrac{1}{\sqrt{2}}(A_0 - A_1) \\
A_2 &= G_{xy} & G_{xy} &= A_2 \; .
\end{aligned} \quad (5)$$

These kernels are in fact the polar separable representation of the cartesian kernels (see e.g. [9]). We can now express all corner operators discussed above in terms of the new kernels A_k. For the corner detectors KIT and DET we find:

$$\text{KIT} = \frac{\frac{1}{\sqrt{2}}(G_x{}^2 + G_y{}^2)A_0 + \frac{1}{\sqrt{2}}(G_x{}^2 - G_y{}^2)A_1 - 2G_xG_yA_2}{G_x{}^2 + G_y{}^2} \quad (6)$$

$$\text{DET} = \frac{1}{2}(A_0{}^2 - A_1{}^2) - A_2{}^2 \; . \quad (7)$$

Here A_k, G_x and G_y stand for the projections of the image to the kernels rather than for the kernels themselves. In 3D, the orthogonal kernels are defined by:

$$\begin{aligned}
A_0 &= \tfrac{1}{\sqrt{3}}(G_{xx} + G_{yy} + G_{zz}) & G_{xx} &= \tfrac{1}{\sqrt{3}}A_0 - \tfrac{1}{\sqrt{2}}A_1 + \tfrac{1}{\sqrt{6}}A_2 \\
A_1 &= \tfrac{1}{\sqrt{2}}(G_{yy} - G_{xx}) & G_{yy} &= \tfrac{1}{\sqrt{3}}A_0 + \tfrac{1}{\sqrt{2}}A_1 + \tfrac{1}{\sqrt{6}}A_2 \\
A_2 &= \tfrac{1}{\sqrt{6}}(G_{xx} + G_{yy} - 2G_{zz}) & G_{zz} &= \tfrac{1}{\sqrt{3}}A_0 - 2\tfrac{1}{\sqrt{6}}A_2
\end{aligned} \quad (8)$$

$$A_3 = G_{xy} \quad A_4 = G_{xz} \quad A_5 = G_{yz} \; .$$

Substituting the right hand side of (8) into equation (3) provides an expression for the corner detector represented in the new basis. Although there is no compact analytical form for the operator, the calculation in practice is straight forward.

4 Localization Properties

As already mentioned in the introduction, in our approach the LoG-kernel A_0 is used to preselect possible corner points. However, the positions of the extrema of the LoG response and the responses of the different corner detectors don't coincide. The mislocalization has already been investigated for corner detectors by Rohr [13], Deriche [4], Guiducci [6] and for the LoG by Berzins [2] and others. None of these authors however, addressed explicitly the extrema of the LoG and the relation between both mislocalizations.

The quantitative knowledge of the mislocalization is essential to derive a suitable scheme for selecting candidate corner points. For this reason we introduce a synthetic 2D corner model to compute analytically the exact position of the extrema of the responses of DET and KIT in order to compare it with the position obtained from the LoG activity detector. Concerning the mislocalization, we are not so much interested in the deviation from the true corner position but in the deviation from the positions given by the corner detectors. The corner model consists of a unit step corner with arbitrary corner angle θ.

4.1 Analytical Study

Mathematically, the corner can be described by

$$p(x, y) = U(mx - y)U(mx + y) \qquad (9)$$

where $U(x)$ is the 1D unit step function and $m = \tan \frac{\theta}{2}$. The convolution integral with the Gaussian smoothing kernel $g_0(x)g_0(y)$ is defined as

$$P(x, y) := \int_{-\infty}^{\infty} \int_{-\infty}^{\infty} g_0(x')g_0(y')p(x - x', y - y')dx'dy' . \qquad (10)$$

To simplify we set $\sigma = 1$ in all our calculations, i.e. the results are given in units of σ. The first and second order partial derivatives of $P(x, y)$ can be calculated from (10) by differentiating under the integral. According to (1), (2) and (5) these derivatives are combined to the operator response functions. For reasons of symmetry the extrema of the operator response are located along the symmetry axis (x-axis) of the corner. Restricting the partial derivatives and hence the operator responses $r\{A_0\}$, $r\{KIT\}$ and $r\{DET\}$ to the symmetry axis $y = 0$ simplifies the results to

$$\begin{aligned}
r\{A_0\} &= -2sxg_0(sx)\Phi(cx) \\
r\{KIT\} &= -2csg_0(sx)(cx\Phi(cx) + sg_0(cx)) \\
r\{DET\} &= -4s_2cg_0(sx)^2 - cs^3x^2\Phi(cx)^2 + (c^2 - s^2)xg_0(cx)\Phi(cx)
\end{aligned} \qquad (11)$$

where $\Phi(x) := \int_{-\infty}^{x} g_0(x')dx'$ is the error function and $s = \sin \frac{\theta}{2}$, $c = \cos \frac{\theta}{2}$. These operator responses along the symmetry axis are depicted in Fig. 1a. The zeros of the derivatives of these equations with respect to x yield the extrema of these responses:

$$A_0: \qquad (1 - s^2 x^2)\Phi(cx) + cx g_0(cx) = 0 \qquad (12)$$

$$\text{KIT:} \qquad (1 - s^2 x^2)c\Phi(cx) + s^2 x g_0(cx) = 0 \qquad (13)$$

$$\text{DET:} \qquad \begin{aligned} &(c^2 - s^2)((1 - 2s^2 x^2)g_0(cx)\Phi(cx) + cx g_0(cx)^2 - c^2 x^2 g_0(cx)\Phi(cx)) \\ &- cs^2((1 - s^2 x^2)2x\Phi(cx)^2 + 2cx^2 g_0(cx)\Phi(cx)) \\ &- 2cx g_0(cx)^2 = 0 . \end{aligned} \qquad (14)$$

The calculation of the zeros has been done numerically.

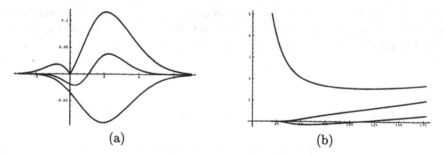

(a)   (b)

Fig. 1. (a) Response function of the corner detectors and $|A_0|$ along the symmetry axis $y = 0$ (upper curve: $|A_0|$, middle curve: DET, lower curve: KIT). Here $|A_0|$ means the absolute value of the response of A_0. (b) Distance (in units of σ) between the maxima of the responses of the corner detectors and the activity detector $|A_0|$ as a function of the corner angle (upper curve: $\text{DET}_{\min}$, middle curve: KIT, lower curve: $\text{DET}_{\max}$). Except for the minimum of DET, the maxima of the corner detectors ly within 1σ to the maximum of $|A_0|$.

4.2 Extended Thresholding

From the results that are summarized in Fig. 1 we can conclude the following. At the true corner position the response of A_0 always has a zero crossing. However, we already said that it is not so much the true corner position that is of interest for our scheme, but the positions of the local maxima of the corner detector responses. Figure 1 shows that the maxima of the responses of $|A_0|$ ly close to the maxima of the corner detectors. Consequently, we take the local maxima of $|A_0|$ as candidates for corner points. Due to the distances in the positions of the extrema of $|A_0|$ to DET and $|A_0|$ to KIT, depicted in Fig. 1b, we cannot only take the thresholded response of $|A_0|$. For $\text{DET}_{\max}$ and KIT we find that for small corner angles the distance is close to zero, while for large angles the distance is bounded by approximately 1σ. For $\text{DET}_{\min}$ the distance is essentially below 2σ.

We therefore apply to the thresholded response of $|A_0|$ a morphological dilation with a circular structuring element with radius 1σ (respectively 2σ in the case that we are interested in detecting DET_{min}). We will call this an *extended thresholding*. The resulting binary image is a mask for the positions where the corner detectors are evaluated. Usually the selected pixels/voxels are only a subset of the whole image so that a significant speed up can be achieved. In 3D the analytical computations even for a simple corner model are rather complex. Symmetry considerations and experiments show that the maximal mislocalization is of the same order as in 2D, i.e. 1σ.

Our proposed hierarchical corner detection scheme based on the set of orthogonal kernels A_k is then composed of the following steps:

1. Activity detection by convolving the image with the kernel A_0 (equation (5) or (8) respectively).
2. Extended thresholding on the activity image provides a binary mask for the corner point candidates.
3. Projection of the remaining kernels $A_i, i \neq 0$, and calculation of the curvature at the selected positions.

5   Experimental Results

Our approach is based on two assumptions. First, that the pixels/voxels selected by the activity detection cover all relevant corner points and second, that the selected pixels/voxels are only a small part of the whole image in practice. Concerning the first point, the theoretical considerations for a synthetic corner model in section 4 motivated the extended thresholding. Real world image structures however, are far more varied and complex than the synthetic corner used. The experiments in this section are to prove that also in real situations both assumptions are satisfied.

The proposed filter scheme was evaluated on CT image data in 2D and 3D in order to use the corner points as robust features for matching contrast enhanced to native CT data. We compare the results of our hierarchical scheme to the results of a direct application of the same corner detector.

Fig. 2b shows the response of KIT to a CT slice of the skull (Fig. 2a) calculated by the direct approach using Gaussian derivative kernels at a scale of $\sigma = 1.2$. About 350 corner points were selected by thresholding the response image (Fig. 2c).

Figures 2d, 2e and 2f show the different steps of the hierarchical approach. There is only one convolution of the whole image with the activity detector $|A_0|$ (Fig. 2d). About the same number of points as for the direct approach (about 350) has to be selected by thresholding the activity image. After the dilation (extended thresholding) we obtain about $4,000$ candidates for corner points (Fig. 2e). The remaining four projections are calculated only for these $4,000$ pixels. This gives a significant speed up compared to the calculation of the projections to all 256^2 pixels. Comparing the resulting curvature image (Fig. 2f) to the direct calculation of the curvature in the whole image (Fig. 2b) reveals differences only at regions with low gray value activity. The corner points selected by thresholding both curvature images are exactly the same for both methods (Fig. 2c).

In 3D a $256^2 \times 43$ CT data set, of which Fig. 2a shows one slice, was used to verify the performance of our hierarchical scheme. About 20,000 points were selected by thresholding the activity image, giving about 120,000 candidates for corner points after dilation. The number of points where the curvature calculations have to be performed can be reduced significantly compared to the direct approach, so that the overall computational costs are essentially given by the costs of the activity detection.

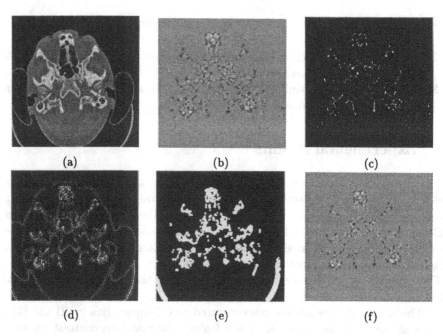

Fig. 2. Comparison of our scheme and the standard scheme for corner detection: (a) shows the input image, a CT slice of the skull. The standard scheme calculates the response of KIT in the whole image (b) and obtains the corner points by thresholding this response image (c). Our scheme in contrast calculates first the response to the activity detector $|A_0|$ (d). (e) shows the extended thresholding of the activity image. (f) shows the application of KIT to the pixels that are masked by (e) (about 4,000 pixels, the whole image is 256×256). Thresholding this image provides the same about 350 corner points (c) as the standard scheme.

6 Summary

The hierarchical approach to corner detection that has been presented in this work offers an important gain in CPU as well as in memory requirements. The whole image is convolved with only one kernel that serves as an isotropic activity detector. More kernels and calculations are only applied to pixels/voxels selected by the first kernel. As long as only a small part of the pixels or voxels are selected by this detection, the performance gain of our approach is approximately a factor of 5 (2D) or 9 (3D).

The main problem with selecting only a small part of the image for further processing is the mislocalization between the maxima of the responses of the LoG filter and the various corner detectors. By theoretical considerations we have shown that the mislocalization can be estimated to be less then one sigma.

Our experiments with real medical image data in 2D and 3D gave proof of the applicability of these considerations to real data. For many types of medical images it turns out to be true that only a small part of the image is selected without loosing relevant corner points.

References

1. Beaudet, P.R.: Rotationally invariant image operators. International Joint Conference on Pattern Recognition (1978) 579-583
2. Berzins, V.: Accuracy of Laplacian edge detectors. Computer Vision, Graphics, and Image Processing **27** (1984) 195-210
3. Drescher, L., Nagel, H.H.: Volumetric model and 3D trajectory of a moving car derived from monocular TV frame sequences of a street scene. Computer Graphics and Image Processing **20** (1982) 199-228
4. Deriche, R, Giraudon, G.: A computational approach for corner and vertex detection. International Journal of Computer Vision **10** (2) (1993) 101-124
5. Florack, L.:. The syntactical structure of scalar images. PHD Thesis, University of Utrecht (1993)
6. Guiducci, A.: Corner characterization by differential geometry techniques. Pattern Recognition Letters **8** (1988) 311-318
7. Kitchen, L., Rosenfeld, A.: Gray-level corner detection. Pattern Recognition Letters **1** (2) (1982) 95-102
8. Krueger, W.M., Phillips, K.: The geometry of differential operators with application to image processing. IEEE PAMI **11** (1989) 1252-1264
9. Koenderink, J., van Doorn, A.: Gerneric neighborhood operators. IEEE PAMI **14** (6) (1992) 597-605
10. Lipschutz, M.: Differential Geometry. McGraw-Hill: New York (1969)
11. Monga, O., Benayoun, S.: Using partial derivatives of 3D images to extract typical surface features. Computer Vision and Image Understanding **61** (2)(1995) 171-189
12. Nagel, H.H.: Displacement vectors derived from second-order intensity variations in image sequences. Computer Vision, Graphics and Image Processing **21** (1983) 85-117
13. Rohr, K.: Localization properties of direct corner detectors. Journal of Mathematical Imaging and Vision **4** (1994) 139-150
14. Thirion, J.P., Gourdon, A.: Computing the differential characteristics of isointensity surfaces. Computer Vision and Image Understanding **61** (2) (1995) 190-202
15. Van den Elsen, P.: Automatic registration of CT and MR brain images using correlation of geometrical features. IEEE Transactions on Medical Imaging **14** (2) (1995) 384-396

Defect Detection on Leather by Oriented Singularities

A. Branca, F.P. Lovergine, G. Attolico, A. Distante

Istituto Elaborazione Segnali ed Immagini - CNR
Via Amendola 166/5, 70126 Bari
ITALY
Phone number: (+39)-80-54419696
Fax number: (+39)-80-484311
E-mail: [branca,loverg,attolico,distante]@iesi.ba.cnr.it

Abstract. This paper presents a system for leather inspection based upon visual textural properties of the material surface. Defects are isolated from the complex and not homogeneous background by analyzing their strongly oriented structure. The patterns to be analyzed are represented in an appropriate parameter space using an optimization approach: in this way a parameter vector is associated to each different textured region in the original image. Finally a filter process, based upon knowledge about the parameter vectors representing the leather without defects, detects and classifies any abnormality.
Subject terms: leather inspection, texture analysis, texture classification, oriented structures, parameter space, vectorial bases.

1 Introduction

The proposed system is intended to address the *receiving inspection* of raw leather with the goal of supplying the manufacturing process with information about location and classification of defects. Inspection decisions are often based on superficial properties, therefore machine vision is considered a natural tool for an automated inspection system. A satisfactory vision inspection system for many contexts can be based on the analysis of texture because most natural surfaces and defects do exhibit rich textural contents.

Leather has not received much attention: its visual appearances may change over a wide range (it is not homogeneous in color, thickness, brightness, and finally in wrinkledness) depending on both the original characteristics of skin and on the subsequent curing processes. The defects too, often due to natural reasons, lack a well-defined classification and description in terms of visual cues. Therefore their separation from background can become a quite hard task. **Directionality and local variation** of intensity are the major elements to rely upon for solving the defect detection and classification problem. A practical solution for this application must be able to estimate at the same time *how* and *how much* the intensities vary in the image, in order to select defects that produces sharp variation points (edge points) with high directionality. The solution is related to an analysis of the derivatives of the image smoothed with a Gaussian filter. The gradient vector of the smoothed image indicates at each point the direction along which the derivative has the largest absolute value. Points where its modulus is maximum are inflection points (sharp variation points) of the smoothed image. Directionality is efficiently estimated through the *coherence image* [2] measuring locally the accord between image gradient orientations. Because we are interested in sharp intensity variations with high directionality, we have to search through local maxima associated to high coherence values for finding defect edges. Since texture is a property strongly tied to the resolution used for the analysis, each image should be analyzed at a number of resolutions (or scales) granting that every image region has been considered at the most appropriate scale. The multiresolution scheme may provide useful support also for the characterization of signal variation. In fact defect edges produce modulus maxima at several scales. This persistence across scales becomes a powerful tool for removing the background. The **wavelet theory** allow the detection of edges as local maxima of a suitable wavelet transform. Moreover, Mallat et al. [3] have proved that local maxima of a wavelet transform can not only detect the singularities of a signal but even to provide information about the local Lipschitz exponents can often be measured from the evolution across scales of these local maxima. We develop an algorithm that removes textural background by discriminating the signal singularities through an analysis of wavelet transform maxima indicating the location of edges in images. The background can be discriminated from the image information from the geometrical properties of the maxima curves and the evolution across scales of the wavelet transform values along these curves. Detected defects are identified by projecting in a suitable parameter space the vector field of gradient vectors corresponding to selected singular points. In this way defects are represented as a set of projection coefficients (texture parameters) on a set of basis vectors (elementary texture vectors) spanning a vector sub-space where the vector fields associated to the considered defects can be defined. Though the method has been proposed to solve the difficult problem of defect identification on leather, it can be considered as a general technique characterizing compositional textures of a complex disordered background and overimposed oriented structures. Actually, the present work integrates an oriented singularity detection framework based on wavelet theory [3] with a previous approach [1] analyzing compositional textures through the vector fields of dominant local gradient orientations [2]. In section 2 the multiscale edge detection approach based on wavelet theory is described, while in section 3 the coherence measure used to select singular points in regions with high directionality is presented. Section 4 presents the segmentation model in the parameter space. Finally, section 5 shows experimental results obtained on several defects on different kinds of leather.

2   Singularity detection through wavelet theory

The first step of our approach consists in estimating an oriented flow field in which at each position is assigned the gradient vector of the corresponding pixel in the original image. Defects on the leather surface produce edges with high directionality. To select defect points we use a multiscale approach based on a suitable wavelet transform that uses the wavelet theory for detecting image singularities. Moreover an algorithm for estimating the coherence of selected edges is applied.

Let $\theta(x, y)$ be the 2D Gaussian smoothing function, and $\psi^1(x, y)$ and $\psi^2(x, y)$ its first partial derivatives with respect to x and y respectively. These two functions can be considered to be wavelets because their integrals are equal to 0. The corresponding wavelet transform $W_s f(x, y)$ is computed by convolving the image function $f(x, y)$ with the two wavelets dilated by a scaling factor s. The wavelet transforms $W_s^1(x, y)$ and $W_s^2(x, y)$ of $f(x, y)$ are proportional to the gradient vectors of the signal smoothed at the scale s.

In particular, at any scale s, the modulus of the gradient vector is proportional to

$$M_s f(x, y) = \sqrt{\|W_s^1 f(x, y)\|^2 + \|W_s^2 f(x, y)\|^2} \tag{1}$$

while the angle of the gradient vector with the horizontal direction is given by

$$A_s f(x, y) = argument(W_s^1 f(x, y) + iW_s^2 f(x, y)) \tag{2}$$

The sharp variation points of $f * \theta_s(x, y)$ are the points (x, y) where the modulus $M_s f(x, y)$ has a local maxima in the direction of the gradient given by $A_s f(x, y)$.

A remarkable property of the wavelet transform is its ability to characterize the local regularity of a function. In mathematics, the local regularity of functions can be measured using Lipschitz exponents.

The function $f(x, y)$ is said to be uniformly Lipschitz α over an interval $]a, b[\times]c, d[$ if and only if there exists a constant A such that for any $(x, y) \in]a, b[\times]c, d[$ and any scale s,

$$\|W_s f(x, y)\| \leq As^{\alpha} \tag{3}$$

Equation 3 is a condition on the asymptotic decay of $\|W_s f(x, y)\|$ when the scale s goes to zero. The **Lipschitz uniform regularity** of $f(x, y)$ is the upper bound of all α such that $f(x, y)$ is uniformly Lipschtz α.

Edges at the scale s are defined as local sharp variation points of $f(x, y)$ smoothed by the Gaussian $\theta_s(x, y)$. For edge or singularity detection, we are only interested in the local maxima of $W_s f(x, y)$. The values of local maxima can be useful for characterizing the Lipschitz exponents of signal irregularities.

All the singularities of $f(x, y)$ can be located by following the maxima lines when the scale goes to 0. It can been proved that the Lipschitz regularity of $f(x, y)$ in (x_0, y_0) can be estimated from the decay of the wavelet transform along one line of maxima that converges towards (x_0, y_0). The equation 3 is equivalent to

$$\log \|W_s f(x, y)\| \leq \log(A) + \alpha \log(s) \tag{4}$$

Lipschitz regularity at (x_0, y_0) is the maximum slope of straight lines that remain above $\log \|W_s f(x, y)\|$, on a logarithmic scale. The Lipschitz exponents are computed by finding the coefficient α such that As^α approximates at best the decay of $\|W_s f(x, y)\|$ over a given range of scales larger than 1. It has been proved [3] that if a signal is singular at a point (x_0, y_0) there exist a sequence of wavelet transform modulus maxima that converge to x_0 when the scale decreases. Hence, we detect all the singularities from the positions of the wavelet transform modulus maxima. Moreover, the decay of the wavelet transform is bounded by the decay of these modulus maxima, and we can thus measure the local uniform Lipschitz regularity from this decay. If $f(x)$ is differentiable at (x_0, y_0), then it is Lipschitz $\alpha = 1$. As the uniform Lipschitz regularity approaches 0, the corresponding singularity becomes less and less regular. If $f(x, y)$ is discontinuous but bounded in the neighborhood of (x_0, y_0), its uniform Lipschitz regularity in the neighborhood of (x_0, y_0) is 0. If the uniform Lipschitz regularity is positive, the amplitude of the wavelet transform modulus maxima should decrease when the scale decreases. If the wavelet transform maxima increase when the scale decreases, singularities can be described with negative Lipschitz exponents, meaning that they are more singular than discontinuities.

In our application context, defects are singularities with positive Lipschitz regularity, while discontinuities on the background are singularities with negative Lipschitz exponents, then the values of the wavelet local maxima located on the background decrease on average, when the scale increases. In the neighborhood of singularities with positive Lipschitz exponents, the local maxima of wavelet transform have an amplitude increasing or remaining constant when the scale increases. In order to remove the background components, we suppress all the maxima that do not propagate along enough scales or whose average amplitude increases when the scale decreases.

3 Orientation Detection

If most of the singularities of the original image have positive L. exponents, we can separate the noise from the signal by measuring the evolution across scales of the wavelet local maxima. We integrate this information with a-priori knowledge on the geometrical properties of the image singularities in the image plane (x, y). Since leather defects exhibit high directionality, we estimate this property by analyzing the *coherence image* computed as described by Rao and Schunck [2].

The coherence is a local measure of the accord between image gradient orientations. It is essentially based on normalizing and projecting the gradient vectors of a given neighborhood on a given orientation. In our case we compute the coherence of the orientation of sharp points (with module maxima) with respect to neighboring gradient orientations. If the vector is coherent, its normalized projections will be close to one; otherwise they will tend to cancel producing results close to zero. Being $\phi(x_0, y_0)$ the gradient direction at point (x_0, y_0), the measure of coherence in that point is given by

$$G(x_0, y_0) \frac{\sum_{(x,y)\in W} \| G(x,y)\cos(\alpha(x_0, y_0) - \alpha(x,y)) \|}{\sum_{(x,y)\in W} G(x,y)} \tag{5}$$

Leather defects are identified between sharp variation image points using also their coherence. The points with gradients characterized by module maxima and argument with high coherence are considered at each scale to estimate the Lipschitz regularity.

4 Parameter Space Modeling

Points selected using their Lipschitz regularity are represented as a vector field by considering the corresponding gradient vector. This particular representation can help in classifying the defects and in segmenting, at same time, the image into different defected areas, moreover points misclassified as defects in the previous step are now removed.

A vector field can be analyzed using its representation as a set of M projection coefficients $\{c_i\}_{i=1,..,M}$ (texture parameters) on N-dimensional basis vectors $\{\psi_i(x,y)\}_{i=1,..,M}$ (elementary texture vectors) spanning a vector space where the vector fields associated to defects can be defined. The coefficients of these projections are the parameters by means of which texture classification can be performed. To represent a vector $V(x, y)$, either exactly or in some optimal sense, by projecting it onto a chosen set of vectors $\{\psi_i(x, y)\}$ we must find the projection coefficients c_i making the linear combination of the basis $(H(x, y))$ either identical to or with minimum distance from $V(x, y)$. The desired set of coefficients is determined minimizing the squared norm of the difference-vector:

$$E = \|V(x, y) - H(x, y)\|^2 \tag{6}$$

This quadratic error function E reaches its global minimum only when its partial derivatives with respect to any of the M coefficients c_i equal zero. Satisfying this condition for each c_i generates a system of M simultaneous equations in M unknowns:

$$\left| \sum_{(x,y)} V(x, y)\psi_i(x, y) = \sum_{(x,y)} \left(\sum_{k=1}^{M} c_k \psi_k(x, y) \right) \psi_i(x, y) \right| \tag{7}$$

A central role in the estimation of this projection is played by the reference system used for expressing the basis orientation fields. Moving the reference system inside a single region with homogeneous textural content may produce, especially if a singular point of the orientation field belongs to that region, different parameter vectors. This makes necessary to *interleave* the classification process with the segmentation process: the evaluation of texture parameters supports the region-growing by providing the measure of similarity between patches while the amount of growth of a seed patch does validate or confute the classification results. The resulting scheme iterates alternatively the steps of *classification* and *segmentation*.

Since small patches belonging to the same large region, with homogeneous textural contents, do not produce necessarily the same coefficient vectors when analyzed locally, each with respect to its own reference system, an acceptable segmentation cannot be obtained merging patches solely on the base of a straight comparison of their parameter vectors evaluated by the classification. A region-growing method has been designed which attempts to extend seed patches if their parameters are able to describe the textural contents of their neighborhoods. The reference system is kept centered on the seed patch during all the growing trials: this provides a meaningful estimation of textural similarity and better final results. The region growing algorithm is quite simple: it starts from a temporary classified patch (seed patch) and considers a set of surrounding patches (having the same dimension of the seed patch and the center lying on its border) that still need to be classified. They are allowed to overlap no more than half of their size. The orientation field in each patch is compared with the linear combination of the basis field vectors (expressed with respect to the reference system of the seed patch) weighted using the texture parameters of the seed patch. A single network iteration allows the merging decision, based on the difference between the orientation field of the patch and its approximation (evaluated with respect to the reference system and the texture parameters of the seed patch).

During each patch-growing process, the network weights coding the projection coefficients (classification vector) are kept fixed to the values c_i estimated during the classification of the seed patch. The number of patches in the final region is used for judging the correctness of texture parameters: in fact wrong classifications do not expand. In those cases the seed-patch is kept unclassified, waiting for its subsequent inclusion into another region. In case of successfully expansion all the patches in the region are marked as classified and excluded from any further processing.

This extension policy increases the computational efficiency of the process: testing the validity of a parameter vector in a neighboring patch does require a single iteration of the neural net instead of the 300-400 iterations required for solving the equations 7.

Furthermore it accounts for the misclassification that a local analysis of orientation fields could introduce keeping the reference system used for the basis vector fields fixed onto the seed patch during all the growing process.

5 Experimental Results

The proposed approach is general and can be applied to analyze any oriented texture choosing appropriately the set of basis vectors $\{\psi_i(x, y)\}$ which spans the vector space of the input patterns.

Generally the defects occurring on leather surfaces are characterized by particular oriented structures representable by orientation vector fields that can be optimally projected onto the basis vectors (Fig.(1)) defined by the following functions:

$$\psi_1(x,y) = \begin{pmatrix} 1 \\ 0 \end{pmatrix} \qquad \psi_2(x,y) = \begin{pmatrix} 0 \\ 1 \end{pmatrix} \qquad \psi_3(x,y) = \begin{pmatrix} -x \\ -y \end{pmatrix}$$

$$\psi_4(x,y) = \begin{pmatrix} -y \\ x \end{pmatrix} \qquad \psi_5(x,y) = \begin{pmatrix} -x \\ y \end{pmatrix}$$

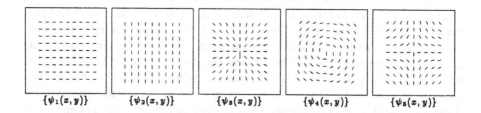

$\{\psi_1(x,y)\}$ $\qquad$ $\{\psi_2(x,y)\}$ $\qquad$ $\{\psi_3(x,y)\}$ $\qquad$ $\{\psi_4(x,y)\}$ $\qquad$ $\{\psi_5(x,y)\}$

Fig. 1. Elementary texture vector fields

For our experiments we have analyzed images (512 × 512) acquired from leather samples of about 10-15 by 10-15 cm using the same geometrical and ligthing setup (the camera was hold perpedicular to the leather sample and ligth sources was at about 45° from the optical axis). The images have been digitized using a COHU b/w camera.

Several samples have been considered with different kinds of defects. Their names are related to the events, often occurred during the animal's life, that have caused them. Each defect produces a different oriented pattern that the proposed method is able to separate from the textural structure of the background.

The following figures show the results obtained on the some samples juxtaposing the original textured image and the final segmented image.

References

1. A.Branca, M.Tafuri, G.Attolico, A.Distante "Directionality Detection in Compositional Textures" *International Conference on Pattern Recognition*, August 1996.
2. R.Rao and B.G.Schunk, Computing oriented texture fields *CVGIP: Graphical Models Image Processing 53* (1991) 157-185.
3. Mallat, S. and Hwang, W.L. "Singularity Detection and Processing with Wavelets", Technical Report March 1991.

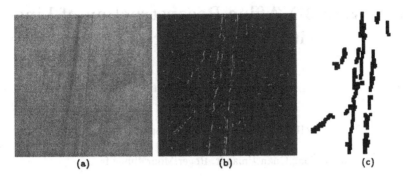

Fig. 2. (a) Original image. (b) Defects selected using Lipschitz exponents decay across scales and coherence of module maxima. (c) Defects selected by the segmentation approach.

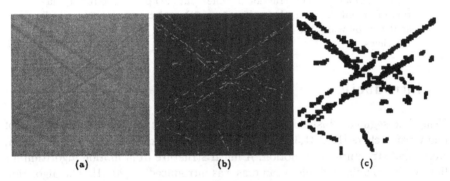

Fig. 3. (a) Original image. (b) Defects selected using Lipschitz exponents decay across scales and coherence of module maxima. (c) Defects selected by the segmentation approach.

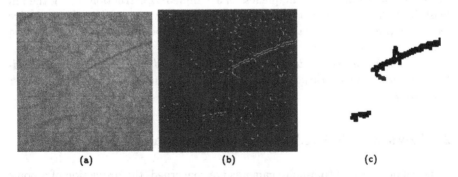

Fig. 4. (a) Original image. (b) Defects selected using Lipschitz exponents decay across scales and coherence of module maxima. (c) Defects selected by the segmentation approach.

Uniqueness of 3D Affine Reconstruction of Lines with Affine Cameras

Long Quan and Roger Mohr

CNRS-GRAVIR-INRIA
ZIRST – 655 avenue de l'Europe
38330 Montbonnot, France
Email: Long.Quan@imag.fr, Roger.Mohr@imag.fr

Abstract. We prove that 3D affine reconstruction of lines with uncalibrated affine cameras is subject to a two way ambiguity. The key idea is to convert 3D affine reconstruction of "lines" into 2D projective reconstruction of "points". Then, the ambiguity of 2D projective reconstruction is analyzed by using the full tensorial representation of three uncalibrated 1D views.

1 Introduction

Using line segments instead of points as features has attracted the attention of many researchers [11,2,31,30,29,1,18] for various tasks such as pose estimation, stereo and structure from motion. A linear structure from motion algorithm for lines with uncalibrated affine cameras was introduced in [20,21]. The algorithm requires a minimum number of seven line correspondences over three views. The key idea is the introduction of a one-dimensional projective camera which converts 3D affine reconstruction of lines into 2D projective reconstruction of points. However, the reconstruction ambiguity was unclear in the previous papers. In this paper, we examine the uniqueness of 3D affine reconstruction. Using the full tensorial representation of three uncalibrated views, we prove that 2D projective reconstruction has always a two way ambiguity, so does 3D affine reconstruction of lines.

Throughout the paper, tensors and matrices are denoted in upper case boldface, vectors in lower case boldface and scalars in either plain letters or lower case Greek.

2 Review of 3D Affine Reconstruction of Lines

As far as perspective (pin-hole) cameras are concerned, the projection of a point $\mathbf{x} = (x, y, z, t)^T$ of $\mathcal{P}^3$ to a point $\mathbf{u} = (u, v, w)^T$ of $\mathcal{P}^2$ can be described by a 3×4 homogeneous projection matrix $\mathbf{P}$:

$$\lambda \mathbf{u} = \mathbf{P}_{3 \times 4} \mathbf{x}. \tag{1}$$

For a restricted class of camera models, by setting the third row of the perspective camera $\mathbf{P}$ to $(0,0,0,\lambda)$, we obtain the affine camera initially introduced by Mundy and Zisserman in [16]

$$\mathbf{A}_{3\times4} = \begin{pmatrix} p_{11} & p_{12} & p_{13} & p_{14} \\ p_{21} & p_{22} & p_{23} & p_{24} \\ 0 & 0 & 0 & p_{34} \end{pmatrix}$$

$$\equiv \begin{pmatrix} \mathbf{M}_{2\times3} & \mathbf{t}_{3\times1} \\ \mathbf{0}_{1\times3} & \end{pmatrix}. \tag{2}$$

This is the uncalibrated affine camera which emcompasses all the uncalibrated versions of the orthographic, weak perspective and paraperspective camera models.

Now consider a line in $\mathbb{R}^3$ through a point $\mathbf{x}_0$ with direction $\mathbf{d}_x$:

$$\mathbf{x}_a = \mathbf{x}_0 + \lambda \mathbf{d}_x.$$

The affine camera $\mathbf{A}_{3\times4}$ projects this to an image line:

$$\mathbf{A}_{3\times4} \begin{pmatrix} \mathbf{x} \\ 1 \end{pmatrix} = (\mathbf{M}\mathbf{x}_0 + \mathbf{t}_0) + \lambda\mathbf{M}\mathbf{d}_x = \mathbf{u}_0 + \lambda\mathbf{M}\mathbf{d}_x,$$

with direction

$$\rho\mathbf{d}_u = \mathbf{M}_{2\times3}\mathbf{d}_x, \tag{3}$$

passing through the image point

$$\mathbf{u}_0 \equiv \mathbf{M}\mathbf{x}_0 + \mathbf{t}_0.$$

Equation (3) describes a linear mapping between directions of 3D lines and those of 2D lines. It can be derived even more directly using projective geometry, by considering that the line with direction $\mathbf{d}_x$ is the point at infinity $\mathbf{x}_\infty = (\mathbf{d}_x^T, 0)^T$ in $\mathcal{P}^3$ and the line with direction $\mathbf{d}_u$ is the point at infinity $\mathbf{u}_\infty$ in $\mathcal{P}^2$.

Comparing Equation (3) with Equation (1) which is a projection from $\mathcal{P}^3$ to $\mathcal{P}^2$, we see that Equation (3) is nothing but a projective projection from $\mathcal{P}^2$ to $\mathcal{P}^1$ if we consider the 3D and 2D directions of lines as 2D and 1D projective points. This means that the affine reconstruction of lines with a two-dimensional affine camera is equivalent to the projective reconstruction of points with a one-dimensional projective camera!

This 2D projective reconstruction turns out the uncalibrated orientation of the camera and the affine direction vector of lines.

To recover the full affine structure of the lines, we still need to find the vector $\mathbf{t}_{3\times1}$ of the affine cameras defined in (2). These represent the image translation and magnification components of the camera. Recall that line correspondences from two views do not impose any constraints on camera motion: The minimum number of views required is three. The recovery of the uncalibrated translations is essentially linear once the uncalibrated rotations have been recovered. A detailed linear algorithm is developed in our previous work [20,21].

The final reconstruction step of lines can be easily formulated as a subspace selection and solved by SVD [20,21].

In summary, the whole algorithm consists of 3 steps:

1. Recovery of affine directions of lines and uncalibrated orientations
2. Recovery of uncalibrated translations
3. Recovery of affine shape

3 Ambiguity of 2D Projective Reconstruction

We see from the above review that the solution to the second step which recovers the uncalibrated translations and affine shape is completely linear and unique provided that the uncalibrated orientations and direction vectors of lines are recovered. The affine reconstruction ambiguity comes from the first step which is equivalent to 2D projective reconstruction. Therefore 2D projective reconstruction ambiguity should be carefully analyzed.

There have been many recent works [3,5,26,13,15,4,6,23,24] on projective reconstruction and the geometry of multi-views of two dimensional uncalibrated cameras. Particularly, the tensorial formalism developed by Triggs [26] is very interesting and powerful. We are now extending this study to the case of the one-dimensional camera.

3.1 Uncalibrated one-dimensional camera

First, rewrite Equation (3) in the following form:

$$\lambda \mathbf{u} = \mathbf{M}_{2 \times 3} \mathbf{x} \tag{4}$$

in which we use $\mathbf{u} = (u_1, u_2)^T$ and $\mathbf{x} = (x_1, x_2, x_3)^T$ instead of $\mathbf{d}_u$ and $\mathbf{d}_x$ to stress that we are dealing with "*points*" in the projective spaces $\mathcal{P}^2$ and $\mathcal{P}^1$ rather than line directions in the vector spaces $\mathbb{R}^3$ and $\mathbb{R}^2$. This exactly describes a one-dimensional projective camera which projects a point $\mathbf{x}$ in $\mathcal{P}^2$ onto a point $\mathbf{u}$ in $\mathcal{P}^1$.

We now examine the matching constraints between multiple views of the same point. There is a constraint only for the case of 3 views.

Let the three views of the same point $\mathbf{x}$ be given as follows:

$$\begin{cases} \lambda \mathbf{u} = \mathbf{Mx}, \\ \lambda' \mathbf{u}' = \mathbf{M'x}, \\ \lambda'' \mathbf{u}'' = \mathbf{M''x}. \end{cases} \tag{5}$$

These can be rewritten in matrix form as

$$\begin{pmatrix} \mathbf{M} & \mathbf{u} & 0 & 0 \\ \mathbf{M'} & 0 & \mathbf{u'} & 0 \\ \mathbf{M''} & 0 & 0 & \mathbf{u''} \end{pmatrix} \begin{pmatrix} \mathbf{x} \\ -\lambda \\ -\lambda' \\ -\lambda'' \end{pmatrix} = 0, \tag{6}$$

which is the basic reconstruction equation for a one-dimensional camera. The vector $(\mathbf{x}, -\lambda, -\lambda', -\lambda'')^T$ cannot be zero, and so

$$
\begin{vmatrix}
\mathbf{M} & \mathbf{u} & 0 & 0 \\
\mathbf{M}' & 0 & \mathbf{u}' & 0 \\
\mathbf{M}'' & 0 & 0 & \mathbf{u}''
\end{vmatrix} = 0. \tag{7}
$$

The expansion of this determinant produces a trilinear constraint of three views

$$
\sum_{i,j,k=1}^{2} T_{ijk} u_i u_j' u_k'' = 0, \tag{8}
$$

or in short

$$
\mathbf{T}_{2\times2\times2}\mathbf{u}\mathbf{u}'\mathbf{u}'' = 0.
$$

where $\mathbf{T}_{2\times2\times2} = (T_{ijk})$ is a $2 \times 2 \times 2$ homogeneous tensor whose components T_{ijk} are 3×3 minors of the following 6×3 joint projection matrix:

$$
\begin{pmatrix}
\mathbf{M} \\
\mathbf{M}' \\
\mathbf{M}''
\end{pmatrix} =
\begin{pmatrix}
1 \\
2 \\
1' \\
2' \\
1'' \\
2''
\end{pmatrix}. \tag{9}
$$

The components of the tensor can be made explicit as

$$
T_{ijk} = [\bar{i}\,\bar{j}'\,\bar{k}''], \text{ for } i, j', k'' = 1, 2.
$$

where the bracket $[ij'k'']$ denotes the 3×3 minor of i-th, j'-th and k''-th row vector of the above joint projection matrix and bar "‾" in $\bar{i}, \bar{j}$ and $\bar{k}$ denotes the mapping

$$
(1, 2) \mapsto (2, -1).
$$

It can be easily seen that any constraint obtained by adding further views reduces to a trilinearity. This proves the uniqueness of the trilinear constraint. Moreover, the $2 \times 2 \times 2$ homogeneous tensor $\mathbf{T}_{2\times2\times2}$ has $7 = 2 \times 2 \times 2 - 1$ d.o.f., so it is a minimal parametrization of three views since three views have exactly

$$
3 \times (2 \times 3 - 1) - (3 \times 3 - 1) = 7
$$

d.o.f., up to a projective transformation in $\mathcal{P}^2$.

Each correspondence over three views gives one linear constraint on the tensor components T_{ijk}. With at least 7 points in $\mathcal{P}^1$, the tensor components T_{ijk} can be estimated linearly.

At this point, we have obtained a remarkable result that for the one-dimensional projective camera, the trilinear tensor encapsulates exactly the information needed for projective reconstruction in $\mathcal{P}^2$. Namely, it is the unique matching constraint, it minimally parametrizes the three views and it can be estimated linearly. Contrast this to the 2D image case in which the multilinear constraints are algebraically redundant and the linear estimation is only an approximation based on over-parametrization.

3.2 2D projective reconstruction by rescaling

According to Triggs [26], the projective reconstruction in $\mathcal{P}^3$ can be viewed as being equivalent to the rescaling of the image points in $\mathcal{P}^2$. We have just proven that recovering the directions of affine lines in 3D space is equivalent to 2D projective reconstruction from one-dimensional projective images. Therefore, a reconstruction of the directions of 3D affine lines can be obtained by rescaling the direction vectors of image lines, viewed as points of $\mathcal{P}^1$.

For each 1D image point through in views (*cf.* Equation (5)), the scale factors λ, λ' and λ''–taken individually– are arbitrary: However, taken as a whole $(\lambda\mathbf{u}, \lambda'\mathbf{u}', \lambda''\mathbf{u}'')^T$, they encode the projective structure of the points $\mathbf{x}$ in $\mathcal{P}^2$. One way to explicitly recover the scale factors $(\lambda, \lambda', \lambda'')^T$ is to notice that the rescaled image coordinates $(\lambda\mathbf{u}, \lambda'\mathbf{u}', \lambda''\mathbf{u}'')^T$ should lie in the joint image, or alternatively to observe the following matrix identity:

$$\begin{pmatrix} \mathbf{M} & \lambda\mathbf{u} \\ \mathbf{M}' & \lambda'\mathbf{u}' \\ \mathbf{M}'' & \lambda''\mathbf{u}'' \end{pmatrix} = \begin{pmatrix} \mathbf{M} \\ \mathbf{M}' \\ \mathbf{M}'' \end{pmatrix} \left(\mathbf{I}_{3\times3} \; \mathbf{x} \right).$$

The rank of the left matrix is therefore at most 3. All 4×4 minors vanish. Expanding by cofactors in the last column gives homogeneous linear equations in the components of $\lambda\mathbf{u}$, $\lambda'\mathbf{u}'$ and $\lambda''\mathbf{u}''$ with coefficients that are 3×3 minors of the joint projection matrix:

$$\mathbf{T}_{\cdot jk}(\lambda\mathbf{u}) - \mathbf{e}_1''(\lambda'\mathbf{u}')^T + \mathbf{e}_1'(\lambda''\mathbf{u}'')^T = \mathbf{0}_{2\times2}, \tag{10}$$

where $\mathbf{T}_{\cdot jk}\mathbf{u}$ is for $\sum_{i=1}^{2} \mathbf{T}_{ijk}\mathbf{u}^i$, a 2×2 matrix.

There are two types of minors: Those involving three views with one row from each view and those involving two views with two rows from one view and one from the other. The first type gives the 8 components of the tensor $\mathbf{T}_{2\times2\times2}$ and the second type gives 12 components of the "*epipoles*" $\mathbf{e}_1, \mathbf{e}_2, \mathbf{e}_1', \mathbf{e}_2', \mathbf{e}_1'', \mathbf{e}_2''$. The epipoles are defined by analogy with the 2D camera case, as the projection of one projection center onto another view.

At present we only know T_{ijk}–the epipoles are still unknown. To find the rescaling factors for projective reconstrucion, we need to solve for the epipoles. One way to proceed is as follows. Taking $\mathbf{x}$ to be the projection center of the second view $\mathbf{o}'$, and projecting into the three views, Equation (10) reduces to

$$\lambda\mathbf{T}_{\cdot jk}\mathbf{e}_2 = -\lambda''\mathbf{e}_1'\mathbf{e}_2''^T.$$

As $\mathbf{e}_1'\mathbf{e}''^T$ has rank 1, so does $\mathbf{T}_{\cdot jk}\mathbf{e}_2$. Its 2×2 determinant must vanish, *i.e.*

$$|\mathbf{T}_{\cdot jk}\mathbf{e}_2| = 0.$$

As each entry of the 2×2 matrix is homogeneous linear in $\mathbf{e}_2 = (u_1, u_2)^T$, the expansion of $|\mathbf{T}_{\cdot jk}\mathbf{e}_2|$ gives a homogeneous quadratic

$$\alpha u_1^2 + \beta u_1 u_2 + \gamma u_2^2 = 0, \tag{11}$$

where α, β, γ are known in terms of T_{ijk}.

Doing the same thing with the projection center of the third view $\mathbf{o}''$ gives

$$\lambda \mathbf{T}_{\cdot jk}\mathbf{e}_3 = \lambda' \mathbf{e}_1'' \mathbf{e}_3'^T.$$

and hence

$$|\mathbf{T}_{\cdot jk}\mathbf{e}_3| = 0.$$

In other words, it leads to exactly the same quadratic equation (11) with $\mathbf{e}_3$ replacing $\mathbf{e}_2$. The two solutions of the quadratic (11) are $\mathbf{e}_2$ and $\mathbf{e}_3$–only the ordering remains ambiguous.

The other epipoles are easily obtained, $\mathbf{e}_1'$ and $\mathbf{e}_2''$ by factorizing the matrix $\mathbf{T}_{\cdot jk}\mathbf{e}_2$ and $\mathbf{e}_1''$ and $\mathbf{e}_1'$ by factorizing $\mathbf{T}_{\cdot jk}\mathbf{e}_3$.

If the first solution set is

$$\{\tilde{\mathbf{e}}_2, \tilde{\mathbf{e}}_1', \tilde{\mathbf{e}}_2'', \tilde{\mathbf{e}}_3, \tilde{\mathbf{e}}_1'', \tilde{\mathbf{e}}_3'\},$$

the reordering gives the second solution set

$$\{\mathbf{e}_3 = \tilde{\mathbf{e}}_2, \mathbf{e}_1'' = \tilde{\mathbf{e}}_1', \mathbf{e}_3' = \tilde{\mathbf{e}}_2'', \mathbf{e}_2 = \tilde{\mathbf{e}}_3'', \mathbf{e}_1' = \tilde{\mathbf{e}}_1'', \mathbf{e}_2'' = \tilde{\mathbf{e}}_3'\}.$$

Once all the epipoles have been recovered, the scale factors of the image *"points"* for 3D direction reconstruction can easily be recovered by solving the linear homogeneous equation (10).

4 Conclusion

In view of the results obtained above, we can establish the following.

For the recovery of affine shape and affine motion from line correspondences with an uncalibrated affine camera, the minimum number of views needed is three and the minimum number of lines required is seven for a linear solution. The recovery is unique up to a re-ordering of the views.

This result can be compared with that of Koenderink and Van Doorn [9] for affine structure with a minimum of two views and five points.

Acknowledgement

This work was partly supported by European project CUMULI which is gratefully acknowledged. The key idea of using the full tensorial approach to study the line reconstruction ambiguity comes from a discussion with B. Triggs who also provided much aid on tensor representation and interesting discussion on his work.

References

1. D. Dementhon and L.S. Davis. Model-based object pose in 25 lines of code. *IJCV*, 15(1/2):123–141, 1995.
2. O.D. Faugeras, F. Lustman, and G. Toscani. Motion and structure from point and line matches. In *ICCV*, June 1987.
3. O. Faugeras. What can be seen in three dimensions with an uncalibrated stereo rig? In *ECCV*, pages 563–578. May 1992.
4. O. Faugeras and B. Mourrain. About the correspondence of points between N images. In *Workshop on Representations of Visual Scenes*, June 1995.
5. R. Hartley, R. Gupta, and T. Chang. Stereo from uncalibrated cameras. In *CVPR*, pages 761–764, 1992.
6. R. Hartley. Lines and Points in Three Views - An Integrated Approach. Technical report, G.E. CRD, 1994.
7. R.I. Hartley. Projective reconstruction from line correspondences. In *CVPR*, 1994.
8. R. Horaud, S. Christy, and F. Dornaika. Object Pose: the Link between Weak Perspective, Para Perspective, and Full Perspective. Technical report, Inria, September 1994.
9. J.J. Koenderink and A. J. Van Doorn. Affine structure from motion. Technical report, Utrecht University, Utrecht, The Netherlands, October 1989.
10. C.H. Lee and T. Huang. Finding point correspondences and determining motion of a rigid object from two weak perspective views. *CVGIP*, 52:309–327, 1990.
11. Y. Liu and T.S. Huang. A linear algorithm for motion estimation using straight line correspondences. *CVGIP*, 44(1):35–57, October 1988.
12. H.C. Longuet-Higgins, A computer program for reconstructing a scene from two projections. in *Nature*, vol. 293, pp. 133–135. XX, September 1981.
13. Q.T. Luong and T. Vieville. Canonic Representations for the Geometries of Multiple Projective Views. In *ECCV*, pages 589–599. May 1994.
14. Ph.F. McLauchlan, I.D. Reid and D.W. Murray. Recursive affine structure and motion from image sequences. In *ECCV*, pages 217–224. May 1994.
15. R. Mohr, L. Quan, and F. Veillon, "Relative 3D reconstruction using multiple uncalibrated images", *The International Journal of Robotics Research*, vol. 14, no. 6, pp. 619–632, 1995.
16. J.L. Mundy and A. Zisserman, editors. *Geometric Invariance in Computer Vision*. MIT Press, Cambridge, Massachusetts, USA, 1992.
17. C. J. Poelman and T. Kanade. A paraperspective factorization method for shape and motion recovery. In *ECCV*, pages 97–108, May 1994.
18. L. Quan and R. Mohr. Affine shape representation from motion through reference points. *Journal of Mathematical Imaging and Vision*, 1:145–151, 1992. also in IEEE Workshop on Visual Motion, New Jersey, pages 249–254, 1991.
19. L. Quan. Self-calibration of an affine camera from multiple views. *IJCV*, 1996.
20. L. Quan and T. Kanade. A factorization method for shape and motion from line correspondences. In *CVPR*, June, 1996.
21. L. Quan and T. Kanade. Affine Structure from Line Correspondences with Uncalibrated Af fine Cameras. To appear in IEEE T-PAMI, 1997.
22. L.S. Shapiro, A. Zisserman, and M. Brady. Motion from point matches using affine epipolar geometry. *IJCV*, 1994.
23. A. Shashua. Algebraic functions for recognition. IEEE *T-PAMI*, 1994. in press.
24. M. Spetsakis and J. Aloimonos. A Unified theory of structure from motion. In *Proceedings DARPA IU Workshop*, 1990.

25. C. Tomasi and T. Kanade. Shape and motion from image streams under orthography: A factorization method. *IJCV*, 9(2):137–154, 1992.

26. B. Triggs. The geometry of projective reconstruction I: Matching constraints and the joint image. In *ICCV*, 1995.

27. S. Ullman and R. Basri. Recognition by linear combinations of models. IEEE *T-PAMI*, 13(10):992–1006, 1991.

28. D. Weinshall and C. Tomasi. Linear and incremental acquisition of invariant shape models from image sequences. In *ICCV*. 1993.

29. T. Vieville and O. Faugeras and Q.T. Luong. Motion of points and lines in the uncalibrated case. *IJCV*, 7(3):211–241, 1996.

30. J. Weng and T.S. Huang and N. Ahuja. Motion and structure from line correspondences: Closed-form solution, uniqueness, and optimization. IEEE *T-PAMI*, 14(3):318–336, 1992.

31. Z. Zhang and O. Faugeras. Three-dimensional motion computation and object segmentation in a long sequence of stereo frames. *IJCV*, 7(3):211–241, 1992.

Distortions of Stereoscopic Visual Space and Quadratic Cremona Transformations

Gregory Baratoff

Center for Automation Research
University of Maryland
College Park, MD 20742, USA

Abstract. When incorrect values for the extrinsic and intrinsic parameters of the two cameras of a stereo rig are used in the reconstruction of a three-dimensional scene from image correspondences, the resulting reconstruction is distorted in a systematic way. We show here that the transformation between the true scene structure and its distorted reconstruction is a *quadratic Cremona transformation* - a rational transformation which is one-to-one almost everywhere, but which does not in general preserve collinearity.

We study the distortion of points in the fixation plane on a global, qualitative, level, and on a local, quantitative level. Both global and local viewpoints provide evidence of severe non-linear distortions in the vicinity of the camera centers, thereby indicating that their consideration is of particularly high relevance for near regions of the stereo rig. This is consistent with experimental evidence from psychophysics, which shows that distortions in the near range can not be adequately described by linear transformations.

Our distortion framework describes and enables the analysis of situations where insufficient information is available to compute even a weak calibration of a stereo rig. It also makes possible a thorough error analysis of systematic errors in computing structure from motion.

1   Introduction

A faithful reconstruction of a scene from image correspondences in two views requires knowledge of the relative orientation (translation and rotation) of the two cameras as well as their intrinsic parameters. Here, we will assume that a set of values for these parameters, henceforth called *apparent* parameter values, has been obtained. When, however, the apparent values are different from the true ones, according to which the image projections are in reality obtained, the resulting *distorted reconstruction* is distorted in a systematic way. Such distortions have been studied before [2, 3] in the case of calibrated cameras, and under the assumption of infinitesimal motion. We show in section 2 that the transformation between the true scene structure in the fixation plane and its distorted reconstruction is a quadratic plane Cremona transformation. One can extend this result to show that the transformation between true and distorted 3D structure is a quadratic space Cremona transformation[1], but in this paper we restrict ourselves to the analysis of the fixation plane. Quadratic distortions have been observed by photogrammetrists [7] in certain stereoscopes. In the Computer Vision literature, (plane) quadratic transformations have been used to describe the mapping between corresponding points in two image planes (e.g. [6]), but have not been applied to the tranformation of space as is done here.

Cremona transformations are rational transformations which are one-to-one almost everywhere. They do not in general preserve collinearity, except when they are of degree 1, i.e. when they are projective transformations, or collineations. Concretely, this means that for example a line - a curve with zero curvature - in general maps to a curve with non-vanishing curvature.

In section 3 we first describe some properties of general quadratic planar Cremona transformations, and then apply them to the distortion of the plane of fixation. This provides a qualitative explanation of some global distortion phenomena. In section 4 we then derive the distortion of local differential-geometric properties of curves, namely their tangent and curvature. The analysis is thus complemented by a quantitative account of the local distortions.

2 The Planar Shape Distortion

2.1 Homogeneous Point and Line Coordinates

The use of homogeneous coordinates for points and lines in the plane simplifies expression considerably. We use bold-face letters for both, upper-case ones for points (e.g. $\mathbf{X}$), and lower-case ones for lines (e.g. $\mathbf{m}$). Homogeneous vectors for points and lines in the plane have three components, and they allow the representation of points and lines at infinity. A useful convention is for finite points $\mathbf{X} = (X \ Z \ W)^T$ to satisfy $W \neq 0$, and for points at infinity to satisfy $W = 0$. The line at infinity $\mathbf{h}_\infty$ is given by the homogeneous vector $(0 \ 0 \ 1)$, so that $\mathbf{h}_\infty \cdot \mathbf{X} = 0 \Rightarrow W = 0$.

The cross-product of homogeneous vectors defines both the joining of two points to a line ($\mathbf{m} = \mathbf{X} \times \mathbf{Y}$), and the intersection of two lines in a point ($\mathbf{X} = \mathbf{m}_1 \times \mathbf{m}_2$). The scalar product $\mathbf{m} \cdot \mathbf{X}$ between a line and a point vector defines their scaled distance. Thus, $\mathbf{X}$ lies on $\mathbf{m}$ when $\mathbf{m} \cdot \mathbf{X} = 0$. It follows from this that the homogeneous coordinate vector of a point or of a line is only given up to scale, i.e. $\lambda \mathbf{X}$ represents the same point as $\mathbf{X}$, for any non-zero scalar λ.

The linear combination $\mu_1 \mathbf{m}_1 + \mu_1 \mathbf{m}_2$ of two lines $\mathbf{m}_1$ and $\mathbf{m}_2$ defines a *line pencil* with *vertex* $\mathbf{m}_1 \times \mathbf{m}_2$, i.e. a one-dimensional family of lines that pass through the intersection point of the two basis lines $\mathbf{m}_1$ and $\mathbf{m}_2$. Any line of the pencil can be written as $\mathbf{m} = M\boldsymbol{\mu}$, where $M = (\mathbf{m}_1 \ \mathbf{m}_2)$ and $\boldsymbol{\mu} = (\mu_1 \ \mu_2)^T$. Here, $\boldsymbol{\mu}$ are the (homogeneous) *line-pencil coordinates* of $\mathbf{m}$ in the basis M.

2.2 Camera Geometry

The introduction of homogeneous coordinates allows us to treat the shape distortion problem in great generality. It makes possible arbitrary positioning and orientation of the cameras with respect to the world coordinate frame, as well as a consideration of all intrinsic parameters of a pinhole camera. A camera projection matrix is a general 2×3 matrix $A = ST_{ext}$, where S is the 2×2 matrix of intrinsic parameters, $T_{ext} = (R \mid -R\hat{O})$ the matrix of extrinsic parameters, with R a 2×2 rotation matrix, and $(\hat{O}^T \mid 1)^T$ the camera center. The first and second row of a camera matrix represent the X-axis and the Z-axis, respectively, of the camera frame. Thus, a one-dimensional camera can be regarded as a line pencil, and the camera projection matrix is simply the (transpose of the) line-pencil matrix.

Let $A = (\mathbf{a}_1 \ \mathbf{a}_2)$ and $B = (\mathbf{b}_1 \ \mathbf{b}_2)$ be the true left and right cameras. The camera centers are given by $\mathbf{O}_L \doteq \mathbf{a}_1 \times \mathbf{a}_2$ and $\mathbf{O}_R \doteq \mathbf{b}_1 \times \mathbf{b}_2$. The image projections of a point $\mathbf{X}$ in the left and right cameras are :

$$x_L = \frac{X_L}{Z_L} = \frac{\mathbf{a}_1 \cdot \mathbf{X}}{\mathbf{a}_2 \cdot \mathbf{X}} \qquad x_R = \frac{X_R}{Z_R} = \frac{\mathbf{b}_1 \cdot \mathbf{X}}{\mathbf{b}_2 \cdot \mathbf{X}} \tag{1}$$

2.3 The Distorted Reconstruction

In order to reconstruct $\mathbf{X}$ as a function of the image coordinates, we first rewrite (1) as $(x_L \mathbf{a}_2 - \mathbf{a}_1) \cdot \mathbf{X} = 0$ and $(x_R \mathbf{b}_2 - \mathbf{b}_1) \cdot \mathbf{X} = 0$. Each of these equations is the equation of a line, and therefore $\mathbf{X}$ is given by their intersection

$$\mathbf{X} = (x_L \mathbf{a}_2 - \mathbf{a}_1) \times (x_R \mathbf{b}_2 - \mathbf{b}_1) \tag{2}$$

If instead of using $\mathbf{a}_i$ and $\mathbf{b}_i$ we use $\mathbf{a}'_i$ and $\mathbf{b}'_i$, we obtain the distorted point $\mathbf{X}'$. When we then replace the image coordinates x_L and x_R by their expressions (1) in terms of the true quantities, we obtain :

Proposition 1. *(Planar Shape Distortion)*
The transformation between the true and the distorted reconstruction is a quadratic plane Cremona transformation given by :

$$\mathbf{T} : \mathbf{X} \mapsto \mathbf{X}' = \mathbf{a}'(\mathbf{X}) \times \mathbf{b}'(\mathbf{X}) = P\mathbf{X} \times Q\mathbf{X} \tag{3}$$

where the direct distorted rays are

$$\mathbf{a}'(\mathbf{X}) \doteq (\mathbf{a}_1 \cdot \mathbf{X})\mathbf{a}'_2 - (\mathbf{a}_2 \cdot \mathbf{X})\mathbf{a}'_1 = (\mathbf{a}'_2\mathbf{a}_1^T - \mathbf{a}'_1\mathbf{a}_2^T)\mathbf{X} \doteq P\mathbf{X} \tag{4}$$
$$\mathbf{b}'(\mathbf{X}) \doteq (\mathbf{b}_1 \cdot \mathbf{X})\mathbf{b}'_2 - (\mathbf{b}_2 \cdot \mathbf{X})\mathbf{b}'_1 = (\mathbf{b}'_2\mathbf{b}_1^T - \mathbf{b}'_1\mathbf{b}_2^T)\mathbf{X} \doteq Q\mathbf{X}. \tag{5}$$

Its reverse is given by

$$\mathbf{T}' : \mathbf{X}' \mapsto \mathbf{X} = \mathbf{a}(\mathbf{X}') \times \mathbf{b}(\mathbf{X}') = P^T\mathbf{X}' \times Q^T\mathbf{X}' \tag{6}$$

where the reverse distorted rays are

$$\mathbf{a}(\mathbf{X}') \doteq (\mathbf{a}'_1 \cdot \mathbf{X}')\mathbf{a}_2 - (\mathbf{a}'_2 \cdot \mathbf{X}')\mathbf{a}_1 = (\mathbf{a}_2\mathbf{a}'^T_1 - \mathbf{a}_1\mathbf{a}'^T_2)\mathbf{X}' = -P^T\mathbf{X}' \tag{7}$$
$$\mathbf{b}(\mathbf{X}') \doteq (\mathbf{b}'_1 \cdot \mathbf{X}')\mathbf{b}_2 - (\mathbf{b}'_2 \cdot \mathbf{X}')\mathbf{b}_1 = (\mathbf{b}_2\mathbf{b}'^T_1 - \mathbf{b}_1\mathbf{b}'^T_2)\mathbf{X}' = -Q^T\mathbf{X}'. \tag{8}$$

The reverse transformation is simply obtained by exchanging the true and the apparent quantities.

2.4 An Example

What happens qualitatively in the fixation plane under such a distortion can be seen in Fig. 1, where a rectangular grid is shown together with two distorted reconstructions corresponding to an under-, respectively an over-estimation, of the fixation angle. From this example, we see that straight line segments are not

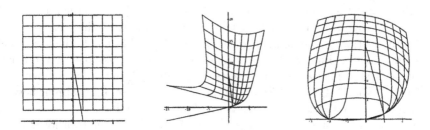

Fig. 1. Distortion of a rectangular grid. From left to right : undistorted grid, distorted grid due to underestimation of the fixation angle, distorted grid due to overestimation of fixation angle. Also shown are the fixation triangles (two camera centers plus fixation point).

preserved, but map into curved segments. In fact, since the transformation (3) is quadratic in (X, Z, W) these curves are of second order, i.e. they are conic sections. Nevertheless, it is clear that the transformation is one-to-one for most points, with the exception of a few regions. We will have more to say about them later.

3 Planar Quadratic Cremona Transformations

Before discussing the particular quadratic transformation (3), we first give some general properties of planar quadratic Cremona transformations. Our discussion is based on *Semple & Roth*[8].

Plane Cremona transformations are rational transformations whose inverse is also rational. This means that they are one-to-one mappings in general, with the exception of a finite number of *base points* at which they are one-to-many, and a finite number of *fundamental curves* where they are many-to-one, i.e. each point of a fundamental curve maps to the same point. Such exceptional elements must exist if the order of the Cremona transformation is greater than one, i.e. if the transformation is not linear. Instead of the inverse, we use the *reverse* transformation. Both map to the same points, but the homogeneous coordinate vectors they use to represent points differ by a multiplicative factor. The reverse transformation has the advantage that it has the same degree as the (forward) transformation, and has the same number of base points and fundamental lines. For *quadratic* transformations, the base and fundamental elements are :

Theorem 2. *(Base Points and Fundamental Lines of Quadratic Cremona Transformations [8, page 48])*

A plane quadratic Cremona transformation **T** *has three base points. If they are all distinct, then* **T** *has three fundamental lines, each one passing through two of the base points.* **T** *and the reverse transformation* **T′** *establish a one-to-one correspondence between the base points of* **T** *and the fundamental lines of* **T′**, *and between the base points of* **T′** *and the fundamental lines of* **T**.

Let $\mathbf{B}_i$, $(i = 1, 2, 3)$ denote the base points of **T**, and let $\mathbf{f}_i$ denote the fundamental lines of **T**, with the meaning that $\mathbf{f}_i$ is the line "opposite" the base point $\mathbf{B}_i$. Similarly, let the corresponding primed symbols $\mathbf{B}_i'$ and $\mathbf{f}_i'$ denote the base points and fundamental lines of **T′**. It will be convenient to assume that the base points and fundamental lines are *consistently indexed*[1], by which we mean that $\mathbf{T}(\mathbf{f}_i) = \mathbf{B}_i'$ and $\mathbf{T}'(\mathbf{f}_i') = \mathbf{B}_i$. Figure 2a) gives an example of the mapping between such consistently indexed base points and fundamental lines.

Let us now consider the transformation of lines. Since the transformation is of second order, we would expect a generic line to map to a conic section, a curve of second order. Indeed, this is usually the case. If, however, the curve passes through a base point, then by Thm. 2 we know that the base point maps to a fundamental line of the reverse. Since a line has order 1, the remaining points of the curve must also map to a line, the order having been reduced by 1 due to the "splitting off" of the fundamental line. Figure 2b) illustrates the three different possibilities for the transformation of lines :

- L_1 : Fundamental line maps to a base point.
- L_2 : Line through one base point maps to a line through one base point.
- L_3 : Generic line maps to a conic through all three base points.

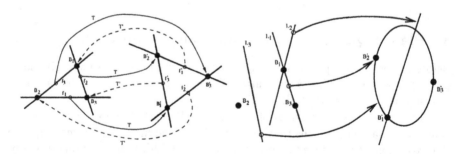

Fig. 2. a) Mapping between base points and fundamental lines. b) Transformation of lines.

3.1 Global Qualitative Distortion Phenomena

The previous section has shown that certain global properties of the distortion of lines can be explained by their location relative to the base points and fundamental lines of a quadratic Cremona transformation. In this section, we apply this discussion to the shape distortion derived in section 2. First, we find the base points of this transformation :

Proposition 3. *The base points of the shape distortion transformation (3) and of its reverse (6) are :*

$$B_1 = O_L \quad B_2 = O_R \quad B_3 = a(O_R') \times b(O_L')$$
$$B_1' = O_L' \quad B_2' = O_R' \quad B_3' = a'(O_R) \times b'(O_L)$$

A proof of this can be found in [1]. The proposition states a most interesting result, namely that *the camera centers are base points*. We can now interpret the transformation of lines by general planar Cremona transformations discussed in section 3 in the context of our distortion. Some consequences are :

- Camera rays remain camera rays, since rays are lines through camera centers, which by Proposition 3 are base points (L_2 in Fig. 2).
- In general, lines map to conic sections passing through both camera centers (L_3 in Fig. 2).

From this last fact that almost all lines map to curves that necessarily pass through both camera centers it follows that *the distortion is particularly severe in the vicinity of the camera centers* where the images of all lines converge. This clearly happens with the distorted grid in Figure 1. This suggests that the investigation of distortions should be of particular relevance for objects in the near range of the stereo rig.

4 Distortion of Curvature

Let a line be given in the apparent space with equation $\mathbf{m}' \cdot \mathbf{X}' = 0$. We define its image under the reverse distortion as : $c(\mathbf{X}) \doteq \mathbf{m}' \cdot \mathbf{T}(\mathbf{X}) = 0$. Its tangent line, given by its gradient, is :

$$\nabla c(\mathbf{X}) = \mathbf{G}(\mathbf{X})\mathbf{m}' \tag{9}$$

where we defined $\mathbf{G}(\mathbf{X}) \doteq (\frac{\partial}{\partial \mathbf{X}}(\mathbf{T}(\mathbf{X})))^T$. We then have

$$\frac{\partial}{\partial \mathbf{X}}(\mathbf{T}(\mathbf{X})) = \frac{d}{d\mathbf{X}}(P\mathbf{X} \times Q\mathbf{X}) = [P\mathbf{X}]_\times \frac{\partial}{\partial \mathbf{X}}(Q\mathbf{X}) - [Q\mathbf{X}]_\times \frac{\partial}{\partial \mathbf{X}}(P\mathbf{X})$$
$$= [P\mathbf{X}]_\times Q - [Q\mathbf{X}]_\times P$$

so that $\mathbf{G}(\mathbf{X}) = -Q^T[P\mathbf{X}]_\times + P^T[Q\mathbf{X}]_\times$.

Next, we present an expression for the curvature of the conic section that is the image of the line $\mathbf{m}'$. In [1], we derive the curvature distortion for general curves. For the special case of lines, the curvature is given by :

$$\kappa_0 = |W| \frac{\mathbf{P}_\infty^T \mathcal{H} \mathbf{P}_\infty}{\|\mathbf{P}_\infty\|^3}, \tag{10}$$

where W is the third component of $\mathbf{X}$, $\mathcal{H} = \frac{\partial^2 c(\mathbf{X})}{\partial \mathbf{X}^2}$ is the Hessian matrix of $c(\mathbf{X})$, and $\mathbf{P}_\infty \doteq \nabla c(\mathbf{X}) \times \mathbf{h}_\infty$ is the point at infinity on the tangent line to $c(\mathbf{X})$ at $\mathbf{X}$.

Before deriving the Hessian, we first rewrite the expression for the tangent line $\nabla c(\mathbf{X}) = \mathbf{G}(\mathbf{X})\mathbf{m}'$:

$$\nabla c(\mathbf{X}) = -Q^T(P\mathbf{X} \times \mathbf{m}') + P^T(Q\mathbf{X} \times \mathbf{m}')$$
$$= Q^T[\mathbf{m}']_\times P\mathbf{X} - P^T[\mathbf{m}']_\times Q\mathbf{X}$$

From this, we obtain for the Hessian : $\mathcal{H} = Q^T[\mathbf{m}']_\times P - P^T[\mathbf{m}']_\times Q$, which allows us to compute the curvature κ_0 of the distortion of the line $\mathbf{m}'$ using formula (10).

The global analysis in the previous section showed that for each line pencil there are three special lines, namely those passing through the three base points. These special lines remain lines, and thus have a zero curvature. Figure 3a) shows the curvature as a function of the orientation of the original line at a point in front of and close to the right camera center. It is clearly seen that the curvature has three roots, corresponding to the lines through the base points.

In order to get a picture of the distortion in the fixation plane, we next plot the maximal distortion (corresponding to the point with the highest peak in Figure 3a)) for the points on a line passing through the right camera center and perpendicular to the baseline connecting the two camera centers. The plot

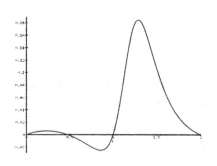
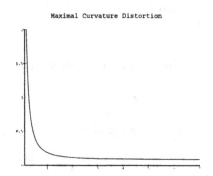

a) Curvature distortion for line pencil   b) Maximal curvature distortion

Fig. 3. a) Curvature of distorted line as a function of line orientation. Parametrization of orientation is such that $\mathbf{m}'(0) = \mathbf{X}' \times \mathbf{O}'_L$ and $\mathbf{m}'(1) = \mathbf{X}' \times \mathbf{O}'_R$.
b) Maximal curvature distortion for points along right camera ray. x-axis corresponds to distance from right camera center.

shows that the curvature increases monotonically with decreasing distance to the camera center. This provides a quantitative account of the qualitative explanation given in section 3.1, where we observed that the distortions are particularly severe in the vicinity of the camera centers.

5   Conclusions

In this paper, we introduced a framework that accounts for systematic distortions occurring in stereo and motion resulting from incorrect calibration of both intrinsic and extrinsic parameters. The full range of distortions for a stereo rig was shown to be exactly modelled by quadratic Cremona transformations. We analyzed the distortion of curves in the fixation plane on a qualitative, global level, as well as on a quantitative, local level. Both approaches indicate that distortions are particularly severe in the near range of the stereo rig. This is consistent with experimental evidence from psychophysics[9], which shows that distortions in the near range can not be adequately described by affine transformations.

Faugeras[4] showed that the *epipolar geometry* needs to be known if a reconstruction up to a projective transformation (a so-called projective reconstruction) is to be obtained. Thus, second-order distortions of the perceived space arise when sufficient information is not available to compute the epipolar geometry (e.g. too few point correspondences), or if the system does not have - or chooses not to invest - the computational resources necessary for the estimation of the epipolar geometry. The consideration of these distortions is also of importance for systems that do compute the epipolar geometry, since it makes possible a thorough error analysis of systematic structure errors that goes beyond a simple first-order sensitivity analysis.

References

1. G. Baratoff, "Distortion of Stereoscopic Visual Space", Tech. Rep. CAR-TR-861, Center for Automation Research, Univ. of Maryland, College Park, USA, May 1997.
2. L. Cheong and Y. Aloimonos, "Iso-distortion contours and Egomotion Estimation," in *Proc. Int. Symposium on Computer Vision*, pp. 55-60, 1995.
3. L. Cheong, C. Fermüller and Y. Aloimonos, "Interaction Between 3D Shape and Motion : Theory and Applications," Technical Report CAR-TR-773, Center for Automation Research, Univ. of Maryland, College Park, USA, June 1996.
4. O. Faugeras, "What can be seen in three dimensions with an uncalibrated stereo rig," in *Proc. 2nd European Conf. on Computer Vision*, G. Sandini, ed., Vol. 538 of Lecture Notes in Computer Science (Springer Verlag, Berlin), pp. 563-578, 1992.
5. C. Fermüller, L. Cheong and Y. Aloimonos, "Explaining Human Visual Space Distortion," Technical Report CAR-TR-833, Center for Automation Research, Univ. of Maryland, College Park, USA, July 1996.
6. S. Maybank, "Theory of Reconstruction from Image Motion", Springer-Verlag, Berlin, 1993.
7. C. C. Slama and C. Theurer and S. W. Henriksen, "Manual of Photogrammetry", Am. Soc. of Photogrammetry, 1980.
8. J.G. Semple and L. Roth, *Introduction to Algebraic Geometry*, Oxford, 1949.
9. J. S. Tittle, J. T. Todd, V. J. Perotti, and J. F. Norman, "Systematic Distortion of Perceived Three-Dimensional Structure From Motion and Binocular Stereopsis", *J. of Experimental Psychology*, vol. 21 (1995), no. 3, pp. 663-678.

Self-Evaluation for Active Vision
by the Geometric Information Criterion

Kenichi Kanatani

Department of Computer Science, Gunma University, Kiryu, Gunma 376 Japan
kanatani@cs.gunma-u.ac.jp

Abstract. We present a scheme for evaluating the "goodness" of camera motion for robust 3-D reconstruction by means of the *geometric information criterion* (*geometric AIC*). The evaluation does not require any knowledge about the environment, the device, and the image processing techniques by which the images are obtained, and we need not introduce any thresholds to be adjusted empirically.

1 Introduction

Suppose a robot is to reconstruct 3-D structures from camera images. The paradigm of *active vision* is to allow a robot to control its motion in such a way that the resulting images make robust 3-D reconstruction possible. In order to do robust 3-D reconstruction, the camera must be displaced over a long distance so that the resulting disparity is sufficiently large. As the disparity increases, however, it is increasingly difficult to match feature points over the two images due to possible occlusions and illumination changes. Feature matching becomes easier as the disparity decreases; for each feature point, only a small neighborhood needs to be searched. It is therefore desirable to keep the camera displacement minimum in such a way that the resulting disparity is sufficient for reliable 3-D reconstruction. This means we need a measure of the "goodness" of the images for evaluating whether or not the disparity is sufficient; without such a measure the active vision paradigm is meaningless. This fact does not have received much attention in the past. In this paper, we show that we can do *self-evaluation* of the images as described above in such a way that (i) the evaluation does not require any knowledge about the environment, the device, and the image processing techniques by which the images are obtained and (ii) we need not introduce any thresholds to be adjusted empirically.

2 Three Models for Structure from Motion

We define an XYZ camera coordinate system in such a way that the origin O is at the center of the lens and the Z-axis is in the direction of the optical axis. With an appropriate scaling, the image plane can be identified with the plane $Z = 1$, on which an xy image coordinate system is defined in such a way that the origin o is on the Z-axis and the x- and y-axes are parallel to the X- and Y-axes,

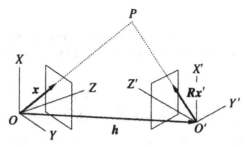

Fig. 1. Geometry of camera motion.

respectively. Suppose the camera moves to a position defined by translating the first camera by vector h and rotating it around the center of the lens by matrix R (Fig. 1); we call $\{h, R\}$ the *motion parameters*.

Let (x_α, y_α), $\alpha = 1, ..., N$, be the image coordinates of a feature point before the motion, and (x'_α, y'_α), $\alpha = 1, ..., N$, those after the motion. We use the following three-dimensional vectors to represent them (the superscript T denotes transpose):

$$x_\alpha = (x_\alpha, y_\alpha, 1)^\mathsf{T}, \qquad x'_\alpha = (x'_\alpha, y'_\alpha, 1)^\mathsf{T}. \tag{1}$$

In order to reconstruct 3-D structures from the two images, we must first detect correspondences of feature points between the two images. However, images are digitized, and image processing operations such as edge detection and template matching may not be accurate. We refer to the resulting inaccuracy of matching, irrespective of its sources, collectively as *image noise* and view it as a random phenomenon. Here, we regard $\{x_\alpha, x'_\alpha\}$ as independent Gaussian random variables and denote their covariance matrices by $V[x_\alpha]$ and $V[x'_\alpha]$. Since the third components of x_α and x'_α are 1, the covariance matrices $V[x_\alpha]$ and $V[x'_\alpha]$ are singular with rank 2.

The absolute magnitude of the image noise is very difficult to estimate a priori, but often its geometric characteristics, such as homogeneity/inhomogeneity, isotropy/anisotropy, and their relative degrees, can be predicted from the characteristics of the imaging device and the image processing algorithm. Here, we assume that the covariance matrices are known *only up to scale* and write

$$V[x_\alpha] = \epsilon^2 V_0[x_\alpha], \qquad V[x'_\alpha] = \epsilon^2 V_0[x'_\alpha]. \tag{2}$$

The constant ϵ, which indicates the average magnitude of the image noise, is assumed unknown; we call it the *noise level*. The matrices $V_0[x_\alpha]$ and $V_0[x'_\alpha]$ indicate in which orientation the deviation is likely to occur; they are assumed known and called the *normalized covariance matrices*.

General model

An optimal solution of the motion parameters $\{h, R\}$ in the sense of *maximum likelihood estimation* (*MLE*) is obtained by minimizing the following function [4, 6]:

$$J = \sum_{\alpha=1}^{N} \frac{|x_\alpha, h, Rx'_\alpha|^2}{(h \times Rx'_\alpha, V_0[x_\alpha](h \times Rx'_\alpha)) + (h \times x_\alpha, RV_0[x'_\alpha]R^\top(h \times x_\alpha))}. \quad (3)$$

In this paper, (a, b) denotes the inner product of vectors a and b, and $|a, b, c|$ denotes the scalar triple product of vectors a, b, and c. The scale of the translation h is indeterminate, so we normalize it into $\|h\| = 1$. Direct minimization of eq. (3) requires numerical search [2, 11], but the computation can simplified by combining the *linearization* technique with a procedure called *renormalization* and an *optimal correction* scheme [4].

Planar surface model

Consider a planar surface in the scene. Let n be its unit surface normal, and d its distance from the origin O (positive in the direction of n). An optimal solution of the surface parameters $\{n, d\}$ and the motion parameters $\{h, R\}$ in the sense of MLE is obtained by minimizing the following function [6, 8]:

$$J_\Pi = \sum_{\alpha=1}^{N} (x'_\alpha \times Ax_\alpha, W_\alpha x'_\alpha \times Ax_\alpha). \quad (4)$$

$$A = R^\top(hn^\top - dI). \quad (5)$$

$$W_\alpha = \left(x'_\alpha \times AV_0[x_\alpha]A^\top \times x'_\alpha + (Ax_\alpha) \times V_0[x'_\alpha] \times (Ax_\alpha)\right)_2^-. \quad (6)$$

We denote by $(\cdots)_r^-$ the *rank-constrained generalized inverse* computed by replacing all the eigenvalues of $\cdots$ other than the r largest ones by zero in the canonical form and computing the (Moore-Penrose) generalized inverse [6]. The product $v \times T$ of a vector v and a matrix T is defined to be the matrix whose columns are the vector products of v and the three columns of T. For a vector v and a matrix T, the symbol $v \times T \times v$ is an abbreviation of $v \times T(v \times I)^\top$. The renormalization procedure can be applied to compute the matrix A (*homography*) that minimizes eq. (4) [6, 8]. The surface parameters $\{n, d\}$ and the motion parameters $\{h, R\}$ are analytically computed from the resulting matrix A [3, 9, 10].

Pure rotation model

No 3-D information can be obtained if the camera motion is pure rotation around the center of the lens; all we can estimate is the amount of the camera rotation R. An optimal solution of R in the sense of MLE is obtained by minimizing the following function [6]:

$$J_R = \sum_{\alpha=1}^{N} (x_\alpha \times Rx'_\alpha, W_\alpha x_\alpha \times Rx'_\alpha), \quad (7)$$

$$W_\alpha = \left((Rx'_\alpha) \times V_0[x_\alpha] \times (Rx'_\alpha) + x_\alpha \times RV_0[x'_\alpha]R^\top \times x_\alpha\right)_2^-. \quad (8)$$

3   Geometric Model Selection

The above three models are generalized as follows. Suppose we observe m-dimensional vectors a_1, ..., a_N constrained to be in an m'-dimensional manifold $\mathcal{A} \in \mathcal{R}^m$, which we call the *data space*. We regard a_α as a Gaussian random variable of unknown mean $\bar{a}_\alpha$ and covariance matrix $V[a_\alpha]$. Since each a_α is constrained to be in $\mathcal{A}$, its covariance matrix $V[a_\alpha]$ is singular. We assume that it is a positive semi-definite symmetric matrix of rank m' whose range coincides with the tangent space $T_{\bar{a}_\alpha}(\mathcal{A})$ to the data space $\mathcal{A}$ at $\bar{a}_\alpha$. We also assume that $V[a_\alpha]$ is known only up to scale, i.e., we decompose it into the noise level ϵ and the normalized covariance matrix $V_0[a_\alpha]$ in the form

$$V[a_\alpha] = \epsilon^2 V_0[a_\alpha], \tag{9}$$

and assume that $V_0[a_\alpha]$ is known but ϵ is unknown.

Suppose the true values $\bar{a}_\alpha$, $\alpha = 1$, ..., N, are known to satisfy L equations parameterized by an n-dimensional vector u. We assume that the domain of the vector u is an n'-dimensional manifold $\mathcal{U} \subset \mathcal{R}^n$, which we call the *parameter space*. Let $F^{(k)}(a, u)$: $\mathcal{R}^m \times \mathcal{R}^n \rightarrow \mathcal{R}$, $k = 1$, ..., L, be smooth functions of arguments $a \in \mathcal{R}^m$ and $u \in \mathcal{R}^n$. We are to estimate $\bar{a}_\alpha \in \mathcal{A}$, $\alpha = 1$, ..., N, and $u \in \mathcal{U}$ that satisfy

$$F^{(k)}(\bar{a}_\alpha, u) = 0, \quad k = 1, ..., L, \tag{10}$$

from the noisy data $a_\alpha \in \mathcal{A}$, $\alpha = 1$, ..., N.

The L equations $F^{(k)}(a, u) = 0$, $k = 1$, ..., L, need not be algebraically independent with respect to the argument $a \in \mathcal{A}$; we call the number r of independent equations the *rank* of the constraint. In order to avoid pathological cases, we assume that each of the L equations defines a manifold of *codimension 1* in the data space $\mathcal{A}$ in such a way that the L manifolds intersect each other *transversally* [6]. It follows that eq. (10) defines a manifold $\mathcal{S} \subset \mathcal{A}$ of codimension r parameterized by $u \in \mathcal{U}$. We call $\mathcal{S}$ the *geometric model* of eq. (10). Our problem is to estimate the model $\mathcal{S} \subset \mathcal{A}$ that passes through N points $\{\bar{a}_\alpha\} \in \mathcal{A}$ from the noisy data $\{a_\alpha\} \in \mathcal{A}$.

The MLE estimators $\{\hat{a}_\alpha\}$ of $\{\bar{a}_\alpha\}$ are computed by minimizing

$$J = \sum_{\alpha=1}^{N} (a_\alpha - \bar{a}_\alpha, V_0[a_\alpha]^-(a_\alpha - \bar{a}_\alpha)), \tag{11}$$

under the constraint that eq. (10) holds for the value u chosen so that the resulting value J is minimum. We denote the minimum value of J by $\hat{J}$ and call it the *residual*. An unbiased estimator of the squared noise level ϵ^2 is obtained in the form

$$\hat{\epsilon}^2 = \frac{\hat{J}}{rN - n'}. \tag{12}$$

Extending the idea of Akaike [1], we define the *geometric information criterion*, or the *geometric AIC* for short, of model $\mathcal{S}$ as follows [6, 7]:

$$AIC(\mathcal{S}) = \hat{J} + 2(pN + n')\epsilon^2. \tag{13}$$

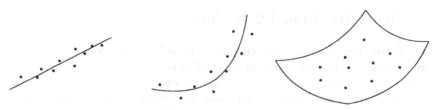

Fig. 2. (Left) A two-dimensional manifold with three degrees of freedom can be fitted. (Middle) A two-dimensional manifold with eight degrees of freedom can be fitted. (Right) A three-dimensional manifold with five degrees of freedom can be fitted.

Let S_1 be a model of dimension p_1 and codimension r_1 with n_1' degrees of freedom, and S_2 a model of dimension p_2 and codimension r_2 with n_2' degrees of freedom. Let $\hat{J}_1$ and $\hat{J}_2$ be their respective residuals. Suppose model S_2 is obtained by adding an additional constraint to model S_1. If model S_1 is correct, the squared noise level ϵ^2 is estimated by eq. (12). Since the geometric AIC estimates the expected sum of squared Mahalanobis distances, the ratio of the deviations from the two models can be evaluated by

$$K = \sqrt{\frac{AIC(S_2)}{AIC(S_1)}} = \sqrt{\frac{r_1 N - n_1'}{(2p_1 + r_1)N + n_1'}\left(\frac{\hat{J}_2}{\hat{J}_1} + \frac{2(p_2 N + n_2')}{r_1 N - n_1'}\right)}. \qquad (14)$$

Using this, we can do the following tests [6, 5, 7]:

Planarity test: The object is judged to be planar if

$$K_\Pi = \sqrt{\frac{N - 5}{7N + 5}\left(\frac{\hat{J}_\Pi}{\hat{J}} + \frac{4N + 16}{N - 5}\right)} < 1. \qquad (15)$$

Rotation test: The camera motion is judged to be a pure rotation if

$$K_R = \sqrt{\frac{N - 5}{7N + 5}\left(\frac{\hat{J}_R}{\hat{J}} + \frac{4N + 6}{N - 5}\right)} < 1. \qquad (16)$$

4 Self-Evaluation for Active Vision

Consider how the information in the images increases as the camera moves, recalling that the structure-from-motion is essentially model fitting in the four-dimensional data space $\mathcal{A}$ (Fig. 2).

- If the camera motion is small, we have $K_R < 1$. The data points are concentrated near a two-dimensional manifold in the four-dimensional data space $\mathcal{A}$ (Fig. 2a). As a result, we can robustly fit the rotation model S_R, but we cannot robustly fit S_Π or S, so we are unable to perceive any 3-D structure of the scene.

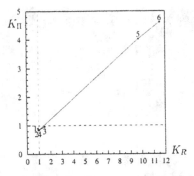

Fig. 3. Left: self-evaluation of the real image sequence. Right: the initial frame (real image).

- As the camera motion increases, the data spread more in $\mathcal{A}$, so we have K_R > 1 but K_Π < 1. We can robustly fit the planar surface model $\mathcal{S}_\Pi$, but we cannot robustly fit $\mathcal{S}$ (Fig. 2b). As a result, the scene we can perceive is merely a planar surface.
- If the camera motion is sufficiently large, we have $K_R > 1$ and $K_\Pi > 1$. The distribution of the data is sufficiently three-dimensional in $\mathcal{A}$ (Fig. 2c), so we can robustly fit the general model $\mathcal{S}$ and thereby perceive the full 3-D structure of the scene.

It follows that in the course of the camera motion we can evaluate the information content of the images by plotting the trajectory of (K_R, K_Π) in the K_R-K_Π plane. Let us call that the data has *2-D information* if $K_R < 1$, *2.5-D information* if $K_R > $ and $K_\Pi < 1$, and *3-D information* if $K_R > 1$ and $K_\Pi > 1$.

Examples. Fig. 3 (left) is the self-evaluation diagram of a real image sequence; the images have 2-D information in positions 1 and 2, 2.5-D information in positions 3 and 4, and 3-D information in positions 5 and 6. Fig. 3 (right) is the initial image; the upper two rows in Fig. 4 shows the subsequent six images. We chose feature points manually as indicated in the figures. In order to simulate real image processing operations, we added independent random noise of mean 0 and standard deviation of 1 pixel to their image coordinates and rounded them into integers. The lower two rows in Fig. 4 are the 3-D shapes reconstructed by referring to the initial frame shown in Fig. 3 (right). The values of (K_R, K_Π) are also listed. The 3-D shapes are viewed from the same position for all the frames. We again see that the 3-D shapes reconstructed from the general model are meaningless if $K_\Pi < 1$ (positions 1, 2, 3, and 4) while a fairly good 3-D shapes are reconstructed if $K_\Pi > 1$ (positions 5 and 6). Thus, we can evaluate the goodness of the images fairly well by computing (K_R, K_Π).

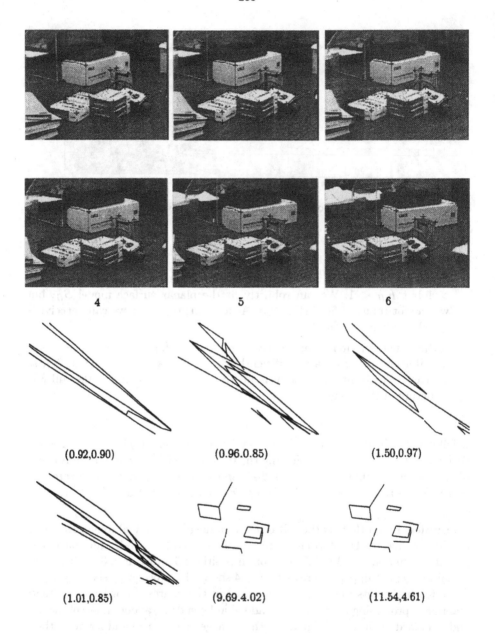

Fig. 4. The first to sixth frames (real images) and the corresponding reconstructed 3-D shapes and the values of (K_R, K_Π).

5 Concluding Remarks

We have presented a scheme for evaluating the "goodness" of camera motion for robust 3-D reconstruction by means of the *geometric information criterion* (*geometric AIC*). We introduced a statistical model of image noise, defined a geometric *model* as a *manifold* determined by geometric constraints in the a four-dimensional data space, and viewed the structure-from-motion problem as *model fitting*. Considering three models, we measured the extent of the "spread" of the data in the four-dimensional data space in term of the geometric AIC. This computation does not require any knowledge about the environment, the device, and the image processing techniques by which the images are obtained, and we need not introduce any thresholds to be adjusted empirically.

This work was in part supported by the Ministry of Education, Science, Sports and Culture, Japan under a Grant in Aid for Scientific Research C(2) (No. 09680352).

References

1. H. Akaike, "A new look at the statistical model identification," *IEEE Trans. Automation Control*, **19**-6 (1974), 176–723.
2. K. Kanatani, "Unbiased estimation and statistical analysis of 3-D rigid motion from two views," *IEEE Trans. Patt. Anal. Mach. Intell.* **15**-1 (1993), 37–50.
3. K. Kanatani, *Geometric Computation for Machine Vision*, Oxford University Press, Oxford, 1993.
4. K. Kanatani, "Renormalization for motion analysis: Statistically optimal algorithm," *IEICE Trans. Inf. & Sys.*, **E77**-D-11 (1994), 1233–1239.
5. K. Kanatani, "Automatic singularity test for motion analysis by an information criterion," *Proc. 4th European Conference on Computer Vision*, April, 1996, Cambridge, U.K., pp. 697–708.
6. K. Kanatani, *Statistical Optimization for Geometric Computation: Theory and Practice*, Elsevier Science, Amsterdam, 1996.
7. K. Kanatani, "Geometric information criterion for model selection," *Int. J. Comput. Vision*, to appear.
8. K. Kanatani and S. Takeda, "3-D motion analysis of a planar surface by renormalization," *IEICE Trans. Inf. & Syst.*, **E78**-D-8 (1995), 1074–1079.
9. H. C. Longuet-Higgins, "The reconstruction of a plane surface from two perspective projections," *Proc. Roy. Soc. Lond.*, **B-227** (1986), 399–410.
10. J. Weng, N. Ahuja and T. S. Huang, "Motion and structure from point correspondences with error estimation: Planar surfaces," *IEEE Trans. Sig. Proc.*, **39**-12 (1991), 2691–2717.
11. J. Weng, N. Ahuja and T. S. Huang, "Optimal motion and structure estimation," *IEEE Trans. Patt. Anal. Mach. Intell.*, **15**-9 (1993), 864–884.

Discrete-Time Rigidity-Constrained Optical Flow*

Jeffrey Mendelsohn[1], Eero Simoncelli[2], and Ruzena Bajcsy[1]

[1] University of Pennsylvania, Philadelphia PA 19104-6228, USA
[2] New York University, New York NY 10003-6603, USA

Abstract. An algorithm for optical flow estimation is presented for the case of discrete-time motion of an uncalibrated camera through a rigid world. Unlike traditional optical flow approaches that impose smoothness constraints on the flow field, this algorithm assumes smoothness on the inverse depth map. The computation is based on differential measurements and estimates are computed within a multi-scale decomposition. Thus, the method is able to operate properly with large displacements (i.e., large velocities or low frame rates). Results are shown for a synthetic and a real sequence.

1 Introduction

Estimation of optical flow, a longstanding problem in computer vision, is particularly difficult when the displacements are large. Multi-scale algorithms can handle large image displacements and also improve overall accuracy of optical flow fields [1, 5, 9, 15]. However, these techniques typically make the unrealistic assumption that the flow field is smooth. In many situations, a more plausible assumption is that of a rigid world.

Given point and/or line correspondences, the discrete-time rigid motion problem has been studied and solved by a number of authors (e.g. [6, 7, 11, 14, 17]). For instantaneous representations, multi-scale estimation techniques have been used to couple the flow and motion estimation problems to provide a direct method for planar surfaces [4, 8]. These methods use the multi-scale technique to capture large motions while significantly constraining the flow with a global model. But the planar world assumption is quite restrictive, and the approach also contains a hidden contradiction; the algorithm can observe large image motions but can only represent small camera motions due to the instantaneous time assumption.

---

* J. Mendelsohn is supported by NSF grant GER93-55018. R. Bajcsy is supported by ARO grant DAAH04-96-1-0007, ARPA grant N00014-92-J-1647, and NSF grants IRI93-07126 and SBR89-20230. E. Simoncelli is supported by NSF CAREER grant MIP-9796040, the Sloan Center for Theoretical Neurobiology at NYU, and ARO/MURI grant DAAH04-96-1-0007 to UPenn/Stanford.

This paper describes an optical flow algorithm for discrete camera motion in a rigid world. The algorithm is based on differential image measurements and estimates are computed within a multi-scale decomposition; the estimates are propagated from coarse to fine scales. Unlike traditional coarse-to-fine approaches which impose smoothness on the flow field, this algorithm assumes smoothness of the inverse depth values.

2 Discrete-Time Optical Flow

The imaging system is assumed to use the following projection model:

$$\mathbf{p}_i = \mathbf{C}\mathbf{x}_i \frac{1}{z_i} \quad \text{where} \quad \mathbf{x}_i \equiv \begin{bmatrix} x_i \\ y_i \\ z_i \end{bmatrix} \quad \text{and} \quad \mathbf{p}_i \equiv \begin{bmatrix} u_i \\ v_i \\ 1 \end{bmatrix} ; \qquad (1)$$

$\mathbf{x}_i$ denotes the point's coordinates in the camera's frame of reference and $\mathbf{p}_i$ the image coordinates. The matrix $\mathbf{C}$ contains the camera calibration parameters and is presumed invertible.[3]

The discrete motion of a point is expressed as:

$$\mathbf{x}'_i = \mathbf{R}\mathbf{x}_i + \mathbf{t} , \qquad (2)$$

where $\mathbf{R}$ is a (discrete-time) rotation matrix, $\mathbf{t}$ is a translation vector, and $\mathbf{x}'_i$ denotes the point's coordinates after the discrete motion.

A classic formulation of this constraint is due to Longuet-Higgins [10]:

$$\mathbf{x}'^T_i (\mathbf{t} \times \mathbf{R}\mathbf{x}_i) = 0 .$$

Using equation (1) to substitute for $\mathbf{x}_i$ gives:

$$z'_i \mathbf{p}'^T_i (\mathbf{C}'^{-1})^T (\mathbf{t} \times \mathbf{R}\mathbf{C}^{-1}\mathbf{p}_i z_i) = 0 . \qquad (3)$$

Let $\mathbf{t}^\times$ represent the skew-symmetric matrix corresponding to a cross-product with $\mathbf{t}$. Using suitable linear algebraic identities, and assuming that $z_i \neq 0$ and $z'_i \neq 0$, leads to the following simplification:

$$0 = z'_i \mathbf{p}'^T_i (\mathbf{C}'^{-1})^T \mathbf{t}^\times \mathbf{R}\mathbf{C}^{-1}\mathbf{p}_i z_i$$
$$0 = \mathbf{p}'^T_i (\mathbf{C}'^{-1})^T \mathbf{t}^\times \mathbf{C}'^{-1} \mathbf{C}'\mathbf{R}\mathbf{C}^{-1}\mathbf{p}_i$$
$$0 = \mathbf{p}'^T_i (\mathbf{C}'\mathbf{t})^\times \mathbf{C}'\mathbf{R}\mathbf{C}^{-1}\mathbf{p}_i$$
$$0 = \mathbf{p}'^T_i \mathbf{L}\mathbf{p}_i , \qquad (4)$$

where $\mathbf{L} \equiv (\mathbf{C}'\mathbf{t})^\times \mathbf{C}'\mathbf{R}\mathbf{C}^{-1}$ is a matrix that depends on the global motion and camera calibration information. Equation (4) provides a constraint on the initial

---

[3] This assumption is valid for any reasonable camera system. For example, the pinhole model is included in this family, as well as more complex models such as that given in [16].

and final image positions assuming rigid-body motion and is, essentially, the fundamental matrix [12].

In addition, it will be useful to develop an expression for the final position, $\mathbf{p}'_i$, given the calibration, motion, and structure parameters. Substituting the inverse of equation (1) into the rigid-body motion constraint of equation (2):

$$\mathbf{C}'^{-1}\mathbf{p}'_i z'_i = \mathbf{R}\mathbf{C}^{-1}\mathbf{p}_i z_i + \mathbf{t} \ .$$

Solving for the image position after the motion:

$$\mathbf{p}'_i = \frac{\mathbf{C}'\mathbf{R}\mathbf{C}^{-1}\mathbf{p}_i + \mathbf{C}'\mathbf{t}\frac{1}{z_i}}{\hat{z}^T\left(\mathbf{C}'\mathbf{R}\mathbf{C}^{-1}\mathbf{p}_i + \mathbf{C}'\mathbf{t}\frac{1}{z_i}\right)} \quad \text{where} \quad \hat{z} = \begin{bmatrix} 0 \\ 0 \\ 1 \end{bmatrix} \ . \tag{5}$$

This rigid-world motion constraint must be connected with measurements of image displacements. Since differential optical flow techniques have proven to be quite robust [2], the formulation is based on the differential form of the 'brightness constancy constraint' [8]:

$$\left.\frac{\partial I}{\partial u}\right|_i \cdot \frac{\partial u_i}{\partial t} + \left.\frac{\partial I}{\partial v}\right|_i \cdot \frac{\partial v_i}{\partial t} + \left.\frac{\partial I}{\partial t}\right|_i = 0 \ .$$

Substituting discrete displacements for the differential changes in image positions[4] and rewriting to isolate $\mathbf{p}'_i$ gives:

$$\begin{bmatrix} \partial I/\partial u \\ \partial I/\partial v \\ \partial I/\partial t \end{bmatrix}_i^T \begin{bmatrix} u'_i - u_i \\ v'_i - v_i \\ 1 \end{bmatrix} = \begin{bmatrix} \partial I/\partial u \\ \partial I/\partial v \\ \partial I/\partial t \end{bmatrix}_i^T \begin{bmatrix} 1 & 0 & -u_i \\ 0 & 1 & -v_i \\ 0 & 0 & 1 \end{bmatrix} \mathbf{p}'_i = 0 \ . \tag{6}$$

This constraint is combined with equation (5), squared, and summed over a local neighborhood, $N(i)$, to produce an error metric:

$$E_i(\mathbf{A}, \mathbf{b}, \frac{1}{z_i}) = \frac{\left(\mathbf{A}\mathbf{p}_i + \mathbf{b}\frac{1}{z_i}\right)^T \mathbf{D}_i \left(\mathbf{A}\mathbf{p}_i + \mathbf{b}\frac{1}{z_i}\right)}{\left(\hat{z}^T \mathbf{A}\mathbf{p}_i + \hat{z}^T \mathbf{b}\frac{1}{z_i}\right)^2} \ ,$$

where $\mathbf{A} \equiv \mathbf{C}'\mathbf{R}\mathbf{C}^{-1}$, $\mathbf{b} \equiv \mathbf{C}'\mathbf{t}$, and $\mathbf{D}_i$ is a matrix constructed from the differential image measurements and known position vectors:

$$\mathbf{D}_i \equiv \begin{bmatrix} 1 & 0 & -u_i \\ 0 & 1 & -v_i \\ 0 & 0 & 1 \end{bmatrix}^T \left(\sum_{j \in N(i)} \begin{bmatrix} \partial I/\partial u \\ \partial I/\partial v \\ \partial I/\partial t \end{bmatrix}_j \begin{bmatrix} \partial I/\partial u \\ \partial I/\partial v \\ \partial I/\partial t \end{bmatrix}_j^T \right) \begin{bmatrix} 1 & 0 & -u_i \\ 0 & 1 & -v_i \\ 0 & 0 & 1 \end{bmatrix} \ ,$$

---

[4] The displacements are assumed to be small; this is reasonable given the multi-scale (coarse-to-fine) framework described in Section 3.

Minimizing $E_i(\mathbf{A}, \mathbf{b}, \frac{1}{z_i})$ with respect to $\frac{1}{z_i}$ gives:

$$\frac{1}{z_i} = -\frac{\mathbf{p}_i^T \mathbf{A}^T \mathbf{D}_i \left(\mathbf{I} - \mathbf{b}\hat{z}^T\right) \mathbf{A}\mathbf{p}_i}{\mathbf{b}^T \mathbf{D}_i \left(\mathbf{I} - \mathbf{b}\hat{z}^T\right) \mathbf{A}\mathbf{p}_i} \; .$$

Substituting back into $E_i(\mathbf{A}, \mathbf{b}, \frac{1}{z_i})$, and noting that $\mathbf{L} = \mathbf{b}^\times \mathbf{A}$:

$$E_i(\mathbf{A}, \mathbf{b}) = -\frac{\mathbf{b}^T \mathbf{D}_i \left(\mathbf{L}\mathbf{p}_i\right)^\times \mathbf{D}_i \mathbf{A}\mathbf{p}_i}{\left(\mathbf{L}\mathbf{p}_i\right)^T \left(\hat{z}^\times\right)^T \mathbf{D}_i \hat{z}^\times \left(\mathbf{L}\mathbf{p}_i\right)} \; .$$

Using a series of linear algebraic manipulations, the numerator may be rewritten as an expression quadratic in $\mathbf{L}$:

$$
\begin{aligned}
-(\mathbf{D}_i \mathbf{b})^T (\mathbf{L}\mathbf{p}_i)^\times \mathbf{D}_i \mathbf{A}\mathbf{p}_i &= (\mathbf{L}\mathbf{p}_i)^T (\mathbf{D}_i \mathbf{b})^\times \mathbf{D}_i \mathbf{A}\mathbf{p}_i \\
&= |\mathbf{D}_i|(\mathbf{L}\mathbf{p}_i)^T \mathbf{D}_i^{-1} \mathbf{b}^\times \mathbf{A}\mathbf{p}_i \\
&= |\mathbf{D}_i|(\mathbf{L}\mathbf{p}_i)^T \mathbf{D}_i^{-1} \mathbf{L}\mathbf{p}_i \; ,
\end{aligned}
$$

where $|\mathbf{D}_i|$ indicates the determinant of the matrix $\mathbf{D}_i$.

To efficiently solve for $\mathbf{L}$, a global metric is formed by summing over the image the weighted numerators of the $E_i(\mathbf{A}, \mathbf{b})$:

$$E(\mathbf{A}, \mathbf{b}) = \sum_i |\mathbf{D}_i|(\mathbf{L}\mathbf{p}_i)^T \mathbf{D}_i^{-1} \mathbf{L}\mathbf{p}_i / w_i \; , \qquad (7)$$

where w_i is the value of the denominator of $E_i(\mathbf{A}, \mathbf{b})$ using the previous estimate of $\mathbf{L}$. This metric is computed iteratively in the coarse-to-fine procedure and, from empirical observations, only one iteration at each scale is necessary.

The algorithm proceeds by first globally minimizing equation (7) to obtain a solution for the nine entries of the matrix $\mathbf{L}$, subject to the constraints $|\mathbf{L}| = 0$ and $\sum_j \sum_i (\mathbf{L}_{j,i})^2 = 1$ (to remove the scale ambiguity). Then, the squared optical flow constraint

$$E_f(\mathbf{p}'_i) = \mathbf{p}_i'^T \mathbf{D}_i \mathbf{p}'_i \; , \qquad (8)$$

is minimized at each pixel, subject to equation (4) with the estimated value of $\mathbf{L}$, to obtain an optical flow field.

3 Multi-Scale Implementation

Since the method is capable of estimating large (discrete) camera motions, it must be able to handle large image displacements. This is accomplished with a coarse-to-fine version of the algorithm on a multi-scale decomposition.

First, a Gaussian pyramid is constructed on the pair of input frames [3]. At the coarsest scale of the pyramid, the algorithm is employed as derived in the previous section to provide an initial coarse estimate of optical flow. This optical flow is interpolated to give a finer resolution flow field, denoted $(\Delta^c u_i, \Delta^c v_i)$. This motion is removed from the finer scale images using warping; the warped images are denoted I^w.

Since the optical flow equation (8) is written only in terms of the final positions p'_i, the constraint on the warped images needs only a slight modification:

$$
\begin{bmatrix} \partial I^w/\partial u \\ \partial I^w/\partial v \\ \partial I^w/\partial t \end{bmatrix}_i^T
\begin{bmatrix} 1 & 0 & -(u_i + \Delta^c u_i) \\ 0 & 1 & -(v_i + \Delta^c v_i) \\ 0 & 0 & 1 \end{bmatrix} p'_i = 0 \ .
$$

The remainder of the algorithm is as before: new matrices D_i are computed from this constraint and these are used to estimate L using equation (7). The weightings, w_i, are computed using the estimate of L from the previous scale. Finally, equation (8) is minimized at each pixel, subject to the constraint of equation (4), to estimate the optical flow.

4   Experimental Results

Experimental results were collected for three different algorithms on two sequences. The first method is a simple multi-scale optical flow (msof) algorithm [15]. The second computes flow for discrete motion of a planar world (planar) [13]. The third is the algorithm presented in this paper (rigid).

The first sequence is the 'Yosemite' sequence.[5] True optical flow vectors were computed from the motion, structure, and calibration data provided with the sequence. The textureless top region was ignored during error calculations. In order to obtain large motions, the computations are performed on the sequence subsampled at different temporal rates.

The second sequence was taken in the GRASP Laboratory from a camera mounted on a tripod. Six markers consisting of seven black disks on a white planar surface were placed in the scene and used to calculate ground-truth. Using knowledge of the individual targets, accurate centroids for the disks were computed. Flow was calculated from the motion of each centroid for a total of 42 flow vectors.

Table 1 shows results. The metric is the mean angular error in degrees [2]:

$$
E_a = \frac{1}{n} \sum_{i=1}^{n} \cos^{-1}(\hat{v}_{t_i}^T \hat{v}_{e_i})
$$

where $\hat{v}_{t_i}$ is a unit three-vector in the direction of the true flow and $\hat{v}_{e_i}$ is a unit three-vector in the direction of the estimated flow. The multi-scale optical flow algorithm did well for small motions but poorly for large ones. Since the range map for the Yosemite sequence is nearly planar, the difference in performance between the discrete algorithms is less significant than those for the real sequence. It is clear that the rigid algorithm provides the best optical flow estimates.

---

[5] This sequence was graphically rendered from an aerial photograph and range map by Lyn Quam at SRI.

sequence →	Yosemite							Real
frame interval →	1	2	3	4	5	6	7	1
msof	6.13°	7.49°	13.10°	21.32°	29.09°	35.82°	42.43°	3.28°
planar	5.82°	6.56°	6.70°	6.70°	6.92°	6.82°	7.63°	2.87°
rigid	5.77°	5.86°	6.18°	6.53°	6.55°	6.77°	6.95°	2.05°

Table 1. Mean angular error in flow vectors for three different algorithms. See text.

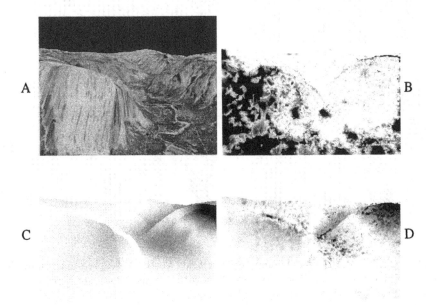

Fig. 1. Sample image from the Yosemite sequence (A) and the angular error metric of the computed flow fields for the Yosemite sequence by the msof (B), planar (C), and rigid (D) algorithms with a frame interval of four. White corresponds to 0° of error and black to 45°.

5 Conclusion

A multi-scale algorithm for estimating optical flow based on an uncalibrated camera moving through a rigid world has been presented. Its implementation is only slightly more complicated and time-consuming than standard multi-scale algorithms. In situations where the camera may be undergoing relatively large motions, the superiority of the rigid model has been demonstrated on both a synthetic and a real sequence.

References

1. P. P. Anandan. A Computational Framework and an Algorithm for the Measurement of Visual Motion. *IJCV*, 2, 283–310, 1989.

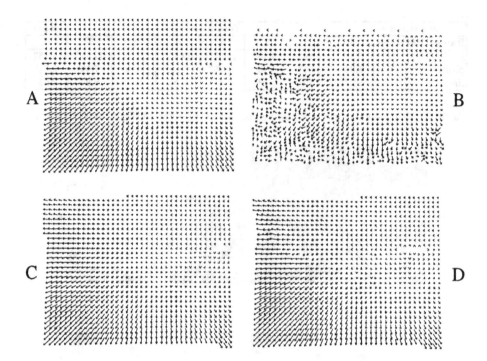

Fig. 2. True flow field (A) and computed flow fields for the Yosemite sequence by the msof (B), planar (C), and rigid (D) algorithms with a frame interval of four.

2. J. L. Barron, D. J. Fleet, and S. S. Beauchemin. Performance of Optical Flow Techniques. *IJCV*, 1992.

3. P. J. Burt. Fast filter transforms for image processing. *Computer Graphics and Image Processing*, 16, 20–51, 1981.

4. J.R. Bergen, P. Anandan, K.J. Hanna, and R. Hingorani. Hierarchical Model-Based Motion Estimation. *Proc. ECCV*, 237–252, Springer-Verlag, Santa Margherita Ligure, Italy, 1992.

5. W. Enkelmann and H. Nagel. Investigation of Multigrid Algorithms for Estimation of Optical Flow Fields in Image Sequences. *Computer Vision Graphics Image Processing*, 43, 150–177, 1988.

6. R. Hartley. Projective Reconstruction from Line Correspondences. *Proc. IEEE CVPR*, 1994.

7. R. Hartley, R. Gupta, and T. Chang. Stereo from Uncalibrated Cameras. *Proc. IEEE CVPR*, 1992.

8. B.K.P. Horn. *Robot Vision*. MIT Press, Cambridge, MA, 1986.

9. M. Leuttgen, W. Karl, and A. Willsky. Efficient Multiscale Regularization with Applications to the Computation of Optical Flow. *IEEE Trans. Im. Proc.*, 1992.

10. H. C. Longuet-Higgins and K. Prazdny. The interpretation of a moving retinal image. *Proc. Royal Society of London B*, 208, 385–397, 1980.

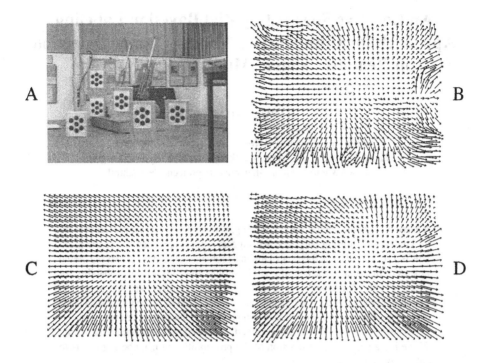

Fig. 3. Sample image from the real sequence (A) and the computed flow fields for the real sequence by the msof (B), planar (C), and rigid (D) algorithms.

11. Q.T. Luong and T. Vieville. Canonic Representation for the Geometry of Multiple Projective Views. *Proc. ECCV*, Stockholm, 1994.

12. Q.T. Luong and O. Faugeras. The Fundamental Matrix: Theory, Algorithms, and Stability Analysis. *IJCV*, 43–76, 1996.

13. J. Mendelsohn, E. Simoncelli, and R. Bajcsy. Discrete-Time Rigid Motion Constrained Optical Flow Assuming Planar Structure. *Univ. of PA GRASP Laboratory Technical Report #410*, 1997.

14. A. Shashua and N. Navab. Relative Affine Structure: Theory and Application to 3D Reconstruction from Perspective Views. *Proc. IEEE CVPR*, Seattle, June 1994.

15. E. P. Simoncelli, E. H. Adelson, and D. J. Heeger. Probability Distributions of Optical Flow. *Proc. IEEE CVPR*, Maui, June 1991.

16. T. Vieville and O. D. Faugeras. The First Order Expansion of Motion Equations in the Uncalibrated Case. *Computer Vision and Image Understanding*, 128–146, July 1996.

17. T. Vieville, C. Zeller, and L. Robert. Using Collineations to Compute Motion and Structure in an Uncalibrated Image Sequence. *IJCV*, 1995.

An Iterative Spectral-Spatial Bayesian Labeling Approach for Unsupervised Robust Change Detection on Remotely Sensed Multispectral Imagery

RAFAEL WIEMKER

Universität Hamburg, II. Institut für Experimentalphysik
D-22761 Hamburg, Germany; wiemker@informatik.uni-hamburg.de
kogs-www.informatik.uni-hamburg.de/projects/Censis.html

Abstract In multispectral remote sensing, change detection is a central task for all kinds of monitoring purposes. We suggest a novel approach where the problem is formulated as a Bayesian labeling problem. Considering two registered images of the same scene but different recording time, a Bayesian probability for 'Change' and 'NoChange' is determined for each pixel from spectral as well as spatial features. All necessary parameters are estimated from the image data itself during an iterative clustering process which updates the current probabilities.

The contextual spatial features are derived from Markov random field modeling. We define a potential as a function of the probabilities of neighboring pixels to belong to the same class.

The algorithm is robust against spurious change detection due to changing recording conditions and slightly misregistered high texture areas. It yields successful results on simulated and real multispectral multitemporal aerial imagery.

1 Introduction

In the field of multispectral remote sensing, change detection is a central task for all monitoring purposes. It uses multitemporal imagery in order to detect land cover changes caused by short-term phenomena such as flooding and seasonal vegetation change, or long-term phenomena such as urban development and deforestation [5,6].

In general, remotely sensed multispectral imagery for monitoring purposes is recorded by overflights over the same land area at two times, T_1 and T_2, say. An appropriate algorithm must then compare the two observed images of the same scene and assist the analyst by designating those areas where the ground cover has apparently changed. For specific applications, certain wavelength bands may be selected, whereas for general purpose monitoring, all spectral bands will be taken into account.

A pre-requisite to pixel-wise change detection is the registration between the two images which are to be compared. The registration can be carried out either by geocoding of both images or by direct image-to-image registration. Imagery from airborne scanners in general requires locally adaptive coordinate transformation functions as described in [8].

The general problem is to compare two images I_1 and I_2 of the same scene recorded at times T_1 and T_2. When using a multispectral sensor with n spectral bands denoted by index $i = 1...n$, we particularly want to compare the spectral values $x_i(T_1)$ and $x_i(T_2)$ for a certain pixel $\mathbf{x}$. Depending on the spectral differences and contextual considerations, a decision is sought whether this pixel shows a ground surface patch which – with respect to its ground cover type – has either changed or has remained unchanged.

2 Known Problems and Here Chosen Approaches

2.1 Iterative, Principal Component Based Change Detection

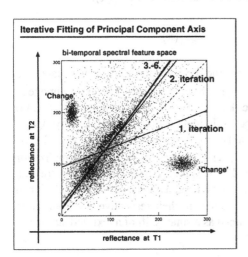

A fundamental problem of comparing two images of the same scene is that the recording conditions may have changed. In particular, the direct solar illumination and the diffuse sky light, the path radiance, and the transmittance of the atmosphere, as well as the dark current and gain setting of the sensor may have changed individually in each spectral band. All these topics can roughly be categorized into influencing the received spectral signal in either a multiplicative or an additive fashion. Thus the relation between the spectral signal $x_i(T_1)$ and $x_i(T_2)$ received from a certain Lambertian reflecting surface at two times T_1 and T_2 is very often modeled approximately as a linear function [5,6].

Let us consider a bi-temporal feature space for a single spectral band i where each pixel $\mathbf{x}$ is denoted by a point $\mathbf{x}_i = [x_i(T_1), x_i(T_2)]^T$. Then, as a consequence of the assumed linear relation between unchanged pixels, we expect all unchanged pixels to lie in a narrow elongated cluster along a principal axis. On the other hand, the pixels which do have experienced 'change' in their spectral appearance are expected to lie far out from this axis [5]. Thus, the amount of 'change' is quantified by the magnitude of the second principal component (PC): $c_i'(\mathbf{x}_i) = g_{1,i}(x_i(T_1) - m_{i,1}) + g_{2,i}(x_i(T_2) - m_{i,2})$ where $\mathbf{g}_i = [g_{1,i}, g_{2,i}]^T$ is the second eigenvector of the overall covariance matrix $\mathbf{C}_i$ (2×2 matrix), and $\mathbf{m}_i = [m_{1,i}, m_{2,i}]^T$ are the mean values of $x_i(T_1)$ and $x_i(T_2)$.

An obvious problem with principal component based change detection is that the principal components are conventionally estimated as the eigenvectors of the covariance matrix $\mathbf{C}_i$ computed from all pixels $\mathbf{x}_i$, including those which have experienced 'change'. Thus, the such found 'NoChange'-axis is prone to error. In our iterative approach the problem is addressed such that the cluster mean $\mathbf{m}_i$ and the covariance matrix $\mathbf{C}_i$ are determined from all pixels $\mathbf{x}$ but weighted with their respective probabilites $P(\mathrm{NC}|\mathbf{x})$ to be 'NoChange'-pixels.

2.2 Avoiding Spurious Change Detection in High Texture Areas

A typical problem of remotely sensed imagery is that the registration of airborne scanner imagery only yields registration accuracies of some pixels at best. In image areas of high texture, even small misregistration errors will cause a large amount of spurious 'Change': If e.g. a test image is shifted by one pixel and subtracted from the original image, we essentially observe a crude edge detector which enhances all contours of the image. In order to avoid this effect, a crucial provision of our approach is to normalize all change components $c_i'(\mathbf{x}_i)$ by the combined local variance $\sigma_{\mathrm{loc},i}^2(\mathbf{x})$:

$$c_i(\mathbf{x}_i) = c_i'(\mathbf{x}_i) \Big/ \sqrt{\sigma_{\mathrm{loc},i}^2(\mathbf{x})} \tag{1}$$

The local variance $\sigma_{\mathrm{loc},i}^2(\mathbf{x})$ is estimated from the spectral variances (mean square scatter) $\sigma_{\mathrm{loc},I_1,i}^2(\mathbf{x})$ and $\sigma_{\mathrm{loc},I_2,i}^2(\mathbf{x})$ in the local neighborhood $\mathcal{N}(\mathbf{x})$ of pixel $\mathbf{x}$ in image I_1 and image I_2 in spectral band i. Here we define the local neighborhood $\mathcal{N}(\mathbf{x})$ as the $k \times k$ window centered around the pixel $\mathbf{x}$, excluding the pixel $\mathbf{x}$ itself. Error propagation yields that after the PC-transformation the errors $\sigma_{\mathrm{loc},I_1,i}^2(\mathbf{x})$ and $\sigma_{\mathrm{loc},I_2,i}^2(\mathbf{x})$ yield a combined error of $\sigma_{\mathrm{loc},i}^2(\mathbf{x}) = g_{1,i}^2\,\sigma_{\mathrm{loc},I_1,i}^2(\mathbf{x}) + g_{2,i}^2\,\sigma_{\mathrm{loc},I_2,i}^2(\mathbf{x})$. The normalization of the change component with the estimated local variance means that the spectral values $x_i(T_1)$ and $x_i(T_2)$ should only then be considered as indicating 'Change' if their difference is large in comparison to the variances in their respective neighborhoods, since otherwise the difference can be explained as a small shift of a textured surface.

2.3 Bayesian Decision on 'Change' vs. 'NoChange'

Conventional change detection depends crucially on the setting of threshold parameters by the analyst. In each spectral band i, it has to be decided just when the change component $c_i(\mathbf{x}_i)$ of a given pixel $\mathbf{x}$ is large enough to be considered 'Change'.

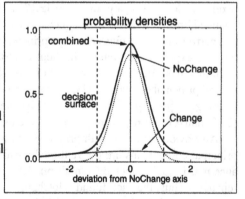

The unsupervised approach presented here considers change detection as a Bayesian labeling problem, where each pixel is assigned to one out of two classes $\omega = \{\mathrm{CH}, \mathrm{NC}\}$: 'Change' or 'NoChange'. This decision is made using *Bayes Rule*, which minimizes the probability of error [2]: The pixel is assigned to the class ω which has the maximum *a posteriori* probability $P(\omega|\mathbf{x})$. This probability is given by the normalized product of the *a priori* probability and the conditional probability density computed from the observation $\mathbf{x}$. We assume that the *a priori* probability for class ω is proportional to the relative abundance $N_\omega/N_{\mathrm{tot}}$, where $N_{\mathrm{tot}} = \sum_\omega N_\omega$ is the total number of pixels and N_ω the sum of the probabilities of class ω over all pixels:

$N_\omega = \sum_{\mathbf{x}} P(\omega|\mathbf{x})$. The normalized conditional probability density derived from *spectral features* in spectral band i is

$$p_i(\mathbf{x}_i|\omega) = \frac{1}{\sqrt{2\pi\sigma_{\omega,i}^2}} \exp\left(-\frac{1}{2}\frac{c_i(\mathbf{x}_i)^2}{\sigma_{\omega,i}^2}\right) \quad ; \quad \text{with} \quad \int_{-\infty}^{\infty} p_i \, dc_i = 1 \qquad (2)$$

and is multiplied with the respective probability densities of the other spectral bands. This tacitly assumes a Gaussian distribution of the 'Change'-components $c_i(\mathbf{x}_i)$. The within-class-variances $\sigma_{\omega,i}^2$ are estimated from the data itself during the iteration. This kind of unsupervised iterative clustering of unlabeled data is described *e.g.* in [2]. The variance of a class ω with respect to the variable $c_i(\mathbf{x}_i)$ is estimated from all pixels $\mathbf{x}$ weighted with their respective probability $P(\omega|\mathbf{x})$: The conditional probability density $p_{\mathrm{con}}(\mathbf{x}|\omega)$ derived from *context features* is described in Section 2.4.

Finally, the *a posteriori* probability is the product of *a priori* probability and the spectral and spatial conditonal probabilities. It is normalized by the sum of the probability densities for all classes ω in order to be constrained to the interval $P(\omega|\mathbf{x}) \in [0\dots1]$ and to add up to unity: $\sum_\omega P(\omega|\mathbf{x}) = 1$. Hence, our overall *a posteriori* probability amounts to

$$P(\omega|\mathbf{x}) = \frac{N_\omega/N \cdot p_{\mathrm{con}}(\mathbf{x}|\omega) \cdot \prod_i p_i(\mathbf{x}_i|\omega)}{\sum_\omega [N_\omega/N \cdot p_{\mathrm{con}}(\mathbf{x}|\omega) \cdot \prod_i p_i(\mathbf{x}_i|\omega)]} \qquad . \qquad (3)$$

2.4 Context Features derived from Markov Random Field Modeling

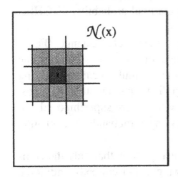

$\mathcal{N}(\mathbf{x})$

The Markovian property means that the probability of pixel $\mathbf{x}$ to belong to class ω is influenced by and only by the classes assigned to the pixels $\mathbf{x}'$ in its neighborhood $\mathcal{N}(\mathbf{x})$ [4]. Here we define $\mathcal{N}$ as the quadrangular $k \times k$ window around the pixel $\mathbf{x}$, excluding $\mathbf{x}$ itself. MRF-based approaches to contextually enhanced multispectral classification have recently been used in remote sensing [3,7]. Unlike these approaches we here use a contextual potential function $U(\mathbf{x}|\omega)$ which not only evaluates identical or different labels in the neighborhood $\mathcal{N}$ with a $\{0,1\}$-Kronecker-function, but which is influenced continuously by the current probabilities of the neighboring pixels:

$$p_{\mathrm{con}}(\mathbf{x}|\omega) = \frac{1}{Z} \exp\left(-\beta U(\mathbf{x}|\omega)\right) \quad \text{with} \quad U(\mathbf{x}|\omega) = \sum_{\mathbf{x}' \in \mathcal{N}(\mathbf{x})} [1 - P(\omega|\mathbf{x}')] \qquad . \qquad (4)$$

Then the context-conditional probability $p_{\mathrm{con}}(\mathbf{x}|\omega)$ is computed from the neighborhood potential $U(\mathbf{x}|\omega)$, where the parameter β defines the magnitude of the contextual influence. For $\beta = 0$ the influence is vanishing, and $\beta = 1$ is a common choice [1,3,7].

3 The Iterative Algorithm

The probabilities and the necessary parameters are estimated during an iterative process using the current conditional probabilities (similar to the recently often used ICM algorithm (iterated conditional mode) [1]). In our case the conditional probabilities depend on the spectral and spatial features as set out above. At each iteration step and for each spectral band i we update the 'NoChange'-cluster center $\mathbf{m}_i$ and its second principal axis $\mathbf{g}_i$, the change components $\mathbf{c}_i$ and probabilities $P(\omega|\mathbf{x})$ for each pixel, and re-estimate the typical scatter widths $\sigma_{\omega,i}^2$ for both the 'NoChange' and the 'Change' cluster, and the probability sums N_ω. The actual algorithm can be outlined as follows (see below Fig. 1):

For a given β, the iteration process converges to a local rather than a global maximum and thus depends on the initial labeling. Therefore we start the iteration with purely spectral features ($\beta = 0$). In our experimental experience we have achieved fast convergence when the context-conditional probabilitites are used with increasing influence. After starting with $\beta = 0$ we proceed eventually towards $\beta = 1$ as the relative number of yet converged pixel probabilities $P(\omega|\mathbf{x})$ increases. So the spatial context information is introduced gradually to the iteration process, and spectral and spatial features are iterated at the same time.

4 Experimental Results on Simulated and Real Multitemporal Multispectral Imagery

The output of our unsupervised iterative algorithm is a binary image showing the locations of 'Change' pixels, and a float-type image giving the final probability $P(\mathrm{CH}|\mathbf{x})$ for each pixel.

The algorithm was verified on simulated test imagery with artificial 'Change' (Fig. 1) and two spectral bands of a real aerial image. These were duplicated and shifted by one pixel ('misregistration'); furthermore, an additive offset and a multiplicative rescaling of the gray values ('varying recording conditons') was applied separately to both bands; additive noise ($\pm$ 5 DC) and multiplicative noise (SNR = 5%) were superimposed; the 'Change' quadrangles were drawn into only one of the two spectral bands ('wavelength specific change').

The results on the simulated test imagery (Fig. 1) clearly show that only the combination of spectral features with local variance assessment and context-conditional probabilities yields optimal results.

We then applied the described change detection strategy to real multispectral image data, which was recorded by a DAEDALUS AADS 1268 line scanner during four campaigns from 1991 to 1995 in cooperation with the German Aerospace Research Establishment (DLR) at flight altitudes of 300 m and 1800 m (nadir ground resolution 70 cm and 4.2 m, respectively).

The multispectral images of 1991 and 1995 in Fig. 2 have $n = 10$ spectral bands and were rectified and registered. The unsupervisedly detected changes were checked by eye appraisal against high resolving aerial photographs and show promising results; most changes are due to industrial construction activity between 1991 and 1995.

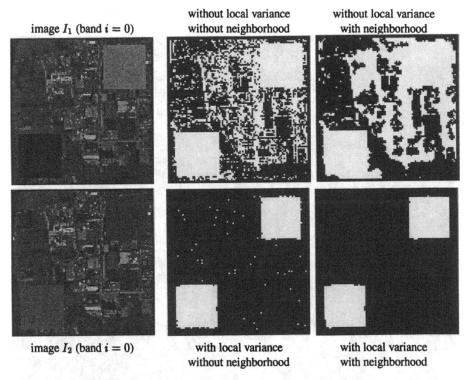

image I_1 (band $i = 0$)	without local variance without neighborhood	without local variance with neighborhood

image I_2 (band $i = 0$)	with local variance without neighborhood	with local variance with neighborhood

Figure 1. Change detection results on a simulated test image. Below: The iterative algorithm.

---

0. register the two images I_1 and I_2 (rectification / geocoding)
1. initialize the probabilities for all pixels to $P(\omega|\mathbf{x}) = 0.5$, where $\omega = \{CH, NC\}$
   pre-compute the local spectral variance estimates $\sigma^2_{\mathrm{loc},I_1,i}(\mathbf{x})$ and $\sigma^2_{\mathrm{loc},I_2,i}(\mathbf{x})$ for all spectral bands i and all pixels $\mathbf{x}$
   2. for all n spectral bands i do:
      2.1 determine principal axes from all pixels weighted with the current probabilities $P(NC|\mathbf{x})$ and the inverse of the combined local variance $\sigma^2_{\mathrm{loc},i}(\mathbf{x})$ in spectral band i
      2.2 compute the 'Change'-components $c_i(\mathbf{x}_i)'$ and normalize them into $c_i(\mathbf{x}_i)$ by dividing by the combined local variance in spectral band i
      2.3 compute the CH- and NC-class-variances $\sigma^2_{\mathrm{CH},i}$ and $\sigma^2_{\mathrm{NC},i}$ from all pixels weighted with the current probabilities $P(CH|\mathbf{x})$ and $P(NC|\mathbf{x})$ respectively
      2.4 compute the spectral probability densities $p_i(\mathbf{x}_i|CH)$ and $p_i(\mathbf{x}_i|NC)$
3. compute current context probabilities $p_{\mathrm{con}}(\mathbf{x}|CH)$ and $p_{\mathrm{con}}(\mathbf{x}|NC)$ from the current probabilities $P(\omega|\mathbf{x})$
4. compute current *a posteriori* probabilities $P(\omega|\mathbf{x})$, the new number of class members N_{CH} and N_{NC}, and the relative number of already having converged probabilities $P(\omega|\mathbf{x})$
5. exit if all probabilities $P(\omega|\mathbf{x})$ have converged, else start over from (2)

1991   **1995**

'Change' coded in white

Figure 2. Change detection between two registered multispectral aerial images recorded in 1991 (top left) and 1995 (top right) of an industrial suburb of Nürnberg (rectified scanner image with 10 spectral bands, 1000×800 pixels).

Bottom: 'Change'-areas are color-coded white on a gray image background. The neighborhood was defined as a 5×5 window.

5 Conclusions

We have addressed the problem of pixel-wise multispectral change detection in the framework of a Bayesian labeling problem, utilizing spectral features and spatial conditional probabilities derived from Markov random field modeling. The labeling decision on 'Change' or 'NoChange' is made with regard to maximum *a posteriori* probability for each pixel. The probabilities and the necessary parameters are estimated during an iterative process using the conditional probabilities (ICM: iterated conditional mode [1]). The here presented formulation of the change detection problem in a Bayesian labeling framework frees the problem from the need of arbitrary *ad hoc* thresholds, and delivers meaningful change probability values as well as error probabilities of first and second kind. Certainly, for a specific monitoring application, particularly useful weights of the various spectral bands may be experimentally found by the experienced analyst and hand-tuned thresholds may prove successful. However, just as unsupervised clustering and labeling of multispectral imagery is used complementarily to costly analyst-trained classifiers, our unsupervised change detection approach can provide a primary and well founded tool for change detection. Our approach uses all spectral information on equal footing and is robust in so far as it particularly addresses a number of well known problems of multispectral change detection:

- varying illumination and recording conditons
- small misregistration errors
- (often unemployed) contextual information
- unknown threshold parameters
- additive and multiplicative noise

Promising results were achieved on simulated test imagery as well as on real multi-temporal, remotely sensed multispectral imagery. We found that the combination of spectral features with local variance and neighborhood potentials delivers a substantial improvement.

References

1. J. Besag. On the statistical analysis of dirty pictures. *Journal of the Royal Statistical Society B*, 48(3):259–302, 1986.
2. R. O. Duda and P. E. Hart. *Pattern Classification and Scene Analysis*. Wiley, New York, 1973.
3. Y. Jhung and Philip H. Swain. Bayesian contextual classification based on modified *M*-estimates and Markov random fields. *IEEE T.o.Geosci.a.Rem.Sens.*, 34(1):67–75, 1996.
4. S.Z. Li. *Markov Random Field Modeling in Computer Vision*. Springer, Tokyo, 1995.
5. J. A. Richards. *Remote Sensing Digital Image Analysis*. Springer, New York, 1993.
6. A. Singh. Review article: Digital change detection techniques using remotely-sensed data. *International Journal of Remote Sensing*, 10(6):989–1003, 1989.
7. A.H.S. Solberg, T. Taxt, and A.K. Jain. A markov random field model for classification of multisource satellite imagery. *IEEE T.o.Geosci.a.Rem.Sens.*, 34(1):100–113, 1996.
8. R. Wiemker, K. Rohr, L. Binder, R. Sprengel, and H.S. Stiehl. Application of elastic registration to imagery from airborne scanners. In *Proceedings of the XVIII. Congress of the International Society for Photogrammetry and Remote Sensing ISPRS 1996, Vienna*, volume XXXI part B4 of *International Archives of Photogrammetry and Remote Sensing*, pages 949–954, 1996.

Contrast Enhancement of Badly Illuminated Images Based on Gibbs Distribution and Random Walk Model

Bogdan Smolka   and   Konrad W. Wojciechowski

Dept. of Automatics Electronics and Computer Science
Silesian University of Technology
Akademicka 16 Str, 44-101 Gliwice, Poland

Abstract. In the paper a new approach to the problem of contrast enhancement of grey scale images is presented. The described method is based on a model, which treats the image as a toroidal, two dimensional lattice, the points of which possess a potential energy. On in this way defined lattice, a regular Markov chain of the positions of this particle can be investigated. The probability of a transition of the virtual particle from a fixed lattice point to a point belonging to its neighbourhood can be determined using the Gibbs canonical distribution, defined on an eight-connectiviy system. The idea of the presented algorithm consists in determininig the stationary probability vector of the Markow chain. The new algorithm registers the visits of the wandering particle and then determines their relative frequencies. It performs especially well in case of images with nonuniform brightness.

1   Introduction

Image enahancement methods consist of procedures, which aim is the improvement of the visual appearance of an image or its conversion to a form better suited for analysis by a human or a machine [1]. The main goal of the image enhancement is the improvement of the detectability of objects represented by the image. Contrast stretching, enhancement of corners and edges, noise filtering and image sharpening are some examples of operations, which allow a better analysis of the image features .

Let us assume that the image spatial domain is the cartesian product of $L_r \times L_c$ where $L_r = \{1, 2, \ldots, N_r\}$ and $L_c = \{1, 2, \ldots, N_c\}$ are spatial domains of the rows and columns of the image array. The digital image is a function $\mathcal{I}$ assigning some value from the set $G = \{g_1, g_2, \ldots, g_Q\}$, $g_k \in [0, 1])$ to every resolution cell (i, j).

$$\mathcal{I} : L_r \times L_c \longmapsto G$$

Let the image matrix be denoted as L and its grey scale values as $L(i, j) \in G$. The total number of image points will be denoted by N. There exists a great variety of techniques of contrast enhancement and a detailed review can be found in [1, 2, 3].

Let us briefly describe some of the basic techniques of contrast enhancement. The simplest kind of contrast modifications are the one-point operations. These are transformations, in which the output value of each point is not dependent on the grey tones of its neighbourhood. A simple example of such an operation is the linear image scaling with clipping

$$T(g) = \begin{cases} 0 & \text{if} \quad g < \alpha \\ 1 & \text{if} \quad g > \beta \end{cases}$$

where g is the original grey scale value, $T(g)$ is the output value and α, β are parameters. This transformation is used for images in which a small number of pixels exceeds the limits α and β.

Another, frequently used procedure, transforms the grey scale values of the image points according to the formula

$$T(g) = g^\delta$$

where δ is the so called power law variable. Choosing the appropraiate value of δ, the dark or bright regions of the image can be enhanced.

Generally, the point modification of the grey scale value of a pixel is a mapping $T : G \mapsto G$ where T may be any function defined on G, for which $\min\{T(g)\} > -\infty$, and $\max\{T(g)\} < +\infty$.

The histogram equalization is an example of a transformation, which uses the global information about an image [1, 4, 5, 6, 7, 8, 9]. In this case the transforming function T is not given but calculated using the information about the grey scale distribution of the image pixels. Let x_k, $k = (1, 2, \ldots, N)$ be the discrete grey scale values of the image points. The histogram equalization is a transformation given by

$$T(x_k) = \sum_{j=0}^{k} p_j$$

where $p_j = \frac{n_j}{N}$ and n_j is the number of points with a grey scale value of x_j. This transformation seeks to create an image, the histogram of which, comprises the whole range of grey scale values.

The local transformations of grey scale values consider its distribution in a specified neighbourhood af a given point. In this way the local histogram equalization is a transformation, which assigns to a given point a new value resulting from the histogram equalization of points belongig to its neighbourhood.

A different approach to the problem of contrast enhancement is based on the mean value and standard deviation of the grey scale values [10, 11]. The mean value is regarded as a measure of image brightness and the standard deviation as its contrast.

$$T(i,j) = A(i,j) \cdot (L(i,j) - m_{i,j}) + m_{i,j} \qquad A(i,j) = \frac{c \cdot \hat{m}}{\sigma(i,j)}$$

where $m_{i,j}$, and $\sigma_{i,j}$ is the mean value and standard deviation of grey scale values of the points belonging to the neighbourhood of the point (i,j) and $\hat{m}$ is the mean grey scale value of all image points and finally $c \in (0,1)$ is a constant. Since $A(i,j)$ is inversely proportional to the local standard deviation, this operation results in the enhancement of image regions with low contrast.

Another method of contrast modification is based on the enhancement of image edges. This can be done, performing the convolution of the image with high-pass convolution masks [1, 2, 3]. Some of the commonly used masks shows figure 1.

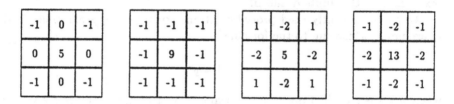

Fig. 1. Typical linear edge crispening masks

The described techniques of image enhancement perform quite well on images with uniform spatial distribution of grey scale values. Dificulties occur when the object and background assume broad range of grey scales or when the background brightness is strongly nonhomogenous. For this reason, a new method of contrast enhancement of badly illuminated images has been developed.

2 Random Walk Model

The presented method is based on a model, treating the image as a two dimensional lattice, the points of which possess a potential energy. On a such defined lattice, a random walk of a virtual Gibbsian particle is simulated. The behaviour of the particle during the random walk offers valuable information about the image features.

Let the distance between points (i,j) and (k,l) be expressed by

$$\rho\{(i,j),(k,l)\} = \max\{|i-k|, |j-l|\}$$

Assuming that the neighbourhood of the point (i,j) consists of points (k,l), which satisfy $\rho\{(i,j),(k,l)\} = 1$ (Fig. 2), the virtual particle can jump to its eight neighbours or stay in its current position.

If we assume that the image points are statistically independent, then the simplest model of a potential V can be used : $V(i,j) = L(i,j)$ and the elements of the probability matrix Π with elements $p[(i,j) \rightsquigarrow (k,l)]$ can be obtained using the Gibbs distribution formula

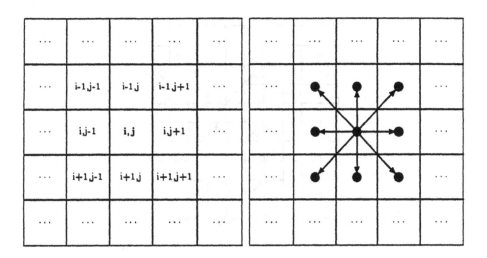

Fig. 2. Transition directions of the virtual particle on the lattice with the eight neighbourhood system

$$
p[(i,j) \rightsquigarrow (k,l)] = \begin{cases} \dfrac{e^{\frac{V(i,j)-V(k,l)}{T}}}{\displaystyle\sum_{m=-1}^{m=+1}\sum_{n=-1}^{n=+1} e^{\frac{V(i,j)-V(i+m,j+n)}{T}}} & : p\{(i,j),(k,l)\} \le 1 \\[20pt] 0 & : p\{(i,j),(k,l)\} > 1 \end{cases}
$$

where $V(i,j)$ is the potential energy, and T plays the role of the temperature of the statistical system. The value of T has to be found experimentally. The experiments have shown that good results can be obtained using $T \in [0.1, 0.2]$.

In order to avoid problems connected with the incomplete neighbourhood of the boundary elements of the image matrix L, an additional matrix has to be introduced, so that the initial lattice acquires the structure of a torus and every boundary point of L has eight neighbours [12]. The imposed toroidal structure of L plays a crucial role in the contrast enhancement of images with nonhomogenous brightness. On a such defined lattice the random walk of the virtual particle can be investigated. Figure 2 shows the way the virtual particle walks around the image points.

Let $p[(i,j) \rightsquigarrow (k,l), n]$ be the probability that the virtual particle starting from the point (i,j) visits after n steps the point (k,l). The matrix of the elements $p[(i,j) \rightsquigarrow (k,l), n]$ will be denoted by $\Pi(n)$.

It is known from the theory of Markow chains, that the matrix $\Pi(n)$, the elements of which denote the probabilities of a transition from (i,j) to (k,l) in n steps, is equal to Π^n.

A matrix R is called regular, if for some integer κ, the matrix R^κ has positive entries [13, 14]. Our matrix Π is regular, because the matrix $\Pi(N_r + N_c) =$

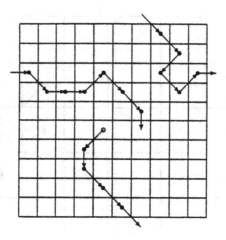

Fig. 3. Random walk on a toroidal lattice

$\Pi^{N_r+N_c}$, where N_r and N_c are the sizes of the image matrix L, has positive elements. It is so, since every point of L can be reached from any other point in fewer than $N_r + N_c$ steps.

If Π is a regular matrix, then there exists a limiting matrix $\mathcal{P}$ satisfying

$$\lim_{n \to \infty} \Pi^n = \mathcal{P}$$

Each row of $\mathcal{P}$ is composed of the same probability vector $\eta = [\eta_1, \eta_2, \ldots, \eta_N]$ which elements satisfy $\eta_i > 0$ and $\sum_{i=1}^{i=N} \eta_i = 1$.

The vector η is the unique probability vector satisfying the eigenvalue equation

$$\eta \mathcal{P} = \mathcal{P}$$

The presented approach seek to determine the vector η. Taking into acoount that the number of image points is N, then the size of the matrix Π is $N \times N$. Solving the eigenvalue equation is connected with a great computational effort and that is why an experimental method of approximation of the η vector was developed.

Let $\psi[(i,j) \rightsquigarrow (k,l)]$ be the expected value of the number of visits in the point (k,l) by a particle, which started from the point (i,j), then

$$\psi[(i,j) \rightsquigarrow (k,l)] = \sum_{n=0}^{n=\infty} p[(i,j) \rightsquigarrow (k,l), n]$$

The expected value of the number of particles, which visit the point (k, l) starting from every point of the image is given by

$$\chi(k, l) = \sum_{i,j} \psi[(i, j) \rightsquigarrow (k, l)]$$

The relative frequency of visits $v(k, l)$ is then

$$v(k, l) = \frac{\sum\limits_{i,j} \psi[(i, j) \rightsquigarrow (k, l)]}{\sum\limits_{k,l}\sum\limits_{i,j} \psi[(i, j) \rightsquigarrow (k, l)]} = \frac{\chi(k, l)}{\sum\limits_{k,l} \chi(k, l)}$$

The algorithm of contrast enhancement assigns to each image point (k, l) its relative frequency of visits $v(k, l)$. Having the probabilities of transitions to the points of the neighbourhood, the random walk on the image can be simulated. Starting from every point of the image, the virtual particle performs λ transitions. Every visit is registered and afterwards the relative frequency $v_\lambda(k, l)$ of visits in (k, l) can be determined. The results of experiments presented in this paper were obtained at $\lambda = 500$ (the image was of the size 125×125). Generally λ must be high enough to ensure a good enough approximation of the real vector η a performance of the contrast enhancement.

The enhanced image V is obtained by transforming the values of v_λ according to

$$V(i, j) = \frac{e^{\gamma v_\lambda(i,j)}}{e^\gamma}$$

where γ is a parameter, which has to be found experimentally. The normalization of V gives the enhanced image.

3 Results

Figure 3 shows four examples of contrast enhancement obtained by means of the presented method. The original images (left) of the size 125×125 are of low contrast resulting from nonuniform image brightness. The enhanced images are presented on the right. The parameters of the algorithm are $\lambda = 500$, $T = 0.1$ and $\gamma = 5$. The experiments have shown that good results can be obtained with $T \in [0.1, 0.2]$ and $\lambda > 100$. This values are independent on the image size. The parameter γ must be chosen experimentally at the end of the algorithm.

4 Conclusion

In the submitted paper, an original method of contrast enhancement of grey scale images is presented. The described method presents a quite novel approach to the old problem of image enhancement. It is based on a model of an image as a two dimensional toroidal lattice, on which a special case of a Markow process, the

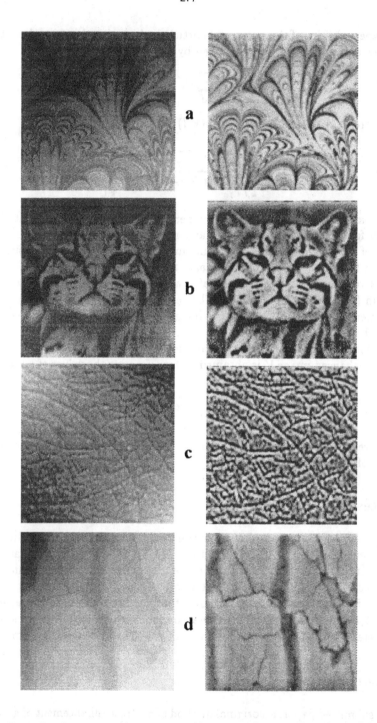

Fig. 4. Example of contrast enhancement of images with nonuniform image brightness
a) hand-made paper b) cat c) human skin d) marble surface

random walk of a virtual particle is performed. The introduced virtual particle is assumed to be able to jump between the neighbouring points and the transition probability is derived from the the Gibbs distribution formula. In this way an analogy between an image and a statistical mechanics approach is established.

The main contribution of this work is the development of a quite new approach to the problem of contrast enhancement. Its essence is the combination of the random walk model and the Gibbsian properties of the virtual particle walking around the image.

Although a very simple model of the potential energy was assumed, the obtained results are encouraging. The presented method works especially well in cases of badly illuminated images. The results obtained with this method are much better than the results got using the methods described briefly in the introduction (histogram equalization, one-point operations, sharpening etc.).

5 Acknowledgements

The research was supported by the Polish Committee of Scientific Research (KBN).

References

1. Pratt W.: Digital image processing. New York, Willey (1991)
2. Habärecker P.: Digitale Bildverarbeitung. Hanser Verlag (1989)
3. Jähne B.: Digitale Bildverarbeitung. Springer Verlag (1989)
4. O' Gorman L.: A note on histogram equalization for optimal intensity range utilization. Computer Vision Graphics and Image Processing 41 (1988) 229-232
5. Pizer S.M.: Adaptive histogram equalization and its variations. Computer Vision Graphics and Image Processing 39 (1987) 355-366
6. Hummel R.A.: Image enhancement by histogram transformation. Computer Graphics and Image Processing 6 (1977) 184-195
7. Otsu N.: A threshold selection method from gray level histograms. IEEE Transactions on Systems, Man and Cybernetics 9 (1979) 62-66
8. Parker J.R.: Gray level thresholding in badly illuminated images. IEEE Transactions on Pattern Analysis and Machine Intelligence 13 (1991) 813-819
9. Weszka J.S., Rosenfeld A.: Histogram modification of threshold selection. IEEE Transactions on Systems, Man and Cybernetics 9 (1978) 38-52
10. Narendra P.M., Fitch. R.C.: Real time adaptive contrast enhancement. IEEE Trans. Pattern Anal. Mach. Intell. 3 (1981), 655-661
11. Gonzalez R.C.: An overview of image processing and pattern recognition techniques. in Handbook of geophysical exploration, Pattern recognition and image processing edited by F. Aminzadeh, Geophysical Press 20 (1987)
12. Hassner M., Sklansky J.: The use of Markov fields as models of texture. Computer Graphics and Image Processing 12 (1980) 357-370
13. Mathar R., Pfeifer D.: Stochastik für Informatiker. B.G. Teubner Verlag Stuttgart (1990)
14. Yokoyama R., Haralick R.M.: Texture pattern image generation by regular Markow chain. Pattern recognition 11 (1979) 225-234

Adaptive Non-linear Predictor
for Lossless Image Compression

Václav Hlaváč, Jaroslav Fojtík

Czech Technical University, Faculty of Electrical Engineering
Center for Machine Perception
121 35 Prague 2, Karlovo náměstí 13, Czech Republic
phone +42 2 24357465, FAX +42 2 290159
{hlavac,fojtik}@vision.felk.cvut.cz

Abstract. The paper proposes the new method for lossless image compression that performs wery well and results can be compared with other methods that we are aware of. We developed further the Slessinger's idea to represent an image as residuals of a special local predictor. The predictor configurations in a binary image are grouped into couples that differ in representative point only. Only residuals that correspond to the less frequent predictor from the couple is stored. An optimal predictor is based on the frequency of predictor configuration in the image. Two main extensions are proposed. (1) The method is generalized for grey-level image or images with even more degrees of freedom. (2) The method that works with addaptive estimator is proposed. The resulting *FH-Adapt* algorithm performs very well and results could be compared with most of the current algorithms for the same purpose. The predictor can learn automatically from the compressed image and cope even with complicated fine textures. We are able to estimate the influence of the cells in a neighbourhood and use only significant ones for compression.

1 Introduction

The main difference between compression of strings and images is the more complex context in the case of images. The stronger is the structure of the available context the more radical can the compression be. The context of a binary image has two degrees of freedom (2-DOF), i.e. in vertical and horizontal direction. The grey-level image context has 3-DOF, the 3rd DOF allows to change intensity values. A colour (e.g. RGB) image context has 4-DOF. The fourth direction points across color image planes.

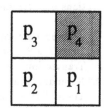

Fig. 1. 2 × 2 testing probe.

One of sources of redundancy in raster images is the fact that neighbouring cells [1] are likely to have the same intensity. That implies that the local context of the current cell is of importance. In this paper, an attempt is

---

[1] The concept *cell* is used instead of common pixel for 2D images. The reason is that the cell will be used for higher dimensions.

made to examine the neighbourhood of the current cell using a small probe. We shall start from a simple probe, the 2×2 window for the case of binary images.

We shall develop further the idea common in predictive coding. The new value in the current cell is estimated using a *predictor* function. The support of the predictor is the context that is encompassed by the probe. In the compression phase, an image is systematically traversed by the probe. The estimated value is compared with the actual value of the cell. There will be a precise match in very many cells and nothing should be written into compressed image. Where there is a discrepancy between estimated and the actual value the x, y coordinates of such cell are stored together with the *residual*. The predictor, residuals and their positions in the image are compressed using a optimal Huffman code.

The proposed method has *two unusual features*. First, the unusual non-linear predictor that was first in its basic form proposed by M. Schlessinger is used. Generalizations of this predictor are proposed. Second, the results of compression are well for available set of test images.

2 Representation and Compression of Binary Images

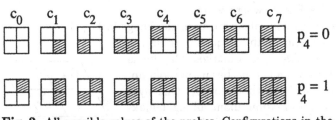

Fig. 2. All possible values of the probes. Configurations in the same column differ in the upper right pixel only.

Our compression method builds on the representation of a binary image proposed by M. Schlessinger from Ukraine [Sch89]. We shall abbreviate it as *FH-2-DOF* in the sequel. We guess that this method is almost unknown in the image analysis community. Thus we explain it first. We use slightly adapted notation to be able to describe newly proposed methods in the same formalism.

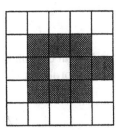

Fig. 3. Hypothetical input binary image

The *FH-2-DOF* method assumes a binary image f as an input. Its support is $T=\{(x,y) : 0 \leq x \leq M; 0 \leq y \leq N\}$. The image is the mapping $f(x,y) \rightarrow \{0,1\}$. Without loss of generality, we assume that the left column and the bottom line of the image have value zero[2], i.e. $f(0,y) = f(x,0) = 0$.

The *FH-2-DOF* method traverses the image f by the probe of 2×2 cells. Its representative cell (x,y) is the upper right corner of the probe. The value of the representative cell is $p_4 = f(x,y)$. Let values of the probe placed in the particular position in the image be $p_1 = f(x,y-1); p_2 = f(x-1,y-1); p_3 = f(x-1,y);$ see Fig. 1.

---

[2] We assume that the leftmost column and bottommost row in the input image is zero for simplicity. It could be nonzero but must be known. Thus some other compression technique should be used for left most column and bottom most row, e.g. RLC coding.

Schlessinger designed the predictor (estimator, probe) e in the following way. All 16 possible combinations of cells within a 2×2 probe are considered, see Fig. 2. Probes are arranged into two rows in such a way that probes in one column differ only in the upper right bit.

The method should consist of two inverse phases, compression and decompression. For all methods proposed in this paper, the compression phase consists of: (1) Optimal predictor design, that is based on the frequency analysis of the probes configurations in the particular image. (2) Data reduction that writes predictor residuals into a sparse matrix. (3) The reduced data are optimally coded and together with the predictor description create the compressed image.

The optimal predictor uses the information about *frequency analysis* of probe configurations in the image. The frequency analysis needs one pass through image and can be performed in parallel. We shall explain algorithms using a small hypothetical input image, see Fig. 3. The frequency of configurations is stored in a table with two rows and eight columns. See Fig. 4, where is an example of the frequency table calculated from the image in Fig. 3.

	c_0	c_1	c_2	c_3	c_4	c_5	c_6	c_7
$p_4 = 0$	0	1	1	2	1	0	0	2
$p_4 = 1$	1	2	0	1	2	1	2	0

Fig. 4. Frequency table for the hypothetical image

Maxima in each column of the frequency table are found and the corresponding probes marked. One bit is needed to distinguish between two probes in one column. The stored bit value equals to the value p_4. One byte is enough to store the information about all eight maxima; e.g. 11001110 for the frequency table in Fig. 4. The principal Schlessinger's idea is that only positions of less frequent configurations in the image are remembered.

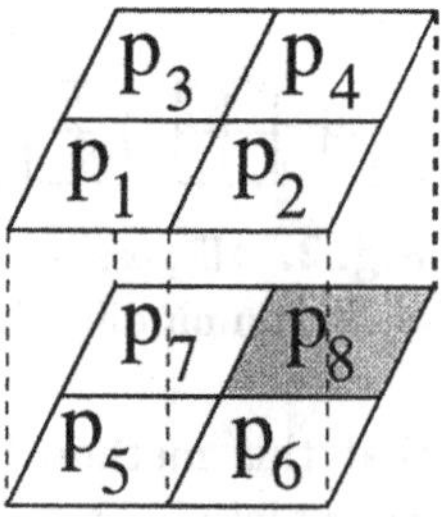

The predictor $\hat{p}_4 = e(p_1, p_2, p_3)$ estimates the value of the image in the representative point. Let define a penalty function g

Fig. 5. 3-DOF predictor

$$g(\hat{p}_4, p_4) = 0 \quad \text{for } \hat{p}_4 = p_4$$
$$= 1 \quad \text{for } \hat{p}_4 \neq p_4 \tag{1}$$

The construction of an optimal predictor can be formulated as an optimization task that minimizes sum of penalties for all positions of the probe in the image

$$\sum_{x=1}^{M} \sum_{y=1}^{N} g(e(p_1, p_2, p_3), p_4) \to \min \tag{2}$$

The image is scanned in the bottom-up manner and from the right to left. In each probe position in the image, it is checked if more frequent or less frequent configuration appeared. In the former case, the value 0 is written to the representative cell. In the latter situation, the value is stored that corresponds to the residual of the predictor. This reduced image is typically sparse as most of the information was captured by the predictor. The description of the predictor (one byte) belongs to the compressed image.

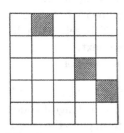

Let us note that Schlessinger codes reduced image as list of x, y coordinates of residuals. He even uses suboptimal coding for special arrangement of probes. In such case, the standard set operation can be performed on reduced images without a need to decompress it.

The *decompression* algorithm has at hand description of the predictor (one byte) and reduced image with residuals. The left most column and bottom most row is known.

Fig. 6. The residuals calculated from Fig. 3

The probe is put to bottom left position first. Its values p_1, p_2, and p_3 are known. This three values uniquely determine one of the eight columns of the predictor. Which of the couple of probes is determined by the value of the stored residual. The new decompressed value p_4 is set.

The probe traverses the image linewise to the right and upward and the whole original image is restored.

3 Generalization for grey-level images

The grey-level image can be treated as stacked bit planes. We generalized the predictor to cope with the current bit plane and the closest higher bit plane, see Fig. 5. The context encompassed by the volumetric probe has 3-DOF. We name the algorithm *FH-3-DOF*. Letters F and H are initials of authors surnames.

The frequency analysis of probes configurations is analogical to the previous case. Here 2^7 bits $= 16$ bytes are needed to store the frequency table information. The image of the bit-plane is reduced and residuals are stored in the same way as in *FH-2-DOF* case. This requires second pass through original image date. The result is a sparse matrix with residuals.

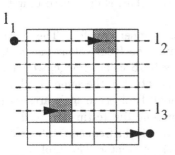

To achieve high compression ratio the matrix with residuals should be effectively coded. Consider Fig. 6 where an example of residuals is shown. The intention is to code residuals using line scans and distances on them between residuals. The counting starts from a hypothetical starting point positioned left from the first row. The distances $l_1, \ldots, l_k$ obtained represent uniquely residuals.

Fig. 7. Lengths to be coded.

The question of effective coding distances $l_1, \ldots, l_k$ is important. The Huffman coding is not suitable as the maximal length $l_{max} = M N +$ 1 is a big number. We have chosen the combination of the Huffman coding and the prefix code with logaritmical growth. The length of the latter code grows logarithmically with the number l_i. The code with logaritmical growth used is shown in Tab. 9. The combination of the Huffman code and the prefix code with logaritmical growth is illustrated in Fig. 8. The number K is the highest number coded by the Huffman code. There are three options: (a) $K = 0$, i.e. only the code with logaritmical growth is used. (b) K is fixed. Experiments have shown that best values are typically $K = 32$ or $K = 64$. (c) The K is set optimally according to data to be compressed. This option is described next.

The length L_1 of the Huffman code table is an increasing function

$$L_1 = 4 + (n + 1) \log_2 S \quad [bits], \tag{3}$$

where S is the size of Huffman code space (that corresponds to the longest coded symbol). $(n+1)$ is number of entries of the table. The $\log_2 S$ term gives the length of the table entry in bits. The length of the coded residuals, roughly speaking[3] monothonically decreases. The difference between these two monotonic function is our function. We are looking for its maximum that corresponds to optimal K.

4 Adaptive Compression

The two previous predictors had fixed number entries (3 in the *FH-2DOF* case and 7 in the *FH-3-DOF* case) in a fixed arrangement. The desire is to exploit more complex context in the predictor. The support set

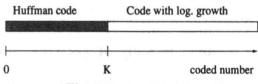

Fig. 8. Coding lengthes.

of the predictor can have more cells. Such predictor allows to encode more complicated structure in the image as, for instance, some fine texture.

The size of the predictor support is limited from the practical point of view by the size of the frequency table. It grows exponentially with the number of cells k in the predictor support set. Its size is 2^k. Say, for 20 cells, the frequency table size is $2^{20} = 1$Mbit. This is too much. Of course, not all predictor cells bear the same amount of information. The trick we propose is to through away some cells from a relatively large support adaptively. This is done according to the content of the particular image.

Predictor probes are constructed in the similar manner as before. They have more cells, e.g. 20, and may span n-DOF. Assume the RGB colour image. If we compress one of the colour components of the RGB colour image, we have 3-DOF due to grey-level image plus 2-DOF if the context in the same bit-plane in other two colour components is taken into account. Probes are again arranged in couples that differ only in the value of the representative (estimated) cell.

The frequency analysis comes next. The result of it is the frequency table with the size 2^{k+1}. k is the number of support cells of the predictor. This table is too big. The aim is to select only those i cells, $i < k$, that are the most informative and thus reduce the length of the frequency table to 2^i. The selection step does not need to pass through original image data any more and thus is independent of image size.

---

[3] The small perturbation are caused namely by the changes of the exponent, e.g. from 2^3 to 2^4.

We may look at cells of the predictor as at features common in statistical pattern recognition. Features contribute to the estimate of the representative point of the probe. Recall that k is the number of probe feature points.

Number	Code
1	00
2	01
3	100
4	101
5	11000
6	11001
7	11010
8	11011
9	1110000
10	1110001

Fig. 9. Code with logaritmical growth

After the frequency analysis, it is known how often a particular predictor probe occurs in the image. Moreover, the features can be ordered according to their importance for estimation of the representative cell. We found an algorithm that allows to dig out influences of individual features from the frequency table. Input data are from the frequency table.

Let us explain decoupling on the specific example with $k = 3$. All configurations of the probe were shown in Fig. 2. Assume for a moment that we like to decouple influence of the feature No. 3 (top left corner of the probe). If we omit this feature couples of probes will merge as depicted in Fig. 10. The frequency of the new probe is the sum of frequencies of its two constituting probes.

A good representation of the frequency table allows very effective bit operations. Recall that the representative cell has the highest index. Let the frequency table be stored in a vector $\mathbf{q}$ with 2^{k+1} elements. First 2^k elements of the vector $\mathbf{q} = (q_0, \ldots, q_k)$ correspond to frequencies of upper row of probe configurations.

Let e_{all} be the number of incidences of all less frequent configurations of the probes. Let b be an index to the first element of the second half of the frequency table, $b = 2^k$.

$$e_{all} = \sum_{i=0}^{k} \min(q_i, q_{i+b}) \tag{4}$$

Let us define the *estimativeness* e_c of the feature c.

$$e_c = e_{all} - \frac{1}{2} \sum_{i=0}^{k} \min\left(q_i + q_{i \oplus c}, q_{i \oplus b} + q_{i \oplus c \oplus b}\right) \tag{5}$$

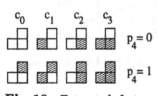

$c_0 \quad c_1 \quad c_2 \quad c_3$

$p_4 = 0$

$p_4 = 1$

Fig. 10. Extracted feature No.3

The factor $\frac{1}{2}$ in front of the sum comes the fact that the estimativeness of the feature c is in fact scanned twice. The estimativeness e_c tells what is the deficit in frequency of the probe occurrence when feature c would be omitted.

The feature with lowest estimativeness is ruled out from the probe. The same algorithms can operate iteratively and throw away second less informative feature, etc.

At last, we have to tell when to stop iterations. The estimativeness e increases monothonically if we rule out features gradually. The number of entries in the frequency table decreases monothonically and thus decreases its length. The difference between these two functions has an extreme that we look for. If the increase of the length of the coded residuals (given by the estimativeness e) is bigger than decrease of the table length or table is empty then the algorithm stops.

5 Experiments

The goal of experiments was to (1) test properties of the proposed method, (2) compare it with the results of others, and (3) find posibilities of further improvements.

Let us show the behaviour of the proposed *FH-3-DOF* compression on the standard grey-level testing image Lena, see Fig. 11 of the size $256 \times 256 \times 8$ bits. Selected bit-planes of the original and compressed image are shown in Fig. 12. Notice that the compression is

Fig. 11. Lena

more effective on bit-planes corresponding to more significant bits as there is more regularity. Notice how well the predictor residuals express the significant features in the image. The mentioned experiment is quantified in tabular form in Tab. 1. Notice how number of residuals increases from the most significant bit-plane to the least significant bit-plane. This result is obvious as in the least significant bit-plane there is much randomnes that is difficult to represent in the estimator. The estimativeness gives a clear insight into the predictor. The feature No. 6 and feature No. 4 from the neigbouring bit plane bear the biggest amount of information.

Bit-plane	bit 7	bit 6	bit 5	bit 4	bit 3	bit 2	bit 1	bit 0
2-DOF residuals	2010	5477	10177	15067	21865	28595	31771	32214
3-DOF residuals	empty	3841	5935	9421	15045	21690	28622	31272
estimativeness e_1	empty	25	160	218	233	91	125	320
estimativeness e_2	empty	616	2165	3108	4930	4919	2521	457
estimativeness e_3	empty	52	180	219	233	91	128	352
estimativeness e_4	empty	1892	4101	5543	6611	6600	2813	447
estimativeness e_5	0	14	123	143	92	26	112	361
estimativeness e_6	1457	2097	3477	4169	5807	5031	2312	467
estimativeness e_7	0	23	118	136	170	85	71	220

Table 1. Quantitative description of the FH-2-DOF and FH-3-DOF methods

The proposed adaptive compression method *FH-Adapt* was compared on several images from the standard JPEG image set [AT94]. Methods and results were taken: SEGMENT from [RA96], Portable Video Research Group at Stanford PVRG-JPEG from [CAJ96], Rounding transform from [JCP96], and Logic coding from [CAJ96].

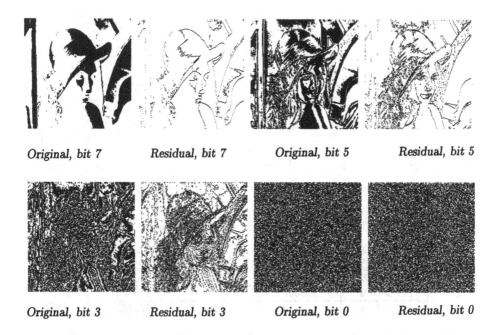

Original, bit 7   Residual, bit 7   Original, bit 5   Residual, bit 5

Original, bit 3   Residual, bit 3   Original, bit 0   Residual, bit 0

Fig. 12. Comparison between bit-planes of the original image and bit-planes storing residuals.

The quantitative measure used for comparison of different methods is *compression efficiency CE* [CAJ96]

$$CE = \frac{\text{total input bytes} - \text{total output bytes}}{\text{total input bytes}} \cdot 100\% \qquad (6)$$

The Table 13 shows how our method compares with others. The value CE shows that our method performs very well. It was beaten by Segment and PNG in the case of an image of a monkey (Baboon). The structure of hair is bigger that the size of the predictor in our method. Segment perform also better on Lena. Unfortunately we were not able to experiment with Segment.

Method/Image	Lena	Baboon	Boats
PVRG-JPEG	22.7-29.4	6.7-13.7	24.9-33.1
Logic	26.1	12.6	29.3
Segment	54.6	36.3	empty
Rounding Tr.	37.6	empty	empty
PNG	20.7	41.7	33.7
FH-3-DOF	**34.0**	**19.1**	**41.2**

Fig. 13. Compression coefficients compared with results of others

The Table 15 shows the comparisson of our method with Lena and a technical drawing, see Fig. 14. This is an image which is perfectly suited for our compression method as can be seen from results.

The last Table 2 shows what is the influence of three method of coding residuals and three different predictors. This allows to understand what individual improvement bring to us. The values in the table are the length of the compressed 512×512 Lena in bytes. The PNG compression is very sim-

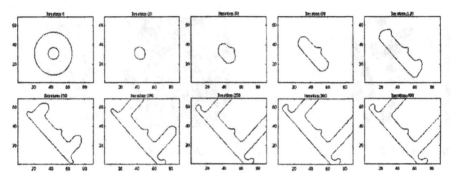

Fig. 14. Image of the scanned technical drawing

iliar to the IBM Q80 coder [ISO94] which is obsolette and its implementetion is inaccessible.

Method/Coding	Exp. code only	Fixed Huffman	Moving Huffman
FH-2-DOF	203000	200758	200599
FH-3-DOF	178655	176021	175922
FH-Adapt.	178130	175541	175432

Table 2. Comparison of proposed 3 predictors and 3 types of coding techniques

Compressing the image Lena $512 \times 512 \times 8$Bits spends 6.2s. Decompression is aproximatelly 2x faster - in Lena case it runs 2.7s. Loading/Storing a data without compression spends 0.1s. All timing tests was done on the machine Pentium 120MHz.

6 Conclusions

We proposed a new lossless image compression method that performs in many respects very well and could be compared with other known methods. Basic idea of the method, i.e. is a bit unusual, i.e. using predictors that differ in one bit only. This idea comes from M.I. Schlessinger [Sch89] and we learned about it during his series of lectures in Prague in Spring 1996.

We propose two significant extensions. First is to have predictor with more degrees of freedom. Second is to use adaptive predictors according to statistics of the particular image to be compressed. The method is namely suitable for images with strong structure as e.g. technical drawings, maps, etc.

We could do better with some images with pallete. The improvement in this respect is under development.

Let us summarize pros and contras of the method. The *advantages* are: (1) The method is fully automatic. No intervention from the user is expected. (2) Method is quick as simple operations are used only. E.g. SHL, XOR, ADD, and selection from the table. (3) The compression results are on the top of current methods. (4) The method does not upper limit on the compression performance. (5) Substantial parts of the compression and decompression can run in parallel. (6) The predictor that constitutes the core of the method is extensible to more degrees of freedom. (7) When the resolution of the image increases the compress ration does not degrade. This happen in the case of pyramidal algorithms, say MLP.

Type of data	Lena	CE Lena	Drawing	CE Drawing
Uncompressed	262656	0	106940	0
ZIP	225007	15.62	10224	90.44
ARJ	228389	14.35	10380	90.29
PCX	274256	-2.85	38624	63.88
GIF	278771	-4.54	17290	83.83
FH-Addapt.	**175434**	**34.21**	**7297**	**93,18**

Fig. 15. Comparison with typical compressions

The *disadvantages* of the method are: (a) The compression needs at least two passes through image data. When the Huffman coding is used even three passes are needed. This passes are fully sequential. (b) The decompression algorithm is very sensitive to errors in data. This is the same with other methods.

The *intentions for the future* are: (i) To improve a compression ratio. (ii) Do experiments with more degrees of freedom (mainly RGB). (iii) To learn if the method will be used more widely in applications.

The reader might want to test the method himself, if so see the www page *http://cmp.felk.cvut.cz/~fojtik/* for the actual version of the compression algorithm.

References

[AT94] R.B. Arps and T.K. Truong. Comparision of international standards for lossless still image compression. *Proceedings of the IEEE*, 82(6):889–899, June 1994.

[CAJ96] A.K. Chaudhary, J. Augustine, and J. Jacob. Lossless compression of images using logic minimization. In P. Delogne, editor, *Proceedings of the International Conference on Image Processing*, volume II, pages 77–80, Lausanne, Switzerland, September 1996. IEEE Signal Processing Society, Louvain-La-Neuve, Belgium.

[ISO94] ISO/IEC 10918-1: Information technology-digital compression and coding of continuous-tone still images: Requirements and guidelines, 1994.

[JCP96] H.Y. Jung, T.Y. Choi, and R. Prost. Rounding transform for lossless image coding. In P. Delogne, editor, *Proceedings of the International Conference on Image Processing*, volume II, pages 65–68, Lausanne, Switzerland, September 1996. IEEE Signal Processing Society, Louvain-La-Neuve, Belgium.

[RA96] K. Ratakonda and N. Ahuja. Segmentation based reversible image compression. In P. Delogne, editor, *Proceedings of the International Conference on Image Processing*, volume II, pages 81–84, Lausanne, Switzerland, September 1996. IEEE Signal Processing Society, Louvain-La-Neuve, Belgium.

[Sch89] M.I. Schlessinger. *Matematiceskie sredstva obrabotki izobrazenij, in Russian, (Mathematic tools for image processing)*. Naukova Dumka, Kiev, Ukraine, 1989.

Beyond Standard Regularization Theory

Zhiyong YANG and Songde MA

National Laboratory of Pattern Recognition
Institute of Automation, Chinese Academy of Sciences
P. O. Box 2728, Beijing 100080, China
Email: yangzy@prlsun6.ia.ac.cn masd@prlsun2.ia.ac.cn

Abstract. A set of local interaction field are suggested to replace the δ error term in usual regularization approaches. These local fields bring some computational and conceptual benefits. A set of local oriented position pinning fields and orientation tuning fields are suggested to use local position and orientation correlations directly in regularization. Some simple experiments show that these generalizations are useful in many cases.

1 Introduction

Regularization theory has been widely used in computational vision[1]. Recently, it has been found to be closely related to another widely used tool in signal and image processing, Markov random fields[2]. Due to their mathematical simplicity, computational power and biological supports, regularization networks have been suggested as working modules of the human brain[3]. In this paper, we use local interaction field to replace the δ interaction in usual regularization. To use local position and orientation correlations directly in regularization, we suggest a set of local oriented position pinning fields and orientation tuning fields. Due to their computational power and neurobiological plausibility, these new theories may be much more relevant to vision problems.

2 Regularization With Local Interaction Fields

To construct a surface $f(x,y)$ from n known points $f_1(x_1,y_1),\ldots,f_n(x_n,y_n)$, usual theory minimizes the following functional

$$H = \sum_{i=1}^{n}(f - f_i(x_i,y_i))^2 + E \tag{1}$$

E is regularization term and can be chosen from several different approaches. In this paper, we suggest two general regularization forms[4]

$$E = \frac{K_1}{2}\int g_1(\nabla f \bullet \nabla f)dxdy + \frac{K_2}{2}\int g_2(\nabla^2 f \bullet \nabla^2 f)dxdy \tag{2}$$

∇, ∇^2 are 1st and 2nd differential operators respectively. $K_1 > 0$ and $K_2 > 0$ are regularization constants. $\frac{\partial g_1(\chi)}{\partial \chi} > 0$, $\frac{\partial g_2(\chi)}{\partial \chi} > 0$. For two dimensional surface reconstruction, we also propose the following regularization

$$E = \frac{K_1}{2} \int g_1((c_1 + c_2)^2) dx dy + \frac{K_2}{2} \int g_2(c_1 c_2) dx dy \qquad (3)$$

c_1, c_2 are two principal curvatures. $\frac{\partial g_1(\chi)}{\partial \chi} > 0$, $\frac{\partial g_2(\chi)}{\partial \chi} > 0$. When $g_1(\chi) = \chi$, $g_2(\chi) = \chi$, $Eq(3)$ is widely used in the study of shapes and interaction of biological membranes[5].

The two terms in $Eq.(1)$ are conceptually asymmetrical: regularization term concerning a neighbor of a point while the error term being a $\delta-$ function form. We try to revise this δ interaction so that it also concerns a neighbor of a point, not an isolated point itself. We propose the interaction between surface $f(x, y)$ and a known data point $f_i(x_i, y_i)$ as

$$h_I(x_i, y_i) = \int T(x_i, y_i, f_i(x_i, y_i), x, y, f(x, y)) dx dy \qquad (4)$$

$T(x_i, y_i, f_i(x_i, y_i), x, y, f(x, y))$ is a kernel function which decreases monotonically with distance $|x_i - x|$, $|y_i - y|$ and increases monotonically with $|f_i(x_i, y_i) - f(x, y)|$. Here we propose a separable form for it

$$T(x_i, y_i, f_i(x_i, y_i), x, y, f(x, y)) = (f(x, y) - f_i(x_i, y_i))^2 \times W(x_i, y_i, x, y) \qquad (5)$$

$(f(x, y) - f_i(x_i, y_i))^2$ is the usual error term. $W(x_i, y_i, x, y) \geq 0$ is a window function which satisfies

$$\lim_{|x - x_i| \to \infty} W(x_i, y_i, x, y) = 0, \quad \lim_{|y - y_i| \to \infty} W(x_i, y_i, x, y) = 0$$

When $W(x_i, y_i, x, y)$ is a $\delta-$ function, the usual error term, $(f(x, y) - f_i(x_i, y_i))^2$ is recovered. We choose window function $W(x_i, y_i, x, y)$ as short range functions of the distance $|x - x_i|$, $|y - y_i|$. Two such short range functions are square well and Gaussian function. The orientation, shape and size of the window can change from point to point and thus adapt to local structures in specific situations. Window shapes can be square, rectangle, sphere, and ellipse. So the total energy functional becomes

$$H = \sum_i \int (f(x, y) - f_i(x_i, y_i))^2 \times W(x_i, y_i, x, y) dx dy + E \qquad (6)$$

We can see two direct benefits from this interaction field at this stage.

- Two terms in total functional are conceptually symmetrical, both concern point neighbors, not isolated points.

- Window orientations, shapes and sizes can be made adaptive to local structures of curves or surfaces. So every data point has its receptive field which is mediated by local structures of surfaces to reconstruct. This fact may bring stability, tolerance and robustness against noises.

At present, we have not found general closed-form solutions to $Eq.(6)$. We will present some examples by gradient descent algorithm.

3 Regularization with Position and Orientation Correlation

For most problems in visual perception, we encounter correlated data, such as line segments or surface patches of certain orientations, discontinuity segments of intensity, depth or motion, line or surface patches moving in certain directions. Line segments or surface patches seem to be much more preferred than points in neurobiology. In the primary visual cortex, most cells can signal simple bars of certain orientations. Most MT neurones exhibit simplest directional selectivity. Present regularization theory does not explicitly use these correlated data. Here, we present a regularization theory to explicitly use these correlated data. Expanded presentation will be published elsewhere[6].

We want to reconstruct surfaces from input $(x_1, y_1, f_1, \rho_1), (x_2, y_2, f_2, \rho_2)$, $\ldots, (x_k, y_k, f_k, \rho_k)$. (x_k, y_k, f_k) is position data, ρ_k is surface orientation at the point (x_k, y_k, f_k), i.e., the normal of the surface at that position. We propose two kinds of interaction, i.e., position pinning and orientation tuning. In the local coordinate with a known point an orientation as in Fig.1, position pinning is proposed as

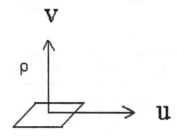

Fig. 1. Local coordinate with a point and orientation

$$U_p = -W \times \int F_p(u)\phi(v)u\,du \qquad (7)$$

And orientation tuning is proposed as

$$U_o = -J \times \int F_o(u,v)\mathbf{n} \bullet \rho u du \tag{8}$$

$W > 0, J > 0$ are constants. $\phi(v) > 0, \frac{\partial \phi(v)}{dv} < 0$. $\mathbf{n}$ is the normal of the surface to be reconstructed. $F_p(u) > 0, F_o(u,v) > 0$ are two shape functions. $\frac{\partial F_p(u)}{\partial u} < 0, \frac{\partial F_o(u,v)}{\partial u} < 0, \frac{\partial F_o(u,v)}{\partial v} < 0$. Due to the symmetry with respect to ρ, integral element is udu. Orientation is an emergent property. Position pinning and orientation tuning are anisotropic fields, not $\delta-$type interactions or homogeneous fields. In the experiments presented here, we choose,

$$\phi(v) = exp(-\frac{v^2}{2\sigma^2_{p\perp}}) \tag{9}$$

$$F_p(u) = exp(-\frac{u^2}{2\sigma^2_{p\|}}) \tag{10}$$

$$F_o(u,v) = exp(-\frac{u^2}{2\sigma^2_{o\|}})exp(-\frac{v^2}{2\sigma^2_{o\perp}}) \tag{11}$$

$\sigma_{p\|}, \sigma_{p\perp}, \sigma_{o\|}, \sigma_{o\perp}$ are sizes. $Z_p = \frac{\sigma_{p\|}}{\sigma_{p\perp}} \geq 1, Z_o = \frac{\sigma_{o\|}}{\sigma_{o\perp}} \geq 1$ are orientation constants. One can choose other forms of shape functions. Then we have total energy

$$H = U_0 + U_1 + E \tag{12}$$

$$U_0 = \sum_i \int (f(x,y) - f_i(x_i,y_i))^2 \times W(x_i,y_i,x,y)dxdy \tag{13}$$

$$U_1 = \sum U_p + \sum U_o \tag{14}$$

U_0 is the total interaction with input points of only position data. U_1 is the total interaction with input points of both position and orientation data.

In $1 + 1$ dimensions, the theory presented here is somewhat similar to the dynamic coverings in [7]. Our theory is intended to include local correlation in regularization. It is quite general and may induce a new kind of neural representation.

4 Some Examples

Surface Reconstruction, Surface Smoothing and Image Restoration

The next step is to find solutions which minimize various functionals given data. We use gradient descent algorithm. Some widely used statistical physics algorithms such as stimulated annealing, fast Monte Carlo algorithms can be also used. As we will see later, even simple gradient descent algorithm is very efficient in many examples.

For surface reconstruction and surface smoothing, $x - y$ plane is two dimensional lattice with discrete index. Suppose we have a data point $f(x_i, y_j)$ at site (i, j), we choose window function W as

$$W(i, i', j, j') = \begin{cases} u_1, & i = i', j = j' \\ u_2, & \text{Either } i = i' \pm 1, or, j = j' \pm 1 \\ u_3, & i = i' \pm 1, and, j = j' \pm 1 \\ 0, & \text{otherwise} \end{cases} \tag{15}$$

$u_1 > u_2 > u_3 > 0$ are constants. We adopt average square root distance as error criteria.

$$SRD = (\sum_{i=1}^{n} (f^\circ - f^*)^2)/n)^{1/2} \tag{16}$$

f° is target. f^* is result.

Fig.2 shows a surface (down) reconstructed from a set of known data point(up). Eq.(2) is used. $g_\sigma(\chi) = \chi$. $K_1 = 0.0002$, $K_2 = 0.0002$, $u_1 = 1$, $u_2 = 0.5$, $u_3 = 0.25$, $l_c = 0.004$(learning rate). $SRD = 0.0036$.

In this and other experiments on surface reconstruction and smoothing, we have found

- When the usual error term is used instead of the local interaction field, it is quite difficult to find converge solutions and the solutions are sensitive to noises and parameters in the algorithm.
- Learning rate can be adjusted in updating. It can be set as bigger value at small SRD and as smaller value at big SRD. Big learning rate at big SRD can lead to divergence or oscillation.
- Regularization constants can be also adjusted in updating.

Surfaces have a few types of local structure, $c_1 = c_2 = 0$; $c_1 \neq 0, c_2 = 0$; $c_1 > 0, c_2 > 0$; $c_1 < 0, c_2 < 0$; $c_1 c_2 < 0$. c_1, c_2 are two principal curvatures. Simple gradient descent algorithm is effective when windows are adaptive to neighbors with big enough curvatures.

We have also used the regularization with local interaction fields for image smoothing and restoration. Experiments at this stage show that the simple method can effectively smooth images and filter salt & pepper noises, gaussian noises, and speckle noises.

Fig.3 and Fig.4 are two examples. Left to right are original images, images with noises and restored images. Eq.(2) is used as regularization, and $g_1(\chi) = \chi, g_2(\chi) = \chi$. For $Fig.3$, $K_1 = 0.0002$, $K_2 = 0.0002$, $l_c = 0.015$, $u_1 = 1, u_2 = 0.2, u_3 = 0.1$. For $Fig.4$, $K_1 = 0.00025$, $K_2 = 0.00025$, $l_c = 0.04$, $u_1 = 1, u_2 = 0.2, u_3 = 0.1$.

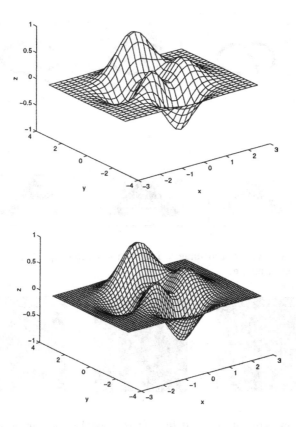

Fig. 2. A surface (down) reconstructed from input(up)

Surface Reconstruction with Position and Orientation Correlation

In the following, we use both position and orientation data to reconstruct surfaces. We choose a proper start surface according to input data set and then update it by the simple gradient algorithm.

Fig.5 is a hat constructed from five points $(1,1,0.5)$, $(1,-1,0.5)$, $(-1,1,0.5)$, $(-1,-1,0.5)$ and $(0,0,0.75)$. Orientations at these points are $(0,0,1)$. It is the surface normal at these points. $Eq.(3)$ is used as regularization, and $g_1(\chi) = \chi$, $g_2(\chi) = \chi$. $K_1 = 0.001$, $K_2 = 0.001$, $l_c = 0.002$. $\sigma_{p\|} = 0.55$, $\sigma_{p\perp} = 0.2$, $\sigma_{o\|} = 0.55$, $\sigma_{o\perp} = 0.2$. At $(0,0,0.75)$, $\sigma_{p\|} = 0.2$, $\sigma_{p\perp} = 0.2$, $\sigma_{o\|} = 0.2$, $\sigma_{o\perp} = 0.2$. $W = 1.5$, $J = 1$.

Fig.6 is a saddle surface constructed from another five points $(1,1,0.65)$, $(-1,-1,0.65)$, $(1,-1,0.35)$, $(-1,1,0.35)$ and $(0,0,0.5)$. Local Orientations at these points are $(0,0,1)$. $Eq.(2)$ is used as regularization, and $g_1(\chi) = \chi$, $g_2(\chi) = \chi$. $K_1 = 0.0002$, $K_2 = 0.0002$, $l_c = 0.0002$. $\sigma_{p\|} = 0.8$, $\sigma_{p\perp} = 0.2$, $\sigma_{o\|} = 0.8$, $\sigma_{o\perp} = 0.2$. $W = 2$, $J = 1$.

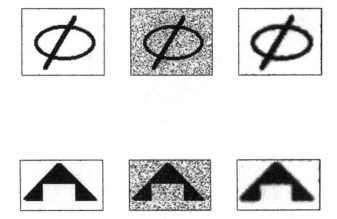

Fig. 3. Results on synthesized images. Left to right are original images, images with noises and restored images.

Fig. 4. Result on a real image. Left to right is original image, image with noises and restored image.

5 Conclusion

We use a set of local interaction field to replace the δ error item in usual regularization theory. These local interaction fields may lead to simple and stable algorithms for converge solutions, tolerance, robustness against noises. Local fields resemble receptive fields in neurobiology, so they may be more favored biologically than usual δ interaction and thus more useful for vision problems. To use local position and orientation correlation directly in regularization, we use a set of local oriented position pinning fields and orientation tuning fields. Anisotropic position and orientation correlation are primary local correlation in visual environment with complex hierarchical correlation and structures. Regularization with local position and orientation correlation may have important meaning for human visual perception.

References

1. Poggio, T., Girosi,F.: Regularization algorithms for learning that are equivalent to multilayer networks. Science **24** (1990)978-982

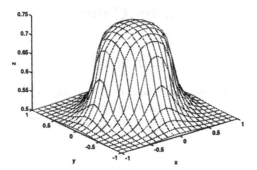

Fig. 5. A hat constructed from five points with both position and orientation data.

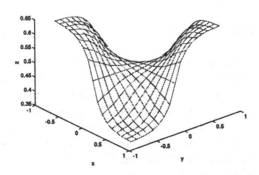

Fig. 6. A saddle constructed from five points with both position and orientation data.

2. Li, S. Z.: Markov Random Field Modeling in Computer Vision. Springer- Verlag 1995
3. Poggio,T.: A theory of how the brain might work. Cold Spring Harbor Sym. on Quantitative Bio. (1990)978-982
4. Zhiyong Yang, Songde Ma: Local Interaction Fields and Adaptive Regularizers For Surface Reconstruction and Image Relaxation. Technical Report, NLPR, Inst. of Automation, Chinese Acad. of Sci.(1996)
5. Safran, S. A.: Statistical Thermodynamics of Surfaces, Interfaces, and Membranes. Addison-Wesley Pub. Com.(1994)
6. Zhiyong Yang, Songde Ma: Combine local oriented position and orientation correlations with regularization. Technical Report, NLPR, Inst. of Automation, Chinese Acad. of Sci. (1996)
7. David, C., S.W. Zucker, S. W.: Potentials, valleys, and dynamic global coverings. Int. J. Computer Vision5(1990) 219 - 238

Fast Stereovision by Coherence Detection

R. D. Henkel

Institute for Theoretical Neurophysics
University of Bremen
henkel@physik.uni-bremen.de

Abstract. A new approach to stereo vision based on similarities between optical flow estimation and disparity computation is introduced. The fully parallel algorithm utilizes fast filter operations and aliasing effects of simple disparity detectors within a coherence detection scheme. It is able to calculate dense disparity maps, verification counts and the cyclopean view of a scene within a single computational structure.

1 Introduction

The structure of 3D-space is hidden in the small relative displacements between the left and right view of a scene (disparities). Beside these small displacements of image parts, numerous other image variations occur between the two stereo views, making the reliable computation of disparities difficult. Noise or different transfer functions of the left and right imaging system introduce stochastic signal variations, and the varying view geometry leads to additional systematic image variations, including occlusion effects and foreshortening.

Most object surfaces in real-world scenes display specular reflection, causing the observed intensities to be not directly correlated with the object surfaces. Image intensities have nearly always a viewpoint-dependent component moving independently of the object surfaces.

Classical approaches to stereo vision try to counteract this set of signal variations with two basic algorithmic paradigms, known as feature- and as area-based approaches [1]. In feature-based stereo algorithms, the intensity data is first converted to a set of features assumed to be a more stable image property than the raw intensities. The matching stage operates only on the extracted image features. In area-based approaches, intensity values within small image patches are directly compared, and a correlation measure between patches is maximized.

Both types of stereo algorithms have intrinsic problems which can be directly attributed to the basic assumptions inherent in these approaches. In the feature-based algorithms, usually only a few discrete feature-classes are used for matching. Therefore most parts of the stereo images end up in the "no feature present"-class, which is not considered further in the matching process.

The tremendous data reduction obtained this way speeds up processing, but in all image regions where not features have been found, no disparity estimate is possible. For dense disparity maps, these missing values have to be interpolated. To complicate matters further, every feature of the left image belonging to one specific class can potentially be matched with *every* feature of the same class in the right image. This is the "false matches"-problem, which is basic to all feature-based stereo algorithms. The problem can only be solved by the introduction of additional constraints to the solution. These are usually reasonable assumptions valid matches have to obey, derived from physical properties of the viewed scene. Classical constraints include the uniqueness of a match, figural continuity and the preserved ordering of matches [2]. In conjunction with the features extracted from the stereo images, these constraints define a complicated error measure which can be minimized by direct search or through cooperative processes.

The problems inherent to feature-based stereo algorithms can simply be reduced by increasing the number of features classes. In the extreme case, one utilizes a continuum of feature-classes, which largely avoids the "false-target"-problem. One example is the locally computed Fourier phase which can be used for classifying local intensity variations into a continuum of feature-classes indexed by the phase value [3].

Using such a continuum of feature-classes derived from some small image area comes very close to area-based algorithms, where the whole vector of image intensities over a small image patch is used for matching. Classical area-based approaches minimize the deviation or maximize the correlation between patches of the left and right view. For stable performance, a minimum patch size is required, which reduces the spatial resolution of the resulting disparity map.

The computation of a correlation measure and the subsequent maximization with regard to a set of parameters describing the mapping between the left and right views turns out to be a computationally expensive process, since extensive search is required in configuration space. This problem is usually solved within hierarchical approaches, where disparity data obtained at coarse spatial scales is used to restrict searching at finer scales.

However, it is not guaranteed for general image data that the disparity information obtained at some coarse scale is valid. This data might be wrong, might have a different value than at finer scales, or might not be present at all. Thus hierarchical approaches may fail under various circumstances.

A third way for calculating disparities is known as phase-based methods [4,5]. These approaches work on derived images, usually the aforementioned Fourier phase. Algorithmically, these approaches are in fact gradient-based optical flow methods [6], with the time-derivative approximated by the difference between the left and right image. The Fourier phase exhibits wrap-around, making it again necessary to employ hierarchical methods — with the already discussed drawbacks. Furthermore, additional steps have to be taken to ensure the exclusion of regions with ill-defined Fourier phase [7].

The new approach to stereo vision presented in this paper is based on similarities between disparity calculations and the estimation of optical flow. It utilizes fast filter operations and aliasing effects of simple disparity detectors in a coherence detection scheme as a new computational paradigm. The algorithm is able to calculate dense disparity maps, verification counts and the cyclopean view of the scene within a single computational structure.

2 Coherence Based Stereo

Estimation of disparity shares many similarities with the computation of optical flow, but has the additional problem that only two discrete "time"-samples are available, namely the images of the left and right view. This discrete sampling of visual space leads to an aliasing limited working range for simple disparity estimators.

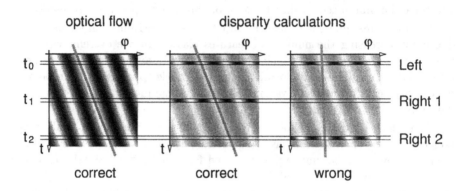

Fig. 1. The velocity of an image patch manifests itself as principal texture direction in the space-time flow field traced out by the intensity pattern in time (left). Sampling such flow patterns at discrete time points only can create aliasing-effects leading to wrong estimates if the flow velocity is too large. If one is using optical flow estimation techniques for disparity calculations, this problem is always present.

For an explanation consider Fig. 1. If a surface patch is shifted over time, the intensity pattern of the patch traces out a corresponding flow pattern in space-time. The principal texture direction of this flow pattern indicates the velocity of the image patch. It can be estimated without difficulty if data for all time points is available (Fig. 1, left). If the flow pattern can not be sampled continuously, but only at some discrete time points, the shift can still be estimated without ambiguity if the shift between the two samples is not too large (Fig. 1, middle). But it becomes impossible to estimate the shift reliably if it exceeds a certain limit (Fig. 1, right). This is caused by simple aliasing in the "time"-direction; an everyday example of this effect is sometimes seen as motion reversal in movies.

To formalize, let $I_L(\varphi)$ be the image intensity of a small patch in the left view of a scene with corresponding Fourier transform $I_L(k^\varphi)$. Moving the left camera on a linear path to the position of the right camera, we obtain a local flow-field very similar to Fig. 1, namely: $I(\varphi, \alpha) = I_L(\varphi + d\alpha)$. The shift-parameter α runs from 0 to 1; d is the disparity of the image patch. The Fourier transform of $I(\varphi, \alpha)$ follows from elementary calculus as $I(k^\varphi, k^\alpha) = I_L(k^\varphi)\delta(k^\alpha - dk^\varphi)$. Now, if the spectrum of $I_L(k^\varphi)$ is bounded by some maximum wavevector $k^\varphi_{\max}$, i.e. if $I_L(k^\varphi) = 0$ for $|k^\varphi| > k^\varphi_{\max}$, we obtain as highest wavevector in α-direction for the flow field $I(k^\varphi, k^\alpha)$: $k^\alpha_{\max} = dk^\varphi_{\max}$. However, the maximal representable wavevector in this direction is given by sampling theory as $k_{\max}^{\alpha} = \pi/\Delta\alpha$. Since sampling in α-direction is done with a step size of $\Delta\alpha = 1$, disparities which can be computed without aliasing effects are restricted by

$$|d| < \frac{\pi}{k_{\max}^{\varphi}} = \frac{1}{2}\lambda_{\min}^{\varphi}. \tag{1}$$

Equation (1) states that the range of reliable disparity estimates is limited by the largest wavevector present in the image data. This size-disparity scaling is well-known in the context of spatial frequency channels assumed to exist in the visual cortex. Cortical cells respond to spatial wavelengths down to about half their peak wavelength λ_{opt}, therefore limiting the range of detectable disparities to less than $1/4\,\lambda_{opt}$. This is known as Marr's quarter-cycle limit [8,9].

Since image data is usually sampled in spatial direction with some receptor spacing $\Delta\varphi$, the highest wavevector $k_{\max}^{\varphi}$ which will be present in the data after sampling is given by $k_{\max}^{\varphi} = \pi/\Delta\varphi$. This leads to the requirement that $|d| < \Delta\varphi$ — without additional processing steps, only disparities *less than the receptor spacing* can be estimated reliably by a simple disparity unit.

Equation (1) immediately suggests a way to extend the aliasing limited working range of disparity detectors: prefiltering the image data before or during disparity calculation reduces $k_{\max}^{\varphi}$, and in turn increases the disparity range. In this way, larger disparities can be estimated reliably, but only with the consequence of reducing simultaneously the spatial resolution of the resulting disparity map. The application of a preshift to the input data of the detectors before disparity calculation is another way to modify the disparity range. However, modification of the disparity range by preshifting requires prior knowledge of the correct preshift to be applied, which is a nontrivial problem. One could resort again to hierarchical coarse-to-fine schemes by using disparity estimates obtained at coarse spatial scales to adjust the processing at finer spatial scales, but the drawbacks inherent to hierarchical schemes have already been elaborated.

The aliasing effects discussed above are a general feature of sampling the visual space with only two eyes; instead of counteracting them, one can utilize them in a new scheme which analyzes the coherence in the multi-unit activity of stacks of disparity detectors tuned to a common view direction.

Assuming that the disparity units i in such a stack have random preshifts or presmoothing applied to their input data, these units will have different, but slightly overlapping disparity ranges $D_i = [d_i^{\min}, d_i^{\max}]$ of valid disparity estimates. An object seen in in the common view direction of the stack with true disparity d will split the whole stack into two disjunct classes: the class $\mathcal{C}$ of detectors with $d \in D_i$ for all $i \in \mathcal{C}$, and the rest of the stack, $\overline{\mathcal{C}}$, where $d \notin D_i$.

All disparity detectors $\in \mathcal{C}$ will code more or less the true disparity, $d_i \approx d$, but the estimates of detectors belonging to $\overline{\mathcal{C}}$ will be subject to the random aliasing effects discussed, depending in a complicated way on image content and disparity range D_i. Thus, we will have $d_i \approx d \approx d_j$ whenever units i and j belong to $\mathcal{C}$, and random values otherwise.

A simple coherence detection within each stack, i.e. searching for all units with $d_i \approx d_j$ and extracting the largest cluster found, will be sufficient to single

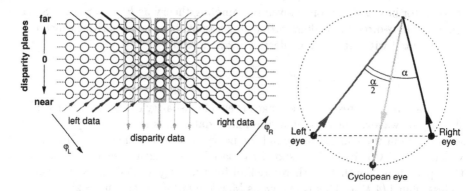

Fig. 2. *The network structure for calculating stereo by coherence detection along a horizontal scan-line (left). Disparity units are arranged in layers having slightly overlapping disparity ranges. The links defining the disparity stacks run perpendicular to these disparity layers. Image data is fed into the network along diagonally running data lines. This network structure causes three-dimensional space to be analyzed in a specific way. Reprojecting the network structure, i.e., disparity stacks and data lines, back into 3d-space shows that the disparity stacks sample data exactly along the cyclopean view directions (right).*

out $\mathcal{C}$. The true disparity d can than be estimated as an average over the detected cluster:

$$d \approx \langle d_i \rangle_{i \in \mathcal{C}} \ .$$

Repeating this coherence detection scheme in every view direction leads to a fully parallel network structure for disparity calculation. Neighboring disparity stacks responding to different view directions will estimate disparity values independently from each other, and within each stack, every disparity unit operates independently from all other units in that stack.

3 Results and Discussion

An appropriate neural network for coherence-based stereo follows very simple. Fig. 2 depicts the network structure along a single horizontal scan-line. Disparity units are arranged in layers stacked behind each other, and image data is fed into the network along diagonal data lines. This leads automatically to different, slightly overlapping disparity ranges for neighboring layers. Links running perpendicular to the disparity layers mark the disparity stacks.

Coherence detection is performed independently within each stack. In the simulations reported below, responses of two disparity units i and j in a given stack were marked as coherent with each other whenever $|d_i - d_j| < \epsilon = 0.3$. The exact value of this coherence threshold is not critical; results were obtained for values ranging from 0.05 to 1.5 .

As basic disparity unit, a detector based on energy motion filters was used [10–12]. Various other circuitry is possible, including units based on standard

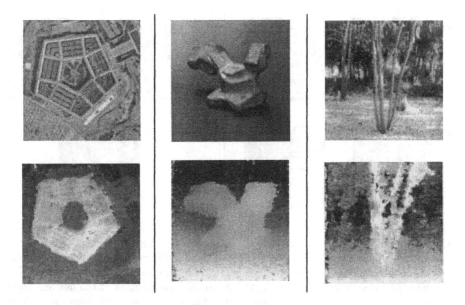

Fig. 3. *Disparity maps for several test images, calculated by coherence-based stereo.*

optical flow estimation techniques or algorithms for the estimation of principal texture direction. In any case, these algorithms can be realized with local spatial filter operations combined with simple arithmetic operations like squaring or dividing values.

The structure of the proposed network resembles superficially earlier cooperative schemes used to disambiguate false matches [8,13], but dynamic and link structures are quite different. In the new scheme, no direct spatial facilitation takes place, which is a central ingredient of most other cooperative algorithms. In addition, units interact here only along the lines defining the disparity stacks. This interaction is of the excitatory type, while usual cooperative schemes have either none or inhibitory interactions along such lines. Furthermore, coherence detection is a fast and non-iterative process, whereas classical cooperative algorithms require many iterations before reaching some steady state.

Since coherence detection is in addition an opportunistic scheme, extensions of the neuronal network to multiple spatial scales and different types of disparity detectors is trivial. Fig. 3 displays some results obtained with the new algorithm by using disparity units operating at several different spatial scales.

Additional units are simply included in the appropriate coherence stacks at the smallest spatial scale. The coherence detection scheme will combine only the information from coherently acting units and ignore the rest of the data. In this way, image areas where disparity estimates are missing will be filled-in with data from some coarser or finer scales having valid estimates. Adding units with asymmetric receptive fields [14] can give more precise disparity estimates close to object boundaries.

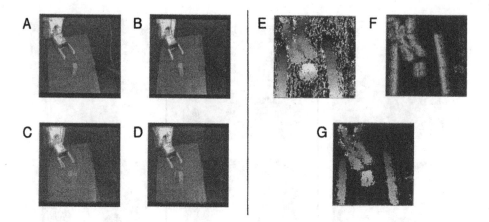

Fig. 4. Several maps are available within the network structure for coherence-based stereo. The disparity map (E) is estimated with subpixel precision in all image areas where enough detail present. Simultaneously, a verification count (F) is available, which can be used to mask unreliable data (G). A simple superposition of the left and right input pictures displays double contours (C), but coherence-based stereo is able to combine both stereo pictures into a cyclopean view (D). Note the change of perspective between either the left (A) or right (b) input image and the constructed cyclopean image.

As Fig. 3 shows, the results calculated with coherence-based stereo are generally comparable to results of classical area-based approaches. A dense disparity map is obtained, but without the complicated hierarchical search structures immanent in these approaches.

Parallel to the disparity data, a simple verification count (Fig. 4F) can be derived from the relative number of coherently acting disparity units in each stack, i.e. by calculating the ratio $N(\mathcal{C})/N(\mathcal{C} \cup \overline{\mathcal{C}})$, where $N(\mathcal{C})$ is the number of units in class $\mathcal{C}$. Using this validation count for masking unreliable disparity estimates (Fig. 4G), the new scheme reduces to something similar to a standard feature-based algorithm when confronted with difficult data. However, none of the usual constraints basic to feature-based stereo algorithms are explicitly build into the new computational scheme.

The coherence-detecting links defining the disparity stacks split the angle between the left and right data lines of the network in half. Reprojecting data lines and coherence-detecting links back into 3D-space shows that the disparity stacks actually analyze space along the cyclopean view directions (Fig. 2). Thus any data which is output by the algorithm is neither aligned with the left nor the right input image, but with the cyclopean view of the scene.

It is very easy to obtain this cyclopean view of the scene itself. Let I_i^L and I_i^R be the left and right input data of disparity-unit i respectively. Than a simple summation over all coherently coding disparity units, i.e.,

$$I^{\mathcal{C}} = \left\langle I_i^L + I_i^R \right\rangle_{i \in \mathcal{C}} ,$$

gives the image intensity in the cyclopean view direction (Figure 4C). Interestingly, even in areas where the disparity estimates are poor, the cyclopean image seems not to be severely distorted. Note that it is not a simple interpolation between the left and right images, as can be seen by noting the difference in perspective between the cyclopean view and the left or right input images.

In general, the cyclopean view will differ from a correct perspective projection of the scene; this depends on the geometry of the imaging system. It will be a proper perspective projection for flat sensor arrays, and a more complicated projection for an imaging situation similar to the human eye.

Acknowledgements

Fruitful discussions and encouraging support from H. Schwegler is acknowledged. Image data courtesy of G. Medoni, UCS Inst. for Robotics & Intelligent Systems, B. Bolles, AIC, SRI International, and G. Sommer, The Kiel Cognitive Systems Group, Christian-Albrechts-University of Kiel. Thanks also to unknown referees for valuable comments.

References

1. U. R. Dhond and J. K. Aggarwal, *Structure from Stereo - A Review*, IEEE Trans. Syst., Man and Cybern. **19**, 1489–1510, 1989.
2. J. P. Frisby, *Stereo Correspondence and Neural Networks*, in "Handbook of Brain Theory", ed. M. A. Arbib, 937–941, MIT Press 1995.
3. J. Weng, *Image Matching Using the Windowed Fourier Phase*, Int. J. of Comp. Vis. **3**, 211–236, 1993.
4. T.D. Sanger, *Stereo Disparity Computations Using Gabor Filter*, Biol. Cybern. **59**, 405–418, 1988.
5. M. R. M. Jenkin and A. D. Jepson, *Recovering Local Surface Structure through Local Phase Difference Methods*, CVGIP **59**, 72–93, 1994.
6. J. L. Barron, D.J. Fleet and S. S. Beauchemin, *Performance of Optical Flow Techniques*, Int. J. Comp. Vis. **12**, 43–77, 1994.
7. D.J. Fleet and A. D. Jepson, *Stability of Phase Information*, IEEE Trans. Patt. Anal. Mach. Intel. **15**, 1253–1268, 1993.
8. R. Blake and H. R. Wilson, *Neural Models of Stereoscopic Vision*, TINS **14**, 445–452, 1991.
9. D. Marr and T. Poggio, *A Computational Theory of Human Stereo Vision*, Proc. R. Soc. Lond. **B 204**, 301–328, 1979.
10. N. Qian and Y. Zhu, *Physiological Computation of Binocular Disparity*, to appear in VISION RESEARCH '97.
11. R. D. Freeman, G. C. DeAngelis, and I. Ohzawa, *Depth is Encoded in the Visual Cortex by a Specialized Receptive Field Structure*, Nature **11**, 156–159, 1991.
12. E. H. Adelson and J. R. Bergen, *Spatiotemporal Energy Models for the Perception of Motion*, J. Opt. Soc. Am. **A 2**, 284–299, 1985.
13. K. Prazdny, *Detection of Binocular Disparities*, Biol. Cybern **52**, 93–99, 1985.
14. A. Fusiello, V. Roberto, and E. Trucco, *A Symmetry-based Stereo Algorithm*, Research report UDMI/53/96/RR, submitted to CVPR 97, Machine Vision Lab, University of Udine, Italy, 1996.

Stereo Matching Using M-Estimators*

Christian Menard and Aleš Leonardis**

Pattern Recognition and Image Processing Group,
Vienna University of Technology, Treitlstraße 3/183/2, Vienna, Austria
e-mail: men@prip.tuwien.ac.at

Abstract. Stereo computation is just one of the vision problems where the presence of outliers cannot be neglected. Most standard algorithms make unrealistic assumptions about noise distributions, which leads to erroneous results that cannot be corrected in subsequent processing stages. In this work the standard area-based correlation approach is modified so that it can tolerate a significant number of outliers. The approach exhibits a robust behaviour not only in the presence of mismatches but also in the case of depth discontinuities. Experimental results are given on synthetic and real images.

1 Introduction

Stereo computation is one of the classical problems of computer vision. It involves the search for the correct match of the points in the two images, which is called the "correspondence problem"—one of the central and most difficult parts of the stereo problem [8, 16]. Several algorithms have been developed to compute the disparity between images, e.g., the correlation methods [7, 15], the correspondence methods [3], or the phase difference methods [5]. If the same point in scene can be located in both stereo images then, given that the relative orientation between the cameras is known, the three dimensional world coordinates can be computed.

Stereo computation is just one of the vision problems where the presence of outliers cannot be neglected. Most standard algorithms make unrealistic assumptions about the noise distributions leading to erroneous results which can not be corrected in the subsequent post-processing stages. It is becoming clear that the algorithms should be designed such that they can cope both with well-behaved noise and outliers, which are either large measurement errors (incorrect matches) or data points belonging to other distributions (models); in the case of stereo— to different depths. It is very important that a vision system incorporates the properties of robustness to the stereo-vision process.

Robust window operators [1, 6, 9, 14] have been brought into computer vision as an answer to the problems encountered when using standard least-squares

* This work was supported by a grant from the Austrian National Fonds zur Förderung der wissenschaftlichen Forschung (No. S7002MAT) and (No. P9110-SPR).

** Also at the University of Ljubljana, Faculty of Computer and Information Science.

method in windows containing outliers or more than one statistical population. However in most cases, the robust filtering techniques are used as a post-processing step. As an example, W. Luo and H. Maitre [7] tried to use a surface model to correct and fit disparity data in stereo-vision as a post-processing step after non-robust stereo computation. By that time the result may already contain errors that cannot be removed. Our approach differs in that the principle of the robust estimators is integrated *directly* in the computation of stereo. The approach not only tolerates a significant number of outliers but also gives robust results in the presence of depth discontinuities.

The paper is organized as follows: In the next section basic notations regarding the area-based stereo matching are given. A robust version of the stereo algorithm is explained in section 3 and experimental results are described in section 4. The work is concluded with a summary and an outlook to future work.

2 Area-based Stereo Algorithm

The knowledge of the epipolar geometry is the starting point for this method. For a given pair of stereo images, the corresponding points are supposed to lie on the epipolar lines. As a similarity measure the correlation of gray-level intensities between left and right windows is used. The correlation $C(x_L, x_R, w)$ between two regions of window size w can be written in the one-dimensional continuous case with a rectangular function $\delta_{1/w}$ [11, 10] as:

$$C(x_L, x_R, w) = \frac{\int_{\xi=-\infty}^{\infty} \delta_{1/w}(\xi)[I_L(x_L + \xi) - \mu_L(x_L, w)][I_R(x_R + \xi) - \mu_R(x_R, w)]d\xi}{\sqrt{\int_{\xi=-\infty}^{\infty} \delta_{1/w}(\xi)[I_L(x_L + \xi) - \mu_L(x_L, w)]^2 d\xi \int_{\xi=-\infty}^{\infty} \delta_{1/w}(\xi)[I_R(x_R + \xi) - \mu_R(x_R, w)]^2 d\xi}}, \quad (1)$$

with

$$\mu(x, w) = I(x) * \delta_{1/w}(x) , \text{ where } \delta_{1/w}(x) = \begin{cases} \frac{1}{w} & |x| \leq \frac{w}{2} \\ 0 & elsewhere \end{cases} . \quad (2)$$

The correlation $C(x_L, y_L, x_R, y_R, w)$ for two-dimensional regions can easily be extended from Eq. (1). For each point in the left intensity image the disparity information for $D(x_L, y_L)$ is computed using the correlation function C as

$$D(x_L, y_L) = \begin{cases} x_L - x_R & x_R = argmax\{C(x_L, y_L, x_R, y_R, w)\} > T \\ -1 & otherwise \end{cases}, \quad (3)$$

where T defines the threshold accepting the corresponding point. In this case the position of the distinct maximum defines the corresponding point in (x_R, y_R).

The major problem with the standard stereo computation based on correlation (1) is that due to occlusions, highlights, discontinuities, and noise, one can not avoid erroneous matches leading to incorrect estimation of disparities and consequently shape. The reason is that the standard correlation follows the least-square argument which is very sensitive to outliers in the data set. The outliers are not only large measurement errors, but also points that belong to a different model (surface) than the one that is currently under consideration. For example, all the points in a window used in standard correlation should have the same disparity, otherwise the estimate will be incorrect. In Schunck [13] it is argued that visual perception is fundamentally a problem in discrimination: data must be combined with like data (having the same property) and outliers must be rejected. In other words, vision algorithms must be able to combine data while simultaneously discriminating between data that should be kept distinct, such as outliers (errors) and data from other regions.

3   Robust Computation of Correlation

When depth maps are obtained by stereo techniques, any false correspondences produced by the matching algorithm will induce outliers in the disparity space. This poses a challenging problem to any surface reconstruction algorithm. It is becoming clear that the algorithms should be designed such that they can cope with various types of noise. Thus, the idea is to incorporate the principles of robustness directly into the computation, in this case—the computation of correlation. The approach being developed not only tolerates a significant number of outliers but also gives robust results in the presence of depth discontinuities.

The estimators that remain stable in the presence of various types of noise and can tolerate a certain portion of outliers are known under the generic name of *robust estimators* [4, 12, 2]. *M-estimators* (maximum likelihood type estimate), for example, tackle the problems by either rejecting outliers from the calculation or down-weighting their influence on the final result. This is achieved by first computing an initial estimate and then refining it by repeated re-weighting of the data points. Some specific weighting functions τ are known as Huber [4]

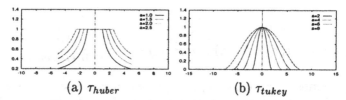

(a) τ_{huber} (b) τ_{tukey}

Fig. 1. τ functions for Huber and Tukey.

and Tukey [17], which are shown in Fig. 1. The functions are defined by

$$\tau_{huber}(x) = \begin{cases} 1 & |x| < aS \\ \frac{aS}{|x|} & |x| > aS \\ aS & x > aS \end{cases} \qquad \tau_{tukey}(x) = \begin{cases} [1 - (\frac{x}{aS})^2]^2 & |x| \leq aS \\ 0 & |x| > aS \end{cases}, \qquad (4)$$

where a is a tuning constant. S is a scale estimator which is usually MAD (median of absolute deviation).

3.1 Robust Correlation Algorithm

Since for most weighting functions τ a closed form solution for the estimation problem has not been found, an iterative method is used. In order to define the correlation as minimization problem in the one-dimensional case, Eq. (1) and (2) are used to modify the correlation such that

$$C = \frac{2 - C_{LS}}{2} , \tag{5}$$

where

$$C_{LS}(x_L, x_R, w) = \int_{\xi=-\infty}^{\infty} \delta_{1/w}(\xi) \ \varepsilon^2(x_L + \xi, x_R + \xi, w)d\xi . \tag{6}$$

The residual error ε is defined as

$$\varepsilon^2(x_L, x_R, w) = [\{\frac{I_L(x_L) - \mu_L(x_L, w)}{\sigma_L(x_L, w)}\} - \{\frac{I_R(x_R) - \mu_R(x_R, w)}{\sigma_R(x_R, w)}\}]^2 . \tag{7}$$

Minimizing C_{LS} is equivalent to maximizing the correlation C. Using the re-weighted least squares (RLS) technique, Eq. (6) can be written in an iterative way as:

$$C_{\Omega_n}(x_L, x_R, w) = \int_{\xi=-\infty}^{\infty} \delta_{1/w}(\xi) \ \Omega_n(x_L, x_R, w) \ \varepsilon_{\Omega_n}{}^2(x_L + \xi, x_R + \xi, w)d\xi , \tag{8}$$

with the residual error

$$\varepsilon_{\Omega_n}{}^2(x_L, x_R, w) = [\{\frac{I_L(x_L) - \mu_{L,\Omega_n}(x_L, w)}{\sigma_{L,\Omega_n}(x_L, w)}\} - \{\frac{I_R(x_R) - \mu_{R,\Omega_n}(x_R, w)}{\sigma_{R,\Omega_n}(x_R, w)}\}]^2 . \tag{9}$$

The weights from $\Omega_n \to \Omega_{n+1}$ are defined by

$$\Omega_{n+1}(x_L, x_R, w) = \tau(\varepsilon_{\Omega_n}{}^2(x_L, x_R, w)) , \tag{10}$$

where τ is the weighting function. The standard deviation for $k = L, R$ can be written as

$$\sigma_{k,\Omega_n}(x, w) = \frac{1}{w} \int_{\xi=-\infty}^{\infty} \delta_{1/w}(\xi) \ \Omega_n(x_L + \xi, x_R + \xi, w)[I_k(x + \xi) - \mu_{k,\Omega_n}(x, w)]d\xi \tag{11}$$

with

$$\mu_{k,\Omega_n}(x, w) = \frac{1}{w} \int_{\xi=-\infty}^{\infty} \delta_{1/w}(\xi) \ \Omega_n(x_L + \xi, x_R + \xi, w)I_k(x + \xi)d\xi. \tag{12}$$

The iteration starts with the initial condition $\Omega_0 = [1, \dots, 1]$. The correlation function can easily be extended from the one- to the two-dimensional case. The function C_Ω (Eq. 8) is used for the robust correlation technique. Now the complete algorithm for robust computation of the correlation can be outlined:

1. Compute the initial disparity value using the standard correlation C.
2. Based on the disparity $D = x_L - x_R$, compute weights Ω_n.
3. Calculate a new correlation C_{Ω_n} using the weights Ω_n and determine new disparities D.
4. Update the weights with $\Omega_{n+1} = \tau(\varepsilon_{\Omega_n}{}^2)$.
5. Iterate steps 2, 3, and 4 until convergence, or until a maximal number of iterations is reached[1].

4 Experimental Results and Comparisons

We tested the approach on synthetic and real data. The synthetic stereo pair (Fig. 2 (a) and (e)) depicts a box on a plane with natural texture added on the surface. As a real object we use an archaeological fragment, depicted in

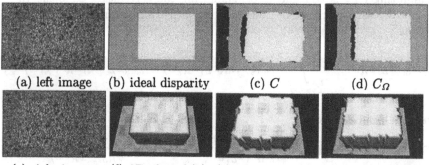

(a) left image   (b) ideal disparity   (c) C (d) C_Ω

(e) right image   (f) 3D plot of (b)   (g) 3D plot of (c) (h) 3D plot of (d)

Fig. 2. Comparison of the results: (a) and (e) stereo pair, (b) ideal disparity map, (c) disparity maps computed with the standard, and (d) with the robust stereo method using τ_{tukey}. (f-h) show 3D plots of the computed disparity maps.

Fig. 3 (a) and (b). In order to compare the accuracy of the stereo matching, the disparity values for the synthetic stereo pair are known. A dense disparity map is computed for both the box image and for the fragment using our method. For the robust technique Tukey's weighting function is used. Fig. 2 shows the ideal disparity map, the result of the standard stereo matching algorithm and the result with the robust correlation method. It can be seen that in the regions near the depth discontinuity, edge points can be detected more exactly with the robust stereo technique than with the standard correlation approach. In Fig. 3 (c) the result for the fragment is shown. It can be seen that the surface of the fragment is recovered. Together with both gray-level and disparity information three-dimensional models of the objects with the texture added on the surface can be created, which are illustrated in Fig. 3 (d).

---

[1] Experiments have shown that a few iterations (2–4) are needed for convergence.

Fig. 3. Archaeological fragment.

4.1 Noise Test

Images are sometimes corrupted by gross errors. In such a case, a random value replaces the value of the pixel. A commonly used model for this behavior is Salt & Pepper noise, in which some percentage of the pixels of the observed area is randomly replaced by white or black pixels. Another type of noise which is also tested is additive Gaussian noise with a standard deviation varying from $\sigma = [0..500]$ with pixels clipped for > 255 and < 0.

For these noise conditions the disparity maps for the box image are computed for both the standard and the robust stereo method. The disparity maps are compared to each other by using the *MSE* (Mean Square Error). The *MSE* is determined between the ideal disparity map (Fig. 2 (b)) and the computed results for both methods under different noise conditions. For this test the thresh-

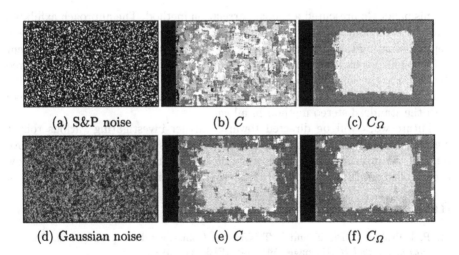

| (a) S&P noise | (b) C | (c) C_Ω |
| (d) Gaussian noise | (e) C | (f) C_Ω |

Fig. 4. (a) 30% Salt & Pepper (d) Gaussian noise, $\sigma =300$ (b),(e) Disparity maps computed with C. (c),(f) Disparity maps computed with C_Ω using τ_{tukey}. The size of the search window is constant for both methods with $w = 7$.

old T is set to zero to accept also wrong matches. The disparity maps for the standard and the robust technique are determined and compared to each other. The results are visualized in Fig. 4, where the first column depicts the box image introduced with Salt & Pepper and Gaussian noise, the second column depicts the disparity maps computed with the standard correlation method and the last column with the robust method using Tukey's biweight. Fig. 5 shows the MSE for the two noise conditions for both weighting functions. The dotted curve represents the error for the standard correlation method. The robust approach shows for both weighting functions better results than the standard method.

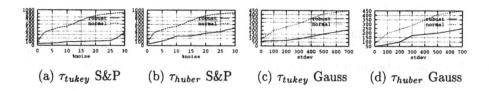

(a) τ_{tukey} S&P   (b) τ_{huber} S&P   (c) τ_{tukey} Gauss   (d) τ_{huber} Gauss

Fig. 5. The $MSE(C[noise] - ideal)$ is plotted as dotted curve and the $MSE(C_\Omega[noise] - ideal)$ as solid curve for τ_{tukey} and τ_{huber}.

5 Conclusion and Outlook

In this paper we presented a robust correlation method. The approach exhibited a robust behaviour not only in the presence of mismatches but also in the case of depth discontinuities. The method was tested on synthetic images under different noise conditions and compared to the results computed with the standard stereo method. For the robust technique the tests were also performed for some specific weighting functions. With the robust approach we obtained better results than with the standard stereo method in all cases.

Future work will be directed towards further investigating various robust methods. In addition, the development of a method is planned that would remove some of the problems with window operators such that the size of the window is controlled using gray-level and disparity information.

References

1. P. J. Besl, J. B. Birch, and L. T. Watson. Robust window operators. In *Proceedings of the 2nd ICCV*, pages 591–600. IEEE, Dec 1988.
2. D.N. Bhat and S.K. Nayar. Ordinal measures for visual correspondence. Technical Report CUCS-009-96, Columbia University,Dept. of Computer Science, February 1996.
3. H. Helmke, R. Janssen, and G. Saur. Automatische Erzeugung dreidimensionaler Kantenmodelle aus mehreren zweidimensionalen Objektansichten. In Grabkopf R.E., editor, *Mustererkennung 1990, 12. DAGM Symposium*, pages 617–624, 1990.

4. P. J. Huber. *Robust Statistics*. Wiley, New York, 1981.
5. M.R.M. Jenkin, A.D. Jepson, and J.K. Tsotsos. Techniques for disparity measurements. *CVGIP*, 53:14–30, 1991.
6. D. Y. Kim, J. J. Kim, P. Meer, D. Mintz, and A. Rosenfeld. Robust computer vision: A least-median of squares based approach. In *Proceedings of the Image Understanding Workshop*, pages 1117–1134, Palo Alto, CA, May 1989. DARPA.
7. W. Luo and H. Maitre. Using surface model to correct and fit disparity data in stereo vision. In *Tenth International Conference on Pattern Recognition*, pages 60–64, Atlantic City, NJ, June 16-21 1990.
8. D. Marr and T. Poggio. A computational theory of human stereo vision. *Proc. R. Soc. Lond. B.*, 204:301–328, 1979.
9. P. Meer, D. Mintz, A. Rosenfeld, and D. Y. Kim. Robust regression methods for computer vision: A review. *International Journal of Computer Vision*, 6(1):59–70, 1991.
10. C. Menard. *Robust Stereo and Adaptive Matching in Correlation Scale-Space*. PhD thesis, TU Wien, Institut für Automation, PRIP, Wien, 1996.
11. Azriel Rosenfeld and Avinash C. Kak. *Digital Picture Processing Volume 2*. Academic Press, Inc., 1982.
12. P. J. Rousseeuw and A. M. Leroy. *Robust Regression and Outlier Detection*. Wiley, New York, 1987.
13. B. G. Schunck. Robust computational vision. In *Proc. of the IWRCV*, Seattle, WA, Oct 1990.
14. S. S. Sinha and B. G. Schunck. A two-stage algorithm for discontinuity-preserving surface reconstruction. *IEEE Trans. on PAMI*, 14(1):36–55, Jan 1992.
15. J. Subrahmonia, J. Hung, and D.B. Cooper. Model-based segmentation and estimation of 3d surfaces from two or more intensity images using markov random fields. *IEEE Conf. on Pattern Recognition*, 1:390–397, 1990.
16. D. Terzopoulos, A. Witkin, and M. Kass. Stereo matching as constrained optimization using scale continuation methods. In *Opt. Dig. PR/ SPIE 754*, pages 92–99, 1987.
17. J.W. Tukey. Some advanced thoughts on the data analysis involved in configural polysampling directed toward high performance estimates. Technical Report 189, Series 2, Department of Statistics, Princeton University, Princeton,N.J., 1981.

Robust Location Based Partial Correlation

Zhong-Dan LAN and Roger MOHR

Laboratoire Gravir, Projet Movi,
Inria, 655 avenue de l'Europe, 38330 Montbonnot France
zhong-dan.lan@imag.fr

Abstract. The visual correspondence problem is a major issue in computer vision. Correlation is a common tool for this problem. Most classical correlation methods fail near the disparity discontinuities, which occur at the boundaries of objects. In this paper, a *partial correlation* technique is proposed to solve this problem. *Robust location* methods are used to perform this partial correlation. Comparisons are made with other techniques and experimental results validate the approach.

1 Introduction

The visual correspondence problem is a major task in computer vision [10, 2, 5]. The use of correlation as a similarity measure between two signals is well known. It is commonly used in stereo vision for the visual correspondence problem and extensive comparisons have been made evaluating different correlation criteria [1].

As indicated in [4, 6], occlusion plays an important rule in stereo matching. In [6], Intille uses Ground Control Points (GCP) to construct the best disparity path. This method gives the best results if GCP are very close to the occlusion boundaries as these are the regions in which all classical methods fail.

We feel that only the use of robust methods can overcome the matching problems that every method encounters at the occluding contours. In [8], a *partial correlation* method is described, based on *robust regression*. It provides good matches at occlusion contours.

In this paper, another partial correlation method is developed, based on *robust location*. It uses a minimum volume estimator (MVE), and is also called the MVE correlation. It can be treated as a robustification of the ZNCC correlation (Zero Mean Cross Correlation), in that sense it is the ZNCC on the *good* parts of the signals.

This paper is organized as follows: after an introduction to related work, we give a short analysis of classical ZNCC correlation from the location view point. Then the MVE estimator is presented, followed by a *robust location* based correlation, which is also called MVE correlation. At the end, some experimental results are shown followed by a conclusion.

2 Related work

The visual correspondence of two vectors is usually defined by correlation. There are many different versions of this (Table 1 in [1]). Experimental comparisons are reported in [1] and [3].

Among existing methods, we have chosen zero mean normalized cross correlation (ZNCC) because of its good experimental performance with respect of varying lighting conditions:

$$ZNCC(X, X + dX) = \frac{cov(I_1(X), I_2(X + dX))}{\sigma_{I_1(X)}\sigma_{I_2(X+dX)}},$$

with $X = (x, y)$ the pixel to be matched, $dX = (dx, dy)$ the disparity, $cov(I_1(X), I_2(X + dX))$ the covariance between $I_1(X)$ and $I_2(X + dX))$, $\sigma_{I_1(X)}$ and $\sigma_{I_2(X)}$ the standard deviations of two image windows $I_1(X)$ and $I_2(X + dX)$.

Other correlations as ZSSD (*Zero mean Sum of Squared Difference*) [1] can also been used. Alternative approaches have been proposed based on transformed versions of the original signal. A common standard transformation is the Laplacian of the image. This has a similar effect to normalizing the image as in the ZSSD method.

All of these methods are very sensitive to outliers, e.g. strong noise or the presence of partial occlusions. In order to reduce such effects, some authors limit the influence of the numerical values by only taking into account the sign of the Laplacian [15, 11], or the orientation of the gradient [13]. In these cases, the perturbations of the correlation are in direct ratio to the size of the perturbed area. A similar idea is exploited in the rank and census transform [14]. Lan [8] proposed a *partial correlation* technique based on *robust regression*.

3 Analysis of ZNCC: location view point

The function ZNCC of two n-dimensional lists $X = (x_i)$ and $Y = (y_i)$, is defined as:

$$\frac{cov(X, Y)}{\sigma_X \sigma_Y} = \frac{\sum_{i=1}^{n} \frac{(x_i - \bar{X})(y_i - \bar{Y})}{n-1}}{\sqrt{\sum_{i=1}^{n} \frac{(x_i - \bar{X})^2}{n-1} \sum_{i=1}^{n} \frac{(y_i - \bar{Y})^2}{n-1}}}$$

which can also be expressed as the radio of the covariance $cov(X, Y)$ between X and Y and the geometrical mean of their variances σ_x^2 and σ_y^2.

This procedure is composed of three steps:

1. Estimation of the data center.
2. Estimation of the covariance matrix, *i.e.* the variances of X, Y and their covariance.
3. Estimation of the correlation coefficient by computing the ratio of the two variable covariance and the geometrical mean of their variances.

In the partial occlusion case, there are some outliers. The estimation of the center and the covariance matrix should be robust so that the estimation of the correlation coefficient is reliable. In the next section, we discuss the robust estimation of location and dispersion.

4 Robust estimation of location and dispersion

We discuss the estimation of the "center" of a set of points and the dispersion around the center. The dispersion is usually expressed by the covariance matrix of this set of points.

Given a set of data: $X = \{x_1, \ldots, x_n\} = \{(x_{11}, x_{12}, \ldots, x_{1p})^t, \ldots, (x_{n1}, x_{n2}, \ldots, x_{np})^t\}$ of n p-dimensional points. we have to estimate its center, and also its dispersion around the center, expressed by a covariance matrix C.

More exactly, we are looking for an estimator T, such that $e_i = x_i - T(X)$ follows a hypothesized distribution and has some useful properties. If some outliers among the e_i do not perturb the result, the estimator is said to be *robust*.

Rousseeuw [12] has introduced a robust location estimator, which has the theoretically maximal breakdown point 50%, defined as: the "center of the ellipse of minimal volume containing at least h points of X" denoted as MVE. Here h can be taken as $[\frac{n}{2}] + 1$. This is called Minimal Volume Ellipse (MVE). The ellipse itself gives an estimate of the covariance matrix, multiplied by a suitable scale factor.

A random sampling technique is used to get the initial solution which can be refined using all inliers [12].

The MVE estimator has been already used in vision, as a classification method for thresholding, analysis of Hough space and image segmentation [7]. In this paper, we will use it for the image matching problem to deduce a correlation function, which is robust to partial occlusions (section 5).

5 Robust location based partial correlation

We begin with a discussion of correlation and location, deduce the procedure for partial correlation and apply it to image matching problem.

- **Analysis of correlation and location**
  Given a list of n p-dimensional points: $X = (x_i)_{i=1}^n$, where $x_i = (x_{i1}, \ldots x_{ip})^t$, we look for an estimator $T = T(X)$, such that:

$$x_i = T + e_i \tag{1}$$

where $e_i = (e_{i1} \ldots e_{ip})$ are some vectorial random variables.

The location problem is the search for such an estimator. The correlation coefficient between e_{ij} and e_{ik} can be computed by the aid of e_i's covariance matrix, as the ratio of the covariance between e_{ij} and e_{ik} and the geometrical mean of e_{ij} and e_{ik}'s variances.

In image matching problem, $p = 2$, $\mathbf{x}_i = (I_{1i}, I_{2i})$ represents a pair of corresponding intensities, $\mathbf{T}$ represents the "center" of $(\mathbf{x}_i)_{i=1}^n$ and it can be estimated as the gravity of points $\mathbf{x}_i$ representing matched pixels, $\mathbf{e}_i = \mathbf{x}_i - T$ represents the "joined" texture of I_1 and I_2. The correlation coefficient between $\mathbf{e}_{i1}$ and $\mathbf{e}_{i2}$ reflexes the association between I_1 and I_2, and it can be estimated by estimating the covariance matrix of $\mathbf{e}_i$.

– **Dealing with the partial occlusion problem**

Most correlation techniques have difficulties near disparity discontinuities, or in places where highlights occur, as the window under consideration is locally partially but severely corrupted (Figure 1). The pixels in such a region represent scene elements from two distinct intensity populations. Some of the pixels correspond to the template under consideration and some to other parts of the scene. This leads to a problem for many correlation techniques, which are usually based on standard statistics methods, and best suited to a single population. We shall call this phenomenon *partial occlusion* and propose the *partial correlation* idea to overcome it.

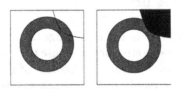

Fig. 1. Occlusion.Left: I_1, right: I_2

– **Procedure of partial correlation**

When partial occlusions occur, relation (1) holds only for the non-occluded part. So we should compute the correlation measure only on this part.

We propose the partial correlation procedure as follows:

1. Estimate robustly T and the associated covariance matrix C using MVE (section 4).
2. Find the occluded part according to a statistical test (section 4).
3. Compute the correlation measure on the non occluded part.

– **Estimating correlation between two image windows**

In case of image matching, we take $\mathbf{x}_i = (I_{1i}, I_{2i})^t$, where I_{1i} (I_{2i}) are intensities of the ith point of the first (second) image window respectively.

Let $I_1 = (I_{1i})$, $I_2 = (I_{2i})$ and $\mathbf{C}_1(X) = \begin{pmatrix} \sigma_1^2 & cov(I_1, I_2) \\ cov(I_1, I_2) & \sigma_2^2 \end{pmatrix}$ be the covariance matrix of $(\mathbf{x}_i)$ estimated by the MVE, we estimate the correlation coefficient between I_1 and I_2 as $\frac{cov(I_1, I_2)}{\sigma_1 \sigma_2}$. This estimate is the generalization of the classical correlation coefficient (called ZNCC [1] in vision) if $\mathbf{C}_1(\mathbf{X})$ is replaced by the classical covariance matrix (section 3).

A real example is shown in Figure 2, where each point represents a pair of corresponding grey levels (I_{1i}, I_{2i}). The ellipse is similar to the uncertainty ellipse of the data set.

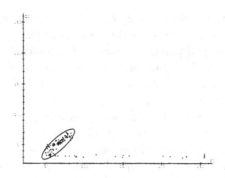

Fig. 2. Robust location by MVE estimator of (I_1, I_2). The points outside the ellipse represent pairs of pixels that are not matched.

6 Experimental results

Select a template of fixed size in the first image. Suppose that there are several candidate matches of the same size. Below we use the *region of interest* constraint and the *epipolar constraint* to reduce the number of the candidates [1].

For comparison with ground truth, experiments were conducted on planar patches. In such cases the exact match can be computed by estimating the homography mapping from the matched targets between two images. We have tested the MVE correlation, and compared it with ZNCC, RZSSDC (*robust zssd with center point constraint*) [8], the standard ZSSD, the rank and the census methods [14]. Owing to space limit, only its comparison with ZNCC is shown, more experimental results can be found in [9].

The results for two pair of images called *Benetton* and *World* are displayed in this paper. Lack of space, only one pair is displayed in Figure 3.

Both images have a planar background but the background textures are different. Two tests on points near occlusion contours were designed for each pair. Figure 4 displays two sets of points for the *Benetton* (Figure 3 (a)) background for which the corresponding points were sought.

The different methods were run on these four tests. The candidates were constrained to stay on the epipolar line and to have a disparity of less than 200. For each method, we compare the results found with the exact value provided by the homography.

The results are reported in the Tables 1, 2. 3 and 4. For two methods MVE and ZNCC are indicated:
- the number of accurate matches: up to one pixel error (good match),

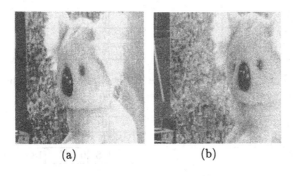

(a)   (b)

Fig. 3. Stereo pair 1: *Benetton.*

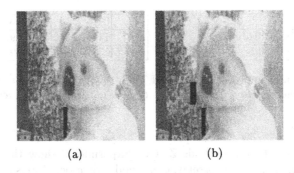

(a)   (b)

Fig. 4. Background *Benetton*. Points selected (in black). (a): example corresponding to Table 1; (b): example corresponding to Table 2.

- the number of matches in a distance between one and two pixels error (near miss),
- the number of matches at a distance of two to three pixels (bad matches),
- the matches which lie more than three pixels away from their exact position (false matches).

From these examples, we see that the MVE correlation performs better than the ZNCC correlation for points near occlusion contours.

The results on general points are displayed in Tables 5 and 6, which show that a compromise between *similarity* and *completeness* is needed [8].

7 Conclusion and future work

Occlusions carry important information in the visual correspondence problem, because they indicate disparity discontinuities and produce useful cues for the shape and location of objects. However, occlusions show the limitations of most of the standard correlation methods.

In this paper, we discussed the occlusion problem and proposed a new robust approach called *partial correlation* to overcome it. We compared our method with

	0–1	1–2	2–3	3–∞
MVE	270	79	0	50
ZNCC	183	37	0	179

Table 1. First example on image *Benetton* (see text).

	0–1	1–2	2–3	3–∞
MVE	730	108	0	56
ZNCC	656	38	1	205

Table 2. Second example on image *Benetton* (see text).

	0–1	1–2	2–3	3–∞
MVE	94	10	1	35
ZNCC	56	0	0	84

Table 3. First example of *World* (see text).

	0–1	1–2	2–3	3–∞
MVE	173	49	7	37
ZNCC	78	0	0	188

Table 4. Second example of *World* (see text).

	0–1	1–2	2–3	3–∞
MVE	12378	2183	40	169
ZNCC	12771	1550	0	449

Table 5. An example for the general case on image *Benetton* (see text).

	0–1	1–2	2–3	3–∞
MVE	11102	2512	414	1697
ZNCC	13216	1970	80	459

Table 6. An example for the general case on image *World* (see text).

the standard correlation methods ZNCC. Experiments show that our method works better than classical correlation methods when occlusions arise. However, when no occlusions are present, our method performs less well. Owing to space limit, comparison results with rank, census [14], ZSSD and RZSSDC [8] are omitted, which can be found in [9].

We think that the *partial correlation* represents only the beginning of the solution to the problem. The fundamental limitation of this technique is the necessity for a tradeoff between the *similarity* and the *completeness* criteria : *Given a template, and several candidates, which one is the most similar to the template ?*

The most similar candidate may have few active pixels hence not be very complete. The straightforward solution could be to take the one which is more similar when occlusions occur and the one which is more complete otherwise. This is an important issue not only in the problem of stereo matching by correlation, but also in the more general template matching problems like recognition. A possible perspective is the application of the *partial correlation* technique to the recognition problem where partial occlusion often occurs.

References

1. P. Aschwanden and W. Guggenbühl. Experimental results from a comparative study on correlation-type registration algorithms. In Förstner and Ruwiedel, editors, *Robust Computer Vision*, pages 268–282. Wichmann, 1992.
2. D.H. Ballard and C.M. Brown. *Computer Vision*. Prentice Hall, 1982.

3. P. Fua. A parallel stereo algorithm that produces dense depth maps and preserves image features. *Machine Vision and Applications*, 1990.

4. D. Geiger, B. Ladendorf, and A. Yuille. Occlusions and binocular stereo. In G. Sandini, editor, *Proceedings of the 2nd European Conference on Computer Vision, Santa Margherita Ligure, Italy*, pages 425–433. Springer Verlag, 1992.

5. B.K.P. Horn. *Robot Vision*. The MIT Press, 1986.

6. S.S. Intille and A.F. Bobick. Disparity-space images and large occlusion stereo. In *Proceedings of the 3rd European Conference on Computer Vision, Stockholm, Sweden*, pages 179–186. Springer-Verlag, 1994.

7. J.M. Jolion, P. Meer, and S. Bataouche. Robust clustering with applications in computer vision. *PAMI*, 13(8):791–802, August 1991.

8. Z. D. Lan, R. Mohr, and P. Remagnino. Robust matching by partial correlation. In *Proceedings of the sixth British Machine Vision Conference, Birmingham, England*, pages 651–660, September 1995.

9. Z.D. Lan and R. Mohr. Robust location based partial correlation. Technical report, Institut National Polytechnique de Grenoble, Movi, Gravir, Inria, 655 avenue de l'Europe 38330 Montbonnot, 1997. to appear.

10. D. Marr. *Vision*. W.H. Freeman and Company, San Francisco, California, USA, 1982.

11. H.K. Nishihara. PRISM, a pratical real-time imaging stereo matcher. Technical Report Technical Report A.I. Memo 780, Massachusetts Institute of Technology, 1984.

12. P.J. Rousseeuw and A.M. Leroy. *Robust regression and outlier detection*, volume XIV of *Wiley*. J.Wiley and Sons, New York, 1987.

13. P. Seitz. Using local orientational information as image primitive for robust object recognition. In *SPIE proceedings*, pages 1630–1639, 1989.

14. R. Zabih and J. Woodfill. Non-parametric local transforms for computing visual correspondance. In *Proceedings of the 3rd European Conference on Computer Vision, Stockholm, Sweden*, pages 151–158. Springer-Verlag, May 1994.

15. J. Zhao. *Extraction d'information tri-dimensionnelle par stéréovision*. Thèse de doctorat, Université Paul Sabatier, Toulouse, July 1989.

Optimization of Stereo Disparity Estimation Using the Instantaneous Frequency

M. Hansen, K. Daniilidis, G. Sommer

Institut für Informatik und Prakt. Mathematik
Christian-Albrechts Universität Kiel
Preusserstr. 1-9, 24105 Kiel, email: mha@informatik.uni-kiel.de

Abstract. The use of phase differences from Gabor filter responses is a well established technique for the computation of stereo disparity. It achieves the subpixel estimation of disparity without applying a correspondence search. However, the problem of scale or central frequency selection is still unsolved. Here, we study the effects of varying filter frequency on the disparity estimation and we compare it to the use of the instantaneous frequency. The analytical results on several models of intensity and disparity variation show robustness of the disparity estimates against variations in the filter-wavelength.

1   Introduction and related work

The computation of stereo disparity is a necessary step before depth reconstruction from two views of the environment. The majority of the approaches is based on searching for the maximum correlation between an area in the left image and the area in the second image that corresponds to the projection of the same scene point (cf. [2] for an extensive review). If the cameras are calibrated and the epipolar geometry is known this search is one-dimensional along the epipolar line in the right image corresponding to the point in the left image. Area-based approaches find the best correlation between two areas in the original or band-pass filtered images. Feature-based approaches first extract characteristic gray-value structures in both images which are then matched by applying a similarity criterion. Both groups suffer under the necessity of a search which increases the complexity and makes an explicit analysis intractable.

The paradigm of active vision introduced new prospects in stereoscopic vision. Real-time constraints on the complexity of the disparity computation are necessary in order to achieve a reactive behavior. Thus, techniques requiring a search for every image position are not appealing for real-time responses. On the other hand, active vision enables the control of the mechanical degrees of freedom of a stereo set-up. The vergence control of a camera yields a decrease on the disparity magnitude in a considerable area around the fixation point. Small disparities can now be computed with local techniques using only the responses of appropriate filters. It was shown that the local disparity variation can be computed from the intensity derivatives. However, a more general filter-based approach exhibited the most robust performance until now: the phase-based approach and studied in [3,4,7,8,5].

The phase-based disparity computation may be apparently justified by the shift-theorem for the Fourier-transform of an image. However, the shift-theorem is exact only for a globally constant disparity. To capture varying disparity, all approaches apply the local spectral representation of the complex-valued Gabor-filter responses. As we will describe later, the shift-theorem for the phase difference of the Gabor responses is not valid even in the case of simple sinusoidal input. A more plausible formulation of the shift-theorem [6,3] assumes the preservation of the local phase in the left and the right image. Application of the Taylor expansion to the local phase yields the disparity as the ratio of phase-difference and the first spatial derivative of the phase if the higher order disparity terms and derivatives are neglected. The first derivative of the local phase is well known as the definition of the instantaneous frequency of a signal [1].

The phase matching aspect of the local shift-theorem still involves the selection of the appropriate frequency to compute the local phase as well as its derivative. The Gabor-filters applied have coupled bandwidth and central frequency like wavelets so that the filter support is small for high central frequencies and large for low ones. Thus, frequency tuning is equivalent to the scale selection in the scale-space framework. Applying a bank of Gabor-filters with different scales yields a disparity estimate for each scale. We could choose the central frequency of the maximal energy response and compute the phase from the response of the Gabor filter with the same central frequency. However, this is not valid for multi-component signals and it is computationally very expensive to densely sample the scale (or frequency) space. Furthermore, the left and the right image can have different frequency contents due to perspective effects so that the search for the maximum response must be done for both images.

The main contribution of this study is the result that the instantaneous frequency remains almost the same for a wide range of scales. This fact is analytically shown for prototype signals like a sinusoid and an edge. The practical implications are extremely useful: the instantaneous frequency can be computed from a small filter-bank by sparsely sampling the frequency-space. In a second step the disparity can be computed with a filter tuned to the instantaneous-frequency. We show, that even if the second step is omitted the disparity error is still low if the instantaneous frequency and a phase difference from an almost arbitrary tuned Gabor-filter are used. We analyze the behavior of disparity with respect to the relative bandwidth and we show that the above mentioned strategy is valid for the sizes of the filter-supports used in real-time constrained applications.

The dependence of local phase and instantaneous frequency on scale has been extensively studied by Fleet and Jepson [3] where the effects of spatial deformations between the left and the right image on the preservation of scale were studied. Moreover, they found out that sensitivity of disparity to position and scale appears due to singularities in phase as a function of position and scale. Maki et al. [5] studied the discrepancy between actual phase of the signal and the local phase computed from Gabor-responses. However, they did not derive an analytical expression for a simple signal like the one in our work and in [4]. The focus of the study is the effect of narrow or wide bandwidth on the discrepancy factor. Langley et al. [4] found the same formula for the above mentioned discrepancy with a sinusoidal input. However, they delved into the quadrature approximation without studying the relation between scale, disparity and instantaneous frequency.

2 Phase-based disparity estimation

Stereo disparity has a horizontal and a vertical component but only one of them suffices to compute the depth. As it is usual and justified in active binocular systems we will use only the horizontal disparity. The filters we apply have a two-dimensional support with a smoothing component in the vertical direction. Therefore we restrict our analysis to disparity estimation in 1D-signals.

Let us suppose that the disparity d is constant over the image. Then, according to the shift-theorem of the Fourier transform a spatial domain shift $d(x)$ is mapped to a modulation in frequency domain and can be recovered from the phase difference:

$$f(x) \;\circ\!\!\!-\!\!\!-\!\!\!\bullet\; F(\omega) \quad f(x+d) \;\circ\!\!\!-\!\!\!-\!\!\!\bullet\; F(\omega)e^{i\omega d} \Rightarrow d(x) = \frac{\phi_l(x) - \phi_r(x)}{\omega} \tag{1}$$

Unfortunately disparity is hardly ever globally constant so that some sort of local spectral representation must be extracted. We apply here complex Gabor-filters with coupled bandwidth and central frequency. The phases are denoted $\phi_r(x)$ for the right and $\phi_l(x)$ for the left image. Since a bank of Gabor-filter responses samples the frequency space the question arises whether the shift theorem still holds for the local phase extracted from the complex Gabor responses. In this case, the mean frequency of the left and the right image could be used in order to apply (1). A slight variation of the shift-theorem can be obtained if we assume that local phase is preserved in the left and the right image [4,3]

$$\phi_l(x+d) - \phi_r(x) = 0. \tag{2}$$

A Taylor series expansion of $\phi_l(x+d)$ leads to : $\phi_l(x+d) = \phi_l(x) + \phi_l'(x) \cdot d + \mathcal{O}(d^2)$ The spatial shift $d(x)$ of the signal in the left image is defined in a first order approximation as:

$$d(x) = \frac{\phi_l(x) - \phi_r(x)}{\phi_l'(x)} \tag{3}$$

The derivative $\phi_{l/r}'(x)$ of the local phase is well known [1] as *instantaneous* or *local* frequency of the signal. We will next elucidate the importance of the phase derivative in the task of disparity estimation. We use the following definition of the Gabor impulse response

$$G(x; \sigma, \omega_g) = \frac{1}{\sqrt{2\pi}\sigma} e^{-x^2/2\sigma^2} e^{-i\omega_g x} \tag{4}$$

The results in disparity estimation with a phase based approach using Gabor filters depends strongly on the instantaneous frequencies of the signals due to the bandpass characteristic of the Gabor filter. We treat here analytically the case of a sinusoid with a linear ground-truth disparity $d_{tr}(x) = Ax + B$. The case of an edge modeled as a superposition of sinusoidal functions is studied in the full paper (http://www.informatik.uni-kiel.de/~mha/meinpub-www.html).

3 Analysis of a sinusoidal pattern

In the sinusoidal case the left and right signals read

$$T_l(x) = \sin(\omega_0(x + d_{tr})), \qquad T_r(x) = \sin(\omega_0 x). \tag{5}$$

The local phase $\phi_{l/r}(x)$ and its derivative $\phi'_{l/r}(x)$ are functions of the Gabor-filter parameter ω_g, t, the signal frequency $\omega_0 = 2\pi/\lambda_0$ and the parameters A, B of the disparity variation. Our goal is to study the effects of the above parameters on the instantaneous frequency ω_{inst} and the disparity estimate $d_{est}(x)$.

The convolutions of the functions $T_{r/l}(x)$ with the Gabor filter $G(x; \omega_g, t)$ yield responses with phases $\phi_{r/l}(x) = arg[G * T_{r/l}(x)]$ and the corresponding first derivative $\phi'_{r/l}(x)$ with respect to x:

$$\phi_r(x) = \arctan\left(-\frac{\tanh(\sigma^2\omega_0\omega_g/2)}{\tan(\omega_0 x)}\right), \quad \phi'_r(x) = \omega_0 \frac{\tanh(2\sigma^2\omega_0\omega_g)}{(1 - \frac{\cos(2\omega_0 x)}{\cosh(2\sigma^2\omega_0\omega_g)})} \tag{6}$$

The same expressions are obtained for the left signal with $\tilde{\omega}_0 = \omega_0(A+1)$ instead of ω_0 and the phase shifted by a constant offset $\omega_0 B$.

To improve the intuitive understanding of the above expressions we use wavelengths measured in pixels instead of frequencies with $\lambda_0 = 2\pi/\omega_0$ for the signal and $\lambda_g = 2\pi/\omega_g$ for the Gabor-filter, and the bandwidth factor $t = 1/(\omega_g\sigma)$ respectively. Without loss of generality, we assume that $\lambda_g^{max} \geq \lambda_0^{max}$ for the case of a sinusoid, which means that the signal wavelength is smaller than the filter wavelength. For the disparity computation in eq. (3) we use the average of the left and right signal $\lambda_{inst}(x) = \frac{4\pi}{(\phi'_l(x) + \phi'_r(x))}$ as the instantaneous wavelength $\lambda_{inst}(x)$ of the Gabor filter response [3].

We observe in eq. (6) and in Fig. 1 that the instantaneous wavelength spatially oscillates with a frequency $2\omega_0$ due to the fraction $\frac{\cos(2\omega_0 x)}{\cosh(2\sigma^2\omega_0\omega_g)}$ in eq. (6). However, the spatial average over a local region (Fig. 1, left) is still equal to λ_0, as expected according to the theory on instantaneous frequency.

Additionally we observe in eq. (6) that ω_{inst} is equal to the signal frequency ω_0 only if $\tanh(2\sigma^2\omega_0\omega_g)$ is approximately one. Then the instantaneous wavelength represents the signal wavelength $\lambda_{inst} \approx \lambda_0$ of the signal over a wide region in the Gabor parameter space (λ_g, t). This condition can be achieved if $\lambda_g \geq \lambda_0 t^2$.

In Fig. 1 (right) we see as expected that the maximum of the local energy achieves its maximum if the Gabor-filter wavelength λ_g at a given pixel x coincides with the signal wavelength λ_0. The practical effect is crucial: If we rely

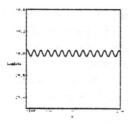 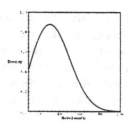

Fig.1. Behavior of the instantaneous frequency $\lambda_{inst}(x)$ as a function of the spatial position (left) and of the local energy as a function of the Gabor wavelength λ_g (right) for a sinusoidal signal with $\lambda_0 = 30$ pixel.

on a maximum response analysis in order to find the optimal scale we have to filter with different wavelengths to capture the variation of the responses. If we compute the instantaneous frequency then just one filter response suffices in order to obtain the optimal scale with a negligible error. The computation of the instantaneous wavelength leads directly to an estimate of the optimal scale without any maximum search over different filter responses which is computational expensive especially for real time implementations.

Accuracy of disparity estimation in sinusoids

We address the accuracy of the disparity estimation for two models: the *constant* model using the filter frequency in eq. (1) and the *instantaneous* model using eq. (3). We choose the true disparity $d_{tr}(x) = Ax + B$ with $A = 0.1$ and $B = 1$. The choice of the disparity gradient $A \neq 0$ yields a scaling by $\frac{1}{(A+1)}$ of the signal wavelength λ_0 in the left view. We analytically compute the disparity and use

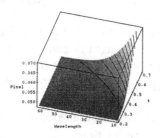 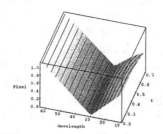

Fig.2. Absolute error of disparity estimation at x=0 as a function of bandwidth factor t and Gabor-wavelength λ_g. The signal wavelength is $\lambda_0 = 30$ pixel and the true disparity is one pixel (left). The error increases for the instantaneous model, if $\lambda_g \approx \lambda_0 t^2$. (right) The error for the constant model is generally higher besides when $\lambda_0 \approx \lambda_g$.

the absolute error $\Delta d_{abs}(x) = |d_{tr}(x) - d_{est}(x)|$ to compare both models for a given position $x = 0$ where $d_{tr}(0) = 1$. Fig.2 shows the variation of the absolute disparity error at $x = 0$ as a function of the Gabor-filter parameters (λ_g, t).

The range (λ_g, t) of these parameters is according to the implementation in a real-time architecture.

In Fig.2 (left) we observe that the absolute disparity error is very low if the instantaneous frequency is used and condition $\lambda_g \geq \lambda_0 t^2$ is satisfied. The phase difference and the instantaneous wavelength are nearly independent of the Gabor-filter wavelength λ_g and strongly dependent on the signal wavelength λ_0 for a wide variation of the bandwidth factor t. The constant model Fig.2 (right) is not able to deal with phase differences independent of λ_g and this leads directly to an error $\Delta d_{abs}(x) \sim |\lambda_g - \lambda_0|$ in disparity estimation.

The use of the instantaneous model (3) definitely leads to better results than the constant model (1). The relative error varies from 5% to 7% in the instantaneous and 0.1% to 100% in the constant model. For the majority of filter wavelengths the instantaneous model outperforms the constant model. However, we notice that the absolute error in the constant model is very low if $\lambda_g \approx \lambda_0$. The error in the instantaneous model increases with the disparity gradient A and can be approximated by $\Delta d_{abs}(x) \sim \frac{A}{A+1}$.

4 Simulations with perspective foreshortening

In this section we will study the perspective effects of a viewed slanted plane. We will again find out that the expressions concerning the simple sinusoid in section 3 are still useful if we assume nonlinear disparity. Fixating to a point of a flat surface leads to a disparity model $d(x) = \frac{Ax+B}{Cx+D}$. By fixating the disparity near the center vanishes and the wavelengths in both signals (left/right) are similar. The obtained stereo signals are shown on the left of Fig. 3. The measured

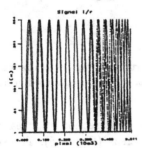
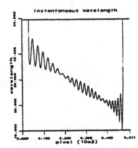

Fig.3. The left figure showed the superimposed signal of the left (dotted line) and of the right view (continuous line). The reader can observe the vanishing disparity in the center. On the right we show the instantaneous wavelength λ_{inst} as a function of spatial position x with a Gabor-filter of wavelength $\lambda_g = 63$ pixel, computed in a simulation.

instantaneous wavelength λ_{inst} for one of the Gabor-filter wavelengths is shown in Fig. 3 (right). The decrease of λ_{inst} is overlayed by an oscillation in x due to the small violation of the inequality constraint $\lambda_g \geq \lambda_0 t^2$.

This oscillation affects also the error in the disparity plotted on the left of Fig.4 obtained by both models for a Gabor-filter with wavelength $\lambda_g=63$ pixels. We repeat the same curves on the right of Fig.4 but this time obtained from the responses to a Gabor-filter with wavelength equal to the instantaneous

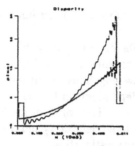 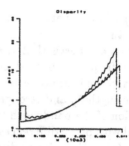

Fig.4. True disparity (continuous line), disparity for the instantaneous (dotted line) and disparity for the constant model (continuous line far from the ground truth) for a Gabor-filter λ_g=63 pixel are shown on the left. On the right we show the same curves but for a Gabor-filter tuned to the instantaneous frequency $\lambda_{inst}(x = 256)$ of the center.

wavelength at the center. We observe that the constant model performs better but does not reach the performance of the instantaneous model which is an order of magnitude closer to the ground-truth at almost every position in the image. The results support the fact that the instantaneous model provides good estimates without a special filter tuning.

Fig.5. Stereo images of our lab with some structured regions at the wall. The system fixates the black cross. The white line indicates the 1D-signal used in further plots.

We will apply the methods of the previous section at a *real* stereo image pair which contains edges and textured areas (Figure 5). The stereo system fixates the black cross at the wall. The *true* disparity in Fig. 6 is computed from the simulation by calibrating the extrinsic and intrinsic camera parameter in relation to the wall. Due to the errors resulting from the erroneous calibration an exact quantitative comparison is not possible. In the following we will use the calibration result as ground-truth.

The results in Fig. 6 for this row (and others) are nearly the same as in the simulation. The instantaneous model provides a good approximation of the *true* disparity with both Gabor-filters (Fig. 6 middle). The constant model (right) has to be tuned to the measured instantaneous wavelength $\lambda_{inst}(256)$=55 of the center of the row in order to get a good local approximation near the center. The zero disparities outside the center are regions of unreliable phase information.

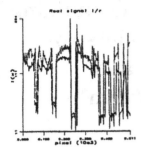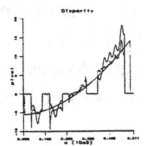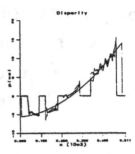

Fig.6. Signals (l/r) of the real images with different amplitudes at row 260 through the image center (left). True and estimated disparity for a Gabor-filter (λ_g=63 pixel, t=0.66) (middle) and a Gabor-filter tuned (λ_g=55 pixel, t=0.33) to the instantaneous wavelength $\lambda_{inst}(x = 256)$ (right). The smooth line represents the true disparity.

5 Conclusions

We have analytically studied the behavior of the constant and the instantaneous model for disparity estimation using complex Gabor- filters. The instantaneous model has a superior performance over the entire spectrum of filter-wavelengths whereas the constant model results are not useful at all over the most wavelengths. Until now the procedure was to optimize by tuning near the frequency of maximal response. This procedure is computationally expensive and fails if such a unique maximum does not exist. Tuning the filters to the instantaneous frequency improves their performance, a step that in most cases is not even necessary. Taking the phase differences over the instantaneous frequency performs perfectly for most filter-wavelengths.

References

1. B. Boashash. Estimating and interpreting the instantaneous frequency of a signal-part 1: Fundamentals. *Proceedings of the IEEE*, 80:520–539, 1992.
2. U.R. Dhond and J.K. Aggarawal. Structure from stereo - a review. *IEEE Trans. Syst. Man. and Cyber.*, 19, 1489-1510 1989.
3. D. J. Fleet, A. D. Jepson, and M. R. M. Jenkin. Phase-based disparity measurement. *CVGIP: Image Understanding*, 53(2), 3 1991.
4. K. Langley, T.J. Atherton, R.G. Wilson, and M.H.E. Larcombe. Vertical and horizontal disparities from phase. In *Proc. First European Conference on Computer Vision*, pages 315–325. Antibes, France, Apr. 23-26, O.D. Faugeras (Ed.), Lecture Notes in Computer Science 427, Springer-Verlag, Berlin et al., 1990.
5. A. Maki, L. Bretzner, and J.-O. Eklundh. Local fourier phase and disparity estimation: An analytical study. In *V. Hlavac et al. (Ed.), Proc. Int. Conf. Computer Analysis of Images and Patterns CAIP, Prag*, pages 868–874, 1995.
6. T.D. Sanger. Stereo disparity computation using gabor filters. *Biol.Cybernetics*, 59:405–418, 1988.
7. W. M. Theimer and H. P. Mallot. Phase-based binocular vergence control and depth reconstruction using active vision. *CVGIP: Image Understanding*, 60(2):343–358, 12 1994.
8. C.J. Westelius, H. Knutsson, J. Wiklund, and C.F. Westin. Phased-based disparity estimation. In J.L. Crowley and H.I.Christensen, editors, *Vision as Process*, pages 157–178. Springer Verlag, Berlin et al., 1994.

Segmentation from Motion: Combining Gabor- and Mallat-Wavelets to Overcome Aperture and Correspondence Problem[*]

Laurenz Wiskott[†]

Institut für Neuroinformatik

Ruhr-Universität Bochum

D–44780 Bochum, Germany

http://www.neuroinformatik.ruhr-uni-bochum.de

Abstract

A segmentation-from-motion algorithm is presented, which is designed to be part of a general object recognition system. The key idea is to integrate information from Gabor- and Mallat-wavelet transform to overcome the aperture and the correspondence problem. The assumption is made that objects move fronto-parallel. Gabor-wavelet responses allow precise estimation of image flow vectors with low spatial resolution. A histogram over this image flow field is evaluated, its local maxima providing *motion hypotheses*. These serve to reduce the correspondence problem on the Mallat-wavelet transform, which provides the required high resolution. The segmentation reliability is improved by integration over time. The system can segment even small, disconnected, and openworked objects of arbitrary number, such as dot patterns. Several examples demonstrate the performance of the system and show, that the algorithm behaves reasonably even if the assumption of fronto-parallel motion is strongly violated.

1 Introduction

The segmentation algorithm presented here has been motivated in context of a larger effort to build a general system for object recognition [6]. It is an attempt to develop a segmentation algorithm which follows similar concepts and joins smoothly with the general object recognition system with respect to the

[*]Supported by grants from the German Federal Ministry for Science and Technology (413-5839-01 IN 101 B9).

[†]Current address: Computational Neurobiology Laboratory, The Salk Institute for Biological Studies, San Diego, CA 92186-5800, http://www.cnl.salk.edu/CNL, wiskott@salk.edu

underlying image representations. This is the main reason why Gabor- and Mallat-wavelet transform are used here. For a pure segmentation system, less expensive representations with similar properties could be used.

Segmentation from motion algorithms can be divided into two classes, the filter based methods and the matching methods. The typical example of a filter based method is the gradient method, where the motion constraint equation [2] is used to estimate image flow. This method suffers from the aperture problem. Matching techniques on the other hand usually suffer from the correspondence problem, i.e. the ambiguity as to which feature point in one frame has to be matched to which feature point in the next frame. In this work I integrate these two types of techniques, filters and matching, in order to overcome the aperture problem and reduce the correspondence problem significantly. This work is described in more detail in [5].

2 The System

The algorithm can roughly be described in five stages:

1. Based on a Gabor wavelet transform, the image flow field between the two images is computed. Since the Gabor wavelets are large, this method does not have a serious aperture or correspondence problem, but the spatial resolution of the computed flow field is low; see Fig. 2.

2. The assumption is made that all objects in the scene move mainly translationally, i.e. fronto-parallel. A histogram over the flow vectors can then be evaluated, its peaks providing precise information on the motion vectors of the different objects, called *motion hypotheses*. The spatial information of the image flow field is disregarded; see Fig. 3.

3. For each edge pixel each of the motion hypotheses is checked. The local grey-value gradient at a pixel in one frame is compared with the gradient in the next frame, but taken from a pixel which is displaced according to the considered motion hypothesis. This comparison provides an *accordance* value for each hypothesis, which is high if the gradients are similar and low if they are not. The accordance values of the whole image yield one accordance map for each motion hypothesis.

4. The accordance maps are integrated over time. No motion continuity is assumed. Objects may jump back and forth from one frame to the other. This is possible by identifying objects of successive pairs of frames through the spatial domain based on the fact that objects cover the same space in the second frame of a pair of frames as they cover in the first frame of the next pair (in fact these two frames are the same). See Fig. 6–9.

5. Finally each pixel is categorized into the motion hypothesis class for which it has achieved the highest accordance value; see Fig. 7. This is done for each pixel individually. No object model is used and no regularization is applied.

2.1 Image Flow Estimation

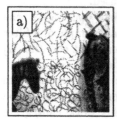

Figure 1: **a)** Frame 14 and **b)** Frame 15 of the moving-animals sequence. The zebra moves fast to the right, the elephant slowly to the left. The images have a resolution of 128×128 pixels with 256 grey values.

The first stage of the algorithm computes an image flow field. The algorithm presented here is based on a Gabor wavelet transform and adopted from a stereo algorithm [4].

A *jet* $\mathcal{J}$ is defined as a set... A Gabor wavelet transform [1, 6] is defined as a convolution with a family of Gabor wavelets in the shape of plane waves with wave vectors $\vec{k}_j$, restricted by Gaussian envelope functions. The wavelets have five different frequencies and eight orientations. A *jet* $\mathcal{J}$ is defined as the set $\{\mathcal{J}_j\}$ of 40 complex coefficients obtained for one image point from this convolution. It can be written as $\mathcal{J}_j = a_j \exp(i\phi_j)$ with amplitudes $a_j(\vec{x})$, which slowly vary with position, and phases $\phi_j(\vec{x})$, which rotate with a rate set by the spatial frequency of the wave vector $\vec{k}_j$ of the respective wavelet.

Consider two jets, $\mathcal{J}$ and $\mathcal{J}'$, taken from the same pixel position in two successive frames. The underlying object in the scene may have moved by a vector $\vec{d}$. The phases of the jet coefficients $\mathcal{J}_j$ then vary mainly according to their corresponding wave-vectors $\vec{k}_j$, yielding phase differences $\Delta\phi_j = \phi_j - \phi'_j \approx \vec{d}\vec{k}_j$. Vice versa, since the phases of $\mathcal{J}$ and $\mathcal{J}'$ are known, the displacement $\vec{d}$ can be estimated by matching the terms $\vec{d}\vec{k}_j$ with the phase differences $\Delta\phi_j$, which can be done by maximizing the function

$$S_\phi(\mathcal{J}, \mathcal{J}') = \sum_j a_j a'_j \cos(\Delta\phi_j - \vec{d}\vec{k}_j) \tag{1}$$

in its Taylor expansion. Fig. 2 shows the image flow field for Frames 14 and 15 shown in Fig. 1. It can be seen that the flow vectors are fairly constant within regions, indicating high precision of the estimated displacements. However, the spatial resolution of this image flow algorithm is low.

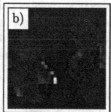

Figure 2: **a)** horizontal and **b)** vertical component of the image flow field for Frames 14 and 15 shown in Fig. 1. Negative values are light and positive values dark.

2.2 Motion Hypotheses

The second stage of the algorithm is concerned with the extraction of motion hypotheses from the image flow field. This can be simply done by detecting local maxima in the image flow histogram, a histogram over the flow field vectors; see Fig. 3. (Actually, the histogram is restricted to edge pixels, as will be motivated later.) In order to avoid detecting too many irrelevant maxima, a low-pass filter is applied to the histogram and only maxima of a certain minimal height are accepted. The result of this stage is usually a small number of displacement vectors $\vec{v}_n$ representing frequently occurring flow vectors.

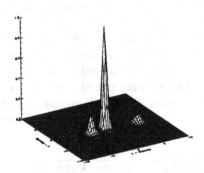

Figure 3: Image flow histogram for Frames 14 and 15 of the moving-animals sequence. It shows three prominent local maxima. The largest one corresponds to the resting background. The smaller ones on the left and right side correspond to the elephant and zebra, respectively. The zebra moves faster to the right than the elephant to the left. In addition there is a fourth local maximum, too small to be visible here and not corresponding to any object.

A displacement $\vec{d}$ can be compared with a motion hypothesis $\vec{v}_n$ by the displacement similarity function

$$S_d\left(\vec{v}_n, \vec{d}\right) = \max\left\{1 - \frac{\left(\vec{v}_n - \vec{d}\right)^2}{r^2}, 0\right\} \tag{2}$$

with a parameter r to set its sensitivity.

The advantage of the motion hypotheses is that they drastically reduce the correspondence problem for an image flow estimation algorithm based on more localized features, such as edges. Instead of testing all possible displacements within a certain distance range, only a few of them suggested by the motion hypotheses need to be taken into consideration; see Fig. 4.

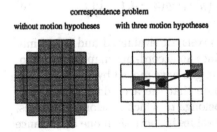

correspondence problem

without motion hypotheses   with three motion hypotheses

Figure 4: Illustration of the reduced correspondence problem: Without motion hypotheses, a whole region within a certain diameter has to be tested for possibly corresponding pixel locations with a similar grey value gradient. With motion hypotheses, the regions to consider reduce to few small spots.

2.3 Edge Valuation

The third stage of the algorithm is based on the Mallat-wavelet transform [3], which can be thought of as the grey-value gradient at different levels of resolution. For simplicity only the highest level of the Mallat-wavelet transform will be used in the following considerations and the terms gradient and edges will be used instead of wavelet transform and modulus maxima, respectively. However it is important from a conceptional point of view that the modulus maxima of the wavelet transform provide enough information to reconstruct images [3].

This stage is similar to a matching algorithm and is therefore faced with the correspondence problem, which is particularly severe for such a simple feature as the local gradient. Two methods are used to bypass the correspondence problem. Firstly, the evaluation of the Mallat-wavelet transform is restricted to edges, as defined by the modulus maxima [3]. Edges have a particularly high information content and are less ambiguous than gradients in general. Secondly, the edges are not matched between two frames, but only the *accordances* with the different motion hypotheses obtained in stage two are determined. This reduces the correspondence problem significantly, but leaves it partially unresolved at the same time, because the final decision between the small number of motion hypotheses is not yet made.

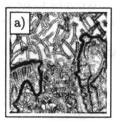 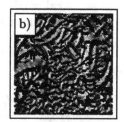 

Figure 5: Local gradient as an edge representation. a) the amplitude of the gradient from zero (white) to maximum (black), b) its direction angle between 0 (black) and 2π (white), and c) the edges as defined by the modulus maxima (black).

Given a gradient similarity function $S_g(\vec{g}, \vec{g}')$ comparing direction and magnitude of two gradients $\vec{g}$, the accordance of a gradient $\vec{g}(\vec{x}, t)$ with a motion hypothesis $\vec{v}_n$ is defined as the maximum over

$$\mathcal{A}(\vec{x}, t, \vec{v}_n) = \max_{\vec{d}} \left\{ S_d\left(\vec{v}_n, \vec{d}\right) S_g\left(\vec{g}(\vec{x}, t)\vec{g}(\vec{x} - \vec{d}, t - 1)\right) \right\}, \qquad (3)$$

found by varying $\vec{d}$. Notice that only edge pixels are evaluated and taken into consideration. This accordance function defines an accordance map for each motion hypothesis indicating whether an edge might have moved by the respective flow vector $\vec{v}_n$ or not. It is important to note that the accordance maps may still contain ambiguities as to which edge belongs to which motion hypothesis. This would be indicated by high accordance values in more than one accordance map for a single edge pixel.

2.4 Integration over a Sequence of Frames

The quality of the accordance maps can be significantly improved by integrating them over a sequence of frames. The main difficulty is to determine which motion hypothesis in one pair of frames corresponds to which motion hypothesis in the next pair of frames. In order to avoid assumptions about the type of movements objects may perform, I associate the motion hypothesis via the spatial overlap in the intermediate frame, i.e. the one that is common to both pairs. This is possible because the first pair of frames already provides a crude segmentation result. The integrated accordance map is the geometric mean of the previous integrated accordance map and the current instantaneous accordance map; see Fig. 6.

Figure 6: Accordance maps for Frame 15 with respect to the three relevant motion hypotheses, disregarding the spurious one, as extracted from the image flow histogram shown in Fig. 3. These accordance maps were integrated over time and take into account the results from preceding frames. High values are shown dark. **a)** corresponds to the zebra, **b)** to the background, and **c)** to the elephant.

2.5 Segmentation

The last stage of the algorithm finally performs the segmentation. Each edge pixel is classified as belonging to that motion hypothesis for which it has the highest accordance value. One can think of it as a pixel-wise winner-take-all competition between the accordance maps; see Fig. 7.

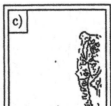

Figure 7: **a)-c)** Segmentation result based on the accordance maps shown in Fig. 6 a)-c), respectively. **d)** shows the pixels which were not categorized, because their accordance values were too low.

3 Examples

One strength of the system is that it generates motion hypotheses on a coarse level, but segments on a single pixel level. This allows the system to segment even small, disconnected, or openworked objects. A second strength is that the motions need not be continuous in order to do the integration over a sequence of frames. Objects may jump back and forth, and the algorithm will still be able to track them. This is demonstrated for dot-patterns in Fig. 8, where random shifts of the frames against each other have been introduced artificially.

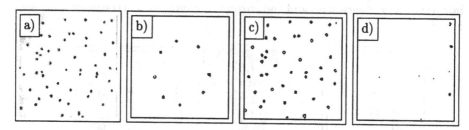

Figure 8: **a)** Seventh frame of the dot-pattern sequence, showing a stationary circle with eight points and a dotted background moving left. Each frame in the sequence is additionally shifted randomly up to $(\pm 2, \pm 2)$ pixels, resulting in an additional relative displacement of up to ± 4 pixels in each dimension. **b) c)** Segmentation result for the circle and the background respectively. **d)** Uncategorized pixels. One can see three dots entering the image on the right and having not yet accumulated enough evidence for segmentation.

The most restrictive assumption of the system is that objects move fronto-parallel. An extreme example of violating the fronto-parallel-motion assumption is shown in Fig. 9. The plant is rotating in depth such that all leaves move against each other. The sequence treats this inappropriate sequence by breaking it down into parts which can be approximated as moving fronto-parallel. The leaves moving fast to the left are grouped together and those which move slowly are grouped with the background. A similar result was found for objects rotating in the image plane.

Figure 9: **a)** Fifth frame of the rotating-plant sequence. **b) c)** Segmentation result for the fast and the slowly moving leaves respectively. **d)** Uncategorized pixels.

4 Conclusions

The system presented performs segmentation from motion on a sequence of frames. It is unconventional in several respects. It integrates two different techniques based on Gabor- and Mallat-wavelets to overcome the aperture and the correspondence problem and it integrates over time for an additional improvement of the segmentation result. Segmentation is only performed on edges and it is argued that edges are the most appropriate representation for segmentation, because they provide enough motion evidence and the complete grey-value distribution could be reconstructed from edges. Since no object model is used, the system can even segment small, disconnected, and openworked objects. The motion model used is pure translation and obviously inappropriate for objects in arbitrary motion. However, the traditional segmentation from motion paradigm that strives to segment individual objects globally on the basis of given motion models may be inappropriate in any case. I propose to do the segmentation locally and integrate the local fragments on a higher level. For a gesticulating person, for instance, the hand may move in front of the body and could be connected with the body and locally segregated from the body at the same time. In such a locally segmenting system, the translational motion model may be the most appropriate one, because of its simplicity, and the system presented could serve as a local mechanism.

Acknowledgement: Many thanks go to Prof. von der Malsburg for his support and helpful comments.

References

[1] DAUGMAN, J. G. Complete discrete 2-D Gabor transform by neural networks for image analysis and compression. *IEEE Transactions on Acoustics, Speech and Signal Processing 36*, 7 (July 1988), 1169–1179.

[2] HORN, B. K. P., AND SCHUNCK, B. G. Determining optical flow. *Artificial Intelligence 17* (1981), 185–203.

[3] MALLAT, S., AND ZHONG, S. Characterization of signals from multiscale edges. *IEEE Transactions on Pattern Analysis and Machine Intelligence 14*, 7 (1992), 710–732.

[4] THEIMER, W. M., AND MALLOT, H. A. Phase-based binocular vergence control and depth reconstruction using active vision. *CVGIP: Image Understanding 60*, 3 (Nov. 1994), 343–358.

[5] WISKOTT, L. Segmentation from motion: Combining Gabor- and Mallat-wavelets to overcome aperture and correspondence problem. Tech. Rep. IR-INI 96-10, Institut für Neuroinformatik, Ruhr-Universität Bochum, D-44780 Bochum, Germany, 1996.

[6] WISKOTT, L., FELLOUS, J.-M., KRÜGER, N., AND VON DER MALSBURG, C. Face recognition by elastic bunch graph matching. IEEE Transactions on Pattern Analysis and Machine Intelligence, accepted for publication., 1997.

Contour Segmentation with Recurrent Neural Networks of Pulse-Coding Neurons[1]

L. Weitzel, K. Kopecz, C. Spengler, R. Eckhorn, H.J. Reitboeck

Dept. of Neurophysics, University of Marburg, Germany

Abstract. The performance of technical and biological vision systems crucially relies on powerful processing capabilities. Robust object recognition must be based on representations of segmented object candidates which are kept stable and sparse despite the highly variable nature of the environment. Here, we propose a network of pulse-coding neurons based on biological principles which establishs such representations using contour information. The system solves the task of grouping and figure-ground segregation by creating flexible temporal correlations among contour extracting units. In contrast to similar previous approaches, we explicitly address the problem of processing grey value images. In our multi-layer architecture, the extracted contour features are edges, line endings and vertices which interact by introducing facilatory and inhibitory couplings among feature extracting neurons. As the result of the network dynamics, individual mutually occluding objects become defined by temporally correlated activity on contour representations.

1 Introduction

Vision systems are faced with the general problem of constructing representations emphasizing regularities and coincidences which have been recognized by experience to be related to coherent physical structures in the world. They have to construct a "meaningful" representation, i.e. to overcome irregularities and corruptions of coherent structures. The mechanisms which solves these problems has often been referred to as *segmentation, perceptual organization*, or *figure/ground segregation*. Although visual memory is known to affect the condtruction of percepts, built-in grouping principles are manifested in the "Gestalt-rules" of perceptual psychology [8], which has been adopted also recently for computer vision ([5]). Whereas these algorithms are essentially symbolic in nature, we define networks with subsymbolic operations, in which the network dynamics evolves in time towards stable states representing "computational" results. The mapping of desired system behaviors onto states with dynamic stability provides a means of endowing the system with robust processing capabilities.

To label two features on a fragmentary representation as belonging to the same perceptual group, the concept of *temporally correlated activity* has been proposed [4,6] and supporting neurophysiological evidence has been accumulated (e.g. [1]). These studies showed, that neurons with similar feature extracting characteristics can synchronize their activity partially. The correlation hypothesis also provides an elegant way of representing multiple object candidates or hypothesis by decorrelating activity among neurons which belong to the representation of different objects. Several theoretical studies based on binary input information demonstrated the feasibility of this approach (e.g. [2,7]). Here, we

[1] We acknowledge support by the BMBF, grant no. 01 M 3013D

address the problem of image segmentation by activity correlations based on contour information extracted from real-world scenes. A prototypical test image is depicted in Fig. 7. There, we perceive a triangle which occludes a possibly rectangular object. From the Gestalt point of view, the triangle is defined as an entity by the straight contours which are connected by corners (vertices). The impression of occlusion is induced by the two T-junctions at the locations where triangle and rectangle meet. Note that the triangle is easily perceived as such although the contrast across its contours vary and, in particular, vanishs at the center of its left side.

In the processing modules of our system (Fig. 1) the image is fed through a preprocessing stage, where it is hexagonally subsampled and local intensity contrast is extracted by "Mexican-Hat"-shaped filters. This analogue contrast signal is converted into a pulse sequence by our model neurons. The basic processing element of the edge processing layers is a neuron which is sensitive to a local contrast edge oriented along a certain direction. Coaxially aligned units are mutually coupled to provide a grouping mechanism which emphasizes the representation of extended edges. From these edge sensitive units, representations of line endings ("end-stops") and vertices are derived. These vertex primitives cooperate with a separate contour synchronization stage to generate correlated activity according to the built-in grouping tendencies. Further, the inhibitory feedback from the synchronization stage to the vertex representation separates different object candidates in time.

2 The Dynamic Pulse-Coding Neuron

Our basic neural unit produces a sequence of pulses ("spikes") in response to an input membrane potential. The pulses are generated by a spike encoder characterized by a dynamic threshold subject to internal feedback from the neuron's output [2]. In brief, to each neuron k three distinct potentials are assigned. The *feeding potential* $F_k(t)$, the *inhibitory potential* $I_k(t)$ and the *linking potential* $L_k(t)$. The total membrane potential $M_k(t)$ is a nonlinear combination of these contributions:

$$M_k(t) = F_k(t)\left[1 + L_k(t)\right] - I_k(t) \quad , \text{ or: } \quad M_k(t) = F_k(t)L_k(t) - I_k(t) \quad (1)$$

Both variants are used within our network. Thus, the linking potential modulates the feeding contribution. Correspondingly, each neuron k can receive spike input $A_j(t)$ from other neurons j by three types of weights, w_{kj}^F, w_{kj}^L and w_{kj}^I. The weighted spike input is finally low-pass filtered:

$$\tau_{P_k}\dot{P}_k(t) = -P_k(t) + V_{P_k}\sum_j w_{kj}^P A_j(t) \quad (2)$$

where P can be F, L, or I, and τ_{P_k} and V_{P_k} are time constants and gain factors, respectively. Whenever the membrane potential $M_k(t)$ exceeds a resting threshold a sequence of pulses is produced. The frequency increases approximately linearly with $M_k(t)$ up to a saturation value.

The functional role of the feeding weights, w_{kj}^F, is to define filter characteristics by forward connections, and thus the feature extracting property of the

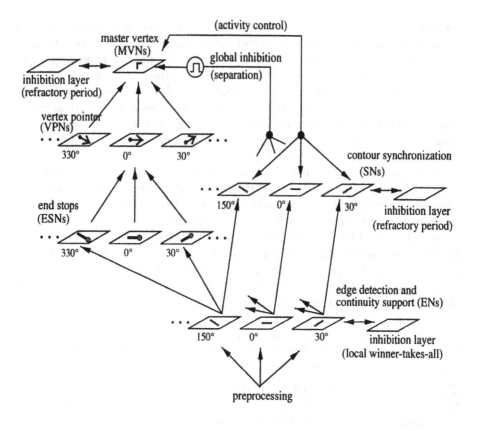

Fig. 1. Processing modules of the contour segmentation system.

neurons. The linking couplings, w_{kj}^L, can be effectively used to quickly establish temporal correlations among neurons [2], which codes for the unity of individual features. Further, the susceptibility of neurons can be changed by linking couplings according to the context in which they are embedded.

3 Edge Detection and Continuity Support

Based on the activity delivered by the preprocessing stage, in a first step neurons specific for an oriented contrast edge (CENs) are constructed by convergent connections from preprocessing neurons. The superposition of the "Mexican-Hat" shaped receptive fields yields wavelet-like filter characteristics of CENs. CENs are arranged in twelve layers (orientation preferences: $0°$ to $150°$ in steps of $30°$ for the two contrast polarities). Within one layer, neurons are coaxially coupled in by linking connections. This coupling amplifies the representation of extended edges and can raise the input into a neuron over its threshold if it is embedded in a matching spatial context. This is demonstrated in Fig. 2 for a layer of CENs which are sensitive for a $0°$ orientation. Neural activity is visualized by superimposing the receptive fields of active neurons in the image domain.

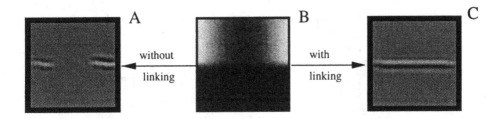

Fig. 2. Linking connections transmit context information.

Fig. 2B shows the grey value distribution used as the test image, a noisy contrast edge with reduced contrast in the center region. Panels A and C show the receptive fields of active neurons with 0° preferred orientation without and with mutual linking connections, respectively. Note that through linking the center neurons are activated by the low contrast part of the edge. This becomes possible because these neurons are embedded in the context of coaxially aligned high contrast edges, so that adjacent neurons can increase the sensitivity of the center neurons by the modulatory linking connections. Thus, each neuron acts as a context dependent adaptive filter.

To become invariant of the polarity of contrast edges, edge detecting neurons (ENs) are constructed by the convergence of two CENs with opposite contrast polarity. ENs are mutually coupled by two types of connections. First, through inhibitory interneurons arranged in a separate inhibition layer (cf. Fig. 1), so that at each spatial location neurons compete by a winner-take-all mechanism. Second, within one layer again ENs are coupled in iso-orientation by linking connections to support also the representation of extended edges along which the contrast polarity change.

4 End-Stop Detection

End-stop detecting neurons (ESNs) are constructed by convergent feeding connections from the edge detecting layers. This wiring is illustrated in Fig. 3, where the construction of an ESN with 0° preferred direction is shown. At a certain location the ESN is connected with adjacent edge detectors in the 0°-edge layer in 0° direction by positive feeding weights. Further, it is connected to a patch extending in 180° direction by negative weights. Thus, a continuous edge passing through the ESN's receptive field (inset in Fig. 3) will not evoke any response, because excitation is cancelled by inhibition. The detector will respond only if the inhibitory subfield is not covered by the edge. The patchy structure of the inhibitory subfield strongly decreases the susceptibility to perturbations of the edge. For instance, if an extended edge is not exactly aligned along 0° orientation (as will generally be the case), the edge is represented by staggered segments in the 0° edge layer. Thus, staggered representations must not be interpreted as an end-stop which requires a spatial tolerance in the inhibitory subfield.

5 Vertex Representations

We define a vertex as being composed of at least two line endings at roughly the same location. Thus, corners of two-, or three-dimensional objects will be

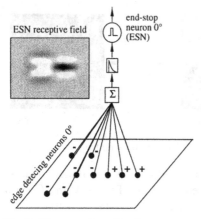

end-stop
neuron 0°
(ESN)

ESN receptive field

edge detecting neurons 0°

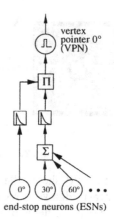

vertex
pointer 0°
(VPN)

0° 30° 60° • • •

end-stop neurons (ESNs)

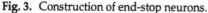

Fig. 3. Construction of end-stop neurons.

Fig. 4. Synthesis of a VP neuron

considered as vertices. This definition excludes the detection of T-junctions as vertices, which are interpreted here as being indicative of object occlusions.

Two different representations of vertex information are introduced, the *vertex pointer layers* and a *master vertex layer* (cf. Fig. 1). Vertex pointers are individual directed parts of a vertex and are detected by convergent connections from the end-stop neurons. Thus, we do not construct an explicit representation of all possible vertices in all possible compositions and orientations, but a vertex is represented implicitly by the concurrent activation of several vertex pointer neurons (VPNs) at the same location. For a 0° VPN the response of the 0° ESN at the same location as the VPN is multiplied with the summed response of the remaining ESNs at that location but with different direction preferences (Fig. 4). Thus, the 0° VPN will fire in the presence of at least two end-stops, one of them being the 0° ESN. It turns out that for real-world scenes one has to relax the strict spatial one-to-one correspondence between ESNs and VPNs, so that here also spatially adjacent ESNs converge onto the VPN.

The second type of vertex representation constitutes the master vertex layer. By spatial one-to-one connections from the vertex pointer layers, a master vertex neuron (MVN) signals the presence of a vertex regardless of its detailed composition. For each MVN there are spatially one-to-one corresponding inhibitory partner neurons arranged in a separate inhibition layer. Through these inhibitory local feedbacks, MVNs become refractory after their activation.

6   Temporal Contour Binding and Separation

Contour binding and separation is now achieved by a dynamic interplay between master vertex layer and the synchronization layers (cf. Fig. 1). At this point it becomes essential that we are using networks of dynamic neurons which are able to produce temporal patterns. These patterns will be designed to reflect the figural unity by synchronous activation of figural parts within a certain temporal window.

The synchronization layers are essentially a copy of the edge detecting layers with corresponding one-to-one feeding connections. Importantly, neurons in the synchronization layers (SNs) cannot be activated by input from the edge detecting neurons (ENs) alone, but SNs become presensitized by edge information. The actual activation of SNs is induced by activity originating in the master vertex layer. A part of the couplings among ENs, SNs and MVNs is depicted in Fig. 5. The connections to the right-hand-side SN are shown. ENs are coupled to SNs at the same spatial location by links which become only effective if the MVN at that location is activated. Further, SNs are laterally coupled to adjacent neurons in iso-orientation (only one of the lateral couplings is depicted in the figure). Consequently, once a SN has been activated by a vertex, this activity can propagate in iso-orientation within the synchronization layers, but only as long as SNs are presensitized by input from the edge detecting neurons. Thus, the emerging "linking wave" propagates along detected edges. Through the introduction of a separate inhibition layer (cf. Fig. 1), SNs enter a refractory period of a certain period once they have been active. There are two possible configu-

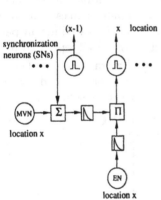

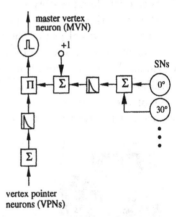

Fig. 5. Couplings from MVN and ENs to SNs   **Fig. 6.** Coupling from SNs to MVN

rations in which a linking wave terminates. First, the detected edge simply ends at a certain location. Second, the edge traced by the linking wave can be connected to a vertex. Then, the linking wave must be directed around the vertex to continue along further edges which constitutes the vertex. This "linking around corners" is established by the linking feedback from SNs to MVNs as depicted in Fig. 6. Whenever a linking wave arrives at a vertex, it will activate the corresponding MVN, provided the MVN receives sufficiently strong support from the vertex pointer neurons and is not in its refractory period due to a previous activation. The activation of the MVN in turn activates non-refractory SNs by the coupling shown in Fig. 6, which again can initiate a linking wave along further edges constituting the vertex. Importantly, linking waves are not diverted along T-junctions, because these do not activate the vertex detecting system. Note that usually several linking waves starting at different vertices trace an object in parallel.

By the above introduced couplings, linking waves trace a contour consisting of line segments connected by vertices within a limited amount of time, which signals the unity of the detected collection of contour segments. Temporal separation of mutually unconnected collections of contours or contours which appear to be occluded due to the presence of T-junctions, is established by introducing a global inhibitory feedback from the synchronization layers to the master vertex layer (cf. Fig. 1). Thus, an inhibitory neuron is introduced which collects signals from all SNs and distribute it to all MVNs. By this inhibitory feedback, contours which are not mutually facilitated due to their connectedness become separated in time (see [2] for details).

7 Simulation Results

We present the processing results for two sample images. The first is an artificial image consisting of a triangle defined by a variable intensity gradient which appears to occlude a possibly rectangular shape (Fig. 7A). After about 400 time steps, the network has entered a cyclic state, where all contour elements become aggregated into two objects. The temporally accumulated spikes of all neurons in the synchronization layers are depicted in Figs. 7B,C for two different periods. As obvious, the triangle and the occluded object become well separated in time.

Similarly, Fig. 8 demonstrates the processing of a traffic scene. There, the contour of the foreground car becomes separated from background structures, which are not resolved in detail due to the chosen spatial resolution.

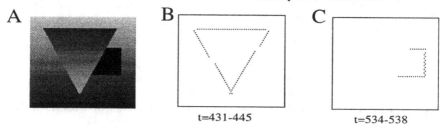

A   B   C

t=431-445   t=534-538

Fig. 7. Binding and segregation of two seemingly overlapping objects.

Discussion and Conclusion:

We presented a recurrent neural network of pulse-coding neurons performing image segmentation based on contour information. Our model neuron and the network architecture are motivated by neurophysiological findings on stimulus dependent correlations in cat and monkey visual cortex. Following the previously developed correlation hypothesis, image segmentation is performed by coding grouping relations through correlated activity and separation through decorrelation. In contrast to previous comparable approaches, our network analyses grey value images of real-world scenes and is not restricted to binary input obtained by thresholding. Instead, the sensitivity of edge detectors is locally modulated by

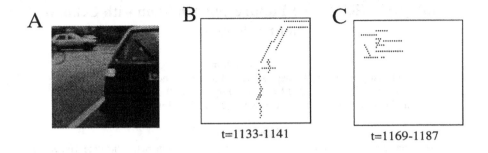

Fig. 8. Binding and segregation of contour elements in a traffic scene.

spatial context information. This is important in the light of coding *relevance* of signals (here given by the intensity contrast and the spatial context) to be used at succeeding, knowledge-based decision stages. Information about relevance is carried in our network by pulse firing rates and is thus not lost at these early stages of visual processing.

The presented results are based on the processing of 44 by 50 image sampling points. The total number of neurons needed is about 10^5 and the number of connections amounts to about $4 \cdot 10^6$. Thus, simulation cycles on usual workstations are far from reaching video rate. However, our network is well suited for non time critical applications and, more important, gains significance also for real time applications by the development of digital pulse-processing hardware [3].

References

1. R. Eckhorn, R. Bauer, W. Jordan, et al. Coherent oscillation: A mechanism of feature linking in the visual cortex. *Biol. Cyb.*, 60:121–130, 1988.
2. R Eckhorn, HJ Reitboeck, M Arndt, et al. Feature linking via synchronization among distributed assemblies: Simulations of results from cat visual cortex. *Neural Comp.*, 2:293–307, 1990.
3. G Frank and G Hartman. An artificial neural network accelerator for pulse-coded model neurons. In *Proc. ICNN 1995*, (CD–ROM). IEEE, 1995.
4. HJ Reitboeck. A multi-electrode matrix for studies of temporal signal correlations within neural assemblies. In Basar et al., (ed.), *Synergetics of the Brain*, 174–181. Springer, Berlin, 1983.
5. S Sarkar and KL Boyer. A computational structure for preattentive perceptual organization. *IEEE Trans Sys Man Cybern*, 24:246–266, 1994.
6. C von der Malsburg. The correlation theory of brain function. Technical Report Internal Report 81-2, MPI für Biophysikalische Chemie, Göttingen, 1981.
7. C. von der Malsburg and J Buhmann. Sensory segmentation with coupled neural oscillators. *Biol. Cybern.*, 67:233–242, 1992.
8. M Wertheimer. Untersuchungen zur Lehre von der Gestalt II. *Psychologische Forschung*, 4:301–350, 1923.

Multigrid MRF Based Picture Segmentation with Cellular Neural Networks

László *Czúni*

University of Veszprém, Dep. of Image Processing and Neurocomputing
H-8200 Veszprém, Egyetem u. 10.
czuni@silicon.terra.vein.hu, Fax: +36-88-422022/605, Tel: +36-88-422022/605

Tamás *Szirányi*

Analogical and Neural Computing Laboratory, Computer and Automation Institute, Hungarian Academy of Sciences
H-1111 Budapest, Kende u. 13-17, Hungary,
sziranyi@lutra.sztaki.hu, Fax: +36-1-2095264, Tel:+36-1-2095265

Josiane *Zerubia*
INRIA, Sophia-Antipolis,
BP 93, 2004 route des Lucioles - 06902 - Sophia-Antipolis Cedex - France
zerubia@sophia.inria.fr, Fax: +33 49365 7865, Tel: +33 49365 7643

Abstract

Due to the large computation power needed in image processing methods based on Markovian Random Field (MRF) [6], new variations of basic MRF models are implemented. The Cellular Neural Network [5,14,15] (CNN) architecture, implemented in real VLSI circuits, is of superior speed in image processing. This very fast CNN can implement the ideas of existing MRF models, which would result in real-time processing of images. On the other hand this VLSI solution gives new tasks since the CNN has a special local architecture [4], but it is already shown that a type of MRF image segmentation with Modified Metropolis Dynamics (MMD [9]) can be well implemented in the CNN architecture [18]. In this paper, we address the improvement of the existing CNN method [17]. We have tested different multigrid models and compared segmentation results. The main reason for this research is to find proper implementation of the CNN-MRF technique on CNNs taking into consideration the abilities of today's and future's VLSI CNN systems.

1 Introduction

Herein, we present segmentation techniques based on a classical MRF model described in [6]. Each process is implemented in the CNN architecture [4,14]. Our improvement can help to produce high-level parallel VLSI imaging chips in the CNN environment.

First, we give some details about the general MRF segmentation model and the optimisation method we used (MMD [9]). Then, we explain how this model was modified to meet the constraints of the CNN. Although the existing CNN-MRF model already gives good segmentation results, as shown in some previous papers [18,17], it needs 2nd order neighbourhood connectivity system what can be technologically costly to be implemented in VLSI hardware. As it is in the forthcoming, reducing neighbourhood connectivity results in a lost of precision. To achieve good results with smaller connectivity, multiscale techniques are introduced. We also use anisotropic diffusion [13] to improve segmentation quality as reported in a previous paper [16]. The implementation of anisotropic diffusion can also be done within the framework of the CNN Universal Machine (CNN-UM [14]).

2 Image Segmentation with a MRF Model

Many of the early vision tasks can be considered as a *labelling* process during which each pixel should be marked with a new, optimal value according to a cost function constructed from the observed data, a priori information and some well defined constraints. Such an approach is usually conducted within a Bayesian framework. The conventional criteria are Maximum A Posteriori, Marginal Posteriori Mode etc. The use of a MRF model is well suited for many image segmentation problems.

2.1 The MRF Segmentation Model

The task of segmentation can be decomposed in the following way: we are looking for the labelling of pixels (MRF) which is the most probable taking into account that there are initial image and some a priori conditions (e.g. to obtain homogeneous regions in the image).

Considering Bayes theorem and conditionally independence hypothesis, the most likely labelling ($\hat{\omega}$) is given by the MAP estimator:

$$\hat{\omega} = \arg \max_{\omega \in \Omega} \prod_{s \in S} P(f_s | \omega_s) \prod_{C \in \mathcal{C}} \exp(-V_c(\omega_c)),$$

Here f is our observation, $s \in S$ is a site and S is the set of all sites, while Ω is the set of possible configurations, C is a clique and $\mathcal{C}$ is the set of all cliques. $\exp(-V_c(\omega_c))$ comes from the Hammersley-Clifford theorem establishing connection between probability distribution and the energy of local configurations (for more details see [2]). The energy of a configuration is expressed in the following way [6,10]:

$$U(\omega, f) = U_1(\omega, f) + U_2(\omega)$$

$$U_1(\omega, f) = \sum_{s \in S} \left(\ln(\sqrt{2\pi}\sigma_{\omega_s}) + \frac{(f_s - \mu_{\omega_s})^2}{2\sigma_{\omega_s}^2} \right)$$

$$U_2(\omega) = \sum_{c \in \mathcal{C}} V_2(\omega_c),$$

where $V_2(\omega_C) = V_{\{s,r\}}(\omega_s, \omega_r)$ is equal to $-\beta$ if $\omega_s = \omega_r$, and it is $+\beta$ if $\omega_s \neq \omega_r$. β is a model parameter controlling the homogeneity of the regions. μ_{ω_s} and σ_{ω_s} are respectively the mean and the variance of a class ω_s. If these parameters are known, then the segmentation process is supervised. There are also techniques for estimating these parameters in case of unsupervised segmentation.

In Section 3, we will see that our CNN model uses local characteristics (variance and mean in a pixel's neighbourhood) of the image enabling unsupervised segmentation. During optimisation, ω_s is changed to look for labelling values corresponding to classes which satisfy our criterion leading to an equilibrium state.

2.2 Optimisation: Modified Metropolis Dynamics (MMD, [9])

The result of [9,10] supports the use of a MRF in a parallel, simple-cell VLSI architecture. To solve the optimisation problem, we have chosen the MMD algorithm [9]. The most important feature of MMD is that it is a pseudo-stochastic process with two phases:

- a pseudo-stochastic phase: similarly to SA [11], energy increase is allowed during optimisation
- a deterministic phase: if temperature reaches a certain limit, only configurations of lower energy are accepted.

This gives us a useful feature: by setting the initial temperature and the number of iteration steps, both deterministic and pseudo-stochastic phases can be performed. In our model, we have chosen MMD optimisation because of this favourable feature and its relative ease of implementation in the CNN environment. More details can be found in [17,18].

3 Adaptation of MRF Models to CNN Architecture [18]

Since during MRF based image segmentation a huge amount of simultaneous computations are needed, a possible architecture of CNN-UM [14] is suitable to carry out these calculations.

The CNN-UM model helps to implement an unsupervised image segmentation method. The main point of this CNN-MRF model is that local image properties are calculated (σ_s, μ_s) instead of the statistical characteristics of the different classes (σ_{ω_s}, μ_{ω_s}). To calculate these local features a smoothed image is generated from the original one. In our model, we suppose that the original pixel values are probably around the average of the observed (noisy) and their smoothed values. The expected value (μ_s) will be the average (or somewhere in-between) of the smoothed and observed pixel, while the standard deviation (σ_s) will be the half of the difference of these two values.

This model requiring a few type of operations fits the CNN-UM architecture and has already been tested with MMD optimisation; a more detailed description can be found in [18]. The high speed of VLSI CNN implementation can be exploited by the parallelism of the optimisation process.

According to [1], we know that the algorithm will converge as long as less than 100% of the pixels are updated at the same time. Experiments show that this criterion is accomplished with high probability and leads to a highly parallel system.

4 The Need for Multigrid Models

Nowadays multiresolution, multiscale and hierarchical approaches are widely applied in the field of image processing. Markov Random Fields modelling is one typical area where the advantages of such techniques seem to be tremendous. A review of this widespreading area is given in [7].

Multigrid models are interesting for MRF modelling, because during the optimisation process they smooth local minima, resulting in faster convergence and becoming less sensitive to initial configurations.

A common feature of all multigrid models is the representation of image models on several levels with decreasing resolution. There can be significant differences in the definition of cliques and energy functions in the different approaches.

As we have already mentioned, the CNN-MRF model with 2nd order neighbourhood system can be technologically costly to be implemented in VLSI and the first order systems do not give satisfactory results. That is why we tried out multigrid methods to overcome this problem with a system of first order neighbourhood connectivity. This solution has still the general advantages mentioned above.

To demonstrate the different segmentation abilities we show some segmentation results [16,17,18] on the artificial noisy image of [10]. In Fig. 1 (a) the input noisy image, in (b), (c) and (d) the different segmented images can be seen. In [9,10] the segmentation error was 1.0% using the original Metropolis algorithm [12] and it was 1.3% using the MMD method.

We have tested some multigrid models appropriate for the CNN implementation. In the followings, the Multiscale MRF model will be sketched up briefly, with its CNN adaptation examined. Two implementations of the Multiscale MRF Model will be shown.

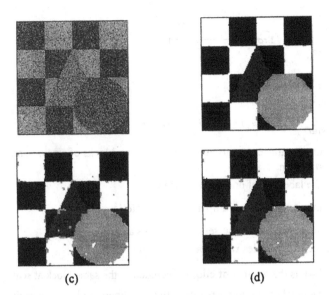

(c)   (d)

Fig. 1. Segmentation with different neighbourhood systems in monogrid CNN structures

(a) observed image, (b) segmented image using second order connectivity (12 neighbours, misclassification error:1.5%), (c) first order connectivity (4 neighbours, error: 5.4%), (d) first order connectivity with anisotropic diffusion pre-processing (error: 3.7%) (see [16])

4.1 Multiscale MRF Model

The following multiscale model has been introduced by Perez at al. in [8]. Now, we are going to give only a very short description of this model, but still to be able to apply its main idea.

In this model the different layers are built up by reducing the resolution by 2 at each level of the label-pyramid. During optimisation there are no interactions between the scales except for the fact that the initial state of one level is always generated from the final state of the corresponding coarser scale. Also, an important feature of this model is the calculation of clique potentials: at a given scale it is done through the cliques of the finest scale: always the clique potentials of the finest scale are computed but with the restriction that there are distinct blocks of n^{2i} sites with the same value (n is the re-scale ratio, i is the level). Cliques of the finest scale (corresponding to a certain level) are partitioned into two sets: one set contains the cliques which are included in any block of the level (i) currently under investigation whereas the other set of cliques contains those which sit astride any two neighbouring blocks of the actual level (i). The clique potential of the given level is the sum of these clique potentials of the finest scale not forgetting about the fact that one site at the actual scale (level i) is equivalent to n^{2i} sites at level 0 (since $n*n$ sites build up a block on the level above).

The following form of energy components at level i represents this idea:

$$U_1^i(\xi^i, F) = \sum_{s^i \in S^i} V_1^i(\xi_{s^i}^i, F)$$

where

$$V_1^i(\xi_{s^i}^i, F) = \sum_{s \in b_{s^i}^i} (\log(\sqrt{2\pi}\sigma_{\omega_s}) + \frac{(f_s - \mu_{\omega_s})^2}{2\sigma_{\omega_s}^2}) - p^i \beta$$

and

$$U_2^i(\xi^i) = \sum_{C^i = (r^i, s^i) \in \mathbf{C}^i} V_2^i(\xi_{C^i}^i)$$

$$V_2^i(\xi_{C^i}^i) = \sum_{(r,s) \in D_{C^i}} V_2(\omega_r, \omega_s) = -q^i \beta \text{ if } \omega_r = \omega_s \text{ and } +q^i \beta \text{ if } \omega_r \neq \omega_s$$

Some explanations with respect to notations assuming blocks of size $n*n$ and a first order neighbourhood system:

- $\xi_{s^i}^i$ means the labelling of one block at scale i

- $s \in b_{s^i}^i$ means sites which build up a block at scale i

- $\mathbf{C}^i$ is one clique, while $\mathbf{C}^i$ is the set of all cliques at scale i

- D_{C^i} is the set of those sites which build up clique C^i

- $p^i = 2n^i(n^i - 1)$ is the number of cliques included in the same block at scale i

- $q^i = n^i$ is the number of cliques between two neighbouring blocks at scale i

See Fig. 2 representing the 0th and 1st scale for easier understanding.

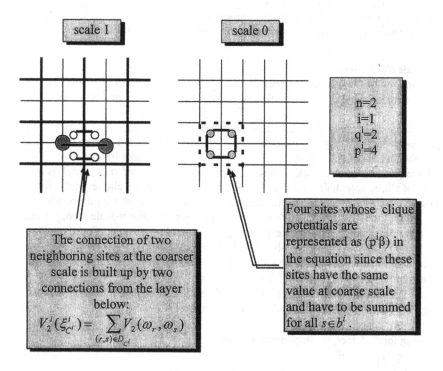

scale 1   scale 0

n=2
i=1
q^i=2
p^i=4

The connection of two neighboring sites at the coarser scale is built up by two connections from the layer below:

$$V_2^i(\xi_{C^i}^i) = \sum_{(r,s) \in D_{C^i}} V_2(\omega_r, \omega_s)$$

Four sites whose clique potentials are represented as (p^i β) in the equation since these sites have the same value at coarse scale and have to be summed for all $s \in b^i$.

Fig. 2. The two kinds of cliques in the Multiscale Model

The segmentation algorithm consists of two main steps:
- energy optimisation of a layer
- initialisation of the next layer from the coarser grid above.

5 The Adaptation of Multigrid Models to the CNN

Multigrid models can be implemented in a one-layer CNN-UM VLSI structure using global cell-organising, local grouping and memory-transfers. The main problem is how to compress the multigrid effects into a complex local (small-neighbourhood) process.

First we tested a simplified version of the Causal Hierarchical Model [3]. Here a label-pyramid of decreasing number of sites is built up and sequential MAP estimation is used during optimisation. In case of 4 levels and only vertical neighbourhood. we can achieve a very fast and low-complexity architecture. However, this simplified structure results in higher segmentation error.

In the followings we show two implementations of a multiscale model. They need higher VLSI complexity but smaller neighbourhood than in the case of the monogrid model [17,18].

5.1 CNN-Multiscale Model

In the above simplified Causal Hierarchical Model, no neighbourhood system was taken into account in the energy computation. The following multiscale algorithm uses a first order neighbourhood system. Now, there is no direct interaction between two scales, the connection is kept by the initialisation step and by clique potentials as explained before: clique potentials and the energy component representing the observation are always calculated considering the cliques of the finest scale.

There are two possible algorithmic implementations of this model. We have tested both of them:
1. One is considering the model as a system with several scales where cliques are originating from the finest layer. Only the current image scale is represented in the memory.
2. In the other implementation, we tried to implement what is behind the formulae: we had only one scale and instead of building up a coarser (smaller) one, we restricted the values so as to be the same in each 2x2 block.

Since both are based on the same theory, we got similar results during testing. However. they have different complexity and different computational time. Table 1 contains characteristic properties of the tested models (Multiscale 1 and 2). Besides these parameters, there are more constraints to satisfy when implementing the models on the CNN-UM and those structural limits of the CNN-UM may determine which model to use. In both models, MMD optimisation was used with several adjustable parameters. Fig. 3 demonstrates the process of the segmentation. In Fig. 4, we can see a real-image segmentation. The input image (a) is a part of an image from Airborne Multisensor Pod System Data Access Catalogue (http://info.amps.gov:2080/). Segmentation was done with a 2 level multiscale method at 2*80 iteration steps.

5.2 Characteristics of Implemented Models

Table 1 roughly summarises the capacities needed by the two multiscale and the monogrid models. To sum up the results, multiscale models need more memories per cell but use smaller neighbourhood connectivity than the monogrid model.

351

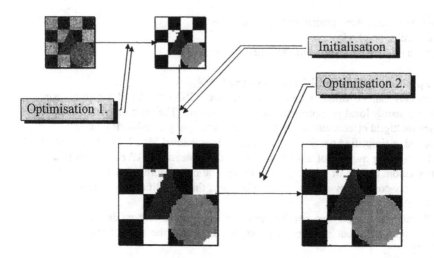

Fig. 3. The results of a segmentation process of two levels: Error: 1.86%

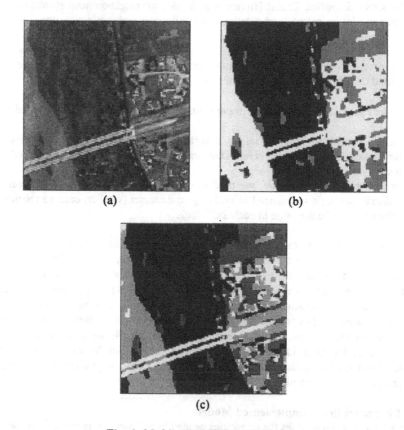

(a)

(b)

(c)

Fig. 4. *Multiscale MRF segmentation*
(a) Air view of a scene with river, bridge, forest, green area and town (from left to right).
Rio Grande, New Mexico. Segmented into: (b) 3 regions, (c) 4 regions.

	Monogrid	Multiscale 1.		Multiscale 2.
Memories per cell	8	15		10
Operations/Iteration	8	coarser: 24: finest scale: 8		13
Processed image size	N*N	c.s.: (N/2)x(N/2); f.s.: N*N		N*N

Table 1. Some characteristics of the models

6 Conclusion

Our main expectation about multigrid implementations has been fulfilled: to reduce the necessary neighbourhood connectivity of the CNN-MRF model at comparative segmentation precision and operations/iteration ratio.

All these trials (the combination of probabilistic, anisotropic and multiscale techniques with CNN) may lead to a new approach to solve general early vision problems in real VLSI architectures.

References.

1. R. Azencott: Parallel Simulated Annealing: Parallelization techniques, Wiley, **1992**.
2. J. Besag: Spatial interaction and the statistical analysis of lattice systems, *Jl. Roy. Statis. Soc. B. 36.*, **1974**.
3. C. A. Bouman, M. Saphiro: A Multiscale Random Field Model for Bayesian Image Segmentation, *IEEE Trans. Image Proc.*, Vol. 3, pp. 162-177, **1994**.
4. L.O. Chua, L. Yang: Cellular Neural Networks, *IEEE Trans. on Circuit and Systems*, Vol. 35, pp. 1257-1295, **1988**.
5. S. Espejo, R. Carmona, R. Dominguez-Castro, A. Rodriguez-Vazquez: A CNN Universal Chip in CMOS Technology, *Int. J. of Circuit Theory and Appl.*,Vol. 24, pp. 93-110, **1996**.
6. S. Geman and D. Geman: Stochastic relaxation, Gibbs distribution and the Bayesian restoration of images, *IEEE Trans. on Pattern Analysis and Machine Intelligence*, Vol.6, pp.721-741, **1984**.
7. C. Graffigne, F. Heitz, P. Pérez, F. Preteux, M. Sigelle, J. Zerubia: Hierarchical Markov random field models applied to image analysis, SPIE Conf., San Diego, July 10-11 **1995**.
8. F. Heitz, P. Perez, P. Bouthemy: Multiscale minimisation of global energy functions in some visual recovery problems, *CVGIP: Image Understanding*, Vol.59. pp. 125-134, Jan. **1994**.
9. Z. Kató, J. Zerubia, M. Berthod: Satellite image classification using a modified Metropolis dynamics, *Proc. ICASSP*, San Francisco, March **1992**.
10. Z. Kató, J. Zerubia, M. Berthod: A Hierarchical Markov Random Field Model and Multitemperature Annealing for Parallel Classification. *Graphical Models and Image Processing*, Vol. 58, No. 1, pp. 18-37, January, **1996**.
11. S. Kirkpatrick, C. Gellatt, M. Vecchi: Optimisation by simulated annealing, *Science* 220, pp. 671-690, **1983**.
12. N. Metropolis, A.W. and M.N. Rosenbluth, A.H. and E. Teller: Equation of State Calculations by Fast Computing Machines, *J. of Chemical Physics*, Vol.21, No.6, pp.1087-1092, **1953**.
13. P. Perona, J. Malik: Scale space and edge detection using anisotropic diffusion, *IEEE Tr. Pattern Analysis amd Machine Int.*,Vol.12,No.7, pp. 629-639,**1990**.
14. T. Roska, L. O. Chua: The CNN Universal Machine: An Analogic Array Computer. *IEEE Trans. on Circuits and Systems-II*, Vol. 40, pp. 163-173, March **1993**.
15. T. Roska, T. Szirányi: Classes of analogic CNN algorithms and their practical use in complex image processing tasks, *Proc. IEEE Nonlinear Signal and Image Proc.*, pp.767-770, **1995**.
16. T. Szirányi, L. Czúni: Picture Segmentation with introducing an Anisotropic Preliminary Step to an MRF Model with Cellular Neural Networks, *Proc. of the ICPR*, IEEE, Vienna, **1996**.
17. T. Szirányi, J. Zerubia: Markov Random Field Image Segmentation using Cellular Neural Network, IEEE Tr. Circuits and Systems I., Vol. 44, pp. 86-89, January, **1997**.
18. T. Szirányi, J. Zerubia, D. Geldreich, Z. Kato: Cellular Neural Network in Markov Random Field Image Segmentation, *CVNA '96, IEEE, Seville*, **1996**.

Computing Stochastic Completion Fields in Linear-Time Using a Resolution Pyramid

Lance Williams*   Tairan Wang†   Karvel Thornber*

*NEC Research Institute
4 Independence Way
Princeton, NJ 08540

†Department of Physics
Massachusetts Institute of Technology
Cambridge, MA 02139

Abstract. We describe a linear-time algorithm for computing the likelihood that a completion joining two contour fragments passes through any given position and orientation in the image plane. Our algorithm is a resolution pyramid based method for solving a partial differential equation characterizing a distribution of short, smooth completion shapes. The PDE consists of a set of independent advection equations in (x, y) coupled in the θ dimension by the diffusion equation. A previously described algorithm used a first-order, explicit finite difference scheme implemented on a rectangular grid. This algorithm required $O(n^3 m)$ time for a grid of size $n \times n$ with m discrete orientations. Unfortunately, systematic error in solving the advection equations produced unwanted anisotropic smoothing in the (x, y) dimension. This resulted in visible artifacts in the completion fields. The amount of error and its dependence on θ has been previously characterized. We observe that by careful addition of extra spatial smoothing, the error can be made totally isotropic. The combined effect of this error and of intrinsic smoothness due to diffusion in the θ dimension is that the solution becomes smoother with increasing time, i.e., the high spatial frequencies drop out. By increasing Δx and Δt on a regular schedule, and using a second-order, implicit scheme for the diffusion term, it is possible to solve the modified PDE in $O(n^2 m)$ time, i.e., time linear in the problem size. Using current hardware and for problems of typical size, this means that a solution which previously took one hour to compute can now be computed in about two minutes.

1 Introduction

The problem of computing the shape of the contour joining a pair of boundary fragments was first examined by Ullman in 1976. It has since become conventional wisdom that the shape of this contour is described by the curve of least energy, i.e., the curve which minimizes a functional of the form, $E = \int_{\Gamma} (\alpha \kappa^2(s) + \beta) \, ds$ (see Horn[4]). The curve of least energy can be regarded as a maximum likelihood estimate of the shape of the completion. However, humans do not experience a sharply defined, well localized illusory contour when presented with a stimulus such as the Ehrenstein figure or Kanizsa Triangle (see Figure 1). In a recent paper, Williams and Jacobs[9] argue that our perception more closely resembles the distribution of possible shapes and not simply the most likely shape. According to this view, the degree of sharpness is related to the variance of this distribution. In an earlier paper, Mumford[5] proposed that the distribution of completion shapes could be modeled as the set of paths followed by particles traveling with constant speed in directions described by Brownian motions. He shows that the maximum likelihood paths followed by such particles are curves of least energy and gives a partial differential equation (PDE) which describes the evolution in time of the probability density function describing a particle's position and direction, i.e., a *Fokker-Planck equation*. Williams and Jacobs[10] subsequently proposed a neural model[1] of illusory contour shape, salience and sharpness based on a finite difference scheme for integrating this PDE. Although the dynamics of this model are consistent with known human visual psychophysics, the algorithm is fairly slow, requiring over an hour for a problem of typical size on a modern workstation. Being able to solve this PDE efficiently will allow it to be applied profitably to problems in computer vision, e.g., to the problem of identifying smooth, closed shapes amidst background clutter.

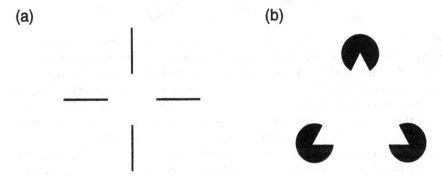

Fig. 1. (a) Ehrenstein Figure. **(b)** Kanizsa Triangle.

2 Stochastic Completion Fields

Given 1) a set of position and orientation constraints (a.k.a. *keypoints*)[2] representing the beginning and ending points of a set of contour fragments; and 2) a probability distribution of completion shapes; the magnitude of the *stochastic completion field* at (u, v, ϕ) is the probability that a completion from this distribution will pass through (u, v, ϕ) on a path joining two of the contour fragments.

---

[1] Other models of illusory contour formation are described by Grossberg and Mingolla[1], Guy and Medioni[2] and Heitger and von der Heydt[3].

[2] We adopt this term used in [3]. A keypoint can represent either of the two orientations at a corner or the normal orientation at a line termination. These are points where it is likely that one surface occludes the boundary of another.

The probability distribution of completion shapes is modeled as the set of paths followed by particles traveling with constant speed in directions described by Brownian motions. Williams and Jacobs[9] showed that the stochastic completion field could be factored into a *source field* and a *sink field*. The source field, $p'(u, v, \phi)$, represents the probability that a contour beginning at (x_p, y_p, θ_p) will pass through (u, v, ϕ) and the sink field, $q'(u, v, \phi)$, represents the probability that a contour beginning at (u, v, ϕ) will reach (x_q, y_q, θ_q) (where p and q are keypoints).

Given a probability density function describing a set of particles' positions and directions at time zero, $p(u, v, \phi; 0)$, the probability density function describing their positions and directions at time t is computed by integrating the Fokker-Planck equation described by Mumford[5]:

$$p(u, v, \phi; t') = p(u, v, \phi; 0) + \int_0^{t'} dt \, \frac{\partial p(u, v, \phi; t)}{\partial t}$$

$$\frac{\partial P}{\partial t} = -\cos\theta \frac{\partial P}{\partial x} - \sin\theta \frac{\partial P}{\partial y} + \sigma^2/2 \frac{\partial^2 P}{\partial \theta^2} - 1/\tau P$$

where $P = p(u, v, \phi; t)$. This PDE can be viewed as a set of independent *advection* equations in (x, y) (the first and second terms) coupled in the θ dimension by the *diffusion* equation (the third term). The effect of the advection equations is to translate probability mass in the θ direction with unit speed. The diffusion term models the Brownian motion in direction. Finally, the effect of the fourth term is that a constant fraction of particles decay per unit time. This represents our prior expectation on the length of gaps—most are quite short.

The algorithm for computing the source field described by Williams and Jacobs is based on a first order, explicit scheme for integrating the Fokker-Planck equation:

STEP 1: $p_{x,y,\theta}^{t+1/4} = p_{x,y,\theta}^t - \cos\theta \cdot \begin{cases} p_{x,y,\theta}^t - p_{x-\Delta x,y,\theta}^t & \text{if } \cos\theta > 0 \\ p_{x+\Delta x,y,\theta}^t - p_{x,y,\theta}^t & \text{if } \cos\theta < 0 \end{cases}$

STEP 2: $p_{x,y,\theta}^{t+2/4} = p_{x,y,\theta}^{t+1/4} - \sin\theta \cdot \begin{cases} p_{x,y,\theta}^{t+1/4} - p_{x,y-\Delta y,\theta}^{t+1/4} & \text{if } \sin\theta > 0 \\ p_{x,y+\Delta y,\theta}^{t+1/4} - p_{x,y,\theta}^{t+1/4} & \text{if } \sin\theta < 0 \end{cases}$

STEP 3: $p_{x,y,\theta}^{t+3/4} = \lambda \, p_{x,y,\theta-\Delta\theta}^{t+2/4} + (1 - 2\lambda) \, p_{x,y,\theta}^{t+2/4} + \lambda \, p_{x,y,\theta+\Delta\theta}^{t+2/4}$

STEP 4: $p_{x,y,\theta}^{t+4/4} = e^{-\frac{1}{\tau}} \cdot p_{x,y,\theta}^{t+3/4}$

where $\lambda = \sigma^2/2(\Delta\theta)^2 < 1/2$, and $\Delta t = \Delta x = \Delta y = 1$. The four steps correspond to the four terms of the PDE. The first two steps employ *upwind differencing* to ensure stability (see [6]). The third step is stable when $\lambda < 0.5$ and the fourth step is unconditionally stable. By repeating the above four steps, $p(x, y, \theta; t)$ can be computed for increasing values of t. The source field,

Fig. 2. Demonstration of the anisotropic nature of the advection error and its correction. Left: The initial condition consists of eight impulses uniformly spaced around the circumference of a circle and in directions tangent to the circle. For illustration purposes, there is no diffusion in θ and no decay. After 12 time-steps, the mass traveling in the 45°, 135°, 225° and 315° directions is noticeably dispersed. The mass traveling in the 0°, 90°, 180° and 270° directions remains concentrated. Right: Advection error after correction.

$p'(x, y, \theta) = \int_0^\infty dt\, p(x, y, \theta\,;t)$, is computed using the following recurrence equation:

$$p'(x, y, \theta) \leftarrow p'(x, y, \theta) + p(x, y, \theta; t)$$

Since the amount of remaining probability mass, $\int \int \int dx\, dy\, d\theta\, p(x, y, \theta; t) = e^{-t/\tau}$, is typically sufficiently small for $t > n$, the time complexity of this method is $O(n^3 m)$.

Due to the way in which the advection equations are finite-differenced on a rectangular grid, the above method introduces additional non-isotropic, spatial smoothing (see Figure 2). It is straightforward to show that after one time-step, the expected values and variances of a particle's position (with respect to its previous position) are given by the following expressions:

$$< x(\theta) > = \cos\theta, \quad < y(\theta) > = \sin\theta$$
$$\sigma^2_{xerr}(\theta) = \cos\theta(1 - \cos\theta), \quad \sigma^2_{yerr}(\theta) = \sin\theta(1 - \sin\theta)$$

Note that this error is highly non-isotropic—σ^2_{xerr} has a minimum value of 0.0 when $\theta = 0°$ and a maximum value of 0.25 when $\theta = 45°$. This means that the PDE which Williams and Jacobs actually solve more closely resembles:

$$\frac{\partial P}{\partial t} = -\cos\theta\frac{\partial P}{\partial x} - \sin\theta\frac{\partial P}{\partial y} + \sigma^2/2\frac{\partial^2 P}{\partial \theta^2} - 1/\tau P + \sigma^2_{xerr}(\theta)/2\frac{\partial^2 P}{\partial x^2} + \sigma^2_{yerr}(\theta)/2\frac{\partial^2 P}{\partial y^2}$$

This non-isotropic advection error in the source and sink fields leads to visible artifacts in the completion fields.

3 A Linear-Time Method

The basic idea underlying our new algorithm is to selectively increase spatial smoothing to make the advection error isotropic and then to increase Δt and Δx on a regular schedule, i.e., as the high spatial frequencies drop out of the solution. Undersampling in this manner will lead to a linear-time algorithm. The modified PDE is:

$$\frac{\partial P}{\partial t} = -\cos\theta\frac{\partial P}{\partial x} - \sin\theta\frac{\partial P}{\partial y} + \sigma^2/2\frac{\partial^2 P}{\partial\theta^2} - 1/\tau P + \sigma_x^2/2\frac{\partial^2 P}{\partial x^2} + \sigma_y^2/2\frac{\partial^2 P}{\partial y^2}$$

where $\sigma_x^2 = \sigma_{xerr}^2(\theta) + \sigma_{xcorr}^2(\theta)$ and $\sigma_y^2 = \sigma_{yerr}^2(\theta) + \sigma_{ycorr}^2(\theta)$. The variances, $\sigma_{xcorr}^2(\theta)$ and $\sigma_{ycorr}^2(\theta)$, are the correction factors needed to make the advection error isotropic. Their values are given by the following expressions:

$$\sigma_{xcorr}^2(\theta) = \frac{1}{4} - \cos\theta(1 - \cos\theta), \quad \sigma_{ycorr}^2(\theta) = \frac{1}{4} - \sin\theta(1 - \sin\theta)$$

which were derived by linear interpolation of the variances for the maximum and minimum error directions. The modified PDE can be solved using the following finite-differencing scheme:

STEP 1: $p_{x,y,\theta}^{t+1/6} = p_{x,y,\theta}^t - \cos\theta \cdot \begin{cases} p_{x,y,\theta}^t - p_{x-\Delta x,y,\theta}^t & \text{if } \cos\theta > 0 \\ p_{x+\Delta x,y,\theta}^t - p_{x,y,\theta}^t & \text{if } \cos\theta < 0 \end{cases}$

STEP 2: $p_{x,y,\theta}^{t+2/6} = p_{x,y,\theta}^{t+1/6} - \sin\theta \cdot \begin{cases} p_{x,y,\theta}^{t+1/6} - p_{x,y-\Delta y,\theta}^{t+1/6} & \text{if } \sin\theta > 0 \\ p_{x,y+\Delta y,\theta}^{t+1/6} - p_{x,y,\theta}^{t+1/6} & \text{if } \sin\theta < 0 \end{cases}$

STEP 3: $p_{x,y,\theta}^{t+3/6} = \lambda_x p_{x-\Delta x,y,\theta}^{t+2/6} + (1 - 2\lambda_x) p_{x,y,\theta}^{t+2/6} + \lambda_x p_{x+\Delta x,y,\theta}^{t+2/6}$

STEP 4: $p_{x,y,\theta}^{t+4/6} = \lambda_y p_{x,y-\Delta y,\theta}^{t+3/6} + (1 - 2\lambda_y) p_{x,y,\theta}^{t+3/6} + \lambda_y p_{x,y+\Delta y,\theta}^{t+3/6}$

STEP 5: $-\lambda p_{x,y,\theta+\Delta\theta}^{t+5/6} + (1 + 2\lambda) p_{x,y,\theta}^{t+5/6} - \lambda p_{x,y,\theta-\Delta\theta}^{t+5/6} =$

$$\lambda p_{x,y,\theta-\Delta\theta}^{t+4/6} + (1 - 2\lambda) p_{x,y,\theta}^{t+4/6} + \lambda p_{x,y,\theta+\Delta\theta}^{t+4/6}$$

STEP 6: $p_{x,y,\theta}^{t+6/6} = e^{-\Delta t/\tau} \cdot p_{x,y,\theta}^{t+5/6}$

where $\lambda = \sigma^2\Delta t/4(\Delta\theta)^2$, $\lambda_x = \sigma_{xcorr}^2(\theta)\Delta t/2(\Delta x)^2$ and $\lambda_y = \sigma_{ycorr}^2(\theta)\Delta t/2(\Delta y)^2$. As long as $\Delta x/\Delta t = \Delta y/\Delta t = 1$, the first two steps will be stable. Since our undersampling schedule ensures that $\Delta x = \Delta t$, this condition will always hold. Because the maximum value of $\sigma_{xcorr}^2(\theta)$ and $\sigma_{ycorr}^2(\theta)$ is 0.25, it follows that $\lambda_x < 0.5$ and $\lambda_y < 0.5$. Consequently the third and fourth steps will always be stable. In order to implement the diffusion in the θ dimension, we have switched from a first-order explicit scheme to a second-order implicit scheme, i.e., the

Crank-Nicholson method (see [6]). With the first-order, explicit scheme, λ was limited to values of less than one-half. However, the Crank-Nicholson method remains stable for the larger values which λ reaches as Δt is increased.

After c time-steps are performed at a given resolution, Δx and Δt are increased by a factor of two. This is accomplished by means of the following projection operation:

$$
\begin{aligned}
p_{i+1}(x,y,\theta;1) \leftarrow [\; & p_i\,(2x-1,2y+1,\theta\,;c) + 2\,p_i\,(2x,2y+1,\theta\,;c) + \\
& p_i\,(2x+1,2y+1,\theta\,;c) + 2\,p_i\,(2x-1,2y,\theta\,;c) + \\
& 4\,p_i\,(2x,2y,\theta\,;c) + 2\,p_i\,(2x+1,2y,\theta\,;c) + \\
& p_i\,(2x-1,2y-1,\theta\,;c) + 2\,p_i\,(2x,2y-1,\theta\,;c) + \\
& p_i\,(2x+1,2y-1,\theta\,;c)]/\,16
\end{aligned}
$$

where i and $i+1$ are successive levels in a resolution pyramid. Using the above strategy, we can efficiently simulate the evolution of the PDE for a sufficient length of time. However, to compute the source field, we must still compute the integral of the probability over all time. Within a given level, the probability can be accumulated as before, using the following recurrence equation:

$$
p_i'(x,y,\theta) \leftarrow p_i'(x,y,\theta) + p_i(x,y,\theta;t)
$$

After c time-steps, $p_i'(x,y,\theta)$ will hold the partial sum for the i-th level:

$$
p_i'(x,y,\theta) = \sum_{t=1}^{c} p_i(x,y,\theta\,;t)
$$

The only thing that remains is to combine the partial sums for each level. This is accomplished using repeated projection operations to push the partial sums down the resolution pyramid:

for $i = L$ downto 1 do

$$
\begin{aligned}
p_{i-1}'(x,y,\theta) \leftarrow p_{i-1}'(x,y,\theta) + [&\tfrac{1}{4}p_i'(\tfrac{x}{2}-1,\tfrac{y}{2}+1,\theta) + \tfrac{1}{2}p_i'(\tfrac{x}{2},\tfrac{y}{2}+1,\theta) + \\
&\tfrac{1}{4}p_i'(\tfrac{x}{2}+1,\tfrac{y}{2}+1,\theta) + \tfrac{1}{2}p_i'(\tfrac{x}{2}-1,\tfrac{y}{2},\theta) + p_i'(\tfrac{x}{2},\tfrac{y}{2},\theta) + \\
&\tfrac{1}{2}p_i'(\tfrac{x}{2}+1,\tfrac{y}{2},\theta) + \tfrac{1}{4}p_i'(\tfrac{x}{2}-1,\tfrac{y}{2}-1,\theta) + \tfrac{1}{2}p_i'(\tfrac{x}{2},\tfrac{y}{2}-1,\theta) + \\
&\tfrac{1}{4}p_i'(\tfrac{x}{2}+1,\tfrac{y}{2}-1,\theta)]/4
\end{aligned}
$$

endo

Some care is required. We have found that simpler projection schemes (e.g., uniform weighting and/or non-overlapping windows) produced noticeable pixelization effects. Stochastic completion fields for a Kanizsa triangle computed using the method of Williams and Jacobs[10] and the linear-time method described in this paper are shown in Figure 3.

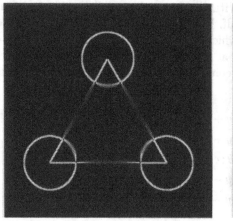 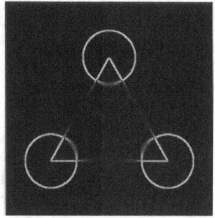

Fig. 3. Left: Stochastic completion field for Kanizsa Triangle computed using method of Williams and Jacobs[9]. The completion field magnitude is summed over 36 discrete orientations and superimposed on the brightness gradient magnitude image for illustration purposes. The image is of size 256×256, $\sigma^2 = 0.005$ and $\tau = 20$. Right: The same, but computed using the linear-time method described in this paper. Twelve steps were performed at each level of the resolution pyramid, i.e., $c = 12$.

4 Time-Complexity

Given an $n \times n$ image with m discrete orientations, i.e., a problem of size $n^2 m$, the number of levels in the resolution pyramid is:

$$L = 1 + \log_2(n/c)$$

and the cost at the i-th level to perform c time-steps is:

$$c \cdot \frac{n^2 m}{4^{i-1}}$$

It follows that the total cost of running the PDE forward in time is:

$$c \cdot (1 + \frac{1}{4} + ... + \frac{1}{4^{L-1}}) \cdot n^2 m$$

The total cost of pushing the partial sums for each level down the resolution pyramid is:

$$(\frac{1}{4^{L-1}} + ... + \frac{1}{4} + 1) \cdot n^2 m$$

Consequently, the total amount of time required by both stages is:

$$(c+1) \cdot (1 + \frac{1}{4} + ... + \frac{1}{4^{L-1}}) \cdot n^2 m < \frac{4}{3}(c+1) \cdot n^2 m$$

so that the time-complexity of the new algorithm is $O(n^2 m)$, i.e., linear in the problem size. In practice, we have observed that problems which previously took over an hour to finish ($n = 256$ and $m = 36$), can now be computed in about 2 minutes.

5 Conclusion

In two recent papers, Williams and Jacobs[9, 10] have described a representation of visible and occluded image contours called a stochastic completion field. The stochastic completion field is based on the idea that the distribution of contour shapes can be modeled by the paths of particles traveling with constant speed in directions described by Brownian motions. The algorithm described in the more recent paper[10] is based on a simple finite-differencing scheme for integrating the partial differential equation given by Mumford[5]. The time-complexity of the algorithm described in [10] is $O(n^3 m)$ (for an $n \times n$ image with m discrete orientations). Their solution also introduced unwanted, anisotropic smoothing which resulted in noticeable artifacts in the completion fields. In our approach, we carefully add smoothing so as to make the error isotropic. We then decrease the spatial and temporal sampling rate as the high spatial frequencies drop out of the evolving solution. The result is an $O(n^2 m)$ method. In practical terms, the previous algorithm took over an hour to produce an answer (for a problem of size $256 \times 256 \times 36$) and the new method takes about 2 minutes.

Acknowledgments The authors wish to thank Achi Brandt for suggesting the use of a multi-resolution method. Joachim Weickert also offered helpful comments.

References

1. Grossberg, S., and Mingolla, E., Neural Dynamics of Form Perception: Boundary Completion, Illusory Figures, and Neon Color Spreading, *Psychological Review* **92**, pp. 173-211, 1985.
2. Guy, G. and G. Medioni, Inferring Global Perceptual Contours from Local Features, *Intl. Journal of Computer Vision* **20**, pp. 113-133, 1996.
3. Heitger, R. and von der Heydt, R., A Computational Model of Neural Contour Processing, Figure-ground and Illusory Contours, *Proc. of 4th Intl. Conf. on Computer Vision*, Berlin, Germany, 1993.
4. Horn, B.K.P., The Curve of Least Energy, MIT AI Lab Memo No. 612, MIT, Cambridge, Mass., 1981.
5. Mumford, D., Elastica and Computer Vision, *Algebraic Geometry and Its Applications*, Chandrajit Bajaj (ed.), Springer-Verlag, New York, 1994.
6. Press, W.H., Flannery, B.P., Teukolsky, S.A., and W.T. Vetterling, *Numerical Recipes in C*, Cambridge University Press, 1988.
7. Strang, G. *Introduction to Applied Mathematics*, Wellesley-Cambridge Press, Cambridge, MA, 1986.
8. Ullman, S. Filling-In the Gaps: The Shape of Subjective Contours and a Model for their Generation, *Biological Cybernetics* **21**, pp. 1-6, 1976.
9. Williams, L.R., and D.W. Jacobs, Stochastic Completion Fields: A Neural Model of Illusory Contour Shape and Salience, *Neural Computation* **9**(4), pp. 837-858, 1997.
10. Williams, L.R., and D.W. Jacobs, Local Parallel Computation of Stochastic Completion Fields, *Neural Computation* **9**(4), pp. 859-881, 1997.

A Bayesian Network for 3d Object Recognition in Range Data

B. Krebs, M. Burkhardt and F. M. Wahl
e-mail:{B.Krebs, M.Burkhardt, F.Wahl}@tu-bs.de

Institute for Robotics and Computer Control,
Technical University Braunschweig,
Hamburger Str. 267, D-38114 Braunschweig, F.R.G.

Abstract. Introducing general CAD descriptions in object recognition systems has become a major field of research called CAD based vision (CBV). However, the major problem using free-form object descriptions is how to define recognizable features which can be extracted from sensor data. In this paper we propose new methods for an extraction of 3d space curves from CAD models and from range data. Object identification is performed by correlating feature vectors from significant subcurves. However, features relying on differential surface properties tend to be very vulnerable with respect to noise. To cope with erroneous data we propose to model the statistical behavior of the features using a Bayesian network. Thus, providing a robust and powerful CAD based 3d object recognition system.

Key Words:

3d Object Recognition – CAD Based Vision (CBV) – B-Spline Space Curves – Bayesian Networks

1 Introduction

In recent years a new field called *CAD Based Vision* (CBV) introduces usual CAD object representations in 3d recognition systems (e.g. [9, 12]). Extraction of significant features from CAD descriptions and mapping those onto sensor data has become the major problem in this research field. Usually high computational effort is spent to achieve reliable results in sensor data analysis (e.g. [2]). Especially, CAD definitions using free-form surface representations have become a challenge because features relying on differential surface properties tend to be very vulnerable with respect to noise.

To cope with these problems we propose to model the statistical behavior of the features within the recognition process. In the field of uncertain reasoning in expert systems *Bayesian networks* has become very popular (e.g. [6]). In this paper we propose a new approach of CAD based 3d object recognition using Bayesian networks to handle uncertain reasoning about object hypotheses. Bayesian networks do not assume independence among features; rather, they encode the dependencies among features. Hence, not only the statistical behavior of the feature values but also the influence of an observed feature on other features can be modeled.

In our approach we compute features from 3d space curves representing *object rims* separating two adjacent surfaces at a discontinuity. A Bayesian network allows to combine even very noisy features.

2 Related Work

In the most object recognition systems using Bayesian networks the nodes represent subparts of an object [1, 8]. However, observable features need not to coincide with object subparts, i.e. features from the same subpart are not statistical independent as it is assumed by constructing objects from subparts. Another disadvantage is that the significance of a subpart to distinguish between different models within the whole model data base can not be specified.

Furthermore, CAD objects using free-form descriptions can not be constructed from a simple set of primitives. Thus, we propose to build Bayesian networks from various spatial features to form composite hypotheses. Not only the observations of features but also relations between different features provide meaningful clues for object recognition.

Sakar and Boyer proposed the Perceptual Inference Network (PIN) which is a special Bayesian network which propagates not only belief values but also position information [11]. But their methods involve highly complex propagation algorithms. Hence, the PIN is reduced to a tree-like structure and is used for reasoning in 2d gray level images only. We propose to model only relational properties of objects and features in a Bayesian network. The nodes represent subsets of object's hypotheses and the links point from more discriminative to more general subsets. This allows to use efficient well-known propagation algorithms [6, 10]. Since the network represents relational properties of observed features only, it only is used for object hypotheses generation. Objects with a high evidence to be present in the scene are validated in a subsequent step.

3 Bayesian Networks from CAD Models

A Bayesian network is a *Directed Acyclic Graph* (DAC). Each node describes a detected observation in the scene or a hidden cause which may have *caused* the observation, i.e. a specific object. These dependencies between nodes are described by the directed links. A link points from the cause to the dependent node. The probability of a node to have caused the dependent node is computed using Bayes Rule. For each root node, i.e. a node with no parents, *prior probabilities* and for each link *conditional probabilities* have to be specified. Our aim is to design a Bayesian network enabling to integrate information about a set of *elementary features* to hypothesize about a set of target object hypotheses.

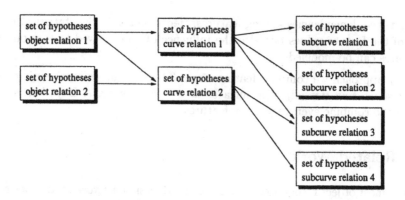

Fig. 1. A part of the two layer Bayesian network is depicted. A root node representing a set of object hypotheses has a link to all curve nodes if the objects share the curve. A curve node represents those object hypotheses which contain a specific curve and are linked to a subcurve node if the subcurve occurs in the curve.

The elementary features serve as an abstract concept to describe observations of object properties in sensor data. Features extracted from geometric models should have strong geometric meaning. Furthermore, they should bind all degrees of freedom to allow efficient computation of an object's location. In real world scenes objects generally overlap. Hence, a feature has to be derived from a local portion of an object's surface. This so called *local feature focus* has been proved to be a robust method in several applications (e.g. [4]).

Using common CAD descriptions 3d objects are defined by a set of surface patches which enclose the entire *interior* of the object. A patch may be a polygon or a free-form Bezier/NURBS surface [3]. The object's topology is described by an $\underline{A}$*ttributed* $\underline{A}$*djacent* $\underline{G}$*raph* (AAG). Each surface represents a node in the AAG and a link occurs where different surface patches share points (e.g. at rims, edges, corners). Thus, a feature can be described as an attributed, named relation and an object as a pair $O = (S, R)$ where S is the set of surfaces $S = \{S_0, \ldots, S_n\}$ and R is the set of relations $R = \{R_0, \ldots, R_m\}$.

Definition: A *feature* is an attributed, named relation with name n:

$$R_i = (n, \{t_{ij}\}, f). \tag{1}$$

The function f provides the values of the relation's attribute which may be either a simple value, a vector or set of values. The set of tuples $\{t_{ij}\}$ contains the elements which are related. Each tuple

$$t_j = (\{S_0, \ldots, S_k\}, T) \in \{t_{ij}\} \tag{2}$$

contains the k adjacent surface patches and a transformation T describing the local position in the model object. The transformation is used in the recognition process to determine the object's location in the scene.

The structure of the Bayesian net can be derived directly from the set of relations R. Relations describing properties which are observable features in the sensor data become leaf nodes, i.e. nodes with no children. These nodes are instantiated when a feature is detected. Objects having this feature as a relation become the parent node of a leaf node. Higher layers of the network are constructed from relations discriminating the sets of object hypotheses. The root nodes, i.e. nodes with no parents, represent the sets of hypotheses which are extracted for the hypotheses validation step. To optimize the internal structure of the net intermediate nodes are constructed by *divorcing* [6].

In industrial applications objects have several corners, edges and rims. Our feature extraction exploits these properties to enable simple and fast matching. A relation between two adjacent surfaces defines a 3d rim curve which separates different surfaces at a discontinuity. All rim curves are subdivided into *significant subcurves*. All subcurves belonging to a curve and all curves belonging to an object define a *belong-to*-relation. Each subcurve and each set of object hypotheses is mapped onto a node in the Bayesian network. The relational dependencies form the links in the network (see Figure 1).

For each subcurve a probability $P(S|C) \in [0,1]$ that S is observable given the presence of C called the *likelihood* $L(C|S)$ must be specified:

$$L(C|S) = P(S|C) = \begin{cases} 1.0 & S \text{ derived from } C \\ \in [0,1] & S \text{ similar to } C \end{cases} \tag{3}$$

Thus, 3d object recognition can be achieved by finding the subcurves in the sensor data and propagating the dependencies in the Bayesian network to get the set of the most likely object hypotheses.

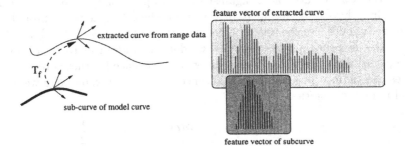

Fig. 2. The correlation of the feature vectors provides the correspondences between Frenet-Frames. Subsequently, The transformation T_f from the subcurve onto the extracted curve can be computed.

4 Determining the Network Probabilities

To complete the network specification we also need to determine the conditional probabilities at the nodes with links to parents. The conditional probability at a node represents the conditional belief that a child node is caused by the parent node, i.e. a set of hypotheses causes the observation of a specific subcurve. Given the likelihood $L(C|S)$ of a subcurve S we have to determine the conditional probabilities for the links to the parent nodes. Since S may occur in several curves $C_S = \{C_0, \ldots, C_n\}$, the node representing S has several parents and the conditional probabilities $P(S|C_S^*)$ for each configuration of subsets from C_S has to be determined. If a subcurve occurs only once in the database the probability of observing the curve equals the probability of observing the subcurve $P(S|C_S^*) = P(S|C)$. But, if the subcurve occurs in many other curves then the impact of the subcurve to select a specific set of hypotheses is small. Nevertheless, if evidence enters the net decreasing the probability of one parent node the probabilities of the other parent nodes are increased. This behavior is known as the "noisy or"-operation [6].

The curve nodes are binary variables $C_0 \ldots C_n$ listing all causes of the binary variable S. Each event $C_i = present$ causes $S = present$ unless an inhibitor prevents it, and the probability for that is

$$q_i = 1 - L(C_i|S). \tag{4}$$

That is, $P(S = not\ present|C_i = present) = q_i$ assuming all inhibitors to be independent. The conditional probabilities are defined by:

$$P(S = not\ present|C_0, \ldots, C_n) = \prod_{j \in Y} q_j \tag{5}$$

where Y is the set of indices for variables in the state present [6].

One of the common arguments against Bayesian probability theory is that prior probabilities have to be specified. In our case we have to specify the probabilities of the root nodes, i.e. the set of object hypotheses $O = \{O_0, \ldots, O_l\}$ which can not be distinguished any further. But we don't consider this to be a drawback since prior knowledge of the probabilities of occurrence for these objects in the scenes can be incorporated. In the absence of any prior knowledge we assume equal prior probabilities:

$$P(O_i = not\ present) = P(O_i = present) = \frac{1}{2} \tag{6}$$

The conditional probabilities $P(C|O^*)$ are computed like $P(S|C_S^*)$.

5 3d Curves for Object Identification

All rim curves of the CAD models are subdivided into significant subcurves to achieve a more robust curve matching. A significant subcurve is extracted around a curvature extremum and extends to the next inflection points. Each subcurve is equidistantly sampled, i.e. $ds \approx \Delta s = \|p_{i+1} - p_i\| = const$. At each sample point $p_i \in [p_o, \ldots, p_{n_s}]$ the *Frenet-Frame* F_i is computed [3]:

$$F_i = \left[\mathbf{t}_i = \frac{\dot{p}_i}{\|\dot{p}_i\|} \ , \ \mathbf{b}_i = \frac{\dot{p}_i \times \ddot{p}_i}{\|\dot{p}_i \times \ddot{p}_i\|} \ , \ \mathbf{m}_i = \frac{\dot{p}_i \times \ddot{p}_i}{\|\dot{p}_i \times \ddot{p}_i\|} \right] \tag{7}$$

The curvature $\kappa = \frac{d\alpha}{ds}$ and the torsion $\tau = \frac{d\beta}{ds}$ at each sample point p_i are the angular velocities of $\mathbf{t}$ and $\mathbf{b}$ with

$$d\alpha_i = arccos \left(\frac{< \mathbf{t}_i, \mathbf{t}_{i+1} >}{|\mathbf{t}_i||\mathbf{t}_{i+1}|} \right) \quad , \quad d\beta_i = arccos \left(\frac{< \mathbf{b}_i, \mathbf{b}_{i+1} >}{|\mathbf{b}_i||\mathbf{b}_{i+1}|} \right) \tag{8}$$

each subcurve is uniquely represented by two feature vectors

$$f_{s_\kappa} = [d\alpha_0, \ldots, d\alpha_{n_s}] \quad , \quad f_{s_\tau} = [d\beta_0, \ldots, d\beta_{n_s}] \tag{9}$$

since $ds \approx \Delta s = const$. Range data from an acquired range image is sorted into several scan planes. In each scan plane line segments are approximated. Adjacent end points define the rim points of an object. Through each set of adjacent rim points a 3d B-spline curve is approximated. The 3d B-spline curves extracted from the range data have to be smoothed with respect to the curvature by a knot-shifting algorithm proposed in [3]. Each 3d B-spline curve is sampled like the model curves and the feature vectors are computed with (9). The *similarity* s between a feature vector of a subcurve $f_s = [f_0, \ldots, f_{n_s}]$ and a feature vector of an extracted curve $f_c = [f_0, \ldots, f_{n_c}]$ at a specific index i is defined by

$$s(i, j) = \begin{cases} 1 & (f_s[i] - f_c[i+j])^2 \leq \varepsilon_{max} \\ 0 & else \end{cases} \tag{10}$$

$$s(i) = \sum_{j \in n_{min}} s(i, j) \tag{11}$$

with the threshold ε_{max} for the maximal error between two features. Subsequently, subcurves are identified in the extracted curves by correlating their feature vectors. A subcurve is fitted into an extracted curve by finding the index i_{max} at the maximum of similarity:

$$i_{max} = \max_{i \in [0, n_c - n_s]} s(i) \tag{12}$$

The maximal similarity $s(i_{max})$ is used to determine the *evidence* e with which a subcurve is recognized in the sensor data by

$$e = \frac{s(i_{max})}{\max(n_c, n_s)} \in [0, 1]. \tag{13}$$

Using the correspondences between Frenet-Frames the transformation T_f from the model subcurve onto the extracted curve can be computed by a least square minimization providing the object's location [5] (see Figure 2).

6 Recognition Results with the Bayesian Net

Our 3d object recognition is an iterative process which collects new sensor data until no further evidence can be computed. A robot depths-eye-in-hand-configuration supplies range information of different views based on the coded light approach which is a well-known active triangulation method. Significant

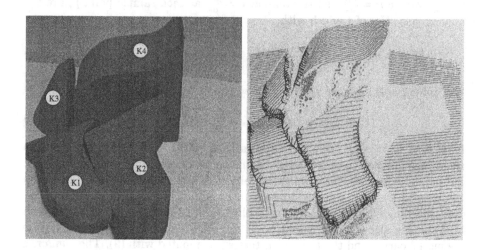

Fig. 3. An example view of a scene with four objects (left Figure) and the ext racted 3d spline curves in the corresponding range image (right Figure).

subcurves are identified in the extracted curves and the evidence is propagated in the Bayesian net implemented in the HUGIN system [10]. If the evidence of a set of hypotheses is high each hypotheses is validated with a *Fuzzy ICP* algorithm [7]. In the right Figure 3 several 3d-space curves are extracted from an

Table 1. Propagation results for a single view

Object	single *belong-to* evidence	no. subcurves	matched
$P(K1 = y)$	0.7634	16	5
$P(K2 = y)$	0.8673	24	6
$P(K3 = y)$	0.6735	17	4
$P(K4 = y)$	0.4534	8	1

range image of a single view. The space curves are divided into significant sub-curves and feature vectors are calculated for each subcurve. This feature vectors are correlated with the feature vectors of the model-subcurves. The correlation evidence e from (13) is instantiated for the corresponding subcurve nodes in the Bayesian net (Figure 1). Table 1 shows the propagation results for the first view (right Figure 3). The object "K2" has the highest evidence subcurves are found because many correct subcurves are found for this object. However, not only the number of matched subcurves but also the quality of the match influences the evidence values, i.e. "K1" and "K2" have high evidence too. A evidence about 0.5 is not significant, i.e. the object "K4" can only be identified after further views.

This clearly shows that modeling the statistical behavior with a Bayesian network can handle even very noisy features. The proposed methods provide a very efficient way for CAD based recognition in range data. All computed features carry 3d information which allows to compute object positions very easily. Hence, features from 3d rim curves provide sufficient information to allow identification and pose estimation of industrial CAD models. The algorithms have been used and tested elaborately to validate their robustness in real world environments.

References

1. D. M. Chelberg. Uncertainty in interpretation of range imagery. In *Proc. International Conference on Computer Vison, Osaka, Japan*, pages 634–657, 1990.
2. C. Dorai and A. K. Jain. Recognition of 3-d free-form objects. In *Proc. International Conference on Pattern Recognition, Vienna, Austria*, 1996.
3. G. E. Farin. *Curves and Surfaces for Computer Aided Geometric Design, a Practical Guide 3rd. ed.* Academic Press, New York, 1993.
4. P. J. Flynn and A. K. Jain. 3d object recognition using invariant feature indexing of interpretation tables. *Computer Vision, Graphics, and Image Processing*, 55(2):119–129, 1992.
5. B. K. P. Horn. Closed-form solution of absolute orientation using unit quaternions. *J. Opt. Soc. of America*, 4(4):629–642, 1987.
6. F. V. Jensen. *An Introduction to Bayesian Networks*. UCL Press, 1996.
7. B. Krebs, P. Sieverding, and B. Korn. A fuzzy icp algorithm for 3d free form object recognition. In *Proc. International Conference on Pattern Recognition, Vienna, Austria*, pages 539–543, 1996.
8. W. B. Mann and T. O. Binford. An example of 3d interpretation of images using bayesian networks. In *DARPA Image Understanding Workshop*, pages 793–801, 1992.
9. J. Mao, A. K. Jain, and P. J. Flynn. Integration of multiple feature groups and multiple views into an 3d object recognition system. In *Proc. CAD-Based Vision Workshop, Champion, Pennsylvania*, pages 184–192, 1994.
10. K. G. Olesen, S. L. Lauritzen, and F. V. Jensen. Hugin: A system creating adaptive causual probabilistic networks. In *Proc. International Conference on Uncertainty in Artificial Intelligence*, pages 223–229, 1992.
11. S. Sakar and K. L. Boyer. *Computing Perceptual Organization in Computer Vision*. World Scientific, 1994.
12. L. G. Shapiro, S. L. Tanimoto, and J. F. Brinkley. A visual database system for data experiment management in model-based computer vision. In *Proc. CAD-Based Vision Workshop, Champion, Pennsylvania*, pages 64–74, 1994.

Improving the Shape Recognition Performance of a Model with Gabor Filter Representation

Peter Kalocsai

University of Southern California
kalocsai@selforg.usc.edu

Abstract

A recognition model which defines a measure of shape similarity on the direct output of multiscale and multiorientation Gabor filters does not manifest qualitative aspects of human object recognition of contour-deleted images in that: a) it recognizes recoverable and nonrecoverable contour-deleted images equally well whereas humans recognize recoverable images much better, b) it distinguishes complementary feature-deleted images whereas humans do not. Adding some of the known connectivity pattern of the primary visual cortex to the model in the form of *extension fields* (connections between collinear and curvilinear units) among filters increased the overall recognition performance of the model and: a) boosted the recognition rate of the recoverable images far more than the nonrecoverable ones, b) increased the similarity of complementary feature-deleted images, but not part-deleted ones, more closely corresponding to human psychophysical results. Interestingly, performance was approximately equivalent for narrow (±15°) and broad (±90°) extension fields.

1 Introduction

A task that both biological and artificial vision systems have to solve is to recover boundaries of objects from many times imperfect, noisy input. The Gestalt grouping principles of co-curvilinearity, proximity, constancy of curvature can help recovering meaningful information under these circumstances. There is considerable evidence from neuroscience [4] and psychophysics [3] that these grouping principles are built into the mammalian visual system in the form of connectivity patterns among processing units. There is both anatomical and physiological evidence that cells with approximately collinear orientation are interconnected mainly by excitatory connections [4]. Psychophysical results seem to suggest a broader field of connections between not only collinear units, but also curvilinear ones [3]. For either the narrow or the broad fields, the excitatory connections reveal smoothly decreasing strength with increasing distance and curvature differences [3,11]. There is also evidence for facilitation (increase in sensitivity for detecting Gabor patches) when local and global orientations are 90 degrees offset (the virtual line connecting two segments is perpendicular to their orientation) which is not modeled here [3,11]. The smoothly decaying excitatory field around an oriented segment is referred to as an *extension field* in this paper. The terms 'association field' or 'stochastic completion field' can be found in the literature to refer to similar constructs. These terms are generally applied to fields considered to manifest broad tuning. The term 'horizontal connections' has been employed to refer to the more narrowly tuned excitatory connections documented for neural units. To compare the effects of both narrow (collinear) and broad (collinear and curvilinear) connectivity patterns among processing units we decided to implement two versions of the extension field: a narrow and a broad one. In the absence of precise neurophysiological data for the strength of connections between collinear and curvilinear units we choose the algorithmic definition of narrow and broad extension fields to be an excitatory gradient +/-15 and +/-90 degrees respectively centered on an oriented segment.

The goal of the present study was to investigate the consequences of adding extension fields to a recognition model that computed shape similarity based on representations of V1 hypercolumn activity. Specifically, we studied whether the extension fields would increase the resemblance of the recognition performance of the model to that shown by humans.

1.1 Brief comparison with previous work

Several previous computer vision models have used extension field type algorithms to guide the grouping process [5,6,7,10,12]. The main contribution in the present effort is the implementation of such a scheme on a biologically plausible multiscale and multiorientation filter representation, roughly similar to that of a lattice of V1 hypercolumns. This representation allowed a measure of shape similarity based on the combined activity produced by both the input image and the grouping process (although this does not necessarily mean that grouping results in activity that is indistinguishable from that produced by the original image). The previously cited efforts did not result in a measure of shape similarity.

Other differences distinguishing the present effort from prior ones was that the latter studies used only one scale as opposed to our multiscale approach. Since our test images were line-drawings, only one scale size--chosen to be the width (frequency) of the lines--could have very well been used, but a multiscale representation better resembles the sampling properties of biological vision systems. Many of the studies in the grouping literature [10,12] used an iterative relaxation algorithm as opposed to the more biologically plausible one-pass operation which was implemented here. An additional feature of the current study is that it directly compared the recognition performance of a grouping model to that of humans on a large number of test images, which is relatively rare in the literature. In the following we will describe two experiments on object recognition and compare human data to the performance of our baseline model.

1.2 Human experimental results and comparison of performance with a baseline model

In a psychophysical experiment [1] equal amount of contour was deleted from line drawings in such a way that the parts were either *recoverable* or *nonrecoverable* as illustrated in Figure 1. Subjects were able to recognize recoverable versions, but not nonrecoverable ones. A model [9] based on the direct output of a number of columns of multiscale and multiorientation Gabor filters (each column is roughly analogous to the simple cells in a V1 hypercolumn) was tested on the same images (this recognition system was originally developed for face recognition and it has achieved high accuracy in recognizing faces from several face databases). The model recognized the nonrecoverable images as well as the recoverable ones, a result that does not correspond to human data (see results later).

In another task, subjects named briefly presented contour-deleted images [2]. For each image, two sets of complementary pairs were created by deleting every other vertex and edge from each simple part in the first set (feature-deleted) and by deleting approximately half the components from each image in the second set (part-deleted) (Figure 2). If the members of the complementary feature-deleted pair or the part-deleted pair were superimposed they would provide an intact image without any overlap in contour.

Members of a complementary feature-deleted image pair (Figure 2, left) were equivalent to each other for human subjects as tested with the priming paradigm [2], but not for the model since the similarity of members of a pair was markedly lower than similarity of one of the images from the pair to itself. Part-deleted complementary images (Figure 2, right) were not equivalent neither to humans nor to the model.

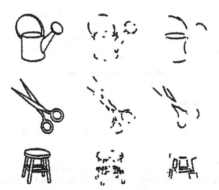

Figure 1. Examples of intact (left), recoverable (middle) and nonrecoverable (right) test images [1].

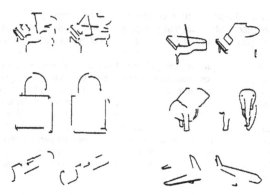

Figure 2. Examples of feature-deleted complementary image pairs (left) and part-deleted complementary image pairs (right). Each member contains approximately half the parts of the object.

2 Additions to the baseline model

The extension field is essentially a probability directional vector-field describing the contribution of a single unit-length edge element to its neighborhood in terms of direction and strength [6]. In other words, it describes the preferred direction and the probability of existence of every point in space to share a curve with the original segment. The field direction at a given point in space is chosen to be tangent to the osculating circle passing through the edge element and that point, while its strength is proportional to the radius of that circle (Figure 3). Also, the strength decays with distance from the origin (the edge segment). The decay of extension field strength is set to be Gaussian for both the proximity and curvature constraints:

$$\overline{EF}(x,\rho) = e^{-Ax^2} e^{-B\rho^2} \qquad (1)$$

where x is the distance along the circular arc and ρ is the curvature of the given arc. Recently, Williams and Jacobs [13] described a very similar type of prior probability distribution of boundary completion based on computing the probability that a particle following a random walk will pass through a given position and orientation on a path joining two edge segments.

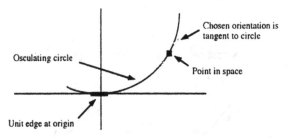

Figure 3. Field direction for every point in space is chosen to be the tangent to the osculating circle passing through the edge segment and the given point.

From each end of an edge segment, an extension field expanded to define a triangular area as shown in Fig. 4. The maximum orientation difference spanned by the broad extension field was ±90°, which were at the ±45° boundaries of the extension field (Figure 4). Beyond those values, the Gaussians for orientation were set to zero so the broad extension fields had zero values above and below the main diagonals, as illustrated in Fig. 4. The narrow extension field is a subset of the broad extension field in that it uses the same direction and strength fields except that the excitation area is limited to ±15° orientation difference. The absence of grouping activity in the regions outside of the extension field merely means that additional information is needed to reconstruct curves between such pairs.

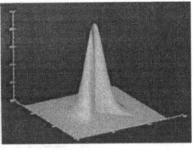

Figure 4. Left panel: The brightness coded directional map of the broad extension fields given a horizontal edge element in the middle. Within the butterfly shaped extension fields, black refers to horizontal and white to vertical orientations. The regions above and below the edge element have no assigned orientation and are shown in black to clearly deliniate the ±45° boundaries of the extension field. Right panel: The strength map of the extension field for locations and directions shown in the left figure. Strength declines with increasing orientation differences and distance from the edge element. There is no strength assigned above and below the diagonals (right).

The extension fields were incorporated into the baseline model by allowing a field to operate on each of the 24 activation fields created by convolving the 24 kernels with an image. Because there were 8 orientations for the activation fields there were also 8 orientations for the extension fields. The additional excitation as provided by the extension field was distributed to the activation fields in such a manner that only the corresponding orientations of the activation fields and extension fields were convolved:

$$(EFWI)(\bar{l},\bar{x}_0) = \int EF_{\bar{l}}(\bar{x}_0 - \bar{x})WI_{\bar{l}}(\bar{x})d^2x = EF_{\bar{l}} * WI_{\bar{l}} \qquad (2)$$

where $\bar{l}$ gives the orientation of both the extension and activation fields. For the broad extension field model the activation fields not only get excitation from the extension field with the same orientation, but also from all the other orientations. For computational ease the excitation fields were divided into 8 subregions based on orientation and only the corresponding range of orientations were applied to an activation field with a given orientation. For the broad extension field model the overall excitation applied to an activation field is then given by summing up the excitation coming from: (a) the extension field with the preferred orientation of the given activation field and (b) the excitation from all the other extension fields. In the narrow extension field model the activation field with a given orientation was only convolved with the excitation field having the same orientation.

Figure 5 provides a direct visual comparison of the workings of the three different model types. The top row displays three versions of the 'boat' image from the set: intact, recoverable and nonrecoverable in left, middle and right columns respectively. Below the 3 x 9 blocks of images show the cumulative activation patterns induced by the three images in the three examined models: baseline, with narrowly tuned extension fields, with broadly tuned extension fields (from top nine image to bottom nine). In each of the three nine image blocks the first row represents the cumulative activation patterns of the kernels at the highest scale and at all 8 orientations. The second row represents the cumulative response at the highest and medium scale and the last row shows the 'total' of the activation for all scales and all orientations as well . This visualization of model activation also shows that for the second and third block of nine images (model with narrow and broad extension fields) the activation patterns for intact and recoverable images are much more similar than for the baseline model (first block of nine images). The visual representation the activation values of the model are normalized to integer values between 0-255 for 8-bit graphical display.

3 Simulations

In the recoverable-nonrecoverable experiment the similarity of 36 intact images with the recoverable and nonrecoverable versions (altogether 108 images) was calculated and compared to each other. In the feature-deleted vs. part-deleted experiment the similarity of the feature-deleted complementary image pair was compared to the similarity of the part deleted complementary image pair for 18 images (altogether 72 used).

3.1 Result of the simulations

The results of the simulations are displayed on Figures 6 and 7. The addition of narrowly tuned extension fields between similarly oriented kernels increased the similarity of both the recoverable and nonrecoverable versions to the original intact image. although it increased the similarity of the recoverable version more. Whereas for the baseline model there was no difference between the similarity of recoverable and nonrecoverable images the addition of narrow extension fields significantly increased the difference between the similarity of recoverable and nonrecoverable types compared with the original images. The addition of broad extension fields further improved similarity for recoverable images, but it

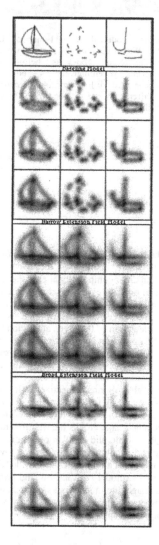

Figure 5. The top row displays the intact. recoverable and nonrecoverable versions of an image respectively. The 3 x 9 block of images below the top row display the activation patterns of the three model types (Baseline, Narrow Extension Fields, and Broad Extension Fields) to these images. The first row in each three blocks represents the cumulative activation of the highest frequency kernels at all eight orientations to the three images. The second row in each three blocks shows the cumulative activation of the highest and medium frequency kernels at all orientations. Finally. the third row in each blocks represents the cumulative activation of all three kernel sizes at all orientations (all 24 kernels).

did not improve similarity for the nonrecoverable ones compared to the narrow extension fields. Consequently, the broad extension field model further increased the difference between the similarity of recoverable and nonrecoverable images compared with the intact versions.

The addition of narrow and broad extension fields significantly increased the similarity of feature-deleted complementary images pairs, but did not improve the similarity of the part-deleted pairs. The similarity of two complementary feature-deleted images was already significantly higher than of two complementary part-deleted ones, but the addition of horizontal connections further improved this difference, just as did the addition of extension fields. The fact that similarity did not improve for part-deleted image pairs was expected considering that there was no any global knowledge provided that could relate the two different part structures in the pair to each other. However, the significant increase of similarity for the feature-deleted pairs was not an obvious outcome of the simulation. The addition of the broad extension field did not improve similarity for feature-deleted images compared to the narrow extension fields.

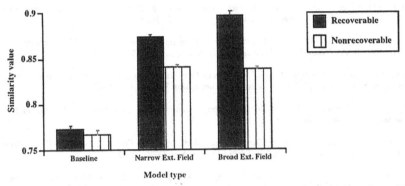

Figure 6. Average similarity values for matching the original intact images with the recoverable and nonrecoverable versions in the three model types.

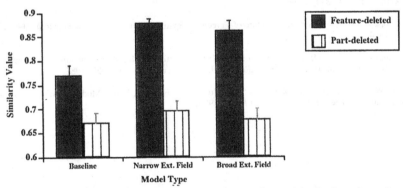

Figure 7. Average similarity values for matching complementary feature-deleted pairs and complementary part-deleted pairs in the three model types.

4 Conclusions

The addition of extension fields to a baseline model of object recognition that operates on the output of multiscale and multiorientation Gabor filters improves its overall recognition performance (at minimum for the given set of images) and brings its performance significantly and qualitatively closer to that of human object recognition. Interestingly, adding broad extension fields to the original model did not improve its performance significantly beyond the improvement already achieved by narrow extension fields.

An obvious direction for future development of the model is to incorporate inhibition and endstopping into the connectivity pattern, both well known characteristics of biological low level vision systems. We might mention though that even without these additions the model's performance could significantly be improved just based on the addition of excitatory connections. The extension field model is currently being tested on 8-bit gray-scale images.

5 References

[1] I. Biederman, "Recognition-by-components: A theory of human image understanding," *Psychological Review*, 94, 115-147, 1987.

[2] I. Biederman and E. E. Cooper, "Priming contour-deleted images: Evidence for intermediate representations in visual object recognition." *Cognitive Psychology*, 23, 393-419, 1991.

[3] D. J. Field, A. Hayes, and R. F. Hess, "Contour Integration by the Human Visual System: Evidence for a Local Association Field," *Vision Research*, 33(2), 173-193, 1993.

[4] C. D. Gilbert and T. N. Wiesel, "Columnar Specificity of Intrinsic Horizontal and Corticocortical connections in Cat Visual Cortex,"*The Journal of Neuroscience*, 9(7), 2432-2442, 1989.

[5] S. Grossberg and E. Mingolla, "Neural Dynamics of Perceptual Grouping: Textures, Boundaries, and Emergent Segmentations," Perception and Psychophysics 38, 141-171, 1985.

[6] G. Guy and G. Medioni, "Inferring Global Perceptual Contours from Local Features," *International Journal of Computer Vision*, 20(1/2), 113-133, 1996.

[7] F. Heitger and R. von der Heydt, "A Computational Model of Neural Contour Processing: Figure-Ground Segregation and Illusory Contours," In *Proceedings of the ICCV*, 32-40, 1993.

[8] M. Lades, J. C. Vorbrüggen, J. Buhmann J. Lange, C. von der Malsburg, R. P. Würtz, and W. Konen, "Distortion Invariant Object Recognition in the Dynamic Link Architecture," *IEEE Transactions on Computers*, 42. 300-311, 1993.

[9] P. Parent and S. W. Zucker, Trace Inference, Curvature Consistency, and Curve Detection, *IEEE Trans. PAMI*, 11(8), 823-839, 1989.

[10] U. Polat and D. Sagi, "The Architecture of Perceptual Spatial Interactions,"*Vision Research*, 34(1), 73-78, 1994.

[11] A. Sha'ashua and S. Ullman, "Structural saliency: the detection of globally salient structures using locally connected network," In *Proceedings of the ICCV*, Tampa Fl., 321-327, 1988.

[12] L. R. Williams and D. W. Jacobs, "Stochastic Completion Fields: A Neural Model of Illusory Contour Shape and Salience," In *Proceedings of ICCV*, 1995.

Bayesian Decision versus Voting for Image Retrieval

R. Mohr, S. Picard, C. Schmid

GRAVIR-IMAG & INRIA Rhône-Alpes
655 avenue de l'Europe
38330 Montbonnot Saint-Martin France

Abstract. Image retrieval from image databases is usually performed by using global image characteristics. However the use of local image information is highly desirable when only part of the image is of interest. An original solution was introduced in [9] using invariant local signal characteristics. This paper extends this contribution by extending the set of invariants considered to allow illumination change. Then it is shown that the invariant distribution is far from uniform and a probabilistic indexing scheme is proposed. Experimental results validate the approch and the different methods are discussed.

1 Introduction

In this paper we address the problem of retreiving images from a large database based on local measures of similarity. Similarity may be based on colour, texture, shape or gray level properties of the image under query . However, variations in viewpoint, scale or illumination complicate the problem significantly. Therefore, it is imperative that the properties used as the basis of any indexing scheme be invariant with respect to these changes. Although it has been pointed out by several authors that there are no generally invariant features [2], our goal is to find realistic simplified assumptions for computing stable signal properties.

Database query on the basis of color histograms [11], of texture [5], greylevel processed information (gradient, etc) [7] or Principal Component Analysis [12] have all achieved some success. However, techniques based on the global analysis and comparison of images and therefore are not well suited to matching when images of an object with large occlusion for instance. Image database query techniques need to be able to locale a pattern in a subimage when that pattern is obtained from different viewing conditions: partial visibility, different viewing angles, complex scene with complex background,many potential reference pattern. Under such conditions, data have to extracted locally, and the general process will have to cope with missing and spurious data.

In this paper we begin by describing the method by which locally invariant characteristics can be extracted from image (section 2.1), and how an indexing scheme based on these features can provide fast access to the relevant corresponding images. This section extends mainly the work presented by Schmid in [9] by considering not only invariance in geometry but also in illumination. The next section provides the Bayesian analysis of the retrieval process and, based on an estimated density of the characteristic in the image, provides a Bayesian scheme for image retrieval. This leads to experiments which compare the new scheme to the basic scheme and its extension with semi local characteristics presented in [9].

2 Indexing with greylevel invariants

2.1 Geometric greylevel invariants

For shape oriented towards the observer, groupe of similarities absorbs the first order of geometric changes with respect to camera motion [1]. Therefore computing invariants for this group of image displacements would address the problem of change in camera position. However, in order to be robust in the presence of occlusion and clutter the computation of global surface invariants [10] must be abandoned in favour of locally invariant features.

As explained in [9] we employ differential invariants that have been theorically studied by Kœnderink [4], Romeny *et al* [3,6]. These greylevel invariants are computed at points of interest, typically corners, ensuring that they are located in regions where the signal is informative.

Following [3] we compute the complete set of differential invariants up to the third order, and stack them in a vector, denoted by $\mathbf{X}$. The first part of this vector contains the complete and irreducible set of differential invariants up to 2nd order (see eq. 1). It is given in the so-called Einstein notation:

$$
\mathbf{X}[0..4] = \begin{bmatrix} L \\ L_i L_i \\ L_i L_{ij} L_j \\ L_{ii} \\ L_{ij} L_{ji} \end{bmatrix} = \begin{bmatrix} L \\ \frac{\partial L^2}{\partial x} + \frac{\partial L^2}{\partial y} \\ \frac{\partial L^2}{\partial x}\frac{\partial^2 L}{\partial x^2} + 2\frac{\partial L}{\partial x}\frac{\partial L}{\partial y}\frac{\partial^2 L}{\partial x \partial y} + \frac{\partial L^2}{\partial y}\frac{\partial^2 L}{\partial y^2} \\ \frac{\partial^2 L}{\partial x^2} + \frac{\partial^2 L}{\partial y^2} \\ \frac{\partial^2 L}{\partial x^2}^2 + 2\frac{\partial^2 L}{\partial x \partial y}\frac{\partial^2 L}{\partial y \partial x} + \frac{\partial^2 L}{\partial y^2}^2 \end{bmatrix} \tag{1}
$$

L represents a Gaussian kernel convolved with the image and the L_i are the first derivatives of the luminance according to i, which represents either x or y coordinates. The derivatives are computed by convolving the image with derivatives of the Gaussian kernel.

The second part of the vector contains a complete set of invariants of third order. Equation 2 presents this set limited to the Einstein notation, with ε_{ij} the 2D antisymmetric Epsilon tensor defined by $\varepsilon_{12} = -\varepsilon_{21} = 1$ and $\varepsilon_{11} = \varepsilon_{22} = 0$.

$$
\mathbf{X}[5..8] = \begin{bmatrix} \varepsilon_{ij}(L_{jkl}L_i L_k L_l - L_{jkk}L_i L_l L_l) \\ L_{iij}L_j L_k L_k - L_{ijk}L_i L_j L_k \\ -\varepsilon_{ij}L_{jkl}L_i L_k L_l \\ L_{ijk}L_i L_j L_k \end{bmatrix} \tag{2}
$$

At this stage we have, for each point of interest, a vector $\mathbf{X}$ of 9 values which are invariant to rotation and translation.

2.2 Illumination invariance

Illumination changes, both in direction and intensity produce strong changes in an image. Although there is no invariant which can be computed under such general changes, we can consider the simplified case in which we ignore problems as shadows by considering that the surface around the point of interest is locally flat. For ideal camera, the Lambertien illumination model is a good first order approximation except close to highlights. With such a model the intensities of the local patch vary by a scaling factor, depending on position and illumination.

However, in real case camera transfer functions look more like sigmoid function, thus a linear assumption can be made almost out of the extremal values. Therefore resulting intensity displayed in each pixel will be an affine version of the real one: $I_{image} = \alpha I_{input} + \beta = \alpha' \rho_{object} + \beta'$.

The two factors $\alpha' \beta'$ have to be cancelled out of our measures. The offset β' is eliminated from all of the measures except the first one since they are based on spatial derivatives. The uniform scaling β' is eliminated by considering ratio's of the remaining differential invariants. Illumination invariance is thus achieved by replacing the vector of differential invariants $\mathbf{X}[0..8]$ with $\mathbf{Y}[2..8]$:

$$
\begin{aligned}
&\mathbf{Y}[2] = \frac{\mathbf{X}[2]^{\frac{2}{3}}}{\mathbf{X}[1]}; \ \mathbf{Y}[3] = \frac{\mathbf{X}[3]}{\mathbf{X}[1]}; \ \mathbf{Y}[4] = \frac{\mathbf{X}[4]}{\mathbf{X}[1]}; \\
&\mathbf{Y}[5] = \frac{\mathbf{X}[5]}{\mathbf{X}[1]^2}; \ \mathbf{Y}[6] = \frac{\mathbf{X}[6]}{\mathbf{X}[1]^2}; \ \mathbf{Y}[7] = \frac{\mathbf{X}[7]}{\mathbf{X}[1]^2}; \ \mathbf{Y}[8] = \frac{\mathbf{X}[8]}{\mathbf{X}[1]^2};
\end{aligned}
\tag{3}
$$

Images can be matched under very different illumination conditions using these invariants. Graph of Fig 1 illustrates the matching performance under illumination changes using using $\mathbf{X}$, just when canceling the β and when using the illumination invariant features $\mathbf{Y}$.

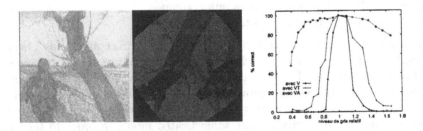

Fig. 1. Example of a change of illumination and %of successful matches when using $\mathbf{X}$ (curve V), when canceling the β (VT) and using $\mathbf{Y}$ (VA).

2.3 A Basic indexing algorithm

Local invariant vectors provide useful information that can be used for indexing image objects. However, the local points of interest from which these vectors are computed cannot be robustly detected in the presence of significant changes in the observed images. Typically only about 50% of the points of interest will be detected in any pair of images of an object. Therefore an index alone is meaningless; moreover, when a large number of such points occurs in an index table (we have more than 150,000 such vectors for our data base), ambiguity may arise. For this reason, a voting technique is used. It is similar to the Hough transform in which multiple feature matches are used to form a consensus of which database images provides the best match to a query.

The basic algorithm is therefore:

- compute the vector of invariants at each point of interest;
- compare them to invariants stored in the database (matching) using a fast access data structure and record the number of matches found for each database image;
- retrieve the image with largest number of matches.

3 A probabilistic model for image retrieval

3.1 Invariant distribution

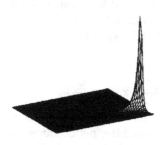

Fig. 2. Distribution of invariants 5 and 6 in the data base

The basic indexing algorithm assumes that all vectors are equally probable, therefore according equal importance to all matches. However, this is not necessarily the case. Following a suggestion made by Schiele [8], we observed the distribution of the invariants collected in our image data base, which contains over 150,000 9-dimensional vectors. Figure 4 represents the repartition of invariants 5 and 6, showing the nature of the distribution. It is clear the the information associated with each point is highly different and should be taken into account.

3.2 The basic Bayesian model

In a study on the distribution of the measure in some "receptive fields", Schiele [8] derived a Bayesian model which allowed the distribution of the measures to be taken into account. In this section we take a similar approach, deriving a model which uses the a priori knowledge of the distribution of the invariants in the matching process. But, as this work is concerned with matches or potential matches, we have to derive a more sophisticated model in order to take into account the matching process.

Let Q be the query image and R be a database image being considered as a retrieval candidate. It is assumed that Q and R will have a large number of features in common if they are correctly matched. Q has $n(Q)$ interest points, and R has $n(R)$ such points. From the $n(Q)$ features of Q, $\{m_i\}, i \in I$ is the set of the features which are matched with features in R. Each m_i is a feature vector of Q that has a matching feature vector f_k appearing in R.

We want to evaluate $P(R|\{m_i\})$. Using Bayes formula and assuming that the individual matches are independent, we get

$$P(R|\{m_i\}) = \frac{P(\{m_i\}|R).P(R)}{P(\{m_i\})} = \frac{\prod_{i \in I} P(m_i|R)P(R)}{\prod_{i \in I} P(m_i)} \quad (4)$$

$P(m_i)$ is the probability that the i-th feature of Q has one match with $n(R)$ random features. This approach considers only the effect of matches, but the fact that many features fail to match can also be integrated in a straitforeward way.

3.3 Probabilistic interpretation of direct voting

Counting the number of feature matches between a query and each database images and selecting the image(s) R for which the number of matches is maximum is an effective strategy for matching. Impressive results are explained by the fact that this counting approach is robust to outlier matches [9]. In this section we consider this matching in a a probabilistic framework.

Let us consider each feature in Q as the result of a uniform random variable whose values can be one of the values observed in all the possible Rs. The probability p_i^R that the i-th feature of Q matches one of the feature knowing that Q is part of R is the probability of repeatability of the feature in R when viewing conditions varies. Such a probability can be considered as a constant α; experimentally, depending of the change of illumination and of viewing conditions, α has been estimated to be in the range $[0.2..0.5]$.

Similarly, the probability $P(m_i)$ will be some constant β. $\beta \ll \alpha$ as the number of features possible in all the data base is much larger than the one appearing in R. This leads to :

$$P(R|\{m_i\}) = \frac{\prod_{i \in I} \alpha}{\prod_{i \in I} \beta} P(R) \tag{5}$$

As the logarithm is a monotic function, the largest probability over R is the largest logarithm.

$$\log(P(R|\{m_i\})) = \sum_{i \in I} (\log \alpha - \log \beta) + \log(P(R)) \tag{6}$$

As $\log \alpha - \log \beta > 0$ and assuming all $P(R)$ equal, this clearly shows that the maximum is obtained when $|I|$ is maximum, i.e. the number of matches is maximum.

This means that assigning a match based on the maximum vote corresponds to a maximum likelihood decision under the assumption of a uniform distribution of features on the object and on matches.

3.4 Posterior Probability of Retrieved Images

Matching also occurs randomly, inducing false matches, and this was not taken into account in the previous model. The model introduced here will consider only correct matches, i.e. corresponding to the good image. Note the false matches, i.e. corresponding to a random process, can also be introduced.

If we assume that Q is a subimage of R under some new viewing condition, the k-th feature of Q might be a feature of R with the previously defined probability α. It could be also a feature that occured due to some random process with the density probability of R.

Let p_R^k be the probability of the k-th random feature vector of Q to appear in image R. p_R^k is estimated using an histogramm on image R. The probability that it might miss all the $n(R)$ features in R is $(1 - p_R^k)^{n(R)}$. Thus the corresponding probability of matching one of these features is $1 - (1 - p_R^i)^{n(R)}$. Combining this two events which are exclusive, the likelihood of match (a correct or a false one) becomes:

$$P(m_i|R) = \alpha + (1 - \alpha)(1 - (1 - p_R^i)^{n(R)}) \qquad (7)$$

Similarly the probability of occurence of the j-th feature vector of Q will be p_B^k (B for data base) and the a priori probability of match with $n(R)$ feature vector is then $P(m_j) = \beta + (1 - \beta)(1 - (1 - p_B^i)^{n(R)})$ with β defined in the previous subsection. Substituting this into equation 4 results in:

$$P(R|\{m_i\}) = \frac{\prod_{i \in I}(\alpha + (1 - \alpha)(1 - (1 - p_R^i)^{n(R)}))}{\prod_{i \in I}(\beta + (1 - \beta)(1 - (1 - p_B^i)^{n(R)}))} P(R) \qquad (8)$$

At this stage one may notice that if p_R^i is very low, then equation 7 simplifies to $P(m_i \mid R) = \alpha$, i.e. to the case when false matches are not considered. One might increase the local information by using much more information, leading to very low probability of false matches and to the simplified model defined by equation 5.

4 Experiments

The experiments reported here are based on querying aerial images of a city. These are the very difficult images: they are similar in texture and shape, since roofs of old houses look very similar seen from the sky.

The query and model images were taken from different positions of observation (about 20 degrees change in the almost vertical viewing direction). Due to the change of viewing direction, the images differ: some facades are visible, illumination has changed, etc. Figure opposite represents a part of a query and model images.

4.1 Experimental process

For each image, the probability p_R^i has been estimated using a histogram, as an image does not provide enough data for estimating the probability density in the high dimensional space we are considering. The value for α in the Bayesian model was set to 0.5 and 0.35 (which is closer to experimental repeatability for this kind of image with 3D texture). The value for β was set to $\frac{\alpha}{1020}$ since 1020 images were used for estimation of a priori probabilities.

The experiments were conducted in the following way: one hundred images corresponding to the first view point are stored in the database, half of the hundred image of the second view point are used for the query. For each of them a subimage which represents x% of the search image from the first view was taken for the query. x was varied from 100 to 9. The rank of the right answer was measured as the output.

In [9] the authors used semi-local constraints for their voting technique: a point was considered as matched only if the invariant descriptor matched, and also if two additional matches were observed in the neighbourhood of the five closest points of interest. This semi local constraint was necessary for obtaining good results in the case of matching complex aerial images. Therefore three different matching strategies were investigated: simple count of the number of matches, count of the number of matches using the semi-local constraint,

4.2 Results

Fig. 3 displays the behavior of the different strategies. The abcissa represents the size of the subwindows considered for the request (percentage of the image surface). The ordinate shows the mean rank of the correctly matching image.

The five curves displayed correspond to the simple voting on invariant, the use of semi local context,the Bayesian model with $\alpha = 0.5$ and $\alpha = 0.35$, and Bayesian model with semi local context.

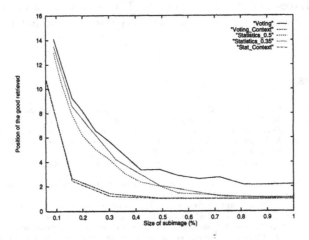

Fig. 3. Behavior of voting *vs* posterior probability

The results show that there is an advantage to using the repeatability factor for the kind of image used, but that the behavior of the Bayesian decision rule is not too much affected by this choice. The use of the a priori knowledge distribution offers a clear gain for large windows, by about a factor 2. This gain is more limited for smaller windows where the number of matched features decreases largely. An additional study of the influence of possible outliers on this result is ongoing.

As forcast in the discussion in 3.4, the semi-local constraint improves the results. It is almost perfect for large windows (the right answer is always the first for up to half of the window size) as is the Bayesian decision rule. For very small window, this method outperforms the Bayesian decision rule by a factor 2.

5 Discussion and conclusion

We have presented a Bayesian model for improving the image retrieval procedure using invariant feature vectors. This model has been compared to the previous indexing technique using experimental data set of very similar images, taken in different viewing conditions.

Bayesian decision definitely improves the voting procedure. The gain is substantial when enough features can be matched, and becomes small otherwise. This gain is achieved at the costs of estimating the a priori density of the possible value for the invariant vectors used for indexing the image data base and estimating the repeatability of the point of interest extraction. For large query images, the method finds the correct answer almost always first.

One could argue that a factor of 2 in the image retrieval process is not enough for the corresponding cost. One clear reason that the gain is not better is that the invariant used are already very discriminant and the probability that a random match will occur is low. In such a case the counting decision rule becomes very close to the Bayesian decision rule (cf 3.4). This is typically what happens with the use of semi-local constraints: they are very discriminant and almost no false matches occur.

References

1. T.O. Binford and T.S. Levitt. Quasi-invariants: Theory and exploitation. In *Proceedings of* DARPA *Image Understanding Workshop*, pages 819–829, 1993.
2. J.B. Burns, R.S. Weiss, and E.M. Riseman. The non-existence of general-case view-invariants. In J.L. Mundy and A. Zisserman, editors, *Geometric Invariance in Computer Vision*, chapter 6, pages 120–131. The MIT Press, Cambridge, MA, USA, 1992.
3. L. Florack. *The Syntactical Structure of Scalar Images*. PhD thesis, Universiteit Utrecht, November 1993.
4. J.J. Koenderink and A.J. van Doorn. Representation of local geometry in the visual system. *Biological Cybernetics*, 55:367–375, 1987.
5. A. Pentland, R.W. Picard, and S. Sclaroff. Photobook: Content-based manipulation of image databases. *International Journal of Computer Vision*, 18(3):233–254, 1996.
6. Romeny. *Geometry-Driven Diffusion in Computer Vision*. Kluwer Academic Publishers, 1994.
7. B. Schiele and J.L. Crowley. Object recognition using multidimensional receptive field histograms. In *Proceedings of the 4th European Conference on Computer Vision, Cambridge, England*, pages 610–619, 1996.
8. B. Schiele and J..L. Crowley. Probabilistic object recognition using multidimensional receptive field histogram. In *Proceedings of the 13th International Conference on Pattern Recognition, Vienna, Austria*, pages 50–54, 1996.
9. C. Schmid and R. Mohr. Combining greyvalue invariants with local constraints for object recognition. In *Proceedings of the Conference on Computer Vision and Pattern Recognition, San Francisco, California, USA*, June 1996. ftp://ftp.imag.fr/pub/MOVI/publications/Schmid_cvpr96.ps.gz.
10. H. Schulz-Mirbach. Constructing invariant features by averaging techniques. In *Proceedings of the 12th International Conference on Pattern Recognition, Jerusalem, Israel*, pages 387–390, 1994.
11. M.J. Swain and D.H. Ballard. Color indexing. *International Journal of Computer Vision*, 7(1):11–32, 1991.
12. M. Turk and A. Pentland. Face recognition using eigenfaces. In *Proceedings of the Conference on Computer Vision and Pattern Recognition, Maui, Hawaii, USA*, pages 586–591, 1991.

A Structured Neural Network Invariant to Cyclic Shifts and Rotations

Sabine Kröner

Technische Informatik I
Technische Universität Hamburg-Harburg
D–21071 Hamburg

Phone: +49 [40] 7718 2539
Fax: +49 [40] 7718 2911
E-mail: kroener@tu-harburg.d400.de

Shift and rotation invariant pattern recognition is usually performed by first extracting invariant features from the images and second classifying them. This poses the problem of not only finding suitable features but also a suitable classifier.

Here a class of structured invariant neural network architectures (SINN) is presented that performs adaptive invariant feature extraction and classification simultaneously. The special characteristic of the pyramidal feedforward architecture of the SINN is sparse connectivity and the use of shared weight vectors. This guarantees the invariance of the network output with respect to cyclic shifts and rotations of the input image. In experiments the recognition ability of the SINN is shown on a database of textile images. Without any preprocessing of the images and without the need to choose an appropriate classifier the SINN achieves similar or even better results than standard pattern recognition methods.

1   Introduction

Most image processing systems for shift and rotation invariant pattern recognition achieve their invariance property by first calculating invariant features and second classifying them with either standard classifiers or neural networks. For the feature extraction usually general methods like moment invariants or geometric invariants are used.

However, the computation of the invariants with these methods can turn out to be very costly since the calculated features are not adapted to the patterns in the present application. Often the separation quality of each feature for the underlying pattern set is not known in advance or can only be estimated heuristicly. This may lead to the calculation of numerous unsuitable features having only poor separating ability for the special application.

The problem can be overcome by combining the feature extraction process with the classification in an adaptive system. Here a structured invariant neural network architecture (SINN) is proposed that simultaneously extracts and classifies adaptive shift and rotation invariant features from the input images. The invariance property is directly built into the network architecture by use of shared weight vectors and a sparse connection structure.

The paper is organized as follows: First the architecture of the structured invariant neural network for the calculation of shift and rotation invariant image features is presented. Then the realization of the node functions is explained together with a learning algorithm for the weights. Experiments show the recognition ability of the invariant neural network compared to a standard feature extraction method using geometric invariants and a higher order neural network on a database of textile images.

2 The architecture of a structured invariant neural network

The characteristic of structured architectures is the use of shared adaptable weights. Examples of such networks are hierarchical feature extraction networks [8] and the neocognitron [2, 7]. Promising results have been achieved with these networks in the field of invariant pattern recognition. However, it has been shown that their invariance property mathematically is only a tolerance with respect to shifts and slight image distortions [1]. In contrast to these networks the structured invariant neural network presented here is designed to be definitely invariant with respect to cyclic shifts and rotations of the input pattern.

First the architecture of the SINN is described for the cyclic shift invariant recognition of a one-dimensional input vector of $a_0 = K$, $K \in I\!N$ elements. These input elements can be grey values directly taken from a line of image pixels or higher order combinations of pixel values.

Definition 1. A *structured invariant neural network (SINN)-architecture* is defined as a n-layer feedforward architecture with $a_0 = K$, $K \in I\!N$ input nodes and $a_n = 1$ output node. K is factorized as $K := p_0 \cdot p_1 \cdot p_2 \cdots p_n$, $p_i \in I\!N$, with $p_0 := 1$ and $1 < p_i \leq K$ for $1 \leq i \leq n$. The cardinality of the sets $H^j := \{h_1^j, \ldots, h_{a_j}^j\}$, of nodes in the layers $j = 0, \ldots, n$ is given as $a_j = \prod_{i=1}^{n-j} p_i$. The layer H^0 is also referred to as input layer, and the layer H^n as output layer. The outdegree of all nodes of the sets $H^j := \{h_1^j, \ldots, h_{a_j}^j\}$, $j = 0, \ldots, n$, is defined as

$$fan_{out}(h_{\alpha_j}^j) = 1 , \quad \text{for} \quad \alpha_j = 1, \ldots, a_j .$$

For the indegree of all nodes of the sets $H^j := \{h_1^j, \ldots, h_{a_j}^j\}$, $j = 1, \ldots, n$,

$$fan_{in}(h_{\alpha_j}^j) = p_{n-j+1} , \quad \text{for} \quad \alpha_j = 1, \ldots, a_j .$$

holds. The connectivity between the sets H^{j-1} and H^j, $j = 1, \ldots, n$, of nodes of two successive layers is described as

$$\bigcup_{\alpha_j=1}^{a_j} \{h_{\alpha_j+i \cdot a_j}^{j-1} | 0 \leq i < p_{n-j+1}\} \times \{h_{\alpha_j}^j\} .$$

The node functions G are defined to be identical for all nodes in one layer:

$$G_1^j = \ldots = G_{a_j}^j \quad \text{for} \quad j = 1, \ldots, n ,$$

and produce the same result for every cyclic permutation $\pi(x_{\alpha_j})$ of the elements of the input vector x_{α_j} into a node $h^j_{\alpha_j}$:

$$G^j(\pi(x_{\alpha_j})) = G^j(x_{\alpha_j}) .$$

Fig. 1 shows the parameter assignments in a SINN-architecture. Depending on the factorization of K different SINN-architectures can be constructed to a given dimension of the input vector.

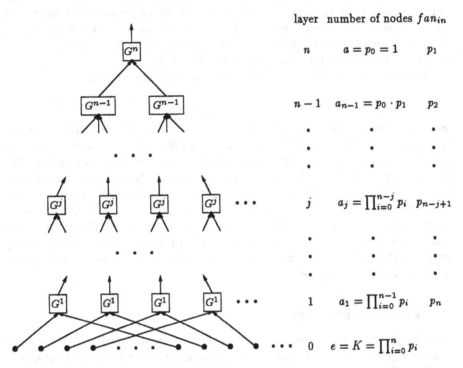

Fig. 1. SINN-architecture

With a group theoretic approach it can be proved that a SINN produces an output that is invariant with regard to cyclic shifts of the input elements [6]. Due to the shared weights and the network architecture on each succeeding layer invariance with respect to a subgroup whose size is related to the factorization of the input elements of the group of cyclic shifts of K elements is achieved. This finally leads to invariance with respect to the full group.

The network architecture for one-dimensional patterns can be extended to an architecture for the recognition of two-dimensional patterns using a special connection structure described in [4]. It allows the integration of additional invariance with respect to discrete rotations of multiples of 90 degrees into the same architecture without any extra costs.

For invariance with respect to shifts and general rotations the network has to be implemented in several different angular orientations so that the networks cover regularly the full circle of 360 degrees. Due to the generalization ability of neural networks not every discrete orientation needs to be covered. The weights are shared between the oriented networks. After the input has been processed in parallel by the oriented SINNs the results are combined to the final output again with an architecture as in Definition 1.

3 Realization of the node function

Since there are no further constraints on the realization of the node functions unless those mentioned in Definition 1 a large variety of networks belongs to the class of SINNs. The incoming values in a node are connected via a connecting function. This can be a weighted sum or a multiplication, thus including sigma-pi neurons. The transfer functions can be any function.

A possible realization of a node function with indegree two is

$$G = \Theta\left(w_3\left(t(w_1 x_1 + w_2 x_2 + w_0)\right) + w_3\left(t(w_2 x_1 + w_1 x_2 + w_0)\right)\right) \quad (1)$$

where w_1 and w_2 are the weights on the connectional links, w_3 and w_0 are the thresholds of the two nonlinear transfer functions Θ and t, respectively. This node function is symmetric and discriminative between the two values x_1 and x_2. The shared weights are applied in such a way that completeness of the mapping of the node function with respect to cyclic shifts of the input elements is achieved.

If the transfer functions in the SINN are of hard limiter type the final network output is binary, permitting only the separation of two classes. If there are more classes to be separated several SINNs have to be processed. The result is a binary output vector with a one in that entry that corresponds to the recognized pattern and zero in all others.

4 The learning algorithm

The weights in a structured neural networks are usually trained with a standard backpropagation algorithm modified for shared weights as it was proposed by Rumelhart [10] or layerwise with a perceptron-like algorithm [7]. However, due to the strong weight sharing in SINNs global training algorithms like backpropagation and also heuristic ones like threshold accepting often lead to the learning of a wrong target or they do not converge at all [5]. Although not much is reported in literature, similar observations have been made for the training of reimplementations of LeCuns networks [9]. On the other hand, layerwise training in multilayer networks poses the problem of finding suitable error functions for the hidden layers so that their minimization eventually leads to a minimum of the global error function.

Therefore for SINNs with hard limiter transfer functions for the separation of two classes a new learning algorithm has been developed [5]. It works layerwise and minimizes an optimization criterion for shared weights on every layer that is derived from the error function of the final layer

$$E = \left(1 - \frac{\|y_A\|}{A} + \frac{\|y_B\|}{B}\right). \quad (2)$$

Each component of y_A describes the output of one of the A patterns of class $\mathcal{A}$ which are to be mapped on the value 1, and y_B the output of the B patterns of class $\mathcal{B}$ which are to be mapped on the value 0, respectively. With $\|\cdot\|$ as the euclidean metric $\|y_A\|$ ($\|y_B\|$) gives the number of patterns of class A (B) that are actually mapped to the value 1 by the SINN. The effect of the use of shared weights means that a pattern on all but the final layer of a SINN is represented by several points in the low-dimensional weight space. The number of points is reduced on every layer due to the pyramidal network architecture. The idea of the learning algorithm is to map as many points of as many patterns of class $\mathcal{A}$ as possible with a linear transfer function to the value 1 on every layer, and to 0 for class $\mathcal{B}$ respectively. It is shown that this sequential minimization of the error function on every layer finally leads to the minimization of the error function (2) as well. Moreover the learning algorithm converges for separable patterns after one training cycle only. This algorithm is used for the training of the SINN in the following experiments.

5 Experimental results

In this experiment the recognition ability of a SINN is compared with the standard feature extraction method of invariant integration [12] and a shift and rotation invariant higher order neural network (HONN), which both have shown convincing results in invariant pattern recognition tasks [3, 12, 13]. The experiment reflects a task that often occurs in vision based quality control: faultless samples, here textiles, are to be distinguished from defect samples. A database of textile samples (see [11] for a detailed description) is used, consisting of four different types of fabrics with different regularity of their surface structure. Examples of the four classes C1 to C4 are shown in Fig. 2.

Fig. 2. Grey-value images of the faultless fabric samples of the classes C1, C2, C3 and C4, size 512×512 pixels

The database contains faultless samples and samples with various defects for each class. The defects are distinguished in five categories of holes and cuts of different size, spots, flaws, loose threads, and crinkles. Fig. 3 shows examples of the defects for the textile classes C1 and C3.

For the experiments 873 samples are used: 50 samples for each class of the faultless fabrics, and 197, 183, 168, and 125 samples for the defect fabrics of the classes C1 to C4, respectively. For the calculation of the recognition rates the different types of defect samples are assumed to be equally likely. The probability for the occurrence of any defect is arbitrarily set to 0.5 as no further information

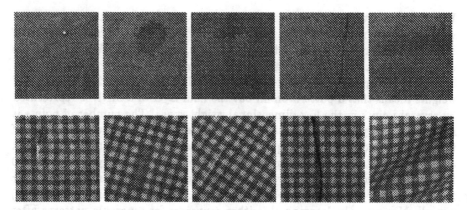

Fig. 3. Examples of grey-value images of defect fabric samples of class C1 and C3, size 512×512 pixels; from left to right: holes and cuts, spots, flaws, loose threads, crinkles

from industrial processes was available. With all methods the classes C1 to C4 are trained separately.

The patterns are processed with SINNs with indegree two of the nodes on all layers, and hard limiter transfer functions. On the first layer a polynomial decision curve of second order is applied, whereas the decision curves on all other layers are linear. The SINNs are trained with the layerwise training algorithm [5] directly on the grey values of the image pixels without any preprocessing. Due to the variety of defects five different SINNs are trained each for separating the faultless samples from one of the defect categories. For the training only five samples are used from the faultless and from the defect patterns. The recognition rates then are obtained according to a maximum decision of the outputs of the five SINNs for the remaining samples which form the test set.

For a comparison the standard feature extraction method of invariant integration is used. The shift and rotation invariant features are based on monomials of up to second order which are calculated from the grey values of the pixels with a filter kernel of maximum size 7×7 pixels. In this experiment five features are calculated and then classified with a Maximim-Likelihood classifier. For the training of the classifier 10 samples of the faultless and also of each category of the defect samples of each class are used. The recognition rates are evaluated on the remaining test patterns.

The method of invariant integration is also used in HONNs for feature extraction. The classification in this network type is then performed with a multilayer perceptron. Therefore the five features in this experiment are also classified with one two-layer perceptron for each class to get the recognition rates for a HONN. The training set consists of 10 patterns of the faultless samples and two for each of the defect categories for each class. The training is performed with a backpropagation algorithm with momentum term and adaptive learning rate.

The upper bound of the computational costs for the recognition of the test images is of order $\mathcal{O}(n)$ for n input pixels with all three methods. However, for images of size $n = 512^2$ pixels the constant factor in this estimation is about ten times smaller for the SINN than for the method of invariant integration and the

HONN where additional costs arise from the size of the filter kernel, the number of features calculated, and the cost for the classifier. The recognition rates for the four classes of fabrics which are obtained with the three different methods are shown in Table 1.

	faultless samples	defect samples	average recognition rates
C1			
SINN	93.2 %	72.3 %	82.8 %
inv. I. + Max.Lik.	32.5 %	85.7 %	59.1 %
HONN	70.0 %	63.3 %	66.6 %
C2			
SINN	97.7 %	71.6 %	84.7 %
inv. I. + Max.Lik.	70.0 %	96.0 %	83.0 %
HONN	95.0 %	85.0 %	90.0 %
C3			
SINN	88.9 %	80.0 %	84.5 %
inv. I. + Max.Lik.	70.0 %	100.0 %	85.0 %
HONN	97.5 %	65.3 %	81.4 %
C4			
SINN	51.2 %	78.0 %	64.6 %
inv. I. + Max.Lik.	35.0 %	90.6 %	62.8 %
HONN	55.0 %	80.0 %	67.5 %

Table 1. Recognition rates for faultless and defect fabric samples, and the averaged recognition rate ($0.5 \times$ (faultless + defect)) for four classes of fabrics based on the processing with an SINN, the method of invariant integration followed by a Maximim-Likelihood classifier, and a HONN

Except for class C4 with lower recognition rates for all methods due to strong variations in lightning conditions in the images itself the SINN yields uniform recognition results of around 83-84%. In contrast to this the recognition rates for the other methods show differences between the classes C1 to C3 of up to 23% points. Such variations are usually undesirable in a recognition system for several classes. One reason for the variations is the different suitability of the feature extraction method for the images of the different classes. Obviously the extraction of five features based on local monomials of second order is not appropriate for the class C1. Another reason is the suitability of the classifier. The comparison of the HONN with the invariant integration and Maximum-Likelihood classifier shows for class C1 and C2 by 7% points higher recognition rates for the HONN. Since both methods are evaluated on the same features this difference is only due to the different classification process. This shows clearly the difficulties one is encountered with when using standard feature extraction methods for invariant pattern recognition. With the SINN, however, stable recognition rates are achieved for all classes without choosing a suitable feature extraction method or classifier. The adaptivity of the network leads to good classification results for all classes.

6 Conclusion

A class of structured invariant neural networks has been presented for the calculation and simultaneous classification of cyclic shift and rotation invariant image features. The invariance property is achieved by a sparsely connected architecture with shared weights on the different layers. For the training of the free parameters in the SINN a layerwise operating learning algorithm is applied which converges after one training cycle only for separable patterns. During the learning process the whole network is adapted to the characteristics of the pattern set in a given application. This means that the usual problem of finding a suitable feature extraction method and a suitable classifier to achieve good recognition results in standard pattern recognition systems is unnecessary for SINNs. Nevertheless the recognition rates which are achieved on a database of textile images with an SINN show stable results similar or even considerably better, depending on the type of fabric, than that achieved with standard feature extraction methods of the same computational expenses.

References

1. E. Barnard, D. Casasent, *Shift Invariance and the Neocognitron*, in Neural Networks, Band 3, S. 403-410, Pergamon Press, 1990
2. K. Fukushima, N. Wake, *Handwritten Alphanumeric Character Recognition by the Neocognitron*, IEEE Trans. on Neural Networks, Vol. 2, No. 3, May 1991
3. T. Kanaoka et al., *A higher-order neural network for distortion invariant pattern recognition*, Pattern Recognition Letters, Vol. 13, pp. 837-841, Dec. 1992
4. S. Kröner, *A Neural Network for Calculating Adaptive Shift and Rotation Invariant Image Features*, Proc. of EUSIPCO'96, 8th European Signal Proc. Conf., G. Ramponi et al. (Eds.), Vol. II, pp. 863-866, Trieste, Italy, Sept. 1996
5. S. Kröner, *A Learning Algorithm for Structured Invariant Neural Networks*, Technical Report 5/96, Technische Informatik I, TU Hamburg-Harburg, Sept. 1996.
6. S. Kröner, H. Schulz-Mirbach, *Fast adaptive calculation of invariant features*, Tagungsband Mustererkennung 1995, (17. DAGM Symposium), Sagerer, G. et al. (Eds.), S. 23-35, Bielefeld, Sept. 1995. Reihe Informatik aktuell, Springer Verlag
7. C. Li, and C.-H. Wu, *Introducing rotation invariance into the Neocognitron model for target recognition*, Pattern Recognition Letters, Vol. 14, pp. 985–995, Dec. 1993
8. Y. le Cun: *Generalization and Network Design Strategies*, Connectionism in Perspective, R. Pfeiffer, Z. Schreter, F. Fogelman-Soulié, L. Steels (Eds.), Elsevier Science Publishers B.V. 143-155, North-Holland, 1989
9. D. de Ridder: *Shared weight neural networks in image analysis* M.Sc. Thesis, TU Delft, March 1996
10. D. Rumelhart, J. McClelland, (Ed.), *Parallel Distributed Processing*, Vol. 1, MIT Press, Cambridge, Massachusetts, 1986
11. H. Schulz-Mirbach, *Ein Referenzdatensatz zur Evaluierung von Sichtprüfungsverfahren für Textiloberflächen*, Tech. Report 4/96, Tech. Inf. I, TU Hamburg-Harburg, Sept. 1996
12. H. Schulz-Mirbach, *Anwendung von Invarianzprinzipien zur Merkmalgewinnung in der Mustererkennung*, VDI-Fortschrittbericht, Reihe 10, Nr. 372, 1995
13. L. Spirkovska, M.B. Reid, *Coarse-Coded Higher-Order Neural Networks for PSRI Object Recognition*, IEEE Trans. Neural Networks, Vol. 4, No. 2, 1993

Morphological Grain Operators
for Binary Images

Henk J.A.M. Heijmans

CWI, P.O. Box 94079, 1090 GB Amsterdam, The Netherlands
email: henkh@cwi.nl, URL: http://www.cwi.nl/~henkh/

Abstract. Connected morphological operators act on the level of the flat zones of an image, i.e., the connected regions where the grey-level is constant. For binary images, the flat zones are the foreground and background grains (connected components) of the image. A grain operator is a special kind of connected operator that uses only local information about grains: grain operators do not require information about neighbouring grains. This paper discusses connected morphological operators for binary images, with an emphasis on grain operators and grain filters.

1 Introduction

An important new development in the area of morphological image processing is the concept of a connected operator. Such operators are defined on the level of the foreground and background grains (by 'grain' we mean 'connected component') rather than on the level of individual pixels. This means that a connected operator cannot break connected components of the foreground or the background.

A grain operator is a particular instance of a connected operator. Characteristic for grain operators is that they can be evaluated grain by grain, without any knowledge about other parts of the image. Every binary image induces a partition of the underlying space whose zones are the foreground and background grains; see Section 2. One of the simplest nontrivial (binary) grain operators is the *area opening* which preserves all zones of X with area (number of pixels) larger than or equal to a given threshold.

Connected operators cannot introduce new discontinuities and as such they are eminently suited for applications where contour information is important. Image segmentation is such an application. The morphological approach towards segmentation, the watershed algorithm usually gives rise to a dramatic oversegmentation. To circumvent this problem, one usually modifies the image using an appropriate set of markers. Connected operators have proved to be useful for determining such markers automatically [2,4,13]. The goal of this paper is to introduce the reader into the relatively new area of connected operators and grain operators. We are exclusively concerned with binary images on the two-dimensional square grid provided with 8-connectivity. A more comprehensive discussion can be found in [8].

2 Connectivity, Reconstruction, and Partitions

By $\mathcal{P}(\mathbf{Z}^2)$ we denote the power set of $\mathbf{Z}^2$, i.e., the collection of all subsets of $\mathbf{Z}^2$. The collection of all subsets of $\mathbf{Z}^2$ that are connected (in the sense of 8-connectivity) is denoted by $\mathcal{C}$. A connected component C of set X is called a *grain of X*, and we write $C \Subset X$. The operator γ_h defined by

$$\gamma_h(X) = \begin{cases} \text{grain of } X \text{ that contains } h, & \text{if } h \in X \\ \varnothing, & \text{if } h \notin X \end{cases}$$

is called *connectivity opening* (indeed, it is not difficult to show that γ_h is an opening [6,14]). Now we can define the *reconstruction operator ρ* as follows:

$$\rho(Y \mid X) = \bigcup_{h \in Y} \gamma_h(X). \tag{1}$$

If $X \cap Y = \varnothing$ then $\rho(Y \mid X) = \varnothing$. The sets X and Y in $\rho(Y \mid X)$ are called the *mask (image)* and *marker (image)*, respectively. Observe that $\gamma_h(X) = \rho(\{h\} \mid X)$. The reconstruction $\rho(Y \mid X)$ (and hence the opening $\gamma_h(X)$) can be computed easily by means of a propagation algorithm: see Fig. 1.

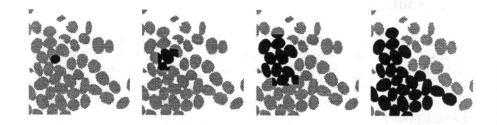

Fig. 1. Computation of $\rho(Y \mid X)$. From left to right: the mask image X (grey) and the marker image Y (black); 15 iterations; 50 iterations; final result $\rho(Y \mid X)$.

Recall that a *partition* of the space $\mathbf{Z}^2$ is a subdivision of this space into disjoint zones. A partition can be represented by a function $P : \mathbf{Z}^2 \to \mathcal{P}(\mathbf{Z}^2)$ that satisfies $x \in P(x)$, for every $x \in \mathbf{Z}^2$, and $P(x) = P(y)$ or $P(x) \cap P(y) = \varnothing$, for any two points $x, y \in \mathbf{Z}^2$. The partition is called connected if every zone is connected. Thus, $P(x)$ is the zone of the partition that contains the point x. The following relation defines a partial ordering on the collection of partitions of $\mathbf{Z}^2$:

$$P \sqsubseteq P' \text{ if } P'(h) \subseteq P(h), \text{ for every } h \in \mathbf{Z}^2.$$

We say that P is *coarser* than P'.

With every binary image $X \subseteq \mathbf{Z}^2$ there corresponds a unique partition $P(X)$ of $\mathbf{Z}^2$ composed of the foreground and background grains of X. Writing $P(X, h) = P(X)(h)$, we have $P(X, h) = \gamma_h(X)$ if $h \in X$ and $P(X, h) = \gamma_h(X^c)$ if $h \in X^c$.

3 Connected Operators

There exist several equivalent definitions of a connected operator. The most intuitive one says that a connected operator is an operator that coarsens the partition corresponding with a binary image.

Definition 1. An operator ψ on $\mathcal{P}(\mathbf{Z}^2)$ is *connected* if the partition $P(\psi(X))$ is coarser than $P(X)$, for every binary image $X \subseteq \mathbf{Z}^2$.

A connected operator acts on the zones of $P(X)$ in an all-or-nothing way: either a zone is left untouched or deleted altogether. This means in particular that contours in the image cannot be broken or changed; they can only be deleted. An opening (closing, filter) that is connected is called *connected opening* (*closing, filter*).

Recall that the *negative* of an operator ψ is defined by $\psi^*(X) = \psi(X^c)^c$, where X^c is the complement of X. It is not difficult to show that an operator is connected if and only if its negative is such as well. Furthermore, compositions, suprema and infima of connected operators are connected, too [8]. Here we will emphasize one other result in particular. Refer to [8] for a proof.

Proposition 2. *Given a Boolean function b of n variables and n connected operators $\psi_1, \psi_2, \ldots, \psi_n$, then the operator $\psi = b(\psi_1, \psi_2, \ldots, \psi_n)$ is connected as well.*

Here the expression $\psi = b(\psi_1, \psi_2, \ldots, \psi_n)$ is defined by means of the expression:

$$\psi(X)(h) = b(\psi_1(X)(h), \ldots, \psi_n(X)(h)),$$

where $X(h)$ equals 1 if $h \in X$ and 0 otherwise. For example, if $b(u_1, \ldots, u_n) = u_1 \cdot u_2 \cdots \cdots u_n$, then $b(\psi_1, \ldots, \psi_n) = \psi_1 \wedge \cdots \wedge \psi_n$, the infimum of $\psi_1, \ldots, \psi_n$ [6].

A general way to construct connected openings is to perform an arbitrary opening α first and to do a reconstruction afterwards:

$$\check{\alpha}(X) = \rho(\alpha(X) \mid X). \tag{2}$$

The following result holds.

Proposition 3. *If α is an opening, then $\check{\alpha}$ is a connected opening. Moreover, α is a connected opening if and only if $\alpha = \check{\alpha}$.*

In Fig. 2 we show an example; here $\alpha(X) = X \circ B$, where B is a square.

4 Grain Operators

Now we come to the main subject of this paper, namely grain operators (called *connected component local operators* by Crespo and Schafer [3]).

Definition 4. A connected operator ψ is called a *grain operator* if it has the following property: if $h \in \mathbf{Z}^2$ and $X, Y \subseteq \mathbf{Z}^2$, are such that $X(h) = Y(h)$ and $P(X, h) = P(Y, h)$, then $\psi(X)(h) = \psi(Y)(h)$.

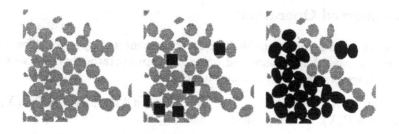

Fig. 2. Opening by reconstruction: the original opening is an opening by a square. From left to right: X, $\alpha(X)$, and $\check{\alpha}(X)$.

Thus a grain operator acts on the level of the individual zones of a partition and does not require information about adjacent zones. A grain operator which is also an opening (closing, filter) is called a *grain opening (closing, filter)*. A simple example of a grain opening is the connectivity opening γ_h: to obtain the new colour at a vertex C, we need to check only whether $h \in C$. Another, more interesting, example is the *area opening* α_S. This is the operator that deletes all grains $C \Subset X$ with area less than S:

$$\alpha_S(X) = \bigcup \{C \mid C \Subset X \text{ and } \mathrm{area}(C) \geq S\}. \tag{3}$$

It is obvious that α_S is a grain operator, and that it is increasing, anti-extensive, and idempotent, and therefore an opening. Fig. 3 depicts the area opening with $S = 10$. Area openings and closings have become very popular since Vincent [15]

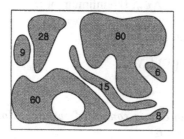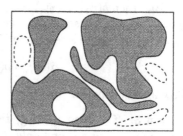

Fig. 3. Area opening α_{10}. The numbers printed inside the grains denote their area.

invented a fast implementation of this operator. The next result states some basic properties of grain operators. Refer to [8] for a proof.

Proposition 5. (a) *If ψ_i is a grain operator for every $i \in I$, then $\bigvee_{i \in I} \psi_i$ and $\bigwedge_{i \in I} \psi_i$ are grain operators.*
(b) *If $\psi_1, \psi_2, \ldots, \psi_n$ are grain operators and b is a Boolean function of n variables, then $b(\psi_1, \psi_2, \ldots, \psi_n)$ is a grain operator.*

(c) *If ψ is a grain operator, then ψ^* is a grain operator.*

However, we emphasize that the composition of two grain operators is not a grain operator in general; see [8] for a counterexample.

5 Grain Criteria

The area opening is determined by a criterion $u : C \to \{0,1\}$ on the foreground grains, namely $u(C) = [\text{area}(C) \geq S]$. Here we use the following convention: if S is a statement then $[S]$ denotes the Boolean value (0 or 1) indicating whether S is true or false. It turns out that every grain operator is uniquely determined by two criteria, one for the foreground and one for the background. This is the contents of the following result. All results stated in this section have been proved in [8].

Proposition 6. *An operator ψ is a grain operator if and only if there exist foreground and background criteria u and v such that $\psi = \psi_{u,v}$, where $\psi_{u,v}$ is given by*

$$\psi_{u,v}(X) = \bigcup \{C \mid (C \Subset X \text{ and } u(C) = 1) \text{ or } (C \Subset X^c \text{ and } v(C) = 0)\}. \quad (4)$$

The criteria u and v are given by

$$u(C) = [C \subseteq \psi(C)] \text{ and } v(C) = [C \subseteq \psi^*(C)]$$

The following properties can easily be established.

Proposition 7. (a) $\psi^*_{u,v} = \psi_{v,u}$.
(b) *Given grain operators ψ_{u_i,v_i}, for $i \in I$, then*

$$\bigwedge_{i \in I} \psi_{u_i,v_i} = \psi_{\bigwedge_{i \in I} u_i, \bigvee_{i \in I} v_i} \text{ and } \bigvee_{i \in I} \psi_{u_i,v_i} = \psi_{\bigvee_{i \in I} u_i, \bigwedge_{i \in I} v_i}.$$

(c) *Let ψ_{u_i,v_i} be grain operators for $i = 1, 2, \ldots, n$, and let b be a Boolean function of n variables, then*

$$b(\psi_{u_1,v_1}, \ldots, \psi_{u_n,v_n}) = \psi_{b(u_1,\ldots,u_n),b^*(v_1,\ldots,v_n)}$$

Here b^* is the negative of b given by $b^*(u_1, \ldots, u_n) = 1 - b(1 - u_1, \ldots, 1 - u_n)$.

We present some examples of grain criteria. Other criteria were given by Breen and Jones [1].

1. $u(C) = [C \ominus B \neq \varnothing]$, which gives the outcome 1 if some translate of B fits inside C. If B is connected, then $\psi_{u,1} = \breve{\alpha}$, where $\alpha(X) = X \circ B$; cf. Proposition 3. However, if B is not connected, then $\psi_{u,1}$ is an opening that is smaller than $\breve{\alpha}$, i.e., $\psi_{u,1} \leq \breve{\alpha}$.
2. $u(C) = [\text{perimeter}(C) \geq S]$, where $\text{perimeter}(C)$ equals the number of boundary pixels in C.

3. $u(C) = [\,\text{area}(C)/(\text{perimeter}(C))^2 \geq k\,]$. Note that this criterion provides a measure for the circularity of C.

A criterion u is called *increasing* if $C, C' \in \mathcal{C}$ and $C \subseteq C'$ implies that $u(C) \leq u(C')$. Increasingness of u and v does not yet guarantee increasingness of $\psi_{u,v}$.

Proposition 8. *The grain operator $\psi_{u,v}$ is increasing if and only if both u and v are increasing criteria, and the following condition holds:*

$$u(\gamma_h(X \cup \{h\})) \vee v(\gamma_h(X^c \cup \{h\})) = 1, \tag{5}$$

if $X \subseteq \mathbb{Z}^2$ and $h \in \mathbb{Z}^2$.

In [8] we present an example that shows that condition (5) is essential here. The notation $v \equiv 1$ means that $v(C) = 1$ for every grain C. In that case we write $\psi_{u,1}$ for $\psi_{u,v}$. Similarly, $\psi_{1,v}$ denotes the grain operator for which the foreground criterion u is identically 1. It is obvious that the grain operator $\psi_{u,v}$ is extensive iff $u \equiv 1$, and that it is anti-extensive iff $v \equiv 1$. The composition of two grain operators is not a grain operator in general. However, composing two (anti-) extensive grain operators yields an (anti-) extensive grain operator. Or more precisely:

$$\psi_{u_2,1}\psi_{u_1,1} = \psi_{u_1,1}\psi_{u_2,1} = \psi_{u_1 \wedge u_2,1} = \psi_{u_1,1} \wedge \psi_{u_2,1} \tag{6}$$

$$\psi_{1,v_2}\psi_{1,v_1} = \psi_{1,v_1}\psi_{1,v_2} = \psi_{1,v_1 \wedge v_2} = \psi_{1,v_1} \vee \psi_{1,v_2} \tag{7}$$

Taking $u_1 = u_2 = u$ in (6), we find that $\psi_{u,1}^2 = \psi_{u,1}$. This implies the following result.

Theorem 9. (a) *Every anti-extensive grain operator is idempotent and of the form $\psi_{u,1}$ for some foreground criterion u. This operator is an opening if and only if u is increasing.*
(b) *Every extensive grain operator is idempotent and of the form $\psi_{1,v}$ for some background criterion v. This operator is a closing if and only if v is increasing.*

6 Grain Filters

In this section we formulate a simple condition on the grain criteria u and v under which $\psi_{u,v}$ is a grain filter. Recall that $\mathcal{C}$ comprises all subsets of $\mathbb{Z}^2$ that are connected. Refer to [8] for additional results.

Definition 10. We write $C \approx D$ if there exists a set X such that $C \in X$, $D \in X^c$ and $C \sim D$.

Recall that $C \sim D$ means that $C \cup D$ is connected. It is evident that

$$C \approx D \iff D \approx C.$$

Consider the following property for the criteria u, v:

$$C \approx D \text{ implies } u(C) \vee v(D) = 1. \tag{8}$$

In [8] it has been shown that for increasing grain criteria u, v, (8) implies (5). However, it is not difficult to give examples that show that the converse is not true: condition (8) is essentially stronger than (5). Now one can prove the following result [8].

Proposition 11. *Let u, v be increasing grain criteria for which (8) holds. Then $\psi_{u,v}$ is a grain filter, and*

$$\psi_{u,v} = \psi_{u,1} \psi_{1,v} = \psi_{1,v} \psi_{u,1} \tag{9}$$

The equalities in (9) means that the grain filter $\psi_{u,v}$ is the composition of a grain opening and a grain closing.

We conclude with an example. Let a_S be the area criterion defined before. If $C \approx D$, then at least one of these grains must contain eight pixels or more. Therefore, if $S, T \leq 8$, the pair a_S, a_T satisfies condition (8). Thus ψ_{a_S, a_T} is a grain filter. If $S = T$, then $\omega_S = \psi_{a_S, a_S}$ is a self-dual grain filter. The filter ω_2 is called *annular filter* [7,9]. Refer to Fig. 4 for an illustration.

Fig. 4. From left to right: a binary image X and the results after filtering with ω_S for $S = 1, 4, 7$, respectively.

Observe that noise pixels adjacent to borders are not affected by the filters ω_S; as we have seen this is a general property of connected operators.

7 Conclusions

In this paper we have discussed the relatively new concept of a grain operator, which is a special kind of connected morphological operator. We have formulated a simple condition on the foreground and background criteria associated with a given grain operator under which this operator is a filter (increasing and idempotent). For the sake of exposition, we have restricted ourselves to 2-dimensional discrete binary images. However, many of our results can be generalised to various other cases: continuous, grey-scale, higher dimensional, other connectivities, etc.

So far, connected operators have found various applications, for example in image filtering [5,11,13], segmentation [4,2,16], and motion tracking [10,12].

References

1. E. Breen and R. Jones. An attribute-based approach to mathematical morphology. In P. Maragos, R. W. Schafer, and M. A. Butt, editors, *Mathematical Morphology and its Application to Image and Signal Processing*, pages 41–48. Kluwer Academic Publishers, Boston, 1996.
2. J. Crespo. *Morphological connected filters and intra-region smoothing for image segmentation*. PhD thesis, Georgia Institute of Technology, Atlanta, 1993.
3. J. Crespo and R. W. Schafer. Locality and adjacency stability constraints for morphological connected operators. *Journal of Mathematical Imaging and Vision*, 7(1):85–102, 1997.
4. J. Crespo, J. Serra, and R. W. Schafer. Image segmentation using connected filters. In J. Serra and P. Salembier, editors, *Mathematical Morphology and its Applications to Signal Processing*, pages 52–57. Universitat Politècnica de Catalunya, 1993.
5. J. Crespo, J. Serra, and R. W. Schafer. Theoretical aspects of morphological filters by reconstruction. *Signal Processing*, 47(2):201–225, 1995.
6. H. J. A. M. Heijmans. *Morphological Image Operators*. Academic Press, Boston, 1994.
7. H. J. A. M. Heijmans. Self-dual morphological operators and filters. *Journal of Mathematical Imaging and Vision*, 6(1):15–36, 1996.
8. H. J. A. M. Heijmans. Connected morphological operators for binary images. Research Report PNA-R9708, CWI, 1997.
9. H. J. A. M. Heijmans and C. Ronse. Annular filters for binary images. Research report BS-R9604, CWI, 1996. Submitted to IEEE Trans. Image Proc.
10. M. Pardàs, J. Serra, and L. Torres. Connectivity filters for image sequences. In *Image Algebra and Morphological Image Processing III*, volume 1769, pages 318–329. SPIE, 1992.
11. P. Salembier and M. Kunt. Size-sensitive multiresolution decomposition of images with rank order based filters. *Signal Processing*, 27(2):205–241, 1992.
12. P. Salembier and A. Oliveras. Practical extensions of connected operators. In P. Maragos, R. W. Schafer, and M. A. Butt, editors, *Mathematical Morphology and its Application to Image and Signal Processing*, pages 97–110. Kluwer Academic Publishers, Boston, 1996.
13. P. Salembier and J. Serra. Flat zones filtering, connected operators, and filters by reconstruction. *IEEE Transactions on Image Processing*, 4(8):1153–1160, 1995.
14. J. Serra, editor. *Image Analysis and Mathematical Morphology. II: Theoretical Advances*. Academic Press, London, 1988.
15. L. Vincent. Morphological area openings and closings for grey-scale images. In Y.-L. O, A. Toet, D. Foster, H. J. A. M. Heijmans, and P. Meer, editors, *Proceedings of the Workshop "Shape in Picture", 7–11 September 1992, Driebergen, The Netherlands*, pages 197–208, Berlin, 1994. Springer.
16. L. Vincent and E. R. Dougherty. Morphological segmentation for textures and particles. In E. R. Dougherty, editor, *Digital Image Processing Methods*, pages 43–102. Marcel Dekker, New York, 1994.

A Parallel 12–Subiteration 3D Thinning Algorithm to Extract Medial Lines[*]

Kálmán Palágyi and Attila Kuba

Department of Applied Informatics, József Attila University
H-6701 Szeged P.O.Box 652, Hungary

Abstract. Thinning is a frequently used method for extracting skeletons of objects in discrete spaces. It is a layer by layer erosion of the objects until only their skeletons are left. This paper presents an efficient parallel thinning algorithm which directly extracts medial lines from 3D binary objects (i.e., without creating medial surface). According to the selected 12 deletion directions, one iteration step contains 12 subiterations (instead of the usual 6). Our algorithm provides good results, preserves topology and makes easy implementation possible.

1 Introduction

Skeletonization provides significant geometric features that are extracted from binary image data. It is a common preprocessing operation in raster–to–vector conversion, in pattern recognition or in shape analysis. Its goal is to shrink elongated binary objects to thin–line patterns. In the 3D Euclidean space, the skeleton can be defined as the locus of the centers of all the maximal inscribed spheres that touch the boundary of the object at least in two points. Thinning is a frequently used method for extracting skeleton in discrete spaces. Border points of the binary object that satisfy certain topological and geometric constraints are deleted in iteration steps. Deletion means that object elements are changed to background elements. The outmost layer of an object is deleted during an iteration step. The entire process is repeated until only the skeleton is left.

In this paper, a new 3D thinning algorithm is proposed that directly extracts medial lines. Each iteration step contains 12 successive subiterations according to the selected 12 deletion directions instead of the usual 6 ones. Subiterations are executed in parallel. It means that the object elements that satisfy the actual deletion conditions are to be changed to background elements simultaneously. The points to be deleted are given by a set of fairly simple deletion configurations of $3 \times 3 \times 3$ lattice points. Our algorithm preserves topology and according to the experiments made on synthetic images, provides a good approximation to the "true" Euclidean skeleton.

---

[*] This work were supported by OTKA T023804 and FKFP 0908/1997 grants.

2 Basic definitions

A *3D binary (26,6) digital picture* $\mathcal{P}$ is a quadruple $\mathcal{P} = (\mathbf{Z}^3, 26, 6, B)$ where each element in $\mathbf{Z}^3$ is called *point* of $\mathcal{P}$. A point in $B \subseteq \mathbf{Z}^3$ is called *black point*; a point in $\mathbf{Z}^3 \backslash B$ is called *white point*. Picture $\mathcal{P}$ is called *finite* if B is a finite set. Value 1 is assigned to each black point; value 0 is assigned to each white point. Two black points are *adjacent* if they are *26–adjacent*; two white points are *adjacent* if they are *6–adjacent* (see Fig. 1).

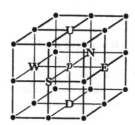

Fig. 1. Adjacencies in $\mathbf{Z}^3$. Points marked **U**, **N**, **E**, **S**, **W**, and **D** are 6–adjacent to point p; points 6–adjacent to p and points marked "•" are 26–adjacent to p.

Transitive closure of the 26–adjacency is called *26–connectivity* relation. Equivalence classes of B under 26–connectivity are called *black 26–components* or *objects*. A singleton black component is called *isolated point*. A black point is called *border point* if it is 6–adjacent to at least one white point. A border point p is called **U**–*border point* if the point marked **U** in Fig. 1 is white. (**N**–, **E**–, **S**–, **W**–, and **D**–border points can be defined in the same way.)

Iterative thinning algorithms delete border points that satisfy certain conditions. The entire process is repeated until there are no more black points to be changed. Each thinning algorithm should satisfy the requirement of topology preservation. Kong and Rosenfeld [1] discussed the definitions of topology preservation of 3D thinning operations.

Ma [2] established sufficient conditions for 3D parallel thinning operations to preserve topology. Topology preservation is not the only requirement to be complied with. Additional conditions have been stated to supply the skeleton.

CRITERION 2.1 [3, 4]. A 3D operation is a *thinning algorithm* if it fulfils the following three conditions.

1. Topology is preserved.
2. Isolated points and curves remain unchanged.
3. After applying the algorithm a black cube of size $2 \times 2 \times 2$ is changed.

Most of the existing 3D thinning algorithms result in *medial surfaces* [5]; a few produce *medial axes* or *medial lines* [4] (see Fig. 2).

A few 3D thinning algorithms have been developed so far. Most of them are *border sequential* [4, 7, 8]. Each iteration step is divided into more successive subiterations. Only a certain type of border points can be deleted during a subiteration. The subiterations are executed in parallel (i.e., all points that satisfy the actual deletion conditions are deleted simultaneously) and the deletion conditions depend on the $3 \times 3 \times 3$ neighbourhood of the points.

Fig. 2. A 3D synthetic picture containing a character "A" (left), its medial surface (centre), and its medial lines (right). (Cubes represent black points.)

3 The new thinning algorithm

In this section, a new algorithm is described for thinning 3D binary $(26, 6)$ pictures.

Our 12–subiteration border sequential 3D thinning algorithm directly creates medial lines without extracting medial surface. The new value of each voxel depends on its $3 \times 3 \times 3$ neighbourhood. This dependence is given by a set of configurations of $3 \times 3 \times 3$ lattice points. Only deletion conditions are used (i.e., black points can be changed but white points remain the same). The deletion condition assigned to a subiteration is described by a set of masks (or matching templates). A black point is to be deleted if and only if its $3 \times 3 \times 3$ neighbourhood matches at least one element of the given set of masks.

The masks are constructed according to the 12 deletion directions **US, NE, DW, UW, SE, DN, UN, SW, DE, UE, NW,** and **DS**. It means that these directions correspond to the 12 non–opposite pairs of 6–neighbours.

The masks assigned to the direction **US** give the condition to delete certain U–border points or S–borders points. The masks of **NE** are to delete certain N– or E–border points, etc. The set of masks $\mathcal{M}_{US}$ are assigned to the direction **US**, see Fig. 3. Deletion conditions of the other 11 subiterations can be derived from the appropriate rotations and reflections of masks in $\mathcal{M}_{US}$.

Let $\mathcal{P} = (\mathbb{Z}^3, 26, 6, B)$ be the finite 3D picture to be thinned. Since the set B is finite, it can be stored in a finite 3D binary array X. (Outside of this array every voxel will be considered as 0 (white).) Suppose that a subiteration results picture $\mathcal{P}' = (\mathbb{Z}^3, 26, 6, B')$ from picture $\mathcal{P}$. Denote $\mathcal{T}(X, \mathcal{D})$ the binary array representing picture $\mathcal{P}'$, where $\mathcal{D}$ is one of the 12 deletion directions **US,...,DS**. We are now ready to present our thinning algorithm formally.

ALGORITHM 3.1 for thinning picture $\mathcal{P} = (\mathbf{Z}^3, 26, 6, B)$

    Input: binary array X representing picture $\mathcal{P}$

    Output: binary array Y representing the thinned picture

       begin

          $Y = X$;

          repeat

             $Y = \mathcal{T}(Y, \mathbf{US})$;

             $Y = \mathcal{T}(Y, \mathbf{NE})$;

             $Y = \mathcal{T}(Y, \mathbf{DW})$;

             $Y = \mathcal{T}(Y, \mathbf{SE})$;

             $Y = \mathcal{T}(Y, \mathbf{UW})$;

             $Y = \mathcal{T}(Y, \mathbf{DN})$;

             $Y = \mathcal{T}(Y, \mathbf{SW})$;

             $Y = \mathcal{T}(Y, \mathbf{UN})$;

             $Y = \mathcal{T}(Y, \mathbf{DE})$;

             $Y = \mathcal{T}(Y, \mathbf{NW})$;

             $Y = \mathcal{T}(Y, \mathbf{UE})$;

             $Y = \mathcal{T}(Y, \mathbf{DS})$;

          until no points are deleted;

       end.

Note that choosing another order of the deletion directions yields another algorithm. We proposed a sequence of directions, in which **U**-, **N**-, **E**-, **S**-, **W**-, and **D**-border points uniformly occur.

Masks in $\mathcal{M}_{\mathbf{US}}$ and all their rotations and reflections can be regarded as a "careful" characterization of simple points for 3D thinning. (A border point is *simple* if its deletion does not change the topology of the picture [6].) This is a Boolean characterization since each of our masks can be described by a Boolean condition. The deletion condition belonging to a subiteration can be given by a Boolean function of 27 variables requiring 199 logical "or" and "and" operations and 123 "not" operations (without optimization) [9]. It has low computational cost and it makes easy implementation possible.

It have been shown in [9] that Algorithm 3.1 is a 3D thinning algorithm in the sense of Criterion 2.1.

4 Discussion

Algorithm 3.1 goes around the object to be thinned according to the 12 deletion directions. One can say that the new 12–subiteration algorithm seems to be slower than a 6–subiteration solution. It is not necessarily true. An iteration step of Algorithm 3.1 is approximately equivalent to four iterations of any 6–subiteration algorithms [4, 7, 8] (see Fig. 4).

Algorithm 3.1 have been tested for several synthetic pictures. Here we present some examples. Thinning eight synthetic 3D objects can be seen in Figs. 5 and 6. However, our algorithm is border-sequential, it creates nearly regular and symmetric skeletons from regular and symmetric objects. (Note that the skeleton of the object forming a $2 \times 2 \times 2$ cube is not symmetric.)

Also, we have tested Algorithm 3.1 in the case of noisy pictures in order to demonstrate its noise sensitiveness (see Fig. 7). This example shows that our algorithm is stable even in the case of noise added to the boundary of the object. Fig. 7 shows also that direct extraction of medial lines from the original object is more practical than the generally used two-phase method (i.e., creating medial surface from the original object and then extracting medial lines from the medial surface).

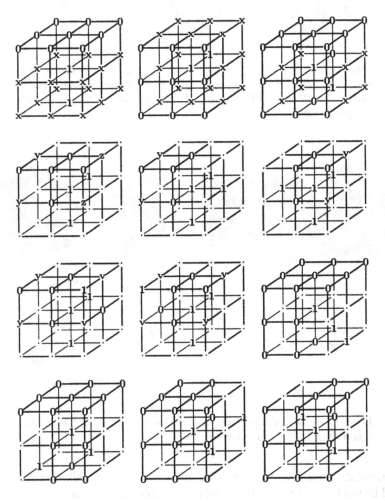

Fig. 3. The set of masks $\mathcal{M}_{\text{US}}$ assigned to the deletion direction **US**. Notations: "." ("don't care") matches either **0** or **1**; at least one position marked "x" is **1**; at least one position marked "y" is **0**; at least one position marked "z" is **0**; two positions marked "v" are different (one of them is **0** and the other one is **1**).

Lee et al. [8] discussed different methods of characterization of 3D simple points in $(26, 6)$ pictures. There are $2^{26}(=67\,108\,864)$ different possible configurations of voxels around a point p in $N_{26}(p)$. Point p is simple in $25\,984\,552$ cases of these configurations. It has been shown that masks in $\mathcal{M}_{\mathbf{US}}$ can delete more than $2^{20}(=1\,048\,576)$ kinds of simple points [9].

Fig. 4. Comparison of directional strategies. The original object is a black cube of size $16 \times 16 \times 16$ (left). A $14 \times 14 \times 14$ cube is generated by one iteration step of 6–subiteration algorithms [4,7,8] (centre). A cube of size $8 \times 8 \times 8$ is created by one iteration step of Algorithm 3.1 (right).

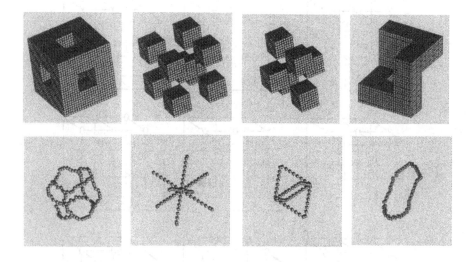

Fig. 5. Thinning of four synthetic pictures of size $24 \times 24 \times 24$. The original objects (top row) and their medial lines (bottom row).

One might think that the algorithm finding and deleting all the simple points would be the best. However, the number of simple points recognized as points to be deleted does not mean the goodness of a border sequential thinning algorithm. We believe that if we delete "too many" kinds of simple points, then we get skeletons having several parasitic parts. In that case the unwanted line segments are to be removed, skeletons are to be "cleaned up" by postprocessing. When we carefully designed masks in $\mathcal{M}_{US}$, we tried to avoid making this mistake (see Fig. 7).

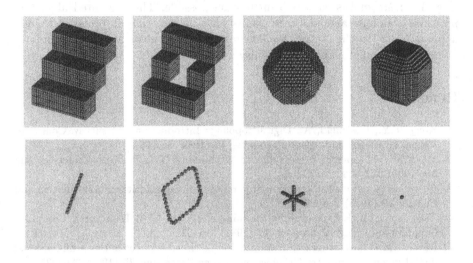

Fig. 6. Thinning of four synthetic pictures of size $24 \times 24 \times 24$. The original objects (top row) and their medial lines (bottom row).

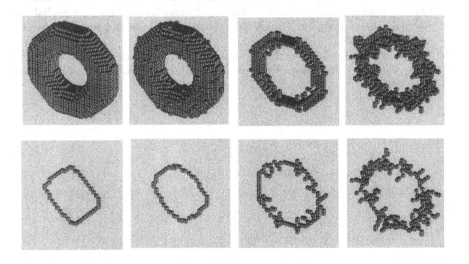

Fig. 7. Thinning of a 3D digital doughnut of size $32 \times 32 \times 9$ and its noisy version (top left two pictures). Their medial surfaces are extracted by the algorithm of Gong and Bertrand [7] (top right two pictures), and medial lines extracted from the objects above them by using Algorithm 3.1 (bottom row).

5 Conclusions

A parallel algorithm has been presented for thinning 3D binary objects. It directly extracts medial lines by using iteration steps containing 12 subiterations. The algorithm makes easy implementation possible. The presented algorithm preserves topology and our test results have shown, that it provides a good approximation to the "true" Euclidean skeleton. It gives stable and uniform (symmetric) medial lines, even in noisy cases.

References

1. Kong, T.Y., Rosenfeld, A.: Digital topology: Introduction and survey. Computer Vision, Graphics, and Image Processing **48** (1989) 357–393
2. Ma, C.M.: On topology preservation in 3D thinning. CVGIP: Image Understanding **59** (1994) 328–339
3. Rosenfeld, A.: A characterization of parallel thinning algorithms. Information and Control **29** (1975) 286–291
4. Tsao, Y.F., Fu, K.S.: A parallel thinning algorithm for 3–D pictures. Computer Graphics and Image Processing **17** (1981) 315–331
5. Reed, G.M.: On the characterization of simple closed surfaces in three–dimensional digital images. Computer Graphics and Image Processing **25** (1984) 226–235
6. Morgenthaler, D.G.: Three–dimensional simple points: Serial erosion, parallel thinning and skeletonization, Technical Report TR–1005, Computer Vision Laboratory, Computer Science Center, University of Maryland, 1981
7. Gong, W.X., Bertrand, G.: A simple parallel 3D thinning algorithm. In Proceedings of the 10th International Conference on Pattern Recognition. (1990) 188–190
8. Lee, T., Kashyap, R.L., Chu, C.: Building skeleton models via 3–D medial surface/axis thinning algorithms. CVGIP: Graphical Models and Image Processing **56** (1994) 462–478
9. Palágyi K., Kuba, A.: A topology preserving 12-subiteration 3D thinning algorithm, Internal report, Department of Applied Informatics, József Attila University, 1996

Architectural Image Segmentation Using Digital Watersheds

Volker Rodehorst

Technical University of Berlin, Photogrammetry and Cartography
Straße des 17. Juni 135, Sekr. EB 9, D-10623 Berlin, Germany

vr@fpk.tu-berlin.de

Abstract. The use of digital image analysis in architectural photogrammetry allows the automation of monotonous low-level tasks and makes the interpretation of facades much easier. Different segmentation methods for a computer assisted mapping of joint contours are discussed. Exemplary the application of the watershed transform is described and a combined watershed-method will be presented, which is well suited for the extraction of irregular stones.

1. Introduction

The purpose of Photogrammetry is to obtain geometric information of objects by taking measurements from images. One of the earliest applications of photogrammetry is the documentation of cultural monuments. A major objective is to document and to archive historically important objects in such a way, that the actual situation can be reconstructed at any time. Today, the acquisition of geometric information is mainly done by human operators who analyze images using classical photogrammetric equipment. Architectural Photogrammetry deals with spatial structured objects, which are in principle suitable for an automatic evaluation [1].

Fig. 1a: Photograph of the northern facade of the brewery house in the Chorin monastery

Fig. 1b: CAD-Model of the analytical measured facade

Fig. 2a: Section of the digital image from the southern facade in Chorin

Fig. 2b: Photograph of the Towers of the Church St. Nicolai

Fig. 2c: Selected area of the digital image from the stone facade

The use of special feature extraction algorithms and the development of interpretation models for these tasks are in a preliminary stage until now. The focus of the described work is the computer assisted mapping of joint contours in architectural images (see Fig. 1a). The aim to automate the creation of a facade map (see Fig. 1b) from gray-level photographs can only be achieved using knowledge based image analysis.

For these studies, historical objects of Brandenburg are selected as test material. The Fig. 2b and 2c show the towers from the Church St. Nikolai in Jüterbog. The stones have different sizes, they have approximately circular but irregular shapes and they are arbitrary distributed in the image. This example of an image class is characterized by mainly dark stones, which stand out against the bright background of the extensive joints.

The second example shows a section of the brewery house facade in the Chorin monastery (see Fig. 2a). Here, a definite separation between the objects and a background is not obvious. Differences in the illumination, shadows and weathering make the analysis tasks difficult. The advantages for an automatic processing are given by the simple shape and the regular and symmetric arrangement of the objects.

2. Image Segmentation

The segmentation based on *discontinuities* searches for areas, where the intensity of the image values changes significant locally. The segmentation based on *homogeneity criteria's* joins pixel to areas of maximum size using a homogeneity predicate. It is not sufficient to search for areas with the same photometric characteristic. The recognition of semantic units, e.g. bricks and joints, is more important.

2.1 Point orientated Segmentation

A simple method to separate objects for a segmentation is the usage of *thresholding*. A lot of approaches are discussed in detail by WAHL [11]. Another approach for a point based segmentation is the usage of a *classificator*, which tries to form groups inside a set of feature vectors to fulfill an appropriate target criterion. The dependence on old gray-level images makes the use of a point orientated segmentation difficult.

 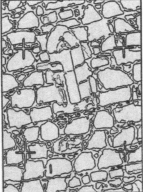 

Fig. 3a: Superposition of an image with the result of a line extraction algorithm

Fig. 3b: Result of an edge detection *(black)* followed by a contour tracing *(gray)*

Fig. 3c: Result of an area based segmentation without preprocessing

2.2 Edge-/Line based Segmentation

The edge based segmentation tries to find borders between objects under the assumption that there is an abrupt change in the gray-levels *(optical edge)*. Edge filters have different characteristics for the localization and the determination of the amplitude and direction. The resulting contours should be closed, straight and should have a small width. Additional criteria for the quality are the handling of contour crosses and the noise sensivety of the filter. More information can be found by MARR-HILDRETH[6] and CANNY[3].

The detection of line structures requires neighbored edges with opposite direction. The result of a *line extraction* algorithm by matching of 2D-functions proposed by BUSCH [2] is shown in Fig. 3a. Problems may occur, because the extracted edge or line contours are not closed. The use of *contour tracing* approaches, which are able to close contour gaps using the gradient information, is only successful if the distance is very small (see Fig. 3b).

2.3 Area based Segmentation

The region growing techniques are based on similarities between neighbored pixel to find a connected area. Beginning from a starting point, features are compared with the neighbors to find pixel with the same characteristic (see Fig. 3c). An investigation for knowledge based image analysis of facades using *hierarchical region aggregation, region growing* and *quadtrees* is described by XU [12]. An advantage of this approach is, that edges which are defined by segment borders have closed contours. A problem is the automatic determination of the segment size. Often, some important details merge to one segment while other uniform areas are still splitted.

2.4 Segmentation Using Morphology

An important approach is to find objects using their shape, which requires that the shape is a-priori known. An application for the *matching* of bricks was described by TRAUTHAN [8]. This approach depends on the absolute gray-level intensity and is not very flexible, especially when the bricks are distorted by perspective projection.

Fig. 4a: High-pass filtered image with the Butterworth-Filter

Fig. 4b: Result of the Dynamic Histogram-Equalization

Fig. 4c: Segmentation improvements after the pre-processing stage

3. Preprocessing

3.1 Homomorphic Filtering

Variations of the lighting or weathering in architectural images are characterized by slow oscillations in the digital image $f(x,y)$. The basic approach for enhancing an image via frequency domain techniques is to compute the *fast fourier transform* (FFT), multiply the result by a filter function $H(u,v)$ and take the inverse transform (IFFT). For sharpening we used the *Butterworth-Filter* of order n, which is defined as

$$H(u,v) = \frac{1}{1 + [D_0 / D(u,v)]^{2n}}.$$

$D(u,v)$ is the distance from the origin of the frequency plane and D_0 is the value, where $H(u,v) = 0.5$. The basic idea behind *homomorphic filtering* [7] is to operate separately on the frequency components for illumination and reflectance. A logarithm operation provides the measure of separation between the low- and high-frequency components. Finally, the filtering process is defined as follows

$$g(x,y) = \exp[\text{IFFT}(H(u,v)\text{FFT}(\ln f(x,y)))].$$

The enhanced image $g(x,y)$ obtained by homomorphic filtering is shown in Fig. 4a.

3.2 Dynamic Histogram-Equalization

A local contrast manipulation can be achieved by modifying the histogram of intensities. The objective is to produce a transformation function $T(i)$ to take advantage of the full range of intensity levels $i = 0,...,G_{MAX}$. Let $n_L,...,n_H$ denote the given discrete intensities and $G(n_i)$ the frequency of intensity i in the image. The mapping function for a *histogram equalization* [11] is given by

$$T(i) = \frac{G_{MAX} \cdot \left(\sum_{j=n_L}^{n_i} G(j) \right) - \frac{G(n_i) + G(n_L)}{2}}{n},$$

Fig. 5a: Over-segmentation using the watershed transform   **Fig. 5b:** Erroneous merging of segments   **Fig. 5c:** Result of the watershed smoothing

where n is the number of pixels in the image. Using the *dynamic histogram equalization* [10] the entire image is divided in several regions and the histogram equalization is computed for every region. In order to get a smooth transition at the boundary between two regions the final intensities are bilinear interpolated. Depending on the region size an optimal local contrast can be achieved (see. Fig. 4b).

4. Watershed Transform

The watershed transform is an approach based on mathematical morphology to divide an image due to discontinuities. In contrast to a classical area based segmentation, the watershed transform is executed on the gradient image. We obtain the gradient information using the algorithm proposed by DERICHE [4]. A digital watershed is defined as a small region that can not assigned unique to an influence zones of a local minima in the gradient image.

For applying the watershed transform, an algorithm proposed by VINCENT [9] was used. The intensity values are sorted and are stored in a table to achieve an efficient execution. The transform can be calculated without the gradient image using the table and neighborhood relations. A result of the watershed transform is shown in Fig. 5a.

4.1 Watershed Transform Using a Smoothed Gradient

It is evident that the segment sizes are very small and that uniform areas are splitted. The reason for the so-called *over-segmentation* is the sensifety of the gradient to noise. An obvious approach is the smoothing of the gradient image, e.g. using a gauss filter. The noise will be reduced and the segments will grow depending on the *standard deviation* σ. The choice of a very large σ may lead to erroneous merged segments (see Fig. 5b). A segmentation result of the watershed transform using a smoothed gradient with an appropriate σ for the bricks is shown in Fig. 5c.

A problem introduced by the smoothing process are the shifted positions of the contours. For simple object structures a description by graphic primitives using a form factor may be sufficient (see Fig. 6a). The correction of object contours with an arbitrary shape is more difficult.

 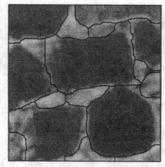 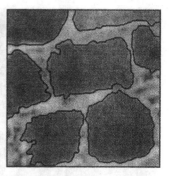

Fig. 6a: Description of the segments using graphic primitives

Fig. 6b: Shifted contours of the watershed transform using a smoothed gradient

Fig. 6c: Result of the optimization process using Active Contours

4.2 Shape Analysis Using Active Contours

In order to find the real contour of a stone using the active contour model it is important to specify an initial curve. A rough approximation to the desired contour in the image is supplied by the watershed algorithm that also defines an outer boundary (see Fig. 6b). An active contour with N points is represented by $s_i = (x_i, y_i)$, $i=1..N$. The optimization will minimize the contour's energy based on the *energy function*

$$E_{TOTAL} = \sum_{i=1}^{N} E_{INT}(s_i) + E_{IMG}(s_i) + E_{CON}(s_i)$$

The *internal energy* E_{INT} consists of the term α, which affects the rigidity of the active contour and the maximum change in the tangent vectors of the curve. Additionally, the term β minimizes the distance between contour points and favors the shortest contour length. The *external* forces are defined by the *constraint energy* E_{CON} and the *image energy* E_{IMG} of a contour. E_{CON} attracts points on the active contour to points in the image and E_{IMG} describes how the contour conforms to the gray-level gradient within the image. Details concerning active contours can be found by KASS et al. [5].

The result obtained by the energy minimization using a dynamic programming technique is shown in Fig. 6c. A major advantage of the active contour model is the retention of the viscosity feature of the watersheds. On the other hand, it is not easy to find appropriate parameters and the optimization process is not very robust.

4.3 Merging of Watershed Regions

Another approach to prevent the over-segmentation is the merging of small segments using a fusion criterion. By merging of neighbored segments, erroneous watersheds should be eliminated (see Fig. 7a). After applying the watershed transform this criterion is calculated for every region. The numbers are normalized to the maximum gray level G_{MAX} and the regions are filled with these values.

The resulting *mosaic image* can be used for another watershed transform during an iterative process (see Fig. 7b). The merging criteria is based on region properties (i.e. texture features) and the gradient amplitude between the regions. Using the region mean value, the quality is comparable with the results obtained by a classical region-growing approach (see Fig. 7c).

Fig. 7a: Over-segmented result of the watershed transform

Fig. 7b: Merged watershed regions using the region mean value

Fig. 7c: Thresholding of the merged watershed regions

4.4 Combination of Watershed Algorithms

Using the watershed smoothing, most of the stones are separated but with shifted segment contours (see Fig. 8a). The mosaic image of the watershed merging process contains the correct contours but several objects are joined erroneously (see Fig. 7c). A combination of these approaches is able to separate objects from a background. First, both algorithms have to be applied to the input image. Using a logical operation the erroneous joined results of the watershed merging can be splitted by the result of the watershed smoothing. A superposition of the joined segments with the shifted watersheds is shown in Fig. 8b. After the separation of the stones a postprocessing stage is necessary. The elimination of remains may be done by a *morphological opening*, which is defined as an erosion followed by a dilation. The final result of the combined watershed algorithm applied to irregular stones is shown in Fig. 8c.

5. Conclusions

Different approaches were presented to support the construction of a facade map by using digital image analysis. We can summarize, that the watershed transformation is well suited for the segmentation of facades. The independence on the absolute gray-levels by using the gradient image and the viscosity while expanding the segments are major advantages. For the known over-segmentation problem different solutions were shown. The smoothing of the gradient image leads to a good separation of relevant structures. For an exact determination of the contour the shifted segment borders have to be revised. Exemplary the active contour method was introduced, which seemed not robust enough for practical use. By merging small watershed regions the original contour position is preserved. The choice of a fusion characteristic may cause a dependence on the absolute gray-levels, which leads to erroneous results. Therefore, appropriate preprocessing techniques for architectural images were shown to improve area based techniques. Finally, the presented combination of the watershed algorithms is able to separate objects from a background and is well suited for the segmentation of irregular stones.

415

 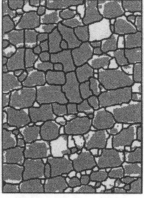 

Fig. 8a: Separation of stones using the watershed transform on a smoothed gradient

Fig. 8b: Superposition of the merged segments *(gray)* with the watersheds *(black)*

Fig. 8c: Segmentation result for irregular stones using the combined technique

Acknowledgement

This work was funded by the German Research Foundation *(DFG)*.

References

1. ALBERTZ, J.: *Invarianten und ihre Bedeutung für Photogrammetrie und Fernerkundung*, Vermessung, Photogrammetrie, Kulturtechnik 83, pp. 295-299, 1985.

2. BUSCH, A.: *Fast Recognition of Lines in Digital Images Without User-Supplied Parameters*, Spatial Information from Digital Photogrammetry and Computer Vision, International Archives of Photogrammetry and Remote Sensing 30, Part 3/1, pp. 91-97, 1994.

3. CANNY, J.: *A Computational Approach to Edge Detection*, Pattern Analysis and Machine Intelligence 8, No. 6, pp. 679-698, 1986.

4. DERICHE, R.: *Using Canny's Criteria to Derive a Recursively Implemented Optimal Edge Detector*, Int. Journal of Computer Vision 1, No. 2, pp. 167-187, 1987.

5. KASS, M., WITKIN, A. UND TERZOPOULUS, D: *Snakes: Active contour models*, Int. Journal of Computer Vision, Band 1, No. 4, pp. 321-331, 1988.

6. MARR, D. AND HILDRETH, E.C.: *Theory of Edge Detection*, Proc. of the Royal Society of London B207, pp. 187-217, 1980.

7. STOCKMAN, T.G.: *Image processing in the context of a visual model*, Proc. IEEE 60, No. 7, pp. 828-842, 1972.

8. TRAUTHAN, F.: *Versuche zur automatischen Erfassung von Steinen einer Hausfassade*, Hrsg.: F.K. List, Publikation zur 13. DGPF-Jahrestagung: Geoinformation durch Fernerkundung, Band 2, Augsburg, pp. 221-228, 1993.

9. VINCENT, L. AND SOILLE, P.: *Watersheds in Digital Spaces: An Efficient Algorithm Based on Immersion Simulation*, IEEE Trans. on Pattern Analysis and Machine Intelligence 13, No. 6, pp. 583-598, 1991.

10. VOSSEPOEL, A.M., STOEL, B.C., MEERSHOEK, A.P.: *Adaptive Histogram Equalization Using Variable Regions*, 9. Int. Conf. on Pattern Recognition, IEEE, pp. 351-353, 1988.

11. WAHL, F.M.: *Digitale Bildsignalverarbeitung*, Nachrichtentechnik 13, Springer, 1984.

12. XU, YONGLONG: *Untersuchung der Bildsegmentation zwecks der nachfolgenden wissensbasierten Bildanalyse*, ISPRS Spatial Information from Digital Photogrammetry and Computer Vision, Munich, pp. 931-938, 1994.

Morphological Iterative Closest Point Algorithm

C. P. Vavoulidis and I. Pitas

Department of Informatics, University of Thessaloniki, Thessaloniki 540 06, GREECE

Abstract. This paper describes a method for accurate and computationally efficient registration of 3−D shapes including curves and surfaces. The method is based on the iterative closest point (ICP) algorithm. The real strength of our algorithm is the use of Morphological Voronoi tessellation method to construct the Voronoi regions around the seed points with respect to a certain distance metric. The tessellated domain can be used to avoid the most computationally expensive step in the ICP, namely to find the closest points. Thus the proposed algorithm is much faster than classical ICP.

1 Introduction

Mathematical morphology [1,4] can be used to construct the Voronoi diagram. Algorithms from mathematical morphology can be found in [2]. Performance analysis of morphological Voronoi tessellation and an implementation of Voronoi diagram based on a region growing method can be found in [6]. The classical ICP algorithm is given in [5]. An accelerated version of the classical ICP algorithm is also given in this paper.

Our algorithm proposes a solution to a key registration problem in computer vision: Given a model 3−D shape and and a data 3−D shape, estimate the optimal rotation and translation that registers the model shape and the actual data shape by minimizing the mean square distance between the shapes. The algorithm requires no extracted features, no curve or surface derivatives and no preprocessing of 3−D data.

An important application of this method is to register actual data sensed from a 3−D object with an ideal 3−D model. Another application is multimodal 3−D image registration in medical imaging (e.g. between NMR and CT volumes). The described method is also useful in the shape equivalence problem as well as in estimating the motion between point sets where the correspondences are not known.

2 The Iterative Closest Point Algorithm

2.1 ICP Algorithm Description

In the ICP algorithm, a data shape P is registered to be in best alignment with a model shape X. In our algorithm the data and the model shape must be decomposed first into point sets if they are not already in this form. The number of 3–D points in the data shape will be denoted N_p and the number of points in the model shape will be denoted N_x.

The distance metric d between an individual data point $\mathbf{p}$ and a model shape X will be denoted $d(\mathbf{p}, X) = \min_{\mathbf{x} \in X} \| \mathbf{x} - \mathbf{p} \|$. We use the Euclidean distance as the distance metric. Let C be the closest point operator and Y denote the resulting set of closest points: $Y = C(P, X)$. Given the resultant corresponding point set Y, the least squares registration is computed as: $(\mathbf{q}, d) = Q(P, Y)$. The positions of the data shape point set are then updated via $P = \mathbf{q}(P)$.

The ICP algorithm is given below:

1. The data point set P with N_p points $\mathbf{p}_i$ and the model point set X with N_x points $\mathbf{x}_i$ are given.
2. The iteration is initialized by setting $P_0 = P, \mathbf{q}_0 = [1, 0, 0, 0, 0, 0, 0]^t$ and $k = 0$. The registration vectors are defined relative to the initial data set P_0 so that the final registration represents the complete registration. The four steps are applied until convergence within a tolerance τ.

   (a) Compute the closest points: $Y_k = C(P_k, X)$.
   (b) Compute the registration vector $(\mathbf{q}_k, d_k) = Q(P_0, Y_k)$.
   (c) Apply the registration: $P_{k+1} = \mathbf{q}_k(P_0)$.
   (d) Terminate the iteration when the change in mean-square error falls bellow a threshold $\tau > 0$ specifying the desired precision of the registration $(d_k - d_{k+1} < \tau)$.

If a dimensionless threshold is desired we can replace τ with $\tau\sqrt{tr(\Sigma_x)}$, where the square root of the trace of the covariance of the model shape indicates the rough size of the model shape.

If someone compute the closest point set in each step of the algorithm then the computational complexity of this step is $O(N_p N_x)$. If N_t is the number of initial translation states and N_q is the number of initial rotation states then the cost of local matching is $O(N_t N_q N_p N_x)$ and the cost of global matching is $O(N_q N_p N_x)$. In our algorithm we compute the Voronoi diagram by using an Euclidean distance metric and then we use it in order that the closest point sets are specified in every step of the algorithm by simple reference. The Voronoi regions can be computed as a preprocessing step independently of the rest of the algorithm. The major drawback of our method is that it is space consuming because it is necessary to keep the entire volume containing the Voronoi region labels in RAM. Another problem is the integer arithmetic used when we assign a point to a Voronoi region instead of the floating point arithmetic. However, according to our experimental results, the error caused by this approximation is not significant.

The unit quaternion is a four vector $\mathbf{q}_R = [q_0 q_1 q_2 q_3]^t$ where $q_0 \geq 0$ and $q_0^2 + q_1^2 + q_2^2 + q_3^2 = 1$. The 3×3 rotation matrix generated by a unit quaternion is the one below:

$$R = \begin{bmatrix} q_0^2 + q_1^2 - q_2^2 - q_3^2 & 2(q_1 q_2 - q_0 q_3) & 2(q_1 q_3 + q_0 q_2) \\ 2(q_1 q_2 + q_0 q_3) & q_0^2 + q_2^2 - q_1^2 - q_3^2 & 2(q_2 q_3 - q_0 q_1) \\ 2(q_1 q_3 - q_0 q_2) & 2(q_2 q_3 + q_0 q_1) & q_0^2 + q_3^2 - q_1^2 - q_2^2 \end{bmatrix} . \quad (1)$$

Let $q_T = [q_4 q_5 q_6]^t$ be a translation vector. The complete registration state vector q is denoted $q = [q_R \mid q_T]^t$. Let $P = \{p_i\}$ be a measured data point to be aligned with a model point set $X = \{x_i\}$, where $N_x = N_p$ and where each point p_i corresponds to the point x_i with the same index. The mean square objective function to be minimized is:

$$f(q) = \frac{1}{N_p} \sum_{i=1}^{N_p} \| x_i - R(q_R)p_i - q_T \|^2 . \quad (2)$$

The center of mass μ_p of the measured point set P and the center of mass μ_x of the model point set X are given by:

$$\mu_p = \sum_{i=1}^{N_p} p_i \quad \text{and} \quad \mu_x = \sum_{i=1}^{N_x} x_i . \quad (3)$$

The cross-covariance matrix Σ_{px} of the sets P and X is given by:

$$\Sigma_{px} = \frac{1}{N_p} \sum_{i=1}^{N_p} [(p_i - \mu_p)(x_i - \mu_x)^t] = \frac{1}{N_p} \sum_{i=1}^{N_p} [p_i x_i^t] - \mu_p \mu_x^t . \quad (4)$$

The cyclic components of the anti-symmetric matrix $A_{ij} = (\Sigma_{px} - \Sigma_{px}^T)_{ij}$ are used to form the column vector $\Delta = [A_{23} A_{31} A_{12}]^T$. This vector is then used to form the symmetric 4×4 matrix $Q(\Sigma_{px})$:

$$Q(\Sigma_{px}) = \begin{bmatrix} tr(\Sigma_{px}) & \Delta^T \\ \Delta & \Sigma_{px} + \Sigma_{px}^T - tr(\Sigma_{px})I_3 \end{bmatrix} \quad (5)$$

where I_3 is the 3×3 identity matrix. The unit eigenvector $q_R = [q_0 q_1 q_2 q_3]^t$ corresponding to the maximum eigenvalue of the matrix $Q(\Sigma_{px})$ is selected as the optimal rotation. The optimal translation vector is given by $q_T = \mu_x - R(q_R)\mu_p$. The least squares quaternion operation is denoted as $(q, d_{ms}) = Q(P, Y)$ where d_{ms} is the mean square point matching error. The notation $q(P)$ is used to denote the point set P after transformation by the registration vector q.

2.2 The Set of Initial Registrations

The converge of the algorithm in a local minimum depends on the initial registration vector. The only way to be sure that we reach the global minimum is to find all the local minima and to select the smallest one.

Consider a 6–D state space Ω where the quaternion component q_0 is determined from the other quaternion components as $q_0 = \sqrt{1 - (q_1^2 + q_2^2 + q_3^2)}$. Thus, the state space Ω is a subset of the space $\Omega' = [-1, 1]^3 \times \mathbb{R}^3$, where $\mathbb{R}$ is the real line. If one have fixed the data and model shapes then there is a finite number of local minima $N_m(X, P)$. Let $\Psi(X, P)$ be the set of all local minima: $\Psi(X, P) = \{\psi_n\}_{n=1}^{N_m}$. This induces a natural partitioning of Ω into equivalence classes Ψ_n, where every value of q that converges to ψ_n is a member of the class Ψ_n. More typically it is stated as:

$$\Omega = \bigcup_{n=1}^{N_m} \Psi_n \quad \text{and} \quad \Psi_n \cap \Psi_m = \emptyset \text{ if } n \neq m . \tag{6}$$

To guarantee that the global minimum will be reached, we must use an appropriate set of initial registration states so that transforming P by at least one initial registration will place the point set into the equivalence class of registrations that leads to global minimum.

From an implementation point of view, we can use all the normalized combinations of $q_0 = \{1, 0\}$, $q_1 = \{+1, 0, -1\}$, $q_2 = \{+1, 0, -1\}$ and $q_3 = \{+1, 0, -1\}$. For more complicated shapes all the normalized combinations of $q_0 = \{1, 0.5, 0\}$, $q_1 = \{+1, 0.5, 0, -0.5, -1\}$, $q_2 = \{+1, 0.5, 0, -0.5, -1\}$ and $q_3 = \{+1, 0.5, 0, -0.5, -1\}$ provides an adequate set of 312 initial rotation states. If we want to achieve local matching, then we have to combine all the initial rotation with all the initial translation vectors in order to construct the set of all the initial registration vectors.

3 The Morphological Voronoi Tessellation

3.1 The Region Growing Approach

The Voronoi diagram partitions the Euclidean space $\mathbb{R}^3$ into a number of regions. Let $X = \{x_1, x_2, ..., x_m\}$ be a set of m points in $\mathbb{R}^3$. These points are called *seeds*. The Voronoi tessellation algorithm assigns a region of the space $\mathbb{R}^3$ to each seed. The points in a certain Voronoi region $V(i)$ have the following property: they are closer to the seed point $x_i \in X$ than to any other seed point $x_j \in X, j \neq i$, according to a specified distance metric. The Voronoi diagram of X is the union of Voronoi regions for all $x_i \in X$. The process of constructing the Voronoi diagram of a set $X \subseteq \mathbb{R}^3$ is called *Voronoi tessellation* of $\mathbb{R}^3$.

Voronoi tessellation is a classical topic in computational geometry. There are several algorithms that based on a divide and conquer strategy and find intersections of half planes. Although the computational complexity of this method is relatively low, $O(N log N)$ where N is the number of seeds, its computation is rather complicated. Furthermore this algorithm poses practical problems in the construction of on the Voronoi diagram on the Euclidean grid $\mathbb{Z}^3$.

The approach we present in the following based on a region growing algorithm from mathematical morphology. The construction of the Voronoi diagram in $\mathbb{Z}^3$ could be implemented with a growth mechanism of the points of X. At each step n new points will be appended to each $x_i \in X$. The set of new points appended to x_i at each step n is called the $n-$Voronoi neighborhood and denoted $D_n(i)$.

$$D_n(i) = \{\mathbf{x} \in J : d(\mathbf{x}, \mathbf{x}_i) = n, \quad n \in \mathbb{N}, \quad d(\mathbf{x}, \mathbf{x}_i) \leq d(\mathbf{x}, \mathbf{x}_j), \quad j \neq i\} \qquad (7)$$

where $J \subseteq \mathbb{Z}^3$, $n \in \mathbb{N}$ is the distance of a point $\mathbf{x} \in J$ from $\mathbf{x}_i$ and $d()$ is a distance function. The n-Voronoi region $X_n(i)$ is defined as:

$$X_n(i) = \bigcup_{k=0}^{n} D_k(i), \quad n \in \mathbb{N} \qquad (8)$$

where n is the growth step.

The proposed algorithm consists of an iterative region growing of all the n-Voronoi regions $X_n(i)$ that corresponds to every seed point $\mathbf{x}_i$. When we have collisions, then the growth stops and the collision points are labeled as border. The algorithm stops when all the Voronoi regions cannot grow any more. If we assume that each growth step has the same computational cost, the computational complexity of this method depends only on the number of growth steps executed.

3.2 Implementation of the Morphological Voronoi Tessellation

The input to the algorithm is the seeds points. We choose a structuring element and in each growth step we evaluate the new Voronoi regions. The structuring element corresponds to an approximation of the Euclidean distance in $\mathbb{Z}^3$. This process is repeated until all the points are assigned to a certain Voronoi region or are specified as borders. The dilation mechanism is implemented by scanning all the volume points at each step. A window that represents the structuring element is center on each point. If the center is zero labeled, that is not yet specified, we check the visible points through the window and label this point with a value that depends on the values of the observed points. The center point can be labeled with:

1. a zero value, when all the observed points are zero labeled.
2. A negative value that represents the border value when two or more points labeled with different positive values are observed from the window.
3. A certain positive value when one or more visible points from the window are labeled with the same value.

4 Experimental Results

4.1 3−D Curve Matching

In this section we demonstrate the ability of morphological ICP algorithm in 3−D curve matching. We take 11 points from a curve in a volume, we rotate them 20° around Z axis and 40° around the X axis and translate all the points with the translation vector (56,55,52). The initial points and the points after rotation and translation are given in Fig.1.

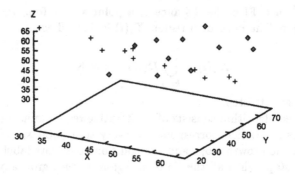

Fig. 1. Model points and Data points after rotation and translation

After the computation of the morphological Voronoi diagram by using the generalization of octagonal distance function in $\mathbb{Z}^3$, it takes only 4 iterations in order the ICP algorithm to converge to the global minimum. The initial points and the points after the registration are given in Fig.2.

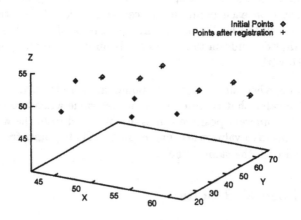

Fig. 2. Model points and Data points after registration

4.2   3–D Surface Matching

In this section we demonstrate the ability of morphological ICP algorithm in surface matching. We take 119 points from the CT frames of a human skull. We rotate the points in a similar way with the curve matching example and translate all the points with the translation vector (81,80,77). The initial points and the points after rotation and translation are given in Fig.3. First we calculate the

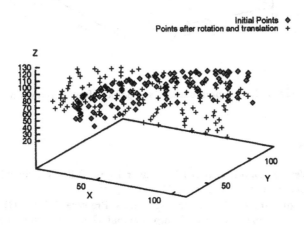

Fig. 3. Model points and Data points after rotation and translation

morphological Voronoi tessellation. It takes only 13 iterations in order the ICP algorithm to lock in the global minimum. The initial points and the points after registration are given in Fig.4.

5   Conclusions

The Morphological Iterative Closest Point algorithm is a strong and fast method for the registration of actual data sensed from a 3–D object with an ideal 3–D model. The Voronoi tessellation is used to eliminate the time consuming step of computing the closest point set in each step of the algorithm as the classical ICP algorithm does. Thus, our method is much faster than classical ICP. This method is useful in 3–D registration, registration of medical images, shape equivalence problems and motion estimation between point sets.

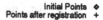

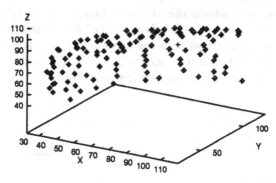

Fig. 4. Model points and Data points after registration

References

1. Pitas, I., Venetsanopoulos, A.N.: Nonlinear Digital Filters: Principles and Applications. Boston: Kluwer Academic (1990)
2. Pitas, I.: Digital Image Processing Algorithms. Prentice Hall (1993)
3. Preparata, F.P., Shamos, M.I.: Computational Geometry. New York: Springer-Verlag (1985)
4. Haralick, R.M., Shapiro, L.G.: Computer and Robot Vision. Addison-Wesley (1992)
5. Besl, P.J., McKay, N.D.: A method for Registration of 3−D Shapes. IEEE Trans. on Pattern Analysis and Machine Intelligence **14** (1992) 239–256
6. Pitas, I.: Performance Analysis and Parallel Implementation of Voronoi Tessellation Algorithms Based on Mathematical Morphology. IEEE Trans. on Pattern Analysis and Machine Intelligence (Submitted)
7. Kotropoulos, C., Pitas, I., Maglara, A.: Voronoi Tessellation and Delaunay Triangulation Using Euclidean Disk Growing in $\mathbf{Z}^2$. Int. Conf. on Acoust. Speech and Signal Processing (1993)

Planning Multiple Views for 3–D Object Recognition and Pose Determination *

Franjo Pernuš, Aleš Leonardis** and Stane Kovačič

University of Ljubljana
Tržaška 25, 1001 Ljubljana, Slovenia
E-mail: feri@.fe.uni-lj.si

Abstract. We present a method for automatic recognition and pose (orientation) determination of 3-D objects of arbitrary shape. The approach consists of a learning stage in which we derive a recognition and pose identification plan and a stage in which actual recognition and pose identification take place. In the learning stage, the objects are observed from all possible views and each view is characterized by an extracted feature vector. These vectors are then used to structure the views into clusters based on their proximity in the feature space. To resolve the remaining ambiguity within each of the clusters, we designed a strategy which exploits the idea of taking additional views. We developed an original procedure which analyzes the transformation of the individual clusters under changing viewpoints into several smaller clusters. This results in an optimal next-view planning when additional views are necessary to resolve the ambiguities. This plan then guides the actual recognition and pose determination of an unknown object in an unknown pose.

1 Introduction

One of the major tasks of computer vision is the identification of an unknown 3-D object as a member of a set of known objects. In certain tasks, the identity of an object is known but the object has to be inspected (observed) from a particular view (direction), e.g., to detect defects or to measure dimensions, or it has to be re-oriented for further fine manipulation. In such cases, only the pose of an object, defined as the orientation and the position within a predefined coordinate system, has to be determined. Therefore, vision systems must be able to reason about both identity and pose. Different approaches to the recognition and pose determination have been proposed, which mainly vary in the representation of 3-D objects and in the search techniques for matching data to models [3]. *Viewer-centered* or *appearance-based* representations use a set of images of an object, obtained from different views, as its implicit description [3]. The approaches based on appearance may be classified according to the degree of structure that

---

* This work was supported in part by The Ministry for Science and Technology of Slovenia (Projects J2-8829, J2-7634 and L2-7691) and by the Austrian national Fonds zur Förderung der wissenschaftlichen Forschung under grant S7002MAT.
** Also with the Technical University Vienna, Austria.

is extracted from the image data and the number of views from which an unknown object must be observed in order to be recognized and its pose determined. The recognition phase very much depends on these two parameters.

Concerning the structure, an extreme approach is to use explicit collections of 2-D images as object models. A similarity measure, like correlation, can then be used to determine how well the unknown image data matches the stored images of the models. Such an approach is very time consuming and requires large amounts of storage. The situation gets worse as the number of objects and views increases. Therefore, images have to be somehow compressed. A well-known image compression technique is the Karhunen-Loeve transform which is based on principal component analysis [1]. Murase and Nayar [8] used this transform to compress the image set to a low-dimensional subset, called the eigenspace, in which an object is represented as a manifold. Higher compression can be achieved by extracting different features from images. This approach models the objects as vectors of characteristic features, each of which corresponds to a point in the multidimensional feature space. Object identification and pose determination is then achieved by using pattern recognition techniques for finding the best match between an unknown object feature vector and the model feature vectors.

Thus far we have assumed that the recognition is based on a single image of an unknown object. The problem with this approach is that it implicitly assumes that the extracted features are sufficient to determine the identity and the pose of an object. In general, however, an image obtained from a single view does not contain sufficient information and is therefore inadequate to uniquely solve the 3-D object- and/or pose-identification task. The reason is that some of the objects are inherently ambiguous, i.e., they differ only in the sequence of observations rather than in the value of any given observation. A natural way to resolve the ambiguities and to gain more information about an object is to observe it from additional views [2, 6, 7]. In this respect, the closest related approach to ours was reported by Gremban and Ikeuchi [2].

The questions that arise in this context are: How many additional views do we need to take? Observing an unknown object from all possible views would be very time consuming and also in most cases unnecessary; thus, only a smaller number of additional views should be used. The next question is therefore: which views contain more information and enable a more reliable recognition? Those views that reduce the ambiguities more than other views should be applied first, which leads us to the last question: How do we determine the sequence of the additional views? This paper is an attempt to answer these questions.

2 Problem formulation

Let us suppose that we are dealing with M different 3-D objects, and that each object m is represented by a set V_m of 2-D images obtained from an ordered set of N different views.

$$V_m = \{v_{m,1}, v_{m,2}, \ldots, v_{m,n}, \ldots, v_{m,N}\}, \quad (m = 1, 2, \ldots, M) \ ,$$

where $v_{m,n}$ is the image of object m from view n. The overall set V then contains $M \times N$ available images

$$V = \cup_{m=1}^{M} V_m = \{v_{m,n}; \ (m = 1, 2, \ldots, M; \ n = 1, 2, \ldots, N)\}.$$

Assume that each 2-D image $v_{m,n}$ is described by I distinguishing characteristics or features, which form the feature vector $\mathbf{f}_{m,n}$:

$$\mathbf{f}_{m,n} = [{}^{1}\!f_{m,n}, {}^{2}\!f_{m,n}, \ldots, {}^{I}\!f_{m,n}].$$

In principle, the feature vector $\mathbf{f}_{m,n}$ can represent the entire image. The set F, containing all feature vectors, is therefore a representation of the set V of available 2-D images:

$$F = \{\mathbf{f}_{m,n}; \ (m = 1, 2, \ldots, M; \ n = 1, 2, \ldots, N)\}.$$

Each feature vector can be represented as a point in the I-dimensional feature space. Because our task is twofold, to identify an unknown object and to determine its pose, let each object m, being observed from view n represent a class. Now, suppose that we obtain an image w of an unknown object in an unknown pose and represent this image by a feature vector $\mathbf{x}$, $\mathbf{x} = [{}^{1}\!x, {}^{2}\!x, \ldots, {}^{I}\!x]$. Based on the information contained in the feature vector $\mathbf{x}$, we want to determine to which class these data belong, i.e., we want to reveal the identity of the object and its pose. Different techniques can be used to tackle this problem [1]. However, in general it is not always possible for recognition system to obtain a correct classification for certain views. There are two reasons for this: some views are inherently ambiguous, and the extracted features do not possess enough discriminative power. Therefore, some points in the feature space will tend to lie very close one to another, causing the recognition to be at least unreliable if not even impossible. Nevertheless, in many cases this problem may be solved by using multiple views, which is exactly the approach we are exploring.

3 View planning and recognition

In this section we show how the next views are planned for a reliable object identification and pose determination by cluster analysis. Feature vectors are first used to structure the views into equivalence classes or clusters based on their proximity in the feature space [5]. Depending on the set of objects, different types of features that we choose to characterize the views, and the required reliability of the recognition and pose determination process, these clusters may contain one, a few, or many views. To resolve the remaining ambiguity within each of the clusters, we designed an original strategy which exploits the idea that by taking an additional view, feature vectors within each cluster will be transformed into feature vectors which may form more than one cluster.

The clustering of a data set F, $F = \{\mathbf{f}_{m,n}\}$ is a partition C of F into a number of subsets, i.e., clusters $C_1, C_2, \ldots$, such that the distance or dissimilarity $d(C_i, C_j)$ between any two clusters is larger than some predefined value δ,

$d(C_i, C_j) > \delta; \ \forall i, j, \ i \neq j$. The clusters satisfy the requirements that each subset must contain at least one feature vector and that each feature vector must belong to exactly one subset. Suppose that a clustering method applied to partition the feature set F, $F = \{\mathbf{f}_{m,n}\}$ resulted in K clusters or equivalence classes $C_1, C_2, \ldots, C_j, \ldots, C_K$. Let each cluster C_j contain the feature subset F_j, such that $F = \cup_{j=1}^{K} F_j$ and $F_i \cap F_j = \emptyset; \ \forall i, j, \ i \neq j$. Depending on the set of objects, different types of features that we choose to characterize the views, and the clustering method, the number of clusters will vary and each cluster may contain one, a few, or many views. Fig. 1 (a) shows 30 feature vectors in a 2-D feature space representing 15 views of two objects, and Fig. 1 (b) shows the clusters ($K=9$) formed by the feature vectors shown in Fig. 1 (a).

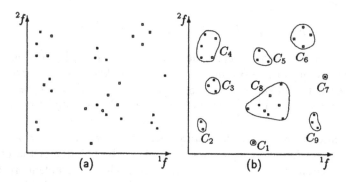

Fig. 1. An example: (a) Feature vectors in a 2-D feature space representing $N = 15$ views of each of the two objects ($M = 2$). Views of the first object are depicted by open squares and views of the second object by closed squares; (b) Clusters formed by the feature vectors.

To reliably recognize an object and its pose we have to resolve the ambiguity within each of the clusters containing more than one feature vector. We apply the strategy which exploits the idea of taking additional views. Let the subset $F_j = \{\mathbf{f}_{o,p}\}$ form a cluster C_j containing more than one feature vector. We can observe the objects represented by these feature vectors from the k-th next view, where $k = 1, 2, \ldots, N-1$. By each new observation the feature vectors in F_j will be transformed to feature vectors forming the subset $F_j^k = \{\mathbf{f}_{o,p+k}\}$:

$$F_j = \{\mathbf{f}_{o,p}\} \longrightarrow F_j^k = \{\mathbf{f}_{o,p+k}\}, \quad (k = 1, 2, \ldots, N-1),$$

where o and p run over ambiguous objects and views and addition is taken modulo N. Each of the $N - 1$ transformations of F_j under changing viewpoint can be analyzed by clustering F_j^k to derive the "best" next-view. For that purpose, the feature space containing the transformed subset F_j^k is further partitioned into a number of clusters, as shown in Fig. 2 for subset F_8 and $k = 1$; see also Fig. 1.

Among the $N - 1$ possible next views, the one yielding the largest number of clusters can be selected as the "best" one. If two or more views produce the same number of clusters, the one with the largest average intercluster dissimilarity is selected. We recursively apply this strategy on all clusters containing more

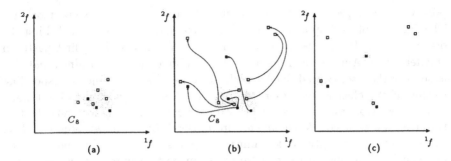

Fig. 2. An example of mapping the subset F_8 under changing viewpoint. (a) The subset F_8 forming cluster C_8 (see also Fig. 1). (b) Transformation of F_8 for $k = 1$. (c) Clustering of the transformed subset F_8^1. Shaded areas represent the clusters.

than one feature vector until only clusters with one feature vector remain, or a transformed cluster can not be further partitioned into more clusters. This procedure results in a recognition-pose-identification (RPI) plan which has the form of a tree. The root of the tree contains the set F and with each non-terminal node a subset of feature vectors and the "best" relative next view are associated. Each terminal node (leaf) may contain one or more views. In the ideal case each leaf in the tree encompasses only a single view and with each leaf, a class label indicating the object and its pose is associated. However, if a leaf contains a set of views, the identity of these views can not be determined by the selected features and clustering method.

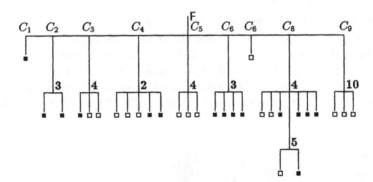

Fig. 3. Recognition-pose-identification plan for the feature vectors shown in Figure 1. Each bold number gives the "best" next view relative to the current one.

Fig. 3 shows the RPI plan for the feature vectors depicted in Fig. 1 (a). This plan guides the actual recognition and pose determination of an unknown object in an unknown pose. The RPI plan reveals that two poses can be identified from one view, that for the majority of poses an additional view is required, and that two poses can be determined after taking two additional views.

Once the RPI plan is built, an image w of an unknown object in an unknown pose which is represented by the feature vector **x** can be recognized on-line. This is accomplished by traversing the RPI tree until a leaf is reached. First the distance of the feature vector **x** to each of the K clusters of the first partition is determined. Among these distances the minimal one is determined and if it is smaller than some predefined threshold, **x** is assigned to the corresponding cluster. If the cluster contains more than one feature vector, the object must be observed from the view associated with the corresponding node and the tree traversal is continued. When we end up in a leaf, the unknown object and its current pose can readily be determined. If necessary, the original pose can be revealed by summing up the relative moves made to acquire new views on the path from the root to the leaf and subtracting (modulo N) the sum from the final view.

If the features do not possess enough discriminating power, feature vectors will lie close one to the other in the feature space and clusters will only be formed for small dissimilarities δ. Because of the small intercluster distances the recognition of an unknown object in an unknown pose will not be very reliable. On the other hand, if the selected features are highly discriminative, the value of δ can be higher, the intercluster distances will be larger and this will result in a more reliable recognition. Obviously, there is a trade-off between the dissimilarity δ, which influences the *reliability* of the recognition and the number of necessary additional views, which influences the *complexity* of the recognition-pose-identification plan.

4 Experiments

The procedure for building the RPI plan has been tested on various objects; four of them are shown in Fig. 4. One object at a time is placed on a motorized turntable and its pose is varied about a single axis, namely, the axis of rotation of the turntable. To explicitly show the power of the approach, we have *intentionally* chosen a very *restricted* set of features with which we model individual views. The features used in this experiment were based on the object silhouettes (Fig. 5).

A   B   C   D

Fig. 4. Four objects used in the experiments.

Fig. 5. Silhouettes of different poses of object A depicted in Fig. 4.

After the silhouettes are segmented, the object contours get extracted and the two moment invariants [4] are computed:

$$^1f = \mu_{20} + \mu_{02}, \quad ^2f = \sqrt{(\mu_{20} - \mu_{02})^2 + 4\mu_{11}^2},$$

where $\mu_{ij}, (i+j=2)$ denote the central moments of second order. Fig. 6 shows the feature vectors $\mathbf{f}_{m,n}$, $\mathbf{f}_{m,n} = \left[{}^1f_{m,n}, {}^2f_{m,n} \right]$, $(m = 1, \ldots, 4; n = 1, 2, \ldots, 13)$ representing the objects from Fig. 4. The outlined clusters were obtained by forming a minimal spanning tree and deleting the edges longer than a predefined distance δ, $\delta = \delta_0$. Fig. 7 shows the RPI plan for the feature vectors depicted in Fig. 6. This plan then guides the actual recognition and pose determination of an unknown object in an unknown pose.

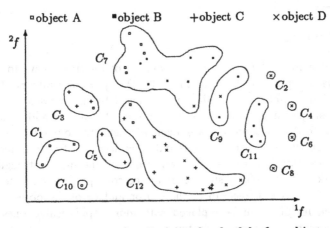

Fig. 6. Feature vectors representing 13 views of each of the four objects. The outlined clusters were obtained by a minimal spanning tree clustering procedure. Edges longer than some prescribed distance δ_0 were deleted.

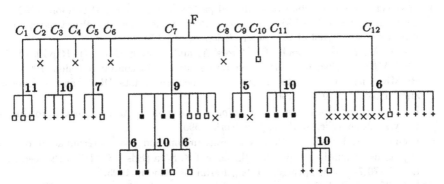

Fig. 7. Recognition-pose-identification plan for δ_0. Bold numbers give the "best" relative next views.

We performed the additional experiments to demonstrate the tradeoff between the reliability (defined by the dissimilarity measure δ) and the speed of recognition (number of necessary additional views). If δ is increased/decreased the recognition is more/less reliable, but more/fewer additional views must be taken. The experiments with 3 different δ-s are summarised in Table 1. As δ increases, in general, less poses can be identified from one view. To identify the more ambiguous ones, additional views must be taken. For instance, for $\delta = 1.15\delta_0$, three additional views are needed to identify the 7 most ambiguous poses.

δ	Additional views			
	0	1	2	3
$0.85\,\delta_0$	9	36	7	0
δ_0	5	37	10	0
$1.15\,\delta_0$	5	15	25	7

Table 1. Number of poses which can be recognized as a function of additional views for different values of δ.

5 Conclusions

We presented a method for automatic planning of multiple views for recognition and pose (orientation) determination of 3-D objects of arbitrary shape. We demonstrated that ambiguities, inherent or caused by the selection of features, present in a single image can be resolved through the use of additional sensor observations. To demonstrate the approach we restricted ourselves to a rather simple situation where the objects have only one degree-of-freedom and extracted features are simple features. However, we view the major strength of the proposed work in its generality since there are no conceptual barriers to increase the number of degrees-of-freedom of the sensor and/or of the object. Besides, currently used features can be replaced with more sophisticated ones without affecting other components of the system.

References

1. Fukunaga K: Introduction to statistical pattern recognition. AP London, 1990.
2. Gremban KD and Ikeuchi K: Planning multiple observations for object recognition. International Journal of Computer Vision 12:137–172, 1994.
3. Hebert M, Ponce J, Boult TE, Gross A, and Forsyth D (Eds): Report of the NSF/ARPA workshop on 3D object representation for computer vision. 1994.
4. Hu MK: Visual Pattern Recognition by Moment Invariants. IRE Trans. on Information Theory 8:179-187, 1962.
5. Kaufman L and Rousseeuw PJ: Finding groups in data. An introduction to cluster analysis. John Wiley & Sons, New York, 1990.
6. Leonardis A, Kovačič S, Pernuš F: Recognition and pose determination of 3-D objects using multiple views. V. Hlaváč and R. Šára (Eds.), CAIP'95 Proceedings, LNCS 970:778–783, Springer-Verlag, Berlin Heidelberg, 1995.
7. Liu CH and Tsai WH: 3D curved object recognition from multiple 2D camera views. Comput. Vision, Graphics, Image Processing 50:177-187, 1990.
8. Murase H and Nayar SK: Visual learning and recognition of 3-D objects from appearances. IJCV 14:5-24, 1995.

Fast and Reliable Object Pose Estimation from Line Correspondences *

Stéphane Christy and Radu Horaud

GRAVIR-IMAG & INRIA Rhône-Alpes
655 avenue de l'Europe
38330 Montbonnot Saint-Martin FRANCE

Abstract. In this paper, we describe a fast object pose estimation method from 3-D to 2-D line correspondences using a perspective camera model. The principle consists in iteratively improving the pose computed with an affine camera model (either weak perspective or paraperspective) to converge, at the limit, to a pose estimation computed with a perspective camera model. Thus, the advantage of the method is to reduce the problem to solving a linear system at each iteration step. The iterative algorithms that we describe in detail in this paper can deal with non coplanar or coplanar object models and have interesting properties both in terms of speed and rate of convergence.

1 Introduction and background

The problem of object pose from 2-D to 3-D correspondences has received a lot of attention for the last years. The perspective camera model has associated with it, in general, non linear object pose computation techniques. This naturally leads to non linear minimization methods which require some form of initialization [3,7,9] . If the initial "guess" is too faraway from the true solution then the minimization process is either very slow or it converges to a wrong solution. Moreover, such resolution techniques does not take into account the simple link that exists between the perspective model and its linear approximations.

Recently, Dementhon and Davis [2] proposed a method for determining the pose of a 3-D object with respect to a camera from 3-D to 2-D point correspondences. The method consists of iteratively improving the pose computed with a weak perspective camera model to converge, at the limit, to a pose estimation computed with a perspective camera model. Although the perspective camera model involves, in general, non linear techniques, the advantage of the method is to reduce the problem to solving, at each iteration step, a linear system – due to the weak perspective model.

In [5], Horaud and al. have extended the method of Dementhon and Davis to paraperspective by establishing the link between paraperspective and perspective models for object pose computation from point correspondences.

---

* This work has been supported by "Société Aérospatiale" and by DGA/DRET.

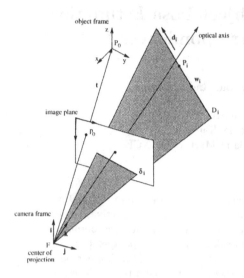

Fig. 1. This figure shows the general setup. The point $\mathbf{P_0}$ is the reference point of the object frame and its projection is $\mathbf{p_0}$. A 3-D line is represented by a reference point $\mathbf{w}_i$ and a direction vector $\mathbf{d}_i$ expressed in the object frame ($\mathbf{w}_i \perp \mathbf{d}_i$). The projection of the 3-D line $\mathbf{D}_i$ is δ_i.

However, it may be useful in some context to be able to deal with other primitives than points – lines for example. The advantage of lines is that they can be more accurately located than points and the robustness of their matching with respect to occlusions. For real time applications, it's generally easier to consider lines, particularly to do the tracking along a sequence of images, and the matching with a 3-D model. In some cases where edges are difficult to extract, the points may be erroneous due to missing segments and thus bring wrong matches which could be solved if we consider the lines described by each edge, without the end points.

In this article, we describe two linear methods to compute object pose from 3-D to 2-D line correspondences. We establish the link between affine camera models (weak perspective and paraperspective), and the perspective model in the case of lines. The proposed algorithms are iterative and suppose, at each iteration step, an affine camera model in order to have linear equations, but converge to the perspective camera model.

2   Camera model

We denote by $\mathbf{P}_i$ a 3-D point belonging to a 3-D line i represented by a reference point $\mathbf{w}_i$ and a direction $\mathbf{d}_i$. We have: $\mathbf{P}_i = \mathbf{w}_i + \lambda_i \mathbf{d}_i$. The origin of the 3-D frame is $\mathbf{P_0}$. A point $\mathbf{P}_i$ projects onto the image in $\mathbf{p}_i$ with camera coordinates x_i and y_i and we have:

$$x_i = \frac{\mathbf{i} \cdot \mathbf{P}_i + t_x}{\mathbf{k} \cdot \mathbf{P}_i + t_z} \qquad y_i = \frac{\mathbf{j} \cdot \mathbf{P}_i + t_y}{\mathbf{k} \cdot \mathbf{P}_i + t_z} \qquad (1)$$

We divide both the numerator and the denominator of eqs. (1) by t_z. We introduce the following notations:

$$\mathbf{I} = \frac{\mathbf{i}}{t_z} \qquad \mathbf{J} = \frac{\mathbf{j}}{t_z} \qquad x_0 = \frac{t_x}{t_z} \qquad y_0 = \frac{t_y}{t_z} \quad \text{and} \quad \varepsilon_i = \frac{\mathbf{k} \cdot \mathbf{P}_i}{t_z}$$

$\mathbf{I}$ and $\mathbf{J}$ represent the first two rows of the rotation matrix scaled by the z-component of the translation vector, and x_0 and y_0 are the camera coordinates of $\mathbf{p}_0$ which is the projection of $\mathbf{P}_0$.

We may now rewrite the perspective equations as:

$$x_i = \frac{\mathbf{I} \cdot \mathbf{P}_i + x_0}{1 + \varepsilon_i} \qquad y_i = \frac{\mathbf{J} \cdot \mathbf{P}_i + y_0}{1 + \varepsilon_i} \qquad (2)$$

with

$$\mathbf{P}_i = \mathbf{w}_i + \lambda \mathbf{d}_i \quad \text{and} \quad \varepsilon_i = \frac{\mathbf{k} \cdot \mathbf{P}_i}{t_z} = \frac{\mathbf{k} \cdot \mathbf{w}_i}{t_z} + \lambda \frac{\mathbf{k} \cdot \mathbf{d}_i}{t_z}$$

By introducing

$$\eta_i = \frac{\mathbf{k} \cdot \mathbf{w}_i}{t_z} \quad \text{and} \quad \mu_i = \frac{\mathbf{k} \cdot \mathbf{d}_i}{t_z}$$

we finally have:

$$\varepsilon_i = \eta_i + \lambda \mu_i$$

3 From weak perspective to perspective

In this section, we derive the pose equations in the case of line correspondences and we show how to find the perspective solution from incremental weak perspective approximations.

Weak perspective assumes that the object points or lines lie in a plane parallel to the image plane passing through the origin of the object frame, i.e., $\mathbf{P}_0$.

The equation of the line i in an image may be written as:

$$a_i x + b_i y + c_i = 0 \qquad (3)$$

a_i, b_i and c_i are the parameters of the image line. If $\mathbf{p}_i$ belongs to this line, we obtain, by substituting x and y (eqs. (2)) in equation (3):

$$a_i \frac{\mathbf{I} \cdot (\mathbf{w}_i + \lambda \mathbf{d}_i) + x_0}{1 + \eta_i + \lambda \mu_i} + b_i \frac{\mathbf{J} \cdot (\mathbf{w}_i + \lambda \mathbf{d}_i) + y_0}{1 + \eta_i + \lambda \mu_i} + c_i = 0$$

or by eliminating the denominator and grouping terms:

$$a_i \mathbf{I} \cdot \mathbf{w}_i + b_i \mathbf{J} \cdot \mathbf{w}_i + a_i x_0 + b_i y_0 + c_i(1 + \eta_i) + \lambda(a_i \mathbf{I} \cdot \mathbf{d}_i + b_i \mathbf{J} \cdot \mathbf{d}_i + c_i \mu_i) = 0$$

This equation is verified for all points $\mathbf{P}_i$ of the 3-D line, i.e., for all values of λ. So we obtain two equations:

$$a_i \mathbf{I} \cdot \mathbf{w}_i + b_i \mathbf{J} \cdot \mathbf{w}_i + a_i x_0 + b_i y_0 + c_i(1 + \eta_i) = 0 \qquad (4)$$
$$a_i \mathbf{I} \cdot \mathbf{d}_i + b_i \mathbf{J} \cdot \mathbf{d}_i + c_i \mu_i = 0 \qquad (5)$$

In these equations, $\mathbf{w}_i$ and $\mathbf{d}_i$ (which represent the 3-D line) are known, and a_i, b_i, c_i are also known (they may be computed from the image). The unknowns are the 3-D vectors $\mathbf{I}$ and $\mathbf{J}$ (which represent the orientation of the camera), the projection of the reference point (x_0, y_0), η_i and μ_i (the perspective effect). However, we don't need any point correspondences, but only line correspondences.

In order to solve the pose problem, we notice that if we take $\eta_i = \mu_i = 0$, eqs. (4) and (5) are similar to the equations obtained for a weak perspective camera model [1,2,5]. We therefore conclude that:

- Whenever the η_i and μ_i are fixed (not necessarily null) the pose equations (4) and (5) become linear in $\mathbf{I}$, $\mathbf{J}$, x_0 and y_0.
- It is possible to solve eqs. (4) and (5) *iteratively* by successive linear approximations.

In this case the pose algorithm starts with a weak perspective camera model and computes an approximated pose. This approximated pose is improved iteratively as follows:

1. For all i, $i \in \{1...n\}$, $\eta_i = \mu_i = 0$;
2. Solve the overconstrained linear system of equations (4) and (5) which provides an estimation of vectors $\mathbf{I}$ and $\mathbf{J}$, and the projection of the origin of the frame (x_0, y_0).
3. Compute the position and orientation of the object frame with respect to the camera frame:

$$t_z = \frac{1}{2}\left(\frac{1}{\|\mathbf{I}\|} + \frac{1}{\|\mathbf{J}\|}\right) \quad t_x = x_0 \, t_z \quad t_y = y_0 \, t_z \quad \mathbf{i} = \frac{\mathbf{I}}{\|\mathbf{I}\|} \quad \mathbf{j} = \frac{\mathbf{J}}{\|\mathbf{J}\|} \quad \mathbf{k} = \mathbf{i} \times \mathbf{j}$$

4. Enforce the orthogonality constraint, i.e., compute a *true* rotation matrix $\mathbf{R}_\perp$ which verifies: $\mathbf{R}_\perp \begin{pmatrix} \mathbf{i}^T \\ \mathbf{j}^T \\ \mathbf{k}^T \end{pmatrix}^T = \mathbf{I}_{3\times 3}$

5. For all i, compute: $\eta_i = \mathbf{k}\cdot\mathbf{w}_i/t_z$ and $\mu_i = \mathbf{k}\cdot\mathbf{d}_i/t_z$. If the η_i and μ_i computed at this iteration are equal to the η_i and μ_i computed at the previous iteration then stop the procedure, otherwise go to step 2.

4 From paraperspective to perspective

In this section, we provide a generalization of the above algorithm to deal with the paraperspective case.

We consider again the perspective equations (2) and we make explicit the paraperspective approximation of the camera model. We obtain [1,5,8]:

$$x_i = \frac{\mathbf{I}_p \cdot \mathbf{P}_i}{1 + \varepsilon_i} + x_0 \quad \text{and} \quad y_i = \frac{\mathbf{J}_p \cdot \mathbf{P}_i}{1 + \varepsilon_i} + y_0 \tag{6}$$

with:

$$\mathbf{I}_p = \frac{\mathbf{i} - x_0 \, \mathbf{k}}{t_z} \quad \text{and} \quad \mathbf{J}_p = \frac{\mathbf{j} - y_0 \, \mathbf{k}}{t_z} \tag{7}$$

By substituting x and y (eqs. (6)) in equation (3) and by grouping terms:

$$a_i \mathbf{I}_p \cdot \mathbf{w}_i + b_i \mathbf{J}_p \cdot \mathbf{w}_i + (a_i x_0 + b_i y_0 + c_i)(1 + \eta_i)$$
$$+ \lambda[a_i \mathbf{I}_p \cdot \mathbf{d}_i + b_i \mathbf{J}_p \cdot \mathbf{d}_i + (a_i x_0 + b_i y_0 + c_i)\mu_i] = 0$$

which yields, as above:

$$a_i \mathbf{I}_p \cdot \mathbf{w}_i + b_i \mathbf{J}_p \cdot \mathbf{w}_i + (a_i x_0 + b_i y_0 + c_i)(1 + \eta_i) = 0 \tag{8}$$
$$a_i \mathbf{I}_p \cdot \mathbf{d}_i + b_i \mathbf{J}_p \cdot \mathbf{d}_i + (a_i x_0 + b_i y_0 + c_i)\mu_i = 0 \tag{9}$$

It is worthwhile to notice that when all the η_i and μ_i are null, the perspective equations above become identical to the paraperspective equations obtained with a paraperspective camera model.

The iterative algorithm is similar to the one described in the previous section. The only difference is step 2: we now compute $\mathbf{I}_p$ and $\mathbf{J}_p$ (instead of $\mathbf{I}$ and $\mathbf{J}$) by solving an overconstrained linear system of equations (8) and (9). Then (step 3), $\mathbf{i}, \mathbf{j}$ and $\mathbf{k}$ are deduced from $\mathbf{I}_p$ and $\mathbf{J}_p$ [1,5,8].

5 Solving the linear equations

Both the weak perspective and paraperspective iterative algorithms need to solve an overconstrained linear system of equations, namely eqs. (4), (5) (weak perspective) and eqs. (8), (9) (paraperspective). In matrix form these equations can be written as:

$$\underbrace{\mathbf{A}}_{2n \times 8} \underbrace{\mathbf{x}}_{8 \times 1} = \underbrace{\mathbf{b}}_{2n \times 1} \tag{10}$$

- More explicitly, we have in weak perspective case:

$$\begin{pmatrix} \cdots & \cdots & \cdots \cdots \\ a_i \mathbf{w}_i^T & b_i \mathbf{w}_i^T & a_i & b_i \\ a_i \mathbf{d}_i^T & b_i \mathbf{d}_i^T & 0 & 0 \\ \cdots & \cdots & \cdots \cdots \end{pmatrix} \begin{pmatrix} \mathbf{I} \\ \mathbf{J} \\ x_0 \\ y_0 \end{pmatrix} = \begin{pmatrix} \cdots \\ -c_i(1+\eta_i) \\ -c_i \mu_i \\ \cdots \end{pmatrix}$$

- In paraperspective case,

$$\begin{pmatrix} \cdots & \cdots & \cdots & \cdots \\ a_i \mathbf{w}_i^T & b_i \mathbf{w}_i^T & a_i(1+\eta_i) & b_i(1+\eta_i) \\ a_i \mathbf{d}_i^T & b_i \mathbf{d}_i^T & a_i \mu_i & b_i \mu_i \\ \cdots & \cdots & \cdots & \cdots \end{pmatrix} \begin{pmatrix} \mathbf{I}_p \\ \mathbf{J}_p \\ x_0 \\ y_0 \end{pmatrix} = \begin{pmatrix} \cdots \\ -c_i(1+\eta_i) \\ -c_i \mu_i \\ \cdots \end{pmatrix}$$

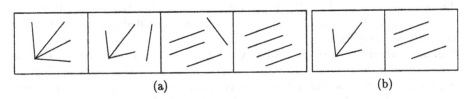

(a)   (b)

Fig. 2. Configurations of image lines which defeat the algorithm for non coplanar object (a) and coplanar object (b).

5.1 Non coplanar object lines

The linear system has 8 unknowns and thus must have at least 8 independent equations. Each line gives us 2 equations, one depending of the translation $\mathbf{w}_i$ and one depending of the direction $\mathbf{d}_i$. We must have at least 4 independent interpretation planes, i.e. we must have a subset of 4 lines in the image with no more than 2 lines intersect in the same point (see figure 2a). Then, the solution for $\mathbf{x}$ is simply given by: $\mathbf{x} = (\mathbf{A}^T \mathbf{A})^{-1} \mathbf{A}^T \mathbf{b}$.

5.2 Coplanar object lines

If the object lines are coplanar then the rank of A is only 6 and the above solution cannot be envisaged anymore. Let's consider the plane formed in this case by the object lines and let u be the unit vector orthogonal to this plane. Vectors I and J (and equivalently I_p and J_p) can be written as a sum of a vector belonging to this plane and a vector perpendicular to this plane, namely: $I = I_0 + \alpha\, u$ and $J = J_0 + \beta\, u$.

By substituting these expressions for I and J into eqs. (4) and (5) (respectively eqs. (8) and (9) for paraperspective), and by noticing that $w_i \cdot u = 0$ and $d_i \cdot u = 0$, the system may be rewritten as: $A'x' = b'$.

The rank of A is 6 if we have at least 3 independent interpretation planes, i.e. we must have a subset of 3 lines in the image with no more than 2 lines intersect in the same point (see figure 2b). The vector u is orthogonal to I_0 and J_0, thus $u \cdot I_0 = 0$ and $u \cdot J_0 = 0$ are two more independent equations and the matrix A' is of rank 8. Thus we obtain solutions for I_0 and J_0: $x' = (A'^T A')^{-1} A'^T b'$.

In order to estimate I and J (and equivalently I_p and J_p) one is left with the estimation of two scalars, α and β.

- In weak perspective case, Oberkampf, Dementhon, and Davis [6] provided a solution using the constraints $\|I\| = \|J\|$ and $I \cdot J = 0$.
- In paraperspective case, Horaud and al. [5] provided a solution using the constraints onto I_p and J_p.

5.3 Enhancements

Solving the linear system: One may notice that in the case of the weak perspective iterative algorithm, the matrix A (or A' for coplanar model) doesn't depend of η_i and μ_i. Thus the pseudo inverse of this matrix remains the same at each iteration step, and therefore needs to be computed only once. It's not the case for paraperspective.

Normalization: The two iterative algorithms solve a linear system at each iteration step. For numerical reasons, we have to normalize each line of the matrix system because the norm of the vectors d_i and w_i may be very different between each line. Moreover, for a coplanar model, the vector u must also be normalized.

6 Experiments

In the first class of experiments, we study the performance of the iterative weak and paraperspective algorithms on simulated images. Two types of performances are studied:

- the precision of pose as a function of position and orientation of the object with respect to the camera in the presence of image (pixel) noise, and
- the convergence of the iterative pose algorithms.

The intrinsic camera parameters are $u_c = v_c = 256$, $\alpha_u = \alpha_v = 1000$. Gaussian noise with standard deviation $\sigma = 1$ pixel has been added to the image measurements and there are 500 trials for each experiments (the object is rotated at 500 random orientations). Finally, the 3-D model represents a simulated house and is made of 18 lines.

Convergence: In all these cases, both algorithms have converged in 100% of the configurations, between 3 and 5 iterations.

Execution time: With no optimization at all, execution time is 0.04 second per iteration on a UltraSparc 1/170 (for 18 lines).

Comparison: According to figure 3, the two iterative algorithms have the same accuracy. The only difference between weak perspective and paraperspective is the number of iterations which is generally one or two less with the paraperspective algorithm (see figure 4).

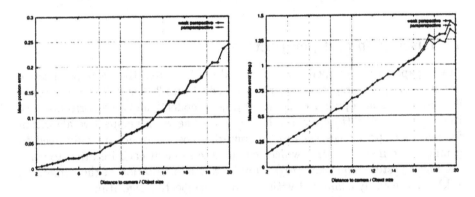

Fig. 3. On the left, error in position as a function of depth in presence of image Gaussian noise – On the right, error in orientation as a function of depth in presence of image Gaussian noise.

An example of application of object pose is the grasping of a polyhedral object by a parallel-jaw gripper (figure 5a). Both the image and the model are described by a network of straight lines and junctions which are matched using a method described in [4]. The following figures represent the quality of object pose computation obtained with two different algorithms. The 3-D object has been reprojected into the image to see the error of the pose.

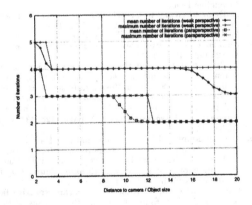

Fig. 4. Speed of convergence as a function of depth.

Figure 5b shows the result obtained with the iterative paraperspective algorithm in the case of lines, described in this paper; In figure 5c, we used a non linear method (Levenberg-Marquardt algorithm) from line correspondences.

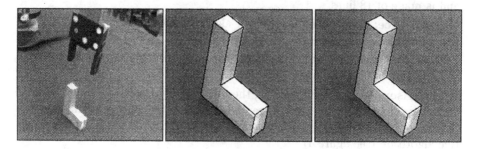

Fig. 5. (a–c) An example of applying the object pose algorithms.

7 Discussion and perspective

In this paper, we developed two iterative algorithms (iterative weak perspective and iterative paraperspective algorithms) to deal with 3-D to 2-D line correspondences, for a non coplanar or coplanar object model. The two algorithms have the same accuracy. However, it's better to use the weak perspective algorithm for line correspondences because computation cost is lower. We only need to compute one inverse matrix which remains the same at each iteration to solve the linear system (equation (10)). Moreover, The orientation of the camera ($\mathbf{i}$, $\mathbf{j}$, $\mathbf{k}$) is more easily computed with the weak perspective algorithm.

References

1. S. Christy and R. Horaud. Euclidean shape and motion from multiple perspective views by affine iterations. IEEE *Transactions on Pattern Analysis and Machine Intelligence*, 18(11):1098–1104, November 1996.
2. D. Dementhon and L.S. Davis. Model-based object pose in 25 lines of code. *International Journal of Computer Vision*, 15(1/2):123–141, 1995.
3. M. Dhome, M. Richetin, J.T. Lapresté, and G. Rives. Determination of the attitude of 3D objects from sigle perspective view. IEEE *Transactions on Pattern Analysis and Machine Intelligence*, 11(12):1265–1278, 1989.
4. P. Gros. Matching and clustering: Two steps towards automatic model generation in computer vision. In *Proceedings of the AAAI Fall Symposium Series: Machine Learning in Computer Vision: What, Why, and How?, Raleigh, North Carolina, USA*, pages 40–44, October 1993.
5. R. Horaud, S. Christy, F. Dornaika, and B. Lamiroy. Object pose: Links between paraperspective and perspective. In *Proceedings of the 5th International Conference on Computer Vision, Cambridge, Massachusetts, USA*, pages 426–433, Cambridge, Mass., June 1995. IEEE Computer Society Press.
6. D. Oberkampf, D.F. Dementhon, and L.S. Davis. Iterative pose estimation using coplanar feature points. *Computer Vision, Graphics and Image Processing*, 63(3):495–511, May 1996.
7. T.Q. Phong, R. Horaud, A. Yassine, and P.D. Tao. Object pose from 2D to 3D point and line correspondences. In *International Journal of Computer Vision*, pages 225–243. Kluwer Academic Publishers, July 1995.
8. C.J. Poelman and T. Kanade. A paraperspective factorization method for shape and motion recovery. In J.O. Eklundh, editor, *Proceedings of the 3rd European Conference on Computer Vision, Stockholm, Sweden*, pages 97–108, May 1994.
9. J.S.C. Yuan. A general phogrammetric solution for the determining object position and orientation. IEEE *Transactions on Robotics and Automation*, 5(2):129–142, 1989.

Statistical 3–D Object Localization Without Segmentation Using Wavelet Analysis

Josef Pösl and **Heinrich Niemann**

Lehrstuhl für Mustererkennung (Informatik 5)
Universität Erlangen–Nürnberg
Martensstr. 3, D–91058 Erlangen, Germany
email: {poesl,niemann}@informatik.uni-erlangen.de

Abstract. This paper presents a new approach for statistical object localization. The localization scheme is directely based on local features, which are extracted for all image positions, in contrast to segmentation in classical schemes. Hierarchical Gabor filters are used to extract local features. With these features statistical object models are built for the different scale levels of the Gabor filters. The localization is then performed by a maximum likelihood estimation on the different scales successively. Results for the localization of real images of 2–D and 3–D objects are shown.

1 Introduction and Motivation

When analysing a 2–D gray–level image with multiple 3–D objects two major problems have to be solved: The pose estimation and classification of each object in the scene. In this paper we focus on the localization of one individual object. All other objects are considered as belonging to the background. We define a statistical density function for pose estimation.

Recognition results in speech understanding were strongly enhanced by the idea of incorporating statistical methods in the recognition task. Despite the obvious success in this area a statistical framework for image object recognition has not been investigated widely up to now. Nevertheless, recent results prove this approach as promising [3].

Most publications in the field of statistical object modeling use geometric information of segmentation results as random variables. Lines or vertices, for example, can be used for the construction of statistical models. There are two major disadvantages of using solely segmentation results. When restricting the recognition process to this level of abstraction a lot of information contained in an image is lost. Another disadvantage are the errors made by the segmentation.

One way to cope with the problem is to avoid segmentation. Instead of that the gray–level information of an image can be used. Correlation is the simplest method of doing this. Another approach focuses in singular value decompositions of vector spaces composed of the gray–level data of several images (appearance based modeling) [1, 8]. Maximization of the mutual information between an object model and an object in a scene is a further possibility [10]. A similar technique is described in [5]. [6] describes a method based on mixture densities of the gray level values of object images. With a focus on the distributions of image pixel values rather than object location values and without an hierarchical solution this approach tends to be very complex.

The cited papers either use only probabilistically chosen model points for matching [10] or use pose restrictions [5] to reduce the complexity of the estimation. In this paper a new approach for the localization of 3–D objects in single gray–level images is presented. The pose of the object is not restricted and the complete image data is considered after hierarchical filtering. Local features are modeled statistically. We demonstrate a new way of formulating a statistical model with a functional basis decomposition of probability density parameters.

2 System overview

The aim of the presented system is the pose estimation of a rigid 3–D object in a single 2–D gray–level image. The parameter space is six-dimensional for this task. Let R_x, R_y and R_z denote the 3D–rotation matrices with rotation angle ϕ_x, ϕ_y and ϕ_z round the x–, y– and z–axis respectively. The 3D–transformation consists of the rotation $R = R_z R_y R_x \in \mathbb{R}^{3 \times 3}$ and the translation $t = (t_x, t_y, t_z)^T \in \mathbb{R}^3$. The parameter space can be split into a rotation R_z with angle $\phi_{int} = (\phi_z)$ and a translation $t_{int} = (t_x, t_y, 0)^T$ inside the image plane and orthogonal components $R_y R_x$ $(\phi_{ext} = (\phi_y, \phi_x))$ and $t_{ext} = (0, 0, t_z)^T$ for the transformations outside. For this work it is assumed that the object does not vary in scale: $t = t_{int}$.

In a first step of the localization process a multiresolution analysis of the image is used to derive feature values on different scales $s \in \mathbb{Z}$ and resolutions (sampling rates) $\Delta x_s = \Delta y_s = r_s \in \mathbb{R}^+$ at the locations of rectangular sampling grids $(r_{s,q+1} < r_{s,q})$. The image $f(x, y)$ is transformed to signals $h_s = (h_{s,0}, \ldots, h_{s,N-1})^T$ by local transformations $\mathcal{T}_{s,n}$ for scale s: $h_{s,n}(x, y) = \mathcal{T}_{s,n}\{f\}(x, y)$. Feature vectors $c_{s,k,l} = (c_{s,k,l,0}, \ldots, c_{s,k,l,N-1})^T$ at discrete locations are obtained by subsampling: $c_{s,k,l,n} = \mathcal{T}_{s,n}\{f\}(kr_s, lr_s)$. Possible definitions of this transformation are described in section 3.

We define a statistical measure for the probability of those features under the assumption of an object transformation. The complexity of the pose estimation is high if all features on the different scale levels are combined into one measure function. Therefore, a hierarchical solution is used. Measures are defined for each scale. The analysis starts on a low scale and a rough resolution. The resolution is then decreased step by step. The transformation estimation becomes more exact with each step. Let $\bar{c}_s$ be the vector of the concatenated feature vectors detected in an image on scale s, B_s the model parameters of an object class and R, t be parameters for rotation and translation. The model parameters B_s consist of geometric information like probability density locations and other density parameters. The density $p(\bar{c}_s | B_s, R, t)$ is then used for localization. The maximum likelihood estimation results in $(\widehat{R}_s, \widehat{t}_s) = \text{argmax}_{(R, t)} p(\bar{c}_s | B_s, R, t)$.

Given a descending sequence $(s_q)_{q=0,\ldots,N_s-1}$ of scales, the analysis begins with the roughest scale s_0. The parameter space is searched completely at this level. The best H transformation hypotheses on scale level s_q are then used to perform a search with restricted parameter range on scale level s_{q+1}. The optimum search on all levels consists of a combination of a grid search for the given parameter range and a successive local search. The grid search on level s_0 is a global search of the full parameter range while the grid search on each other level evaluates the optimum only in the neighbourhood of the previous local optimum. The grid resolution thereby decreases with the scale levels (Fig. 1).

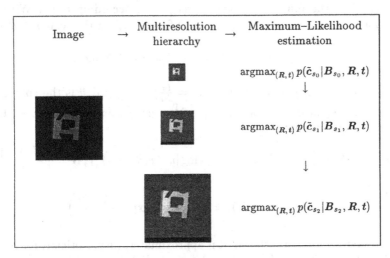

Fig. 1. System overview: Probability density maximization on multiresolution hierarchy of images.

3 Features

Features used for object recognition and localization must have at least two properties. First, they should be robust with respect to noise and different lighting conditions in a scene. Second, they should be invariant to the transformations of the object in the image plane. The feature values at certain object locations should not change if the object is translated or rotated in the image plane.

We define local features derived from Gabor filters which fulfil the requirements of robustness and invariance. By suppressing high frequencies at different scales Gabor filters are a means to compensate for noise. Gabor functions are Gaussians modulated by complex sinusoids. The 2–D Gabor function can be written as:

$$g(\boldsymbol{x}, \boldsymbol{\omega}) = \exp\left(-\left[\frac{x^2}{2\sigma_x{}^2} + \frac{y^2}{2\sigma_y{}^2}\right] + 2\pi i\left[\boldsymbol{\omega}^{\mathrm{T}}\boldsymbol{x}\right]\right).$$

$\boldsymbol{\sigma} = (\sigma_x, \sigma_y)^{\mathrm{T}}$ defines the width of the Gaussian in the spatial domain and $\boldsymbol{\omega}$ is the frequency of the complex sinusoid. These functions achieve the minimum possible joint resolution in space and frequency [2]. They furthermore form a complete but nonorthogonal basis for the set of two-dimensional functions. In order to derive suitable features from the filter results, Gabor wavelets are defined [9]. They allow the analysis of the signal at different scale levels and vary spatial and frequency windows accordingly. The *basic wavelet* of the Gabor wavelet transform with circular spatial windows ($\sigma_x = \sigma_y$) is defined as:

$$g(x, \theta) = \exp\left(-x^2 + p\pi i x'\right),$$

where θ is the orientation and $x' = (x', y')^T = R_\theta x$. R_θ denotes the 2-D rotation matrix with angle θ. The constant p which specifies the ratio of spatial extent and period of the functions is chosen as $p = 1$ according to [7]. With these definitions the wavelet transform of a 2-D function f on the scale $s \in \mathbb{Z}$ is:

$$w_s(x, \theta) = \int f(x_0)\bar{g}\left(d^{-s}(x - x_0), \theta\right) dx_0,$$

with $d \in \mathbb{R}$, $d > 1$, $s \in \mathbb{Z}$ and $\theta \in \{\theta_l = \frac{l\pi}{N_l}\}_{l=0,\ldots,N_l-1}$. $\bar{g}$ is the conjugate of g. A feature vector $c_s = (c_{s,0}, \ldots, c_{s,N-1})^T = c_s(x)$ can now be defined for the locations x on different scales s:

$$c_{s,n} = \left|FT_{k=1\ldots N-1}\left\{|FT_{l=0\ldots N-1}\left\{\log|w_s(x, \theta_l)|\right\}_k|\right\}_n\right|, N = \left\lfloor \frac{N_l + 1}{2} \right\rfloor,$$

where

$$FT_{k=k_0\ldots k_1}\{f_k\}_l = \sum_{k=k_0\ldots k_1} f_k \exp\left(-\frac{2\pi i k l}{N}\right)$$

is the discrete Fourier transform. It is approximately (asymtotically for $N \to \infty$) rotationally invariant and robust to changes in lighting conditions.

As already stated in Sect. 2 the localization is performed hierarchically on different scale levels. The resolution r_s (sampling distance) of the analysis on scale s is connected to the spatial filter width by a constant λ: $r_s = \lambda d^s$.

4 Statistical model

4.1 Model formulation

This section shows the definition of a probability density function on each of the scale levels of the analysis. To simplify the notation the index s indicating the scale level is omitted.

The model object is covered with a rectangular grid of local feature vectors (see Fig. 2). The grid resolution is the same as the image resolution on the actual scale. Let $A \subset X$ be a small region (e.g. rectangular) which contains the object projection onto the image plane for all possible rotations ϕ_{ext} and constant ϕ_{int} and t (see Fig. 2). Let $X = \{x_m\}_{m=0,\ldots,M-1}$, $x_m \in \mathbb{R}^2$ denote the grid locations and $c(x)$ the observed feature vector at location $x = (x, y)^T$. In this paper we choose the grid locations as the sampling locations of the image transformation: $X = \left\{x_m = (kr_s, lr_s)^T\right\}$. The local features $c_n(x_m)$ are the components of the image feature vector $\bar{c}$ if the object is not transformed in the image plane: $c_n(x_m) = c_n(kr_s, lr_s) = c_{k,l,n} = c_{m,n}$.

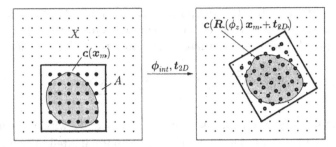

Fig. 2. Object covered with grid for feature extraction.

The local features $c_{m,n}$ are interpreted as random variables. The randomness thereby is, among others, the consequence of noise in the image sampling process and complex changes in environment (e.g. lighting) conditions. Assuming the densities $p(c_m)$ of the local features as stochastically independent leads to:

$$p(\tilde{c}) = \prod_{x_m} p(c(x_m)) = \prod_{x_m \in A} p(c(x)|x = x_m) \prod_{x_m \notin A} p(c(x)|x = x_m).$$

If a uniform distribution for the features outside the model area A (which belong to background) is assumed the second product in the above equation is constant. So it is sufficient to consider

$$p(c_A) = \prod_{x_m \in A} p(c(x)|x = x_m),$$

where c_A is the subvector of $\tilde{c}$ which belongs to A, for pose estimation.

We will first derive the density for the two-dimensional case. We use linear interpolation for reconstruction of $c_n(x)$ from the image feature vector $\tilde{c}$.

The grid positions and A are part of the model parameters B. If the model is transformed by (ϕ_{int}, t) in the image plane the density can be written as:

$$p(c_A|B, \phi_{int}, t) = \prod_{x_m \in A} p(c(x)|x = R(\phi_z) x_m + t_{2D}), \qquad t_{2D} = (t_x, t_y)^{\mathrm{T}}.$$

The feature vectors are assumed to be normally distributed with independent components. Let $\mathcal{N}(c|\mu_m, \Sigma_m)$ denote the normal densities, where μ_m is the mean vector and Σ_m the covariance matrix of the feature vector c_m. In the case of independence, Σ_m is a diagonal matrix diag $(\sigma_{m,0}^2, \dots, \sigma_{m,N-1}^2)$. This results in:

$$p(c_A|B, \phi_{int}, t_{int}) = \prod_{x_m \in A} \mathcal{N}(c(R(\phi_z) x_m + t_{2D})|\mu_m, \Sigma_m).$$

For 3–D objects there are two additional degrees of freedom. They allow an object rotation $\phi_{ext} = (\phi_y, \phi_x)$ perpendicular to the image plane. With the same density model for all possible rotations ϕ_{ext} the density parameters are functions of these additional parameters, so that:

$$p(c_A|B, R, t) = \prod_{x_m \in A} \mathcal{N}(c(R(\phi_z) x_m + t_{2D})|\mu_m(\phi_y, \phi_x), \Sigma_m(\phi_y, \phi_x)).$$

Assuming continuous functions $\mu_m(\phi_y, \phi_x)$, $\Sigma_m(\phi_y, \phi_x)$, they can be rewritten using a basis set for the domain of two-dimensional functions $\{v_r\}_{r=0,\ldots,\infty}$ on the domain of (ϕ_y, ϕ_x) with appropriate coordinates $a_{m,n,r}, b_{m,n,r} \in \mathbb{R}$ $(r = 0, \ldots)$:

$$\mu_{m,n} = \sum_{r=0}^{\infty} a_{m,n,r} v_r, \qquad \sigma_{m,n}^{-2} = \sum_{r=0}^{\infty} b_{m,n,r} v_r.$$

Possible basis functions are $v_r = v_{st}(\phi_y, \phi_x) = \phi_x^s \phi_y^t$ with the enumeration $r = \frac{1}{2}(s+t)(s+t+1) + t$. The functions can be approximated by using only part of the complete basis set $\{v_r\}_{r=0,\ldots,L-1}$. The approximation error can be made as small as possible by choosing L large enough. If ϕ_x is constant, as in our experiments, the v_r are only one-dimensional polynomial basis functions ϕ_y^r.

4.2 Parameter estimation

The model parameters are estimated by a maximum likelihood estimation. Under the assumption of N_ρ independent observations ${}^\rho c_A$ this leads to the estimation

$$\left\{ (\widehat{a}_{m,n}, \widehat{b}_{m,n}) \right\} = \operatorname*{argmax}_{\{(a_{m,n}, b_{m,n})\}} \prod_\rho p({}^\rho c_A | x_m, \{(a_{m,n}, b_{m,n})\}, {}^\rho R, {}^\rho t),$$

with the assumption of known transformation parameters ${}^\rho R, {}^\rho t$ and a predefined number L of basis functions.

The optimization of this function is rather complex. In order to reduce the complexity by providing an analytical solution, $\sigma_{m,n}$ is assumed to be constant.

Solving the equations for the parameters to be estimated results in:

$$\widehat{a}_{m,n} = Q^{-1} \left(\sum_\rho {}^\rho c_{m,n} v({}^\rho \phi_{ext}) \right), \quad \widehat{\sigma}_{m,n} = \frac{1}{N_\rho} \sum_\rho ({}^\rho c_{m,n} - \widehat{\mu}_{m,n}({}^\rho \phi_{ext}))^2,$$

with

$$Q = \sum_\rho v({}^\rho \phi_{ext}) v^{\mathrm{T}}({}^\rho \phi_{ext}),$$

${}^\rho c_{m,n} = {}^\rho c_n(R({}^\rho \phi_z) x_m + {}^\rho t_{int})$ and $v = (v_0, \ldots, v_{L-1})^{\mathrm{T}}$.

5 Results

Fig. 3 shows the objects used in this work. The images are 256 pixels in square. Several real images containing one object at different positions have been used for experiments. The localization was performed on two scale levels: $s_0 = 4$, $s_1 = 3$ ($d = 2$ pixels) with resolution $r_s = 0.5 d^s$ ($r_{s_0} = 8$, $r_{s_1} = 4$ pixels) and constant $\sigma_{m,n}$. The Gabor filters were calculated for 16 equidistant angles from $0°$ to $180°$, resulting in nine-dimensional feature vectors. Only the best localization result of level s_0 was used for further refinement at s_1. The Downhill Simplex algorithm was used for the local parameter search following the global grid search. The

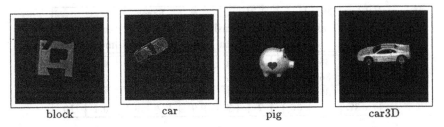

| block | car | pig | car3D |

Fig. 3. Examples for objects used for 2–D (left) and 3–D (right) experiments.

computation time on a SGI Impact is about 45 seconds for feature extraction on both scale levels and 30 seconds for localization of object *pig* with its four-dimensional parameter space (level s_0: 100 grid locations, s_1: 400).

Training and test sets are disjoint. For each of the 2–D objects *halfcircle*, *car* and *plug* one image sequence with a complete object rotation in the image plane in 36 equidistant steps was available. The correct object positions were determined manually with two reference points. The average accuracy of the object positions available for training and testing is about half a pixel with respect to translation and half a degree with repect to rotation. Object *block* was available in three such sequences with different lighting conditions. Training images were taken out of two sequences, the rest was used for testing. The correct positions for *block* were determined by the algorithm described in [3]. The sequences of the 3–D objects *pig* and *car3D* are taken from the Columbia Object Image Library (COIL). They consist of images for different object positions of one rotation axis ϕ_y of ϕ_{ext} and fixed ϕ_x, ϕ_z, t. The range of ϕ_{ext} was treated as one-dimensional in the experiments. The range of ϕ_z, t was searched completely, resulting in a four-dimensional search. Tables 1 and 2 show the results of the tests. Experiments with a translation error of more than ten image pixels were categorized as failure.

Object	Number		Error			
	Train	Test	Transl. (Pix)		Rot. (^{o})	
			mean	max	mean	max
block	40	66	0.3	1.4	0.5	1.6
halfcircle	18	18	0.8	1.8	0.5	1.6
car	18	18	1.3	3.8	1.7	4.0
plug	18	18	0.9	2.0	0.9	1.7

Table 1. Localization results for 2–D objects.

6 Conclusion

A new approach for object localization using statistical models was presented. The localization scheme works without segmentation of the input images. Gabor filters are used to extract local features. The local features are transformed in order to be rotationally invariant and robust to changes in lighting conditions.

| Object | Number | | L | Fail | Error | | | | | |
| | Train | Test | | | Transl. (Pix) | | int.Rot. (^{o}) | | ext.Rot. (^{o}) | |
					mean	max	mean	max	mean	max
pig	36	36	6	2	1.2	2.9	1.5	4.7	4.8	16
car3D	36	36		4	1.5	7.4	3.0	9.4	6.6	22
pig	36	36	8	0	1.1	2.1	1.5	5.6	3.7	14
car3D	36	36		2	1.7	9.9	2.7	9.4	5.4	16

Table 2. Localization results for 3–D objects.

The positions of the local features with respect to the object provide the information for localization. The object model consists of the distributions of the local feature vectors. Assuming stochastic independence the densities are combined to a complete model for the object. The localization itself is performed on different scale levels to reduce the computational complexity.

Several experiments with real images of 2–D and 3–D objects were carried out. The results show that the approach is capable of localizing objects. Future research will focus on the investigation of alternative and rotationally variant features. Statistical dependence will be considered to a certain degree.

References

1. M. Black and A. Jepson. Eigentracking: Robust matching and tracking of articulated objects using a view-based representation. In B. Buxton and R. Cipolla, editors, *Computer Vision — ECCV '96*, volume I of *Lecture Notes in Computer Science*, pages 329–341, Heidelberg, 1996. Springer.
2. J. G. Daugman. Uncertainty relation for resolution in space, spatial frequency, and orientation optimized by two-dimensional visual cortical filters. *Journal of the Optical Society of America A*, 2(7):1160–1169, 1985.
3. J. Hornegger and H. Niemann. Statistical learning, localization, and identification of objects. In ICCV 95 [4], pages 914–919.
4. *Proceedings of the 5th International Conference on Computer Vision (ICCV)*, Boston, June 1995. IEEE Computer Society Press.
5. H. Kollnig and H.-H. Nagel. 3D pose estimation by fitting gradients directly to polyhedral models. In ICCV 95 [4], pages 569–574.
6. V. Kumar and E. S. Manolakos. Unsupervised model–based object recognition by parameter estimation of hierarchical mixtures. In *Proceedings of the International Conference on Image Processing (ICIP)*, pages 967–970, Lausanne, Schweiz, September 1996. IEEE Computer Society Press.
7. B. S. Manjunath, C. Shekhar, and R. Chellappa. A new approach to image feature detection with applications. *Pattern Recognition*, 29(4):627–640, 1996.
8. H. Murase and S. K. Nayar. Visual learning and recognition of 3–D objects from appearance. *International Journal of Computer Vision*, 14(1):5–24, January 1995.
9. A. Shustorovich. Scale specific and robust edge/line encoding with linear combinations of gabor wavelets. *Pattern Recognition*, 27(5):713–725, 1994.
10. P. Viola and W. Wells III. Alignment by maximization of mutual information. In ICCV 95 [4], pages 16–23.

A Real-Time Monocular Vision-Based 3D Mouse System

Sifang Li, Wynne Hsu, Pung H. K.
Department of Information Systems and Computer Science
National University of Singapore
Singapore 119260

Abstract

Speed, robustness, and cost are three important factors that determine the success of a vision-based human-computer interaction system. In our system, we introduce a scheme that uses only one camera (together with a mirror) to derive the 3D coordinates of a target object. To ensure a more natural interaction between the human and the computer, our system allows the user to use his/her bare fingers to point to any position in the 3D space. In addition, robustness and speed is achieved through the use of the chain-code algorithm. The accuracy of the results is improved through the use of suitable post-processing filters. The performance of our system is thoroughly evaluated.

1 Introduction

With the increased popularity of 3D applications, researchers are focusing more and more on efficient yet cost effective 3D interaction solutions. Most existing 3D interaction solutions are electromagnetic-based. One example of an eletromagentic-based 3D input device is the DataGlove. Compared to electromagnetic 3D input devices, the chief advantage of a vision-based solution is that it frees the user from any form of physical attachment that may constraint his/her movement. The vision-based approach has attracted much effort in computer vision fields [1]-[7]. However, current vision-based solutions are too sensitive and computationally too expensive for practical applications, and many of them impose unacceptable constraints on the user. In this paper, we propose a novel 3D vision-based solution using only one camera and a mirror. A number of experiments have been performed to show that our solution is *robust to movement*; has a *fast pick-up* rate so as to allow the user's hand to move in and out of the scene just as s/he tends to pick up a normal 2D mouse now and then; and has a *low computational cost* – thus resulting in fast response time. In addition, it is also *robust to false recognition*. This is important because no algorithm is able to achieve 100% accuracy rate. Thus, recovery from false recognition is critical. Finally, since our solution uses only one camera and one mirror, it satisfies the *low cost* criterion.

2 Related Work

Vision-based 3D mouse was mentioned a number of times in recent publications. Most of them rely on 2D shape models [2,3] (Kalman Snakes [8]) or 3D anatomic models [4]-[6] to guide in deriving the measurements of an image. Movement models such as smoothness, slowness, affinity or periodic properties have also been used to predict the next model state. However, when it comes to applying these models in practical human computer interaction (HCI) applications, a number of problems surface. The first problem is the difficulty in initiating the model. For practical HCI application, the user's hand should be allowed to re-enter the scene easily. This implies that model initiation needs to be done each time the user re-enters the scene. Such initiation is very expensive without introducing additional constraints such as fixed background [4]-[7] or arranged initialization area [2,3]. The second problem is the inability to restore from false recognition. In the model-based approach, estimation of the image is directly computed from the previous result. If the result is false due to the inaccuracy of the feature measurement, lock-loss is possible. Once the lock is lost, regaining the lock is difficult because the initiation of the model is expensive. In [2], learning is used to prevent lock losing, but this only applies to those movements which are used to train the system. Given the fact that the false recognition in a vision-based system is inevitable and the user's movement cannot be constrained too much, model-based approach is not suitable for our application.

To reduce hardware cost, people try to use only one camera to extract 3D position instead of stereo camera systems [3,7]. Monocular 3D position detection has been addressed in the literature [2,4,5,6]. Their approaches fall into two broad categories:

i. Using depth cues to restore 3D depth information indirectly

The cues can be used are shape size, shading and texture. In [2] the area within the hand contour is used as the depth cue. This approach assumes the hand doesn't turn around before the camera. We can not expect that using size, shading and texture as depth cue can provide accurate outputs.

ii. Using relative 3D structure to restore the depth

[4]-[6] are examples which try to restore the detailed hand or arm gestures using 3D hand or arm models. After the 3D model being successfully restored, one part of the model's 3D position can be determined by introducing some constraints (static should [4]; palm on a plane [6]). The constraints introduced in these works are too strong from the user's point of view. Actually in [4], the author also mentioned that users could not maintain their shoulder static at all and the shoulder's movement caused significant error in the final 3D coordinate outputs.

3 System Setup and Architecture

3.1 System Setup

To accurately determine the 3D position of an object, we need at least two views. In our system, a novel method using one camera and a mirror is introduced. The physical setup is shown in Figure 1. A video camera (VC) is placed at an angle of 45° to the mirror. We build up XYZ and UVW coordinate systems in the 3D space. XY plane is actually VC's image plane and UW plane is parallel to the mirror. The output of our 3D mouse refers to the UVW system, which is more natural for the user, instead of XYZ. A (real) object is presented somewhere in the space, we are going to derive the 3D coordinates of it.

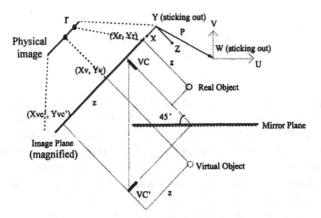

Figure 1. The System Setup.

On the XY plane, let the projection of the real object be (x_r, y_r) and the projection of the mirror image be (x_v, y_v). We call this mirror image the corresponding *virtual object* of the real object. Let's assume that we also know the projection $(x_{vc'}, y_{vc'})$ of the virtual camera VC'. We can then decide the 3D position of the real object by using these three projections - (x_r, y_r), (x_v, y_v) and $(x_{vc'}, y_{vc'})$.

It's straightforward that the X and Y positions of the real object are x_r and y_r respectively. From the symmetric property provided by the mirror, we have

$z = x_{vc'} - x_v.$

By a simple geometry transform, we transform the real object's XYZ co-ordinates into the UVW coordinate system (see Figure 1) where P is the translation vector that maps the origin of the XYZ coordinate system to the origin of the UVW coordinate system.

$$
\begin{bmatrix} u \\ v \\ w \end{bmatrix} = \begin{bmatrix} -\sin\dfrac{\pi}{4} & 0 & \cos\dfrac{\pi}{4} \\ -\cos\dfrac{\pi}{4} & 0 & -\sin\dfrac{\pi}{4} \\ 0 & 1 & 0 \end{bmatrix} \bullet \begin{bmatrix} x \\ y \\ z \end{bmatrix} + P = \begin{bmatrix} -\sin\dfrac{\pi}{4} & 0 & \cos\dfrac{\pi}{4} \\ -\cos\dfrac{\pi}{4} & 0 & -\sin\dfrac{\pi}{4} \\ 0 & 1 & 0 \end{bmatrix} \bullet \begin{bmatrix} x_r \\ y_r \\ x_{vc'} - x_v \end{bmatrix} + P \tag{1}
$$

Consider the starting point $O(0,0,0)$ of UVW system. We denote the XY coordinates of the projection of O and the projection of its corresponding virtual object as (x_{r0}, y_{r0}) and (x_{v0}, y_{v0}), respectively.

From (1), we have (2).

$$
\begin{bmatrix} 0 \\ 0 \\ 0 \end{bmatrix} = \begin{bmatrix} -\sin\dfrac{\pi}{4} & 0 & \cos\dfrac{\pi}{4} \\ -\cos\dfrac{\pi}{4} & 0 & -\sin\dfrac{\pi}{4} \\ 0 & 1 & 0 \end{bmatrix} \bullet \begin{bmatrix} x_{r0} \\ y_{r0} \\ x_{vc'} - x_{v0} \end{bmatrix} + P \tag{2}
$$

Subtract (2) from (1), we get (3).

$$\begin{bmatrix} u \\ v \\ w \end{bmatrix} = \begin{bmatrix} -\sin\dfrac{\pi}{4} & 0 & \cos\dfrac{\pi}{4} \\ -\cos\dfrac{\pi}{4} & 0 & -\sin\dfrac{\pi}{4} \\ 0 & 1 & 0 \end{bmatrix} \bullet \begin{bmatrix} x_r - x_{r0} \\ y_r - y_{r0} \\ -(x_v - x_{v0}) \end{bmatrix} \tag{3}$$

Now we can restore the 3D position of the object using (3) provided we know the XY projections of the following points: UVW origin O and its mirror image, the object and the virtual object. Because the XY plane is actually the video camera image plane, we can consider a physical video image frame is miniature of the projections of all things before the camera (A orthographic camera model is assumed here) . So any point's projection can be found in the video image, if the camera can "see" the point. After we set up the system, (x_{r0}, y_{r0}) and (x_{v0}, y_{v0}) are fixed. As long as the camera can "see" the object (x_r, y_r) and its virtual object (x_v, y_v), we know the object's 3D co-ordinates in UVW system using (3). With this method, we merely need one camera to obtain the accurate 3D position. This results in substantial cost saving.

3.2 Overview of the System Architecture

Figure 2 shows an overview of the system architecture. The initialization module is responsible for building up the background models describing the intensities properties of each background pixel. Next, a target object is introduced into the scene and an object model is built for this target. When all models have been built, the chain-code algorithm is invoked to detect the contour of the target object. This contour is then analyzed and interpreted. Finally, post-processing is performed to obtain a more accurate 3D position of the target object. In our implementation, the target object is a human hand. The index finger is used to specify the desired 3D position while the thumb is used to specify the desired operation. For example, an erected thumb signifies the releasing of the mouse button while a bent thumb signifies the pressing of the mouse button (see Figure 3).

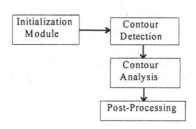

Figure 2. Overview of System Architecture   **Figure 3**. The Two Thumb States

3.3 Initialization Module

Background knowledge is probably the most powerful knowledge in video segmentation. In our system, we assume a fixed but complex background, and instead of using the hand shape knowledge, we use prior background and hand intensity knowledge to perform feature measurement, which guarantees the robustness for a practical 3D mouse. The background and hand intensity knowledge are modeled using simple Gaussian models. To learn the background models, an initial startup time of n continuous frames is required to build the Gaussian models:

$$m(x,y) = \frac{1}{n}\sum_{k=0}^{n-1} i_k(x,y), \quad \sigma^2(x,y) = \frac{1}{n}\sum_{k=0}^{n-1} i_k(x,y)^T i_k(x,y) - m(x,y)^T m(x,y) \tag{4}$$

where i is the intensity vector expressed in RGB space.

Based on the background Gaussian models, a single Gaussian model describing the hand intensity is built up by placing the hand before the learned background. Hand pixel segmentation is achieved by:

$$f(x,y) = \begin{cases} i(x,y), & if \ |m(x,y) - i(x,y)| > \tau\sigma(x,y) \\ 0, & if \ |m(x,y) - i(x,y)| \le \tau\sigma(x,y) \end{cases} \tag{5}$$

where τ is the threshold value.

After the segmentation, the Gaussian model for all hand pixels can be built in a similar way:

$$m_h = \frac{1}{n_h}\sum_{f(x,y)\neq 0} f(x,y), \quad \sigma_h^2 = \frac{1}{n_h}\sum_{f(x,y)\neq 0} f(x,y)^T f(x,y) - m_h^T m_h \tag{6}$$

where n_h is the total number of non-zero pixels in segmented image f.

Using these models, we can classify all the pixels in an image by thresholding (7).

$$\frac{P(I|H)}{P(I|B)} = \frac{\sigma}{\sigma_h}\exp(\frac{|i-m|^2}{2\sigma^2} - \frac{|i-m_h|^2}{2\sigma_h^2}) \tag{7}$$

where H, B and I denote the following events respectively: the pixel belongs to the hand (H); the pixel belongs to the background (B) and the pixel intensity is I (I).

Figure 4 shows the results when we compute (7) over all image pixels. The pixel brightness reflects the magnitude of the results (value of the right side of (7)). The bright area in the figure is the hand and its mirror image with some noise caused by the shade and illumination. Some pixels in the hand is not very bright because they are too similar to the local background.

3.4 Detection of Hand Contour

The chain code algorithm [10] does not require any prior shape knowledge is thus robust to fast and drastic changes in shape. Note that the chain code generation are usually carried out on a grid that is imposed on the image. In [11], Scholten et. al proved theoretically that a hexagonal grid can describe a curve more accurately as compared to the triangle or square grid. The chain coding algorithm on hexagonal grid is given below.

Algorithm chain-code

Let N and V be two functions where

$N_j(C)$ returns the jth neighbor of vertex C, and

$V(C_i,C_j) = t$, if C_i is the tth neighbor of C_j;

$V(C_i,C_j) = -1$, otherwise;

Given the initial contour queue "C_0C_1," the algorithm is:

1. k=1;
2. $t=V(C_k, C_{k-1})$;

3. $j=(t+3)$ mod 12 ;
4. $C_{k+1}= N_j(C_k)$;
    if (C_{k+1} in the boundary) goto 6;
    else $j=(j-1)$ mod 12;
5. goto 4
6. add C_{k+1} to the contour queue, $k=k+1$;
    if ($C_k=C_0$) then halt;
7. goto 2

In Figure 4, we make all pixels above the dotted line hand pixels, thus we apply the above algorithm on the image and constrain the algorithm to search the whole image from left to right instead of terminating too early. The detected contour is shown in black solid line in Figure 4.

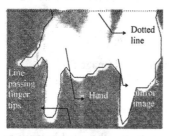

Figure 4. The Hand Contour.

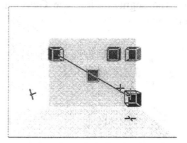

Figure 5. A Simulated 3D Room

4 Contour Analysis

To achieve real-time performance, the detected contour is not analyzed in an elaborated manner. To obtain the position of the index finger tip, we constrain the user's index finger tip to be farthest point from the dotted line. After the rough position of the two index finger tips (the real finger and its mirror image) are detected from Figure 4, we still use the similar chain coding algorithm on a denser square grid to search for more accurate finger tip positions in the neighborhood based on the rough finger tip positions. Because only a small neighborhood needs to be searched, the search is very fast.

After that, the thumb state detection is done by measure finger tip features. Using the discovered index finger tip position as reference, we "guess" a range within the contour whereby the thumb must lie. If the contour bends significantly within the selected range, we say that the thumb has been found.

5 Post-processing

After the finger tips are fixed, (3) is used to restore the 3D coordinates of the finger tip thus a stream of 3D coordinate signals is generated from the video stream. Due to the incorrect detection or noise, the signals generated are not stable thus need to be post-processed.

5.1 Screening the False Signals

From the geometry constraint provided by our setup, we find all straight lines, that pass through the real finger (R') and its mirror image (V') (the white solid line in Figure 4), converge to single point in the image plane [12]. This properties is used to filter out most false signals. In our experiment, less than 5% of the raw signals are filtered by this constraint. When a signal has been filtered, there is a "hole" in the signal stream. Such holes are filled by the predictions of Kalman filters (see below).

5.2 Kalman Filtering and Adaptive Filtering

A set of linear Kalman filters are used both to suppress the Gaussian noise in the signal and to make predictions for the holes in the signal. To ensure a more stable signal with short delay, a set of adaptive filters are also used to smooth the signal.

6 System performance

A 3D mouse application (Figure 5) is built to demonstrate the final performance of our system on a SGI Indy workstation. A simulated 3D room with several cubes inside the room is displayed on the screen. User is asked to manipulate the cubes in 3D space and arrange the cubes in some pre-defined patterns.

- *Sampling rate* -- Because only the boundary pixels are analyzed, the computation cost is low and we are able to achieve a rate of 25 frames per second (each frame is 640X480).
- *Robustness* -- We tested the system with different backgrounds, hand sizes, hand colors, and even with hands holding objects. The test results indicate that our system is robust to dramatic shape changes, fast movements, and frequent re-entry. A experiment in which a hand is doing fast (1-2 Hz) circular movement before a clustered background is shown in Figure 6. The results indicate that our system are able to follow the finger tip in real-time and the output are relatively stable and smooth. Since false recognition has little effect on the later frames recognition, it doesn't have lock loss problem which is quite normal in strong model approach (see results in [2]).
- *Resolution* -- We map the 3D coordinates into a 320x240x200 space. Because of finger thickness, the hand shaking and noise, we achieved an 80x60x100 effective resolution, that is to say the user can effectively specify 80x60x100 positions in 3D space. This result is good enough for most entertainment and education applications.
- *Lag* -- The lag in our system is mainly caused by the Kalman filters and adaptive filters which are used to stabilize the signal. This lag in our system is 0.08 - 0.16 sec. The lag is acceptable from the user's point of view.

7 Conclusions

We have proposed a novel low cost system setup to achieve accurate 3D position and address the constraint provided by the setup. Using the setup, a practical 3D mouse is designed and implemented. Successful post-processing utilizing the constraint provided by the setup is applied to stabilize the outputs. Experimental results show that our system satisfies the requirements for a robust and user friendly 3D input device.

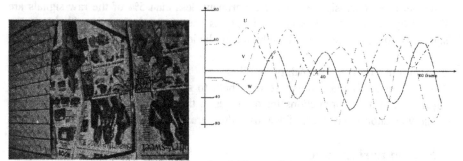

Figure 6. An Example of Clustered Background.

REFERENCES

1. F. K. H. Quek, "Eyes in the Interface," *Image and Vision Computing,* Vol. 13, No. 6, Aug. 1995. pp. 511-525.
2. A. Blake, M. Isard and D. Reynard, Learning to Track the Visual Motion of Contours, *Artificial Intelligence,* Vol 78, 1995, pp. 101-133.
3. R. Cipolla and N.J. Hollinghurst. Human--robot interface by pointing with uncalibrated stereo vision. *Image and Vision Computing,* 14(3):171--178, 1996.
4. L. Goncalves and E. D. Bernardo, Monocular tracking of the human arm in 3D, *Proc. Of 15th International Conference on Computer Vision,* 1995, pp 764-770.
5. J. Kuch and T. Huang, Virtual Gun: A Vision Based Human Computer Interface Using the Human Hand, *Proc. IAPR Workshop on Machine Vision Application,* Tokyo, 1994, pp. 196-199.
6. J. Rehg, DigitEyes: Vision-Based Human Hand Tracking for Human-Computer Interaction, *Proc. of the 1994 IEEE Workshop on Motion of Non-rigid and Articulated Objects.* 1994, pp. 16-22.
7. A. Utsumi Real-time Hand Gesture Recognition System, *Proc. of ACCV'95,* 1995, pp. 249-253.
8. M. Kass, A. Witkin, and D. Terzopoulos Snakes: Active Contour Models, *Proc. Ist Int. Conf. On Computer Vision,* 1987, pp. 259-268.
9. C. R Wren., A. Azarbayejani, and A. Pentland, "Pfinder: Real-time Tracking of the Human Body," *Proc. of the 2nd International Conference on Automatic Face and Gesture Recognition,* 1996.
10. Freeman, H., "Computer Processing of Line-drawing Data," *Comput. Surveys,* v. 6, Mar. 1974, pp. 57-59.
11. D. K. Scholten and S. G. Wilson Chain Coding with Hexagonal Lattice, *IEEE Trans. PAMI,* No. 5, Sept. 1983, pp. 526-533.
12. Sifang, Li, A Vision-based 3D Mouse, MSc. Thesis, National University of Singapore, 1997.

Face Recognition by Elastic Bunch Graph Matching*

Laurenz Wiskott[1†], Jean-Marc Fellous[2‡],
Norbert Krüger[1], and Christoph von der Malsburg[1,2]

[1] Institut für Neuroinformatik
Ruhr-Universität Bochum
D-44780 Bochum, Germany
http://www.neuroinformatik.ruhr-uni-bochum.de

[2] Computer Science Department
University of Southern California
Los Angeles, CA 90089, USA

Abstract

We present a system for recognizing human faces from single images out of a large database with one image per person. The task is difficult because of image variation in terms of position, size, expression, and pose. The system collapses most of this variance by extracting concise face descriptions in the form of *image graphs*. In these, fiducial points on the face (eyes, mouth etc.) are described by sets of wavelet components (*jets*). Image graph extraction is based on a novel approach, the *bunch graph*, which is constructed from a small set of sample image graphs. Recognition is based on a straight-forward comparison of image graphs. We report recognition experiments on the FERET database and the Bochum database, including recognition across pose.

1 Introduction

The system presented here is based on a face recognition system described in [3]. Faces are represented by labeled graphs based on a Gabor wavelet transform. A

*Supported by grants from the German Federal Ministry for Science and Technology (413-5839-01 IN 101 B9) and from ARPA and the U.S. Army Research Lab (01/93/K-109).

†Current address: Computational Neurobiology Laboratory, The Salk Institute for Biological Studies, San Diego, CA 92186-5800, http://www.cnl.salk.edu/CNL, wiskott@salk.edu.

‡Current address: Volen Center for Complex Systems, Brandeis University, Waltham, MA 02254-9110.

matching process allows finding image graphs of new faces automatically. Recognition is then based on a simple similarity measure between the image graph and a gallery of model graphs. We have made three major extensions to the preceding system in order to handle larger galleries and larger variations in pose, and to increase the matching accuracy. Firstly, we use the phase of the complex Gabor wavelet coefficients to achieve a more accurate location of the nodes of the graph. Secondly, we employ object adapted graphs, so that nodes refer to specific facial landmarks, called *fiducial points*. The correct correspondences between two faces can then be found across large viewpoint changes. Thirdly, we have introduced a new data structure, called the *bunch graph*, which serves as a generalized representation of faces by combining jets of a small set of individual faces. This allows the system to find the fiducial points in one matching process, which eliminates the need for matching each model graph individually. This reduces computational effort significantly. A more detailed description of this system is given in [8].

2 The System

2.1 Preprocessing with Gabor Wavelets

The representation of local features is based on a Gabor wavelet transform. It is defined as a convolution of the grey value image with a family of *Gabor kernels*

$$\psi_j(\vec{x}) = \frac{k_j^2}{\sigma^2} \exp\left(-\frac{k_j^2 x^2}{2\sigma^2}\right) \left[\exp(i\vec{k}_j\vec{x}) - \exp\left(-\frac{\sigma^2}{2}\right)\right] \tag{1}$$

in the shape of plane waves with wave vector $\vec{k}_j$, restricted by a Gaussian envelope function. We employ a discrete set of 5 different frequencies, index $\nu = 0, ..., 4$, and 8 orientations, index $\mu = 0, ..., 7$, yielding wave vectors $\vec{k}_j = (k_{jx}, k_{jy}) = (k_\nu \cos\varphi_\mu, k_\nu \sin\varphi_\mu)$ with $j = \mu + 8\nu$, $k_\nu = 2^{-\frac{\nu+2}{2}}\pi$, and $\varphi_\mu = \mu\frac{\pi}{8}$. This sampling evenly covers a band in frequency space. The width σ/k of the Gaussian is controlled by the parameter $\sigma = 2\pi$. The second term in the bracket of (1) makes the kernels *DC-free*, i.e. the integral $\int \psi_j(\vec{x})d^2\vec{x}$ vanishes. One speaks of a wavelet transform since the family of kernels is self-similar, all kernels being generated from one *mother wavelet* by dilation and rotation.

A jet $\mathcal{J}$ is defined as the set $\{\mathcal{J}_j\}$ of 40 complex coefficients obtained for one image point by this convolution. It can be written as

$$\mathcal{J}_j = a_j \exp(i\phi_j) \tag{2}$$

with amplitudes $a_j(\vec{x})$, which slowly vary with position, and phases $\phi_j(\vec{x})$, which rotate with a rate set by the spatial frequency or wave vector $\vec{k}_j$ of the kernels.

Due to phase rotation, jets taken from image points with a relative distance $\vec{d}$ of few pixels have very different coefficients although representing almost the same local feature. This can cause severe problems for matching. We therefore either ignore the phase or compensate for its variation explicitly. The former leads to the similarity function

$$S_a(\mathcal{J}, \mathcal{J}') = \frac{\sum\limits_j a_j a'_j}{\sqrt{\sum\limits_j a_j^2 \sum\limits_j a_j'^2}}, \tag{3}$$

already used in [1, 3]. Compensating phase variation leads to the similarity function

$$S_\phi(\mathcal{J}, \mathcal{J}') = \frac{\sum\limits_j a_j a'_j \cos(\phi_j - \phi'_j - \vec{d}\vec{k}_j)}{\sqrt{\sum\limits_j a_j^2 \sum\limits_j a_j'^2}}. \tag{4}$$

In order to compute it, the displacement $\vec{d}$ has to be estimated. This can be done by maximizing S_ϕ with respect to $\vec{d}$ in its Taylor expansion. If applied iteratively from low to high frequencies, displacements of up to eight pixels can be estimated [8], which is useful to guide the matching process.

2.2 Face Representation

For faces, we have defined a set of *fiducial points*, e.g. the pupils, the corners of the mouth, the tip of the nose, the top and bottom of the ears, etc. A *labeled graph* $\mathcal{G}$ representing a face consists of N nodes on these fiducial points at positions $\vec{x}_n, n = 1, ..., N$ and E edges between them. The nodes are labeled with jets $\mathcal{J}_n$. The edges are labeled with distances $\Delta \vec{x}_e = \vec{x}_n - \vec{x}_{n'}, e = 1, ..., E$, where edge e connects node n' with n. Hence the edge labels are two-dimensional vectors. (When wanting to refer to the geometrical structure of a graph, unlabeled by jets, we call it a *grid*.) This face graph is *object-adapted*, since the nodes are selected from face-specific points (fiducial points); see Fig. 2.

Graphs for different head pose differ in geometry and local features. Although the fiducial points refer to corresponding object locations, some may be occluded, and jets as well as distances vary due to rotation in depth. To be able to compare graphs for different poses, we manually define pointers to associate corresponding nodes in the different graphs.

In order to find fiducial points in new faces, one needs a general representation rather than models of individual faces. This representation should cover a wide range of possible variations in the appearance of faces, such as differently shaped eyes, mouths, or noses, different types of beards, variations due to gender, age, and race etc. It is obvious that it would be too expensive to cover each feature combination by a separate graph. We instead combine a representative set of individual model graphs into a stack-like structure, called a *face bunch graph* (FBG) (see Fig. 1). Each model has the same grid structure and the nodes refer to identical fiducial points. A set of jets referring to one fiducial point is called a *bunch*. An eye bunch, for instance, may include jets from closed, open, female, and male eyes etc. to cover these local variations. During the location of fiducial points in a face not seen before, the procedure described in the next section selects the best fitting jet, called the *local expert*, from the bunch dedicated to each fiducial point. Thus, the full combinatorics of jets in the bunch graph is available, covering a much larger range of facial variation than represented in the constituting model graphs themselves.

Assume for a particular pose there are M model graphs $\mathcal{G}^{Bm}$ ($m = 1, ..., M$) of identical structure, taken from different model faces. The corresponding FBG $\mathcal{B}$ then is given the same structure, its nodes are labeled with bunches of jets $\mathcal{J}_n^{Bm}$ and its edges are labeled with the averaged distances $\Delta\vec{x}_e^B = \sum_m \Delta\vec{x}_e^{Bm}/M$.

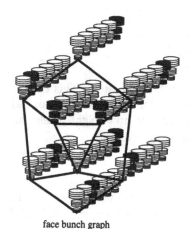

face bunch graph

Figure 1: The face bunch graph (FBG) serves as a representation of faces in general. It is designed to cover all possible variations in the appearance of faces. The FBG combines information from a number of face graphs. Its nodes are labeled with sets of jets, called bunches, and its edges are labeled with averages of distance vectors. During comparison to an image the best fitting jet in each bunch is selected independently, indicated by grey shading.

2.3 Generating Face Representations by Elastic Bunch Graph Matching

So far we have only described how individual faces and general knowledge about faces are represented by labeled graphs and the FBG, respectively. How can these graphs be generated? The simplest method is to do so manually. We have used this method to generate initial graphs for the system, one graph for each pose, together with pointers to indicate which pairs of nodes in graphs for different poses correspond to each other. Once the system has an FBG (possibly consisting of only one manually defined model), graphs for new images can be generated automatically by elastic bunch graph matching. Initially, when the FBG contains only few faces, it is necessary to review and correct the resulting matches, but once the FBG is rich enough (approximately 70 graphs) one can rely on the matching and generate large galleries of model graphs automatically.

A key role in elastic bunch graph matching is played by a function evaluating the *graph similarity* between an image graph and the FBG of identical pose. It depends on the jet similarities and the distortion of the image grid relative to the FBG grid. For an image graph $\mathcal{G}^{\mathcal{I}}$ with nodes $n = 1, ..., N$ and edges $e = 1, ..., E$ and an FBG $\mathcal{B}$ with model graphs $m = 1, ..., M$ the similarity is defined as

$$S_B(\mathcal{G}^{\mathcal{I}}, \mathcal{B}) = \frac{1}{N}\sum_n \max_m \left(S_\phi(\mathcal{J}_n^{\mathcal{I}}, \mathcal{J}_n^{Bm})\right) - \frac{\lambda}{E}\sum_e \frac{(\Delta\vec{x}_e^{\mathcal{I}} - \Delta\vec{x}_e^B)^2}{(\Delta\vec{x}_e^B)^2}, \quad (5)$$

where λ determines the relative importance of jets and metric structure. $\mathcal{J}_n$ are the jets at node n and $\Delta\vec{x}_e$ are the distance vectors used as labels at edges e. Since the FBG provides several jets for each fiducial point, the best one is selected and used for comparison. These best fitting jets serve as *local experts* for the image face.

The goal of elastic bunch graph matching on a probe image is to find the fiducial points and thus to extract from the image a graph which maximizes the similarity with the FBG as defined in Equation (5). In practice, one has to apply a heuristic algorithm to come close to the optimum within reasonable time. We use a coarse to fine approach in which we introduce the degrees of freedom of the FBG progressively: translation, scale, aspect ratio, and finally local distortions. Phase information is used only in the latter steps. The resulting graph is called the *image graph* and is stored as a representation for the individual face of the image.

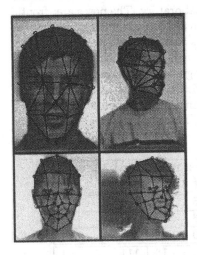

Figure 2: Object-adapted grids for different poses. The nodes are positioned automatically by elastic graph matching against the corresponding face bunch graphs. The two top images show originals with widely differing size and grids as used for the normalization stage with many nodes on the outline. The other images are already rescaled to normal size. Their grids have more nodes on the face, which is more appropriate for recognition (The grids used in the experiments had about 14 additional nodes which are not shown here for simplicity).

To minimize computing effort and to optimize reliability we extract a face representation in two stages. The first stage, called *normalization stage*, has the purpose of estimating the position and size of the face in the original image, so that the image can be scaled and cut to standard size. The second stage takes this image as input and extracts a precise image graph appropriate for face recognition purposes. The two stages differ in emphasis. The first one has to deal with greater uncertainty as to size and position of the head and has to optimize the reliability with which it finds the face, but there is no need to find fiducial points with any precision or extract data important for face recognition. The second stage can start with little uncertainty as to position and size of the head but has to extract a detailed face graph with high precision. The two stages took about 20 plus 10 seconds on a SPARCstation 10-512 with a 50 MHz processor.

2.4 Recognition

After having extracted model graphs from the gallery images and image graphs from the probe images, recognition is possible with relatively little computational effort by comparing an image graph with all model graphs and picking the one with the highest similarity value. The similarity function is based on the average similarity between corresponding jets. For image and model graphs referring to different pose, we compare jets according to the manually provided correspondences. Here we use the similarity function without phase (Eq. 3). A comparison against a gallery of 250 individuals took slightly less than a second.

3 Experiments

For the experiments we used image galleries taken from the ARPA/ARL FERET database provided by the US Army Research Laboratory. The poses are: frontal view, half-profile right or left (rotated by about 40-70°), and profile right or left. For most faces there are two frontal views with different facial expression. Apart from a few exceptions there are no disguises, variations in hair-style or in clothing. The background is always a homogeneous light or grey, except for smoothly varying shadows. The size of the faces varies by about a factor of three (but is constant for each individual, information which we could have used to improve recognition rates, but didn't). The format of the original images is 256×384 pixels.

Model gallery	Probe images	First rank #	First rank %	First 10 ranks #	First 10 ranks %
250 fa	250 fb	245	98	248	99
250 hr	181 hl	103	57	147	81
250 pr	250 pl	210	84	236	94
249 fa + 1 fb	171 hl + 79 hr	44	18	111	44
171 hl + 79 hr	249 fa + 1 fb	42	17	95	38
170 hl + 80 hr	217 pl + 33 pr	22	9	67	27
217 pl + 33 pr	170 hl + 80 hr	31	12	80	32

Table 1: Recognition results for cross-runs between different galleries (f: frontal views; a, b: expression a and b; h: half profiles; p: profiles; l, r: left and right). Each gallery contained only one image per person; the different compositions in the four bottom lines are due to the fact that not all poses were available for all people. The table shows how often the correct model was identified as rank one and how often it was among the first 10 (4%).

We used various model and probe galleries with faces of different pose. Each model gallery contained 250 faces with just one image per person. Recognition results are shown in Table 1. The recognition rate is very high for frontal against frontal images (first row). This is mainly due to the fact that in this database two frontal views show only little variation, and any face recognition system should perform well under these circumstances (cf. results on the Bochum database below for a more challenging example). Before comparing left against right poses we flipped all right pose images over. The recognition rate is high for right profile against left profile (84%, third row). The sharply reduced recognition rate of 57% (second row) when comparing left and right half-profiles is probably due to the poor control in rotation angle in the database — inspection of images shows that right and left rotation angles differ by up to 30°. When comparing half profiles with either frontal views or profiles another reduction in recognition rate is observed (although even a correct recognition rate of 10% out of a gallery of 250 is still high above chance level, which would be 0.4%!).

We also did experiments on the Bochum database, which is described in greater detail in [3]. In this database the differences in facial expression are stronger than in the FERET database, some faces were even half covered by hair or a hand. First rank recognition rates against a model gallery of 108 frontal views were 91% on frontal views of different facial expression, 94% on faces rotated in depth by 11°, and 88% on faces rotated by 22°. The results are similar to those in [3] and demonstrate robustness against variations in facial expression and rotation in depth up to 22°.

4 Discussion

The system presented is general and flexible. It is designed for an *in-class recognition* task, i.e. for recognizing members of a known class of objects. We have applied it to face recognition but the system is in no way specialized to faces and we assume that it can be directly applied to other in-class recognition tasks, such as recognizing individuals of a given animal species, given the same level of standardization of the images. In contrast to many neural network systems, no extensive training for new faces or new object classes is required. Only a moderate number of typical examples have to be inspected to build up a bunch graph, and individuals can then be recognized after storing a single image.

We tested the system with respect to rotation in depth and differences in facial expression. We did not investigate robustness to other variations, such as illumination changes or structured background. The performance is high on faces of same pose. We also showed robustness against rotation in depth up to about 22°. For large rotation angles the performance degrades significantly. Our system performs well compared to other systems. Results of a blind test of different systems on the FERET database were published in [5, 6].

In comparison to the system [3] on the basis of which we have developed the system presented here we have made several major modifications. We now utilize wavelet phase information for accurate node localization. Previously, node localization was rather imprecise. We have introduced the potential to specialize the system to specific object types and to handle different poses with the help of object-adapted grids. The face bunch graph is able to represent a wide variety of faces, which allows matching on face images of unseen individuals. These improvements make it possible to extract an image graph from a new face image in one matching process. Even if the person of the new image is not included in the FBG, the image graph reliably refers to the fiducial points. This accelerates recognition from large databases considerably since for each probe image, correct node positions need to be searched only once instead of in each attempted match to a gallery image, as was previously necessary. We did not expect and the system does not show an improvement in terms of recognition rates compared to the preceding system.

Besides the ability to handle larger galleries and larger rotation in depth, the main achievement of the system presented here is that the increased matching accuracy, the object adapted graphs, and the face bunch graph provide the basis for further developments [2, 4, 7].

Acknowledgements: This work has benefit from many colleagues. We particularly wish to thank Irving Biederman, Michael Lyons, Thomas Maurer, Jonathon Phillips, Ladan Shams, Marni Stewart Bartlett, and Jan Vorbrüggen. For the experiments we have used the FERET database of facial images collected under the ARPA/ARL FERET program and the Bochum database.

References

[1] BUHMANN, J., LANGE, J., VON DER MALSBURG, C., VORBRÜGGEN, J. C., AND WÜRTZ, R. P. Object recognition with Gabor functions in the dynamic link architecture: Parallel implementation on a transputer network. In *Neural Networks for Signal Processing*, B. Kosko, Ed. Prentice Hall, Englewood Cliffs, NJ 07632, 1992, pp. 121–159.

[2] KRÜGER, N. An algorithm for the learning of weights in discrimination functions using *a priori* constraints. accepted for publication in IEEE Transactions on Pattern Analysis and Machine Intelligence, 1997.

[3] LADES, M., VORBRÜGGEN, J. C., BUHMANN, J., LANGE, J., VON DER MALSBURG, C., WÜRTZ, R. P., AND KONEN, W. Distortion invariant object recognition in the dynamic link architecture. *IEEE Transactions on Computers 42*, 3 (1993), 300–311.

[4] MAURER, T., AND VON DER MALSBURG, C. Linear feature transformations to recognize faces rotated in depth. In *Proceedings of the International Conference on Artificial Neural Networks, ICANN'95* (Paris, Oct. 1995), pp. 353–358.

[5] PHILLIPS, P. J., RAUSS, P. J., AND DER, S. Z. FERET (face recognition technology) recognition algorithm development and test report. Tech. Rep. ARL-TR-995, U. S. Army Research Laboratory, 2800 Powder Mill Road, Adelphi, MD 20783-1197, Oct. 1996.

[6] RAUSS, P. J., PHILLIPS, J., HAMILTON, M. K., AND DePERSIA, A. T. FERET (face-recognition technology) recognition algorithms. In *Proc. of the Fifth Automatic Target Recognizer System and Technology Symposium* (1996).

[7] WISKOTT, L. Phantom faces for face analysis. *Pattern Recognition 30*, 6 (1996).

[8] WISKOTT, L., FELLOUS, J.-M., KRÜGER, N., AND VON DER MALSBURG, C. Face recognition by elastic bunch graph matching. Tech. Rep. IR-INI 96-08, Institut für Neuroinformatik, Ruhr-Universität Bochum, D-44780 Bochum, Germany, 1996.

A Conditional Mixture of Neural Networks for Face Detection, Applied to Locating and Tracking an Individual Speaker

Raphaël Feraud *, Olivier Bernier **, Jean-Emanuel Viallet, Michel Collobert, Daniel Collobert

France Télécom C.N.E.T., Technopole Anticipa, 2 av. Pierre Marzin, BP 40
22307 Lannion Cedex France

Abstract. We present a neural network approach to human face detection. Using a modular system, a conditional mixture of networks, we are able to detect front view faces as well as turned faces (up to 50 degrees) with excellent performances. This modular network is integrated into LISTEN, our face tracking system. It enables this system to detect and track in real-time faces in a variety of orientations, extending its previous applicability.

1 Introduction

Transmission of visual information and specifically of video image is of increasing importance in telecommunication applications and complements the traditional audio information. For these applications, an important high level information which can be extracted from a real time video signal is the position of the face of a speaker immersed in a priori unknown scene. In [2], we presented LISTEN (Locating Individual Speaker and Tracking ENvironment), a computer vision system able to locate and track a human face in real time. The core of the face detector module was based on a neural network able to evaluate if a given 15x20 window extracted from an image corresponds to a face [3]. As with most face detectors (see for example [6]), face detection was limited to front view faces.

In the case of a video-phone, this problem is not too critical: the speaker is supposed to look at the screen where his correspondent appears (and consequently to look in a direction not far from the camera). The problem is more critical if we want to extend the system to a (relatively small) room. In this context, a few persons may be present, some of whom may not look directly at the camera at any given time. Even if we only want to track one person at a time (for example the one nearest to the camera), it becomes necessary to be able to detect turned faces. With this goal in mind, we have developed a new face detection sub-system to extend the previous system to turned faces, without loosing its performances: this module must not only feature a good detection rate, but

---

* Email: feraud@lannion.cnet.fr
** Email: bernier@lannion.cnet.fr

also a very small false alarm rate to insure that what is detected as faces are indeed human faces.

2 Face Detection by Neural Networks

Two types of statistical model can be applied to face detection: generative models and discriminant models. Discriminant learning is based on the notion of boundary. To discriminate between two classes, two sets of data, statistically representative of each class, are needed. Obtaining a representative set of faces is possible, but not a representative set of non-face windows, as the number of possible non-face windows is too large. Generative models deal with another approach. The goal of generative learning is to estimate the probability distribution of the data, in this case of the face windows. Consequently, only one probability distribution is needed. The risk with generative models is to overestimate this probability distribution and to obtain an important false alarm rate.

A principal component analysis (PCA) can be used as a generative model to detect faces in an image [8]. A compression multi-layer perceptron (MLP) [9] can also be used. This kind of MLP, with one hidden layer, and linear activation function, simply performs PCA. We use a compression MLP with an additional hidden-layer (see Fig. 1) to obtain a more general non-linear PCA. The standard approach to non-linear PCA is to use four weight layers, but in all our experiments, the use of four layers instead of three did not improve the results, certainly because of the difficulty of back-propagating a gradient through too many layers.

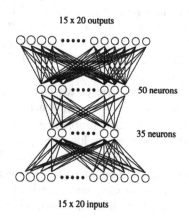

Fig. 1. Structure of a CGM.

Without any additional constraint, the solution obtained by such MLPs is close to PCA. In order to closely fit the probability distribution of the set of faces, and to constrain the algorithm to converge to a non-linear solution (to obtain a non-linear PCA), we add constraints to the generative model using non-face examples: each non-face example is constrained to be reconstructed as its projection in the set of faces (the nearest point in the set of faces). This projection $P(x)$ of a point x of the input space $\mathcal{E}$ in $\mathcal{V}$, the set of faces, is approximated by:

- if $x \in \mathcal{V}$, then $\mathcal{P}(x) = x$,
- if $x \notin \mathcal{V}$:
  - let $V = \min_{y \in \mathcal{V}}(d(x, y))$, where d is the Euclidean distance,
  - let $V_1, V_2, \ldots, V_n \in \mathcal{V}$, be the n nearest neighbors of V,
  - then the projection $\mathcal{P}$ of x in $\mathcal{V}$ is approximated by: $\mathcal{P}(x) = \frac{1}{n} \sum_{i=1}^{n} V_i$.

The goal of the learning process is to approximate the distance $\mathcal{D}$ of a point x of the input space to the set of faces $\mathcal{V}$:

- $\mathcal{D}(x, \mathcal{V}) = \|x - \mathcal{P}(x)\| \approx \frac{1}{M}(x - \hat{x})^2$, where M is the size of the input image x and $\hat{x}$ the image reconstructed by the neural network,
- let $x \in \mathcal{E}$, then $x \in \mathcal{V}$ if and only if $\mathcal{D}(x, \mathcal{V}) \leq \tau$, where $\tau \in \mathbb{R}^+$ is a threshold used to adjust the sensitivity of the model.

In the case of PCA, the reconstruction error is the distance between a point and the principal subspace (the space generated by the principal components) in the input space. A point can be close to the principal subspace and yet far from the set of faces (if the set of faces represents only a portion of this space). With the algorithm proposed, the reconstruction error is related to the distance between a point and the set of faces. As a consequence, if we assume that the learning process is consistent [10], our algorithm is able to evaluate the probability that a point belongs to the set of faces. For example we can approximate the probability of an input x belonging or not to $\mathcal{V}$ by:

$$\Pr(x \in \mathcal{V}) = e^{-\frac{(x-\hat{x})^2}{\sigma^2}} \qquad \Pr(x \notin \mathcal{V}) = 1 - e^{-\frac{(x-\hat{x})^2}{\sigma^2}}.$$

We call Constrained Generative Model or CGM this type of compression MLP trained using counter-examples.

3   The Modular Architecture

In order to extend the response of the network to turned faces, we use a modular approach. The mixture of experts was introduced by Jacobs et al [5]. This architecture is composed of several neural subnetworks and a gating network. During the training phase, each subnetwork specializes itself on a part of the training set and the gating network combines the outputs of the subnetworks.

Considering that to detect a face in an image, there are two subproblems to solve, detection of front view faces and turned faces, we use a similar but different modular architecture, which we call a conditional mixture of networks, using two subnetworks (two CGMs). We separate the training set in two subsets (domains) corresponding to an arbitrary binary label θ: front view faces and the corresponding counter-examples ($\theta = 1$) and rotated faces and the corresponding counter-examples ($\theta = 0$). The first subnetwork (CGM), trained on the first subset, evaluates the probability of the tested image to be a front view face, knowing the label equals 1 ($P(y = 1|x, \theta = 1)$), where y is the category of the input (1 for a face, 0 otherwise). The second (trained on the second subset) evaluates the probability of the tested image to be a turned face, knowing the label equals 0 ($P(y = 1|x, \theta = 0)$). A gating network is trained to evaluate $P(\theta = 1|x)$, supposing that the partition $\theta = 1, \theta = 0$ can be generalized to every input. The modular network then simply evaluates the probability that the input is a face (Fig. 2):

$$P(y = 1|x) = P(y = 1|\theta = 1, x)P(\theta = 1|x) + P(y = 1|\theta = 0, x)(1 - P(\theta = 1|x))$$

This system is different from a mixture of experts: each module is trained separately on a subset of the training set and then the gating network learns to combine the outputs. Since we use prior knowledge to specialize each module, the learning process is more simple and the capacity (as defined by [10]) of the modular system is less than if it was constructed by training the system with all the data at the same time.

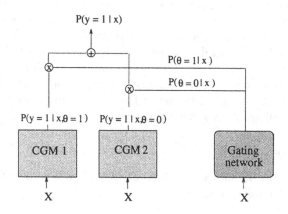

Fig. 2. Each CGM produces the probability that x is a face (supposing it belongs to its domain) and the gate network computes the probability of each domain knowing x.

4 Learning

The size of the training images is 15x20 pixels. The faces are normalized manually in position and scale. The images are enhanced by histogram equalization to obtain a relative independence to lighting conditions, smoothed to remove the noise and normalized by the average face, evaluated on the training set. Two face databases are used, one for the front view face subnetwork B_{f1} and one for the turned face subnetwork B_{f2}. After mirroring, B_{f1} is composed of 3600 different faces with orientation between 0 degree to 20 degrees. B_{f2} is composed of 1600 different faces with orientation between 20 degrees to 60 degrees. All of the training faces are extracted from *Usenix face database*(*), from the test set B of CMU(*), and from 100 images containing faces and complex backgrounds. The non-face databases (B_{nf1}, B_{nf2}, corresponding to each face database), are collected by the following iterative algorithm, similar to the one used in [7] or in [6]:

- 1) $B_{nf} = \emptyset$, $\tau = \tau_{min}$,
- 2) the neural network is trained with $B_f + B_{nf}$,
- 3) the face detection system is tested on a set of background images,
- 4) a maximum of 100 sub-images x_i are collected with $\mathcal{D}(x_i, V) \leq \tau$,
- 5) $B_{nf} = B_{nf} + \{x_0, \ldots, x_n\}$, $\tau = \tau + \mu$, with $\mu > 0$,
- 6) while $\tau < \tau_{max}$ go back to step 2.

After mirroring, the size of the obtained set of non-face examples is respectively 1500 for B_{nf1}, 600 for B_{nf2}.

The gating network is a simple MLP with three layers (300 inputs, 35 hidden units and one output neuron). It is trained using the two already trained CGM and the complete database formed by the concatenation of the previous databases (front view and turned faces, and the corresponding non-face examples).

5 Performances of the Face Detector

To apply the detector to a region of an image, each 15x20 patch of this region, at different scales, is extracted and normalized by enhancing, smoothing and subtracting the average face, before processing by the network. To obtain patches at different scales, the image is sub-sampled. The probability of face presence calculated by the modular network is thresholded. Multiple detections are used to confirm the presence of a face.

To evaluate the performance of this face detection module, three tests are performed. The first consists to test the detection rate with images belonging to *Usenix face database*(*), which are not used in the training process (Table 1).

Since the number of images processed is important (≈ 4200), this test allows the evaluation of the detection rate. Nevertheless, the backgrounds of these images are very simple, the lighting conditions are very stable and the variations in orientation of the faces are small. To evaluate the limits in orientation of the face detector, the *Sussex face database*(*), containing different faces with ten orientations between 0 degree and 90 degrees, is used (Table 2).

The performance of this face detection module in terms of false alarm rate has been tested on a standard database, the test set A (*) [6] of the CMU, containing

Table 1. Results on the *Usenix face database*(*):

System	Detection rate	False alarm rate
Mixture	98 %	1/4,700,000
of CGMs	4233/4320	5/23,500,000

Table 2. Results on the *Sussex face database*(*):

Orientation (degree)	Detection rate
0	100.0%
10	100.0%
20	100.0%
30	100.0%
40	87.5%
50	62.5%
60	37.5%
70	25.0%
80	0.0%
90	0.0%

42 images of various quality (Table 3). This test set enables the comparison of our results with the best result published so far [6]. An example of the detections obtained by the module is given in Fig 3.

6 Application to the LISTEN System

The obtained face detector is integrated into the LISTEN ([2]) system which is aimed at robustly detecting and tracking a face against an unknown background, in real time. This system can be in two mode: detection or tracking. In detection mode, the camera is fixed, and the face detector module is enabled. Two other modules, the movement and skin color detectors, are used to obtain the regions of interest, corresponding to moving skin color zones [2]. These regions of interest are then processed by the face detector. Once a face is detected, the system switch to tracking mode, and the camera follows the detected face (the face detector is disabled).

The whole system, using the new face detector, is tested in the context of an individual communication terminal, in rooms of small to medium size. The system tracks only one person, even if more are present. Automatic framing allows this person to rise from its chair, freely move around, walk to a blackboard, while continuously framed by the camera. The lighting conditions ranges from

Table 3. Results on the CMU test set A. CGM: constrained generative model. SWN: shared weights network.

System	Detection rate	False alarm rate
Mixture of CGMs	**86.5 %**	**1/864,100**
(two CGMs)	142/164 **	39/33,700,000
[Rowley,1995]	84 %	1/123,000
(one SWN)	142/169 **	179/22,000,000
[Rowley,1995]	84.6 %	1/245,000
(two SWNs)	143/169 **	90/22,000,000
[Rowley,1995]	**85.2 %**	**1/469,000**
(three SWNs)	144/169 **	47/22,000,000

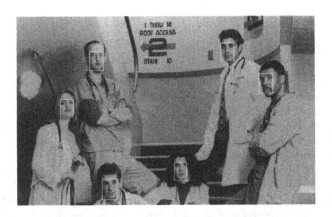

Fig. 3. Sample output of the face detection module.

standard artificial lighting to natural lighting provided by windows. This system operates at the input video rate (25 fr/sec) when no command is sent to the camera (sending a command to the camera causes at most a one frame skip in the video stream). Initialization of the tracking by face detection is typically done in less than half a second. For faces turned up to around 30 degrees, no difference of performance is found compared to front view faces. Up to 50 degrees, the effective detection rate is still good. These results were obtained using a DEC Alphastation 500/333. The captured images, on which the processing modules are working, is in the CIF format (384 by 288 pixels).

7 Conclusion and Future Work

The chosen architecture (conditional mixture of networks) resulted in a very efficient face detecting system, extending the capability of the previous detector to turned faces, (up to 50 degrees). In the future, one possibility of extension of our LISTEN system will be to adapt it to the context of videoconferences. In such conditions, the system must be able to locate and track continuously all the participants, using a wide field camera (the camera must be able to view the entire scene). In this context, the detection of turned faces becomes even more important. More subnetworks may be needed to gives the modular network the ability to detect turned faces up to 90 or 100 degrees, which may be necessary in this case.

(*) *Usenix face database, Sussex face database* and CMU test sets can be retrieved at *www.cs.rug.nl/~peterkr/FACE/face.html.*
(**) we do not take into account drawing faces and non human faces, contrary to Rowley [6].

References

1. R. Chellappa, C. L. Wilson, and S. Sirohey, "Human and Machine Recognition of Faces: A Survey", in *Proceedings of the IEEE*, Vol. 83(5), May 1995.
2. M. Collobert, R. Feraud, G. Le Tourneur, O. Bernier, J. E. Viallet, Y. Mahieux, and D. Collobert, "LISTEN: A System for Locating and Tracking Individual Speakers", in *Proceedings of the second International Conference on Automatic Face and Gesture Recognition* Killington, Vermont, October 1996.
3. R. Feraud, O. Bernier, and D. Collobert, "A Constrained Generative Model Applied to Face Detection", in *Neural Processing Letters*, Vol. 5(2), 1997.
4. M. Hunke, and A. Waibel, "Face Locating and Tracking for Human-Computer Interaction", in *Proceedings of the 28th Asimolar Conf. on Signals, Systems, and Computers*, Pacific Grove, California, November 1994.
5. R. A. Jacobs, M. I. Jordan, S. J. Nowlan, and G. E. Hinton, "Adaptative Mixture of Local Experts", in *Neural Computation*, Vol 3, pp 79-87, 1991.
6. H. Rowley, S. Baluja, and T. Kanade, "Human Faces Detection in Visual Scenes", in *Advances in Neural Information Processing Systems 8*, 1995.
7. K. Sung, and T. Poggio, "Example-based Learning for View-based Human Face Detection", Tech. Report, MIT, 1994.
8. M. Turk, and A. Pentland, "Eigenfaces for Recognition", in *Journal of Cognitive Neuroscience*, Vol. 3(1), pp. 71-86, 1991.
9. D. Valentin, H. Abdi, A. J. O'Toole, and G. W. Cottrell, "Connectionist Models of Face Processing: A Survey", in *Pattern Recognition*, Vol. 27(9), pp. 1209-1230, 1994.
10. V. Vapnik, "The Nature of Statistical Learning Theory", Springer Verlag, 1995.
11. C. Wren, A. Azarbayejani, T. Darrell, and A. Pentland, "Pfinder: Real-Time Tracking of the Human Body", in *SPIE Photonics East*, Vol. 2615, pp. 89-98, 1995.

Lipreading Using Fourier Transform over Time

Keren Yu, Xiaoyi Jiang, and Horst Bunke

Department of Computer Science, University of Bern, Switzerland

Abstract. This paper describes a novel approach to visual speech recognition. The intensity of each pixel in an image sequence is considered as a function of time. One-dimensional Fourier transform is applied to this intensity-versus-time function to model the lip movements. We present experimental results performed on two databases of ten English digits and letters, respectively.

1 Introduction

For decades, most research efforts in automatic speech recognition have focused on the acoustic signal only. One problem in those efforts, however, is that the acoustic recognition rate often decreases significantly in noisy environments such as offices, airports, train station, factory floors, automobile and airplane cockpits, and others. One of the approaches to solving this problem is using the visual signal which is not affected by acoustic noise. The lip movements represented in the visual signal often contain enough information for a categorization of speech. In addition, a combination of both the acoustic and visual signal possesses the potential of capturing the information from two independent sources and thus improving the overall speech recognition performance. As a matter of fact, studies in human perception have shown that visual information allows people to tolerate an extra 4-dB of noise in the acoustic signal [10]. Also in computer speech recognition, fusion of the acoustic and visual signal has been investigated to improve recognition performance [3].

There have been several works [3, 4, 6, 7, 8, 9] on lipreading based on Hidden Markov Model (HMM), neural networks, principal component analysis, a.s.o; see [5] for an overview and [12] for a collection of recent works in visual speech recognition. In these approaches features are usually extracted from individual images, and lip movements are modeled by HMM, neural networks etc. In this paper, we present a novel alternative method where the intensity curve of pixels along the time axis is considered. The one-dimensional Fourier coefficients of the intensity curves encode the motion information in a compact manner and are used as features for matching. The same basic idea of processing intensity-versus-time curves has been successfully applied to medical image analysis [1, 2, 11].

In the next section we describe the Fourier transform over time. After that, we give a description of our model construction and matching method. Then, experimental results are reported. Finally, some conclusions are given.

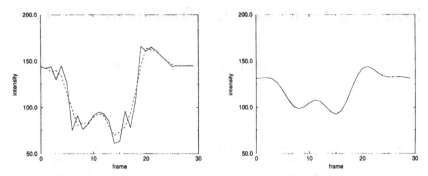

Fig. 1. Intensity-versus-time curve: original and smoothed version (dashed line) (left); reconstruction (right).

2 Fourier Transform over Time

In visual speech recognition, the lip movements are analyzed in a sequence of digitized images of the mouth. The intensity $I(n)$ of each pixel of an image sequence can be considered as a function of time, see Figure 1 (left). Clearly, the complete information of the image sequence is contained in the intensity-versus-time curves if we consider the curve in every pixel. But eventually, we are looking for features that represent the intensity-versus-time curves, and thus the lip movements, in a compact way. For this purpose we perform an one-dimensional Fourier transform of the intensity function $I(n)$,

$$c(k) = \frac{1}{N} \sum_{n=0}^{N-1} I(n)e^{\frac{-j2\pi kn}{N}}, \qquad k = 0, 1, \ldots, N-1,$$

where N is the number of frames of the sequence, and $c(k)$ is a complex coefficient. Using the first few Fourier coefficients we are able to approximately describe the time evolution of the curve.

In order to demonstrate this point, let us choose the intensity-versus-time curve of a pixel on the center of the mouth from a real image sequence. The left curve in Figure 1 shows the intensity of this particular pixel over time. Figure 2 illustrates the real and imaginary parts of the corresponding Fourier coefficients. We observe that indeed only low-frequency Fourier coefficients are significantly different from zero. Therefore, a compact representation of an intensity-versus-time curve $I(n)$ is given by the first k Fourier coefficients ($k = 5$ in our experiments), or alternatively by the their magnitudes. This way we obtain a compact representation of the whole image sequence by means of a $h \times w$ matrix C of k-dimensional vectors

$$C = [c_{ij}]_{hw},$$

where h and w are the height and width of the images, respectively, and c_{ij} is a vector of dimension k, containing the magnitude of the first k Fourier coefficients for pixel (i, j).

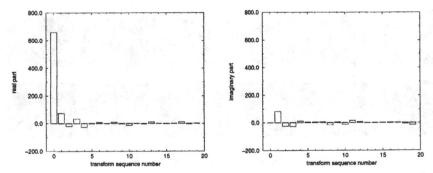

Fig. 2. Fourier coefficients of the original intensity-versus-time curve in Figure 1 (left).

The effectiveness of this representation is demonstrated for the intensity function in Figure 1 (left) by a reconstruction through the inverse Fourier transform, as shown in Figure 1 (right). Obviously, the reconstructed curve has a high similarity to the original curve resp. its smoothed version. Therefore, these Fourier coefficients are able to approximately describe the original signal curve. As an example of this representation for a whole image sequence, Figure 3 shows the matrix C for an image sequence of digit zero, where the magnitudes of the first five Fourier coefficients are represented from left to right. In this visualization a logarithmic transformation has been applied to enhance the visibility of low magnitude values.

3 Model Construction and Matching

Given L training image sequences of a class d and their corresponding Fourier coefficient matrices, $C_k = [c_{ij}^k], k = 1, \ldots, L$, we construct a model matrix $M_d = [m_{ij}^d]$ for the class by averaging the Fourier coefficient matrices,

$$M_d = \frac{1}{L} \sum_{k=1}^{L} C_k.$$

For a test image sequence, its Fourier coefficient matrix $T = [t_{ij}]_{hw}$ is calculated. Then, we match T against all model matrices

$$M_d = [m_{ij}^d], \qquad d = 1, \ldots, D,$$

Fig. 3. The Fourier coefficients of a digit image sequence.

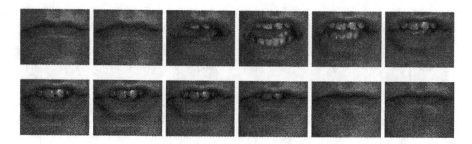

Fig. 4. An image sequence of letter H.

where D is the total number of classes, using

$$\sum_{i=1}^{h} \sum_{j=1}^{w} ||t_{ij} - m_{ij}^d||.$$

The best match r is simply given by

$$r = \arg\min_{d} \sum_{i=1}^{h} \sum_{j=1}^{w} ||t_{ij} - m_{ij}^d||.$$

It is important to notice that the first Fourier coefficient corresponds to the average of the intensity function $I(n)$ along the time axis. Therefore, the first dimension of the model matrices corresponding to the first Fourier coefficient will be approximately identical for all models, as long as the illumination during the training phase is kept constant (but not necessarily uniform). That is, only coefficients other than the first one mainly contribute to the characterization of intensity-versus-time curves and to the matching. Thus, we can potentially ignore the first Fourier coefficient in the matching process.

If the illumination in the test phase differs from that in the training phase, it can be expected that the intensity-versus-time curve of a pixel has a shape similar to that of the same pixel in the corresponding model, although its height may be different. This implies that by ignoring the first Fourier coefficient our lipreading method possesses the potential of illumination invariance.

4 Experimental Results

The algorithm discussed in the previous sections has been implemented in C on a Sun Sparcstation. So far we have performed tests on two image databases. The first one was acquired in our lab. It has a vocabulary of 10 English digits (from zero to nine) spoken by two speakers. From each of the two speakers 20 sets of image sequences were collected, each set containing the ten digits. Images of 190×150 pixels and 8 bits per pixel were collected at a rate of 12 frames per second, centered around the lips, under normal lighting conditions. The second

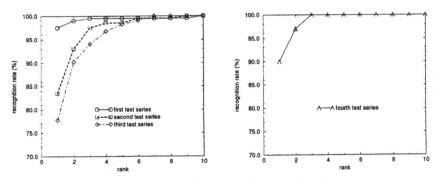

Fig. 5. Recognition rates of four test series.

image database comes from University of Central Florida, Orlando, and has been used in earlier works [7] on visual speech recognition. It consists of 18 sets of lip image sequences, each set containing 10 English letters from A to J. Images of 240 × 200 pixels and 8 bits per pixel were collected at a rate of 15 frames per second under similar conditions as for the first image database. All sequences were supplied by a single speaker. As an example of a lip image sequence, Figure 4 shows one of letter H from the second database.

On the first image database three test series have been carried out. In two test series the image sequences acquired by each of the two speakers were considered separately, while the whole image database was used for the third test series. In addition, a fourth test series was done on the whole second image database. In each test series the jackknife, or so-called "leave-one-out", procedure was applied. That is, leaving one set of image sequences out for testing, the other sets were used for training. Therefore, there were 200 tests in the first two test series respectively, 400 tests in the third, and 180 tests in the fourth test series.

In Figure 5, the recognition rates of the four test series are given. From Figure 5, we can see that the recognition rate for the first test series is better than that for the second. The reason is that the two speakers have made different head movements during the image acquisition. While the first speaker moved his head slightly from right to left when he pronounced the digits, the second speaker moved it slightly downward. Due to the shape of mouth, a vertical head movement changes the intensities of a larger area than the same amount of head movement in the horizontal direction. Thus, our lipreading method is more sensitive to vertical head movements. The third test series brought the lowest recognition rate where the image sequences provided by both speakers were put together. However, a consideration of the second rank already leads to a recognition rate close to 90%. This fact is particularly important in the context of classifier combination.

Our lipreading method also demonstrated promising results in the fourth test series on the second image database. Our recognition rate, nevertheless, is not directly comparable with that reported in [7] on the same image database. In [7] the authors only carried out 20 tests, achieving a recognition rate of 95%. By

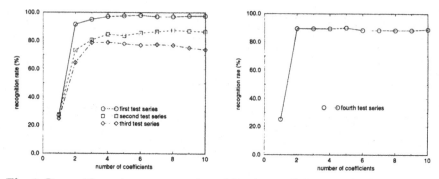

Fig. 6. Recognition rates versus number of Fourier coefficients in the four test series.

contrast, a total of 180 tests were done in our fourth test series.

The only parameter in our lipreading method is the number of Fourier coefficients k used to represent the intensity-versus-time curve. Based on our observations (see Figure 2), k has been set to 5. But we were also interested in the relationship between the recognition rate and the number k in a more fundamental sense. For the four test series, this relationship is shown in Figure 6. Interestingly, the use of only two Fourier coefficients gives already a quite reasonable recognition rate. In general the recognition rate becomes approximately stable after about three coefficients. In the fourth test series this is already the case after only two coefficients, while the recognition rate decreases slightly in the third test series. Overall, the use of four or five coefficients is a good compromise, supporting our earlier choice of k.

As mentioned in Section 3, our lipreading approach has the potential of illumination invariance. For the verification of this property we have simulated a non-uniform illumination on the second test image database. An additional position-dependent intensity is added to the pixels. The intensities are further disturbed by a uniformly distributed noise from the intervals [0,4], [0,8], and [0,12], respectively. As an example, Figure 7 shows the simulated image sequence of noise level 12 originating from that in Figure 4. Using the same leave-one-out procedure, a total of 180 tests has been carried out. In these tests the simu-

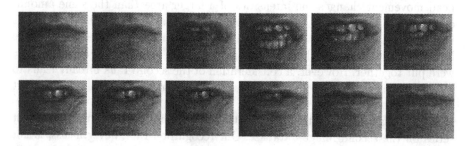

Fig. 7. An image sequence of letter H under non-uniform illumination.

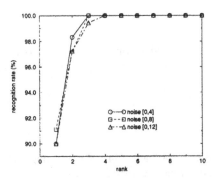

Fig. 8. Recognition rates for non-uniform illumination and different noise levels.

lated lip sequences under non-uniform illumination are matched against models constructed from the original image data. It turned out that for all noise levels essentially the same recognition rates as shown in Figure 5 could be achieved, see Figure 8 for the recognition rates in non-uniform illumination and different noise levels.

5 Conclusions

In this paper we have presented a new method for lipreading. Instead of the usual approach of extracting features from the individual images, we consider the intensity-versus-time curves of the individual pixels. Lip movements are encoded by a few Fourier coefficients of the intensity-versus-time curves. The primary experimental results on two image databases have demonstrated the usefulness of this encoding scheme to lipreading.

There are a number of details of our basic lipreading method to be further investigated. So far only the magnitude of the Fourier coefficients has been considered. But we could incorporate the phase information to our image sequence encoding scheme as well. The relative importance of the individual Fourier coefficients to the overall lipeading performance and thus the need for a weighted matching has not been explored yet. In the two test image databases the image acquisition condition has been kept constant. The potential of illumination invariance of our approach was only verified by simulated image data so far and has to be extended to real image data. These and other aspects will be investigated in the future.

In such complicated tasks as visual speech recognition, reliable classification is difficult to achieve for a single algorithm. Classifier combination is an effective way to improve recognition performance. In [13] we described a combination of the lipreading method proposed in the present paper with some other classifiers. It could be shown that even simple combination concepts can bring a significant improvement of classification accuracy in lipreading.

Acknowledgment

The second image database used in our experiments is supplied by Computer Vision Lab, Computer Science Department, University of Central Florida, Orlando.

References

1. W.E. Adam and F. Bitter, "Advances in Heart Imaging", *Proc. of Int. Symposium on Medical Radionuclide Imaging*, 1980.
2. M. Boehm, U. Obermoeller, and K.H. Hoehne, "Determination of Heart Dynamics from X-Ray and Ultrasound Image Sequences", *Proc. of Int. Conf. on Pattern Recognition*, pp. 403–408, 1980.
3. C. Bregler, S. Manke, H. Hild, and A. Waibel, "Bimodal Sensor Integration on the Example of 'Speech-Reading'", *Proc. of IEEE Int. Conf. on Neural Networks*, pp. 667–671, 1993.
4. A.J. Goldschen, O.N. Garcia, and E. Petajan, "Continuous Optical Automatic Speech Recognition by Lipreading", *Proc. of 28th Annual Asilomar Conference on Signals, Systems, and Computers*, pp. 572–577, 1995.
5. M. Hennecke, D.G. Stork, and K.V. Prasad, "Visionary Speech: Looking Ahead to Practical Speechreading Systems", in *Speechreading by Humans and Machines*, D.G. Stork and M.E. Hennecke (Eds.), pp. 331–350, 1995.
6. M. Kirby, F. Weisser, and G. Dangelmayr, "A Model Problem in the Representation of Digital Image Sequences", *Pattern Recognition*, Vol. 26, No. 1, pp. 63–73, 1993.
7. N. Li, S. Dettmer, and M. Shah, "Lipreading Using Eigensequences", *Proc. of Int. Workshop on Automatic Face- and Gesture-Recognition*, pp. 30–34, 1995.
8. J. Luettin, N.A. Thacker, and S.W. Beet, "Visual Speech Recognition Using Active Shape Models and Hidden Markov Models", *Proc. of IEEE Int. Conf. on Acoustic, Speech and Signal Processing*, 1996.
9. U. Meier, W. Hürst, and P. Duchnowski, "Adaptive Bimodal Sensor Fusion for Automatic Speechreading", *Proc. of IEEE Int. Conf. on Acoustic, Speech and Signal Processing*, 1996.
10. J.R. Movellan, "Visual Speech Recognition with Stochastic Networks", in *Advances in Neural Information Processing System*, G. Tesauro, D. Toruetzky, and T. Leen (Eds.), Vol. 7, MIT Press, Cambridge, 1995.
11. C. Nastar and N. Ayache, "Time Representation of Deformations: Combining Vibration Modes and Fourier Analysis", in *Object Representation in Computer Vision*, M. Hebert, J. Ponce, T. Boult, and A. Gross (Eds.), pp. 263–275, 1994.
12. D.G. Stork and M.E. Hennecke (Eds.), *Speechreading by Humans and Machines*, Springer-Verlag, 1996.
13. K. Yu, X.Y. Jiang, and H. Bunke, "Lipreading: A Classifier Combination Approach", accepted by *Pattern Recognition in Practice V*, 1997.

Phantom Faces for Face Analysis*

Laurenz Wiskott[†]
Institut für Neuroinformatik
Ruhr-Universität Bochum
D–44780 Bochum, Germany
http://www.neuroinformatik.ruhr-uni-bochum.de

Abstract

The system presented here is a specialized version of a general object recognition system. Images of faces are represented by graphs, labeled with topographical information and local features. New graphs of faces are generated by an elastic graph matching procedure comparing the new face with a set of stored model graphs: the *face bunch graph*. The result of this matching process can be used to generate composite images of faces and to determine facial attributes represented in the face bunch graph, such as sex or the presence of glasses or a beard. The performance of the system is comparable to that of other systems.

1 Introduction

The presented work is part of an extended algorithmic system for object recognition. Two dimensional views of objects are represented by image graphs, in which nodes are labeled with jets and edges are labeled with distance vectors. Jets are a robust and flexible representation of local grey value patches based on a Gabor-wavelet transform. In addition to single graphs for representing individual faces a new data-structure is introduced, the *face bunch graph*, to represent faces in general. The face bunch graph is a stack-like structure which combines graphs from a set of model faces and allows the system to represent new faces by combining jets from different models. New image graphs can be generated by elastic bunch graph matching, which is described in greater detail elsewhere [6]. In this work I only focus on analyzing the matching result, generating composite or phantom faces, and determining facial attributes. A more detailed description of this work is given in [5].

*Supported by grants from the German Federal Ministry for Science and Technology (413-5839-01 IN 101 B/9).

†Current address: Computational Neurobiology Laboratory, The Salk Institute for Biological Studies, San Diego, CA 92186-5800, http://www.cnl.salk.edu/CNL, wiskott@salk.edu

2   Face Representation

We use *graphs* with an underlying two–dimensional structure. The nodes are labeled with jets $\mathcal{J}_n$ and the edges are labeled with distance vectors $\Delta\vec{x}_e$. In the simplest case the graph has the form of a rectangular grid with constant spacing between nodes.

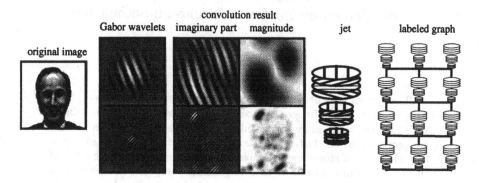

Figure 1: Graph Representation of a Face

The jets are based on a wavelet transform, which is defined as a convolution with a family of complex Gabor kernels

$$\psi_j(\vec{x}) = \frac{k_j^2}{\sigma^2} \exp\left(-\frac{k_j^2 x^2}{2\sigma^2}\right) \left[\exp(i\vec{k}_j\vec{x}) - \exp\left(-\frac{\sigma^2}{2}\right)\right],$$

providing at each location $\vec{x}$ the coefficients

$$\mathcal{J}_j(\vec{x}) = \int I(\vec{x}')\psi_j(\vec{x} - \vec{x}')d^2\vec{x}'$$

given the image grey level distribution $I(\vec{x})$.

Gabor-wavelets are localized in both space and frequency domain and have the shape of plane waves of a wave vector $\vec{k}_j$ restricted by a Gaussian envelope function of width σ. In addition the kernels are corrected for their DC value, i.e., the integral $\int \psi_j(\vec{x})d^2\vec{x}$ vanishes. All kernels are similar in the sense that they can be generated from one kernel by dilation and rotation. We use kernels of five different sizes, index $\nu \in \{0, \ldots, 4\}$, and eight orientations, index $\mu \in \{0, \ldots, 7\}$. Each kernel responds best at the frequency given by the characteristic wave vector

$$\vec{k}_j = \begin{pmatrix} k_\nu \cos\phi_\mu \\ k_\nu \sin\phi_\mu \end{pmatrix}, \quad k_\nu = 2^{-\frac{\nu+2}{2}}\pi, \quad \phi_\mu = \mu\frac{\pi}{8},$$

with index $j = \mu + 8\nu$.

The full wavelet transform provides 40 complex coefficients at each pixel (5 frequencies × 8 orientations). We refer to this array of coefficients at one pixel as a *jet* $\mathcal{J}(\vec{x})$, see Figure 1.

The complex jet coefficients $\mathcal{J}_j$ can be written as $\mathcal{J}_j(\vec{x}) = a_j(\vec{x}) \exp(i\phi_j(\vec{x}))$ with a smoothly changing magnitude $a_j(\vec{x})$ and a phase $\phi_j(\vec{x})$ spatially varying with approximately the characteristic frequency of the respective Gabor kernel. Due to this variation one cannot compare the jets directly, because small spatial displacements change the individual coefficients drastically. One can therefore either use only the magnitudes or one has to compensate explicitly for the phase shifts due to a possible displacement. The latter leads to the similarity function

$$S_\phi(\mathcal{J}, \mathcal{J}') = \frac{\sum\limits_j a_j a'_j \cos(\phi_j - \phi'_j - \vec{d}\vec{k}_j)}{\sqrt{\sum\limits_j a_j^2 \sum\limits_j a'^2_j}} \, ,$$

where $\vec{k}_j$ is the characteristic wave vector of the respective Gabor kernel and $\vec{d}$ is an estimated displacement vector which compensates for the rapid phase shifts. $\vec{d}$ is determined by maximizing S_ϕ in its Taylor expansion around $\vec{d} = 0$, which is a constrained fit of the two–dimensional $\vec{d}$ to the 40 phase differences $\phi_j - \phi'_j$ [6].

This preprocessing was chosen for its technical properties and its biological relevance. Since the wavelets have a limited localization in space and frequency, they are robust to shift, scaling, and rotation. The jets and the similarity function are robust against changes in lighting conditions in two respects. Firstly, since the kernels are DC free, the jets are invariant with respect to general offsets in the image grey values. Secondly, since the similarity function S_ϕ is normalized, it is invariant with respect to contrast variations. Finally the receptive fields of simple cells in the primary visual cortex are of similar shape as the Gabor kernels [2, 4].

New image graphs can be generated by an *elastic graph matching* process, comparing one or a set of stored model graphs with an image. The result is a set of image pixel positions and their attached jets which maximizes the similarity with the model graphs in terms of jet similarity and minimal distortion (see [6] for a detailed description of the elastic graph matching).

3 Face Bunch Graph

Besides the individual image graphs which are obtained automatically by elastic graph matching, a more general face representation is required: a representation which is able to cover a wide variety of different faces not seen before. For this purpose I combine a representative set of individual model graphs into a stack–like structure, the *face bunch graph*, see Figure 2. Each model has the same graph structure and the nodes refer to the same facial landmarks (termed hereafter fiducial points). All the nodes referring to the same fiducial point are bound together and represent various instances of this local face region.

I assume that for each new face and for each fiducial point there is an *expert* jet in the face bunch graph, sufficiently similar to the jet of the new face at that location. These experts are already determined during elastic graph matching

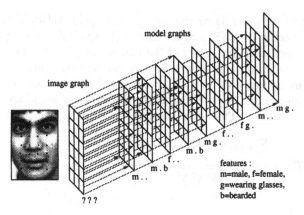

Figure 2: The stack structure of the face bunch graph. It is shown how the individual nodes of an image graph fit best to different model graphs. Each model graph is labeled with known facial attributes, on the basis of which the attributes of the new face can be determined.

and are used to find the fiducial points as precisely as possible. This information — which model provides the expert jet at which node — is important for generating phantom faces and for determining facial attributes.

4 Phantom Faces and Determining Facial Attributes

What can we say about the new face if we discard all of its local feature information, i.e. the original image jets, just keeping the geometry of the matched image graph and the identity of the expert jets at each node?

First I am going to reconstruct the face image on the basis of the matching result; I build a composite or *phantom face* resembling the original. For each node of the graph the local grey level distribution of the respective expert model is copied with a smooth transition between patches of neighboring nodes. This simple method gives a good reconstruction of the original, see Figure 3. Such a phantom face is typically composed of patches from about ten to twenty different models.

The simple and general idea to determine facial attributes now is the following: If, for example, the expert jets are taken mostly from female models, one can expect that the phantom face will look female and consequently that the original face was probably a female as well. This also holds for other attributes, such as facial hair or glasses. If the expert models for the lower half of the image graph are mostly bearded, then the original face was probably bearded as well, and similarly for glasses. One only has to label all models in the face bunch graph with their respective attributes, decide which region of the face is relevant for a particular attribute, and then compare which attribute was most often provided by the experts in that region.

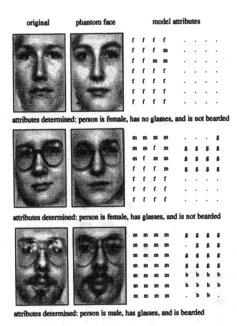

original   phantom face   model attributes

attributes determined: person is female, has no glasses, and is not bearded

attributes determined: person is female, has glasses, and is not bearded

attributes determined: person is male, has glasses, and is bearded

Figure 3: Shown is the original and the phantom face for three different persons. Notice that the phantom image was generated only on the basis of information provided by the match with the face bunch graph; no information from the original image was used. That is the reason why certain details, such as the reflections on the glasses or the precise shape of the lips of the top image are not reproduced accurately. The fields of labels on the right indicate the attributes of the models which were used as experts for the individual nodes; m: male, f: female, b: bearded, g: glasses.

5 Statistical Analysis

This simple idea can be refined by applying a standard Bayesian approach to determine automatically which nodes are most reliable. For each node n a stochastic variable X_n is introduced which can assume the values 1 and 0 depending on whether the respective expert model has a particular attribute (male, beard, glasses) or not (female, no beard, no glasses). X is the random variable for the image face itself. A sample of these stochastic variables is denoted by x_n and x respectively. Given an image with a certain value x of X, one can ask for the probability $P(x_1, ..., x_N | x)$ of a certain combination of node labels. I make the assumption that the probabilities for the individual nodes are independent of each other $P(x_1, ..., x_N | x) = \prod_n P(x_n | x)$. The Bayes a posteriori probability for a new image having the attribute x given the node labels x_n then is

$$P(x|x_1, ..., x_N) = \frac{P(x) \prod_n P(x_n|x)}{P(1) \prod_n P(x_n|1) + P(0) \prod_n P(x_n|0)}.$$

The decision whether the attribute is present or not, i.e. $x = 1$ or $x = 0$, is then based on whether $P(1|x_1, ..., x_N) > P(0|x_1, ..., x_N)$ or not. This can easily be transformed into the more illustrative weights formulation in which the decision is made on the basis of a weighted sum over the nodes of one attribute. If one takes into account the formula for the Bayes a posteriori probability and the fact that the x_n may assume the values 0 and 1 only, one obtains:

$$P(1|x_1, ..., x_N) > P(0|x_1, ..., x_N)$$
$$\iff \sum_n x_n \beta_n > \theta,$$

with

$$\beta_n = \ln\left(\frac{P(1_n|1)P(0_n|0)}{P(1_n|0)P(0_n|1)}\right),$$

$$\theta = \ln\left(\frac{P(0)}{P(1)}\right) - \sum_n \ln\left(\frac{P(0_n|1)}{P(0_n|0)}\right).$$

The weights β_n are shown in Figure 4.

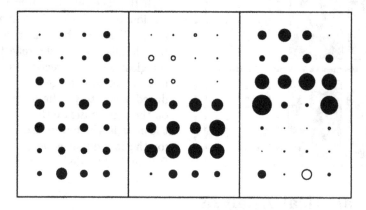

Figure 4: Weights β_n of the nodes. Radii grow linearly with the weights, white circles indicate negative values. From left to right for sex identification, for beard detection, and for glasses detection. The weights are determined on pure sets (see Section 6).

The conditional probabilities $P(x_n|x)$ are not known but have to be estimated on images for which the attributes are known. The absolute probabilities were chosen manually to be $P(1) = P(0) = 0.5$ in order to exclude prejudices about the test set composition derived from the training set.

6 Results

In the test runs a gallery of 111 neutral frontal views was used, 65% of which were male, 19% were bearded, and 28% had glasses. Each of the 111 faces was analyzed while the remaining 110 models constitute the face bunch graph. All faces were normalized in size while keeping the format (height to width ratio) constant and the influence of hairstyle was reduced by a grey frame around the faces (see visible region in Figure 5). The 111 model graphs of 7×4 rectangularly arrayed nodes were positioned by hand; the image graphs were generated automatically. Results are shown in Table 1 for the three attributes sex, beard, and glasses. The sets of faces were always split randomly into a training set and a test set of same size. On the training sets the conditional probabilities were determined and on the test sets the performance was tested. For each figure

Table 1: Rates of Correct Attribute Determination.

	sex [%]	beard [%]	glasses [%]
complete sets (111/111/111)	92 ± 3	94 ± 3	96 ± 2
small sets (68/45/51)	88 ± 5	94 ± 5	92 ± 5
pure sets (68/45/51)	83 ± 5	86 ± 5	89 ± 4

several hundred sample sets were drawn randomly. The standard deviation is shown in the table as well.

Besides the results for the complete sets, results on small and pure sets are given as well. In a pure set, faces differ only in one of the three attributes, i.e., 68 beardless faces with no glasses for sex identification, 45 male faces without glasses for beard detection, and 51 beardless males for glasses detection. The performance degrades significantly. Part of this degradation is due to the reduced gallery size. For comparison, results on small sets randomly selected from the complete set are shown as well. The dependency on gallery size indicates that the system would improve significantly for a face bunch graph larger than 111. Part of the degradation on pure sets is due to the fact that attributes such as sex and beard cannot longer cooperate.

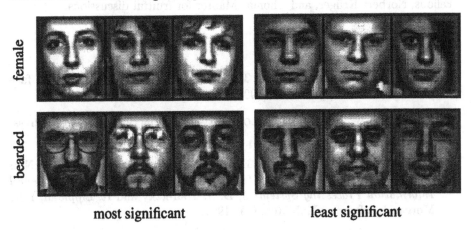

most significant   least significant

Figure 5: Female and bearded sample faces ordered by their attribute significance as judged by the system.

Figure 5 shows several sample faces with respect to their significance for the attributes 'female' and 'bearded'. For the bearded males the order correlates well with contrast and extension of the beard. For the females it is conspicuous that the three least significant females included the two youngest ones in the gallery (last and third last), of an age were facial sex is not yet fully developed. For glasses no such obvious order was obtained.

7 Discussion

I have demonstrated a simple and general principle for determining facial attributes. No extensive training is required. The system generalizes from a collection of single sample faces, the face bunch graph, by combining subparts into new composite or phantom faces. Abstract attributes can be transferred to a new face in a simple manner. The classification performance relies on what is represented in the face bunch graph: one cannot expect that with a Caucasian face bunch graph the system performs well on Asian people, for example. I expect, however, that with an appropriate face bunch graph, other attributes such as age, ethnic group, or facial expression can be detected.

The performance of the system is comparable to others. In [1] a hyper basis function network was trained on automatically extracted geometrical features. Correct sex identification rate was 87.5%. In [3] a back–propagation network was trained on a compressed representation (40 units) of low resolution face images of 30×30 pixels and a performance of 91.9% was achieved. In that system limited hair information was used and the faces were aligned under manual control.

Acknowledgements: Many thanks go to C. von der Malsburg for his support and helpful comments. I would also like to thank Irving Biederman, Jean-Marc Fellous, Norbert Krüger, and Thomas Maurer for fruitful discussions.

References

[1] BRUNELLI, R., AND POGGIO, T. Caricatural effects in automated face perception. *Biol. Cybern. 69* (1993), 235–241.

[2] DEVALOIS, R. L., AND DEVALOIS, K. K. *Spatial Vision*. Oxford Press, 1988.

[3] GOLOMB, B. A., LAWRENCE, D. T., AND SEJNOWSKI, T. J. SexNet: A neural network identifies sex from human faces. In *Advances in Neural Information Processing Systems 3*, D. S. Touretzky and R. Lippman, Eds. Morgan Kaufmann, SanMateo, CA, 1991.

[4] JONES, J. P., AND PALMER, L. A. An evaluation of the two dimensional Gabor filter model of simple receptive fields in cat striate cortex. *J. of Neurophysiology 58* (1987), 1233–1258.

[5] WISKOTT, L. Phantom faces for face analysis. *Pattern Recognition 30*, 6 (1996).

[6] WISKOTT, L., FELLOUS, J.-M., KRÜGER, N., AND VON DER MALSBURG, C. Face recognition by elastic bunch graph matching. Tech. Rep. IR-INI 96-08, Institut für Neuroinformatik, Ruhr-Universität Bochum, D-44780 Bochum, Germany, 1996.

A New Hardware Structure for Implementation of Soft Morphological Filters

A. Gasteratos, I. Andreadis and Ph. Tsalides
Laboratory of Electronics
Section of Electronics and Information Systems Technology
Department of Electrical and Computer Engineering
Democritus University of Thrace
671 00 Xanthi, Greece
e-mail : ioannis@orfeas.ee.duth.gr

Abstract : A new hardware structure for implementation of soft morphological filters is presented in this paper. This is based on the modification of the majority gate technique. A pipelined systolic array architecture suitable to perform real-time soft morphological filtering is presented as an illustrative example. The processing times of the proposed hardware structure do not depend on the data window size and its hardware complexity grows linearly with the number of its inputs.

1 Introduction

Mathematical morphology is a methodology for image analysis and image processing, based on set theory and topology [1]. It offers effective solutions to many computer vision problems including noise removal, feature extraction, texture analysis, skeletonizing etc. Dilation and erosion are the two basic morphological operations. In mathematical morphology primary role has the structuring element (s.e.). Many morphological filters, based on the two basic operations, have been have been studied [2]. Another well studied class of non-linear filters are rank order filters, which have excellent robustness properties and have been applied into many image processing applications [3]. Soft morphological filters are a combination of morphological and rank order filters [4]. They have been introduced to improve the behaviour of traditional morphological filters in noisy environments. They are less sensitive to additive noise and to small variations in object shape than the standard morphological filters.

Soft morphological filters is a relatively new class of non-linear filters, and therefore, only a few designing methods have been reported [4]. Soft morphological filters can be VLSI implemented using the threshold decomposition technique [5]. However, in this approach hardware complexity increases exponentially both with the resolution of the numbers and the size of the data window. Therefore, implementation of filters capable of handling high resolution numbers is not practical. In this paper a pipelined systolic array is presented, capable of implementing soft morphological dilation (s.m.d.) and soft morphological erosion (s.m.e.). This hardware structure is based on the majority gate technique [6]. This technique has been applied for both morphological filtering [7] and rank order filtering [8] and, thus, it is suitable for their combination, i.e. soft morphological filtering. The architecture of the proposed hardware structure is scalable; its hardware complexity grows linearly with the number of inputs and its processing times do not depend on the data window size.

2 Soft Morphological Operations

In soft morphological operations the max/min operations, used in standard morphology, are replaced by weighted order statistics and the s.e. B is divided into two subsets, i.e. the core A and the soft boundary B\A ("\" denotes the set difference). In s.m.d. (s.m.e.) the pixels of the image are combined with the pixels of the s.e. as in standard morphology; the results which are related to the soft boundary of the s.e. and the results which are related to the core of the s.e. are repeated k times and ordered in a descending (ascending) sequence. The kth element of this sequence is the result of s.m.d. (s.m.e.). Let $\{k \lozenge f(x)\}$ denote the k time repetition of $f(x)$; $\{k \lozenge f(x)\} = \{f(x), f(x), \dots f(x)\}$ (k times). The s.m.d. of a gray-scale image f by a soft gray-scale s.e. $[a, \beta, k]$ is defined as [4]:

$$f \oplus [\beta, a, k](z) = k\text{th larger of } \left(\left\{ k\lozenge \big(f(y) + a(z - y)\big) \right\} \cup \left\{ f(b) + \beta(z - b) \right\} \right) \qquad (1)$$

where $(z - y) \in A$ and $(z - b) \in B \backslash A$.

Also the s.m.e. of f by $[a, \beta, k]$ is defined as:

$$f \ominus [\beta, a, k](z) = k\text{th smaller of } \left(\left\{ k\lozenge \big(f(y) - a(z + y)\big) \right\} \cup \left\{ f(b) - \beta(z + b) \right\} \right) \qquad (2)$$

where $(z + y) \in A$ and $(z + b) \in B \backslash A$.

Based on the above definitions (eqns (1) and (2)) a hardware structure that computes s.m.d/s.m.e. can be constructed. This consists of adders/subtractors, followed by a module, which computes the required order statistic of the addition/subtraction results. The aforementioned module is implemented using the algorithm, which was first presented in [8] and it is suitable for rank order filtering. This algorithm applies a median computation algorithm [6] to a sequence of numbers and uses additional dummy inputs. By computing the median value of the expanded sequence and by being able to control the dummy numbers, any order statistic of the original sequence of the numbers can be determined. Figures 1a, 1b and 1c illustrate this concept. The dummy numbers are used for the computation of median, 2nd order statistic and maximum values, respectively. The bold window contains five binary numbers $x_{(1)}$, $x_{(2)}$, ...$x_{(5)}$ in an ascending order (the subscript in the parentheses denotes the rank). The larger window contains nine binary numbers also in an ascending order. By controlling the dummy numbers d_i which are pushed to the top and to the bottom, any order statistic r of the numbers x_i, can be obtained.

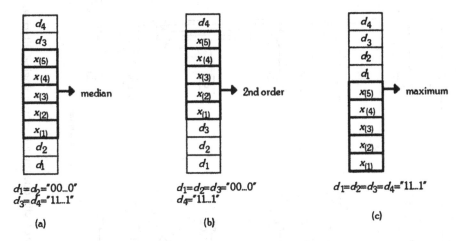

Fig. 1. Illustration of dummy inputs use : (a) median, (b) 2nd order and (c) maximum computation.

3 Systolic Array Implementation for Soft Morphological Filtering

A pipelined systolic array (Figure 2) capable of computing s.m.d./s.m.e. using a four neighbor s.e. and eight bit resolution images is presented in this section. The inputs to this array are five pixels of the image window, five pixels of soft morphological s.e. (this includes the core, the soft boundary and k) and a control signal MODE. The central pixel of the s.e. is its core, whereas the rest four neighbor pixels are the soft boundary. The first five latches (L1) hold the image window, the next five latches (L*1) hold the s.e. and the remaining one (L**1) holds number k. Global signal MODE is used to select the operation; "1" selects the s.m.d., whereas "0" selects the s.m.e. operation. It can be seen from eqns (1) and (2) that in s.m.d. the reflection of the s.e., with respect to the origin, interacts with the image window, whereas in s.m.e. it interacts exactly with the image pixels it overlays. Image data is collected by five multiplexers MUX1, which are controlled by the signal MODE. In the same stage the pixels of s.e. remain either unchanged when the operation of s.m.d. is considered or they are complemented in the operation of s.m.e., by means of XNOR gates controlled by the signal MODE. In the next stage of the pipeline, data is fed to five adders. In the case of s.m.e. the 1's complements of the pixels of the s.e. are added to the image data and the carry in (C_{in}) bit to the adder is "1". Thus, the 2's complements of the pixels of the s.e. are added to the image pixels. This is equivalent to the subtraction operation.

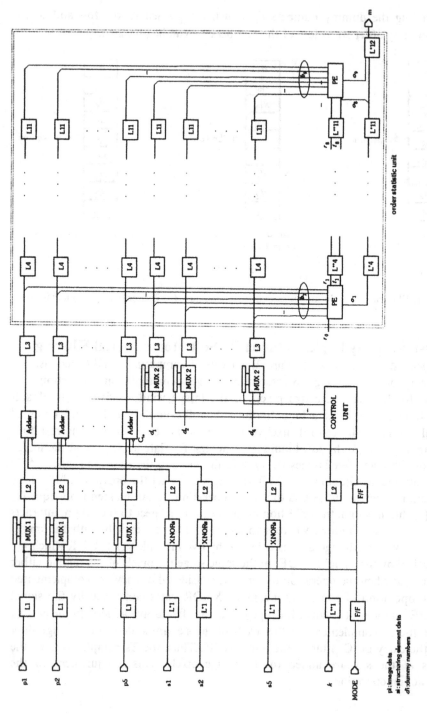

Fig. 2. Systolic array hardware structure implementing the majority gate technique for soft morphological filtering.

In soft morphological filtering, if $k >$ Card(B\A), then the soft morphological operations are reduced and only the core of the s.e. is considered [5]. Therefore in order to preserve the nature of soft morphological operations, the constraint that $k \leq \min\{$Card(B)/2, Card(B/A)$\}$ is used. In this case : Card(B)=5, Card(A)=1, Card(B\A)=4 and $k \leq \min\{2.5, 4\} \Rightarrow 1 \leq k \leq 2$. Figure 3a illustrates the position of the dummy numbers in s.m.d. for k =1 and k=2, whereas Figure 3b illustrates the position of the dummy numbers in s.m.e. For s.m.d. all the dummy inputs are pushed to the top, whereas for s.m.e. they are pushed to the bottom. Thus, the appropriate result is obtained from the order statistic unit. A control unit, the input of which is the number k, controls an array of multiplexers MUX2 and inputs to the order statistic unit either a dummy number or a copy of the addition/subtraction result of the core. The order statistic unit consists of identical Processing Elements (PEs) separated by latches (L**4 to L**11). The block diagram of the PE is shown in Figure 4. In each PE the bits $(b_{1,j}...b_{9,j})$ of the same significance of the numbers of the resulting sequence are processed, starting with the most significant bits in the first stage until the less significant bits in the last stage. The process is based on the majority selection of intermediate signals $(i_{1,j}...i_{9,j})$. The selection is achieved using a majority gate, which operates as follows : its output (o_{j+1}) is "1", if over half of its inputs are "1"; otherwise its output is "0". Flag signals $(r_{1,j}...r_{9,j},t_{1,j}...t_{9,j})$ derived from previous stages are used for further processing in the successive stages, in order to show whether a number has been rejected or not. At the end of the process the median value of the resulting sequence is obtained, which, by appropriate choice of the dummy numbers through the control unit, is the result of the soft morphological operation.

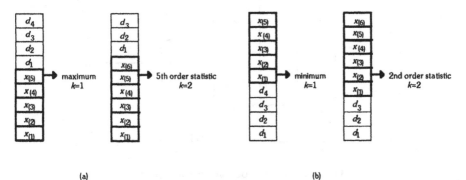

(a)   (b)

Fig. 3. (a) Arrangement of the dummy numbers for s.m.d. and (b) arrangement of the dummy numbers for s.m.e.

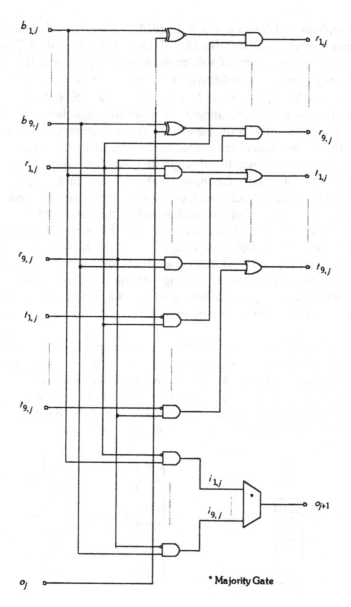

Fig. 4. The basic processing element (PE).

4 Conclusions

A new technique for realization of soft morphological filtering based on the modification of the majority gate has been presented in this paper. A pipelined systolic array architecture based on this technique has been also presented. The processing times of the proposed hardware structure are independent of the data window size and its hardware complexity grows linearly with the number of its inputs.

REFERENCES

[1] Serra, J. (1982). *Image Analysis and Mathematical Morphology, Vol. 1*, Academic Press, N. York.

[2] Dougherty, E. R., Ed. (1994). *Digital Image Processing Methods*, Marcel Dekker Inc., N. York.

[3] Pitas, I. and A. N. Venetsanopoulos (1990). *Non-lineal Digital Filters - Principles and Applications*, Kluwer Academic Publishers, Boston.

[4] Kuosmanen, P. and J. Astola (1995). Soft Morphological Filtering. *Journal of Mathematical Imaging and Vision* 5 (3), 231-262.

[5] Shih, F.Y. and C. C. Pu (1995). Analysis of the Properties of Soft Morphological Filtering Using Threshold Decomposition, *IEEE Trans. Signal Proc.* 43 (2), 539-544.

[6] Lee, C. L. and C. W. Jen (1992). Bit-sliced Median Filter Design Based on Majority Gate, *IEE Proc. G*, 139 (1), 63-71.

[7] Gasteratos, A., I. Andreadis and Ph. Tsalides (1996). Improvement of the Majority Gate Algorithm for Gray Scale Dilation/Erosion, *Electronics Letters*, 32 (9), 806-807.

[8] Gasteratos, A., I. Andreadis and Ph. Tsalides. Realization of Rank Order Filters Based on Majority Gate, To appear in *Pattern Recognition*.

A Method for Anisotropy Analysis of 3D Images

Vassili A.Kovalev and Yaroslav S.Bondar

Institute of Mathematics, Belarus Academy of Sciences
Kirova St., 32-A, 246652 Gomel, BELARUS   *e-mail:* goim@nauka.belpak.gomel.by

Abstract. In this paper we present an extension of anisotropy analysis methods for 3D image volumes. Two approaches based on orientation-sensitive filtering and a 3D version of spatial gray-level difference histograms are compared. The performance of the method is demonstrated on synthetic image volumes and original 3D CT and MRI medical images. The orientation structure of left and right hemispheres of some brain images as well as left and right kidneys are compared.

1   Introduction

The anisotropy (directionality) is a significant property of images. This is supported by two fundamental facts. First, mammalians have orientation selective mechanisms in their visual cortex [1,2] and therefore orientation alongside regularity and structural complexity guides the process of perceptual grouping of textural patterns [3,4]. Second, the anisotropy may be the result of the process by which the imaged object might have been formed [5,6,7]. Thus, on numerous occasions anisotropy reflects properties and determines the behaviour of the textured objects.

The importance of anisotropy in visual perception and object characterization inspired a range of studies for anisotropy analysis. The existing methods can be conditionally categorized in two groups. (Note that we refer here to methods that lie in the image processing and pattern recognition domain. Obviously, in other areas of scientific explorations some specific methods can be employed. For instance, in mechanics certain methods of anisotropy estimation are based on evaluation of the angular scattering of coherent light across the rough surfaces [8], calculation of maximum-to-minimum ratio of radii of the covariance function for several azimuth directions [9], etc.). The first group of methods involves orientation-sensitive filtering of an image, followed by the calculation of the distribution of local orientations. Chetverikov [5] following Zucker et al. [10] utilized local orientation in his edge slope density descriptor to discriminate of different textures. Sasov [11] implemented Sobel masks and applied the method to a wide range of anisotropy analysis tasks. Similar approach was also exploited for studying the anisotropy of engineering surfaces at different scale levels [7] and for detecting tumors in liver ultrasound images [12]. An original idea for the calculation of a local orientation (tissue disorder) measure was reported by Hasegawa and Toriwaki [13] with application to cancer detection in X-ray images. Simoncelli and Farid [14] proposed some generalization of steerable filters for local orientation analysis and examined their behaviour by computing a set of typical "orientation maps". Gorkani and Picard [15] used a low-level orientation and steerable pyramid principle for quick search in a video data base. The result their study proved that a simple measure like "dominant perceived orientation" can successfully be used for the quick coarse classification of certain kinds of image scenes (it is known that pigeons classify cities and countries in a similar way [16]).

The second group of methods evaluates the image anisotropy by calculating suitable measures of image intensity variation for several predetermined directions in an image. Perhaps Longuet-Higgins who proposed a method of evaluating the spatial roughness distribution on solid surfaces [17] can be seen as one of the predecessors of this approach. Chetverikov carefully investigated different aspects of such methods ranging from a simple linear density measure [5,18] to features derived from gray level difference histograms [19].

Let us consider the anisotropy analysis problem in the context of 3D imagery. A number of works dealing with different orientation measures have provided the basis for the development of such 3D image analysis tool. Morgenthaler and Rosenfeld [20] extended for application to multi-dimension arrays the approach that fits a surface (plane, quadric) to a neighborhood of each image point to estimate the magnitude of the gradient. Zucker and Hummel [21] introduced an operator that finds the best oriented plane at each point in a 3D image. This operator complements other approaches that are either interactive or heuristic extensions of corresponding 2D techniques. Liou and Singh [22] generalized 3D gradient operators to image data on anisotropic grids that occur frequently in medical imaging and demonstrated experimental results on many anisotropically sampled CT and MR images.

We shall discuss key details of our method of anisotropy analysis of 3D images in section 2. In section 3 we shall apply the method on real 3D medical images. Finally, in section 4 we shall summarize our results.

2 The Method

The method involves the following sub-problems:
- Homogenous partitioning of direction space to form bins of the orientation histogram. Such partitioning is a simple task for 2D but nontrivial for the 3D case.
- Selection of a way to evaluate image intensity variation in a given direction. The value that characterizes the change of image intensity in some direction is sometimes called orientation descriptor of the image (e.g., [18]).
- Devising of an anisotropy visualization technique.
- Introduction of features to solve the corresponding image analysis problems.

Homogeneous partitioning of direction space. An arbitrary direction in 3D space can be specified in different ways, in particular, by means of two values: "longitude" φ, $0<\varphi<360$ and "latitude" ψ, $0<\psi<180$. It is easy to see, that the uniform partition of φ and ψ into ranges does not produce a homogeneous partition of the direction space. This is caused by the presence of poles in the coordinate system. Every bin of a partition of the direction space refers to a solid angle. The magnitude of a solid angle is determined by the area which it cuts out on the surface of the unit sphere. The center of the sphere coincides with the vertex of the angle. Thus, for the generation of a uniform partition of the whole solid angle, it is sufficient to generate a uniform (with equal area) partition of the surface of the unit sphere. We perform such partition in two steps: (a) Divide the surface of the unit sphere originated in (0,0,0) into M equal spherical digons with solid angle $\alpha = 2\pi / M$ and vertices in points $\delta_1=(0,0,1)$ and $\delta_2=(0,0,-1)$; (b) Divide the surface of the sphere into N spherical layers. These layers are bounded by N-1 planes drawn perpendicularly to the diameter $[\delta_1, \delta_2]$ at distances $H=2/N$ from each other. The area of every layer is $2\pi H$, where H is the height of a spherical layer.

The above procedure results in a partition P_{NM} of the unit sphere into $(N-2)M$ spherical quadrangles and $2M$ spherical triangles with area $S_{bin} = 4\pi / MN$. Let us denote by $P_{NM}[i,j]$ the solid angle formed by the intersection of the j-th digon and i-th layer of the sphere. Then an arbitrary direction defined by a vector (a,b,c) belongs to $P_{NM}[i, j]$ if the following two conditions are satisfied:

$$\frac{2\pi}{M}i \le \varphi < \frac{2\pi}{M}(i+1), \text{ where } \varphi = tan^{-1}(b/a);$$

$$-1+\frac{2}{N}j \le \bar{c} < -1+\frac{2}{N}(j+1), \text{ where } \bar{c} = c\Big/\sqrt{a^2+b^2+c^2}.$$

Visualization of the orientation histogram. In the 2D case visualization of orientation the histograms is usually performed with help of a polar diagrams, called "orientation indicatrix" [5], "orientation diagram" [12], or "orientation map" [14]. In 3D case these techniques should be modified. We compared three ways of orientation histogram representation: by intensity variation, by means of height variation, and with the help of a 3D orientation indicatrix. The first two techniques are widely used in image display and have no need of explanation. We define the 3D orientation indicatrix as the surface formed in two steps: (a) triangulation of the surface of the unit sphere with nodes in the centers of bins of partition P_{NM}; (b) recalculation of node locations so that the distance to the center of the sphere becomes proportional to the magnitude of the corresponding bin of the histogram. Fig. 1 illustrates all the ways of visualization of the orientation histogram with a CT volume image of a liver. As observed in Fig. 1, the third way is the most expressive and thus it will be used later.

Orientation descriptors. We are going to examine here two orientation descriptors for the 3D case: gradient density (GD) and intensity variation (INV) descriptors.

The GD descriptor based on orientation-sensitive filtering of an image volume. Let us say that $GD[i,j]$ represents the number of image voxels where the orientation of the local gradient of intensity belongs to $P_{NM}[i, j]$. We calculate the local gradient $G(x,y,z)$ by convolution of the image intensity $I(x,y,z)$ with three masks of a 3D edge detection operator proposed by Zucker and Hummel [21]:

$$G(x,y,z) = (G_x, G_y, G_z), \text{ where}$$

$$G_x = I(x,y,z) \otimes F_x, \; G_y = I(x,y,z) \otimes F_y, \; G_z = I(x,y,z) \otimes F_z,$$

$$F_z = \begin{pmatrix} 1/\sqrt{3} & 1/\sqrt{2} & 1/\sqrt{3} \\ 1/\sqrt{2} & 1 & 1/\sqrt{2} \\ 1/\sqrt{3} & 1/\sqrt{2} & 1/\sqrt{3} \end{pmatrix} \begin{pmatrix} 0 & 0 & 0 \\ 0 & 0 & 0 \\ 0 & 0 & 0 \end{pmatrix} \begin{pmatrix} -1/\sqrt{3} & -1/\sqrt{2} & -1/\sqrt{3} \\ -1/\sqrt{2} & -1 & -1/\sqrt{2} \\ -1/\sqrt{3} & -1/\sqrt{2} & -1/\sqrt{3} \end{pmatrix}$$

Since these three masks are simple rotations of one another, only one (the one oriented along the z axis) is shown. As pointed out in [21], this operator gives the best approximation of an intensity surface normal.

The INV descriptor measures variations of image intensity along different spatial directions. We calculate such variations by means of a 3D version of the Spatial Gray-

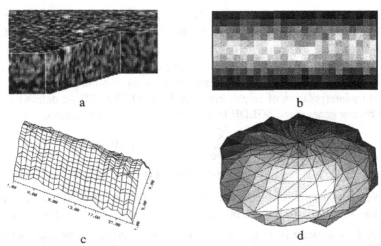

Fig. 1. Different ways of orientation histogram visualization for an example of a 3D medical image. (a) Original CT liver image volume. Orientation histogram of 24×11 bin size displayed by means of intensities (b), height variations (c), and as a 3D orientation indicatrix (d).

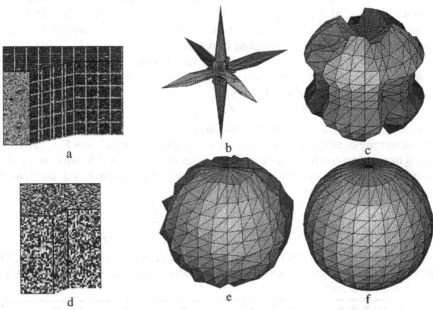

Fig. 2. Comparison of two approaches for 3D anisotropy evaluation on two synthetic image volumes. Left column: (a) A synthetic "lattice" image of 101×101×101 voxels; (d) Pure noise image of 51×51×51 voxels. Middle column: corresponding orientation histograms (b) and (e) calculated by the approach based on orientation sensitive 3D filtering. Right column: orientation histograms (c) and (f) obtained with the help of a 3D version of spatial gray-level dependence histograms.

Level Dependence Histogram (GLDH). For this purpose we generalize Chetverikov's approach [19] for arbitrary spacing in partitioned 3D image space. Let us define the five-dimensional histogram $R=h(i,j,d,\varphi,z)$, where i and j are gray-values of image voxels at distance d apart and the direction (φ,z) is that of the line joining the centers of these voxels. The value of the image intensity in non-integer positions can be obtained by interpolation of neighboring voxels. Much like [23] we define the *inertia* feature for our generalized SGLDH histogram R in the following manner:

$$I(R_{\varphi,z}(d)) = \sum_{i=0}^{Ng-1} \sum_{j=0}^{Ng-1} (i-j)^2 h_{\varphi,z}(i,j|d),$$

where $h_{\varphi,z}(i,j|d)$ is the i,j-th element of $h(i,j,d,\varphi,z)$ for a distance d and direction (φ,z); Ng is the number of image intensity values. Then we use the inertia magnitude $I(R_{\varphi,z}(d))$ for characterization of the anisotropy in (φ,z) direction. As it was established by separate investigation on a range of synthetic and original 3D images, the approach based on the orientation-sensitive filtration (GD descriptor) offers certain advantages in sensitivity and computational cost. Fig. 2 compares the orientation indicatrices derived by both approaches to illustrate the differences on synthetic image volumes.

The problem of anisotropic sampling. The slice thickness and slice separation is usually greater than the internal slice resolution in tomographic data acquisition protocols and some other 3D imaging systems. The natural way to overcome this problem (e.g., [22]) is passing from an anisotropic sampling grid to an isotropic one. Thus the problem of anisotropic sampling can be overcome through the use of sampling rates for x, y, and z sampling directions.

Features. It is commonly known that an important result of the anisotropy analysis is the calculation of rotation-invariant features. In this section we are going to outline similar features for the 3D case.

Anisotropy coefficient: $\quad F_1 = 1 - H_{min}/H_{max}$,

where H_{min} and H_{max} are the minimal and maximal values of a histogram. For example, the anisotropy coefficient of the anisotropic test image "lattice" (see Fig. 2a) for the histograms calculated from the orientation descriptors GD and INV are 0.46 and 0.94 respectively. This is caused by the high sensitivity of the gradient based descriptor for the anisotropy of an image.

Integral anisotropy measure: $\quad F_2 = \sqrt{\dfrac{\sum_{i=1}^{N}\sum_{j=1}^{M}(H[i,j]-H_{med})^2}{NM}}$,

where H_{med} is the arithmetic mean; N and M are the dimensions of the histogram. This anisotropy measure can be considered as a smooth integral feature of the anisotropy.

Local curvature:

$$F_3 = \sqrt{\dfrac{\sum_{i=1}^{N}\sum_{j=1}^{M}(H[i,j] - \frac{1}{4}(H[i-1,j]+H[i+1,j]+H[i,j-1]+H[i,j+1]))^2}{NM}}.$$

Local curvature can be used for evaluation of the "roughness" of the surface of an orientation indicatrix. For example, for histograms (e) and (f) on Fig. 2 the values of the curvature are 0.13 and 0.09 respectively.

3 Examples of application in 3D medical imaging

The goal of the following experiments with medical images is to illustrate our method with real 3D data. Three image volumes of brain referred by BR1, BR2, and BR3 are involved in the first experiment. The third one, BR3 has a large tumor in the right hemisphere. One can see from Fig. 3 that the orientation histograms of the hemispheres of the pathologic brain differ from others. To evaluate the difference we introduced feature F_4 that expresses the volume ratio of left and right parts of a histogram. Feature F_4 was calculated for both hemispheres of all brains. Results are presented in the table. As we can see, feature F_4 for left hemispheres of the normal brains is less than for the right ones. But this is not the case for BR3 brain. We may suppose that this is caused by the tumor.

	BR1	BR2	BR3
Left hemisphere	0.87	0.90	1.06
Right hemisphere	1.06	1.10	0.97

In the next experiment we used pair of symmetric organs of the human body to see whether their corresponding indicatrices reflect their symmetry. For this purpose the histograms of kidneys were calculated (see Fig. 4). We evaluated two distances between them: Euclidean distance D_E and symmetrical distance D_S:

$$D_S = \sqrt{\sum_{i=1}^{N}\sum_{j=1}^{M}(H_1[i,j] - H_2[i,j_s])^2} \,,$$

where $j_s = M/2 - j$, when $j < M/2$, $j_s = 4M/3 - j$, when $j \geq M/2$. For these histograms the distances are $D_E = 5.74$ and $D_S = 3.30$. This shows that our representations preserve the symmetry of the objects.

4 Conclusions

We presented a new method for anisotropy analysis of 3D image volumes. Comparison of two orientation descriptors, gradient density and intensity variation descriptors, allows us to conclude that the first one offers certain advantages in the sensitivity and computational costs. Orientation histograms provided the basis on which a range of features of image spatial structure and its disturbances could be constructed. There may be many examples of such disturbances to be extracted even in the medical imaging field only, exemplified by different invasions, tissue disturbances generated by diseases, different kinds of tumors, etc.

Acknowledgments - The authors wish to acknowledge Belarus Academy of Sciences for financial support of this work.

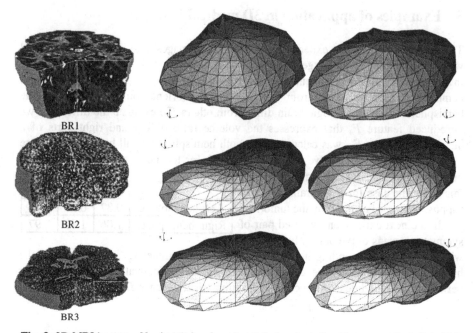

Fig. 3. 3D MRI images of brains (left column) and their orientation histograms for left (middle column) and right (right column) hemispheres.

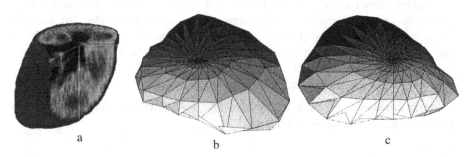

Fig. 4. (a) A section from a CT image of the left kidney of a patient. (b) and (c) The orientation histograms of the left and the right kidneys of the same patient.

502

References

1. D.H.Hubel and T.N.Wiesel, *"Receptive Fields, Binocular Interaction and Functional Architecture in the Cat's Visual Cortex,"* Journal of Physiology, London, 166, pp.106-154, 1962.
2. D.H.Hubel and T.N.Wiesel, *"Receptive Fields and Functional Architecture of Monkey Striate Cortex,"* Journal of Physiology, 195, pp.215-243, 1968.
3. B.Julesz, *"Experiments in Visual Perception of Texture,"* Scientific American, Vol. 232, No. 4, pp. 34-43, 1975.
4. A.R.Rao and G.L.Lohse, *"Identifying High Level Features of Texture Perception,"* CVGIP: GMIP, Vol. 55, pp. 218-233, 1993.
5. D. Chetverikov, *"Textural Anisotropy Features for Texture Analysis,"* Proc. IEEE Conf. on PRIP, pp. 583-588, Dallas, Aug 3-5, 1981.
6. M.Kass and A.Witkin, *"Analyzing oriented patterns,"* CVGIP 37, pp. 362-385, 1987.
7. V.A.Kovalev and S.A.Chizhik, *"On the Orientation Structure of Solid Surfaces,"* Journal of Friction and Wear, Vol. 14, No. 2, pp. 45-54, 1993.
8. E.L. Church, *"The Measurement of Surface Texture and Topography by Differential Light Scattering,"* Wear, 57, pp. 93-105, 1979.
9. B.Bhushun, J.C.Wyant, and J.Meilling, *"A New Three-Dimensional Digital Optical Profiler,"* Wear, pp. 301-312, 1988.
10. S. Zucker, A.Rosenfeld, and L.Devis, *"Picture Segmentation by Texture Discrimination,"* IEEE Trans. Comput., Vol. C-24, pp. 1228-1233, 1975.
11. A.Y. Sasov, *"Development of the Analytical Methods in Scanning Electron Microscopy,"* Ph.D. thesis, Moscow State Univ., Moscow, 1984 (Russ.).
12. V.A.Kovalev, *"Rule-Based Method for Tumor Recognition in Liver Ultrasonic Images,"* 8th Int. Conf. on Image Analysis and Processing, Sanremo, Italy, Sept. 1995, LNCS, Vol. 974, Springer Verlag, pp. 218-222, 1995.
13. J. Hasegawa and J. Toriwaki, *"A new Filter for Feature Extraction of Line Pattern Texture with Application to Cancer Detection,"* Proc. of 11th ICPR, Vol. 3, pp. 352-355, the Hague, the Netherlands, August-September 1992.
14. E.P. Simoncelli and H. Farid, *"Steerable Wedge Filters for Local Orientation analysis,"* IEEE Trans. on Image Proc., Vol. 5, No. 9, pp. 1377-1382, 1996.
15. M.M.Gorkani and R.W. Picard, *"Texture Orientation for Sorting Photos "at a Glance,"* Proc. of 12th ICPR, pp. 459-464, Jerusalem, Israel, Oct. 9-13, 1994.
16. R.J. Herrngstein, D.H. Loveland, and C.Cable, *"Natural Concepts in Pigeons,"* Journal of Experim. Psych.: Animal Behavior Procs., 2:285-302, 1976.
17. M.S. Longuet-Higgins, *"The Statistical Analysis of Random, Moving Surfaces,"* Phil. Trans. Royal Soc., Vol. 249, Series A, pp. 321-384, 1957.
18. D. Chetverikov, *"Experiments in the Rotation-Invariant Texture Discrimination Using Anisotropy Features,"* Proc. 6th ICPR, Munich, pp. 1071-1073, 1982.
19. D. Chetverikov, *"GLDH Based Analysis of Texture Anisotropy and Symmetry: an Experimental Study,"* 12th ICPR, pp. 444-448, Jerusalem, Israel, Oct. 9-13, 1994.
20. D.G. Morgenthaler and A. Rosenfeld, *"Multidimensional Edge Detection by Hypersurface Fitting,"* IEEE Trans. on PAMI, Vol. 3, No 4, pp.482-486, 1981.
21. S.W. Zucker and R.A. Hummel, *"A Three-Dimensional Edge Operator,"* IEEE Trans. on PAMI, Vol. PAMI-3, No. 3, May, pp. 324-331, 1981.
22. S.-P. Liou and A. Singh, "High-Resolution Anisotropic 3D Edge Detector for Medical Images," Medical Imaging 1994, SPIE, Vol. 2167, pp. 315-325, 1994.
23. R.W. Conners and C.A. Harlow, *"A Theoretical Comparison of Texture Algorithms,"* IEEE Trans., PAMI-2, No. 3, pp.204-222, 1980.

Fast Line and Rectangle Detection by Clustering and Grouping

Dmitry Lagunovsky and Sergey Ablameyko

Institute of Engineering Cybernetics, Belarussian Academy of Sciences,
Surganov str, 6, 220012, Minsk, Belarus
e-mail: (dmitry,abl)@newman.basnet.minsk.by

Abstract. Fast algorithms to detect lines and rectangles in grey-scale images are proposed. At first, a contour image is obtained by the modified edge detection scheme. The linear primitives are extracted in the contour image and joined into line segments by cluster analysis method. The line merging algorithm is developed to get straight lines from segments. Algorithm to detect rectangles from the extracted straight lines is suggested. The developed algorithms are fast and permit to get the qualitative result.

1 Introduction

Analysis of grey-scale images usually supposes extraction and recognition of objects contained in the images. The most often encountered problem which is quite important for processing of remote sensing images, integrated circuit (IC) images, and various scenes is detection of straight lines and rectangular objects [10]. Detection of straight lines is a very important step in recognition of roads in remote sensing images. Rectangles are the key elements for buildings' recognition in remote sensing and for recognition of contact areas in integrated circuit quality inspection task.

Various approaches are known to solve these tasks [1,4,5]. A method for runway detection in airport aerial images is described in [2]. To detect runways, straight lines are extracted in the paper [3] by means of the histogram analysis. The runway is represented as a pair of anti-parallel lines.

A method for geometric primitive extraction using implicit description is presented in [4]. The search of geometric primitives is performed as search of local minimum of a cost function. This method permits to detect all geometric primitive types including rectangles though the number of operations is consuming enough for industrial use. Another rectangle detection algorithm is suggested in [1]. It is based on the search of angles as basic units for quadrangle followed by building shadow analysis. These units are more sensitive to gaps in the corners than the units used in our approach.

Rectangle can be described by means of its sides. They can be detected with the help of straight line extraction procedure [6,9]. Being extracted, straight lines usually have some deviations in straightness and collinearity. The use of straight line as the basic unit is more effective in comparison to pixel as the basic unit, suggested in [4]. Rectangle detection performed at a higher level simplifies the whole algorithm. It can be represented as:

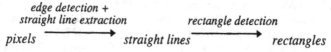

In this paper, we propose extraction of straight lines as the first step[9] with the developed algorithm and then the rectangle-shaped object detection algorithm which is based on the above scheme but allows to improve quality and speed-up the rectangle detection process.

2 Main Notions

There are three types of neighbours in 8- neighbourhood of a pixel. *4-neighbours* of the pixel are its horizontal/vertical neighbours. *D-neighbours* of the pixel are its diagonal neighbours. *8-neighbours* of the pixel are its 4-neighbours and D-neighbours.

The following pixels are considered in a contour image. The *end pixel* has a unique 8-neighbour. A *simple pixel* has two 4-neighbours which are D-neighbours. A *normal pixel* has two 4-neighbours which are not D-neighbours. An *isolated pixel* has no neighbours.

The contour image has the following types of lines. *Linear primitive (LP)* is a set of 4-neighbour pixels in a horizontal/vertical direction. LP has three possible orientations: horizontal, vertical and undetermined. The undetermined orientation is used for one-pixel length LP. A *line segment* is a line formed by LPs with similar features (orientation, length, etc.).

We regard the rectangles to be detected through quadrangles. The *quadrangle* is a closed figure constructed by four straight lines with definitely imposed constraint on opposite side collinearity.

3 Edge Detection

Step edge [7] detector was used for object boundary extraction. The scheme with edge strength and direction calculation was modified to accelerate edge detection process. The modified algorithm contains no convolution or multiplication operations inherent in the scheme and has capability to preliminary detect the edge pixels. It is described in detail in [9].

3.1 Edge Strength Calculation

The edge strength is calculated to locate the edge and to remove weak edges. The required edge strength $L(i,j)$ is calculated for image $U(i,j)$ by the formula:

$L(i,j) = |U(i+1,j)-U(i,j)| + |U(i,j+1)-U(i,j)| + |U(i+1,j+1)-U(i,j)| + |U(i,j+1)-U(i+1,j)|$,

which is rotationally invariant. Such small area operator is rather noise-dependent, but that drawback is eliminated by means of the following operation.

3.2 Edge Direction Calculation

The edge direction is used to evaluate precise location of the edge and to accelerate edge strength analysis. The edge direction is the direction of contour line clockwise tracing. Edge direction, usually discrete, is actually calculated by convolution with a set of masks.

Edge direction can be also evaluated by analysing the surface formed by pixel values in the pixel neighbourhood. We suggested another technique [9] that has a simple implementation, moreover allows to preliminary decide whether analysed pixel is a contour one or not. We use a local binarization method to make necessary decision. It is assumed that there is a unique relation between brightness surface orientation and the result of thresholding of the pixel neighbourhood. Therefore, the edge direction can be evaluated by the obtained binary image fragment.

3.3 Thresholding

To calculate the threshold, the following technique was used: values of fragment pixels are ranked and intervals between neighbouring values are analysed. The required threshold value is placed in the maximum interval between neighbouring values. Such a technique yields the minimal deviation between grey-scale and binary image.

During thresholding procedure pixels having the lower values become black and the higher values do white, respectively. To determine orientation of the brightness surface we use a set of samples with corresponding edge directions. To build it, we thresholded a set of grey-scale fragments containing step edges of different orientations inside. Each binary fragment was assigned orientation of edge creating it.

When edge direction and strength are detected for all image pixels, further steps are the same as in known edge detection scheme and will be not described here. Finally, we remove all simple and isolated pixels from extracted contours. As a result, we obtain one-pixel width contour lines with strict 8-connectivity.

4 Extraction of Line Segments by Cluster Analysis

Straight line in a discrete image can be represented by linear primitives, extracted from detected edges. The straight line can be considered as a set of LPs with the same orientation and NLP direction and approximately the same length. For extraction of line segments it is necessary to extract clusters of connected LPs with the above characteristics.

4.1 Linear Primitive Extraction

Linear primitive extraction is performed on the basis of detected edges. The vector description of LP contains: orientation, length, the first and the last pixel coordinates, and direction to the next LP(NLP). This direction is from last pixel of current LP to the first pixel of the next LP along contour.

4.2 Grouping of linear primitives

Straight lines are formed from the detected LPs by a cluster analysis method. One cluster corresponds to one line segment. An 8-connected fragment of contour image is analysed at a time. First LP of the fragment is considered as a line segment. Neighbouring LPs are joined to the cluster if the following conditions are satisfied:

•it is connected with the cluster of line segment already found;

•it has the same orientation;

•it has the same NLP direction;

•deviation of its length from average cluster length does not exceed the given threshold (it was equal to 3 pixels in our experiments).

If these conditions are not met, then another cluster corresponding to another line is started. Two separate clusters are created when the fork in contour is encountered.

Several clustering conditions were added to reduce the noise sensitivity of the algorithm. As a result, line segments are extracted and described by their end point coordinates.

4.3 Line Merging

Some straight boundaries have gaps due to noise influence. One straight boundary can be represented by a set of collinear and not connected straight lines. They have to be merged. The way to detect spatial relationship between separate lines using a label image [6] is not suitable for large-size image processing unless it is simple, because it requires large computer memory resources to store intermediate values. We implement this process by multiple application of line merging operator. Some of operator parameters are changed between passes, to sequentially increase length of lines to be merged and gaps between them.

4.4 The Line Merging Operator

For any initial line with length greater than *MinLength*, we are looking for the candidate line in following way: the difference between angles of initial and candidate line must be less than *DelAngle* value and candidate line end point must be situated in a search area.

The value of *DelXY* defines the maximal difference in the coordinates of the nearest points of a line pair and *DelBridge* defines the maximal difference between slopes of an initial line and line joining the nearest end points of lines to be merged. The candidate line found, the hypothetical line is formed, which is used instead of initial and candidate lines after merging.

That sequential comparison during candidate line search allows to reduce drastically the number of operations to find the required line. The described configuration of search area allows to choose unique line for deviation calculation. Considering our experiments, it is only approximately 15% of cases, when search area contains more than one candidate line.

The search algorithm has various parameters. We apply it to the image three times with different parameters for local gaps removing, line linking and long line growing. All merging stages are applied repeatedly, i.e. each step is performed until there are no more changes.

5 Rectangle Detection

Most of rectangular objects are represented as a set of almost parallel lines, i.e., quadrangles. To detect rectangle, we are looking for a quadrangle and by its analysis we replace it by a rectangle or not. Only those quadrangles, which are 'potential' rectangles, are considered. To consider lines 'almost' parallel, the difference between their angles must be less than *ParallelDev* value (in our research *ParallelDev*=20°).

To solve a rectangle detection task, it is necessary to find four straight lines, forming a "potential" quadrangle (see [8] for details). Edge detectors do not work well at corners and contour connectivity is usually corrupted there. That is why the extracted straight lines are usually not intersected.

5.1 Fast Preliminary Search and Grouping of Parallel Lines

The direct evaluation of "rectangularity" degree for all possible combinations of lines has relatively high computational cost and the number of possible combinations is enormous. That is why this evaluation is not relevant in rectangle detection.

To reduce the number of operations, we introduce a parametrical space for preliminary detection of parallel lines which can be opposite sides of quadrangle. To build the space, we transform each line into a point in 3-D space having the following coordinates: (S,L,D), where S- slope of the line, L - its length and D - distance from a perpendicular line which passes the line centre to the origin of coordinates. (Fig.1.).

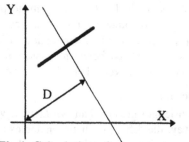

Fig.1. Calculation of space parameter D

The parametrical space is constructed in such a manner that 'almost' parallel lines which are opposite sides of quadrangle will be situated close to each other. That permits to easily detect all pairs of 'almost' parallel lines in the image. When all pairs are detected, it is necessary to extract those which form a quadrangle. It is done by parametrical space because a pair of detected parallel lines unambiguitively defines a location of corresponding perpendicular lines.

To detect the perpendicular line location, we advance a preliminary hypothesis on the basis of extracted parallel lines. This perpendicular line has a slope which differs from the initial line slope in 90^0 and its end points are situated on the parallel lines being initial ones. The required parameters in the search space, namely (S_o,L_o,D_o) can be easily calculated to define a search area for perpendicular lines in parametrical space. The search area can be formed as a small area containing the initial point inside. Presence of the line pair in this area preliminary indicates the perpendicular candidate existence.

The described configuration of the search area is used to detect 'ordinary' rectangles. Some of the aerial image objects can be represented as a set of rectangles with specific spatial relationship (a building with its shadow, for example).

Most of the enclosed quadrangles have common sides and one line can belong to several rectangles. It would be not possible to detect internal rectangles by the suggested search area forming technique due to their different length. To detect them, we consider lines which are longer and shorter than the line length given by the advanced preliminary hypothesis, but still can be used to form quadrangles. For this, we form the search area as shown in Fig.2 where 2D parametrical subspace is presented with fixed value of $S=(S_o,L,D)$ space.

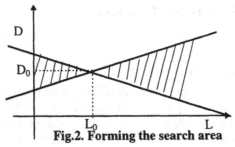

Fig.2. Forming the search area

The area formed in such a manner should be applied for search of all quadrangle lines such as the opposite sides of enclosed quadrangle having different length. These sides will not be situated close to each other in the parametrical space. To remove this, we look for only external quadrangles by using a small search area as the first step and use only extracted quadrangle sides to advance a preliminary hypothesis as the second step with enlarged search area. It also allows to simultaneously divide the detected rectangles into external and internal ones.

5.2 Extraction of Quadrangles

For precise evaluation, we directly analyse the mutual spatial relationship of the detected pairs of lines and calculate the "rectangularity" degree (Dev). To calculate the value of Dev, we build quadrangle based on chosen lines by their growing until intersection.

The intersection points determine the vertices of quadrangle. The $DeviationRectangle$ value is the maximal distance among distances of intersection points from nearest endpoints of lines corresponding to them. If the $DeviationRectangle$ value is less than Dev value, this set of lines is considered as the quadrangle. The choice of Dev value depends on the line representation quality and is defined by the average gap values (in the rectangle corners) and average distance between neighbouring rectangle objects.

When a quadrangle is situated inside another one, the internal quadrangle will have large value of Dev and will be not detected. To take it into account, we calculate the distance value only when the intersection point does not belong to line segments forming the quadrangle. In the opposite case, the distance value corresponding to this line endpoint is assumed to zero. The use of this feature allows to obtain more detail description of image objects.

5.3 Approximation by rectangle

The obtained figure is approximated by rectangle. Two opposite quadrangle sides having a greater length are chosen as the basic ones and two new parallel lines are built instead of them. Their slope is equal to an average slope of the initial pair and they pass through the centre points of the initial pair. The remaining rectangle sides are perpendicular to obtained parallel lines and pass through the centre points of the remaining two. The intersection points of the obtained lines are considered as rectangle vertices.

6 Experimental Results and Conclusion

The proposed algorithms have been developed to process aerial photographs and integrated circuit images.

6.1 Remote Sensed Image Processing

The input grey-scale images are obtained from aerial photographs. The example of the initial image having 256 grey-levels is shown in Fig.3, the detected rectangles are shown in Fig.4. The edge detection takes 2 sec, and the straight line extraction using cluster analysis takes 2 sec. (PLs extraction included). The line merging takes 1 sec. The rectangle detection process takes 3 sec. All measurements of processing time were performed on personal computer with the Pentium processor.

   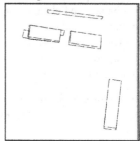

Fig.3.An initial aerial image   **Fig.4. Detected rectangles in aerial image**

The obtained results proved our algorithm being computationally effective. It was found, that the use of adaptive thresholding in edge detection makes implementation of large-sized masks (like 5x5) impossible, because the corners of neighbouring objects have not been detected since their mutual influence. But the possibility to use our edge detector for other edge types, ('convex roof' and 'concave roof', in particular) was discovered. Such modification was implemented in another application described below.

6.2 Integrated Circuit Image Processing

The described technique was also implemented for quality inspection of integrated circuit images.

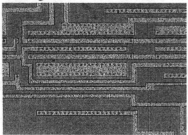   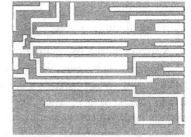

Fig.5.An initial IC image   **Fig.6.Detected IC topology on the basis of initial image**

This task has some special requirements: very low error percentage together with small processing time. The latter is caused by the big number of images to be processed (up to thousand images per one integrated circuit). Each image is 640x442 pixel size with 256 grey-levels.

The initial grey-scale image is shown in Fig. 5 and the result of topology detection (containing straight line extraction, line merging and rectangular contact area detection) is shown in Fig.6. Processing time strictly depends on the number of straight lines in the image. On an average, one image processing with reading/writing operations takes 15-20 sec.

6.3 General Remarks

On the basis of these applications, we investigated, that cluster analysis method gives us many opportunities to become faster, because it allows to use almost all possibilities of the computer while programming. The absence of arithmetic calculations also results in small computational time. The only drawback was found. The cluster analysis method has a higher sensitivity to corruption of line straightness than algorithms based on approximation. This sensitivity was noticeably reduced by introduction of additional cluster forming rules, but it is not as low as we would like. We consider the basic advantage to be the absence of any arithmetical operations which reduces drastically the time of calculation. The rectangle forming technique shows good results which were caused by development of the special strategy of rectangle forming. This strategy reduces to minimum the number of search acts when the candidate line is being searched. Now we investigate the possibility to extend this technique to parallelogram detection. The performed experiments corroborated our theoretical study and allow us to determine some ways to extend and to improve the developed algorithms.

7 References

1. A. Huertas, R. Nevatia, 'Detecting buildings in aerial images', Computer vision, graphics and image processing, Vol. 41, pp. 131-152, 1988.
2. A. Huertas, V. Cole, R. Nevatia, 'Detecting runways in complex airport scenes', Computer vision, graphics and image processing, Vol. 51, pp. 107-145, 1990.
3. R. Nevatia, K. Babu, 'Linear feature extraction and description', Computer graphics and image processing, Vol. 13, pp. 257-269, 1980.
4. G. Poth, M. D. Levine, 'Extracting geometric primitives', CVGIP: Image Understanding, Vol. 58, pp. 1-22, 1993.
5. J. A. Shufelt, D. M. McKeown, 'Fusion of monocular cues to detect man made structures in aerial imagery' , CVGIP: Image Understanding, Vol.57, pp 307-330, 1993.
6. V. Venkateswar, R. Chellapa, 'Extraction of straight lines in aerial images', IEEE Trans. on PAMI, Vol. 14, pp. 1111-1114, 1992.
7. J. Canny, 'A computational approach to edge detection', IEEE Trans. on PAMI, Vol. 8, pp. 679-698, 1986.
8. S.Ablameyko, D.Lagunovsky 'Aerial images: from straight lines to rectangles', SPIE's Proceedings (Visual communications and image processing), Vol.2308.,pp.2040-2048, 1994.
9. D.Lagunovsky, S.Ablameyko, 'Fast straight line extraction in aerial images' Pattern Recognition and Image Analysis, vol.6., pp.627-633, 1996
10. D.Torkar, N.Pavesic, 'Feature extraction from aerial images and structural stereo matching' Proc. of Intern.Conf.on Pattern Recognition, Vol.C, pp.880-884, 1996.

1st and 2nd Order Recursive Operators for Adaptive Edge Detection

Ling-Xiang Zhou   and   Wei-Kang Gu

Department of Information & Electronic Engineering
Zhejiang University, Hangzhou 310027
People's Republic of China
Email: guwk@sun.zju.edu.cn

Abstract — Edge detection is to trade off between noise restraining and edge localization. In this paper we introduce a novel edge detection criterion based on the general gradient edge model. Hence two adaptive edge detectors with optimal properties as well as their recursive implementations are presented. It can be proven that their linear shift-invariant approximations are Shen's and Canny-Deriche's operators respectively. Experimental results involving both 1D and 2D, test and real images are compared for these operators and previous algorithms.

1 Introduction

Edge detection has been one of the most essential problems in image processing for many years. Most of the step-edge detection operators can be concluded to a suitable smoothing method and the gradient or zero-crossing calculation after filtering. There have been many researchers discussing the problem of "optimal" edge detection, or designing the smoothing filter $f(x)$, under all kinds of criteria [1-4].

Generally speaking, the noise smoothing and edge localization abilities are contradictory in these algorithms. Good noise restraining needs a relatively flat filtering mask; however large filtering mask will bring serious localization uncertainty. Non-linear operators as median filters can preserve edge structure while smoothing image noise. For example, Bovik and Huang described ordering statistical filter (OSF), which utilized the linear combination of ordered value in a local window [5]. However OSF is only efficient for small windows because of lacking of fast ordering algorithms. Another way toward image smoothing is adaptive filtering. Nagao and Matsuyama's oriented mask algorithm is perhaps the most typical one [4]. They defined nine oriented masks around the given pixel, measured the uniformity by variance in each mask and selected the most uniformity direction to take the average. This algorithm has a quite good edge preserving and image smoothing effect, but its iterative computations are quite slow.

Jeong and Kim [7], Gökmen and Lee [8] have respectively present some regularization approaches for adaptive edge detection. According to their options, the problem is indeed to estimating ideal signal from noisy data using the regularization method. However these methods have no analytical solutions. Some numerical calculation or relaxation procedures are not avoidable.

We find it is interesting that Shen's operator is an IIR filter that can be implemented recursively [2]. Similarly, Deriche has derived an IIR implementation of

Canny's operator. It is a second order linear recursive system [3]. This is their obvious advantage among all kinds of edge-detection means. In this paper we will present a new algorithm for edge-preserving adaptive image smoothing and edge detection. It uses nonlinear shift-variant operators, but they can be implemented recursively as common IIR filters. The experimental results involving both 1D and 2D, tested and real images are available.

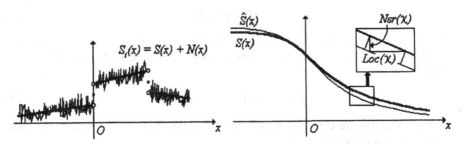

Fig. 1. Gradient edge model   **Fig. 2.** Filtering criterion

2 Criterion for Edge Detection

Let us consider the one-dimensional case. Fig. 1 shows the *gradient edge model* used in this paper. The original signal $S(x)$ is blurred by additional noise $N(x)$, hence the input signal $S_i(x)$. Image signals are regarded as piecewise continuous functions in this model, which is much more general than those used by Canny [1] and Shen [2], etc. The nonlinear image smoothing filter is $f(x;\chi)$, which denotes a filter centered at $x = \chi$. We simplify it to $f(x) = f(x;0)$ when $\chi = 0$. $S(x)$ is the noise-free signal.

As shown in Fig. 2, $S(x)$ with noise $N(x)$ is estimated as $\hat{S}(x)$ after filtering. In previous edge detection methods, $\hat{S}(x) - S(x)$ is always used as the index of signal recovery. However, the difference between $S(x)$ and $\hat{S}(x)$ consists of not only the vertical interval (which is usually measured by the noise-to-signal ratio, $Nsr(\chi)$), but also the horizontal interval (which is usually measured by the localization error, $Loc(\chi)$), where χ is the displacement of the filter. In [10] we have derived that

$$Nsr(\chi) = \frac{n}{A(\chi)} \sqrt{\int_{-\infty}^{+\infty} f^2(x;\chi)dx}, \quad Loc(\chi) = \frac{n\sqrt{\int_{-\infty}^{+\infty} f''^2(x;\chi)dx}}{\left|\int_{-\infty}^{+\infty} S'(x)f''(x;\chi)dx\right|} \quad (1)$$

where $A(\chi) = \left| \lim_{x \to \chi_+} S(x) - \lim_{x \to \chi_-} S(x) \right|$ is the absolute value of step amplitude, n is the white noise power.

As shown in Fig. 1, it is easy to know that $Nsr(f)$ and $Loc(f)$ describes the mathematical expectation of vertical and horizontal error, respectively. They can not represent the distance between two curves independently. They should be projected to the perpendicular direction of signal. In other words, noise-to-signal ratio is the primary index where signal is relatively flat; while localization error becomes the main item in which the signal varies sharply. Therefore we present a new edge-detection criterion as following:

$$C(f) = E\{Nsr^2(\chi)\cos^2\theta\} \cdot E\{Loc^2(\chi)\sin^2\theta\} \to \min \qquad (2)$$

where θ is the inclination angle of signal. We have

$$\tan\theta = K \cdot S'(x) \qquad (3)$$

K is a scalar constant that links the physical means of $Nsr(\cdot)$ and $Loc(\cdot)$, which have different dimensions. Using the Cauchy-Schwarz inequality, $C(f)$ achieves its minimum if and only if

$$Nsr(\chi)\cos\theta = \lambda \cdot Loc(\chi)\sin\theta \qquad (4)$$

where λ is a non-zero constant. Finally Eq. (4) yields the following necessary condition for "optimal" operators

$$\int_{\chi_-}^{\chi_+} f^2(x;\chi)dx = \frac{(\lambda K)^2}{f''^2(x;\chi)} \cdot S'^2(\chi) \cdot \int_{\chi_-}^{\chi_+} f''^2(x;\chi)dx \qquad (5)$$

For more details please refer to [9, 10]. Obviously the filter $f(x;\chi)$ is nonlinear which parameters are modulated by the signal gradient. In practice the signal gradient can not be obtained beforehand, however it can be assessed by any simple linear masks[1]:

$$G(x) \approx |S'(x)| \qquad (6)$$

3 Recursive Implementation

To derive $f(x;\chi)$ directly from the Eq. (5) is difficult and insufficient. Some formal constraints on the shape of $f(x;\chi)$ can be added. Because computational efficiency is primary important in low-level vision, we are sincerely expect such operators that can be implemented recursively. Consider the following recursive systems:

$$\hat{S}(k) = S_-(k) + S_+(k) \qquad (7)$$

where

$$S_-(k) = \sum_{i=-M+1}^{0} \beta_i S_i(k+i) + \sum_{j=-N}^{-1} \alpha_j S_-(k+j)$$

$$S_+(k) = \sum_{i=1}^{M-1} \beta_i S_i(k+i) + \sum_{j=1}^{N} \alpha_j S_+(k+j) \qquad (8)$$

---

[1] It is well known that the gradient estimation in real image is usually error prone. In our algorithm the initial gradient $G(x)$ is only used as an adjutant reference and does not affect the edge detection filters directly. Experimental results have shown that our algorithm is quite robust under different initial gradient estimators [10].

α_i and β_i are recursive coefficients, and $\sum_{i=-\infty}^{+\infty}\beta_i + \sum_{j=-\infty}^{+\infty}\alpha_j = 1$. Let the recursive coefficients vary with respect to the displacement,

$$\alpha_j = \alpha_j(k), \quad \beta_i = \beta_i(k) \tag{9}$$

Hence the basic form of our recursive filters. For the first order case, the equivalent continuous filter is

$$f(x) = a \cdot b^{|x|} \tag{10}$$

where $a = \ln b/2$. Substitute $f(x)$ in Eq. (5) by Eq. (10) we have deduced that [10]

$$b = \exp\{-2\lambda K \cdot G(x)\} \tag{11}$$

where $G(x)$ is the estimated signal gradient. According to Eq. (7), the discrete form of the first order filter is that

$$\begin{aligned} S_-(k) &= \beta_0 S_i(k) + \alpha_{-1} S_-(k-1) \\ S_+(k) &= \beta_1 S_i(k+1) + \alpha_1 S_+(k+1) \end{aligned} \tag{12}$$

Let the recursive coefficients vary with respect to the displacement, finally we have

$$\begin{aligned} \alpha_{-1}(k) &= \alpha(G) \\ \beta_0(k) &= 1 - \alpha(G) \\ \alpha_1(k) &= \alpha^2(G) \\ \beta_1(k) &= \alpha(G) - \alpha^2(G) \end{aligned} \tag{13}$$

and the gradient-related coefficient

$$\alpha(G) = \exp\{-2\lambda K \cdot G(k)\} \tag{14}$$

where $2\lambda K$ is a control constant. We have proven in [10] that this value must be more than 1/15. The shape of $\alpha(G)$ is shown in Fig. 3. A look-up table can be built beforehand to improve the computational efficiency.

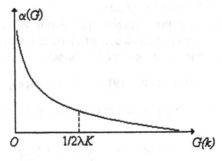

Fig. 3. Relation between the coefficient and estimated gradient

We have also derived the recursive filter of 2nd order [10]. For the limitation of space, only results are available here. The computational format is still Eq. (7), and

$$S_-(k) = w\big[S_i(k) + e^{-\alpha}(\alpha - 1)S_i(k-1)\big] + 2e^{-\alpha}S_-(k-1) - e^{-2\alpha}S_-(k-2)$$
$$S_+(k) = w\big[e^{-\alpha}(\alpha + 1)S_i(k+1) - e^{-2\alpha}S_i(k+2)\big] + 2e^{-\alpha}S_+(k+1) - e^{-2\alpha}S_+(k+2) \tag{15}$$

where w is the normalizing factor. Let α vary with respect to the displacement,

$$\alpha = \alpha(G) = \frac{2}{\sqrt{5}} \cdot 2\lambda K \cdot G(x) \tag{16}$$

However w can not be simply determined, because it is shift-variant too. To solve this problem, we introduce a substitution:

$$S_-(k) = \gamma \cdot \big[S_i(k) + e^{-\alpha}(\alpha - 1)S_i(k-1)\big] + 2e^{-\alpha}S_-(k-1) - e^{-2\alpha}S_-(k-2)$$
$$S_+(k) = \gamma \cdot \big[S_i(k) + e^{-\alpha}(\alpha - 1)S_i(k+1)\big] + 2e^{-\alpha}S_+(k+1) - e^{-2\alpha}S_+(k+2) \tag{17}$$

where

$$\gamma = \frac{(1 - e^{-\alpha})^2}{1 + e^{-\alpha}(\alpha - 1)}$$

is the normalizing factor. The final filtering output is

$$\hat{S}(k) = \frac{S_-(k) + S_+(k)}{2} \tag{18}$$

4 Image Edge Detection

Images are two-dimensional signals. The adaptive filtering procedure is applied to image by doing it for each row and redoing it for each column. When the gradient map is obtained on the smoothed image, an efficient edge detector is introduced by restraining the non-maximum points in the gradient image and take the thresholds. We choose two thresholds, $T_1 > T_2$. If a gradient pixel is brighter than T_1, it will be output immediately. All pixels more than T_2 are labeled. If a labeled pixel is connected to an output pixel, it will be output too.

In real image edge detection experiments, we choose the constant $2\lambda K = 0.10$. Because the digital images are bounded, a $\alpha(G)$ value very closed to 1.0 will bring some unexpected problems. So in practice we choose

$$\alpha(G) = (0.8 \sim 0.9) \cdot \exp\{-2\lambda K \cdot G(k)\}$$

for the 1st order edge detector. For the same reasons, we choose

$$\alpha(G) = \max\left\{0.2, \frac{2}{\sqrt{5}} \cdot 2\lambda K \cdot G(k)\right\}$$

for the 2nd order edge detector.

Fig. 4 illustrates the equivalent first order operators and estimated signal under the a computer generated one-dimensional signal. In this figure position B is a flat area, so the equivalent filter B is also a flat, symmetrical exponential series truncated by the left and right edges. There is an edge left to position C, so the equivalent filter is not symmetric and is relatively sharp on the left side.

Fig. 5 is row 127 sampled from the *Girl* image (Fig. 6). This figure illustrates that a Gaussian smoothing with constant σ is perhaps insufficient or excessive. However our filter can preserve the edge structure while smoothing noise.

Fig. 6 and Fig. 7 are 256×256 real images. The images after smoothing using the 1st or 2nd order adaptive filters are also shown. We compare our means to some previous detectors. The parameters of Shen's, Canny's and Deriche's operators have all been carefully adjusted so that the outputs seem quite well. In Fig. 7 the Deriche's operator has two thresholds. We found that the output of Shen's operator usually generates too many unnecessary small edge pieces in the presence of noise. Canny's and Deriche's operators usually provide fairly good results, however sometimes it maybe lost fine details. The edge detectors presented in this paper lead to best trade-off between noise restraining and edge localization.

In addition, a comment is maybe suitable for interpreting the experimental results. The edge detection criterion Eq. (5) can also be used for designing optimal linear shift-invariant edge detectors. If we take the signal gradient $G(x)$ as constants, the linear forms of our first and second order operators are respectively Shen's and Canny-Deriche's edge detectors. They are unified under our new criterion [10].

5 Conclusions

Marr had indicated that a good edge detection algorithm should use more than one scales [11]. This idea has been extended to the so-called "scale-space filtering" by Witkin [12]. In this paper we have presented a new criterion for optimal edge detection, and have deduced two adaptive image filters. The presented edge detectors have the ability of adaptively selecting filtering scales, however they can be efficiently implemented by recursive operations. It can be proven that their linear shift-invariant approximations are Shen's and Canny-Deriche's operators respectively.

Because the filters in this paper are indeed nonlinear and shift-variant, strict analyses are still not completely available. However experiments on both computer generated and real images have shown that the new detectors are powerful, and have been successfully used in many visual systems.

References

1 J. Canny: A computational approach to edge detection. IEEE Trans. PAMI-8, 679-698 (1986)
2 J. Shen, S. Castan: An optimal linear operator for step edge detection. CVGIP: Graph Model Image Processing, 54, 112-133 (1992)
3 R. Deriche: Optimal edge detection using recursive filtering. Proc. Int'l Conf. Comput. Vision, London (1987) pp. 501-505

517

4   F. van der Heijden: Edge and line feature extraction based on covariance modals. IEEE Trans. PAMI-17, 16-33 (1995)
5   A. C. Bovik, T. S. Huang: A generalization of median filtering using linear combination of order statistics. IEEE Trans. ASSP-31, 1342-1350 (1983)
6   M. Nagao, T. Matsuyama: Edge preserving smoothing, CGIP, 9, 394-407 (1977)
7   H. Jeong, C. I. Kim: Adaptive determination of filter scales for edge detection. IEEE Trans. PAMI-14, 579-585 (1992)
8   M. Gokmen, C. C. Lee: Edge detection and surface recognition using refined regularization. IEEE Trans. PAMI-15, 492-499 (1993)
9   L. X. Zhou, W. K. Gu: Adaptive edge detection: a new criterion and its recursive implementation. Proc. Intern. Conf. Signal Processing, Beijing (1996) pp. 1130-1133
10 L. X. Zhou: Fundamental researches in computer vision, Dissertation, Zhejiang University (Hangzhou, China) (1997)
11 D. Marr, E. C. Hildreth: Theory of edge detection. Proc. Roy. Soc. London, B, 207 (1980) pp. 187-217
12 A. P. Witkin: Scaling-space filtering. Proc. Intern. Joint Conf. Artif. Intel., Karlsruhe, Germany (1983) pp. 1019-1022

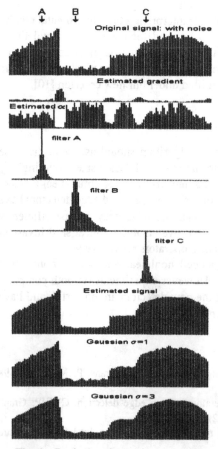

Fig. 4. Equivalent first order smoothing filters and estimated signal: simulated data

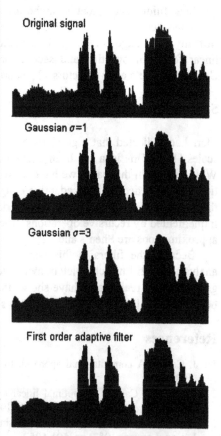

Fig. 5. Smoothed by both *Gaussian* and the first order filter: real data

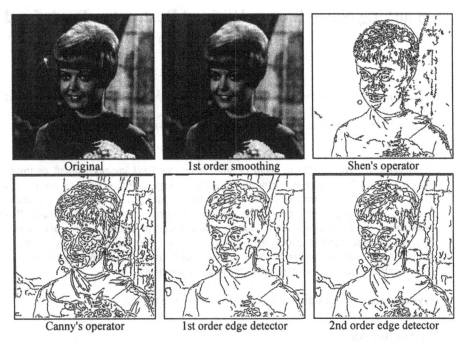

Original   1st order smoothing   Shen's operator

Canny's operator   1st order edge detector   2nd order edge detector

Fig. 6. *Girl* image

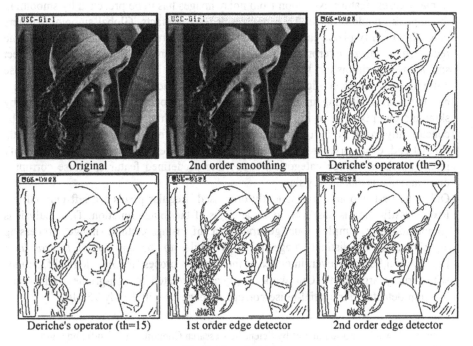

Original   2nd order smoothing   Deriche's operator (th=9)

Deriche's operator (th=15)   1st order edge detector   2nd order edge detector

Fig. 7. *Lena* image

Smoothing Noisy Images Without Destroying Predefined Feature Carriers

Andrzej J. Kasinski[1]

Katedra Automatyki, Robotyki i Informatyki, Politechnika Poznanska, 60965 Poznan,Poland

and

Grupo Vision y Robotica, Universidad de Murcia ,30203 Cartagena, Paseo Alfonso XIII
Spain

Abstract - We address the problem of smoothing gray-level images without destroying feature carriers. Smoothing is performed to suppress high, spatial-frequency noise in the image, whose relevant features contain high spatial-frequency components. The separation is obtained by using a heuristical image-surface geometry criterion over 5x5 mask. Pixel classification results with bit-fields associated with image processing tasks such as noise suppression, edge and/or some 2D-features extraction. We demonstrate the results on standard benchmark image disturbed by uncorrelated gaussian noise. Peformance of some filters applied to feature-less domains of the image is compared.

Keywords - smoothing, feature extraction, segmentation, grouping.

1 Introduction

In many cases, feature extraction from noisy images has to be preceded by smoothing to suppress random noise. This may result with artefacts. Global smoothing filters inevitably modify pixels carrying feature-relevant information. This may result in the reduction of image interpretation cues. A goood example of the destructive effect smoothing algorithms can have, is the median filter which, while eliminating impulse noise, destroy apparent edge junctions and object corners. In the study we propose a non-destructive filtering method. The goal is to smooth selectively the image by skipping the pixels carrying interesting features. This unmodified pixels can be concurrently extracted by using algorithms different from the ones used to smooth. To that end, every pixel is analysed, classified and labelled. In this study, we classify image pixels into three categories: supporting predefined features, noise corrupted ones, and pixels that are regular.

Only the first two are subsequently processed. Regular pixels are left unmodified. Pixel classification is based on local analysis of the image function. This analysis is computationally simple. Our general idea of using distinct image processing algorithms to the distinct image-fields, is similar to the concept of *logical filter* [Kulikowski,96]. However, in our case, the decision process involve logical functions with ternary arguments. Our work is somehow related to non-gradient edge- and junction-detectors for spatial-noise-corrupted images [Smith,Brady 96]. It

---

[1] This work was supported in part by Scientific Research Committee (Poland) under grant KBN-8T11A00510, and in part by the Ministry of Education and Science (Spain) under grant SAB95-0358

has been inspired also by the morphological filtering methods developed for gray images, [Dougherty, Laplante,96]. The experience with classical Sobel operators and the linear edge detectors of Canny, and Deriche [Deriche,87] has contributed to the idea as well.

Our aim is to improve the reliability of real-image analysis registered with relatively strong, high-frequency, spatial-noise. In particular, we look for smoothing methods which could cope with the extraction of linear features (edges, contours), and can guarantee a precise location of features in image coordinates.

The important part of our approach is the segmentation based on local geometrical properties of the image function. Features are not limited to contrast edges, i.e. due to discontinuity of the image-function derivatives. They also include groups of pixels being the locus of the image surface singular points. This generalization can be useful in the analysis of scenes presenting smoothly bounded objects. The results of non-destructive smoothing are demonstrated on gray images, and the image-function considered is the brightness.

2 Local analysis of the image

Reasoning about 3D geometry of the scene from 2D images analysis requires exact location of features. In most cases features are related to surface discontinuities. With smooth-boundary, softly illuminated objects - weaker, intrinsic contours, corresponding to the loci of the first- and second-order derivative zeros are also of interest. They have to be extracted in order to support the image understanding process. Moreover, they have to be classified into the appropriate feature class (as 1D/2D, collinear/curvilinear, concave/convex etc.).

There have been previous studies on extending the optimal edge detection method to ramp edges [Petrou,Kittler,91], and on zero-crossing based edge detectors [Sarkar, Boyer,91]. These works extended the class of detectable features which could be associated with the edge paradigm. Machine extraction of such a general set of features is difficult when the high-spatial-frequency noise is contained in the image.

Most of image smoothing methods are ranked by considering overall image quality indices or subjective visual effects. There is, in general, no consideration directly relating the merits of smoothing methods with their invariance with respect to the predetermined set of features. The exception are the works of [Ben-Ari and Rao,93], in which they introduce the performance measure DSNR (Discriminative Signal-to-Noise Ratio), which accounts not only for noise suppression but also for the suppression of the off-center responses to the feature extraction filter. To some extent, our ideas are similar to their approach. However, we do not parametrize edge templates.

Our idea is to partition the image domain into three pixel subdomains: feature carriers (which can be noisy as well), noisy featureless pixels, and so called regular pixels. Smoothing could then be applied only to the second category of pixels, and could be more radical compared to the global image smoothing methods, which are usually compromised by the postulate of non-damaging features. The approach is thus similar to the idea of DNSR.

Pixel classification criteria are based on the following assumptions, resulting from observations:
- meaningful contours on the image are reasonably distant to each other,
- local geometry of the image function surface can be classified as a result of joint evaluation of short othogonal profiles that intersect at the central pixel of the window under study.

We propose a 5x5 window as local analysis domain. It is an area large enough to make decision based on the second of the above assumptions. At the same time its small size gives quite non-restrictive implementation of the first assumption.

If we consider an 8-bit grey.level representation and a 5x5 window, our observation space can count 256^{25} possible local situations. By the geometrical encoding described below we can reduce it to the much smaller number of cases. Image samples are read in 5x5 sliding blocks. If we give local indices to samples (in row-order), then $x(13)$ is the central pixel of the 5x5 window. Orthogonal profiles, intersecting at $x(13)$, will consist of $x(11)$ to $x(15)$ (in the horizontal direction), and of $x(3)$, $x(8)$, $x(13)$, $x(18)$, $x(23)$ (in the vertical direction, respectively).

We compare neighbour image samples (subsequent elements of the profile vector in the appropriate direction). For each vector there are 4 comparisons. We store the results of each comparison ($<,=,>$), as respectively (-1, 0, $+1$), in registers $\underline{z}^h$ and $\underline{z}^v$. Each 4-cell register is associated with the appropriate profile and holds ternary variables. Using that scheme we encode local directional properties of the image surface such as constancy, monotonicity, unimodality.

At that level of description we can have $3x3x3x3 = 81$ situations for each directional profile. The abstract (qualitative) observation space (the Cartesian product of two "state registers") contains $81x81 = 6561$ cases. As $6561 << .256^{25}$ we get a considerable reduction in the number of diagnostic situations.

To that represenation, we associate our geometry-based, heuristic assumptions in the following way. As regular we define those pixels for which at least three consecutive cells of both $\underline{z}^h$ and $\underline{z}^v$ contain equal values. In that way all pixels belonging to smooth flat or monotonous (convex/concave) patches of the image-surface are regular.If we denote horizontal or vertical "state register" cells by: $[z_1, z_2, z_3, z_4]$, the local regularity condition can be easily expressed as the following logical function:

$$f_1 = ((z_1 = z_2) \cap (z_2 = z_3)) | ((z_2 = z_3) \cap (z_3 = z_4)) \qquad (1)$$

Here we make an implicit use of the first assumption saying that meaningful .features are isilated on the image surface and separated by the feature-less patches.

To carry a feature, the "logical profile" in the neighbourhood of the pixel under study, has to satisfy the alternative of the following functions:

$$f_2 = (z_1 = z_2) \cap (z_3 = z_4) \cap (z_2 \neq z_3) \qquad (2)$$

$$f_3 = (z_1 = 0) \cap (z_4 = 0) \cap (z_2 \neq z_3) \qquad (3)$$

The condition expressed by (2) represents a generalised (convex, concave, step-like, ramp-like) edge through the window center, in one of orthogonal directions. On the other hand, function (3) selects sequences of image signal samples, which jointly satisfy the following relationships: $x(11)=x(12)$ and $x(14)=x(15)$, (in the case of vertical profile we take the appropriare indices), with window central pixel $x(13)$ above or below the value to its left and to its right (or up and down). By adding this particular criterion , we prevent thin isolated features, and also, closely collocated directional features from being eliminated (they can be used in the local texture analysis). As feature-carriers we take all pixels for which condition "f_2 or f_3" is satisfied in at least one orthogonal direction. By that condition we extract edges in noisy image, without explicit numerical differentiation. The remaining pixels , i.e. pixels whose neighbour samples do not satisfy the above conditions are treated as noisy and subject of smoothing with the selected algorithm. In other words, the above procedure is the way of smoothing images only at those pixels, whose horizontal and vertical neighbourhoods are characterized by multiple zero-crossings of the gradient.

3 Noise suppression

The choice of smoothing algorithm depends on assumed noise model. Eliminating part of the image from smoothing affects the statistical properties of the filtered image slightly (the number of feature carriers in most practical cases is small compared to the number of pixels of the image). In our work, we have been comparing standard filters by applying them to selected pixels of well-known test images. For the sake of space limitations here we only present the results for LENA. Equivalent processing took place with the "Golden Gate Bridge" and some "geometric" test images. Among the filters tested were: median, mean, mode and Wiener over 3x3, 5x5, and 7x7 windows. Only the most interesting results are given here. In the case of Wiener filter, to avoid boundary effects due to the window size, we have been using a nominal 3 x 3 filter window, collocated centrally w.r.t. the 5x5 window used for the classification of the central pixel. Model noise used was an additive, white gaussian noise of various strengths and salt & pepper noise. As much worse filtering results have been obtained for the gaussian noise, only this results are presented. The images could be compared directly, however more insight into the effect of restoration was given by the comparison of histograms of the 'clean image', the'noisy image', and the 'selectively smoothed image'. Our tests were performed on images carrying a variety of features such as linear and curvilinear contours and image-surface extrema, as well as corners, cusps, end contour junctions.

In fig.1 we present the 8x256x256 test image - the central part of LENA to which a gaussian noise with zero mean and the standard deviation $\sigma = 10\%$ (of the max signal range) has been added. Classification results of the image in fig.1 are shown in fig.5. As black are marked either pixels classified as noisy or regular. Only edge-hypothesis-supporting pixels were left unmodified and are visible. They can be subsequently processed by the feature extraction algorithms with specialized noise suppression directional means. In fig.2 we give the result of selective median filtering (here only pixels which have been masked as black on fig.5 have been filtered over 5x5 window). Filtering improved the SNR value from 14.5dB to 21.2dB, which is almost optimal result. For comparison, the best so far, selective Wiener filtering over 3x3 window improved the SNR value from 14.5dB to 21.8dB.

In fig.3 there is the result of mean-filtering with 5x5 window. Here the improvement w.r.t SNR value is more than 5.8dB, a slightly worse result as compared to median and Wiener filtering. The directly observable differences between images in fig.2 and fig.3 are not significant. In both cases a substantial growth in the value of SNR as a result of filtering, can be attributed to the suppression of the high-spatial-frequency noise components.The reconstruction property of the method is more visible on histograms. Note that the histogram envelope of the selectively-filtered image in fig.4 has roughly the same pattern as the envelope of the "clean image". However the strips of the original histogram couldn't be restored. Nevertheless, the restoration of the histogram strips via selective filtering can be observed with weaker additive noise (σ=0.02). Anyway, the principal modes of the original histogram have been extracteded. Thus, from the first-order statistics point of view, the method gives relevant results.

4 Features Preservation

It is interesting to check how well in reality features are preserved as a result of selective filtering. We have been checking it with Sobel operator. The results are given in fig. 6 and 7 (in inverse graph. conv.). A sum of absolute horizontal and vertical Sobel mask responses is displayed. In fig.6 the results for noisy image of fig.1 are given. Many short and weak edges are visible on the smooth background patches (cheeks, front). The results of edges extraction for the selectively median-filtered image from fig.2 are in fig.7. It is noteworthy, that significant (strong or long) edges have been preserved. At the same time a significant "cleaning effect" with respect to spurious features could be seen. It has to be noted that the face image, characterised by fine, natural texture is not the best benchmark. Moreover, Sobel operator thickens edges, obscuring feature preservation property. However, still the effect of smoothing, while egdes are preserved, can be observed.

5 Conclusions

A method of splitting image for smoothing and edge-based feature extraction processes has been proposed. It has been shown, by using exemplary noisy images, that even smoothing restricted to some subdomain of the image can visibly reduce the noise, giving a proper statistical restoration of the image. Whether features are effectively preserved as a result of the proposed proceeding has to be studied further. A relatively weak effect of cleaning presented in the paper can be attributed to the choice of Sobel operator. Much better results are expected with specialized algorithms applied to the features-carrying image subdomain. This is the subject of on-going investigation. It has to be pointed out that preliminary pixel classification does not involve the dislocation of potential features. Filters performing pixel classification which preceede smoothing and feature extraction, can be implemented in compact hardware form due to the "logical character" of the pixel labeling process. The method can be also cosidered as a simple way of encoding information about the local geometrical context of the particular pixel. This encoding can be used in image texture analysis for segmentation. This topic has been studied in [Kasinski, Noriega,1997].

6 References

Ben-Arie J. and K.R. Rao: A Novel Approach for Template Matching by Non-Orthogonal Image Expansion, **IEEE Trans. Circ. & Syst. for Video Techn.**, vol.3, no.1, pp.71-84, 1993.

Deriche.R.: Usind Canny's Criteria to Derive a Recursively Implemented Optimal Edge Detector, **Int. J. of Comp. Vision** , 1987, pp. 167-187.

Dougherty E.R and Ph.A.Laplante: Nonlinear Real-Time Image Processing Algorithms, in: **Real-Time Imaging**, P. Laplante & A.Stoyenko Eds., IEEE Press, New York 1996, pp.3-26.

Kasinski A. and L.Noriega. Image Texture Segmentation Using Microstructural Features.1997, **submited for publication.**

Kulikowski J.L.: Basic concepts in the theory and design of logical filters. **Mach. Graph. & Vision Int. J.**, vol.3, no. 3/96, pp.465-482, 1996.

Petrou M. and J.Kittler: Optimal Edge Detectors for Ramp Edges, **IEEE Trans. on Pattern Anal.&Mach. Intell.**, vol.13, no5/May 91, pp.483-495. 1991.

Sarkar S.and K.L.Boyer: Optimal Infinite Impulse Response Zero Crossing Based Edge Detectors, **CVGIP: Image Understanding**, vol.54, Sept.1991, pp.224-243.

Smith S.M. and J.M. Brady: SUSAN - a new approach to low level image processing. (to be published in 1996, in **Int. J. of Comp.Vision**).

Fig.1. Image LENA (256x256 fragment), added white gaussian noise σ=0.1, SNR=14.5dB

Fig.2. Noisy image after selective 5x5 median filtering. Measured SNR=21 dB.

Fig.3. Noisy image after selective 5x5 selective mean filtering . Measured SNR=20dB.

Fig.4 Noisy image from fig.1 after selective 3x3 Wiener filtering. SNR= 21.8 dB.

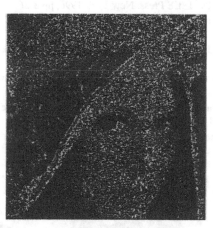

Fig. 5 Noisy and regular pixels in fig.1 are in black, unchanged are left feature-carriers.

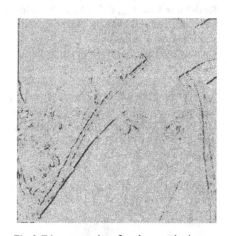

Fig.6 Edge extraction for the rough. image (fig. 1) with $\text{Max}\{\text{Sob}_H ; \text{Sob}_V\}$ operator.

Fig. 7 Edge extraction results for the fig.2 filtered image with $\text{Max}\{\text{Sob}_H;\text{Sob}_V\}$ operat.

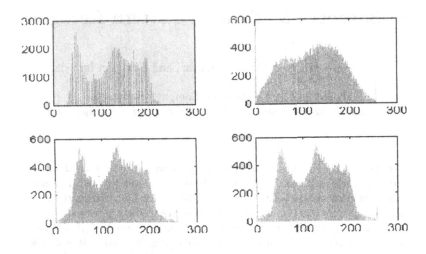

Fig.8 256-box histograms of tested images . Upper-left: "clean" LENA. Upper-right: "noisy" LENA. Lower left: selectively median-filtered image (fig.2). Lower right: selectively mean-filtered image (fig.3).

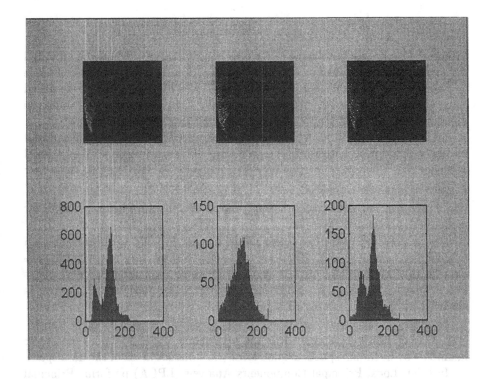

Fig.9. Fragments of "clean" LENA (left), white N(0; 0.03)-noise corrupted LENA (center) , median-filtered LENA from fig.2 (right). Below are the appropriate 256-boxes histograms (not-normalized). Note the qualitative restoration effect of the selective median filtering. Test image has strong, fine texture.

Local Subspace Method for Pattern Recognition

Władysław Skarbek, Miloud Ghuwar, and Krystian Ignasiak

Polish-Japanese Institute of Computer Techniques
Koszykowa 86, 02-008 Warsaw, Poland, email: skarbek@ipipan.waw.pl

Abstract. Local Principal Components Analysis, i.e. Principal Component Analysis performed in data clusters, is discussed and a neural algorithm is developed. This algorithmic tool is used for pattern recognition. A decision function based on the subspace method is generalized by introducing normalization matrix Γ and affine coefficients α, β. Assuming that feature measurements are Gaussian in data clusters, it is shown that the new method is equivalent to maximum likelihood method. However, no explicit knowledge on probability distributions of feature vectors in classes, is required. For handwritten numerals the technique reaches the recognition rate of about 99%.

1 Introduction

Principal Component Analysis (PCA) is a classical statistical modelling method which fits a linear model to experimental data (Hotelling [1]). In the technique an object of interest ω from a class Ω is represented by a data item $\mathbf{x}$ (so called feature vector or measurement vector) from the Euclidean space R^N while the population Ω of objects is modelled by a random, zero mean variable $\mathbf{X}$. A K-dimensional ($1 \leq K \leq N$) linear model is specified by K normalized and mutually orthogonal vectors $\mathbf{w}_1, \ldots, \mathbf{w}_K$ from R^N which span K-dimensional subspace (KLT subspace) of the largest variance for $\mathbf{X}$. It is easy to show that K eigenvectors of the covariance matrix $\mathbf{R}_{xx} = E\left(\mathbf{X}\mathbf{X}^T\right)$, corresponding to K largest eigenvalues, define such a linear model. Further, we identify the linear model with the matrix $\mathbf{W} = [\mathbf{w}_1, \ldots, \mathbf{w}_K]$ with $\mathbf{w}_i$ placed into i-th column.

Despite the optimality of a linear model obtained by PCA, the error of such modelling depends on data distribution, and it can be large. Especially, multimodal random variables are not suitable for linear modelling. Moreover, it is known that for given N, K, and covariance matrix $\mathbf{R}_{xx}$, the least PCA error is achieved by the Gaussian distribution.

In order to reduce the modelling error, nonlinear models must be considered. One of possible approaches is the approximation of nonlinear stochastic manifolds by linear subspaces defined locally in different areas of the data space.

In brief: Local Principal Components Analysis (LPCA) performs Principal Component Analysis in data clusters created in a vector data space. However, we should be aware that clustering process could be affected by PCA, and obviously PCA can depend on current clusters.

In section 2 a generic algorithm is described and next it is specialized to the neural algorithm for LPCA. The neural network approach is based on Oja/RLS learning rule recently introduced by Kung and Diamantaras [9] .

Essentially, PCA was used in pattern recognition for two goals:

- data compression: reducing the dimensionality from N of data vector $\mathbf{x}$ to K of data vector $\mathbf{y} = \mathbf{W}^T(\mathbf{x} - \overline{\mathbf{x}})$ in order to accelerate the classification task;
- subspace method (Watanabe [12], Oja [8]): defining a new proximity function for pattern matching which measures the distance to the principal subspace of the i-th class, i.e. replacing $(\mathbf{x} - \overline{\mathbf{x}}_i)^T(\mathbf{x} - \overline{\mathbf{x}}_i)$ by:

$$\rho_i(\mathbf{x}) \doteq (\mathbf{x} - \overline{\mathbf{x}}_i)^T \left(\mathbf{I} - \mathbf{W}_i \mathbf{W}_i^T\right)(\mathbf{x} - \overline{\mathbf{x}}_i) \tag{1}$$

where $\overline{\mathbf{x}}_i$ is the centroid of the i-th class and $\mathbf{W}_i$ is the linear model of the i-th class.

In section 3, we discuss a generalization of the subspace method which is obtained by introducing a normalization diagonal matrix Γ_i and affine coefficients α_i, β_i :

$$\rho_i(\mathbf{x}) \doteq \alpha_i(\mathbf{x} - \overline{\mathbf{x}}_i)^T \left(\mathbf{I} - \mathbf{W}_i \Gamma_i \mathbf{W}_i^T\right)(\mathbf{x} - \overline{\mathbf{x}}_i) + \beta_i \tag{2}$$

In this generalization we do not require from ρ to have positive values. It is used rather as a decision function than a distance measure in the following classification rule: *assign* $\mathbf{x}$ *to a class* j if and only if:

$$j = \arg\min_i \rho_i(\mathbf{x}) \tag{3}$$

We should emphasize that in our approach we use local analysis LPCA which splits original classes into subclasses corresponding to determined data clusters. Therefore the index i in the above rule goes over all subclasses of all classes.

2 Algorithm for LPCA

Local Principal Components Analysis can be implemented as a result of mutual interaction of two processes: a clustering process and a PCA process.

Though it is relatively easy to modify both, Kohonen's learning scheme [4] for data clustering and Oja's learning rule [7] for PCA, to get an "on line algorithm" of data item level, the resulting computational scheme is not only of higher complexity, but also less stable than epoch level algorithms (see Joutsensalo and Miettinen [2]).

Therefore we advocate LPCA which performs PCA in clusters after each clustering epoch (epoch level communication).

Let N be the dimensionality of the data space, K - the dimensionality of local linear models, and L - the number of data clusters.

LPCA algorithm can be conveniently described using a collection U of L *computational units* $(L = |U|)$. The unit $i \in U$ in a discrete moment has a state $q_i = (Z_i, \mathbf{c}_i, \mathbf{W}_i, N_i)$ where:

- Z_i – current list of data items from training set $X \subset R^N$, assigned to the unit i;
- c_i – reference vector for Z_i (at the end of the epoch it is the centroid of the list Z_i);
- $W_i \in R^{N \times K}$ – $N \times K$ matrix (at the end of epoch K, principal vectors of the i-th linear model are in its columns);
- $N_i \subset U$ – units which are currently recognized as *neighbors* of the unit i; we assume that $i \in N_i$.

In design of any clustering method we have to choose a *proximity measure* $\rho_i(x)$, $i \in U$, $x \in R^N$ which is used when x is to be joined to a cluster. We consider here two proximity measures based on Euclidean norm $\| \cdot \|$:

$$\text{proximity to centroid } \rho_i^c: \quad \rho_i^c(x) \doteq \|x - c_i\|^2 \tag{4}$$

$$\text{proximity to subspace } \rho_i^s: \quad \rho_i^s(x) \doteq \|Q_i(x - c_i)\|^2 \tag{5}$$

where $Q_i = I - W_i W_i^T$. As Q_i is symmetrical and idempotent ($Q_i^2 = Q_i$), we obtain an equivalent form for ρ_i^s :

$$\rho_i^s(x) \doteq (x - c_i)^T Q_i (x - c_i) \tag{6}$$

Note that

$$\rho_i^c(x) = \rho_i^s(x) + \left\| W_i W_i^T (x - c_i) \right\|^2 \tag{7}$$

Therefore if for some reasons W_i is the zero matrix then $\rho_i^s = \rho_i^c$.

Further, it will be convenient to assume that if either c_i or W_i is not defined then $\rho_i(x) = \infty$.

Generic LPCA Algorithm
Input: $N, K, L, X \subset R^N$
Output: $(c_i, W_i), i = 1, \ldots, L$
Method:
  for each $i \in U$:
    $Z_i := \emptyset$; Init c_i, W_i, and N_i;
  loop
    for each $x \in X$:
      $i(x) := \text{Argmin}(x, L, \rho)$;
      for each $i \in U$:
        if $i \in N_{i(x)}$ **then** $Z_i := Z_i \cup \{x\}$;
    for each $i \in U$:
      $c_i := \overline{Z_i}$;
      $W_i := \text{PCA}(N, K, Z_i)$;
    if $\text{Stop}(c, W, X)$ **then** return;
    for each $i \in U : Z_i := \emptyset$; Update N_i;

$\overline{Z}$ denotes here the mean operator which actually calculates the centroid of elements in the given list. Note also that $\cup$ operation joins elements to a list, so duplicates of elements in Z_i are possible.

A concrete LPCA algorithm is obtained from the above generic algorithm by the specification of the following meta variables: ρ, Init, Argmin, PCA, Stop, and Update. Several such specifications with references to the literature are described by the first author in [10]. Here only one possible way is presented.

Our algorithm is not "true SOM" type algorithm as neighborhoods are fixed to $N_i = \{i\}$. Reference vectors c_i are initially chosen randomly from the training data set X and the proximity measure $\rho = \rho^s$.

Initially the matrix $\mathbf{W}_i$ is set to zero, so for the first epoch the proximity measure $\rho_i = \rho_i^s = \rho_i^c$. In our LPCA results of Argmin depend not only on currently computed proximity values ρ_i^s but on the distance to the closest reference vector in the previous epoch, too. In this way we avoid possible cases when a data item $\mathbf{x}$ is close to a subspace centered on a quite distant reference vector. Let $i'(\mathbf{x})$ be the index of the closest reference vector computed in the previous epoch (say $= 1$ if no previous epoch). Then $i(\mathbf{x})$ in the current epoch is computed conditionally:

$$i(\mathbf{x}) \doteq \begin{cases} \arg\min_{i \in U} \rho_i^s(\mathbf{x}) & \text{if } \rho_{i(\mathbf{x})}^c(\mathbf{x}) < \alpha \rho_{i'(\mathbf{x})}^c(\mathbf{x}) \\ i'(\mathbf{x}) & \text{otherwise,} \end{cases}$$

where typically $\alpha = 1.5$.

For PCA in the given cluster, Oja/RLS learning scheme [9] is used which gives a good approximation of the principal component.

3 LPCA Based Pattern Recognition

For the given data set X, LPCA finds local models for existing data clusters in X. Such a model is defined by three elements: the centroid of the cluster $\bar{\mathbf{x}}$, its variance σ^2, and the matrix of the principal components $\mathbf{W} \in R^{N \times K}$. In recognition we deal with classes of objects and separate data sets for each class. Therefore we have as many local models as the number of all data clusters found in all data sets. We can number all of them from one to say L, and identify the result of LPCA stage by L local models $(\bar{\mathbf{x}}_i, \sigma_i^2, \mathbf{W}_i)$, $i = 1, \dots, L$. As a result, the classifier is designed for L new classes Ω_i which are subclasses of the original ones. In practice, original random variable representing measurements for the original class is multimodal while in subclasses found by LPCA it becomes unimodal (usually well approximated by a Gaussian distribution).

The recognition is based on identifying a subclass j for which the minimum of the decision function $\rho_i(\mathbf{x})$, defined by formula (2), is achieved:

$$j = \arg\min_i \left[\alpha_i (\mathbf{x} - \bar{\mathbf{x}}_i)^T \left(\mathbf{I} - \mathbf{W}_i \Gamma_i \mathbf{W}_i^T \right) (\mathbf{x} - \bar{\mathbf{x}}_i) + \beta_i \right] \tag{8}$$

By minimizing the recognition error, we could try to find unknown parameters α_i, β_i, and Γ_i. However, it appears that we can specify explicit formulas for those parameters getting a classifying scheme equivalent to an optimal estimator for maximum likelihood method if probability distributions in subclasses are Gaussian. For this goal additionally to local linear models we will need a probability *apriori* $P(\Omega_i)$ for the subclass Ω_i. It can be easily estimated from relative frequencies of data clusters.

Since further relationships are derived for a fixed subclass, we drop the index i in the formulas below.

Let λ_j be the variance for the j-th component of the vector random variable $\mathbf{Y} \doteq \mathbf{W}^T(\mathbf{X} - \bar{\mathbf{X}})$. Let λ be a *missing variance per component* of $\mathbf{Y}$ in $\mathbf{X}$, i.e.:

$$\lambda_j \doteq E[\mathbf{w}_j(\mathbf{X} - \bar{\mathbf{X}})], \quad \lambda \doteq \frac{\text{var}(\mathbf{X}) - \text{var}(\mathbf{Y})}{N - K} \tag{9}$$

Then the unknown parameters are defined as follows:

$$\alpha = 1/\lambda$$
$$\beta = \sum_{j=1}^{K} \ln \lambda_j + (N - K) \ln \lambda - 2 \ln P(\Omega) \tag{10}$$
$$\Gamma = \text{diag}(1 - \lambda/\lambda_1, \ldots, 1 - \lambda/\lambda_K)$$

We are going to show that the decision function

$$\rho(\mathbf{x}) \doteq \left[\alpha(\mathbf{x} - \bar{\mathbf{x}})^T \left(\mathbf{I} - \mathbf{W}\Gamma\mathbf{W}^T\right)(\mathbf{x} - \bar{\mathbf{x}}) + \beta\right] \tag{11}$$

defined for class Ω is an optimal estimator of the Mahalanobis decision function $\delta(\mathbf{x})$ defined as follows:

$$\delta(\mathbf{x}) \doteq (\mathbf{x} - \bar{\mathbf{x}})^T \mathbf{R}^{-1}(\mathbf{x} - \bar{\mathbf{x}}) + \ln \det \mathbf{R} - 2 \ln P(\Omega) \ ,$$

where $\mathbf{R}$ is the covariance matrix of $\mathbf{X}$. For Gaussian distribution, the Mahalanobis decision function is equivalent to maximum likelihood decision function $p(\mathbf{x}|\Omega)P(\Omega)$. Namely

$$\delta(\mathbf{x}) = -2 \ln(p(\mathbf{x}|\Omega)P(\Omega)) - N \ln(2\pi) \ ,$$

where the density p of multivariate Gaussian distribution has the form:

$$p(\mathbf{x}|\Omega) = \frac{\exp\left(-\frac{1}{2}(\mathbf{x} - \bar{\mathbf{x}})^T \mathbf{R}^{-1}(\mathbf{x} - \bar{\mathbf{x}})\right)}{(2\pi)^{N/2}(\det \mathbf{R})^{1/2}} \ .$$

Let $\mathbf{R} = \mathbf{U}\Lambda\mathbf{U}^T$ be the spectral decomposition of $\mathbf{R}$. Then

$$\Lambda = \mathbf{U}^T \mathbf{R} \mathbf{U} = E\left[\mathbf{Y}\mathbf{Y}^T\right],$$

where $\mathbf{Y} \doteq \mathbf{U}^T(\mathbf{X} - \bar{\mathbf{X}})$.

Hence diagonal elements of $\Lambda = diag(\lambda_1, \ldots, \lambda_N)$ are variances of $\mathbf{Y}$ components and therefore they are identical to elements defined in (9).

The matrix of eigenvectors can be decomposed into $\mathbf{U} = [\mathbf{W}\mathbf{W}']$ where $\mathbf{W}$ consists of the first K principal vectors of the random variable $\mathbf{X}$. Then the diagonal matrix Λ_U of eigenvalues has the form:

$$\Lambda = \begin{bmatrix} \Lambda_W & 0 \\ 0 & \Lambda_{W'} \end{bmatrix} \ .$$

From the spectral decomposition we have:

$$\ln \det \mathbf{R} = \ln \det \Lambda_W + \ln \det \Lambda_{W'} \ .$$

The decomposition of $\mathbf{U}$ and Λ leads to the decomposition of the Mahalanobis function:

$$\delta(\mathbf{x}) = \delta_W(\mathbf{x}) + \delta_{W'}(\mathbf{x})$$
$$\delta_W(\mathbf{x}) \doteq (\mathbf{x} - \bar{\mathbf{x}})^T \mathbf{W}\Lambda_W^{-1}\mathbf{W}^T(\mathbf{x} - \bar{\mathbf{x}}) + \ln \det \Lambda_W - 2 \ln P(\Omega) \tag{12}$$
$$\delta_{W'}(\mathbf{x}) \doteq (\mathbf{x} - \bar{\mathbf{x}})^T \mathbf{W}'\Lambda_{W'}^{-1}\mathbf{W}'^T(\mathbf{x} - \bar{\mathbf{x}}) + \ln \det \Lambda_{W'}$$

We would like to estimate the second term in order to avoid the computation of low magnitude eigenvalues. To this end we introduce a parameter μ and replace $\Lambda_{W'}$ by μI getting a parametric function $\delta_{W'}(\cdot; \mu)$:

$$\delta_{W'}(\mathbf{x}; \mu) \doteq \frac{1}{\mu}(\mathbf{x} - \overline{\mathbf{x}})^T \mathbf{W}' \mathbf{W}'^T (\mathbf{x} - \overline{\mathbf{x}}) + (N - K) \ln \mu .$$

In order to find the best μ, we analyze the average error expression:

$$\epsilon(\mu) = E[\delta_{W'}(\mathbf{X}; \mu) - \delta_{W'}(\mathbf{X})] .$$

It appears that $\epsilon(\mu) \geq 0$ and its unique minimum is achieved for $\mu = \lambda$ which is defined by formula (9). Therefore our estimate of Mahalanobis function has the form:

$$\begin{aligned}
\hat{\delta}(\mathbf{x}) &\doteq \delta_W(\mathbf{x}) + \delta_{W'}(\mathbf{x}; \lambda) \\
&= (\mathbf{x} - \overline{\mathbf{x}})^T \mathbf{W} \Lambda_W^{-1} \mathbf{W}^T (\mathbf{x} - \overline{\mathbf{x}}) + \tfrac{1}{\lambda}(\mathbf{x} - \overline{\mathbf{x}})^T \mathbf{W}' \mathbf{W}'^T (\mathbf{x} - \overline{\mathbf{x}}) \\
&\quad + \textstyle\sum_{j=1}^{K} \ln \lambda_j + (N - K) \ln \lambda - 2 \ln P(\Omega) .
\end{aligned}$$

We can eliminate $\mathbf{W}'$ from the above formula using the following relation:

$$\mathbf{I} = \mathbf{U}\mathbf{U}^T = \mathbf{W}\mathbf{W}^T + \mathbf{W}'\mathbf{W}'^T .$$

Namely:

$$\begin{aligned}
&(\mathbf{x} - \overline{\mathbf{x}})^T \mathbf{W} \Lambda_W^{-1} \mathbf{W}^T (\mathbf{x} - \overline{\mathbf{x}}) + \tfrac{1}{\lambda}(\mathbf{x} - \overline{\mathbf{x}})^T \mathbf{W}' \mathbf{W}'^T (\mathbf{x} - \overline{\mathbf{x}}) \\
&= \tfrac{1}{\lambda}(\mathbf{x} - \overline{\mathbf{x}})^T \mathbf{W} (\lambda \Lambda_W^{-1}) \mathbf{W}^T (\mathbf{x} - \overline{\mathbf{x}}) + \tfrac{1}{\lambda}(\mathbf{x} - \overline{\mathbf{x}})^T (\mathbf{I} - \mathbf{W}\mathbf{W}^T)(\mathbf{x} - \overline{\mathbf{x}}) \\
&= \tfrac{1}{\lambda}(\mathbf{x} - \overline{\mathbf{x}})^T \left[\mathbf{I} - \mathbf{W}(\mathbf{I} - \lambda \Lambda_W^{-1}) \mathbf{W}^T \right] (\mathbf{x} - \overline{\mathbf{x}}) \\
&= \tfrac{1}{\lambda}(\mathbf{x} - \overline{\mathbf{x}})^T \left[\mathbf{I} - \mathbf{W}\Gamma\mathbf{W}^T \right] (\mathbf{x} - \overline{\mathbf{x}}) .
\end{aligned}$$

The last equality proves finally that the decision function ρ specified in (10, 11) is identical to the estimated Mahalanobis function $\hat{\delta}$.

4 Experimental results

We have applied LPCA method for handwritten digits stored in NIST database (see sample input data in Fig. 1(a)) which were collected from zip codes handwritten on envelopes. The database includes about 200 thousands of pictures for already segmented digits. For training and testing random 5% samples are chosen dozen times.

4.1 Results for distance to class subspaces

In these experiments we test the subspace method which is obtained in our program by setting the number of clusters in each class to one. The dimension K of the subspace for each class is set to the same value. From the Table 1, we see that increasing K results in increase of the global recognition rate. However, further increase of K up to maximum possible value N results in worse performance.

The recognition rate in individual classes can achieve its maximum for its individual K but for the design simplicity we have decided in this research not to incorporate this fact into the program.

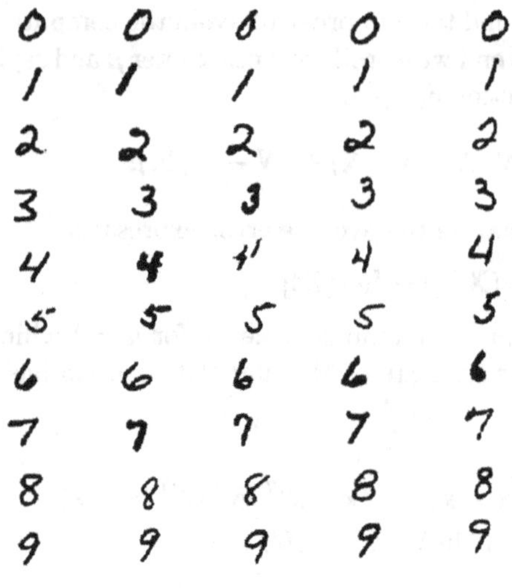

digit	PCA	LPCA
0	99.8	99.5
1	99.8	100.0
2	96.8	99.3
3	97.3	99.6
4	97.6	99.3
5	97.3	98.9
6	98.7	99.3
7	99.7	99.8
8	93.9	98.7
9	97.8	99.6
all	97.87	99.41

(a) (b)

Fig. 1. (a) Sample input data, (b) Recognition rates for Local Subspace Method

4.2 Results for distance to class local subspaces

The local subspace method was tested for different dimensionality K of local subspaces and different number L of clusters in classes. Results are presented in the Table 1. Similarly to the subspace method further increase of K deteriorates algorithm's performance too. The increase of L to values more than three makes results less confident as the number of samples per cluster decreases to the level at which learning of the local model by the neural approach is not possible.

In order to compare the performance of the subspace method versus the local subspace method on the basis of the Table 1 (the rows labeled by $k_{PCA} = 16$), we arrange the Table (b) in the Figure 1.

5 Conclusions

Local Principal Components Analysis, i.e. Principal Component Analysis performed in data clusters, when applied for pattern recognition problem, appears to be a better tool than classical subspace method. The reason is in ability of discrimination for existing subclasses in the given pattern class. Assuming that feature measurements are Gaussian in data clusters, we have very desired property of minimum recognition error like in maximum likelihood method. However now, no explicit knowledge on probability distributions of feature vectors in classes, is required. For handwritten numerals the technique reaches the recognition rate of about 99%. Interestingly they are better than maximum likelihood approach by about 2% that means data clusters are not Gaussian.

k_{PCA}	0	1	2	3	4	5	6	7	8	9	all
PCA											
1	95.5	95.9	92.8	94.9	88.5	91.9	96.5	92.4	81.3	89.8	91.95
2	97.1	98.1	92.8	94.9	89.4	95.6	98.0	94.0	84.7	91.4	93.59
4	99.3	99.8	93.2	94.4	93.5	95.6	98.5	94.3	89.5	96.4	95.45
8	99.3	99.8	96.2	96.6	95.9	95.6	98.3	96.7	93.0	98.0	96.95
16	99.8	99.8	96.8	97.3	97.6	97.3	98.7	99.7	93.9	97.8	97.87
LPCA $(k_{VQ} = 3)$											
1	97.9	99.8	94.5	94.6	93.9	96.3	97.2	90.9	87.3	96.0	94.84
2	98.9	100	94.7	96.5	93.5	97.1	98.3	96.0	89.3	95.7	96.00
4	99.3	100	96.6	97.0	97.2	98.3	99.1	96.4	90.6	97.7	97.20
8	99.6	100	97.5	98.3	97.8	98.1	98.9	99.3	94.6	98.9	98.31
16	99.5	100	99.3	99.6	99.3	98.9	99.3	99.8	98.7	99.6	99.41

Table 1. Recognition rates for the subspace (PCA) and local subspace (LPCA) methods: $N = 20 \times 20$, $k_{PCA} = 1, 2, 4, 8, 16$.

References

1. Hotelling, H.: Analysis of a complex of statistical variables into principal components. Journal of Educational Psychology, 24 1933 417-441
2. Joutsensalo, J., Miettinen, A.: Self-organizing operator map for nonlinear dimension reduction. ICNN'95, 1995 IEEE International Conference on Neural Networks. 1(1) 1995 111-114
3. Kambhatla, N., Leen, T.K.: Fast nonlinear dimension reduction. ICNN'93, 1993 International Conference on Neural Networks. 3 1993 1213-1218
4. Kohonen, T.: The self-organizing map. Proc. of IEEE 78 1990 1464-1480
5. Kohonen, T.: Self-Organizing Maps. Springer, Berlin, 1995
6. Linde, Y., Buzo, A., Gray, R.M.: An algorithm for vector quantizer design. IEEE Trans. Comm. COM-28 1980 28-45
7. Oja E.: Principal components, minor components, and linear neural networks. Neural Networks 5 1992 927-935
8. Oja E.: Subspace methods of pattern recognition. Research Studies Press, England, 1983
9. Diamantaras, K.I., Kung, S.Y.: Principal component neural networks. John Wiley & Sons, New York, 1996
10. Skarbek, W.: Local Principal Components Analysis for Transform Coding. 1996 Int. Symposium on Nonlinear Theory and its Applications, NOLTA'96 Proceedings, Research Society NTA, IEICE, Japan, Oct. 1996 381-384
11. Suen, C.Y., Legault, R., Nadal, C., Cheriet, M., Lam, L.: Building a new generation of handwriting recognition systems. Pattern Recognition Letters, 14 1993 303-315
12. Watanabe, S., Pakvasa, N.: Subspace method of pattern recognition. Proc. of 1st Int. Joint Conf. on Pattern Recognition, 1973

Testing the Effectiveness of Non-linear Rectification on Gabor Energy

George Paschos

Department of Computer Science
Michigan Technological University
1400 Townsend Drive
Houghton, MI 49931-1295
USA

Abstract. The gabor family of filters has received a great deal of attention as it seems to closely approximate the type of low-level processing found in biological vision systems. It has also been widely held that further processing stages involving energy calculation and Non-Linear Rectification (NLR) should be employed. Recent performance evaluation tests of gabor-based features in texture classification [1], where NLR was not used, produced poor results and the hypothesis that use of NLR would improve performance was made.

The work presented here explores the extend to which such a hypothesis can be valid. Two sets of gabor energy features are extracted, before and after an NLR operator is applied to a filtered image; these two sets form the initial texture classification feature set. Feature effectiveness is evaluated using a statistical redundancy test and features are ranked based on the resulting Redundancy Factor (RF). Best features are selected among those with the lowest RF values and subsequently used for classification. Thus, the RF-test provides a strong indicator regarding the effectiveness of features with respect to pattern discrimination. Based on that, the relative effectiveness of pre-NLR and post-NLR features is compared and conclusions on the NLR hypothesis are drawn.

1 Introduction

Recent research in biological vision, as it relates to the initial processing stages along the visual path, has pointed in the direction of spatial frequency/orientation-sensitive forms of filtering ([2, 3]). In response, the gabor family of filters has received a great deal of attention as it seems to closely approximated such type of low-level processing ([5, 7]). It has also been widely held that two additional stages should follow ([2, 4]). First, image energy is calculated over the filter output, and subsequently, a non-linear rectification (NLR) operator is applied to create a map consisting of blob and line-segment types of features. These features could be then readily used for texture segmentation/classification. In a classification study conducted by [1], where gabor filtering was used without NLR, the results were poor and the hypothesis was made that use of NLR would improve classification performance.

The work presented here is part of a wider effort to evaluate the extend to which a hypothesis such as the above can be valid. A set of 25 texture classes (images) of textures of various types (i.e., from uniform to random texture) is used. Class samples are produced by creating sub-images of different sizes and inducing pepper noise of various degrees. A first set of gabor energy features is extracted after filtering, an NLR operator is then applied, and a second set of energy features is obtained after rectification; these two sets form the initial texture classification feature set. In order to evaluate feature effectiveness, a statistical redundancy test is performed and features are ranked based on the resulting Redundancy Factor (RF). Since lower redundancy is equivalent to higher robustness, the RF-test provides a strong indicator for the effectiveness of each feature with respect to classification. Based on this ranking, pre-NLR and post-NLR features can be compared and a number of conclusions on the NLR hypothesis can be drawn.

The following section describes the four-stage processing environment used for feature generation, section 3 presents several test cases and the corresponding results, while in the last section the test results are discussed and suggestions are made.

2 Feature Generation

The energy feature generation system can be viewed as a sequence of four processing stages (Fig. 1) : gabor filtering (G), pre-NLR energy computation (E), NLR, and post-NLR energy computation (E). Three main computational components are involved (filtering, NLR, and energy computation). The details of each for these components are provided in the following.

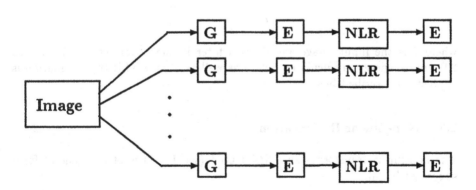

Fig. 1. The Feature Generation System (G:Gabor filter, E:Energy computation, NLR:Non-Linear Rectification).

2.1 Gabor Filters

Image intensity in the CIE XYZ color space is defined as a linear transformation of the RGB space coordinates:

$$Y = 0.299R + 0.587G + 0.114B \tag{1}$$

This information is passed through a set of gabor filters which are defined as follows:

$$G(x,y) = e^{-\frac{(x^2+y^2)}{2\sigma^2}} e^{-2j\pi\phi(x\cos\theta + y\sin\theta)} \tag{2}$$

The filter parameter values used are: $\sigma = 1.0$ (gabor window of size 5x5), spatial frequency(ϕ)=2, 4, 8, 16, and 32 cycles per image size, and orientation(θ)=0^0, 45^0, 90^0, and 135^0. As a result, a set of 20 filters is formed (the real (even-symmetric) part of the complex exponential is utilized).

2.2 Energy Features

The energy for a filtered image F_i of size $N x N$ is measured as follows:

$$E_i = \frac{1}{N^2} \sum_{x=1}^{N} \sum_{y=1}^{N} |F_i(x,y)| \tag{3}$$

$$F_i = I * G_i \tag{4}$$

where I is the input image and G_i is a filter in the gabor set (i= 1, 2, ... , 20). Image energy is calculated before as well as after a NLR transformation is applied, as described below.

2.3 Non-Linear Rectification

The non-linear transformation (NLR) used (see Fig.1) is of the general form suggested by [4]:

$$\psi(x) = \frac{1 - e^{-\alpha x}}{1 + e^{-\alpha x}} \tag{5}$$

where $\alpha = 3.5$ (Fig.2).

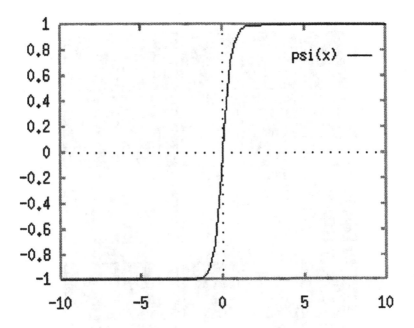

Fig. 2. The Non-Linear Rectification Function.

Table 1. Feature Redundancy Factors Before (pre-RF) and after (post-RF) NLR. Sample Size:64x64, Number of Samples:75 and 300 (left, right).

i	pre-RF	post-RF	pre-RF	post-RF
1	27.835	42.649	117.139	175.542
2	30.524	760.765	132.017	2842.181
3	27.835	42.647	117.139	175.535
4	30.615	759.562	135.903	2846.239
5	27.835	42.647	117.139	175.535
6	27.745	28.887	117.087	120.108
7	27.835	42.646	117.139	175.530
8	27.827	28.942	117.402	120.336
9	27.835	42.646	117.139	175.530
10	28.086	397.289	119.783	1524.469
11	27.835	42.647	117.139	175.535
12	28.390	394.262	123.067	1535.586
13	27.835	42.647	117.139	175.535
14	27.725	144.619	118.060	576.513
15	27.835	42.647	117.139	175.534
16	28.047	143.648	120.333	580.647
17	27.835	42.647	117.139	175.534
18	29.273	652.765	125.369	2451.543
19	27.835	42.647	117.139	175.534
20	29.392	648.551	129.566	2461.485

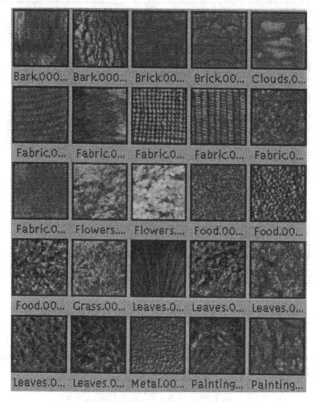

Fig. 3. The Texture Image Set.

3 RF Ranking - Results

The end result of the feature generation system is a set of 40 energy features (pre- and post-NLR energies with 20 filters) for a given image. The original set of 25 texture classes/images of size 512x512 are shown in Fig.3. Image samples for each of them are produced by randomly cutting a number of subimages and inducing various amounts of pepper-type noise.

To test the effectiveness of the energy features in discriminating among the texture classes, the *within-to-between* variance ratio is determined for each of the features based on the sample data set ([6]). For a random variable x, the *within* variance is defined as:

$$\sigma_w^2(k,i) = \sum_{l=1}^{n_k} x_{kl}^2 - \frac{1}{n_k}[\sum_{l=1}^{n_k} x_{kl}(i)]^2 \tag{6}$$

while the *between* variance is given by:

$$\sigma_b^2(i) = \sum_{k=1}^{C} x_k^2(i) - \frac{1}{C}[\sum_{k=1}^{C} x_k(i)]^2 \qquad (7)$$

where $x_{kl}(i)$ is sample l of feature i in class k, $x_k(i)$ is the mean of feature i in class k, n_k is the number of samples in class k, and C is the number of classes. The redundancy of feature i is then determined by the following ratio:

$$RF(i) = \sum_{k=1}^{C} \frac{\sigma_w^2(k,i)}{\sigma_b^2(i)} \qquad (8)$$

Table 2. Feature Redundancy Factors Before (pre-RF) and after (post-RF) NLR. Sample Size:128x128, Number of Samples:75 and 150 (left, right).

i	pre-RF	post-RF	pre-RF	post-RF
1	27.995	42.160	53.611	83.539
2	29.593	814.033	57.354	1561.622
3	27.995	42.160	53.611	83.539
4	28.686	813.925	56.791	1562.952
5	27.995	42.160	53.611	83.539
6	28.006	28.450	53.620	55.702
7	27.995	42.160	53.611	83.538
8	27.991	28.428	53.634	55.719
9	27.995	42.160	53.611	83.538
10	28.341	445.780	54.142	852.630
11	27.995	42.160	53.611	83.539
12	27.876	444.565	54.017	854.595
13	27.995	42.160	53.611	83.539
14	28.176	163.748	53.857	313.993
15	27.995	42.160	53.611	83.539
16	27.910	163.264	53.828	314.291
17	27.995	42.160	53.611	83.539
18	28.955	709.569	55.527	1358.421
19	27.995	42.160	53.611	83.538
20	28.245	708.768	55.220	1361.044

where small RF values indicate low redundancy or, equivalently, high effectiveness and vice versa. Thus, a ranking order can be imposed on the feature set and the best features can be identified as the ones with the lowest RF values.

Several test cases are shown in Tables 1-3 where image samples of size 64x64, 128x128, and 256x256 have been used with noise levels (n_s) ranging from 0% (original subimage) to 45% in increments of $45/n_s$.

Table 3. Feature Redundancy Factors Before (pre-RF) and after (post-RF) NLR. Sample Size:256x256, Number of Samples:75.

i	pre-RF	post-RF
1	24.601	40.823
2	25.266	939.706
3	24.601	40.823
4	25.242	940.809
5	24.601	40.823
6	24.599	25.396
7	24.601	40.822
8	24.599	25.391
9	24.601	40.822
10	24.634	514.287
11	24.601	40.823
12	24.624	516.577
13	24.601	40.823
14	24.609	183.960
15	24.601	40.823
16	24.608	184.698
17	24.601	40.823
18	24.974	822.867
19	24.601	40.823
20	24.925	824.962

4 Discussion

This work addresses the following question: "In practice, is there a potential gain in texture classification performance when gabor energy features are calculated after a non-linear transformation has been applied?" Although such types of transformations have been used in various cases, the hypothetical comparative advantage of post-transformation features over pre-transformation ones has not been explicitly and directly justified apart from any references to models of biological vision. And it is important to evaluate and justify the potential of such processing models in artificial visual systems.

The work presented in this paper provides a significant amount of results from tests performed on a wide collection of texture patterns. Further experiments will be needed with additional types of textural patterns. However, the experimental evidence provided is a first step in the direction of answering the question posed above.

References

1. Ohanian P. P., Dubes R. C.: Performance Evaluation for Four Classes of Textural Features. Pattern Recognition **25(8)** (1992) 819-833.

2. Hess R. F., Badcock D. R.: Metric for Separation Discrimination by The Human Visual System. Journal of The Optical Society of America, A **12(1)** (1995) 3-16.
3. Malik J., Perona P.: Finding Boundaries in Images. Neural Networks for Perception **1** (1992) 315-344.
4. Caelli T. M.: An Adaptive Computational Model for Texture Segmentation. IEEE Transactions SMC **18(1)** (1988) 9-17.
5. Bigün J., Hans du Buf J. M.: , N-folded Symmetries by Complex Moments in Gabor Space and Their Applications to Unsupervised Texture Segmentation. IEEE Transactions PAMI **16(1)** (1994) 80-87.
6. Fugunaga K.: Introduction to Statistical Pattern Recognition. Academic Press 1990.
7. Bovik A. C., Clark M., Geisler W. S.: Multichannel Texture Analysis Using Localized Spatial Filters. IEEE Transactions PAMI **12(1)** (1990) 55-73.

Neural-Like Thinning Processing

Lucius Chudý[1] and Vladimír Chudý[2]

[1] Institute of Measurement Science, Slovak Academy of Sciences, Dúbravská cesta 9, 842 19 Bratislava, Slovakia, email: umerchud@savba.sk
[2] Department of Psychology, Comenius University, Gondova 2, 818 01 Bratislava, Slovakia, email: chudy@fphil.uniba.sk

Abstract. Neural-like approach to the thinning of binary patterns based on local thickness estimate is proposed. Thinning is considered to be a self-organizing process where a binary neuron - contour pixel value is given by the hard limiter transfer function. Input to this function is given by the pattern local thickness estimate from flexible (both in size and shape) neighborhood together with neuron threshold which ensures the preservation of the 4-connectedness and genuine end point. The final skeletons obtained within sequential, hybrid and parallel mode of operation maintain perfectly connectedness and topology. Due to the presence of the more global and structural information about pattern (involved in the local thickness estimate) the enhanced preservation of geometric properties is observed.

1 Introduction

The design of thinning methods is governed mainly by the requirement to produce most efficiently a one-pixel-wide skeleton which maintains connectedness together with topology and preserves geometric properties of the original character. While the first two properties have been resolved by various approaches (see [6]), the last property still remains a difficult problem. The main reason is that more global and structural information is required to capture the essential geometry what is needed in order to prevent the excessive erosion together with avoiding the creation of spurious end points. Two surprising facts can be noticed when scanning the literature on thinning methods. First is the lack of neural network dedicated methods for thinning (we found only three of them from the overall number of more than 300). Second is the absence of any method which would use the local thickness estimate for the design of straightforward thinning strategy based on one-pixel-wide property of the skeleton :

- *delete the contour pixel if the local thickness estimate of the character is greater than one; otherwise, retain it.*

It is clear that every successful thinning algorithm involves this idea implicitly. It is also clear that the reliable estimate of the local thickness may be the task

---

* This work was partially supported by Slovak Grant Agency for Science grant No. 2/2040/95 and grant No. 95/5305/468

of the similar difficulty compared to the thinning problem itself. Hence, the usefulness of this idea will strongly depend on the design of an efficient and simple estimate of the local thickness and its ability to incorporate the relevant geometry of the character. However, the more thorough analysis can discover that the above simple idea may not always be sufficient to ensure the connectedness of the skeleton and preserve genuine end points. Those considerations lead us to the design strategy for thinning which would benefit from the complementarity and trade-off between some simple local thickness estimate procedure and incorporation of some simple test to preserve the connectedness (e.g. [9], or [4]) and genuine end points.

The peculiar feature of the resulting iterative approach is its similarity to the (recall phase) processing within perceptron-type neural networks ([8]). If we understand the pixel value as the output activity of the corresponding binary neuron, then the proposed thinning processing can be characterised by following principles:

- every neuron calculates its input as the sum of nonzero pixel values (until zero pixels on borders are reached) along four main directions (local thickness estimate procedure)
- every neuron calculates its threshold from the neighbourhood pixel values (connectedness test)
- every neuron possesses some proper transfer function which includes the thresholding operation to get the resulting output (pixel) value

It should be stressed from the beginning that the most distinctive feature of the neural network processing — the presence of neural weights and their adaptation — is not involved in our approach. Thus, the more precise characterization of the method could sound "the cellular automaton with the perceptron recall phase type of processing". In this respect the only similar approach is the analog and parallel thinning by means of a cellular neural network proposed in [7]. This iterative method is based on a set of differential equations in which so-called peeling and stopping templates play the crucial role. However, both the structure of the neuron input information and the dynamics of the thinning process are completely different compared to the proposed approach. For the sake of completeness, it is to be noted that within the framework of noniterative thinning methods, recently there appeared two papers based on neural network methodology. Approach proposed in [1] is based on the modification of a self-organizing feature map [5]. The second approach [3] is based on so called Dynamic cell structures [2].

The main motivation of the proposed approach is to investigate the usefulness of the direct use of the local thickness estimate for the thinning purpose and its ability to capture the geometric properties of the binary character. A comparison with the iterative thinning methods (either sequential or parallel) points to the following main differences:

- input information is represented by some proper simple estimate of the local thickness which is extracted from the flexible (both in size and shape) local

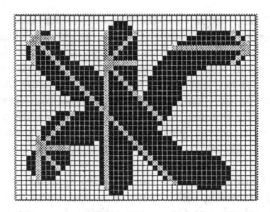

Fig. 1. Thickness estimates at various pixel locations.

neighbourhood compared to the information about pixel values from the fixed (both in size and shape) window (usually 3×3, or generally K×K)
– the decision function for deletion of a pixel is on neural-like basis compared to the "set of expert rules" or logical function on window pixel values

2 The Proposed Approach

The proposed approach belongs to the class of the contour following algorithms. The basic idea of the approach is simple and common to all three modes of operation. *Some proper thickness characteristic (derived from the primary features described below) is calculated for every neuron-pixel and the neuron threshold value is subtracted first. Finally, this neuron activity is thresholded i.e. the neuron output value below (above) some specified value will cause the deletion (retaining) of the pixel, i.e. setting the neuron output to 0 (1).* This holds for both sequential and hybrid algorithm. Parallel algorithm assumes a two-level decision process when particular transfer function can even yield 3 outcomes.

As it was already noted the proposed thinning approach resembles formally the recall phase of the perceptron neural network [8]. Output (pixel) value within proposed approach also uses hard limiter transfer function f()

$$y = f(T(\sum_{vertical} x_k, \sum_{horizontal} x_k, \sum_{diagonal1} x_k, \sum_{diagonal2} x_k) - Thr(x \in Neighbourhood)$$

Input to this function is represented by thickness characteristic T() which is based on magnitude ordering of directional thickness estimates and the threshold Thr() which is a nonlinear function of the neighbourhood pixel values of a current pixel.

Directional thickness estimates can be interpreted as the simple estimates of the local thickness for every contour neuron-pixel in the four basic directions: vertical, horizontal and two diagonals (see Fig.1). They are defined as a weightless sum of (nonzero) pixels along some particular i-th direction (until zero pixels on borders are reached):

$$l_i = l_i^+ + l_i^- = \sum_{\substack{\text{along } i-th \text{ direction}}}^{\text{until zero pixels reached}} x_k$$

It is important to note that the implicitly involved in our considerations chess-board is finally replaced by the Euclidean one. I.e. that both diagonal distances are always recalculated to reflect the Euclidean metrics i.e. $l_3 \to \sqrt{2}l_3, l_4 \to \sqrt{2}l_4$. Hence, two neighbouring pixels along vertical or horizontal have distance equal to 1, whereas two pixels along either diagonal have distance equal to $\sqrt{2}$. It is important to distinguish these two situations within all three proposed modes of operation. All three modes of operation require the knowledge about (ascending) magnitude ordering of the directional thickness estimates which will be denoted as $l^i, i = 1, 2, 3, 4$ (i.e. l^1 stands for the smallest one).

Another calculation common to the all three modes of operation concerns the neuron threshold value. The only motivation for introducing the neuron threshold is to ultimately prohibit the deletion of the pixel if it should result in breaking the connectivity of the pattern or deleting genuine end point. We assume 4-connectedness defined by crossing number introduced in [9]. Based on this crossing number we can design the following expression for the neuron threshold (corresponding to a particular contour pixel) which incorporates also the end point constraint:

$$Thr(p) = \exp\left(\sum_{j=1}^{8} k * (\delta(b) \mid x_{j+1} - x_{j-1} \mid -2)^2\right)$$

Function $\delta(b)$ with argument b of the number of black pixels yields value 1 when having only one black neighbour pixel otherwise gives a zero value. Hence the final value of the threshold assures the threshold value equal to 1 for connectivity and end point safe situations. It yields very large values (controlled by real parameter k) for such situations when deletion of the pixel would break the connectivity or make the undesirable erosion at the genuine end point.

3 Sequential Algorithm

Sequential character of the processing is reflected by the fact that both direc-tional thickness estimates l_i and the neuron threshold $Thr(p)$ are being updated after every contout neuron-pixel examination in predetermined order. Output of every contour neuron-pixel is given by the following transfer function:

$$y(p) = \begin{cases} 0 \text{ if } & l^3 - Thr(p) > 0 \\ 1 \text{ otherwise} \end{cases}$$

where l^3 stands for the third smallest directional thickness estimate and hence it represents the thickness characteristic for sequential mode of operation. This is motivated by following simple observations — heuristics:

(1) Pixels of the final skeleton have always first two smallest directional thickness estimates l^1, l^2 equal to 0 and the greatest one l^4 can have any value.

(2) Skeleton property of the pixel is encoded in the third smallest directional thickness estimate l^3 which can have value either 0 or 1 .

Note that for connectivity and end point safe situations threshold has the value

Fig. 2. Sequential mode of processing in successive iterations.

1 and the transfer function reduces to the simple intuitive decision rule presented in the Introduction.

4 Hybrid Algorithm

Directional thickness estimates are now determined in parallel for all contour neuron-pixels at the beginning of the iteration. The neuron threshold is updated sequentially. Output of every contour neuron-pixel is now given by:

$$y(p) = \begin{cases} 0 \text{ if } & \tilde{l}^1 - Thr(p) + 1 > 0 \\ 1 \text{ otherwise} \end{cases}$$

where $\tilde{l}^1$ stands for so called regularized smallest directional thickness estimate:

$$\tilde{l}^1 = \begin{cases} l^2 \text{ if } & \frac{l^2}{l^1} \geq 2 \\ l^1 \text{ otherwise} \end{cases}$$

The thickness characteristic is now given by the regularized smallest directional

Fig. 3. Hybrid mode of processing in successive iterations.

thickness estimate $\tilde{l}^1$. Introduction of the regularization is motivated by the existence of various corner pixels which may give the smallest directional thickness estimate l^1 equal to 0 and this could result in the spurious tails in the skeleton. Note also that the proper functioning within this mode of operation needs the neuron threshold to be shifted by one.

5 Parallel Algorithm

Parallel algorithm assumes that both the thickness characteristic and the neuron threshold are calculated in parallel for all contour neuron-pixels at the beginning of the iteration. The transfer function accomplishes now a two-level decision process, with 3 possible outcomes at the first level: retain the pixel, delete the pixel and go to the local decision process (for mid range values).

The crucial situations within parallel mode of processing occur when two-pixel-wide lines appear and one needs to be careful to not delete both opposite neighbour pixels and thus creating a hole, i.e. (undesirable) discontinuity in the pattern skeleton. Hence the basic idea behind our parallel algorithm is to proceed by the simple neural-like processing by means of the smallest directional thickness estimate l^1 until its magnitude reaches the value equal 1 and starting from this value to make careful comparisons with the opposite neighbour neuron-pixels whether to make a deletion or not. Hence values of $l^1 > 1$ result in the pixel deletion, values $l^1 \leq 1$ leave the final decision to the local decision process. This picture holds for connectivity and end point safe situations (i.e. threshold value is equal 1, whereas for "unsafe" situations threshold yields great values). Combining both above factors we can formulate following transfer function which applies for the output of every contour neuron-pixel:

$$y(p) = \begin{cases} 0 & \text{if} \quad l^1 - Thr(p) > 0 \\ Ld(p) & \text{if} \quad -1 \geq l^1 - Thr(p) \leq 0 \\ 1 & \text{if} \quad l^1 - Thr(p) < -1 \end{cases}$$

The smallest directional thickness estimate l^1 now represents the local thickness characteristic. Note that the proposed parallel algorithm has no explicit subiterations involved (i.e. single pass algorithm), although it is sometimes at the expense of the more sophisticated local decision computations.

Local Decision — Ld(p).

Local decision process applies in the final stages of the thinning when two-pixel-wide or one-pixel-wide lines occurs. In these situations decision about deletion of the pixel is not possible solely on the basis of the thickness estimate l^1 given by (from the point of view of) the current contour pixel. We have to take into account also some sort of thickness characteristic $ln^1(i), i \in N_c$ given by contour neuron-pixels from below defined "checking neighbourhood" N_c. The basic strategy of the local decision is simple. Compare the thickness characteristics $ln^1(i)$ given by all neuron-pixels from N_c with the thickness characteristic (i.e. l^1) of the current pixel p. If all $ln^1(i)$ are smaller than l^1 then delete the current pixel, and otherwise preserve it. This corresponds to the natural strategy of safe deletion of the pixel i.e. to delete the pixel with the greatest thickness estimate as a first.

Checking neighbourhood N_c of the current pixel neuron p consists of such black pixels along the vertical or horizontal which are in the direction with the thickness l^1 equal 1 (i.e. together with the current pixel they create two-pixel-wide lines). This neighbourhood also includes all white neighbour pixels which are assumed further to coincide with the current pixel p and have thickness characteristic $ln^1(i)$ equal to 0. Hence every neuron-pixel within N_c is specified by its own thickness characteristics which is denoted as $ln^1(i)$ (where i stands for the index within N_c) and by its position (which can play the important role in some special although very rare situations when information about thickness characteristics is not sufficient to decide about deletion in unique way).

Local fecision rule can be described by the following transfer function:

$$Ld(p) = \begin{cases} 0 \text{ if } \forall i \in N_c \text{ holds that } ln^1(i) + 0.5pos(i) < l^1 + 0.5pos(p) \\ 1 \text{ otherwise} \end{cases}$$

Fig. 4. Parallel mode of processing in successive iterations.

6 Experiments

All three modes of operation examine the pixels by contour following. Hence the whole process was divided into the set of iterations at the beginning of each the contour pixels were recognized and ordered in one dimensional array (to speed up the process). The sequence of the processing results in the course of iterations is illustrated for all three algorithms in Fig.2, 3 and 4, respectively. We performed testing of the proposed algorithms on the 45 Microsoft Windows alphabet character sets of various font sizes and some hand-written characters including hand-written signatures. We present for the illustration the selected patterns of brush script letters (see Fig.5). Based on obtained final skeletons we can summarize the following general observations:

(1) The final skeletons preserve perfectly the topology of original patterns.

(2) All observed skeletons were perfectly connected, i.e. no disconnected lines were observed. Note that the final skeleton is 4-connected what is the consequence of using corresponding crossing number for the threshold calculation.

(3) The final skeletons seem to be thinned to maximal degree.

(4) The sensitivity to the bigger perturbations along the outline of the pattern can lead in some cases to spurious tails in the final skeleton. However, in these cases it seems that distinguishing noise from significant contour protrusions or prominences can not be made solely by means of geometric information and supplementary a prior knowledge (e.g. level of the noise present in the pattern) is necessary. Within all three proposed algorithms there were described mechanisms involved, intended for dealing with smaller perturbations (i.e. preserving smaller significant contour protrusions or prominences). Except above mechanisms the enhanced ability to preserve relevant geometric properties of

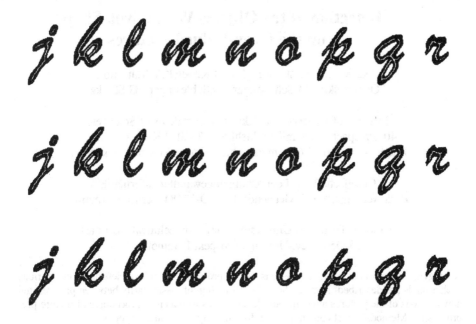

Fig. 5. Final skeletons of selected brush script characters. The three rows (from top to bottom) correspond to sequential, hybrid and parallel mode of processing, respectively.

the original pattern is primarily provided by the more structural and global nature of the local thickness estimates.

References

1. Ahmed, P.: A neural network based dedicated thinning method. Pattern Recognition Letters 16, (1995) 1267-1275.
2. Bruske, J., Sommer G.: Dynamic cell structure learns perfectly topology preserving map. Neural Computation 7, (1995) 845-865.
3. Farkaš, I., Chudý, L.: Application of a growing self-organising map to thinning of binary patterns with noise. Accepted in *Workshop on Self- organising Maps*, Helsinki, June 1997
4. Hilditch, C.J.: Linear skeletons from square cupboards. In: B.Meltzer and D.Michie, Eds., Machine Intell., New York: Amer. Elsevier, vol. 4, (1969) 403-420
5. Kohonen, T.: Self-organizing maps. Springer Verlag (1995).
6. Lam, L., Lee, S.W., Suen, C.Y.: Thinning Methodologies — A Comprehensive Survey. IEEE Trans. Pattern Anal. Mach. Intell. 14, (1992) 869-885.
7. Matsumoto, T., Chua, L.O., and Yokohama, T.: Image Thinning with a Cellular Neural Network. IEEE Trans. Circ. Sys. 37, (1990) 638-640.
8. Rosenblatt, F. The Perceptron: A probabilistic model for information storage and organization in the brain. *Psychological Review* 65, (1958) 386-408
9. Rutowitz, D.: Pattern recognition. J.Roy.Stat.Soc., 129, Series A, (1966) 504-530.

Detection of the Objects With Given Shape on the Grey-Valued Pictures

I.Aksak[1], Ch. Feist[2] ,V.Kijko[1], R.Knoefel[2],V.Matsello[1],
V.Oganovskij[3], M.Schlesinger[1], D.Schlesinger[3], G.Stanke[2]

[1]Institute of Cybernetics, Ukrainian Academy of Sciences,
40, Prospect Akademika Glushkova, 252022 Kiev, Ukraine,
Phone: (44) 266 25 69, email: schles%image.kiev.ua@ts.kiev.ua

[2]Gesellschaft zur Foerderung angewandter Informatik
Rudower Chaussee 5, Gebaeude 13.7, D-12484 Berlin, Germany

[3]National Technical University (Kiev Polytechnical Institute)
252056 Kiev, Ukraine, Prospekt Peremohy, 37

Abstract. The results of investigations of a problem of object detection on various backgrounds are described. New methods for estimation of dissimilarity between pictures and new methods of the pictures matching are developed as well as new procedures of picture pre-processing. Methods of fast computation of dissimilarity criterion are proposed.

1. Introduction

The detection is formulated as a special case of classical best-matching problem. Let e be a picture of some ideal object. The picture depends on its geometrical parameters: scale μ, displacements ξ and η and rotation φ , so that $e = e(\xi, \eta, \mu, \varphi)$. Let ρ be the picture under analysis. Let $R(\rho, e)$ be a function of dissimilarity that shows how well the real picture ρ can be matched with the ideal picture e . The problem consists in finding such displacements ξ^*, η^* , scale μ^* and rotation φ^* of the ideal picture, that provide the best matching of real picture with ideal one in accordance with R . The task consists in calculation of

$$(\xi^*, \eta^*, \mu^*, \varphi^*) = \arg \min_{\xi, \eta, \mu, \varphi} R(\rho, e(\xi, \eta, \mu, \varphi)) \qquad (1)$$

The problem of reasonable dissimilarity function has already a long history (see, for example [1] and referred investigations). In this work the problem is considered in connection with thresholding and noise removing on the thresholded picture.

It is well-known that the threshold for picture binarization can be rarely pre-defined so, that a single value is suitable for every picture. Because of that a lot of adaptive thresholding procedures were developed [2,3]. The new approach, presented here, consists in that a threshold depends on both pictures, whose similarity is to be calculated. It is also necessary to remove some noise in the binary pictures before their similarity is calculated. The most popular methods of such noise removal are based on the detecting of connected components or on the ideas of dilation-erosion [4,5]. This operations must be fulfilled after binarization. As threshold's value cannot be pre-defined, such noise removal must be done for several values of threshold and computational complexity becomes large enough. We propose such procedure that the "noise removing" with subsequent thresholding gives the same result as a thresholding with subsequent commonly used noise removing in the binary picture. This equivalency holds for every threshold's value.

The dissimilarity function is described in section 2. In section 3 the main ideas of the solution of optimisation task are described and then discussed in sections 4,5,6.

2. The Function of Dissimilarity of Ideal and Real Pictures

Let T be a set of pixels (field of vision) and t be a pixel. An ideal picture e is meant as a pair of two subsets T_O - a set of pixels which belong to the object, and T_B - a set of background pixels. A real picture ρ is a function $\varphi: T \to \{0,1,...,255\}$. The dissimilarity function is defined on the base of some assumptions about pictures properties using the following considerations.

2.1. The first step is based on the assumption, that a background is, as a rule, darker than an object. In this case for every two pixels t' and t'', such that the first one belongs to the background and the second - to the object, the inequality

$$\forall t' \in T_B \forall t'' \in T_O \left(\rho(t') < \rho(t'') \right) . \tag{2}$$

would be valid This event could be surely interpreted that the object $e = (T_B, T_O)$ is present on the picture. While the set T is finite, the event (2) is equivalent to

$$\exists \theta \left(\forall t' \in T_B \left(\rho(t') < \theta \right) \right) \& \left(\forall t' \in T_O \left(\rho(t') > \theta \right) \right) . \tag{3}$$

Event (3) means, that there exists such threshold θ, that thresholding of the real picture ρ with θ transforms it into the ideal picture e. It is clear that, firstly, the event (3) gives a good reason for the decision that the object e is present on the picture and, secondly, that this event occurs rather rarely. To weaken this criterion it is assumed that one can neglect the small noisy distortions of the picture.

2.2. Let $B(\rho, \theta)$ be the result of thresholding of picture ρ with θ. Then the event (3) may be represented as $\exists \theta \left(B(\rho, \theta) = e \right)$ (4)

Let us define noise removal operation. Let ρ be a binary picture, T_0 be the set of pixels for which $\rho(t) = 0$ and T_1 be the set of pixels with $\rho(t) = 1$. The connected subsets of small size from T_0 and T_1 are considered as noise. The removing of noise consists in inverting of its brightness. The result of noise removing will be denoted as $S(\varphi)$. Matching criterion: $\exists \theta \left(S(B(\rho, \theta)) = e \right)$ (5)

is more weak, than (3). This criterion means that there exists such threshold θ that after thresholding of the picture ρ and subsequent noise removing the binary picture is obtained, that is equal to the picture e. One can see, that if some picture satisfies (4) then it satisfies (5). Of course, opposite statement is not valid and, consequently, criterion (4) is stronger than (5). At the first sight the computational complexity of (5) is much greater than complexity of (4). Really, for testing of (5) the real picture must be thresholded with every possible value of threshold and then noisy spots must be removed on every obtained binary picture. As for criterion (4), it requires only the testing of simple inequality $\min\limits_{t\in T_O} \rho(t) > \max\limits_{t\in T_B} \rho(t)$. Nevertheless testing of (5) can be essentially accelerated due to the following statement. There exists such transformation S' , that for every ρ and θ

$$B\big(S'(\rho),\theta\big) = S\big(B(\rho,\theta)\big) \qquad (6)$$

is valid. By virtue, (5) can be represented as $\quad \exists\theta\ \big(B(S'(\rho),\theta) = e\big) \qquad (7)$

Testing of (7) requires one-fold transformation of picture ρ into the picture $\rho_1 = S'(\rho)$ and subsequent testing of $\min\limits_{t\in T_O} \rho_1(t) > \max\limits_{t\in T_B} \rho_1(t) \qquad (8)$

The concrete transformation S' that satisfies (6) is described in section 4.

2.3. The next step is based on the assumption that thin lines ("hairs") can also be considered as a noise. Let t be a pixel for which $\rho(t) = 1$ and there exists such pixel t' ,($\rho(t') = 0$), that the distance between t and t' is less then some predefined value. Such pixel t will be referred to as a black border pixel. The dilation consists in inverting of black border pixels. Similarly the erosion is defined. Removing of black lines, whose thickness is not greater than ε pixels, is defined as a sequence of ε erosions and ε dilations. Removing of thin white lines is defined as a sequence of ε dilations and ε erosions.

Let us denote by T the removing of thin distortions and weaken (5) with an account that such distortions can be neglected. As a result the following criterion

$$\exists\theta\ \big(T\big(S\big(B(\rho,\theta)\big)\big) = e\big) \qquad (9)$$

is obtained. For T the same statement is valid as the statement (6). Namely, there exists such transformation T' , that $\quad T\big(B(\rho,\theta)\big) = B\big(T'(\rho,\theta)\big) \qquad (10)$

holds for every threshold θ and every picture ρ . Due to (10) the criterion (9) can be represented as $\quad \exists\theta\ \big(B\big(T'\big(S'(\rho)\big),\theta\big) = e\big) \qquad (11)$

Using (11) one can test the criterion computationally much more effectively. The picture transformation T' , that satisfies (10), is described in the section 5.

Binary criterion $\quad \exists\theta\ \big(B(\rho,\theta) = e\big)$, $\qquad (12)$

can be replaced by the integer-valued function that represents a natural fuzzy variant of (12) and the dissimilarity function $R(\rho,e) = \min\limits_{\theta} d\big(B(\rho,\theta),e\big)$, $\qquad (13)$

where d is a distance by Hemming between two binary pictures. It is evident that if $R(\rho,e) = 0$ then (12) holds. The value $R(\rho,e) \neq 0$ is considered as a degree of dismatching of ρ and e .

As a result, the following formulation of the problem of detection of pre-defined ideal picture on the real picture is obtained. It is necessary to remove some distortions on the real picture ρ . Then transformed picture ρ_1 must be obtained and for real picture such threshold θ^* , such displacements ξ^* and η^* , scale μ^* and rotation φ^* of ideal picture must be found, that the distance by Hemming between thresholded picture $B(\rho_1,\theta^*)$ and transformed ideal picture $e(\xi^*,\eta^*,\mu^*,\varphi^*)$ would be minimal. It means, that it is necessary to calculate

$$(\xi^*,\eta^*,\mu^*,\varphi^*)=\arg\min_{\xi,\eta,\mu,\varphi}\min_{\theta} d\big(B(\rho_1,\theta),e(\xi,\eta,\mu,\varphi)\big) \qquad (14)$$

3. Methods for Acceleration of the Matching

3.1. Let us analyse the innermost part of (14). Number of operations for calculation the distance d under fixed values of $\theta,\xi,\eta,\mu,\varphi$, is of order n - number of pixels in the field of vision. To calculate $\min_{q} d\big(B(r_1,q),e(x,h,m,j)\big)$ the distance d must be calculated for each threshold value. The amount of this values will be denoted by ω. However the total amount of necessary operations is not of order $\omega\cdot n$, as one can suppose at the first glance, but is of order $\omega+n$.

3.2. Computation of (14) requires that the value $\min_{\theta} d\big(B(\rho_1,\theta),e(\xi,\eta,\mu,\varphi)\big)$ must be obtained for every four-tuple ξ,η,μ,φ. Let g be the amount of these values. The total amount of computation is not of order $g\cdot(\omega+n)$, as it could be suspected, but is of essentially less order $g\cdot(\omega+n')$, where n' is the number of pixels of ideal picture at the border of subsets T_B and T_O.

4. Noise Removing in the Grey-Valued Picture

The algorithm for noise removing consists of four steps: 1) sorting of pixels in increasing order of their brightness; 2) "white" noise removing; 3) sorting of pixels in decreasing order of their brightness; 4) "black" noise removing.

Ordering of pixels is fulfilled during two-fold scanning of the pixels using additional array with the length equalled to the number of different brightness.

"White" noise removing is carried out during one-fold scanning of ordered pixels. Let t_i be a pixel having brightness $v(t_i)$. Processing of the pixel consists in checking whether there exists "white" noise region G that contains this pixel. Initially the region contains only t_i. Then the region grows so that. some pixel t' is included into it if: a). t' is a neighbour of the region G; b). $v(t')\geq v(t_i)$. The region grows until one of the following two conditions are satisfied: 1) the size of the obtained region G exceeds the pre-defined size. In this case the region G is not considered as a noise and the next pixel t_{i+1} is processed; 2) no pixel can be included into G. In this case G is a noise and the brightness of all its pixels are equalled to maximal brightness of the neighbouring pixels.Removing of "black" noise is similar to "white" noise removing.

5. Removing of Thin Distortions on the Multilevel Pictures

Let $\rho\,|$ be a multilevel picture, $B(\rho,\theta)$ be a result of thresholding of ρ using θ, $T(B(\rho,\theta))$ be a result of removing of "hairs" on B. It is necessary to construct such transformation T' that the equality $\quad B(T'(\rho),\theta)=T(B(\rho,\theta)) \qquad (15)$

is valid for every multilevel picture ρ and every threshold θ. For the removal of white "hairs" the pair "erosion-dilation" must be fulfilled. So it is necessary to find

out such transformation T' that $$B\big(T'(\rho),\theta\big)= D\Big(E\big(E\big(D\big(B(\rho,\theta)\big)\big)\big)\Big) \qquad (16)$$

is valid for every ρ and every θ, where D and E mean dilation and erosion. Let us consider only D and find the transformation TD, that satisfy the equality
$$B\big(TD(\rho),\theta\big)= D\big(B(\rho,\theta)\big) \qquad (17)$$

Let $T(t)$ be the d-neighbourhood of pixel t, ρ max be the maximal brightness of pixel from $T(t)$. Let us define the picture $TD(\rho)$ such that the brightness of t is ρ max. Then $B\big(TD(\rho),\theta\big)$ is equal 1 if and only if $\theta \le \rho$ max. $\qquad (18)$

The brightness of picture $D\big(B(\rho,\theta)\big)$ is equal 1 if and only if there exists a pixel t' in $T(t)$ for which $B(\rho,\theta)$ is equal 1. It is equivalent to the condition that there exists a pixel t' for which $\rho(t') \ge \theta$ and in its turn is equivalent to
$$\rho\ max \ge \theta. \qquad (19)$$

One can see, that the condition under which brightness $B\big(TD(\rho),\theta\big)$ of the pixel t equals 1 is the same as the condition, under which brightness $D\big(B(\rho,\theta)\big)$ is 1. Consequently, these brightness are always equal. So, transformation TD satisfies (17). Similarly, one can construct a transformation TE such that:
$$B\big(TE(\rho),\theta\big)= E\big(B(\rho,\theta)\big). \qquad (20)$$
This transformation assigns for a pixel t the brightness ρ min -minimal brightness of picture ρ from $T(t)$. The properties (17) and (20) cause that the equality

$$B\Big(TD\big(TE\big(TE\big(TD(\rho)\big)\big)\big),\theta\Big) = D\Big(E\big(E\big(D\big(B(\rho,\theta)\big)\big)\big)\Big)$$ holds also for every θ. The last expression defines explicitly the operator T' that satisfies (16).

6. Fast Matching of Grey-Valued Pictures
6.1. Statement of the Problem

Let $\rho:T \to \{0,1,\dots 255\}$ be a picture under analysis; $\rho(t)$ be a brightness of pixel t; Θ be a threshold; $B(\rho,\Theta)$ be a result of thresholding ρ with Θ; $e =(T_o,T_b)$ be an ideal pattern, T_o be an object, T_b be a background, $T_o \wedge T_b = \varnothing$; ξ,η,μ,φ be values of horizontal and vertical translations, scale and rotation respectively (this four-tuple will be referred to as a variant); $e(\xi,\eta,\mu,\varphi)$ be the ideal pattern under translation ξ and η, scale μ and rotation φ; $d(v,e),e =(T_o,T_b)$ be a dissimilarity function between v and e. Its value is defined as a total quantity of pixels for which

$\left(\left(v(t)=1\right)\wedge\left(t\in T_b\right)\right)\vee\left(\left(v(t)=0\right)\wedge\left(t\in T_o\right)\right)$. The problem is formulated as the optimisation task: $\left(\xi^*,\eta^*,\mu^*,\varphi^*\right)=\arg\min\limits_{\xi,\eta,\mu,\varphi}\min\limits_{\Theta}d\left(B(\rho,\Theta),e(\xi,\eta,\mu,\varphi)\right)$.

6.2. The Problem Analysis and Elimination of Calculations

6.2.1. Let us consider dissimilarity minimisation under fixed variant, i.e. calculation of $F(\rho,\xi,\eta,\mu,\varphi)=\min\limits_{\Theta}d\left(B(\rho,\Theta),e(\xi,\eta,\mu,\varphi)\right)$. It is fulfilled by three steps.

a. Two histograms must be calculated: $G_o(\Theta)$ (respectively $G_b(\Theta)$) being an amount of pixels, for which $\left(t\in T_o\right)\wedge\left(\rho(t)=\Theta\right)$ (respectively $\left(t\in T_b\right)\wedge\left(\rho(t)=\Theta\right)$).

b. Then the cumulative histograms $G'_o(\Theta)=\sum\limits_{i=0}^{\Theta}G_o(i),0\leq\Theta<\Theta_{max}$ and

$G'_b(\Theta)=\sum\limits_{\Theta+1}^{\Theta_{max}}G_b(i),0\leq\Theta<\Theta_{max}$ must be calculated using recurrent expressions:

$G'_o(0)=G_o(0),G'_o(\Theta)=G'_o(\Theta-1)+G_o(\Theta),\Theta=1,2,\ldots,\Theta_{max}-1$;

$G'_b(\Theta_{max}-1)=G_b(\Theta_{max}),G'_b(\Theta)=G'_b(\Theta+1)+G_b(\Theta+1),\Theta=\Theta_{max}-2,\ldots,1,0$

$G'_o(\Theta)$ means the amount of pixels from subset T_o, which would be black after thresholding of picture ρ with Θ. The function $G'_b(\Theta)$ has similar meaning.

c. $F(\rho,\xi,\eta,\mu,\varphi)$ is being calculated. This value is equal $\min\limits_{\Theta}\left(G'_o(\Theta)+G'_b(\Theta)\right)$.

The amount of operations that are necessary to obtain functions $G_o(\Theta),G_b(\Theta)$, is of order n - the amount of pixels in $T_o\cup T_b$. All other calculations are of order Θ_{max} - the amount of brightness values.

6.2.2. Let us consider now dissimilarity minimisation task under fixed scale and rotation, i.e. calculation of $F'(\rho,\mu,\varphi)=\min\limits_{\xi,\eta}F(\rho,\xi,\eta,\mu,\varphi)$, which is necessary to

obtain the values of $F(\rho,\xi,\eta,\mu,\varphi)$ for every ξ and η. It requires to fulfil steps *a*, *b* and *c*. For the first value of ξ and η step *a* is fulfilled by the following algorithm.

For every Θ $G_o(\Theta)=G_b(\Theta)=0$;For every $t\in T_o$ $G_o\left(\rho(t)\right)=G_o\left(\rho(t)\right)+1$;

For every $t\in T_b$ $G_b\left(\rho(t)\right)=G_b\left(\rho(t)\right)+1$;

Calculation time is proportional to the amount of pixels in $T_o\cup T_b$. However, the processing time for all other values of ξ and η can be reduced essentially. Really, let ξ_0 and η_0, ξ_1 and η_1 be two different translations, $e_0=\left(T_{o0},T_{b0}\right)$ and $e_1=\left(T_{o1},T_{b1}\right)$ be corresponding ideal patterns. Let us suppose that histograms $G_{o0}(\Theta)$ and $G_{b0}(\Theta)$ for the ideal pattern e_0 are already available. Then histograms $G_{o1}(\Theta)$ and $G_{b1}(\Theta)$ must be constructed by recalculation of histograms G_{o0} and G_{b0} using the following algorithm.

a. The four-tuple of difference sets
$T_{o01} = T_{o1} \setminus T_{o0}, T_{o10} = T_{o0} \setminus T_{o1}, T_{b01} = T_{b1} \setminus T_{b0}, T_{b10} = T_{b0} \setminus T_{b1}$ must be constructed.

b. $\forall \Theta: \{ G_{o1}(\Theta) = G_{o0}(\Theta); G_{b1}(\Theta) = G_{b0}(\Theta); \}$

$\forall t \in T_{o01} \quad G_{o1}(\rho(t)) = G_{o1}(\rho(t)) + 1; \quad \forall t \in T_{o10} \quad G_{o1}(\rho(t)) = G_{o1}(\rho(t)) - 1;$

$\forall t \in T_{b01} \quad G_{b1}(\rho(t)) = G_{b1}(\rho(t)) + 1; \quad \forall t \in T_{b10} \quad G_{b1}(\rho(t)) = G_{b1}(\rho(t)) - 1.$

As the lists of pixels from the difference sets can be constructed in advance, the time of computations on the step *b* may be proportional to the volume of difference sets. When the difference between the ideal patterns e_0 and e_1 is small, the volume of difference sets is considerably less than the volume of $T_o \cup T_b$. The computation of step *a* may be accelerated. Let the ideal pattern e_1 is obtained by the translation $(\Delta i, \Delta j)$ of the ideal pattern e_0. Let the ideal pattern e_2 is obtained by the same translation of the ideal pattern e_1. It is obvious that the difference sets for e_2 and e_1 can be obtained by the translation $(\Delta i, \Delta j)$ of the difference sets for e_1 and e_0. The complexity of such recalculation is also proportional to the volume of difference sets. Due to this the construction of difference sets is made only once for the first position of ideal pattern. For every other position the difference sets are only recalculated.

7. Application

The developed best-matching method was implemented for searching of geometrical objects in the pictures of shoeprints with the aim of subsequent search of the given shoeprint in the database. Two examples of detecting some geometrical objects are shown in Fig. 1.

Acknowledgement

This research was sponsored the Ministry of Education and Investigations of Germany which is gratefully acknowledged.

References

1. V.Starovoitov, Towards a Measure of Diversity between Grey-Scale Images. In *Comp. anal. of images and patterns*: 6th int. conf; proc./ CAIP'95, Prague, 214-221, 1995.
2. W.Tao, H.Burkhardt, An Effective Image Thresholding Method Using a Fuzzy Compactness Measure. In *Proc. of 12th IAPR Int. conf. on pattern recogn.*, 1, 47-51, 1994.
3. P.K.Sahoo, S.Soltani, A.K.C.Wong, Y.C.Chen, A Survey of Thresholding Techniques. In *Comput. Vision Graphics Image Process.*, 41, 233-260, 1988.
4. Haralick R. M., S. R. Sternberg, and X. Zhuang, 1987. Image analysis using mathematical morphology, *IEEE Trans. on Pattern Anal. and Machine Intell.* 9(4), 532-550.
5. Serra, J. In *Image Analysis and Mathematical Morphology*, Acad.Press, NY, 1982.

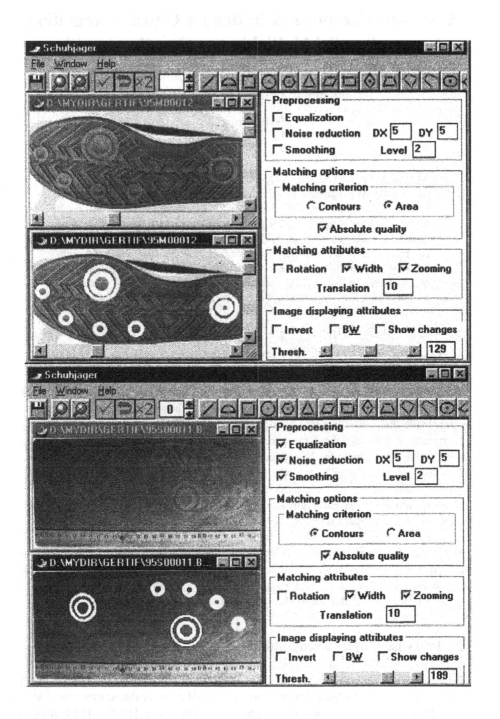

Fig. 1. Detecting of circles on shoeprint images.

Automatic Parameter Selection for Object Recognition Using a Parallel Multiobjective Genetic Algorithm

Frank Aherne, Peter Rockett and Neil Thacker

Dept. EEE, Sheffield University, Sheffield S1 3JD, UK

Abstract. This paper describes the application of a Multiobjective Genetic Algorithm (MOGA) to optimise the selection of parameters for an object recognition scheme: the Pairwise Geometric Histogram (PGH) paradigm. The overall result of the algorithm is to select PGH parameters giving a more compact, efficient histogram representation. The MOGA applied uses Pareto-ranking as a means of comparing individuals within a population over multiple objectives. The significance of this work is that it enables the process of pairwise object recognition to be fully automated so that it can reach the full potential for use on large databases. In future work these 'optimal' histograms will be incorporated into the recognition process.

1 Introduction

Although image features can be characterised to some extent by intrinsic attributes such as local image gradients and curvatures the context of the surrounding shape geometry can provide the basis for a much more powerful descriptor. By selecting appropriate parameters and storing these measurements in the form of a frequency histogram, a concise shape descriptor can be produced which promotes robust feature classification. This histogram is referred to as a *pairwise geometric histogram* because it records geometric measures made between pairs of image features [1], [2].

Built into the PGH paradigm is a set of parameters defining the histogram type, accuracy and quantization of the axes of the histograms. Presently some of these parameters are decided using rules of thumb and trial-and-error and are *constant* for all PGHs used. This paper describes the application of a MOGA to determine these parameters in a principled manner producing 'optimal' PGHs with variable parameters. The significance of this work is that it enables the process of pairwise object recognition to be automatically extended for use on large generic model databases.

2 Pairwise Object Recognition

Pairwise geometric histograms (PGH) are a representation used for the recognition of rigid shape. They have been shown to be a robust solution for the recognition of arbitrary 2D shape in the presence of occlusion and scene clutter [1], [2]. The method is both statistically founded and complete in the sense that a shape may be reconstructed from its PGH representation [3]. The complete algorithm comprises a number of stages:

- Model image data (during training) and scene image data (during recognition) are processed by an edge extraction algorithm and the edge data approximated by line segments (edgels could also be used, and this is mainly done for speed).
- Model histograms (during training) and scene histograms (during recognition) are constructed for *each* line segment (reference line) by comparing the reference line to all other lines (within a circular window of radius d_{max}) and making entries into a histogram according to the measured relative angles and perpendicular distances (see Figure 1). To account for errors in the measurement processes and to encode the variability in the way a shape may be segmented into lines, entries are blurred appropriately when being placed into the PGH. This representation encodes local shape geometry in a manner which is invariant to rotation and translation and is robust to missing data, line fragmentation and clutter.
- Each line in a given scene is treated as a reference line and an associated PGH is constructed, these scene PGHs are then matched to the model database PGHs using the Bhattacharyya metric [7].
- Object classifications are confirmed by finding consistent labelling within a scene using a probabilistic Hough Transform.

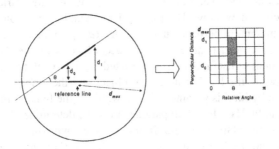

Fig. 1. PGH entry for a Single Line Comparison

2.1 Pairwise Parameters to be Optimised

The optimisation of the representation process can be viewed as a way of matching the stored model data to the difficulty of the recognition task. Thus we can expect to have to construct measures which quantify: repeatability, discriminability and compactness. As each of these cannot be measured using a single unified statistic we can expect to have to use optimisation procedures which operate on non-commensurate objectives. We will now define the motivation and construction of our measures.

Prior to constructing geometric histograms it is necessary to decide on the histogram scale to be used. The choice of maximum perpendicular distance d_{max}, is driven by two conflicting requirements. On one hand d_{max} should be small enough that the PGH represents *local* shape and is robust to missing data and occlusion. On the other hand d_{max} should be large enough so that shape information in each PGH is distinct. Prior to the present work, rule-of-thumb used was to ensure that most of a shape was encoded into each histogram.

The simplest type of histogram is constructed by restricting angles to the range 0 to π and distances to 0 to d_{max}. This histogram is invariant to reflections of the shape data about the reference line and is described as *mirror symmetric*. Mirror reflection invariance is not always desirable and can be removed by using the direction of angles (clockwise or anti-clockwise) to extend the range of angle measurements to $-\pi$ to π. This doubles the number of entries in a histogram and thus the computation needed for histogram matching, but also increases the sparseness so improving robustness of matching in cluttered scenes. Depending on whether the point of intersection between the two line fragments (Figure 1) lies to the right or left of the reference line d_{max} can be signed, thus again doubling the area of the histogram and extending the distance range to $-d_{max}$ to d_{max}.

Typically, there are three *types* of PGHs used: $\{[0, d_{max}], [0, \pi]\}$, $\{[0, d_{max}], [-\pi, \pi]\}$ and $\{([-d_{max}, d_{max}], [-\pi, \pi]\}$, known as 'mirrored', 'rotated' and 'directed' histograms respectively, which encode different degrees of geometric invariance. The local circular window used to make the PGH entries adds robustness to the PGH construction process and is centred around the mid-point of the reference line (Figure 1). The associated histogram is constructed by making entries for each line truncated to the circular window. The radius of this window is a continuous variable and equal to the maximum perpendicular distance (d_{max}) value of the associated histogram. The d_{max} parameter is normally set in accordance with the object size so that the maximum amount of useful data is constructed during matching. Its determination is a trade-off between the local scale of the object and the amount of clutter in the window. Associated with this parameter is the width of the bins on the perpendicular distance axis of the PGHs. The width parameter can be determined from the accuracy of the polygonal segmentation process. In this work we will simultaneously optimise the value of d_{max} and the histogram *type* but the bin width remains at its theoretical value [3].

In order to construct a large recognition system automatically we will require an automated procedure for selecting the parametric representation of a line in terms of the free parameters specifying a PGH - the very nature of this problem requires us to take multiple competing objectives into account.

3 Multiobjective Genetic Algorithms

Several methods for adapting GAs to cope with the simultaneous optimisation of a problem over a number of dimensions have been proposed, including the use of Pareto-ranking, [6]. The MOGA applied in this work uses Pareto-ranking as a means of comparing solutions across multiple objectives. The Pareto-optimal set of a multiobjective optimisation problem consists of all those vectors for which their components cannot be *all* simultaneously improved without having a detrimental effect on at least one of the remaining components. This is known as the concept of Pareto optimality, and the solution set is known as the *Pareto-optimal* set (or *non-dominated* set). GAs have been recognised to be well-suited to multiobjective optimisation as described in [6].

```
for(run=0;run<max_runs;run++)
{
        Initpop();
        for(gen=0;gen<max_gen;gen++)
        {
                Crossover();
                Mutation();
                Pareto();
                Fitness();
                Grouping();
                Selection();
        }
}
```

Fig. 2. Pseudo Code for MOGA

The advantages of GAs have already been used by computer vision researchers, examples of which are described in [4] and [5]. GAs are particularly applicable to the problem addressed in this paper, as our search space is noisy, discontinuous, impractical to search exhaustively and contains no derivative information.

3.1 The Objective Functions

To explain the selection of the objective functions for the MOGA, it is useful to recall the motivation for the application: for a given scene or model line we have a choice of the parameters for the associated PGH. Our algorithm will optimise these parameters producing optimal sets of lines (and their associated PGHs) relative to all others in the database. In what follows B denotes the Bhattacharyya match score between two histograms and is defined as:

$$B = \sum_{x=0}^{n_\theta} \sum_{y=0}^{n_d} \sqrt{H_1(x,y)} \sqrt{H_2(x,y)}$$

where H_1 and H_2 are two PGHs of the *same type*, n_θ the number of bins on the θ axis and n_d the number of bins on the perpendicular distance axis. B is used as a measure of similarity between two histograms - two identical histograms yield $B = 1$ (assuming the histograms are normalised) and two completely dissimilar histograms give $B = 0$.

Suppose we have a set of PGHs, we require a principled technique for determining the relative distinctiveness of each of the histograms in the set. For a given PGH belonging to such a set, this can be achieved by calculating the match scores between our given PGH and all others in the set. We seek to maximise the distance between the identical match $B = 1$ and the *mean* of all the other matches, as the greater this quantity, the more distinct our given PGH will tend to be relative to the rest of the set, hence our first objective, f_1, is:

$$f_1 = \max(1 - B_{mean}) \tag{1}$$

In addition, we want a measure that reflects histogram consistency (or line labelling) across different examples of a given object subject to variable segmentation or fragmentation. We will consider different sets of data, *variant sets*, defined as follows:

– Sets of objects consisting of a fixed known object together with variants, (for example, objects with different degrees of fragmentation).

So for a given histogram this measure will tell us *which* of the variant sets that particular PGH was most likely to have come from. We can achieve this by calculating the matches between our given histogram and all PGHs in each variant set. The best match between our given PGH and all the PGHs for *each* object in *each* variant set is stored (the inner summations of equation 2). The variant set with the highest sum is the winner and the given histogram is deemed to have come from the winning variant set. This avoids the difficult problem of labelling corresponding lines in the database. Thus:

$$f_2 = max\left\{\sum_{i=1}^{n_1} max(B_i), ..., \sum_{i=1}^{n_N} max(B_i)\right\} \tag{2}$$

(where n_j is the number of objects in variant set j and N is the number of variant sets). f_2 is potentially computationally intensive, as it involves computing the matches between all histograms in the variant sets for *every* member of our GA population. Here the ability of the GA to optimise stochastically is of value.

Finally we desire as compact a representation as possible and therefore want to minimise the area of the histograms. For this work the bin width of the histograms is fixed and the only variable parameters for area are the perpendicular distance, d_{max}, and the histogram type. So the third objective function minimises the histogram area. A scale factor of 1, 2 and 4 for mirrored, rotated and directed histograms respectively was used reflecting the associated area. Hence we have:

$$f_3 = min(\text{histogram area}) \tag{3}$$

4 The Experiments

In order to demonstrate the automatic selection of histograms and parameters for inclusion in the object recognition database, we have specifically constructed sets of objects which include representational issues that need to be automatically resolved. For example, when selecting representative histograms, we would like the solution to include examples constructed from noise free data, even though more noisy examples were present. Also, we would wish the representation chosen for each histogram to be driven according to the degree of invariance characteristics in the variant sets provided. Finally, we would want the selected histograms to be based on the most complete version of an object. An automatic object recognition system with these characteristics would then be capable of learning to recognise examples of objects in cluttered scenes (as humans can).

The algorithm was tested on two groups of data containing 1207 lines and 2699 lines respectively. The first group of data consisted of the four data sets shown in Figure 3, (from top-left going clockwise: pterodactyl, stegosaurus, triceratops and dimetrodon).

Fig. 3. Dinosaur Data Sets

The pterodactyl set of Figure 3 consists of variants having the original line length shortened or lengthened by a different factor. The second set of Figure 3 consists of six objects the first is the original stegosaurus, the following two are the first with increasing levels of Gaussian noise added. The last three stegosaurus objects are the first three mirrored around the midpoint of the objects length. The third data set consists of six dimetrodons having increasing levels of Gaussian noise added to the lines. Finally the triceratops set contains objects all having different segmentation accuracy, giving various degrees of a coarse polygonisation effect.

Fig. 4. Duplo Data Sets: Person, Cow and Horse

The second group of test objects consisted of a duplo person, cow and horse and are shown in Figure 4. The persons of Figure 4 were produced by shortening and lengthening the lines of the original duplo person (the first object). The second cow of Figure 4 is the first cow with gaussian noise added and the third and fourth cows are the first two mirrored around the midpoint of the objects length. Finally the horse set consists of horses having different accuracies of segmentation.

5 Results

The MOGA applied to the dinosaur data of Figure 3 consisted of 10 sets of 58 individuals and was run for 100 generations. Figure 5(a) shows a distribution of match scores for an initial and final population of PGHs. These data were generated by calculating the match scores between all histograms having a similar type and then binning the results. The data distribution for the initial population shown in Figure 5(a) has mean = 0.415 and variance = 0.013. Similarly, the rightmost histogram shown in Figure 5(a) shows a typical distribution of match scores for a final population, this distribution has mean = 0.275 and variance = 0.015. The first plot of Figure 6 shows how for each of

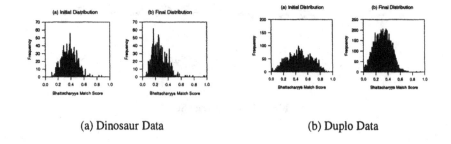

(a) Dinosaur Data   (b) Duplo Data

Fig. 5. Example Initial and Final Distributions

the 10 populations, the histogram representation has become more distinct as is evident by the reduced means of the final populations.

The MOGA applied to the data of Figure 4 consisted of 10 populations containing 107 lines. An example initial and final histogram is shown in Figure 5(b). As for the dinosaur data, the second plot of Figure 6 shows how for each of the 10 populations the histogram representation has become more distinct by the reduced means and greater compactness of the final populations.

By comparing the means of the initial and final populations shown in Figure 6 we can see that the effect of the MOGA is to produce histograms that have, on average, a lower match score across their associated populations, confirmed by the lower means of the final populations. This is due to the first objective function optimising on histogram distinctiveness, and thus the MOGA has produced sets of PGHs containing less redundant information than the initial populations. It can also be seen by comparing the initial and final population distributions of Figure 5(a) and Figure 5(b), that the MOGA has reduced the number of PGHs having high match scores, another indication of how the amount of redundancy in the representation has been reduced and the distinctiveness increased. Thus the three separate objective criteria are jointly optimised as required. Also, an examination of the histograms (line fragments) selected for inclusion in the database shows that:

- histograms are selected from those with the lowest quantity of noise, (as these are the most reliable and on average give the best match scores).
- histograms are constructed from the most complete versions of objects within the variant sets (as these are the most discriminable).
- histograms are selected with the requisite degree of invariance (as these are the representations which have the best recognition repeatability). For example, using mirror symmetric PGHs for objects with mirror symmetry.

Thus the database optimisation process operates as required mimicking our own rule-of-thumb criteria and also performing noise filtering.

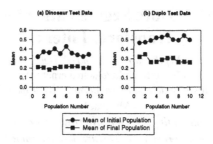

Fig. 6. Plot Showing Population Means For Test Data

6 Discussion and Conclusions

In this work we have applied a Pareto-based MOGA to parameter selection for the PGH object recognition paradigm. This is not a trivial problem as the set of optimal histograms is defined recursively by all others in the database. By carefully selecting our data sets we have shown that the optimisation technique works irrespective of many common problems such as noise, missing data and segmentation. The effect of applying the MOGA is the optimisation of a set of pairwise parameters which until now could only be specified by the user. We are therefore making the construction of PGHs a fully automatic, optimised process. The algorithm produces distinct, efficient histogram definitions for subsets of lines in our database. Potentially it could be used to learn to recognise objects in an environment in which classification feedback is only available for a cluttered (un-segmented) scene.

References

1. Evans, A., Thacker, N. & Mayhew, J., Pairwise Representation of Shape, Proc. 11th Int. Conf. Pattern Recognition (ICPR92), The Hague, Netherlands, (1992) vol 1, pp 133- 136
2. Ashbrook, A., Thacker, N. & Rockett, P., Multiple Shape Recognition Using Pairwise Geometric Histogram Based Algorithms, Proceedings 5th IEE Int. Conf. Image Processing & Applications (IPA95), Edinburgh, 1995
3. Thacker, N., Riocreux, P. & Yates, R., Assessing the Completeness Properties of Pairwise Geometric Histograms, Image & Vision Computing, (1995) 13(5):423-429
4. Roth, G. & Levine, M., Geometric Primitive Extraction using a Genetic Algorithm, IEEE Trans. on Pattern Analysis & Machine Intelligence, (1994) 16(9):901-905
5. Hill, A. & Taylor, C., Model-based Image Interpretation using Genetic Algorithms, Image & Vision Computing, (1992) 10(5):295-300
6. Fonseca, C. & Fleming, P., An Overview of Evolutionary Algorithms in Multiobjective Optimisation, Evolutionary Computing, (1995) 3(1):1-16
7. Thacker, N., Aherne, F. & Rockett, P., The Bhattacharyya Metric as an Absolute Similarity Measure for Frequency Coded Data, Proc. Statistical Techniques in Pattern Recognition, Prague, Czech Republic, June 1997

Unsupervised Texture Segmentation Using Hermite Transform Filters

Atul Negi*, P.Sreedevi and P. Debbarma
(*contact: atulcs@uohyd.ernet.in)
A.I Lab, University of Hyderabad, Gachibowli, Hyderabad-500 046, INDIA.

Abstract:

A Hermite transform multi-channel filtering approach to unsupervised texture segmentation is presented. Texture feature images are obtained by first applying Hermite transform filters of various orders and in two directions to the input images, then next passing filtered components through a nonlinearity and computing local averages. These texture feature images are decimated in a hierarchical pyramid image representation. Segmentation by a square error clustering technique is performed on the decimated feature image with the classification being propagated down the pyramid to obtain pixel labels of the segmented image. Good segmentation results were obtained with about ten channels.

1. Introduction

Image segmentation is a fundamental problem of computer vision research, where global and local properties of an image have a complex interaction. Texture has been recognized as an intrinsic feature of realistic objects, and plays an important role in segmentation of images by computer vision systems. The focus of this paper is on the process of texture segmentation: i.e. identifying regions with similar texture and in segregating regions with different texture. The difficulty in texture segmentation can be traced to the complex interplay of gray level distributions at local and larger regional scales, which prevent a general definitive statement of the nature of texture itself. In the literature [10] several texture segmentation methods have been proposed, for example: use of gray level co-occurence, autocorrelation function analysis, second order spatial averaging, filtering approaches in spatial and frequency domain filtering etc.

More recently with the advances in psychophysical research in biological visual systems, multi-channel filtering approaches to texture segmentation have gained prominence with their ability to discern texture boundaries in arbitrary gray-scale images [9]. Subsequent research by Jain & Farrrokhnia [2] utilized Gabor filters and presented a good exposition of the multi-channel filtering paradigm reporting good results on 16 natural textures and Gaussian Markov Random Field textures.

Essential Characteristics of Multi-channel filtering approaches: Multi-channel filtering approaches succeeded in discerning textures of differing sizes and orientations due to their inherent multi-resolution processing. Further the functional characterization of filtering channels makes use of the gray level statistical information of the images as textural features. Broadly the following are key aspects of multi-channel approach:(a) Characterization of the various filter bank channels in terms of their signal content, and number of channels (b) development of discriminatory features (c) Feature integration and classification. *Motivation* In this paper we adopt Hermite transforms (HT) for multi-channel filter characterization. Martens [1] cites references as to the equivalent accuracy of Derivative of Gaussians (DOG) (which are computed by HT) to the Gabor filters. Further HT filtering has a simpler specification as compared to the radial frequencies, orientation, frequency bandwidth, and orientation bandwidth parameters needed by Gabor filters. Further HT approach allows formation of a DOG receptive field profile due to orthoganality, which is not the case with Gabor elementary expansion functions. Computationally too the approach is also appealing since it involves direct computation of texture features from the gray-level images with low order coefficients proving adequate for classification in

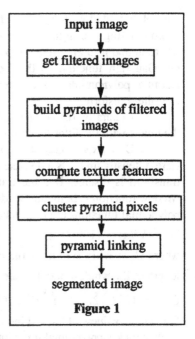

Figure 1

most images. Hermite transform possess the multi-resolution property (hierarchical coding as described by Martens) which is crucial in multi-channel feature based texture segmentation..

The rest of this paper is organized as follows: Section 2 introduces Hermite polynomial filters and gives formulae for computing convolution masks, and obtaining texture feature images, then presents a feature integration and classification method that segments the texture images. In section 3 experimental results on various natural textures are presented, while some concluding remarks are at the end in section 4.

2. Texture Segmentation

The outline of our texture segmentation algorithm is shown in the flowchart of Figure 1 and consists of the following steps: (a) Compute Hermite polynomial coefficients within a small window around each pixel. (Get multi-channel filtered images) (b) Build a pyramid structure for these filter images. (c) Compute the texture features from these pyramided filters by applying a nonlinear transformation followed by an averaging operation (texture energy computation). (d) Perform an unsupervised clustering on the pyramided feature images. (f) Perform pyramid linking by assigning the parent label to its children.

2.1 Hermite Polynomial Transform

Our algorithm uses the Hermite polynomial transform coefficients [1] of different orders of an image to compute texture features. Hermite polynomial transform is an alternative signal decomposition technique (to Gabor methods), in which signals are locally approximated by polynomials. The analysis by a polynomial transform involves two steps. In a first step the original signal $L(x,y)$ is localized by multiplying it by a window function $V(x,y)$. The second step consists of approximating the signal piece within the window by a polynomial $G_{m,n-m}(x,y)$, where m and n-m are orders of polynomials in x and y directions respectively. If we use Gaussian function as the local window then the associated polynomials are Hermite polynomials, so the resulting local decomposition technique is called Hermite transform.

Generally, the polynomial coefficients $L_{m,n-m}(x,y)$ are derived by convolving the image with the filter functions

$$D_{m,n-m}(x,y) = G_{m,n-m}(-x,-y) \, V^2(-x,-y) \qquad (1)$$

Here, intensity image is a discrete function of two variables $L(x,y)$ and the associated transform is discrete Hermite transform. For discrete Hermite transform the local window function is binomial window (Discrete counterpart of Gaussian window) and polynomial functions are Krawtchouk's polynomials.

i.e. $\qquad V^2(x) = (1/2^w) C_w^x \qquad (2)$

where $C_w^x = w!/[x!(w-x)!]$ is the combinatorial computation for $x=0,...,w$, and w is length of the window and further

$$G_n(x) = 1/\sqrt{C_w^x} \sum_{k=0}^{n} (-1)^{n-k} \, C_{w-x}^{n-k} \, C_x^k \qquad (3)$$

For large values of w, binomial window reduces to Gaussian window and Krawtchouk's polynomial turns into Hermite polynomial.

Thus filter function associated to discrete Hermite transform for $x = -w/2,..., w/2$. is,

$$D_n(x) = G_n(w/2-x) \, V^2(w/2 -x) \qquad (4)$$

The filter function is centered on the origin by shifting the binomial window over w/2. The filter function determines which information is made explicit in the coefficients of the Hermite transform. The Hermite polynomial coefficients $L_{m,n-m}(x,y)$ are obtained in two steps as the Gaussian window is a separable function:(a) First convolve $L(x,y)$ with filter $D_m(x)$ in x direction. (b) Convolve the resulting image with filter $D_{m-n}(y)$ in y direction.

The following important points are noted:

• These filters are interpreted as local feature detectors. For example, the filter D_{00} computes the mean intensity value within the window. The filters D_{10} and D_{01} would respond to sudden intensity changes in the x and y directions, respectively.

• The size of the window is important. As the window size gets larger features more global in extent are detected while finer textures would require smaller windows. This suggests that the choice of window size could possibly be tied to the contents of the image. In this paper, we do not suggest any method of selecting the appropriate window size automatically.

Feature Images and Non-Linearity The set of values for each filter over the entire image can be regarded as a new feature image. Let F_k be the kth such image. If we use n filters, then there will be n such filter images. In our experiments, we used filters up to third order.

The filter coefficients alone are not sufficient to obtain good texture features in certain images. Some iso-second order texture pairs which are preattentively discriminable by humans, would have the same average energy over finite regions.

However, their distribution would be different for different textures. One solution suggested by Caelli [5] is to introduce a nonlinear transducer that maps filter coefficients to texture features. Caelli suggests that the nonlinear transducer is "usually logistic, sigmoidal, or power function in form."

We have chosen to use the hyperbolic tangent function as our nonlinear transducer which is logistic in shape. This is followed by measuring Laws texture energy [4] (essentially an averaging process). We obtain the texture feature image T_k corresponding to the filtered image F_k using the following transformation (here both operations are combined into a single formula) where W_{ij} is an M x M averaging window centered at location (i,j) and α controls the shape of the logistic function:

$$T_k(i,j) = 1/M^2 \sum_{(a,b) \in W_{ij}} | \tanh(\alpha(F_k(a,b))) | \qquad (5)$$

2.2 Segmentation Algorithm

If n filtered images are computed over the image, then each pixel will have n feature values associated with it. For a pixel at (i,j), we define a textural feature vector $T_{ij} = <T_1(i,j),...,T_n(i,j)>$ which is a point in an n-dimensional feature space. We perform the texture segmentation by applying a well known clustering algorithm, CLUSTER [3] to the texture features T_{ij}. This partitional clustering technique requires the separation of outliers for proper results.

Decimation by Pyramid and Classification An important fact is that the number of pixels for clustering has to be reduced to a reasonable number for classification. We use a pyramid data structure [7] for the following reasons: (a) allows spatial proximity information to be included in the classification, without the need of arbitrary subsampling such as that by Tuceryan [6] and (b) allows a larger range of scales (greater dynamic range) to be examined at a reasonable computational expense. The linking process we use has the class label of the higher level pixel to be propagated back to the original image pixels which it represents.

The segmentation process operating on the decimated pyramid image consists of two steps:

(a) The pyramided feature images are clustered using partitional clustering algorithm CLUSTER [2,3]. As the squared error clustering methods are sensitive to the initial seed configuration, to improve the initial partition we used a min-max computation procedure [8] to find the farthest point from the existing clusters. This algorithm requires that the user provide the number of clusters. In our experiments, we usually request the correct number of clusters. Thus a segmentation of the feature vectors into a number of clusters (texture classes) is obtained . In case of an oversegmentation (that is a request for more than the correct number of clusters), then some border regions appear as separate clusters. This we found especially in Test image 4.

(b) In second step, as stated earlier is merely propagating the label of the parent pixels to its children in the pyramid. In our experimentation only one higher level than the original image was sufficient.

3. Experimental Results

We tested our segmentation algorithm on natural textures. We scanned the textures of a woven jute fiber (gunny) sack, a woven polymer fiber (polymer) sack, two types of woolen cloth and cotton towel using HP ScanJet 4c. Our algorithm successfully segmented the texture pairs. The results are presented in the following format:

(a) Input image (b) Filtered images (due do space constraint we did not show all filtered images) (c) Feature images of the corresponding filtered images (d) Segmented image. We used the percentage of misclassified pixels as the criterion to observe the performance of our algorithm. We selected α as 0.25 based on the experimental results that gave best classification.

Description of Images and Discussion **Test Image.1.** This image contains textures of *gunny sack* and *cotton towel.*. This gave the best results with a convolution window of size 7X7.: **Test Image.2** This image contains textures of *woolen cloth* and *cotton towel.* **Test Image. 3.** : This image contains textures of *cotton cloth* and *woolen fur.* The size of the image is 234 x 370. Observe that the fur texture (right half) is not such a regular one, and would pose problems to any texture discrimination method. **Test Image. 4.** : This image contains four different textures of *gunny sack, polymer sack; cotton towel, woolen cloth.* Better results were obtained with D_{10}, D_{01}, D_{02}, D_{20}, D_{11}, D_{30}, D_{03} filters. However because of boundary effects from the textural energy at energy stage the results in terms of pixels misclassified obtained were not so good.

4. Concluding Remarks

In this paper we have developed a texture segmentation algorithm based on the Hermite transform of an image. Certain aspects of algorithm need to be further

Table .1

NOTE:* W is the convolution window size to get the filtered images, for all the results texture energy window size is 13x13.

♣ d is the decimation factor if d=2 the image size is reduced to half of its resolution,

			% misclassification		
Image	size (pixel)	Filtered Features	*w=7x7 ♣d=2	w=9x9 d=2	w=11x11 d=2
Image 1	192 x 101	$D_{01}, D_{10\ 20}, D_{02}, D_{03}, D_{30}, D_{11}, D_{12}, D_{13}.$	0.271	1.291	3.208
Image 2	259 x 371	$D_{01}, D_{10}\ D_{20}, D_{02}, D_{03}, D_{30}, D_{11}, D_{12}, D_{21}, D_{31}, D_{13}.$	0.863	0.825	5.526
Image 3	234 x 370	$D_{01}, D_{10\ 20}, D_{02}, D_{03}, D_{30}, D_{11}$	2.785	1.718	2.222
Image 4	201 x 201	$D_{01}, D_{10} D_{20}, D_{02}, D_{03}, D_{30}, D_{11}$	42.68	8.68	12.67

studied carefully. First, the size of the window within which the coefficients are computed may be selected automatically. This window size depends on the content of the image: The window size may possibly be determined by an edge analysis of the image using techniques suggested by Kayargadde [11]. Second, how many filters need to be computed? In our experiments we used filters up to third order and selected those filters having a high energy content which seemed to do a good job in segmenting most textures pairs we tried. However, the usefulness of higher order coefficients need to be studied more carefully. Third, the selection of the window size to find Laws texture energy is also not selected automatically. In this case, the window should cover enough texture elements for the features to be meaningful. Finally, to avoid the border effects due to Laws texture energy one can integrate the region based and boundary based segmentation techniques.

Acknowledgments Atul Negi gratefully acknowledges discussions and initial conceptualization of this paper with Vishwakumara Kayargadde.

References

[1] Jean Bernard Martens, "The Hermite Transform - Theory," IEEE Trans. Acoust., Speech, Signal Processing Vol.38 (1990) pp. 1595-1606.

[2] A. K. Jain and F. Farrokhnia , "Unsupervised texture segmentation using Gabor filters," Pattern Recognition, Vol.24 (1991) pp. .1167-1186.

[3] A. K. Jain and R. C. Dubes, "Algorithms for clustering Data," Prentice Hall, Englewood Cliffs, New Jersey. (1988).

[4] K.I. Laws, "Textured image segmentation," Technical Report USCCIPI-940 , Image Process. Inst., University of Southern California(1980).

[5] T. Caelli and M. N. Oguztoreli. "Some tasks and signal dependent rules for spatial vision," Spatial Vision, Vol.2 (1987), pp.295-315.

[6] Mihran Tuceryan, "Moment Based Texture Segmentation 11th IAPR Conference on Pattern Recognition, The Hague Netherlands (1992) pp. 45-48.,"

[7] P. J. Burt "The Pyramid as a Structure for Efficient Computation," in A. Rosenfeld, Ed. Multi-resolutional Image processing, Springer-Verlag, Berlin, (1984) pp. 8-35.

[8] J. T. Tou and R C. Gonzalez, " Pattern Recognition Principles," Addison-Wesley Publishing Company, (1974).

[9] J. Malik and P. Perona, Preattentive texture discrimination with early vision mechanisms," J. Optical Society. of America. A 7 (1990) pp. 923-932.

[10] Mihran Tuceryan and A. K. Jain, "Texture Analysis," in C.H. Chen, L.F. Pau and P.S.P. Wang (Eds.) "Handbook of Pattern recognition and Computer Vision" World Scientific Publishing Co., Singapore(1993).

[11] Viswakumara Kayargadde "Feature Extraction for Image Quality Prediction, Ph.D. Thesis, Technische Universiteite Eindhoven, (1995).

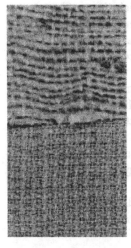

Test Image 1

Top Left : Original Image
Top Right Segmented Image

Centre Left: D_{01} feature
image
Centre Right: D_{03} feature
image

Bottom Left: D_{01} feature
image
Bottom Right: D_{03} feature
image

Test Image 2

Left: Original Image   Right: Segmented Image

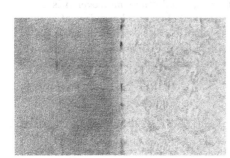

Test Image 3

Left: Original Image   Right: Segmented Image

Test Image 4

Left: Original Image   Right: segmented Image

Decomposition of the Hadamard Matrices and Fast Hadamard Transform

Hakob G. Sarukhanyan

Institute for Informatics and Automation Problems NAS,
P.Sevak 1, 375044, Yerevan, Republic of Armenia,
Tel. (3742) 28 20 62, Fax: (3742) 28 58 12,
E-mail: hakop@ipia.sci.am

Abstract. It is well known, that the classical algorithm of the Walsh-Hadamard fast transform needs only $n \log_2 n$ additions moreover n is a power of two. A problem of decomposition of Hadamard matrices of arbitrary order n, $n \equiv 0 (mod4)$ by orthogonal $(-1, +1)$-vectors of size k is investigated in this paper. An algorithm of the Hadamard fast transform which needs only $n \log_2 k + n(\frac{n}{k} - 1)$ addition operations and in some cases is more efficient than the classical algorithm is proposed.

The square $(-1, +1)$-matrix H_n of order n is called Hadamard matrix, if conditions

$$H_n H_n^T = H_n^T H_n = n I_n, \tag{1}$$

are satisfied, where T is transposition sign, I_n is an identity matrix of order n.

It was proved that if H_n is a Hadamard matrix of order n then $n = 2$ or $n \equiv 0 (mod4)$. The inverse problem has not been yet solved [1].

It is known that the Kronecker product of two Hadamard matrices of orders n and m is a Hadamard matrix of order mn. According to multiplicative theorem of Agayan and Sarukhanyan [1-3] there exists also Hadamard matrix of order $\frac{mn}{2}$. Later in [4], starting with four Hadamard matrices of orders m, n, p and q the existence of Hadamard matrix of order $\frac{mnpq}{16}$ is proved. In [5] for any natural number k, $k \geq 4$ the existence of Hadamard matrices of order $\frac{n_1 n_2 \cdots n_k}{2^k}$, where n_i are orders of Hadamard matrices, is proved.

The analysis of the given results shows that they all are particular solutions of the following problem: what conditions must satisfy $(0, -1, +1)$- and $(-1, +1)$-matrices A_i and X_i, $i = 1, 2, \ldots, k$ with sizes $p_1 \times p_2$ and $q_1 \times q_2$ correspondingly, moreover $p_1 q_1 = p_2 q_2 = n \equiv 0 (mod4)$, for the matrix $H = \sum_{i=1}^{k} X_i \times A_i$ be Hadamard matrix of order n.

Below we will consider a particular case of this problem, when X_i are the orthogonal $(-1, +1)$−vectors of size k. Such Hadamard matrices we will call Hadamard matrices of type $A(n, k)$ or simply $A(n, k)$−matrices.

Theorem 1. Matrix H_n of order n is $A(n, k)$-matrix if and only if there exist $(0, -1, +1)$- matrices A_i, $i = 1, 2, \ldots, k$ of size $n \times \frac{n}{k}$, satisfying the following conditions:

$$A_i \star A_j = 0, \quad i \neq j \quad i,j = 1,2,\ldots,k;$$

$$\sum_{i=1}^{k} A_i \quad \text{is a} \quad (-1,+1) - \text{matrix};$$

$$\sum_{i=1}^{k} A_i A_i^T = \frac{n}{k} I_n; \tag{2}$$

$$A_i^T A_i = \frac{n}{k} I_{\frac{n}{k}}, \quad A_i^T A_j = 0, \quad i \neq j, \quad i,j = 1,2,\ldots,k;$$

where $\star$ is Hadamard product [1].

Proof. Prove the theorem for $k = 4$. Let H_n be $A(n,4)$-matrix, i.e. H_n represents as: $H_n = (+ + + +) \times A_1 + (+ + - -) \times A_2 + (+ - - +) \times A_3 + (+ - + -) \times A_4$, where the sign $+$ means $+1$, the sign $-$ means -1, and $\times$ is a Kronecker product. From definition of $A(n,4)$-matrices it just follows the first tree conditions of (2). Further, from the conditions (1) we obtain a system of equations in the unknowns $A_i^T A_j$, $i,j = 1,2,3,4$. The solution of this system gives two last conditions of (2). The sufficiency is proved by direct verifying the definition of $A(n,4)$-matrices

Note. Any Hadamard matrix of order n is $A(n,2)$-matrix.

Theorem 2. Let there exist $A(n,k)$-matrix and Hadamard matrix of order m ($m = 0(mod\ k)$). Then there also exists Hadamard matrix of order $\frac{mn}{k}$.

Proof. Represent Hadamard matrix H_m of order m as follows:

$$H_m^T = (P_1^T, \ldots, P_k^T),$$

where P_i are $(-1,+1)$-matrices of size $\frac{m}{k} \times m$. It is not difficult to show that the matrix $\sum_{i=1}^{k} P_i \times A_i$ is Hadamard matrix of order $\frac{mn}{k}$, where matrices A_i satisfy the conditions (2).

The following theorem is correct.

Theorem 3. Let there exist Hadamard matrix of order m and $A(n,k)$-matrix. Then for any even numbers r such that $m,n \equiv 0(mod\ r)$, there exist $(0,-1,+1)$-matrices $B_{i,j}$, $i = 1,2,\ldots,\frac{r}{2}$, $j = 1,2,\ldots,k$ of size $\frac{mn}{r} \times \frac{mn}{k}$, satisfying the following conditions:

$$B_{t,i} \star B_{t,j} = 0, \quad i \neq j, \quad i,j = 1,2,\ldots,k, \quad t = 1,2,\ldots,\frac{r}{2};$$

$$\sum_{i=1}^{k} B_{t,i} \quad \text{is a} \quad (-1,+1) - \text{matrix}, \quad t = 1,2,\ldots,\frac{r}{2};$$

$$\sum_{t=1}^{k} B_{i,t} B_{j,t}^T = 0, \quad i \neq j, \quad i,j = 1,2,\ldots,\frac{r}{2}; \tag{3}$$

$$\sum_{t=1}^{r/2} \sum_{i=1}^{k} B_{t,i} B_{t,i}^T = \frac{mn}{2k} I_{\frac{mn}{r}}.$$

Proof. Represent the matrix H_n of $A(n,k)$-type by the following form $H_n = (P_1^T, P_2^T, \ldots, P_r^T)^T$. Here $(-1,+1)$-matrices P_i with dimension $\frac{n}{r} \times n$ represent as $P_i = \sum_{t=1}^{k} v_t \times A_{i,t}$, $i = 1, 2, \ldots, r$, where v_t are the mutually orthogonal k-dimensional $(-1, +1)$-vectors.

It is not difficult to show that $(0, -1, +1)$-matrices $A_{i,j}$ with dimension $\frac{n}{r} \times \frac{n}{k}$ satisfy to the following conditions:

$$A_{t,i} \star A_{i,j} = 0, \quad i \neq j, \quad t = 1, 2, \ldots, r. \quad i, j = 1, 2, \ldots, k;$$

$$\sum_{i=1}^{k} A_{t,i} \quad \text{is a} \quad (-1, +1) - \text{matrix}, \quad t = 1, 2, \ldots, r;$$

$$\sum_{t=1}^{k} A_{i,t} A_{j,t}^{T} = 0, \quad i \neq j, \quad i, j = 1, 2, \ldots, r;$$

$$\sum_{t=1}^{k} A_{i,t} A_{i,t}^{T} = \frac{n}{k} I_{\frac{n}{r}}, \quad i = 1, 2, \ldots, r.$$

Now represent Hadamard matrix H_m of order m as follows:

$$H_m = (Q_1, Q_2, \ldots, Q_r),$$

where $(-1, +1)$-matrices Q_i with dimension $m \times \frac{m}{r}$ evidently satisfy to the condition $\sum_{i=1}^{r} Q_i Q_i^T = m I_m$. Introduce the following matrices

$$U_{2i-1} = \frac{Q_{2i-1} + Q_{2i}}{2}, \quad U_{2i} = \frac{Q_{2i-1} - Q_{2i}}{2}, \quad i = 1, 2, \ldots, \frac{r}{2}.$$

One can show that these matrices satisfy to the following conditions:

$$U_{2i-1} \star U_{2i} = 0, \quad i = 1, 2, \ldots, \frac{r}{2},$$

$$U_{2i-1} \pm U_{2i} \quad \text{is a} \quad (-1, +1) - \text{matrix},$$

$$\sum_{i=1}^{r} U_i U_i^{T} = \frac{m}{2} I_m$$

Consider $(0, -1, +1)$-matrices with the dimension $\frac{mn}{r} \times \frac{mn}{kr}$

$$B_{t,i} = U_{2i-1} \times A_{2t-1,i} + U_{2t} \times A_{2t,i}, \quad t = 1, 2, \ldots, \frac{r}{2}, \quad i = 1, 2, \ldots, k.$$

Now we can show that these matrices satisfy to the conditions (3).

Some useful corollaries follow from this theorem.

Corollary 1. (Agayan-Sarukhanyan [1]). If there exist Hadamard matrices of order m and n then there exists Hadamard matrix of order $\frac{mn}{2}$.

Really, according to theorem 3, for $k = r = 2$ there exist $(0,-1,+1)$-matrices $B_{1,1}$ and $B_{1,2}$ satisfying conditions (3). Now it is not difficult to show that $(++) \times B_{1,1} + (+-) \times B_{1,2}$ is a Hadamard matrix of order $\frac{mn}{2}$.

From theorem 3 it follows also

Corollary 2. (Craigen-Seberry-Zhang [1,4]). If there exist Hadamard matrices of order m, n, p, and q, then there exist Hadamard matrix of order $\frac{mnpq}{16}$.

Direct and inverse Hadamard transforms of vector–column $f = \{f_i\}_{i=1}^n$ are correspondingly determined by formulas [6,7]

$$F = \tfrac{1}{n}H_n f \quad \text{and} \quad f = H_n^T F, \tag{4}$$

For Walsh–Hadamard transform, i.e. when H_n is Hadamard matrix of order $n = 2^k$, there is developed a fast algorithm using only $n \log_2 n$ addition operations [6,7].

Below there is proposed the Hadamard transform fast algorithm for the cases, when Hadamard matrix order is not a power of two.

Let $H_n = v_1 \times A_1 + \cdots + v_k \times A_k$ be $A(n,k)$-matrix. According to the theorem 1 matrices A_i satisfy to the conditions (2) and v_i are the orthogonal $(-1,+1)$–vectors of length k. Represent vector–column f as follows $f = \sum_{i=1}^{n/k} x_i \times P_i$, where x_i is a vector –column of the form: $x_i = \{f_{k(i-1)+j}\}_{j=1}^k$, and P_i is a vector–column of length $\frac{n}{k}$ whose i-th element is equal to 1, and the other elements are zero.

Now the direct Hadamard transform of vector f is represented as follows (normalizing coefficient $1/n$ is omitting).

$$H_n f = \sum_{i=1}^k \sum_{j=1}^{n/k} v_i x_j \times A_i P_j. \tag{5}$$

Consider a part of sum (5) having a form

$$v_1 x_j \times A_1 P_j + v_2 x_j \times A_2 P_j + \cdots + v_k x_j \times A_k P_j. \tag{6}$$

From condition (2) and definition of the matrices P_i it follows that $A_i P_j$ is j-th column of the matrix A_i which has $\frac{n}{k}$ non–zero entries. The product $v_i x_j \times A_i P_j$ means that i-th element of the Walsh–Hadamard transform of vector x_j is located in $\frac{n}{k}$ positions of j-th column of the matrix A_i. Hence, the calculation of all elements of n-dimensional vector (5) needs $k \log_2 k$ addition operations. Note that the expression (5) is the sum of $\frac{n}{k}$ vectors of the form (6). Hence, only $D = n \log_2 k + n(\frac{n}{k} - 1)$ addition operations are required to compute transform of the form (5) by $A(n,k)$–matrix. Now it follows that for $n = k$ the estimate D coincides with the estimate for the fast Walsh–Hadamard transform (n is the power of two).

Note that when in practice there arises a problem of image or signal processing whose dimension is not a power of two, we have to reduce or increase image dimension up to the nearest power of two. This leads to the loss of information in the first case and to the slowing of processing in the second case.

Denote by $D_1 = N \log_2 N$ a number of operations for the fast Walsh-Hadamard transform. (Here N is a power of two). In Table 1 there are given some values of n, k, the nearest orders equal to powers of two and the corresponding numbers of addition operations for classic and new algorithms of the transform.

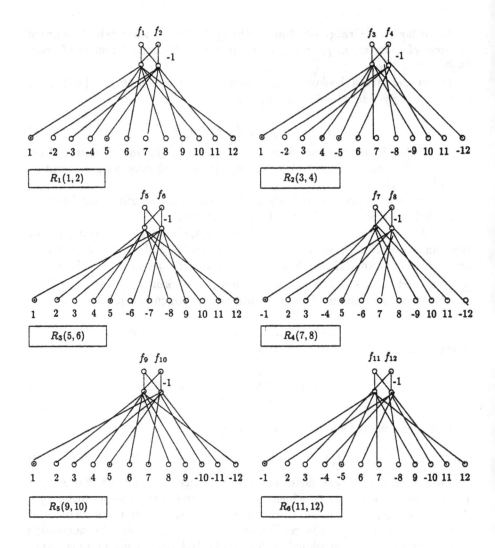

Fig. 1. Vector x_i transform and sending diagram $R_i(j, k)$

From table it follows that new algorithm is more efficient than the classic one already for $n = 96$.

Give the graph of the fast Hadamard transform for the matrix of order 12. Consider the Hadamard matrix of order 12 of the form: $H_{12} = v_1 \times A_1 + v_2 \times A_2$, where $v_1 = (++)$, $v_2 = (+-)$, and A_1 and A_2 are block-circulant matrices whose first block-rows represent as: (R_0, R_1, R_1) and (T_0, T_1, T_1), where

$$R_0 = \begin{bmatrix} 1 & 1 \\ 0 & 0 \\ 0 & 0 \\ -1 & 1 \end{bmatrix}, R_1 = \begin{bmatrix} 0 & -1 \\ 1 & 0 \\ 0 & 1 \\ 1 & 0 \end{bmatrix}, T_0 = \begin{bmatrix} 0 & 0 \\ -1 & -1 \\ -1 & 1 \\ 0 & 0 \end{bmatrix}, T_1 = \begin{bmatrix} 1 & 0 \\ 0 & 1 \\ 1 & 0 \\ 0 & -1 \end{bmatrix}.$$

Represent the vector-colomn f being transformed as: $f = \sum_{i=1}^{6} x_i \times P_i$, where $x_i = \{f_{2(i-1)+j}\}_{j=1}^{2}$, $i = 1, 2, \ldots, 6$. According to the formula (5) we have: $H_{12}f = \sum_{j=1}^{6} (v_1 x_j \times A_1 P_j + v_2 x_j \times A_2 P_j)$.

The computational graphs of the vectors

$$\{F_j(i)\}_{i=1}^{12}, \qquad j = 1, 2, \ldots, 6$$

of the form $v_1 x_j \times A_1 P_j + v_2 x_j \times A_2 P_j$ are given in figure 1.

From graphs mentioned above it is not difficult to calculate that the number of additions is 72.

Introduce the notations: $R_i(j, k)$ is i-th diagram of sending of the components with number j and k (for example, $R_1(1, 2)$ corresponds to the first graph in figure 1), $FHT(x_i)$ is the fast Walsh-Hadamard transform of the vector x_i, and $FHT(x_i)_j$ is j-th component of the transformed vector x_i.

Now give the general algorithm of the Hadamard transform for the vector f of the length $6 \cdot 2^k$. Denote the rows of the Walsh-Hadamard matrix of order 2^k by v_i^k, $i = 1, 2, \ldots, 2^k$. Then Hadamard matrix of order 12 will represent as: $H_{12} = v_1^1 \times A_1^1 + v_2^1 \times A_2^1$, where A_1^1, A_2^1 as above.

It is possible to show that $H_{6 \cdot 2^k} = \sum_{i=1}^{2^k} v_i^k \times A_i^k$ is Hadamard matrix of order $6 \cdot 2^k$, where $A_{2i-1}^k = A_1^1 \times P_i$, $A_{2i}^k = A_2^1 \times P_i$, $i = 1, 2, \ldots, 2^{k-1}$. Here P_i is vector-column of the length 2^k whose i-th element is equal to 1, and the other elements are equal to zero.

From the starting vector f we form six vector-columns by the following formula: $x_i^k = \{f_{2^k(i-1)+j}\}_{j=1}^{2^k}$, $i = 1, 2, \ldots, 6$.

In these notations the Hadamard transform of the vector f represents as: $H_{6 \cdot 2^k}f = \sum_{j=1}^{2^k} \sum_{i=1}^{6} v_j x_i^k \times A_j^k P_i$.

From this formula it follows that to compute $H_{6 \cdot 2^k}f$ it is necessary to fulfil the fast Walsh-Hadamard transforms for six vectors of the length 2^k, and then to form vectors of the form $\left\{ R_i \left(\{FHT(x_i^k)_{2(r-1)+j}\}_{j=1}^{2} \right) \right\}_{r=1}^{2^{k-1}}$, $i = 1, 2, \ldots, 6$, and to add them. One can compute that the general number of additions will equal to $6 \cdot 2^k(k + 5)$.

Table 1

No	n	k	N	D	D_1
1	12	2	16	72	64
2	24	4	32	168	160
3	48	8	64	384	384
4	96	16	128	864	896
5	192	32	256	1920	2048
6	384	64	512	4224	4608
7	768	128	1024	9216	10240

References

1. Seberry J., Yamada M. Hadamard matrices, sequences and block designs. Surveys in Contemporary Designs Theory, Wiley–Interscience Series in Discrete Mathematics. Jhon Wilay, New York, 1992.
2. Agaian S.S., Sarukhanian A.G. Recurrent formulae of the construction Williamson type matrices. Math. Notes, vol. 30, No. 1, 1981, p. 603–617.
3. Agaian S.S., Hadamard matrices and their applications, Lecture Notes in Math., Springer–Verlag, Berlin, Heidelberg, New York, Tokyo, No. 1168,1985, p. 227.
4. Craigen R., Seberry J., Zhang X. Product of four Hadamard matrices. J. Comb. Theory, ser. A, 59, 1992, p. 318–320.
5. Agayan S.S., Sarukhanyan H.G. Hadamard matrices representation by (-1,+1)-vectors. Int. conference dedicated to Hadamard problem centenary, Australia, 1993, 1p.
6. Beauchamp K.R. Walsh functions and their applications. Academic Press, London, New York, Francisco, 1975.
7. Ahmed N., Rao K.R. Orthogonal transforms for digital signal processing. Springer–Verlag, Berlin, Heidelberg, New York, 1975.

A Characterization of Digital Disks
by Discrete Moments

Joviša Žunić and Nataša Sladoje
University of Novi Sad , Faculty of Engineering,
Trg D. Obradovića 6, 21000 Novi Sad, Yugoslavia
ftn_zunic@uns.ns.ac.yu , sladoje@uns.ns.ac.yu

Abstract

In this paper our studies are focused on the digital disks and problems of their characterization (coding) with an appropriate number of bits, and reconstruction of the original disk from the code that is used. Even though the digital disks appear very often in practice of the computer vision and image processing, only the problem of their recognition has been solved till now. In this paper a representation by constant number of integers, requireing optimal number of bits, is presented. One-to-one correspondence between the digital disks and their proposed codes, consisting of:
  – the number of points of the digital disk,
  – the sum of x-coordinates of the points of digital disk,
  – the sum of y-coordinates of the points of digital disk,
is proved.

The efficiency of the reconstruction of the original disk from the proposed code is analysed. It is shown that the errors in estimating the radius of the disk, and the coordinates of its center, tend to zero while the radius of the disk tends to infinity. More precisely, if a disk, having the radius equal to r, is digitized and proposed coding scheme is applied, then the radius and the center position of the original disk can be reconstructed (from the obtained code) with relative errors bounded by $\mathcal{O}\left(\frac{1}{r \cdot \sqrt[3]{r}}\right)$, and absolute errors bounded by $\mathcal{O}\left(\frac{1}{\sqrt[3]{r}}\right)$.

The numerical data strongly confirm the theoretical results. The illustration by several experimental results is given.

Key words: pattern analysis, low level processing and coding, shape representation.

1   Introduction

Among the most important problems considered in computer vision and image processing related to digital pictures analysis, there are the recognition of the studied object, its efficient representing and finding the algorithm for recovering the object from its representation. Digital shapes which appear the most often in practice are digital straight lines and conic sections (in the Euclidean plane) and so-called surfaces of the second order (in the Euclidean space). While the above mentioned problems, connected with the sets of digital points resulting from digitization of the straight line segments, are completely solved ([1],[6],[7]), only the recognition problem is solved for the digital disks ([3],[5],[8],[9]). Statistical parameter estimations of circular arcs are given in [10].

Since the least squares method, which gives efficient representations for certain classes of digital curves, such as digital straight line segments ([7]), digital polynomials ([11]), and for digital plane segments ([4]), is not appropriate for representing disks, where it leads to nonlinear problems which require complex numerical techniques for the solution, in this paper we develop the idea of separating sets for proving one-to-one correspondence between digital shapes and their representations by a constant number of integers, introduced in [4].

In Section 2 we give an asymptotically optimal representation of digital disks by three integer parameters.

An efficient estimation of the radius of the original disk, as well as the coordinates of its center, is given in Section 3. It is shown that the errors in estimating the radius and the center position of the original disk tend to zero while radius of the disk tends to infinity.

The numerical data (Table 1. and Table 2.) strongly acknowledge the theoretical results. Even for the disk (at random position), containing 20 pixels ($r = 2.5$), relative deviation is less than 2%, and absolute error is less than 10^{-1}. For the disk with the radius exceeding 100, relative error is less then 0.0001% (for some examples it is less than 0.0000001%), and the absolute error is less than 10^{-2}.

2 Representation of Digital Disks

Consider a disk K in the Euclidean plane, defined by $(x - a)^2 + (y - b)^2 \leq r^2$. The disk K will be digitized by using the digitization scheme in which all digital points (points with integer coordinates, often referred to as pixels) in the disk are taken. In this way, we get a set of digital points known as digital disk. So, the associated set of digital points for the disk K is defined by

$D(K) = \{(i,j) \mid (i - a)^2 + (j - b)^2 \leq r^2, \quad i, j \text{ are integers}\}.$

The main contribution of this paper is to give an efficient representation for the digital disk. For such representation three integer parameters will be used:

- number of points of the digital disk, denoted by $R(K)$;

- sum of x-coordinates of the points of the digital disk, denoted by $X(K)$;

- sum of y-coordinates of the points of the digital disk, denoted by $Y(K)$.

The values of $R(K), X(K), Y(K)$ can be computed easily and, obviously, are uniquely determined for a certain disk K.

The key question is whether there exist two different digital disks $D(K_1)$, $D(K_2)$ with the same corresponding representation. The answer is negative, which implies that the digital disks and their proposed representations (codes) are in one-to-one correspondence.

Theorem 1 *Let $D(K_1)$ and $D(K_2)$ be two digital disks and let $(R(K_1), X(K_1), Y(K_1))$ and $(R(K_2), X(K_2), Y(K_2))$ be their codes, respectively. Then*
$$(R(K_1) = R(K_2) \quad \wedge \quad X(K_1) = X(K_2) \quad \wedge \quad Y(K_1) = Y(K_2))$$
is equivalent to
$$D(K_1) = D(K_2).$$

Proof: The direction $D(K_1) = D(K_2) \quad \Rightarrow$
$$(R(K_1) = R(K_2) \quad \wedge \quad X(K_1) = X(K_2) \quad \wedge \quad Y(K_1) = Y(K_2))$$
is obvious and follows directly from the definitions.

The opposite direction will be proved by a contradiction. By the assumption, relation $R(K_1) = R(K_2)$ is satisfied. That means that the number of digital points belonging to $D(K_1)$ is the same as the number of digital points of $D(K_2)$. Moreover, considering $D(K_1) \neq D(K_2)$, we have that $\#(D(K_1) \setminus D(K_2)) = \#(D(K_2) \setminus D(K_1)) \neq 0$ holds. This, together with the assumptions $X(K_1) = X(K_2)$ and $Y(K_1) = Y(K_2)$, implies

$$\sum_{(x,y) \in D(K_1) \setminus D(K_2)} x \; = \sum_{(x,y) \in D(K_2) \setminus D(K_1)} x \qquad \text{and} \qquad (1)$$

$$\sum_{(x,y) \in D(K_1) \setminus D(K_2)} y \; = \sum_{(x,y) \in D(K_2) \setminus D(K_1)} y \; . \qquad (2)$$

Then, there exists a straight line $ax + by = c$, which separates sets $D(K_1) \setminus D(K_2)$ and $D(K_2) \setminus D(K_1)$. Without loss of generality, we can suppose that
$$ax + by < c \quad \text{for} \quad (x,y) \in D(K_1) \setminus D(K_2) \qquad \text{and}$$
$$ax + by > c \quad \text{for} \quad (x,y) \in D(K_2) \setminus D(K_1)$$
is satisfied. This, together with (1) and (2), gives:

$$c \cdot \#(D(K_1) \setminus D(K_2)) \; = \sum_{(x,y) \in D(K_1) \setminus D(K_2)} c$$

$$> a \cdot \sum_{(x,y) \in D(K_1) \setminus D(K_2)} x \; + \; b \cdot \sum_{(x,y) \in D(K_1) \setminus D(K_2)} y$$

$$= a \cdot \sum_{(x,y) \in D(K_2) \setminus D(K_1)} x \; + \; b \cdot \sum_{(x,y) \in D(K_2) \setminus D(K_1)} y$$

$$> \sum_{(x,y) \in D(K_2) \setminus D(K_1)} c \; = \; c \cdot \#(D(K_2) \setminus D(K_1)) \; .$$

The contradiction $c \cdot \#(D(K_1) \setminus D(K_2)) > c \cdot \#(D(K_2) \setminus D(K_1))$ finishes the proof. $\square$

For convenience, without loss of generality, it will be assumed that all the digital points that appear have positive coordinates - in other words, the origin is placed in the left lower corner of the observed integer grid.

3   Estimation of the Original Disk from its Code

An important question is how efficiently an original disk K can be recovered from $(R(K), X(K), Y(K))$-code of its digitization $D(K)$. In this section it will be shown that the proposed coding scheme enables an approximate reconstruction of a, b and r with errors tending to zero while $r \to \infty$.

A (k, l)-moment, denoted by $m_{k,l}(S)$ for a shape S, in 2D-space is defined by:

$$m_{k,l}(S) = \iint\limits_{S} x^k y^l \, dx \, dy \quad .$$

If S is the disk K, given by the inequality $(x - a)^2 + (y - b)^2 \leq r^2$, then

$$m_{0,0}(K) = \pi \cdot r^2, \quad m_{1,0}(K) = \pi \cdot a \cdot r^2 \quad \text{and} \quad m_{0,1}(K) = \pi \cdot b \cdot r^2 \quad .$$

Thus, if the moments of the disk K are known, then K can be reconstructed easily. Namely,

$$a = \frac{m_{1,0}(K)}{m_{0,0}(K)}, \quad b = \frac{m_{0,1}(K)}{m_{0,0}(K)} \quad \text{and} \quad r = \sqrt{\frac{m_{0,0}(K)}{\pi}} \quad .$$

Three integers, $R(K)$, $X(K)$ and $Y(K)$, appearing in the proposed characterization of digital disk $D(K)$, can be understood as discrete moments of that discrete shape. So, it is natural to expect that the digitization of the disk, defined by

$$\left(x - \frac{X(K)}{R(K)} \right)^2 + \left(y - \frac{Y(K)}{R(K)} \right)^2 \leq \frac{R(K)}{\pi} \quad ,$$

can be a good approximation for the represented digital disk $D(K)$.

In order to estimate the radius and the coordinates of the center of the original disk, we need asymptotic expressions for $X(K)$, $Y(K)$ and $R(K)$.

Asymptotic behaviour of $R(K)$ is a well-known result of the number theory ([2]):

$$R(K) \quad = \sum_{\substack{i,j \text{ are integers} \\ (a-i)^2 + (b-j)^2 \leq r^2}} 1 = \pi \cdot r^2 + \mathcal{O}\left(r^{\frac{2}{3}} \right) \quad . \tag{3}$$

The following result is derived in [12]:

Theorem 2 *Let real numbers a, b and r, satisfying $r \leq a$ and $r \leq b$, be given. Then the following asymptotic expressions hold:*

$$\sum_{\substack{i,j \text{ are integers} \\ (a-i)^2 + (b-j)^2 \leq r^2}} i = a \cdot \pi \cdot r^2 + \mathcal{O}(a \cdot r^{\frac{2}{3}}) \tag{4}$$

and

$$\sum_{\substack{i,j \text{ are integers} \\ (a-i)^2 + (b-j)^2 \leq r^2}} j = b \cdot \pi \cdot r^2 + \mathcal{O}(b \cdot r^{\frac{2}{3}}) \quad . \tag{5}$$

The optimality of the proposed code is a consequence of Theorem 2:

Corollary 1 *The proposed $(R(K), X(K), Y(K))$-code requires an asymptotically optimal (minimal) number of bits.*

Proof. Under the assumptions of Theorem 2, the number of bits, required for coding $R(K)$, $X(K)$ and $Y(K)$ is

$$\lceil \log R(K) \rceil + \lceil \log X(K) \rceil + \lceil \log Y(K) \rceil = \mathcal{O}(\max\{\log(r^2), \log(a \cdot r), \log(b \cdot r)\}),$$

that is, $\mathcal{O}(\log(\max\{a, b\}))$. On the other side, a trivial lower bound for the number of different digital disks which can be inscribed into an integer grid of size $(a + r) \times (b + r)$ is $a \cdot b$. ($(a + r) \times (b + r)$ is the minimal size of integer grid in which K can be inscribed). So, the required number of bits for unique coding of digital disks is at least $\log(a \cdot b) = \mathcal{O}(\log(\max\{a, b\}))$. Since the lower bound is reached, the proposed code is optimal. $\square$

Possible difference in the order of magnitude of a, b and r, which is not excluded by the assumptions of Theorem 2, affects the order of the absolute error in estimating r, a and b, respectively, by $\sqrt{\frac{R(K)}{\pi}}$, $\frac{X(K)}{R(K)}$ and $\frac{Y(K)}{R(K)}$, if the asymptotic expressions for discrete moments $R(E)$, $X(E)$ and $Y(E)$ are directly substituted. However, as a consequence of the following statement, it can be assumed that a, b and r are of the same order.

Theorem 3 *Let the digital disk $D(K(a, b, r))$ be the digitization of a disk $K(a, b, r): (x - a)^2 + (y - b)^2 \leq r^2$. Then the the numbers $\sqrt{\frac{R(K)}{\pi}}$, $\frac{X(K)}{R(K)} - a$ and $\frac{Y(K)}{R(K)} - b$ are the constants with respect to the translation by the vector having integer components. In other words, the following equalities are satisfied:*

a) $R(K(a, b, r)) = R(K(a + k, b + l, r))$,

b) $\frac{X(K(a,b,r))}{R(K(a,b,r))} - a = \frac{X(K(a+k,b+l,r))}{R(K(a+k,b+l,r))} - (a + k)$,

c) $\frac{Y(K(a,b,r))}{R(K(a,b,r))} - b = \frac{Y(K(a+k,b+l,r))}{R(K(a+k,b+l,r))} - (b + l)$,

where k and l are integers.

Proof. Follows from the definitions of $R(K)$, $X(K)$ and $Y(K)$. $\square$

Now, we can give an upper bound on the precision of estimation of the radius and center of an original disk from its code.

Theorem 4 *Let the disk K, given by $(x - a)^2 + (y - b)^2 \leq r^2$, be digitized. Then the following error estimations hold:*

$$\frac{\sqrt{\frac{R(K)}{\pi}} - 1}{r} = \mathcal{O}\left(\frac{1}{r^{\frac{4}{3}}}\right) , \quad \text{while}$$

$$\frac{\frac{X(K)}{R(K)}}{a} - 1 = \mathcal{O}\left(\frac{1}{r^{\frac{4}{3}}}\right) \quad \text{and} \quad \frac{\frac{Y(K)}{R(K)}}{b} - 1 = \mathcal{O}\left(\frac{1}{r^{\frac{4}{3}}}\right) .$$

Proof. By using (3),(4) and (5), we have:

$$\frac{\sqrt{\frac{R(K)}{\pi}}}{r} = \frac{\sqrt{\frac{\pi \cdot r^2 + \mathcal{O}(r^{\frac{4}{3}})}{\pi}}}{r} = \sqrt{1 + \mathcal{O}\left(\frac{1}{r^{\frac{2}{3}}}\right)} = 1 + \mathcal{O}\left(\frac{1}{r^{\frac{2}{3}}}\right) \quad ,$$

$$\frac{\frac{X(K)}{R(K)}}{a} = \frac{a \cdot \pi \cdot r^2 + \mathcal{O}(a \cdot r^{\frac{4}{3}})}{a \cdot \left(\pi \cdot r^2 + \mathcal{O}(r^{\frac{4}{3}})\right)} = 1 + \mathcal{O}\left(\frac{1}{r^{\frac{2}{3}}}\right) \quad . \quad \square$$

The previous theorem shows that the relative error in estimating the radius of the original disk, as well as its center position, tends to zero while $r \to \infty$. These results are in accordance with the experimental results given in Table 1.

Relative errors for the values of the radius and coordinates of the center, reconstructed from the digitization of the disk $K : (x - a)^2 + (y - b)^2 \leq r^2$

r	a	b	$\left\|\frac{\sqrt{\frac{R(K)}{\pi}}}{r} - 1\right\|$	$\left\|\frac{\frac{X(K)}{R(K)}}{a} - 1\right\|$	$\left\|\frac{\frac{Y(K)}{R(K)}}{b} - 1\right\|$
	7.2	3.6	0.016301848	0.001461988	0.008771929
2.5	272.7	33.3	0.009253008	0.000183351	0.001501501
	2777722.7	33111.1	0.009253008	0.000000072	0.000001511
	47.1	31.2	0.000810481	0.000107204	0.000993347
23.4	4744.7	313.1	0.000520012	0.000000685	0.000022627
	7444774.4	133311.3	0.000642707	0.000000004	0.000000141
	4474.4	311.2	0.000047575	0.000000189	0.000026017
234.3	474477.7	313.1	0.000035978	0.000000022	0.000009612
	7444774.4	333311.3	0.000038877	0.000000001	0.000000034
	4474.7	3111.1	0.000000228	0.000000151	0.000000888
2345.6	474477.7	3313.3	0.000000257	0.000000009	0.000041316
	7444774.4	333311.3	0.000001449	0.000000001	0.000000022

Table 1.

In the next theorem the absolute errors in estimating the radius r and the coordinates of the center (a, b) of the original disk are analysed.

Theorem 5 *Let the disk K, given by: $(x - a)^2 + (y - b)^2 \leq r^2$, be digitized. Then the following error estimations hold:*

$$\sqrt{\frac{R(K)}{\pi}} - r = \mathcal{O}\left(\frac{1}{r^{\frac{1}{3}}}\right) \quad , \quad \text{while}$$

$$\frac{X(K)}{R(K)} - a = \mathcal{O}\left(\frac{1}{r^{\frac{1}{3}}}\right) \quad \text{and} \quad \frac{Y(K)}{R(K)} - b = \mathcal{O}\left(\frac{1}{r^{\frac{1}{3}}}\right) \quad .$$

Proof. As the proof of Theorem 4. Theorem 3 and (3)-(5) are used. □

The next table contains experimental results.

Absolute errors for the values of the radius and coordinates of the center, reconstructed from the digitization of the disk $K : (x - a)^2 + (y - b)^2 \leq r^2$

r	a	b	$\sqrt{\frac{R(K)}{\pi}} - r$	$\frac{X(K)}{R(K)} - a$	$\frac{Y(K)}{R(K)} - b$
	9.4	6.2	-0.04075	-0.03157	0.01052
2.5	282.8	33.3	0.02313	-0.05001	-0.04999
	2444422.4	33222.2	-0.04075	-0.03157	0.01052
	44.7	31.3	-0.00823	0.00041	-0.00041
23.4	29992.3	313.2	0.00536	0.01144	-0.05415
	3111331.3	229992.9	0.01216	-0.00325	0.00708
	2282.4	339.9	-0.00299	0.00751	-0.00053
234.3	229992.9	323.2	0.00855	-0.00413	0.00511
	2882288.8	333322.2	0.00991	-0.00391	0.00391
	4424.2	3888.8	0.00182	-0.00217	0.00217
2345.6	299222.9	6464.4	-0.00068	0.00392	-0.00594
	7444774.7	333311.3	0.00061	0.00436	-0.00436

Table 2.

4 Comments and Conclusion

In this paper our studies are focused on the digital disks and problems of their representation and reconstruction. Even though the digital disks appear very often in practice of the computer vision and image processing, only the problem of their recognition has been solved till now. In previous sections a representation by constant number of integers, requireing optimal number of bits, is presented. One-to-one correspondence between the digital disks and their proposed codes is proved. That enables an approximate, constant time reconstruction of the digital disk from its proposed code. The efficiency of the reconstruction is analysed and it is shown that relative and absolute errors in estimating the radius of the disk and the coordinates of its center tend to zero while the radius of the disk tends to infinity. The illustration by the experimental results is given.

Let us mention that there exists a better estimation for $R(K)$ ([2]):

$$R(K) = \sum_{\substack{i,j \text{ are integers} \\ (a-i)^2+(b-j)^2 \leq r^2}} 1 = \pi \cdot r^2 + \mathcal{O}\left(r^{\frac{7}{11}} \cdot (\log r)^{\frac{47}{22}}\right)$$

The estimation (3), which is sharp enough for our purpose, is used because of simplicity.

References

[1] L. Dorst and A.W.M. Smeulders, "Discrete representation of straight lines", *IEEE Trans. Pattern Analysis and Machine Intelligence*, vol. 6, pp. 450-463, 1984.

[2] A. Ivić, "Introduction in Analytic Number Theory", (in Serbian), Novi Sad, 1996.

[3] C. E. Kim, "Digital disks", *IEEE Trans. Pattern Analysis and Machine Intelligence*, vol. 6, pp. 372-374, 1984.

[4] R. Klette, I. Stojmenović and J. Žunić, "A parametrization of digital planes by least square fits and generalizations", *Graphical Models and Image Processing*, vol. 58, no. 3, pp. 295-300, 1996.

[5] V. A. Kovalevsky, "New definition and fast recognition of digital straight segments and arcs", Proc. of the tenth international conference on pattern recognition, IEEE Proc. 10662, pp. 31-34, 1990.

[6] M. Lindenbaum and J. Koplowitz, "A new parametrization of digital straight lines", *IEEE Trans. Pattern Analysis and Machine Intelligence*, vol. 13, pp. 847-852, 1991.

[7] R. A. Melter, I. Stojmenović and J. Žunić, "A new characterization of digital lines by least square fits", *Pattern Recognition Letters*, vol. 14, pp. 83-88, 1993.

[8] A. Nakamura and K. Aizawa, "Digital circles", *Computer Vision, Graphics Image Processing*, vol. 26, pp. 242-255, 1984.

[9] P. Sauer, "On the recognition of digital circles in linear time", *Computational Geometry: Theory and Applications*, vol. 2, pp. 287-302, 1993.

[10] M. Worring and A.W.M Smeulders, "Digitized Circular Arcs: Characterization and Parameter Estimation", *IEEE Trans. Pattern Analysis and Machine Intelligence*, vol. 17 pp. 587-597, 1995.

[11] J. Žunić and D.M. Acketa, "Least Squares Fitting of Digital Polynomial Segments", *Lecture Notes in Computer Science: Discrete Geometry for Computer Imagery* vol. 1176. pp. 17-23, 1996.

[12] J. Žunić and N. Sladoje, "Efficiency of Characterizing Ellipses and Ellipsoids by Discrete Moments", submitted.

"One-Step" Short-Length DCT Algorithms with Data Representation in the Direct Sum of the Associative Algebras

Vladimir M. Chernov and Marina A. Chicheva

Image Processing Systems Institute of RAS
151 Molodogvardejskaja st., IPSI RAS, Samara, Russia

e-mail: chernov@sgau.volgacom.samara.su

Abstract. A method of synthesizing DCT fast algorithms for short lengths is proposed. The main ideas are data inclusion into a direct sum of real algebras and the transform interpretation in terms of multiplication rules in these algebras. The synthesized algorithms require less arithmetic operations than the known ones. Results of numerical experiments are given for images being coded with FAs of the DCT described in the present work

1 Introduction

Fast algorithms of a discrete cosine transform (FA DCT)

$$C_x(m) = \lambda_m \sum_{n=0}^{N-1} x(n) \cos\left(\frac{\pi m(2n+1)}{2N}\right), \quad (m = 0, ..., N-1) \tag{1}$$

for short lengths are considered in the work (λ_m are normalizing coefficients).

A special attention devoted to this particular case is explained by three reasons.

First, DCT of length $N = 8$ is the fundamental transform in the set of the accepted videoinformation coding standards (JPEG, MPEG, ITU-T and others [1]).

Second, comparatively short transform length allows to obtain "evident, one-step" formulas of a DCT spectrum calculation. Algorithms synthesized in such a manner are more effective then the common algorithms based on reducing the DCT to the complex DFT of the same length [1].

Third, in tasks of videoinformation coding based on the DCT a restoration quality increases with the block size increase [3].

Many papers are devoted to fast algorithms of the DCT for $N = 8$ (for example, [4], [5], [6]). The algorithms of Ref. [6] have the least computational complexity. Usually, all these algorithms are based on heuristically found factorizing identities for DCT matrices. The lack of common factorization theorems for matrices of discrete orthogonal transforms limits the possibilities of such approach by the classification of algorithms synthesized by independent methods.

Fast algorithms for the DCT for $N > 8$ are significantly less studied. Ref. [7] shows that an odd-length DCT can be reduced to a *real* DFT of the same length. In the Ref. [8] this fact was used along with the modification of data representation as elements of the Eisenstein numbers field [9] to synthesize effective DCT algorithms for $N = 3^k$. Ref. [10] generalizes this approach to the two-dimensional case and Ref. [11] generalizes it for DCTs of other types [12].

In this paper the authors work in the framework of their method of the synthesis of FAs of discrete orthogonal transforms. The method is based on the inclusion data and transform parameters into algebraic structures of the special kind (finite-dimensional alternative algebras) with the following interpretation of the associative transform results in the main real or complex field [13], [14].

In this paper we consider an example of FA DCT synthesis for $N = 8$ in details, and we are limited by comments for $N = 10$. The "evident" formulas for DCT-spectra calculation for all considered cases can be given at the presentation.

2   DCT algorithm of length 8

Let us consider the DCT (1) in the non-normalized matrix form

$$\mathbf{X} = \mathbf{Fx} \qquad (2)$$

where

$$\mathbf{X}^t = (X(0),...,X(7)),\ X(m) = \lambda_m^{-1}C_x(m),\ m = 0,...,7,\ \mathbf{x}^t = (x(0),...,x(7)),$$

$\mathbf{F}$ is the DCT matrix, t is the transposition sign.

Rearranging components of the input and output vectors

$$\mathbf{Y}^t = (Y(0),...,Y(7)) = (X(0),X(2),X(4),X(6),X(1),X(3),X(5),X(7)),$$

$$\mathbf{y}^t = (y(0),...,y(7)) = (x(0),x(2),x(4),x(6),x(7),x(5).x(3),x(1)),$$

the matrix representation of the DCT can be represented in the form $\mathbf{Y} = \mathbf{Ty}$, where

$$\mathbf{T} = \begin{pmatrix} h & h & h & h & h & h & h & h \\ g & -g & g & -g & g & -g & g & -g \\ f & e & -f & -e & f & e & -f & -e \\ e & -f & -e & f & e & -f & -e & f \\ a & c & -d & -b & -a & -c & d & b \\ c & d & -b & a & -c & -d & b & -a \\ d & b & a & c & -d & -b & -a & -c \\ b & -a & c & d & -b & a & -c & -d \end{pmatrix}. \qquad (3)$$

$$a = \cos\left(\frac{7\pi}{16}\right),\ b = \cos\left(\frac{\pi}{16}\right),\ c = \cos\left(\frac{5\pi}{16}\right),\ d = \cos\left(\frac{3\pi}{16}\right),$$

$$e = \cos\left(\frac{6\pi}{16}\right),\ f = \cos\left(\frac{4\pi}{16}\right),\ g = \cos\left(\frac{2\pi}{16}\right),\ h = \cos\left(\frac{0\pi}{16}\right),$$

Let us show that mapping an 8-dimensional vector **y** being transformed into different associative algebras (or, the same, into their direct sum), it is possible to calculate the matrix product $\mathbf{Y} = \mathbf{Ty}$ with 16 multiplications.

Proposition 1. *There exists such an eight-dimensional $\mathbb{R}$-algebra $\mathfrak{B}$, a four-dimensional $\mathbb{R}$-algebra $\mathfrak{A}$, with the basis $\{1, \varepsilon_1, \varepsilon_2, \varepsilon_3\}$, an element $\beta \in \mathfrak{B}$ and operators $\mathcal{B}_0, \mathcal{B}_1, \mathcal{B}_2$:*

$$\mathcal{B}_0 : \mathbb{R}^8 \longrightarrow \mathfrak{B} = \mathbb{R} \oplus \mathbb{R} \oplus \mathbb{C} \oplus \mathfrak{A}, \ \ \mathcal{B}_1, \mathcal{B}_2 : \mathbb{R}^8 \longrightarrow \mathfrak{A}$$

that

- *(a) calculation of the matrix product $\mathbf{Y} = \mathbf{Ty}$ is equivalent to the calculation of the function*
$$Y = \beta \mathcal{B}_0 \mathbf{y} + d\mathcal{B}_1 \mathbf{y} + a\mathcal{B}_2 \mathbf{y};$$
- *(b) calculation of element Y requires 16 real multiplications.*

Proof. Let us consider the set of linear operators:

$$\mathcal{R}_1 : \mathbb{R}^8 \longrightarrow \mathbb{R}, \ \ \mathcal{R}_1 \mathbf{y} = y(0) + y(1) + \ldots + y(7);$$

$$\mathcal{R}_2 : \mathbb{R}^8 \longrightarrow \mathbb{R}, \ \ \mathcal{R}_2 \mathbf{y} = y(0) - y(1) + \ldots - y(7);$$

$$\mathcal{C} : \mathbb{R}^8 \longrightarrow \mathbb{C}, \ \mathcal{C}\mathbf{y} = (y(0) - y(2) + y(4) - y(6)) - i(y(1) - y(3) + y(5) - y(7));$$

$$\mathcal{A} : \mathbb{R}^8 \longrightarrow \mathfrak{A}, \ \mathcal{A}\mathbf{y} = (y(0) - y(4)) + \varepsilon_1(y(1) - y(5)) + \varepsilon_2(y(2) - y(4)) + \varepsilon_3(y(3) - y(7));$$

where $\mathfrak{A}$ is the four-dimensional associative $\mathbb{R}$-algebra with the basis $\{1, \varepsilon_1, \varepsilon_2, \varepsilon_3\}$ and multiplication rules of the basic elements:

$$\varepsilon_1^2 = 1, \ \varepsilon_2^2 = -1, \ \varepsilon_3^2 = -1, \ \varepsilon_3 = \varepsilon_1 \varepsilon_2, \ \varepsilon_1 \varepsilon_2 = \varepsilon_2 \varepsilon_1.$$

It is verified immediately that

$$h\mathcal{R}_1 \mathbf{y}^t = Y(0), \ \ g\mathcal{R}_2 \mathbf{y}^t = Y(1).$$

Therefore calculation of the pair $(Y(0), Y(1))$ requires only two real multiplications.

The next relation is verified immediately too:

$$Y(2) + iY(3) = (f + ie)\left(\operatorname{Re}(C\mathbf{y}^t) + i\operatorname{Im}(C\mathbf{y}^t)\right)$$

and calculation of the pair $(Y(2), Y(3))$ requires three real multiplications.

The least trivial part of the algorithm is reducing the calculation of components

$$(Y(4), Y(5), Y(6), Y(7))$$

to the multiplication of elements of the algebra $\mathfrak{A}$.

The next matrix equality is valid:

$$U = \begin{pmatrix} a & c & -d & -b \\ c & d & -b & a \\ d & b & a & c \\ b & -a & c & d \end{pmatrix} = S + (d-a)\,T_1 + (a+d)\,T_2 =$$

$$\begin{pmatrix} a & c & -d & -b \\ c & a & -b & -d \\ d & b & a & c \\ b & d & c & a \end{pmatrix} + (d-a)\begin{pmatrix} 0&0&0&0 \\ 0&1&0&0 \\ 0&0&0&0 \\ 0&0&0&1 \end{pmatrix} + (a+d)\begin{pmatrix} 0&0&0&0 \\ 0&0&0&1 \\ 0&0&0&0 \\ 0&1&0&0 \end{pmatrix}.$$

Calculation of the matrix product Sz, where z is the four-dimensional vector

$$z^t = (z(0), z(1), z(2), z(3))$$

is equivalent to calculation of the product of elements of the algebra $\mathfrak{A}$:

$$K = (a + c\varepsilon_1 + d\varepsilon_2 + b\varepsilon_3)(z(0) + z(1)\varepsilon_1 + z(2)\varepsilon_2 + z(3)\varepsilon_3) \qquad (4)$$

and requires nine real multiplications.

Indeed, let us represent (4) in the form:

$$K = (A + B\varepsilon_2)(P + Q\varepsilon_2) = V + W\varepsilon_2, \qquad (5)$$

where

$$A = a + c\varepsilon_1, \quad B = d + b\varepsilon_1, \quad P = z(0) + z(1)\varepsilon_1, \quad Q = z(2) + z(3)\varepsilon_1.$$

Then the next equalities are valid:

$$V = AP - BQ = (A - B)Q + A(P - Q). \qquad (6)$$
$$W = BP + AQ = (A - B)Q + B(P + Q).$$

Therefore the calculation of V and W by means of formulas (6) requires three multiplications of elements of the type $(\alpha + \beta\varepsilon_1)$. In its turn, every multiplication of such elements is reduced to three real multiplications:

$$\xi + \zeta\varepsilon_1 = (\alpha + \beta\varepsilon_1)(\mu + \nu\varepsilon_1), \quad \xi = (\alpha + \beta)(\mu + \nu) - \alpha\nu - \beta\mu. \quad \zeta = \alpha\nu + \beta\mu.$$

Calculation of the matrix product $[(d - a)\,T_1 + (a + d)\,T_2]\,z$ can be realized with two real multiplications. Indeed,

$$([(d - a)\,T_1 + (a + d)\,T_2]\,z)^t =$$

$$(0, (d - a)\,z(1) + (a + d)\,z(3), 0, (a + d)\,z(1) + (d - a)\,z(3)) =$$

$$(0, d(z(1) + z(3)) - a(z(1) - z(3)), 0, d(z(1) + z(3)) - a(z(1) - z(3))).$$

Adding together multiplicative complexities of the matrix multiplications given above one obtains that the considered algorithm is realized with $1+1+3+9+2 = 16$ real multiplications.

Corollary 1 *There exists an algorithm of the DCT (2) calculation for $N = 8$ requiring 12 real multiplications.*

Proof. The structure of the considered algorithm does not depend on particular values of parameters $a, b, ..., h$. Therefore, assuming

$$h' = g' = f' = a' = 1, \ e' = ef^{-1}, \ c' = ca^{-1}, \ d' = da^{-1}, \ b' = ba^{-1},$$

it is possible to unite four multiplications from 16 ones with the normalization of the DCT-spectrum components (with multiplications by λ_m).

Remark. The best DCT algorithms from ones known to the authors have the next complexity estimations:

- 12 multiplications and 29 additions for FA DCT of Ref. [6];

- 12 multiplications, 29 additions and 5 multiplications by powers of two for FA DCT of Refs. [4], [5].

3   DCT algorithm for $N = 9, 10, 12$

Using of the techniques given above allows us to synthesize effective FAs DCT for $N = 9, 10, 12$ too. In these cases corresponding algebras $\mathfrak{B} = \mathfrak{B}_N$ have the forms:

$$\mathfrak{B}_9 = \mathbb{R} \oplus \mathbb{R} \oplus \mathbb{R} \oplus \mathbb{R}\mathfrak{G}_3 \oplus \mathbb{R}\mathfrak{G}_3,$$

$$\mathfrak{B}_{10} = \mathbb{R} \oplus \mathbb{R} \oplus \mathbb{C} \oplus \mathbb{C} \oplus \mathfrak{A},$$

$$\mathfrak{B}_{12} = \mathbb{R} \oplus \mathbb{R} \oplus \mathbb{R} \oplus \mathbb{R} \oplus \mathbb{C} \oplus \mathbb{C} \oplus \mathbb{H},$$

where $\mathfrak{A}$ is the algebra mentioned above, $\mathbb{H}$ is the quaternion algebra. $\mathbb{R}\mathfrak{G}_3$ is a group $\mathbb{R}$-algebra of the order-3 cyclic group.

Excluding the quite obvious process of the permutating input and output vector components, one obtains the following analogies of the DCT algorithm considered for $N = 8$.

For example, the DCT matrix for $N = 10$ has the form:

$$
\mathbf{T}_{10} =
\begin{pmatrix}
a & b & c & d & -d & -c & -b & -a & \gamma & -\gamma \\
b & d & -a & -c & c & a & -d & -b & -\gamma & \gamma \\
c & -a & d & -b & b & -d & a & -c & \gamma & -\gamma \\
d & -c & -b & a & -a & b & c & -d & \gamma & -\gamma \\
e & f & -f & -e & -e & -f & f & e & 0 & 0 \\
f & -e & e & -f & -f & e & -e & f & 0 & 0 \\
g & -h & -h & g & g & -h & -h & g & -1 & -1 \\
h & g & -g & h & h & g & -g & h & 1 & 1 \\
\gamma & -\gamma & \gamma & \gamma & -\gamma & -\gamma & \gamma & -\gamma & -\gamma & \gamma \\
1 & 1 & 1 & 1 & 1 & 1 & 1 & 1 & 1 & 1
\end{pmatrix},
$$

where

$$\gamma = \cos\left(\frac{\pi}{4}\right); a = \cos\left(\frac{\pi}{20}\right); b = \cos\left(\frac{3\pi}{20}\right); c = \cos\left(\frac{7\pi}{20}\right);$$

$$e = \cos\left(\frac{\pi}{10}\right); f = \cos\left(\frac{3\pi}{10}\right); g = \cos\left(\frac{2\pi}{20}\right); h = \cos\left(\frac{4\pi}{10}\right).$$

The matrix

$$\mathbf{U}_{10} = \begin{pmatrix} a & b & c & d \\ b & d & -a & -c \\ c & -a & d & -b \\ d & -c & -b & a \end{pmatrix}$$

differs from the matrix $\mathbf{U}$ for $N = 8$ only by the permutation of columns and rows and notifications, the multiplication by this matrix is also interpreted as multiplications of elements in the algebra $\mathfrak{A}$. The multiplications by the matrix of the type

$$\begin{pmatrix} e & f \\ f & -e \end{pmatrix}$$

are interpreted as multiplications in the algebra of complex numbers. The multiplications by numbers $\pm\gamma$ can be united with multiplications by normalizing coefficients λ_m.

4 Conclusions

The algorithms of the DCT described in this work are based on new authors' principle of FAs of discrete orthogonal transforms. An expansion of the set of the lengths for which effective DCT algorithms exist provides videoinformation coding with an adaptive choice of sizes of rectangle $(M \times N)$-blocks, for example, as in the Ref. ([2]). Application of the DCT for $N > 8$ in the block coding of videoinformation increases a processing speed and improves a coding quality. Method possibilities are not limited by the case of one-dimensional DCT. However, for two-dimensional case to obtain direct (non-recursive) formulas becomes complicated.

5 Acknowledgment

This work was performed with financial support from the Russian Foundation of Fundamental Investigations (Grant 97-01-009000).

References

1. Rao, K. R., Yip, P.: Discrete Cosine Transform. Academic Press, San Diego, 1990.
2. Krupiczka, A.: Interblock Variance as a Segmentation Criterion in Image Coding. Mashine Graphics and Vision **5**, Nos **1/2** (1996) 229-235
3. Chichyeva, M. A.: On a Specific of Biosignals Block Coding Based on Discrete Trigonometric Transforms. Proceedings of the 13th Biennial International Conference "Biosignal'96", Czech republic, Brno (1996) 122-124
4. Hou, H. S.: A Fast Recursive Algorithm for Computing the Discrete Cosine Transform. IEEE Transactions on Acoustics, Speech and Signal Processing **ASSP-35**, No 10 (1987) 1455-1461
5. Suheiro, N., Hatori, M.: Fast Algorithms for the DFT and Other Sinusoidal Transforms. IEEE Transactions on Acoustics, Speech and Signal Processing **ASSP-34**, No **6** (1986) 642-644
6. Chan-Wan, Y.-H., Siu, C.: On the Realization of Discrete Cosine Transform Using the Distributed Arithmetic. IEEE Transactions on Circuits and Systems - I: Fundamental Theory and Applications **39**, No **9** (1992) 705-712
7. Heideman, M. T.: Computation of an Odd-Length DCT from a Real-Valued DFT of the Same Length. IEEE Trans. Signal Process **40**, No 1 (1992) 54-61
8. Chernov, V. M.: Fast Algorithm of the Odd-Length Discrete Cosine Transform. Automatic Control and Computer Science **3** (1994) 62-70
9. Ireland, K., Rosen M.: A Classical Introduction to Modern Number Theory. Springer, 1982
10. Chernov, V. M.: Algorithms of Two-Dimensional Discrete Orthogonal Transforms Realized in the Hamilton-Eisenstein Codes. The Problem Transmission of Information **31**, No **3** (1995) 38-46
11. Chichyeva, M. A.: Algorithms of Discrete Cosine Transforms with Data Representation in Eisenstein's Codes. Image Processing & Communications (to appear)
12. Wang, Z.: Fast algorithms for discrete W transform and for the discrete Fourier transform. IEEE Trans. Acoust., Speech, Signal Processing **ASSP-32** (1984) 803-816
13. Chernov, V. M.: Arithmetic Methods in the Theory of Discrete Orthogonal Transforms. Proceedings SPIE **2363** (1995) 134-141
14. Chernov, V.M.: Fast Algorithms of Discrete Orthogonal Transforms for Data Represented in Cyclotomic Fields. Pattern Recogn. and Image Anal. **3**, No4 (1993) 455-458

Character Extraction from Scene Image Using Fuzzy Entropy and Rule-Based Technique

Opas Chutatape and Li Li
School of Electrical and Electronic Engineering
Nanyang Technological University, Singapore 639798
E-mail: eopas@ntuvax.ntu.ac.sg

Abstract: This paper presents a segmentation technique based on fuzzy logic and information entropy principle to extract alphanumeric characters from general scene images. In this approach, the scene images are segmented into various regions based on fuzzy entropy method. A coarse searching technique is then implemented to locate potential character regions, followed by a rule-based character detecting technique to accurately extract characters. The x-projection/profile is also applied to search for any missing characters. Test results show that the proposed method is capable of extracting alphanumeric characters from various scene images under badly illuminating condition.

Key words: Thresholding, region grouping, fuzzy entropy, pattern analysis

1. Introduction

Problems related to the replication of human functions are often attractive to researchers. Such an immense effort has been spent on this subject not only because it is a very challenging problem, but also it provides a solution for processing large data automatically. In recent years, great progress has been made in optical character reader (OCR) technology. Most OCRs in current use, however, can only read characters printed on a sheet of paper according to rigid format restrictions and are mainly applied to office automation systems such as document readers. If OCRs could read characters in complex scene images, they would be very useful in more applications such as electronic road pricing system, package and label reading system, automatic monitoring systems in assembly line, and robotic control systems, etc.

Unlike characters in documents, characters in scene images can suffer from a variety of noise components because the scene images are captured under uncontrolled illumination conditions. A number of algorithms [1,2] have been proposed to identify and extract characters by using x-projection/profile to determine the position of the car license plate. However for a complex scene image which is very different from document reading system, it's difficult to find any obvious difference between two regions if one contains characters. Other character extraction algorithms [3,4,5] utilise the grey-level difference between neighbouring pixels to segment individual characters. Yet how to choose the difference of grey levels is perplexing because various scene images are captured under different circumstance. If this value is too low, the image will be subject to the noise effect and if too high, the plate identity may be lost. In Ref.[6,7,8], it was suggested to do local threshold first by dividing the whole image into equal blocks. Using local threshold will help overcome the non-linear illumination condition, however, the binary result can still be influenced by the size of sub-block and it is time consuming.

For segmentation based on threshold, the ideal situation requires that the threshold value, which can separate objects of interest from their background, can be easily found. Yet in a scene image, there are various uncertainties and non-uniformity among various regions such that the conventional global thresholding method will frequently fail. More recently, fuzzy set theory has gained a lot of attention in computer vision particularly in the areas of thresholding to partition the image space into meaningful regions, due to its superior performance in handling uncertainties. In this paper, an algorithm based on fuzzy logic and the information entropy principle is extended to solve the problem of alphanumeric character extraction from scene image. Specific rule-based criteria of character like region identification are used to effectively locate potential characters, then a fine grouping technique is later used to detect extra characters. No knowledge of the approximate location of the characters is assumed and the method is robust for extracting the character object that is either light under dark background or dark under light background.

In Section 2, fuzzy sets, fuzzy entropy and membership function are given. The algorithm outline of the proposed method is discussed in Section 3. The results on scene images which contain alphanumeric character string are shown in Section 4, and finally the conclusion is given in Section 5.

2. Fuzzy Sets, Fuzzy Entropy and Membership Function

An image I of size $M \times N$ with L grey levels can be considered as an array of fuzzy sets [9]. Each pixel has a value of membership denoting its degree of brightness relative to some brightness level l, $l=0, 1, 2, ..., L-1$. The notation of fuzzy sets can be defined as follows:

$$I = \{\mu_l(x_{mn}) = \mu_{mn}/ x_{mn}\} \qquad m =1,2...M, n=1,2... N$$

where $0 \le \mu_l(x_{mn}) \le 1$ denotes the grade of some brightness property possessed by the (m, n)th pixel with the intensity x_{mn}. For simplicity, $\mu_l(x_{mn})$ is replaced by μ_{mn}. The degree of ambiguity in an image I can be measured by the entropy of a fuzzy set [10]:

$$H(I) = \frac{1}{MN\ln 2} \sum_{m=1}^{M} \sum_{n=1}^{N} S_n(\mu_{mn})$$

where $S_n(\mu_{mn})$ is the Shannon's function

$$S_n(\mu_{mn}) = - \mu_{mn} \ln\mu_{mn} - (1-\mu_{mn}) \ln(1-\mu_{mn})$$

Using grey level g instead of the (m, n)th pixel intensity x_{mn}, $H(I)$ can be replaced as:

$$H(I) = \frac{1}{MN\ln 2} \sum_{g} S_n(\mu_l(g)) \cdot h(g) \qquad g= 0,1,... L-1 \qquad (1)$$

where $h(g)$ denotes the number of occurrences of the grey level g. The standard S-function[10] is applied to define the following membership function $\mu_l(g)$:

$$\mu_I(g) = S(g, a, b, c) = \begin{cases} 0 & g \le a \\ 2 \times \left[\dfrac{(g-a)}{(c-a)}\right]^2 & a \le g \le b \\ 1 - 2 \times \left[\dfrac{(g-c)}{(c-a)}\right]^2 & b \le g \le c \\ 1 & g \ge c \end{cases} \tag{2}$$

where $0 \le g \le L\text{-}1$, b is the cross-over point; the interval $[a, c]$ is called a fuzzy region. The portions of image I in $[g_{min}, a]$ and $[c, g_{max}]$ are crisp or non-fuzzy. For a given interval $[a, c]$ if the crossover point b is varied, different values of fuzzy entropy $H(I)$ will be obtained. A set of local maxima of $H(I)$ indicates the occurrence of great uncertainties where the boundary points may locate. Here the maximum value of $H(I)$ rather than the minimum as indicated in [10] will be used since this is in agreement with the maximum information concept used in [12].

3. Algorithm
3.1 Proposed Fuzzy Entropy Algorithm
Given an $M \times N$ image in 256 levels, with minimum and maximum grey level g_{min} and g_{max}, the algorithm to determine the threshold values is stated as follows:

Step 1: Compute the histogram $h(g)$ of the image I which is requantized from 256 to 64 levels. The requantization helps reduce individual noise and save processing time.

Step 2: Construct the membership μ_I, using Eq. (2)

Step 3: Vary b from g_{min} to g_{max}, and compute the fuzzy entropy function $H(I)$ using Eq. (1) for each b.

Increasing the fuzzy region width $w = c\text{-}a$ leads to the possibility of losing some of the weak minima in $H(I)$ as mentioned in [10] as well as its maxima, and as w decreases, some undesirable local maxima will be detected. Based on various testing, it was found that $w = 10$ was suitable for 64 requantized levels used in the images under study.

Step 4: Select those b's, i.e., $b_1, b_2, ..., b_n$ corresponding to local maxima of fuzzy entropy $H(I)$ within g_{min} to g_{max}.

Step 5: Let N_T be the total number of local maxima, t_i be the ith threshold value, set t_i in decreasing order of $H(I)$ magnitude, i.e.,

$t_i = b_i$ given that $H(I)|_{t_i} \ge H(I)|_{t_{i+1}}$, $i=0 ... N_T\text{-}1$

After the above effective thresholding method is employed, the thresholded image $f'(x,y)$ results in N_T binary images given by

$$f'(x, y) = \begin{cases} 0 & \text{for } f(x,y) \le t_i \\ 1 & \text{otherwise} \end{cases} \quad i=1,2,...N_T$$

The image $f(x,y)$ is divided into regions based on the local maximum t_i, the analysis is performed on the image $f'(x, y)$ in which the objects of interest may be present.

3.2 Boundary Detection

After one threshold is selected, pixels whose grey levels are larger than the threshold are labelled as "1", and pixels whose grey levels are smaller than the threshold are labelled as "0". Then the mathematical morphology operations [11] are implemented to smooth the image and get boundary pixels. This is also followed by boundary tracing by chain-code method to obtain geometric specifications of the regions. Due to the grey-level information on characters and their background are not given in advance, whether characters appear as dark or bright is unknown. So both opening and closing operations are applied to obtained smooth image E, and region boundary β can be later extracted.

$$E(A) = (A \circ B) \bullet B \tag{3}$$

$$\beta(A) = E - (E \ominus B) \tag{4}$$

$$B = \begin{bmatrix} * & 1 & * \\ 1 & 1 & 1 \\ * & 1 & * \end{bmatrix} \text{ is a structuring element matrix} \tag{5}$$

where "$\circ$" is opening, "$\bullet$" is closing, "$\ominus$" implies erosion and "$-$" is set *difference*. After the combined morphology operations (Eq.3 and Eq.4) are applied, only those pixels labelled as "1" are boundary pixels. The boundary chain-code method is then used to trace the contours and the top-left and the bottom-right points of the bounding box are then recorded.

3.3 Extraction of the Character Candidates

Because of some general knowledge of objects being known a priori, the rule-based criteria can be applied to identify the character-like regions. The aspect ratio and size testing are the first two steps to carry out to discard the non-character regions. It can be expressed as follows:

if $AR_{min} \leq AR_i \leq AR_{max}$ then   keep region R_i.

if $Height_{min} \leq y_{imax} - y_{imin} \leq Height_{max}$ then   keep region R_i

where $AR_i = \dfrac{x_{i\,max} - x_{i\,min}}{y_{i\,max} - y_{i\,min}}$ $(x_i, y_i) \in R_i$ is *Aspect Ratio*, $Height_{min}$ and $Height_{max}$ are size limits of a potential character, which are obtained from the image system set-up knowledge and object knowledge.

After the first two steps, only those objects that match these two criteria would be kept. The algorithm will continue to compare the relationship and other geometric attributes of the remaining objects using other rules as follows:

Horizontal alignment:

if $\dfrac{\left| (y_{i\,max} + y_{i\,min}) - (y_{j\,max} + y_{j\,min}) \right|}{2} \leq \rho_h \times \max.\{(y_{imax} - y_{imin}), (y_{jmax} - y_{jmin})\}$

then   keep region R_i and R_j

where ρ_h is parameter for horizontal alignment checking

Distance measure:

if $\quad \dfrac{\left|(x_{i\,min}+x_{i\,max})-(x_{j\,min}+x_{j\,max})\right|}{2} \le \rho_d \times \max.\{(x_{imax}-x_{imin}),(x_{jmax}-x_{jmin})\}$

then keep the region R_i and R_j,

where ρ_d is the parameter for distance checking

Neighbouring region similarity:

if $\quad \dfrac{\left|(y_{i\,max}-y_{i\,min})-(y_{j\,max}-y_{j\,min})\right|}{\min\{(y_{i\,max}-y_{i\,min}),(y_{j\,max}-y_{j\,min})\}} \le \rho_s$ then keep region R_i and region R_j

where ρ_s is the parameter for distance checking

Run-length measure:

if $\quad H_{max} \le \rho_r \times [\max(y_i)-\min(y_i)]$ $\qquad\qquad$ then keep the region R_i

where ρ_r is the parameter for run-length measure, H_{max} is maximum number of consecutive rows with $W_{max} > \rho_r \times [\max(x_i)-\min(x_i)]$ and W_{max} is maximum number of consecutive pixels belonging to the object in one row.

If there is no potential character-like region that can be detected after examining the above criteria, it implies the currently setting threshold t_i is not suitable for extracting objects of interest. Thus the threshold value is updated to another local maximum t_{i+1}.

The aligned object regions are grouped together after all the criteria are examined. Because some characters may be missing, it is required to extend the detected character regions by one character-like width each time at both left and right ends. The histogram of x-projection of the extended area is used to check the grey-level frequency variation in order to find more possible character regions till no more is detected. Only those regions that pass through all the criteria checking will still remain for further processing.

3.4 Binarization of the Character Region

A local threshold is selected for each accepted character region in the original grey image based on maximum entropy principle[12]. This method considers two probability distributions: object and background entropy. The sum of the individual entropy of the object and background is then maximised, which means the information of the two clusters in the histogram is maximised. The total entropy of the image is defined as follows:

$$H(X; y=x_t) = H_{(O)} + H_{(B)}$$
$$= -\sum_{i=1}^{T}[p_i/P_1(T)]\log[p_i/P_1(T)] - \sum_{i=T+1}^{n}[p_i/P_2(T)]\log[p_i/P_2(T)]$$

After binarization, some post-processing such as smoothing and normalisation are performed. Thereafter the bi-level extracted characters can be passed on to the next stage for recognition.

4. Experimental Results

The character extraction algorithm presented here has been applied to various images taken in different scenes and lighting conditions. The image is acquired through CCD camera as an input signal and detected character segments are given as output results. Fig.1-3 are three of the experimental results of character extraction. Fig.1 and Fig.2 are car images containing car plate numbers under poor lighting condition. Fig.3 is an image of identification card with pattern background. Table 1 shows the geometric features of character regions in Fig.1a. As shown in Fig.2c, only the regions passing aspect ratio and size checking are kept, and some of these regions will fail to get through after using other criteria checking. In Fig.3c, there is no existing potential character region based on the local maximum of $H(I)$ at 158. The threshold value is therefore updated to another local maximum at 26 and the boundary detection result is obtained and shown in Fig.3d. In Fig.1d, 2d, and 3e, only the character regions are binarized for further processing. The algorithm proposed here is successful in extracting character strings in these cases.

5. Conclusion and Discussion

In this paper, an effective algorithm has been developed to successfully extract alphanumeric characters from scene images. The experimental results show that the proposed algorithm is capable of dealing with various complicated scenes with non-uniform illumination conditions and containing different character sizes and of moderate viewing angle variation.

The experimental results also indicate that the thresholding method based on fuzzy logic and information entropy principle gives satisfactory performance. The use of the fuzzy range can help locate effectively the local maxima of entropy corresponding to the uncertainties among various regions in the image in order to find the optimal partition of image. This approach is suitable for the images whose histograms have no clear peaks and valleys or the number of the segmentation classes is unknown.

References

[1] B.T. Chun, J. Soh and B.W. Min, Locating Number Plates on Vehicles by Using Topographical Features, *Proc. ACCV'93*, 430-433, Japan, 1983

[2] T. K. Lim, J. T. Ong, *et al.*, Image Segmentation and Character Recognition of Vehicle License Plates for Electronic Road Pricing, *Proc. ICARCV'92*, 1663-1667, Singapore, 1992

[3] J. Ohya, A.Shio and S.Akamatru, Recognising Characters in Scene Images, *IEEE Trans. PAMI*, Vol.16, No. 2, 214 - 220, Feb. 1994

[4] M. He and A. L. Harvey, Vehicle Number Plate Location for Character Recognition, *Proc. ACCV'95*, 1425-1428, Singapore, 1995

[5] C. M. Lee and A. Kankanhalli, Automatic Extraction of Characters in Complex Scene Images, *Patt. Recog. Artif. Intell*, 9(1), 67 - 82, 1995

[6] L. Z. Yong and Opas Chutatape, A Method for Car Number Plate Character Extraction, *Proc. IEEE Int. Conf. on SPCS'95*, 256 - 259, Singapore, 1995

[7] K. Lim, J.T. Ong, and M. A. Do, Development of an Electronic Camera Enforcement System for Electronic Road Pricing, *Proc. ICARCV'92*, CV15.3.1- 15.3.6, Singapore, 1992

[8] K. Miyamoto, K. Nagano, *et al.*, Vehicle Licence-Plate Recognition By Image Analysis, *Proc. IEEE IECON'91*, 1734 - 1738, Japan, 1991

[9] S.K. Pal and D. Dutta Majumder, *Fuzzy Mathematical Approach to Pattern Recognition*, John Wiley & Sons, New York, 1986

[10] S.K. Pal and Azriel Rosenfield, Image Enhancement and Thresholding by Optimization of Fuzzy Compactness, *Patt. Recog. Lett.* 7, 77-86, 1988

[11] A.K. Jain, *Fundamentals of Digital Image Processing*, Prentice-Hall, 1989

[12] J.N.Kapur, P.K.Sahoo and A.K.C. Wong, A New Method for Grey-Level Picture Thresholding Using the Entropy of the Histogram, *CVGIP*, V29, 273-285, 1985

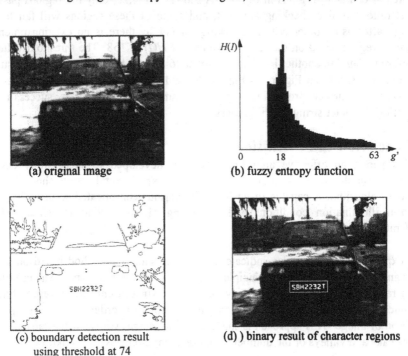

(a) original image   (b) fuzzy entropy function

(c) boundary detection result   (d)) binary result of character regions
using threshold at 74

Fig.1. Testing image of a car

regions	co-ordinate				size		aspect
	left	top	right	Bottom	width	height	ratio
S	249	324	262	341	13	17	0.765
B	267	323	280	342	13	19	0.684
H	283	323	296	341	13	18	0.722
2	302	323	314	340	12	17	0.706
2	319	322	332	341	13	19	0.684
3	336	322	348	340	12	18	0.667
2	353	322	366	340	13	18	0.722
T	372	321	385	339	13	18	0.722

Table 1 geometric features of character regions

604

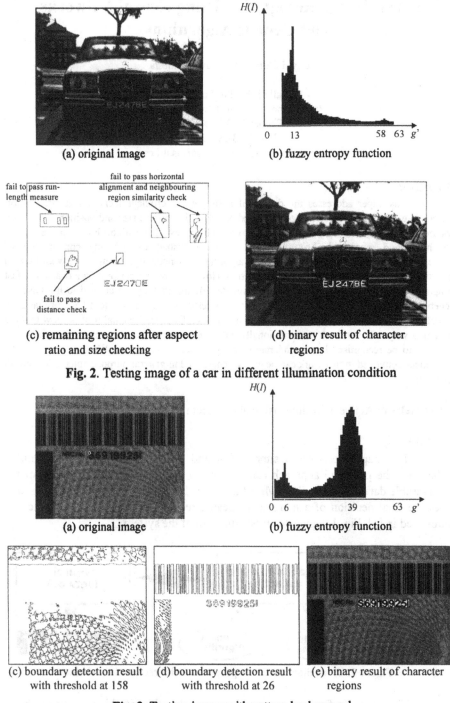

(a) original image

(b) fuzzy entropy function

(c) remaining regions after aspect
ratio and size checking

(d) binary result of character
regions

Fig. 2. Testing image of a car in different illumination condition

(a) original image

(b) fuzzy entropy function

(c) boundary detection result
with threshold at 158

(d) boundary detection result
with threshold at 26

(e) binary result of character
regions

Fig. 3. Testing image with pattern background

Facial Image Recognition Using Neural Networks and Genetic Algorithms

David Carreño and Xavier Ginesta

La Salle School of Engineering
Pg. Bonanova 8, Barcelona 08022. Spain
Phone: 34-3-290-2436
Fax: 34-3-290-2416
e-mails: se03727,xginesta@els.url.es

Abstract:

This paper addresses the design of a simple yet efficient facial image recognition system. We show that a face can be recognised based on the relative size and position of its basic features, i.e., eyes, nose and lips. The key to the efficiency of our algorithm is in the feature search method employed. Feature search is accomplished through the combination of conventional template matching and genetic algorithms. Genetic algorithms alone would take a long time in order to converge to a valid solution. However, by first performing a coarse but fast template matching, we can obtain an approximate solution that can be utilised to initialise the genetic algorithm. The output of the facial feature detection stage is fed to a back-propagation neural network which accomplishes the recognition task. Our experimental results show that the system is very efficient both computationally and in recognition accuracy as long as the facial database to be recognised has a moderate size (16 in our experiments). We also note that the basic ideas conveyed in this work can be easily generalised to general purpose object recognition applications.

CATEGORY: Active, real-time vision and Object recognition.

SUMMARY

This paper describes a simple, fast and efficient facial image recognition technique. The proposed approach can be used in applications where the size of the recognisable database is small or, when this size is large, as an image classifier used to asses the final decision of a more sophisticated recognition system, such as the one introduced in [1]. Fig. 1 depicts the block diagram of the system.

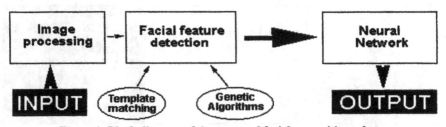

Figure 1: Block diagram of the proposed facial recognition scheme

The image is first pre-processed in order to eliminate high frequency noise and to obtain its gradient. By working on the gradient, the system will be insensitive to contrast and illumination conditions. Next, the facial feature detection stage identifies the relative positions between the main facial features, namely eyes, nose and lips. Facial feature detection has been widely used as a basic building block in several image recognition systems [1], [2], [3].

In this paper we introduce a new technique, based on the combination of pattern matching and genetic algorithms, to efficiently carry out the feature search. In a first step, a template containing the facial features being searched for is matched at different positions and scales against the gradient of the input image. In order to keep the computational complexity down, the matching is performed only for a limited number of positions and scales. This will provide a rough estimate about the position and size of the facial features. The accuracy of both parameters will be refined through the use of genetic algorithms. In fact, the template matching process can be considered as an initialisation method for the genetic algorithm.

The last stage of the system is a back-propagation neural network. This network has been trained with the images that comprise the database that must be recognised. Using the information provided by the feature detection stage, the neural network was always capable in our experiments to correctly identify the input image. This indicates that the relative size and position of the selected facial features are excellent parameters for face recognition.

The image processing stage is discussed in Section 1. Section 2 deals with facials feature detection whereas Section 3 describes the recognition engine. Our experimental results are provided in Section 4, which also draws the conclusions.

1.- Image Pre-processing

In order to eliminate the sensitivity to contrast and illumination variations from image to image our system operates on the gradient of a face. Prior to computing the gradient, however, any possible high frequency noise in the image is filtered out using a 2-D median filter of size 3x3. Next, the gradient is computed using a Sobel mask. Fig. 2 Shows the result of computing the gradient of an image with high frequency noise and the improvement obtained when the image is pre-processed with a median filter.

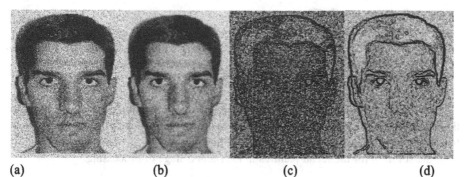

(a)   (b)   (c)   (d)

Figure 2: **(a) Original image; (b) Median filtered image; (c) Gradient of (a); (d) Gradient of (b)**

2.-Facial Feature Detection

The goal of facial feature detection is to identify the relative positions of the essential features of a face, i.e., eyes, nose and lips. This information will be passed along to the recognition engine which, using a back-propagation neural network, will try to recognise the face. Facial feature search proceeds in two consecutive stages. The first is based on a simple template matching approach. The purpose of template matching is to find the position and scale of the face in the image and to provide a rough estimate about the location of its basic features. This initial estimate will be refined in a second phase through the use of genetic algorithms. In the following, a detailed description of both processes, template matching and genetic algorithms, is provided.

(a) Template matching:

The first stage of feature detection is template matching. The key idea is to find the best match between the gradient image and a standard template. This template is a binary pattern, such as the one shown in Fig. 3, obtained by averaging the facial patterns of the same people that will be included in the database of recognisable faces. Since the position and size of the face is not known a priori, the pattern is scaled at different sizes and, for each size, it is tested in a predetermined number of positions. The actual number of positions checked depends on the scale of the pattern and on the admissible tolerance (in pixels) for the alignment error. Since the result of template matching will be further refined by a genetic algorithm, a relatively small number of positions is checked at each scale This significantly contributes to keep down the computational complexity of the system. For each position tested, the *quality* of matching is measured by the following matching factor:

$$\text{MF} = \sum_i \sum_j Gradient(i,j) * Pattern(i,j) \quad \text{(Eq. 1)}$$

where *gradient(i,j)* and *Pattern(i,j)* are the gradient and pattern images, respectively. Since the pattern is binary, the matching factor will be largest at the position for which the degree of coincidence between pattern and gradient is highest.

Figure 3: Binary facial template used for pattern matching. Notice that it contains all the basic features being searched for

Fig. 4 shows the result of template matching for a sample image. Notice that the matching is not perfect. However, the main goal of finding the position and scale of the whole face has been accomplished. Genetic algorithms will provide a more accurate solution for the facial features in the second stage of processing.

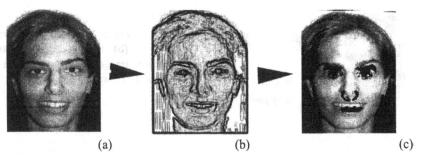

Figure 4) First stage of feature detection. (a) original image; (b) gradient (c) result of template matching

(b) Genetic Algorithms:

Genetic algorithms imitate nature's evolution process in order to find the best solution to a problem [5]-[7]. A genetic algorithm is initialised with any given set of possible solutions, where each solution is characterised by a chromosome (a certain binary value). Chromosomes evolve from generation to generation according to the rules of evolution: (1) *Selection*: only the fittest survive; (2) *Mutation*: The value of a chromosome can randomly change (mute); and (3) *Crossover*: the cross of any two chromosomes may give rise to a brand new chromosome. Due to the random features of genetic algorithms, they can often get around local minima in the quest for the optimum solution. Another interesting characteristic of this process is that it is non-deterministic, i.e., different runs of the algorithm will provide different results.

In the context of facial feature detection, genetic algorithms will help in refining the solution provided by the template matching algorithm. It is assumed that the relative positions and scaling of the basic facial features found in the previous search are approximately correct. Thus, we now concentrate on refining the position and scaling of only the eyes. Once these are found, the exact position and scaling of nose and lips will be implicitly determined. Fig. 5 illustrates the basic chromosome structure used to represent each possible solution. As shown, a chromosome is a 32-bit binary value divided into four 8-bit fields. From left to right, the fields represent the (x,y) position and the x and y scaling of the *eyes* pattern shown in Fig. 6.

00001001001....011001001

X	Y	X Scale	Y Scale
8 bits	8 bits	8 bits	8 bits

Chromosome binary representation   Chromosome basic structure

Figure 5: Chromosome structure and its different fields

The algorithm is initialised with a population of 20 chromosomes, each being a zero-mean Gaussian perturbation of the position and scaling values obtained through template matching. The rules of evolution are iteratively applied to the initial population as explained next.

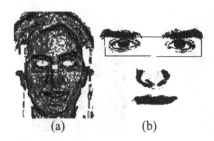

Fig 6: The pattern within the box in (b) is matched against the gradient image in (a) to find the right position and scaling of the eyes.

Selection: A matching factor is computed for each chromosome according to eq. (1). All matching factors are added up to obtain the *fitness* level of the overall population. The probability of a chromosome to survive to the next generation is defined as the ratio of its matching factor to the overall fitness factor. At each iteration, 20 new chromosomes are selected according to these probabilities, so that only the fittest will have a god chance of making it to the next generation.

Mutation: A randomly chosen chromosome can be muted by randomly changing the binary value of some of its bits.

Crossover: The crossover operation results from the combination of two randomly selected chromosomes, which are crossed as follows:

$$(b_1 b_2 ....b_p b_{p+1} ....b_m) \text{ and } (c_1 c_2 ....c_p c_{p+1} ....c_m)$$

will be replaced by:

$$(b_1 b_2 .....b_p c_{p+1} .....c_m) \text{ and } (b_1 b_2 .....b_p c_{p+1} .....c_m)$$

The operations of mutation and crossover prevent the genetic algorithm from getting stuck in local minima. After the algorithm has evolved for a sufficiently long number of generations, the fittest chromosome from the last generation is selected as the solution to the pattern matching problem. Notice that this solution gives us only the right position and scaling of the eyes in the gradient image. The position and scaling of nose and lips can be inferred by properly combining this information with the results obtained in the template matching process.

3.- Recognition

The pattern matching block discussed in the previous section had the mission of identifying the relative position and size of the basic features in a face. This information will be used by a back-propagation neural network for recognition purposes. Basic acquaintance with this type of network is assumed and we do not dwell with its intricacies here. The interested reader is referred to the numerous literature on this topic [4], [8]- [10]. We simply emphasise that the network has as many outputs as images it has been trained with (each output is *linked* to a given image). The number of input and hidden neurones is a design parameter. We used 168 input and 60 hidden neurones, respectively, in our experiments. An important issue is how the data are fed to the network. The output of the facial feature detection stage is a scaled pattern consisting of real values. These values are first normalised to unity and, next, the pattern is re-scaled so that the number of nonzero pixels equals the number of input neurones. Each nonzero pixel will be fed to an input neurone following a raster scan order.

Upon presenting an input pattern to the network, each of its outputs will supply a value which represents the probability that the input is the image linked to that output. The network always provides an answer regardless of whether the input image was in the training set or not, Fig. 7 provides a graphical description of all the processes involved in the proposed recognition system.

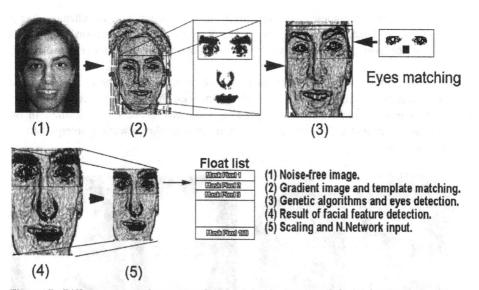

Figure 7: Different processing stages involved in the proposed facial image recognition scheme.

4.- Results and Conclusions

In order to test the performance of the proposed image recognition system we trained it using a database of 16 facial images. The back-propagation neural network had 168 input neurones, 60 hidden neurones, and 16 output neurones, each linked to an image in the database. The output value of each neurone can be interpreted as the degree of confidence that the input is the image linked to that neurone. We run numerous tests, many of them using highly distorted images, and the system always correctly recognised the input. Fig. 8 illustrates some examples showing the level of confidence output by the winner neurone.

One of the main attributes of the system is its simplicity: It can provide the right answer based only on the relative size and position of eyes, nose and lips. This translates into a processing time of about 15 seconds on a 120 MHz Pentium-based PC. Notice too that the system has been simulated on an inherently sequential machine. A parallel implementation can significantly speed up the algorithms involved in the facial feature detection and recognition processes. One limitation is the size of the image database that can be recognised. This size is a function of several system parameters: the resolution of the input image, the size of the pattern that is fed to the neural network, and the number of input, hidden and output neurones. The problem is that all of these parameters have a significant impact on system complexity. For example, a large database requires a large number of output neurones, causing the complexity of the neural network to grow unwieldy. For large databases, an interesting application of the proposed system could be as a classifier used in conjunction with another recognition technique, such as that of [1]. We are currently exploring this line of research.

Another limitation is the sensitivity of our system to pose change. A 3-D rotation of the head, such as the one in Fig. 9, causes an identification failure. This problem, however, can be easily resolved using image warping techniques as in [1].

An important contribution of our work is the use of genetic algorithms for feature search. Initialising the genetic algorithm with the result of a coarse template matching process can significantly speed up the convergence of the former. In our experiments, in just a few iterations the genetic algorithm always converged to the right solution.

   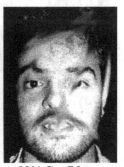

40% Confidence   56 % Confidence   83% Confidence   92% Confidence

Figure 8: Some examples of positive recognition on highly distorted input images.

Figure 9: Face misidentification
due to 3-D rotation of the head

References

[1] B. Moghaddam and A. Pentland, "Face Recognition Using View-Based and Modular Eigenspaces for Face Recognition", Automatic Systems for the Identification and Inspection of Humans, SPIE vol. 2277. 1994.

[2] X. Jia and M.S. Nixon, "Extending the Feature Vector for Automatic Face Recognition", IEEE Trans. on Pattern Analysis and Machine Intelligence, vol. 17, no 12, pp. 1167-1176, Dec. 1995.

[3] D. Carreño, "Sistema de Reconeixement d' Imatges Facials a través de Xarxes Neuronals i Algorismes Genètics", Bachelor's thesis, La Salle School of Engineering, Barcelona, Spain

[4] Madan M. Gupta and George K. Knopf, Neuro Vision Systems, IEEE Press, 1994

[5] J. Stender, Parallel Genetic Algorithms: Theory and Applications. IOS Press, 1993

[6] W. Lewicz, Genetic Algorithms + Data Structures = Evolution Programs. Springer Verlag

[7] D.E. Goldberg, Genetic Algorithms in Search, Optimization and Machine Learning. Addison Wesley, 1989.

[8] C. Torras and G. Wells, "An Introduction to Neural Networks", Institut de Cibernètica internal report, Univ. Politècnica de Catalunya, Barcelona, Spain

[9] G.A. Carpenter and S. Grossberg, Neural Networks for Vision and Processing, MIT Press, 1993

[10] J.L.McCLelland, D. E.Rumelhart, Parallel Distributed Processing , "Explorations in the Microstructure of Cognition". Vol II: *Psychological and Biological models*.

An Energy Minimisation Approach to the Registration, Matching and Recognition of Images

Vassili Kovalev[1] and Maria Petrou[2]

[1] Institute of Mathematics, Belarus Academy of Sciences, Kirova St., 32-A, Gomel 246652, Belarus
[2] School of Electronic Engineering, Information Technology and Mathematics, University of Surrey, Guildford GU2 5XH, United Kingdom

Abstract. This paper describes an image registration method for deformable objects like objects in medical images, in face or fingerprint recognitions tasks, etc. The method uses a global optimisation approach where a cost function is defined, capable of incorporating all available data and knowledge concerning the task. The method is demonstrated within the context of face recognition.

1 Introduction

The problem of registering images that depict deformable objects has attracted considerable interest recently. It pertains to several important application areas, like medical image registration, fingerprint identification, face recognition etc. In all these cases the objects depicted are subject to minor, but possibly significant, for the recognition process deformations, in their different instantiations. Methods that have been proposed in the past include correlation techniques (eg [9]), landmark points based methods (eg [4,6,7]), deformable super quadrics (eg[10,8]), finite element methods (eg [2]), locally adaptive scaling (eg [1]) and flexible contour matching extracted after the segmentation of images (eg [3]).

In this paper we propose an approach that is based on the optimisation of a global energy function. This approach allows the incorporation of any available source of information, including constraints expressing the model we want to impose on the data, and any measurements that are available. Our work is similar to the global optimisation approach presented in [5], where the membrane model of regularization is imposed on the data, and new matching states are generated by hypothesising a local displacement vector from which the likelihood of similarity between the two points is inferred based on the measurements at the two points. In our approach, although the constraint imposed by the membrane model is also used, the new states are generated by an "elastic" local deformation in a randomly chosen direction, controlled by a parameter expressing the elasticity of the image, and the quality of each registration is judged by the value of the global cost function that incorporates the global correlation coefficient between the two images, and the overlapping achieved.

We demonstrate our method in the context of the problem of face recognition. To that effect, we make use of 39 images from the Manchester face image database [6,7]. The results presented here are preliminary results aimed at demonstrating a generic approach rather than exhaustive tests. Thus, we make no attempt at this stage to compare the results with those reported in [6,7].

In section 2 we shall present the method proposed, in section 3 we shall describe some experimental results, and we shall conclude in section 4.

2 The proposed global optimisation approach

Let us assume that we have two images, I_1 and I_2. A pixel i at location (x_i, y_i), is supposed to have grey value p_i^1 in the first image and p_i^2 in the second image. To register the two images, we create a sequence of gradually deformed images, starting from I_2, and denoted by $I_3, I_4, \ldots$. The grey values, p_i^l, of each pixel i in I_l have to be chosen from those of I_{l-1} in order to increase the similarity with I_1. To achieve that, we need:
- A mechanism to distort I_{l-1} in order to form I_l.
- A cost function that measures how bad the registration between I_1 and I_l is.
- A mechanism to search the minimum of the cost function.

It is clear that the distortions observed in the types of image pairs we are interested in, are inhomogeneous, anisotropic and non-linear. Therefore, they cannot be modelled by a global transformation function. Instead, one has to adopt a series of local deformations, which collectively model the overall deformation. We adopt the following process: We choose at random a pair of pixel positions (x_i, y_i) and (x_j, y_j), within a certain distance from each other. All remaining pixels of image I_{l-1} will shift location according to the following law: A pixel k will move in the direction of the vector defined from i to j, and by a distance given by

$$d_k = d_{ij} e^{-s d_{ik}}, \tag{1}$$

where d_{ij} is the distance between j and i, d_{ik} is the distance between k and i, and s is the "springiness" parameter that controls the severity of the distortion. Figure 1 demonstrates this model of distortion for various values of the parameter s. These new positions of the pixels of image I_l are clearly non-integer positions. The grey values at the integer positions of the image grid are calculated using the nearest neighbour interpolation rule. These interpolated values are only used for the comparison of image I_l with the target image. Once this comparison is made, these values are discarded, and the next image in the sequence is formed from the non-integer positions of pixels in image I_l. Image I_l is going to be accepted as the next one in the sequence, provided that it reduces the cost function of the quality of registration with image I_1. Then in the new image a pixel i is picked at random, and a new position (x_j, y_j) chosen also at random, and the process is repeated, to create image I_{l+1}, and so on. The sequence of images created this way is characterised by linearly decreasing values of the "springiness" parameter

s, while parameter d which determines how far away from pixel i the randomly chosen pixel j is allowed to be, is kept constant throughout the process.

The cost function that expresses the quality of registration between images I_1 and I_l is defined as follows:

$$U = \alpha U_1 + \beta U_2 + \gamma U_3. \tag{2}$$

In this expression, α, β and γ are parameters controlling the relative importance of each term. The three terms combined are the following:

$$U_1 \equiv 1 - R(I_1, I_l), \tag{3}$$

where $R(I_1, I_l)$ is the correlation coefficient between the two images defined as

$$R(I_1, I_l) \equiv \frac{\sum_{k \in A_{1,l}} \left(p_k^l - \overline{p}^l\right)\left(p_k^1 - \overline{p}^1\right)}{\sqrt{\sum_{m \in A_{1,l}} \left(p_m^l - \overline{p}^l\right)^2 \sum_{n \in A_{1,l}} \left(p_n^1 - \overline{p}^1\right)^2}} \tag{4}$$

where $A_{1,l}$ is the overlapping part between images I_1 and I_l, and $\overline{p}^1$ and $\overline{p}^l$ are the mean grey values of the overlapping parts of the two images I_1 and I_l respectively. This term of the cost function expresses the dissimilarity between the registered images.

The next term expresses the desire for image I_l to be distorted as little as possible to fit image I_1. If all the pixels in image I_l are subject to the same shifting from their original position in image I_2, then the value of this distortion term must be zero, as the grid is subject only to a mere translation. Thus, this term is a purely geometric term that does not involve any grey values. It simply relies on keeping account on the overall grid distortion, by preserving at all steps the current position of a grid point as it moves about in the continuous 2D domain. If there were no grid distortion, we would expect two pixels which are next to each other along the x axis, to have x coordinates that differ by 1, and y coordinates that differ by 0. We denote these two neighbouring pixels k and $k + 1$, and calculate the square of the deviation of their coordinate differences from the above expected values. Similarly, two neighbouring pixels along the y axis, are expected to have 0 difference in their x coordinates, but difference equal to 1 in their y coordinates. In the single index pixel identification system we use, two such pixels have indices that differ by the size of the image along the x axis, call it N_x. So, two such neighbours are indexed as k and $k + N_x$. The term then that measures the overall grid distortion is:

$$U_2 \equiv \frac{1}{N_{2,l}} \sum_{k \in A_{2,l}} \left((x_{k+1} - x_k - 1)^2 + \right.$$

$$\left. (y_{k+1} - y_k)^2 + (x_{k+N_x} - x_k)^2 + (y_{k+N_x} - y_k - 1)^2\right) \tag{5}$$

where $A_{2,l}$ is the overlapping part between images I_2 and I_l, and $N_{2,l}$ is the number of pixels in this overlapping part. Finally, the third term of the cost function expresses the desire for maximum overlap between images I_2 and I_l:

$$U_3 \equiv 1 - \frac{N_{2,l}}{N} \tag{6}$$

where N is the total number of pixels in each image.

3 Experimentation

For the experiments we used the Manchester's face images lifted from the PEIPA archive. This database contains 39 images, the first 10 of which are of the same person, named agj. We treated image 0.agj01 as the reference image, and kept the remaining 38 images for testing. There are two issue to be considered here: First the registration of the 38 images with the reference image and second, the classification of the registered image into either of the two classes: "identical to agj", or "different from agj". We addressed the problem of registration by using the method described in the previous section. The problem of classification was addressed by using the final values of the U_1 and U_2 terms of the cost function as indicators of the similarity between the two images. In particular, we expect that if the two images belong to the same person, the dissimilarity term U_1 will be low and the stretching term U_2 also to be low. On the other hand, images from different persons are expected to have to be significantly stretched to be brought to registration, and even then to be dissimilar.

We performed three series of experiments:

(1) Recognition of full images.
(2) Recognition of the facial parts of the images.
(3) Recognition of the eye parts of the images.

Figure 2 shows some typical registration result from the second series of experiments: The first face is the reference image, the second is the original tested image, the third is the tested image after registration, and the fourth panel shows the two registered images superimposed. Figure 3 shows the original and the registered image with their corresponding grids, to show the amount and type of stretching used. Figure 4 shows how U_1 and U_2 terms of the cost function evolve with the number of iterations used.

Figures 5a, 5b and 5c show the recognition results for the three series of experiments. In all figures black dots are used to indicate the images of the reference person agj, while triangles are used for all other images. The best results are obtained for the registration of the whole images (figure 5a) where all images of the same person are quite distinctly clustered near the origin of the axes. These results were obtained with parameter values $\alpha = 0.95$, $\beta = 0.05$ and $\gamma = 0.01$, while the springiness parameter varied slowly from $s = 0.03$ down to $s = 0.0005$. The maximum allowed shift was $d = 10$ pixels and typically sequences of 1500 images were also used for the results shown in figure 5b, concerning the recognition of the facial parts only, except that in this case the maximum allowed shift was $d = 5$ pixels. There is one image out of step in these results. It is image agj08 which depicts person agj with a very broad smile with the teeth showing. This probably contributes significantly to the dissimilarity value. Note that the stretch value in this set of results is one order of magnitude smaller that for the whole faces. The results shown in figure 5c concern the identification of the person using the eye region only. Although the there is no overlap between the two classes (the dissimilarity value of the topmost dot is 0.042 and that of the lowest-most triangle is 0.043), there is no really a significant enough gap between the two classes to allow their unambiguous discrimination. These results were obtained with exactly the same parameter values as those for the whole images.

4 Conclusions

We have presented a method for the flexible registration of images. The method is based on the global optimisation of a cost function that allows the incorporation of various constraints, measurements and items of prior knowledge concerning the problem. We demonstrated the proposed algorithm using a database of face images, although the proposed algorithm is applicable to the registration of many types of images, including remote sensing images and particularly medical images. From the limited number of experiments reported here, the results obtained by the method are quite good and encourage the further development of the approach.

Acknowledgements: This work was supported by a grant from the British Royal Society.

References

1. Bani-Hashemi, A., Krishnan, A. and Samaddar, S.: Warped matching for digital subtraction of CT-angiography studies. Medical Imaging 1996: Image Processing, SPIE **2710** (1996) 428–437.
2. Cohen, L. D. and Cohen, I.: Finite element methods for active contour models and baloons for 2D and 3D images. PAMI, **15** (1993) 1131–1147.
3. Cuisenaire, O., Thiran, J. P., Macq, B., Michel, C., DeVolter, A. and F Marques, F.: Automatic registration of 3D MR Images with a computerised brain atlas. Medical Imaging 1996: Image Processing, SPIE **2710** (1996) 438–448.
4. Fang, S., Raghavan, R. and Richtsmeier, J. T.: Volume morphing methods for landmark based 3D image deformation. Medical Imaging 1996: Image Processing, SPIE **2710** (1996) 404–415.
5. Gee, J. C. and Haynor, D. R.: Rapid coarse-to-fine matching using scale-specific priors. Medical Imaging 1996: Image Processing, SPIE **2710** (1996) 416–427.
6. Lanitis, A., Taylor, C, J. and Cootes, T. F.: An automatic face identification system using flexible appearance models. British Machine Vision Conference 1994, 65–74.
7. Lanitis, A., Taylor, C. J. and Cootes, T. F.: A unified approach to coding and interpreting face images. ICCV'95, IEEE Computer Society Press, (1995) 368–373.
8. Pentland, A.: Perceptual organization and the representation of natural form, AI J **28** (1986) 1–38.
9. Richardson, D. B. and Bury, E. A.: 1996. Correlative techniques for cross-modality medical image registration. Medical Imaging 1996: Image Processing, SPIE **2710** (1996) 368–375.
10. Terzopoulos, D. and Metaxas, D.: Dynamic 3D models with local and global deformations: Deformable superquadrics, PAMI, **13** (1991) 703–714.

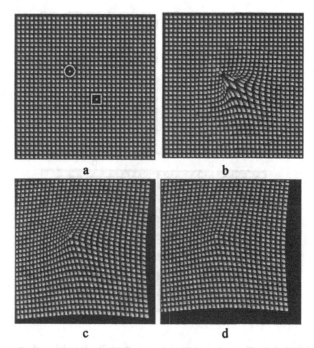

Fig. 1. The effect of a local deformation: (a) An original 256×256 lattice. It is deformed by choosing pixel i (150,150) (marked by a square) and pixel j (100,100) (marked by the circle). (b)-(d) The results of the deformation for s=0.05, s=0.01 and s=0.005 respectively.

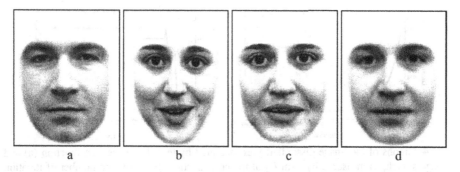

Fig. 2. Example of facial image registration. (a) Target image. (b) Original image to be registered with target one. (c) Result of deformations of original image (α=0.5, β=0.5, γ=0.005). (d) Superposition of the resultant image on the target one by the pixel-chess-order method.

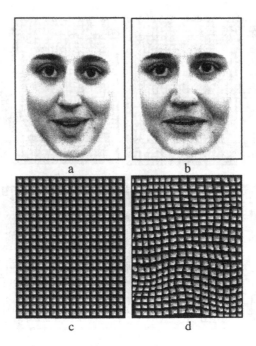

Fig 3. Deformation map of the image shown in Figure 2b. (a) Original image. (b) Result of deformations. (c) Initial grid of image (a). (d) Deformed grid.

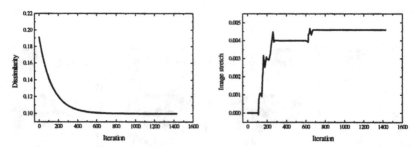

Fig 4. Graphs of the image dissimilarity as measured by term U_1 of the cost function (a) and image stretch, as measured by term U_2 of the cost function (b) versus the number of iteration steps.

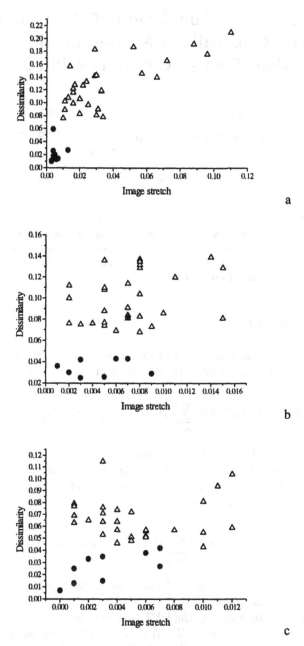

Fig. 5. Face recognition results. Feature scatter plots in space of image stretch and dissimilarity measurements for (a) whole face images; (b) images of facial part only; (c) eyes images. Circles denote images of the same person "agj" and triangles represent images of other persons.

"Error-Free" Calculation of the Convolution Using Generalized Mersenne and Fermat Transforms over Algebraic Fields

Vladimir M.Chernov and Maria V.Pershina

Image Processing Systems Institute of RAS
151 Molodogvardejskaja st., IPSI RAS, Samara, Russia

e-mail: chernov@sgau.volgacom.samara.su

Abstract. A method of the calculation of a discrete convolution via number-theoretic transforms realized without multiplications is described. It is shown that the data representation over algebraic fields allows to generalize the known Mersennse and Fermat transforms onto a wider set of periods of transformed sequences.

1 Introduction

One of universal problems of digital signal processing is the calculation of the discrete cyclic convolution of N-periodical functions:

$$z(t) = (x * h)(t) = \sum_{n=0}^{N-1} x(n)h(t-n), \quad (t = 0, ..., N-1). \tag{1}$$

An effective method of fast obtaining the array $z(t)$ involves the discrete Fourier transform (DFT) for which fast algorithms exist. If the functions $x(n)$ and $h(n)$ are integer, then "error-free" calculation of (1) may be provided using modular analogues of the DFT:

$$\hat{x}(m) = \sum_{n=0}^{N-1} x(n)\omega^{mn} \pmod{p}, \tag{2}$$

where the multiplicative order $Ord_p(\omega)$ of the element $\omega \in \mathbb{F}_p$ is equal to N, p is a reasonable large prime [1].

Such modular analogues of the DFT (number-theoretical transforms (NTTs) or Fourier-Galois transforms (FGTs)) were introduced by Strassen and Shoenhage in [2] and reintroduced by Rader in [3]. The last work has become the most known and has been the ground for different NTT modifications and generalizations [4, 5]. In particular, in Ref. [6] NTTs over algebraic fields were considered. This branch had the most popularity in 1970s, that can be explained mainly by calculating advantages of the integer arithmetics for the digital means which were used. The revival of interest to NTTs obvious in last years is connected with the development of a new VLSI devices generation. They use modular calculations for arithmetic operations realizations [7-9].

The divisibility $N \mid (p-1)$ is a specific defect of FGTs, besides the operation of obtaining residues (mod p) can not be regarded elementary computer operations. By special choice of a prime p (for example, if $p = 2^q - 1$ is a Mersenne prime) arithmetic operations in the finite field $\mathbb{F}_p$ can be significantly easier realized [5]. If, moreover, ω is equal to two, then multiplications in Eq. (2) are replaced by cyclic shifts of binary represented vectors of integer data. Prime Mersenne numbers (as other primes having a "good" realization of modular operations) are infrequent in the series of positive integers.

Since $Ord_p(2) = q$, calculation of Eq. (2) in Mersenne's arithmetics with $\omega = 2$ is possible only for $N = q$ (that is for $q = 3, 5, 7, 13, 17, 31, \dots$ etc. [4]).

The use of the transform (2) modulo a composite Mersenne number:

$$p = 2^q - 1 = p_1 p_2 \dots p_d \tag{3}$$

for calculation of the convolution (1) has some principal difficulties, in particular:

- basis functions of the transform (2) are not orthogonal due to the presence of zero divisors in the quotient-ring $\mathbb{Z}/_{p\mathbb{Z}}$.

- the prime cofactors p_j in (3) has not to be Mersenne numbers in case of parallel calculation of the convolution (1) using systems of the transform (2) (mod p_j) (and further reconstruction of the result using the Chinese remainder theorem (CRT)).

2 The main ideas

In this paper methods of parallel calculation of the convolution (1) using discrete transforms realized without multiplications and associated with data representations in "non-traditional" number systems are considered. It is important, that specific calculation advantages of Mersenne's arithmetics are preserved for some other lengths. The following simple ideas are a basis of the introduced methods of convolution calculation.

1. The ring $\mathbb{Z}$ is included into an algebraic extension $\mathbb{S}$ of rational field $\mathbb{Q}$.

2. A quotient-ring $\mathbb{Z}/_{m\mathbb{Z}}$ (modulo, in general, a composite number m) is included into quotient-ring $\mathbb{S}/_{(m)} = \mathbb{K}$.

3. The extension $\mathbb{S}$ is chosen so that the factorization (3) over the ring $\mathbb{S}$ contains only cofactors of the type $p_j = \alpha_j^{k_j} \pm 1$.

4. The convolution (1) is calculated using the usual parallel scheme with a set of NTTs.

5. Basis functions $h_m^j(n)$ of NTTs are chosen in the form $h_m^j(n) = \alpha_j^{mn}$, where $\alpha_j \in \mathbb{S}/_{(m)}$ are not to be elements of the ring $\mathbb{Z}/_{m\mathbb{Z}}$.

3 Some definitions and theorems

Definition 1 *Let $< X >$ be a $(K + 1)$-bits vector:*

$$< X >= (X_K, X_{K-1}, ..., X_0).\qquad(4)$$

then an operator $\mathcal{M}$ defined by the equality

$$\mathcal{M} < X >= (X_{K-1}, ..., X_0, X_K)\qquad(5)$$

is called an operator of the Mersenne left shift. An inverse operator of the Mersenne right shift is defined similarly:

$$\mathcal{M}^{-1} < X >= (X_0, X_K, ..., X_1)\qquad(6)$$

Definition 2 *An operator $\mathcal{F}$ defined by the equality*

$$\mathcal{F} < X >= (X_{K-1}, ..., X_0, -X_K)\qquad(7)$$

is called an operator of the Fermat left shift. An inverse operator of the Fermat right shift is defined similarly:

$$\mathcal{F}^{-1} < X >= (-X_0, X_K, ..., X_1).\qquad(8)$$

Remark 1. If $p = p_\mu = 2^q - 1$ is a Mersenne prime or $p = p_f = 2^q + 1$ is a Fermat prime, then the operators $\mathcal{M}$ and $\mathcal{F}$ introduced above are the same as transforms associated with multiplications in quotient-ring ${}^{\mathbb{Z}}\!/_{p\mathbb{Z}}$ by the element $\omega = 2$. They are transforms of binary representations of elements of quotient-rings modulo a Mersenne number or a Fermat one, respectively.

Remark 2. It is easy to verify that for any $< X >$ the sequences:

$$\{\mathcal{M}^n < X >: n \in \mathbb{Z}\} \quad \text{and} \quad \{\mathcal{F}^n < X >: n \in \mathbb{Z}\}$$

are periodical with periods (may be not the least) which are equal to $T_\mu = K$ and $T_f = 2K$, respectively.

Definition 3 *Let $\mathbb{K}$ be a commutative associative ring with unit and let $\gamma \in \mathbb{K}$. For the binary vector (4), let the element $X \in \mathbb{K}$ be defined by the equality:*

$$< X >_\gamma = X_K \gamma^K + X_{K-1} \gamma^{K-1} + ... + X_0 \gamma^0.\qquad(9)$$

The binary vector $< X >=< X >_\gamma$ is called a code of the element X in the γ-based number system.

Ring $\mathbb{K}$ elements' addition and multiplication in the form (9) induces obvious rules of actions the codes.

If, for example, $\gamma = \sqrt{2}$, then $\gamma + \gamma = 2\gamma = \gamma^3$, etc.

Definition 4 *Let $T = T_\mu$ (or $T = T_f$); $\mathcal{Y}$ be one from the operators $\mathcal{M}$ or $\mathcal{F}$. Let us define a Mersenne (Fermat) shift-transform of the sequence $< x(n) >_\gamma$ by the equality:*

$$< X_\rho(m) >_\gamma = \sum_{n=0}^{T-1} \mathcal{Y}^{mn} < x(n) >_\gamma \quad (m = 0, ... T - 1), \quad (10)$$

where ρ is one from the symbols μ or f, respectively.

For the transform (10) the inverse transform exists.

Theorem 1 *If $\mathbb{K}$ is a field, then the next equality is valid:*

$$< x(n) >_\gamma = T_\rho^{-1} \sum_{m=0}^{T-1} \mathcal{Y}^{-mn} < X_\rho(n) >_\gamma . \quad (11)$$

If $\mathbb{K}$ is a quotient-ring (a field) modulo a Mersenne prime or a Fermat one and γ is equal to two, then the transforms (10) and (11) are the same as well-known Mersenne or Fermat NTTs interpreted in terms of the shift-transforms introduced above.

If $\mathbb{K}$ is not a field, then the shift-transforms theory is significantly more complicated. It is connected, in general, with two factors: first, with a possible absence of the inverse to T_ρ^{-1} element in (11) and, therefore, with an absence of the inverse to (10) transform; second, with a possible absence of inverse to elements $(1 - \gamma^s)$ for some s (what means that the transform (10) is non-orthogonal).

4 Shift-NTT modulo composite Mersenne numbers

Let $m = 2^{2t+1} - 1$ be such a (composite) Mersenne number that:
(A) for all $2 \leq s \leq 2t$ the numbers m and $2^s - 1$ are coprime;
(B) the element $2t + 1$ has an inverse one in $\mathbb{Z}/_{m\mathbb{Z}}$.
Let us consider a ring $\mathbb{S} = \mathbb{Q}(\sqrt{2})$ of integers over the field $\mathbb{Q}$. In $\mathbb{S}$ the number m has the representation in the form

$$m = (2^t\sqrt{2} + 1)(2^t\sqrt{2} - 1) = -((-\sqrt{2})^{2t+1} - 1)((\sqrt{2})^{2t+1} + 1) \quad (12)$$

as well as the representation using integer rational numbers.

Proposition 1 *Elements $q_1 = (2^t\sqrt{2}+1)$, $q_2 = (2^t\sqrt{2}-1) \in \mathbb{S}$ and are coprime in $\mathbb{S}$.*

From the Proposition 1 it follows, in particular, that the ring $\mathbb{Z}/_{m\mathbb{Z}}$ is included isomorphically into the direct sum

$$\mathbb{Z}/_{m\mathbb{Z}} \rightarrow \mathbb{S}/_{(q_1)} \oplus \mathbb{S}/_{(q_2)} = \mathbb{K}_1 \oplus \mathbb{K}_2$$

where (q_1) and (q_2) are main ideals in $\mathbb{S}$ generated by the elements q_1 and q_2, respectively. The representation (12) and the Proposition 1 allow to reduce the convolution calculation using arithmetics of the ring $\mathbb{Z}/_{m\mathbb{Z}}$ to the parallel calculation of two convolutions over the rings $\mathbb{K}_1$ and $\mathbb{K}_2$ and to the further reconstruction of the result via CRT [1]. Specific advantages of Mersenne's arithmetics are not to be preserved for similar parallel calculation of the convolution in quotient-rings associated with an integer rational factorization of the number m. The introduced method allows to calculate the convolution via shift-transforms realized without multiplications.

Proposition 2 *For any* $X \in \mathbb{Z}/_{m\mathbb{Z}}$ *there exists such effectively defined elements* $Y, Z \in \mathbb{K}$ *and constants* $a, b \in \mathbb{Z}/_{m\mathbb{Z}}$ *that:*

(i) the following equalities are valid:

$$X = aY q_1 + bZ q_2, \quad X \equiv Y \ (\mathrm{mod} \ (q_1)), \quad X \equiv Z \ (\mathrm{mod} \ (q_2)); \qquad (13)$$

(ii) Y *and* Z *are represented in the form:*

$$Y = A_1 + A_2\sqrt{2}, \quad Z = B_1 + B_2\sqrt{2};$$

(iii) the elements A_1, B_1, *require* $(t+1)$ *bits for binary representations, the elements* A_2, B_2, *allow representations of length* t.

Let $a_j^i = 0, 1$ and

$$A_1 = 2^t a_t^1 + 2^{t-1} a_{t-1}^1 + ... + 2^0 a_0^1, \quad A_2 = 2^{t-1} a_{t-1}^2 + ... + 2^0 a_0^2.$$

Then, taking the notations into account, one obtains the $(2t+1)$-bits representation for Y in the "number system with irrational base $\alpha = \sqrt{2}$":

$$Y = \alpha^{2t} a_t^1 + \alpha^{t-1} a_{t-1}^2 + ... + \alpha^1 a_0^2 + \alpha^0 a_0^1. \qquad (14)$$

Similarly, an element Z is represented in the "number system with irrational base $\beta = -\sqrt{2}$":

Remark 3. Surely, we use this terminology terrible for non-mathematician only to explain the analogy between the classical Mersenne transforms and generalized ones introduced in this work. No "approximate modular calculations" are executed really. All calculations use either element codes or the original Mersenne arithmetics.

Further for the element Y represented in the form (14) let us call the elements A_1 and A_2 rational and irrational parts of Y and note as the following:

$$A_1 = Rat(Y), \quad A_2 = Irr(Y): \quad Y = Rat(Y) + Irr(Y)\sqrt{2}. \qquad (15)$$

As usually, representation of X in the form (13) is associated with expansion of the quotient-ring (mod m) in a direct sum of rings. Therefore, it can be allowed that $Y \in \mathbb{K}_1$ and $Z \in \mathbb{K}_2$. Multiplications of the element Y by α lead to the left cyclic shift of the vector $< Y >_\alpha$ in (14). Multiplications by two of the elements A_1 and A_2 lead to the left cyclic shift of the vectors $< A_1 >_\alpha$ and $< A_2 >_\alpha$.

Products of the elements represented in the form (15) are reduced to the well-known "binary" interpretation of Mersenne's arithmetics [1]:

$$Rat(Y_1 Y_2) + Irr(Y_1 Y_2)\sqrt{2} =$$

$$(Rat(Y_1)Rat(Y_2) + 2Irr(Y_1)Irr(Y_2)) + (Rat(Y_1)Irr(Y_2) + Rat(Y_2)Irr(Y_1))\sqrt{2}.$$

Proposition 3 *If for* $m = 2^{2t+1} - 1$ *the conditions* (A)-(B) *are valid, then functions*

$$h_l^\alpha(n) = \alpha^{ln} \in \mathbb{K}_1 \quad and \quad h_l^\beta(n) = \beta^{ln} \in \mathbb{K}_2$$

generate the orthogonal sets:

$$(h_l^\mu, h_k^\mu) = \sum_{n=0}^{2t} h_l^\mu(n)h_k^\mu(m-n) = (2t+1)\delta_{lk} \quad (\mu = \alpha, \beta).$$

The algorithm of the convolution calculation using shift-NTTs is similar structurally to the well-known algorithm of the parallel convolution calculation modulo a composite number.

Theorem 1. *If the conditions* (A)-(B) *are valid, then calculation of the convolution of length* $(2t + 1)$ *requires:*

- five Mersenne shift-NTTs (four direct and one inverse transforms);

- calculating products of shift-NTT spectra components using the original Mersenne arithmetics;

- reconstruction of convolution values using CRT.

Direct numerical testing using the table of factorization of Mersenne numbers [11] shows that numbers m of the type

$$m = 2^{2t+1} - 1 = 2^T - 1, \quad T = 11, 23, 29, 37, 41, 43, 47$$

meet to the conditions (A)-(B). For other primes $2 \le T \le 50$, numbers m are Mersenne primes. Thus, using NTTs realized without multiplications, it is possible to calculate the length-T convolution, at least, for $2 \le T \le 50$.

5 Some remarks on the shift-NTTs modulo composite Fermat numbers

If $f = 2^b + 1$, $(b = 2^r = 3t + \nu; \ \nu = 1, 2)$ is a Fermat number, then, since b is an even, direct implementation of the results mentioned above leads to the known algorithm [6]. New results are produced using a cube extension of the field $\mathbb{Q}$.

Then it is possible to represent f in the form

$$f = (\gamma^0 2^t \sqrt[3]{2^\nu} + 1)(\gamma 2^t \sqrt[3]{2^\nu} + 1)(\gamma^2 2^t \sqrt[3]{2^\nu} + 1) = f_0 f_1 f_2,$$

(where γ is a primitive root of unit of third order) as well as in the form of product of integer rational numbers.

Proposition 4 *Let* $\mathbb{S} = \mathbb{Q}(\gamma\sqrt[3]{2^\nu})$, *then elements*

$$f_0 = (\gamma^0 2^t \sqrt[3]{2^\nu} + 1), \quad f_1 = (\gamma 2^t \sqrt[3]{2^\nu} + 1), \quad f_2 = (\gamma^2 2^t \sqrt[3]{2^\nu} + 1)$$

are coprime to each in $\mathbb{S}$.

That the ring $\mathbb{Z}/f\mathbb{Z}$ is included isomorphically into the direct sum

$$\mathbb{Z}\Big/ f\mathbb{Z} \to \mathbb{S}\Big/ (f_0) \oplus \mathbb{S}\Big/ (f_1) \oplus \mathbb{S}\Big/ (f_2).$$

Further, an elements of $f\mathbb{Z}$ are represented in the "number systems with irrational (non-real) bases $\sqrt[3]{2^\nu}$, $\gamma\sqrt[3]{2^\nu}$, or $\gamma^2\sqrt[3]{2^\nu}$", etc.

Theorem 2. *Calculation of the convolution (1) of length 2b requires:*
-nine Fermat shift-NTTs;
-calculations of products of shift-NTT spectra components in the original Fermat arithmetics;
-reconstruction of convolution values using CRT.

The results obtained in this chapter allow to calculate the convolution of lengths 64 and 128 modulo fifth Fermat number and lengths 128 and 256 modulo sixth Fermat number.

6 Conclusion

Thus, the method of synthesis of the algorithm for convolution calculation considered in this work is based on two independent and widely used ideas:
 - including the quotient-ring modulo a composite number m into a direct sum of some finite rings;
 - representing the data in "non-conventional" number systems.
Significantly news in the considered method are the related choice of finite rings, the speciality of including a quotient-ring into the direct sum of these rings and the choice of number systems with an "irrational base" having the properties of binary number system in a quotient-ring modulo Fermat and Mersenne numbers. Advantages of this method, in authors' opinion, are not limited by the problem of "error-free" convolution calculation. It is perspective to use this method together with other methods for fast parallel powering in finite fields, for calculating products of high-order polynomials, etc.

7 Acknowledgment

This work was performed with financial support from the Russian Foundation of Fundamental Investigations (Grants 95-01-00367, 97-01-009000).

References

1. Nussbaumer, H.J.: Fast Fourier Transform and Convolution Algorithms. Springer. 1982
2. Schoenhage A., Strassen V.: Schnelle multiplikation grosser Zahlen. Computing. 7 (1966) 281-292
3. Rader C.M.: Discrete convolution via Mersenne transform. IEEE Trans. Comp. C-21 (1972) 1269-1273
4. Blahut, R.E.: Fast Algorithms for Digital Signal Processing. Addison-Wesley. 1985
5. Mc. Clellan, J.H., Rader, C.M.: Number Theory in Digital Signal Processing. Englewood Cliffs. New York. 1979
6. Dubois, E., Venetsanopoulos, A.N.: The generalized discrete Fourier transform in ring of algerbaic integers. IEEE Trans. ASSP-28. 2 (1980) 169-175
7. Alfredson, L.-I.: VLSI architectures and arithmetic operations with application to the Fermat number transform. Linköping Studies in Sci. and Technology. Dissertation No.425. (1996)
8. Boussakta, S., Holt, A.G.J.: Calculation of the discrete Hartley transform via Fermat number transform using VLSI chip. IEE Proc. 135 (1988) 101-103
9. Towers, P.J., Pajayakrit, A., Holt A.G.J.: Cascadable NMOS VLSI circuit for implementing a fast convolver using the Fermat number transform. 135 (1987) 57-66
10. Van der Waerden, B.L.: Algebra. 7th ed. Springer. 1966
11. Brillhart, J., Lehner, D.H., Selfridge, J.L., Tuckerman, B., Wagstass, S.S.: Factorizations of $b^n \pm 1$, $b = 2, 3, 5, 6, 7, 10, 11, 12$ up to high powers. Contemporary Mathem. 22. 2nd ed. AMS. 1988.

A New Method of Texture Binarization

Bogdan Smolka   and   Konrad W. Wojciechowski

Dept. of Automatics Electronics and Computer Science
Silesian University of Technology
Akademicka 16 Str, 44-101 Gliwice, Poland

Abstract. In the paper a new method of binarization of grey scale regular textures is presented. The approach is based on a model considering the image points as a two dimensional lattice, on which a random walk of a virtual particle is investigated. The model assumes that the image points possess a potential energy equal to their grey scale values and the particle changes its position according to the Gibbs distribution. The described method assigns to each lattice point, the probability that the particle stays in its current position. The elements of in this way obtained matrix are then transformed and treated like the original image, and the same procedure is repeated. The successive iterations lead to a matrix, which is the binary representation of the initial texture. The main advantage of this method is its ability to binarize regular textures with nonuniform brightness.

1 Introduction

One of the basic operations in the preprocessing of images is their binarization. It can be viewed as the simplest case of segmentation of images into two classes - object and background [1, 2, 3]. Binarization of textures is performed in order to increase and simplify their analysis. It is often used as a preliminary step of image processing by means of the methods of mathematical morphology and therby making it an important part of the algorithms [4]. Very often binarization is performed because of the lack of fast enough morphological methods for grey scale images [5].

Let the image be represented by a matrix L of size $N_r \times N_c$, $L = \{L(i,j), i = 1, 2, \ldots N_r, j = 1, 2, \ldots, N_c\}$. The binarization is a mapping, which assigns the value 0 or 1 to the elements of the image matrix L.

There exist many methods of binarization of images [3, 6, 7, 8, 9, 10]. They all perform well in case of images with a bimodal histogram. Diffuculties arrise when the luminance of the object or background is nonuniform [8] or when the object and background assume some broad range of gray scales.

The concept of texture is not precisely defined. The notion *texture* is commonly used in order to describe an image, which possesses some regularity of its structure [11]. Texture can be described as a pattern comprised of repeated texture primitives, which can be viewed as regions of local order [12, 13].

2   New Method of Binarization of Regular Textures

2.1   One Dimensional Algorithm

Let $X = \{X(i), i = 1, 2, \ldots, N\}$, $X(i) \in [0, 1]\}$ be a vector, which shows some local repeating arrangement of its elements. Let the distance between the elements of X be determined by a metric $\rho(i, j) = |\, i - j\,|$ and let the neighbours of a given point (i) be a set of points (j), for which $\rho(i, j) \leq \lambda$, $\lambda \in \mathbf{N}$.

The features of the vector X can be investigated be means of a virtual particle, which can jump betwen its elements. Let us presume, that the particle can visit their neighbourhood, that is the points of X, which distance to the initial point is equal or less than λ. Assuming, that the probability of the transition between point (k) and a point (l), which belongs to the neighbourhood of of (k) is given by

$$P(k \rightsquigarrow l) = \frac{e^{\frac{X(k) - X(l)}{T}}}{Z} \qquad \text{where} \qquad Z = \sum_{i=-\lambda}^{i=+\lambda} e^{\frac{X(k) - X(k+i)}{T}}$$

(Z - normalizing factor, T - temperature of the Gibbs ansamble) we obtain the probability $P(k)$ of the event, that the virtual particle will not abandon its current position (k)

$$P(k) = \frac{e^{t[X(k) - X(k)]}}{\sum\limits_{i=-\lambda}^{i=+\lambda} e^{t[X(k) - X(k+i)]}} = \frac{e^{-tX(k)}}{\sum\limits_{i=-\lambda}^{i=+\lambda} e^{-tX(k+i)]}} \qquad \text{where} \quad t = \frac{1}{T}$$

Introducing a new variable $\Psi(k) = e^{-tX(k)}$ we obtain

$$P(k) = \frac{\Psi(k)}{\sum\limits_{i=-\lambda}^{i=+\lambda} \Psi(k + i)}$$

Because the first and last element of X have incomplete number of neighbours we introduce an auxiliary vector $\tilde{X}$

$$\tilde{X} = [X_\lambda, \ldots, X_2, \mathbf{X_1}, \mathbf{X_2}, \ldots, \mathbf{X_N}, X_{N-1}, \ldots, X_{N-\lambda}]$$

In this way all elements of X posess neighbours from $\tilde{X}$. Assigning to each point (k) the probability $P(k)$ we obtain the probability vector $P_1^{(1)}$

$$P_1^{(1)} = \{P(1), P(2), \ldots, P(N)\}$$

The values of $P_1^{(1)}$ are now transformed, and we obtain a transformed vector $P_2^{(1)}$

$$P_2^{(1)} = e^{-\frac{P_1^{(1)}}{\mu}} \quad \text{where} \quad \mu = \frac{1}{N} \sum_{k=1}^{k=N} P_1^{(1)} \text{ is the mean value of the elements of}$$

$P_1^{(1)}$. After the normalization of $P_2^{(1)}$ we obtain the vector $P^{(1)}$.

The vector $P^{(1)}$ is now treated as the original vector X and the whole procedure is repeated . As a result of the iterations we obtain a sequence of vectors $P^{(1)}, P^{(2)}, P^{(3)} \ldots$.

Extensive experiments and investigations of the behaviour of this sequence performed by the authors, show that if the parameter λ is greater than the distance between the local arrangements of texture, then at $T > 0$ the sequnce of vectors converges to a binary vector, the elements of which are either 0 or 1. The tamperature T plays a thresholding role and the rate of convergence diminishes with increasing T.

Figure 1 shows three examples of binarization of one dimensional regular textures. The upper image shows the values of each vector by means of 32 grey scale values (black line represent the value 0 and white 1), below the graph of the vectors value and at the bottom the result of binarization at a specified parameters of λ and T after N_{it} iterations is shown. Each of the vectors consists of $N = 200$ elements. Figure 2 shows an example of the effect of the parameters λ and T on the results of binarization with the new method. As can be seen, the temperature plays the role of a threshold and λ decides about the resulting binary pattern.

2.2   Two Dimensional Algorithm

The algorithm of binarization of texture images is an extension of the one dimensional case. Let the elements of the image matrix L have the regular textural structure and let $\rho\{(i,j),(k,l)\} = \max\{|\,i-k\,|,|\,j-l\,|\}$ be the distance between the points (i,j) and (k,l). Morover let the neighbours of the point (i,j) be the points (k,l), for which $\rho\{(i,j),(k,l)\} \leq \lambda$.

Assuming that the virtual particle performs a random walk on the points of the image lattice (Fig. 3), the probability $P(i,j)$ that the virtual particle, which can visit its neighbours, will stay at its current position (i,j) is derived from the Gibbs distribution formula

$$P(i,j) = \frac{e^{t[L(i,j)-L(i,j)]}}{\sum\limits_{m=-\lambda}^{m=+\lambda}\sum\limits_{n=-\lambda}^{n=+\lambda} e^{t[L(i,j)-L(i+m,j+n)]}} = \frac{1}{e^{tL(i,j)}\sum\limits_{m=-\lambda}^{m=+\lambda}\sum\limits_{n=-\lambda}^{n=+\lambda} e^{-tL(i+m,j+n)}}$$

If we introduce a new variable $\Psi(i,j) = e^{-tL(i,j)}$ then

$$P(i,j) = \frac{\Psi(i,j)}{\sum\limits_{m=-\lambda}^{m=+\lambda}\sum\limits_{n=-\lambda}^{n=+\lambda} \Psi(i+m,j+n)}$$

Let us assume that the probability $P(i,j)$ is computed. The probability $P(i,j+1)$ can be calculated in the following way

$$P(i,j+1) = \frac{\Psi(i,j+1)}{\sum\limits_{m=-\lambda}^{m=+\lambda}\sum\limits_{n=-\lambda}^{n=+\lambda} \Psi(i+m,j+1+n)} =$$

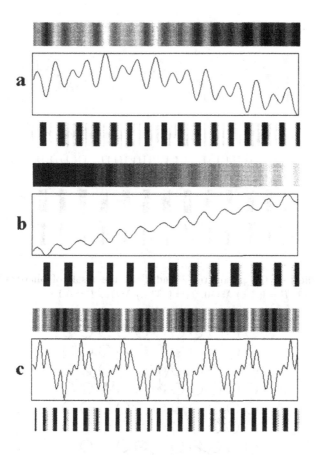

Fig. 1. Binarization of one dimensional regular textures at $T=0.33$, $N_{it}=20$ and a)$\lambda=10$ b)$\lambda=10$ c) $\lambda=5$

$$= \frac{\Psi(i,j+1)}{\Delta(i,j+1) + \sum\limits_{m=-\lambda}^{m=+\lambda}\sum\limits_{n=-\lambda}^{n=+\lambda} \Psi(i+m,j+n)}$$

where

$$\Delta(i,j+1) = \sum_{m=-\lambda}^{m=+\lambda} (\Psi(i+m,j+\lambda+1) - \Psi(i+m,j-\lambda))$$

Since

$$\sum_{m=-\lambda}^{m=+\lambda}\sum_{n=-\lambda}^{n=+\lambda} \Psi(i+m,j+n) = \frac{\Psi(i,j)}{P(i,j)}$$

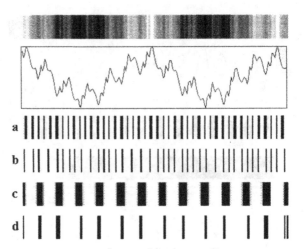

Fig. 2. The effect of the parameter λ and T on the result of binarization a) $\lambda=3$, $T = 0.33$ b) $\lambda=3$, $T = 0.1$ c) $\lambda=10$, $T = 0.33$ d) $\lambda=10$, $T = 0.1$

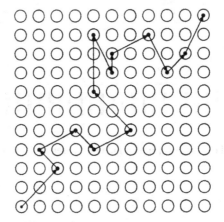

Fig. 3. The random walk on the image lattice

therefore

$$P(i, j+1) = \frac{\Psi(i, j+1) \cdot P(i, j)}{\Psi(i, j) + P(i, j) \cdot \Delta(i, j+1)}$$

This method of calculation of the prabability values leads to an enormous reduction of the computational effort.

Since the boundary elements of L, have incomplete neighbourhood an auxiliary matrix $\tilde{L}$ of size $(N_r + 2\lambda) \times (N_c + 2\lambda)$ has to be introduced. This matrix is created by extending all columns and rows in the way the vector $\tilde{X}$ in the one dimensional algorithm was introduced.

Figure 4 and 5 show some examples of the binarization of regular textures. As can be seen, the presented method performs well in case of images with strongly nonhomogenous image brightness.

3 Conclusion

A new algorithm of binarization of textures is presented. It is based on a new approach to to the problem of image features evaluation. The presented method performs especially well in case of regular textures with nonuniform brightness. It can also be applied to images, which are not characterized by a regularity of its structure.

4 Acknowledgements

The research was supported by the Polish Committee of Scientific Research (KBN).

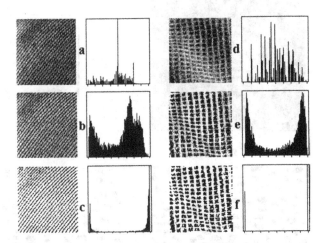

Fig. 4. Binarization of regular textures a) feather b) $P^{(2)}$ - second iteration at $T = 0.2, \lambda = 10$ c) $P^{(10)}$ d) fabric e) $P^{(20)}$ at $T = 0.66, \lambda = 20$ f) $P^{(50)}$, on the right the appropriate histogram

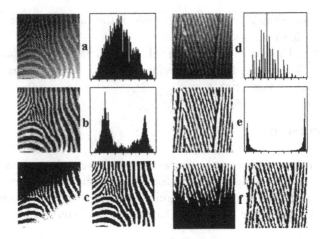

Fig. 5. Binarization of regular textures with nonuniform brightness a) zebra's skin b) $P^{(3)}$ - third iteration at $T = 0.5, \lambda = 20$, on the right the histogram c) original image thresholded at 0.5 and $P^{(10)}$ d) spooled wire e) $P^{(25)}$ at $T = 0.66, \lambda = 10$, on the right the histogram f) original image thresholded at 0.5 and $P^{(25)}$

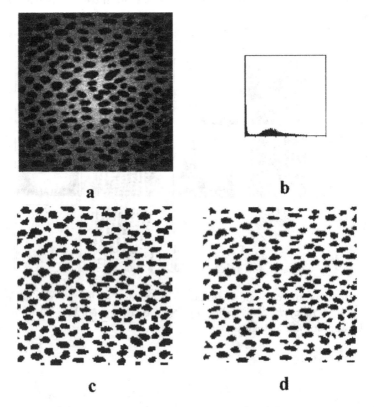

Fig. 6. Thresholding effect of T at $\lambda = 10$, lopard's fur (top), below histogram, $P^{(50)}$ at $T = 0.5$ and $P^{(20)}$ at $T = 0.2$ (bottom)

References

1. Weszka J.: A survey of threshold techniques. Computer Vision, Graphics and Image Processing **7** (1989) 259-265
2. Fu K.S., Mui J.K.: A survey on image segmentation. Pattern Recognition **13** (1981) 3-16
3. Sahoo P.K., Soltani S., Wong A.K.C.: A survey of thresholding techniques. Computer Vision Graphics and Image Processing **41** (1989) 233-260
4. Dougherty E.R., Newell J.T., Pelz J.B.: Morphological texture based maximum-likelihood pixel classification based on local granulometric moments. Pattern Recognition **25** (1992) 1181-1198
5. Iversen H., Lonnestad T. An evaluation of stochastic models for analysis and synthesis of grey scale textures. Pattern Recognition Letters **15** (1994) 575-585
6. Brink A.D.: Gray level thresholding of images using a correlation criterion. Pattern Recognition Letters **9** (1989) 335-341
7. Otsu N.: A threshold selection method from gray level histograms. IEEE Transactions on Systems, Man and Cybernetics **9** (1979) 62-66
8. Parker J.R.: Gray level thresholding in badly illuminated images. IEEE Transactions on Pattern Analysis and Machine Intelligence **13** (1991) 813-819
9. Weszka J.S., Rosenfeld A.: Histogram modification of threshold selection. IEEE Transactions on Systems, Man and Cybernetics **9** (1978) 38-52
10. Lee S.U., Chung S.Y., Park R.H.: A comparative performance study of several global thresholding technique. Computer Vision Graphics and Image Processing **52** (1990) 171-190
11. Haralick R.M., Shanmugan K., Dinstein I.: Texture features for image classification. IEEE Transactions on Systems Man and Cybernetics **3** (1973) 610-621
12. Gonzalez R., Wintz P.: Digital image processing. Addison-Wesley, Reading, Massacusetts 1987
13. Nevatia R.: Machine perception., Prentice-Hall, Englewood Cliffs, New Yersey 1982

Parameter Optimisation of an Image Processing System using Evolutionary Algorithms

Dr. Bertram NICKOLAY, Dr. Bernd SCHNEIDER, Stefan JACOB

Fraunhofer-IPK Berlin
Pascalstraße 8-9, 10587 Berlin, Germany
e-mail: {bernd.schneider|nickolay}@ipk.fhg.de
phone number: +49 (0) 30 39006-201

Abstract

The automatic surface inspection, and thus the determination of quality features, is favourably done by image processing. In the past, numerous methods to improve the performance of image processing systems have been discussed and tested. Unfortunately, new image processing tools and techniques often increase the number of parameters to be adjusted. In order to use the potentiality offered by these techniques, fine-tuned settings of system parameters need to be provided for each image processing task, thus for each object to inspect. Furthermore, in industry the possibility of adjustment is restricted by the lack of image processing knowledge of the operators. In this paper, we propose a method to automatically optimise the parameters of a machine vision system for surface inspection by using specific Evolutionary Algorithms (EAs). Especially, we pay attention onto the following items: first, the optimisation of the systems parameters in order to increase the efficiency, and second, the automatic adjustment of these parameters to simplify the using of the system. The specifications for using Evolutionary Algorithms to optimise the parameters of an image processing system entail modifications of the standard EAs. Two modifications are proposed in this paper: selection switch and σ-comparison.

1 Introduction

Surface properties are an essential feature of the product quality. In spite of the well developed state of automation technology, often the surface inspection is still done by persons. Human evaluations, however, are subjective and therefore only limited reproducible. For this task, machine vision systems are very useful. Nevertheless, it is necessary to improve the processing capacity of computers as well as the functionality and user interface of the complex image processing procedures. This paper is mainly concerned with the last two points.

Numerous opportunities to increase the performance of image processing systems have been intensively researched. This concerns the pre-processing of images by look up table operations [1] as well as the analysis of grey value histograms or co-occurrence matrices (grey-tone spatial dependence matrices) [2, 3]. Other research deals with multiresolution image processing [4]. Also, strategies to find only few but most suitable features to classify textural images have been a main field of research [5]. All these solutions have improved the performance of vision systems. Unfortunately, new tools and algorithms cause new system parameters. In order to make use of these tools, many system parameters must be set properly. This adjustment requires specific knowledge, takes a long time, and sometimes even has to be done in an experimental manner. Therefore, a strategy or method for the automated adjustment and optimisation of the systems parameters is needed. Looking for a suitable search and optimisation strategy for this purpose, one quickly discovers Evolutionary Algorithms (EAs) [6, 7, 8]. Modelling the natural evolution, these algorithms work reliable and robust and yield very good optimisation results [9].

We propose to combine the application of new image processing techniques with the automated adjustment of the processes parameters. This makes it possible to use all the advantages that are given by these techniques. At the same time, the disadvantages of a difficult parameter adjustment is avoided. Furthermore, the handling of the whole system is improved. Even persons that are not familiar with specific image processing tools will be enabled to operate the system, which is important for industrial applications.

The implementation of several new image processing techniques is described in section 2 of this paper. Section 3 gives an overview of Evolutionary Algorithms. Also, the required modifications in design and implementation of these algorithms to allow their use for the proposed purpose are specified there. In section 4, the results of some experiments using EAs for parameter optimisation are presented. Finally, section 5 offers conclusions of this paper.

2 Machine vision system for surface inspection

The main processing stages of a machine vision system for surface inspection are shown in figure 2.1.

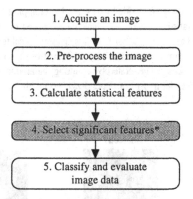

Fig. 2.1: Main steps of the machine vision system
(* step 4 is only performed during learning)

The pre-processing of images is used to compress the image data without loss of information. Therefore it is necessary to increase the contrast before summarising certain grey value sections to grey value classes. To reach this aim, certain grey values have to be faded out and other grey values must be amplified. Figure 2.2 shows the transfer function, that has successfully been used for image pre-processing.

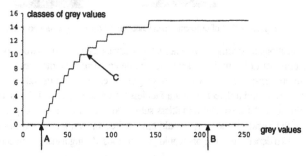

Fig. 2.2: Adaptable transfer function for image pre-processing

The formula of the transfer function contains three parameters, that allow the adaptation to various pre-processing problems:
- Parameter A defines the lowest grey value that is transferred to the new image. All grey values below that specific value are excluded from the further image processing.
- Parameter B, similarly to parameter A, defines the largest grey value.

- Parameter C defines the bending of the transfer function and therefore determines the ranges of grey values that are either faded out or amplified.

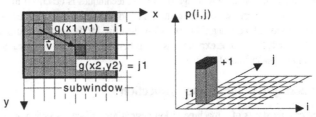

Fig. 2.3: Generation of a co-occurrence matrix

After pre-processing the image is divided into several overlapping or non-overlapping subwindows. Within each section now statistical features are calculated. Based on these feature vectors, the image section later will be assigned to a specified texture class. The statistical features as mean value or standard deviation can be calculated from a grey value histogram or from a co-occurrence matrix of the image sections. Co-occurrence matrices are generated by counting the amount of certain combinations of grey values. The combinations are defined by a dependence vector that is moved over the subwindow as shown in figure 2.3.

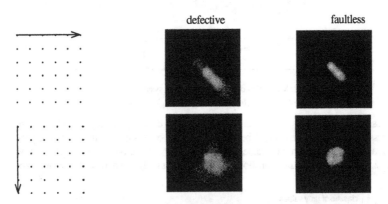

Fig. 2.4: Effect of different dependence vectors on the co-occurrence matrix

In many applications it becomes apparent that it is advantageous to use two dependence vectors in succession. The effect of different dependence vectors on co-occurrence matrices is illustrated in figure 2.4. The co-occurrence matrices are based on an image of a textile with a thread fault and calculated for subwindows which surround a defective and a faultless textile surface area. It is obviously, that the co-occurrence matrices of a defective and a faultless subwindow differ extremely depending on the length and orientation of this vector. If the fabric would have defects in its filling threads as well as in its warp threads, the use of two dependence vectors would give the most distinguishable co-occurrence matrices.

3 Evolutionary Algorithms

Evolutionary Algorithms (EAs) are probabilistic optimisation algorithms based on the model of natural evolution. The three major types are: Evolution Strategies (ESs), Evolutionary Programming (EP) and Genetic Algorithms (GAs). All EAs are based on a learning process within a population of individuals, each of them representing a single point in the space of solutions to a given problem. The outline of a basic EA is shown in figure 3.1.

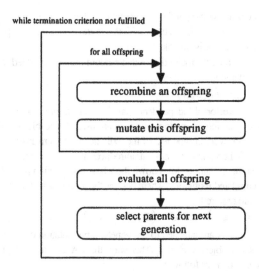

Fig. 3.1: Outline of a basic Evolutionary Algorithm

The population is expected to reach better regions of the search space by means of recombination, mutation, and selection. EAs vary in the use of these evolutionary operators. The recombination mixes the genetic information of the parents which have been chosen to produce a new offspring. The mutation operator changes all components of an individual and therefore introduces new genetic material into the population. The evaluation of an offspring yields a specific fitness value that gives a quality information of the search point represented by this individual. Then, the selection operator determines the best individuals of the current generation to become the parents for the next generation.

EAs seem to be suitable tools to optimise the described system parameters. Classical deterministic and non-deterministic optimisation techniques will in these cases only yield insufficient results because of missing mathematical optimisation functions and time-consuming function evaluations. On the contrary, EAs combine deterministic optimisers (e.g. the selection operator) and non-deterministic optimisers (e.g. the mutation operator) and by this means examine the search space from different points simultaneously. The following parameters of the vision system are supposed to be adapted by EAs:

- The parameters A, B, and C of the transfer function (see figure 2.2). (Contrary to other methods to adjust the contrast, EAs also take the unknown interactions to all other system parameters into consideration.)
- The size and orientation of the two dependence vectors that are used to generate the co-occurrence matrix (see figure 2.4).
- The selection of textural features for classification.

Within EAs, the transfer function parameters as well as the dependence vectors have to be represented by real numbers. This is necessary for the application of evolutionary operators. The selection of statistical features is represented by Boolean values, where true means that the corresponding feature is selected for classification and false means that it is not selected. Thus, two EAs are needed, referred to as EA1 and EA2. Each individual for the EA1 consists of seven components, and the individuals for the EA2 contain 22 components, because the machine vision system allows for computation of 22 statistical features which might be used for the classification.

Different recombination strategies have been utilised in EAs. We found out that either a discrete or an intermediate recombination should be used. During a discrete recombination, the components of the new individual are taken from randomly selected components of two or more parents which are randomly chosen from the parent population. For an intermediate recombination, each component of the offspring is just the mean value of the corresponding components of the parent individuals. In an EA1, both types of recombination may be used, whereas in an EA2 due to the Boolean data type implementation only the discrete recombination operator is suitable.

Generally, for mutation a normally distributed random number is added to every component of the individual under consideration:

$$x_{neu} = x_{alt} + N(0;\sigma)$$

where $N(0;\sigma)$ denotes a vector of independent Gaussian random numbers with zero mean and standard deviation σ. Normally distributed random numbers are used due to biological evidence. The standard deviation σ may vary for every component of the individuals, but remains fixed during the evolutionary procedure. Mutation is necessary, because it introduces innovation into the population. In practical applications, however, the mutation slows down the whole optimisation procedure. It is therefore a good compromise to use correlated mutations, where besides the object variables additional strategy vectors (of standard deviations) are required:

$$x =(x_1,x_2,...x_n) \quad \text{and} \quad \sigma =(\sigma_1,\sigma_2,...\sigma_n) .$$

The strategy parameters are attached to the genetic representation of an individual and modified in the same manner as the object variables. This way, the EA can also adapt the standard deviations. Consequently, the mutation is as follows:

$$\sigma_{neu} = \sigma_{alt} \cdot e^{N(0;\Delta)} \quad \text{and} \quad x_{neu} = x_{alt} + N(0;\sigma_{neu}) .$$

Using EAs for parameter optimisation within a machine vision system, each individual represents a certain setting of parameters of the system. In order to compute the fitness value of an individual, it is necessary to solve the classification problem using the system parameters represented by this individual. Then, the fitness value describes the ability of the parameter setting to yield good classification results. This evaluation process for the EA1 is shown in figure 3.2. For the EA2, it is not necessary to calculate a new vector of statistical features for every individual. Instead, this vector is reduced according to the selection of features which is represented by the individual.

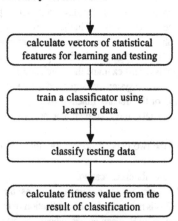

Fig. 3.2: Calculation of a fitness value

The new parent individuals are selected as the best individuals either from the set of current offspring individuals (variant A) or from the union of parent and offspring individuals (variant B). For B, the mean fitness of the population can only increase from generation to generation. This involves the danger of stagnation, if at an early stage of the optimisation process no better individuals are found. On the other hand, in variant A the mean fitness of a population may become worse during the evolutionary process. None of these variants is applicable for the EA2. Using B, the evolutionary progress stagnates very early. This stagnation is caused by the small deteriorations of fitness values occuring when the number of selected features is reduced. Using A, the EA tends to decrease the number of chosen features to 2, because the evolutionary optimisation progresses with no respect to the deterioration of the fitness. This is not acceptable, because more selected textural features may yield considerably better clasification results. Therefore, this paper proposes the combination of both basic types of a selection operator.

In this new strategy we generally use B as selection mechanism. If the progress stagnates, which is verified by the fitness of the parent population, we allow a switch to A for one generation. This „*selection switch*" is controlled by two parameters; the transition may only take place if:
- the mean fitness of the parent population has not increased for X generations *and*
- the fitness of the best offspring is below Y% worse than the fitness of the best paren.

To understand the idea of the selection change, consider the following example. If the mean fitness value of the parent individuals stagnated for X generations, and if the fitness value of the current offspring is below 10% worse than the best parent individual, then the parent will be replaced by the offspring. The parameters X and Y give the user the possibility to put the emphasis of the optimisation either on the reduction of the number of classification features or on the selection of the most significant combination of classification features. The selection switch was successfully applied.

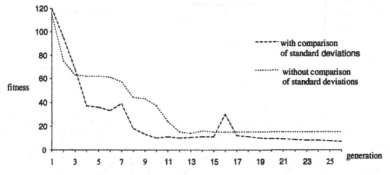

Fig. 3.3: Runs of an EA with and without σ-comparison

For EA1, another problem concerning the selection operator was to solve. The parameter of the machine vision system are integer numbers, e.g. the lowest and largest grey value within the transfer function or the elements of the dependence vectors. Nevertheless, in the algorithms they are implemented as real numbers to ensure the use of evolutionary operators. However, for the evaluation of the individuals fitness these real values need to be transformed to integer numbers. The problem occurs, if the differences between some individuals are getting smaller. Then the transformation to integer numbers yields practically the same individuals which get the same fitness value. In this case, the selection operator can not work properly and the process will stagnate. Thus, we propose to compare the standard deviations that are related to these object vectors. The object vector with the higher standard deviations in its strategy vector is chosen as the better one. This σ-*comparison* will the strategy allow for breaking a stagnation and for further improving the mean fitness. Figure 3.3 shows the mean fitness value of two EAs, which are are run with and without a σ-comparison. The EA without σ-comparison starts to stagnate at generation 14, whereas the EA with σ-comparison shows a fitness deterioration at generation 16. This is caused by the selection of an individual with higher standard deviations. Now, this individual is mutated using higher standard deviations and therfore reaches worse regions of the search space. This fitness deterioration is temporary and restores the slection operator. Hence, the σ-comparison was experimentally found to be useful.

4 Experimental results

We experimentally examined if a specific setting of system parameters as a result of an Evolutionary Algorithm would be reproducible. We generated test data that were independent from the learning data and optimised the system parameters using the adopted EA. This experiment was performed four times. In all four test runs, the resulting parameter settings of the machine vision system were similar. Hence it was proved, that the implemented algorithms are robust and reliable. The results of the four test runs are illustrated in figure 4.1. The table shows the outlines of the transfer functions as they were optimised by the EA. One can see, that the four graphs are nearly identical. Also, it shows the resulting independence vectors of each run. Again, these vectors are similar to each other. Finally, the last line of the table shows the lists of statistical features that were found to be appropriate for classification. We repeated this experiment using four different images and obtained comparable results.

test run	results			
	1	2	3	4
visualisation of the resulting transfer function	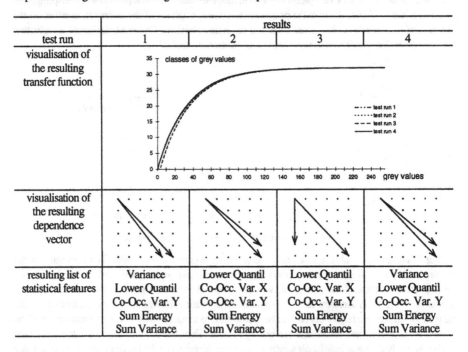			
visualisation of the resulting dependence vector				
resulting list of statistical features	Variance Lower Quantil Co-Occ. Var. Y Sum Energy Sum Variance	Lower Quantil Co-Occ. Var. X Co-Occ. Var. Y Sum Energy Sum Variance	Lower Quantil Co-Occ. Var. X Co-Occ. Var. Y Sum Energy Sum Variance	Variance Lower Quantil Co-Occ. Var. Y Sum Energy Sum Variance

Fig. 4.1: Results of different test runs of Evolutionary Algorithms

The second experiment investigated the benefits to the vision system due to the application of EAs for the adjustment of its parameters. We used test data from six different images, all of them representing another kind of surface, e.g. various textiles, a grey casting part, and a polymer part. We executed all steps of the machine vision system for two cases. In case one, we used the following system parameters:
- a linear, non-adaptable transfer function without parameters,
- just one dependence vector and
- the full selection of all textural features.

  In the second case, all new features of the vision system were used:
- a non-linear, adaptable transfer function, of which the outline was optimised using an EA,
- two dependence vectors that were adjusted by an EA and
- an optimised selection of textural features.

In both cases, the performance of the system was measured by determining the accuracy of classification on a test set of some image data. All other parameter of the machine vision system, e.g. the classificators learning parameters, had been fixed for all experiments. Figure 4.2 shows the results of these experiments. In each case, the portion of test vectors that were classified to the wrong class is given.

image	1	2	3	4	5	6
test case 1	4.5%	8.6%	5%	4.9%	0%	4%
test case 2	0%	0%	0%	0%	0%	0%

Fig. 4.2: Experimental results

As one can see, the use of new system parameters which are adjusted automatically by an EA yields better classification results.

5 Conclusions

Comprising our arguments, we see three main advantages due to the use of new image processing techniques (e.g. the adaptable transfer function and the second dependence vector) and, by the same time, the automatic adjustment and optimisation of the corresponding parameters by Evolutionary Algorithms:
- The accuracy of classification increases, subsequently the reliability of the vision system is highly improved by the use of optimised system parameters.
- The machine vision system is flexible for changes in its conditions (e.g. in lightning).
- By incorporating EA into the system, it became ready to use for industrial applications.

References

[1] B. Jähne, „Digitale Bildverarbeitung", Berlin: Springer-Verlag 1989
[2] Haralick, R.M.; Shanmugam, K.; Dinstein, I.: „Textural Features for Image Classification". IEEE Vol. SMC-3 (1973) 6, pp. 610-621
[3] Zucker, S.-W.; Terzopoulos, D: „Finding Structure in Co-Occurrence Matrices for Texture Analysis". Computer Graphics and Image Processing (1980) 12, pp. 286-308
[4] A. Rosenfeld (Editor), „Multiresolution Image Processing and Analysis", Springer Series in Information Sciences, Volume 12, Berlin Heidelberg: Springer-Verlag 1984
[5] Vigarion, R.N.; Oja, E.: „Signal Seperation and Feature Extraction by Nonlinear PCA Network", Proceedings of the 9th Scandinavian Conf. on Image Analysis, 1995, pp. 811-818
[6] Beasley, D..; Bull, D.R.; Martin, R.R.: „An Overview of genetic Algorithms: Part 1, Fundamentals". in University Computing, 1993, 15(2), pp. 58-69
[7] Fogel, D.B.: „Evolving Artificial Intelligence". Dissertation, University of California, San Diego, 1992
[8] Rechenberg, I.: „Evolutionstrategie '94". Stuttgart: Friedrich Frommann Verlag, 1994
[9] Schwefel, H.-P.; Bäck, T.: „An Overview of Evolutionary Algorithms for Parameter Optimization". Technical Report University Dortmund, 1996

Analysis of Learning Using Segmentation Models *

Anthony Hoogs

University of Pennsylvania and Lockheed Martin Corporation
Bldg. 10, Rm. 1954; P.O. Box 8048, Philadelphia, PA 19101 USA
hoogs@mds.lmco.com

Abstract. A method for change detection using a combination of view-based and model-based representations is presented. Coarse geometric models are enhanced with appearance-based information using a small number of training images, resulting in a hybrid representation that maps local appearance characteristics onto object subparts to provide model-based appearance prediction. The learning behavior of the change detection system is studied as a function of the number and characteristics of the images in the training set. Results are presented showing the level of improvement of change detection performance as images are added to the training set, in comparison with using a purely geometric approach.

1 Introduction

View-based object recognition systems have demonstrated significant capabilities in recognizing complex 3D objects in simple scenes [5, 2]. These systems operate by learning appearance characteristics of objects from training imagery without recovering geometry. This data-driven approach allows view-based systems to learn visually complex objects with many surface features, but often requires an extensive training set spanning the range of all parameters affecting object appearance. In effect, the segmentation problem is circumvented by enumerating many possible imaging conditions in advance. Also, figure-ground discrimination and occlusions can cause difficulties because features are computed over the complete object or image.

In previous work [4, 3] we describe an alternative approach to object representation, called *segmentation modeling*, that uses learning from training images, but is also model-based. The learning paradigm is designed to incrementally improve its performance as training images are added, beginning with no training images at all. The system also relies on coarse 3D geometric models of objects, such as CAD models, but the objects themselves may have complex surface features that would cause difficulties for most model-based methods because the projected model does not effectively match the objects' true appearance. The geometric model provides a spatial framework for localizing these surface appearance characteristics learned from training data, and it also allows the system to operate successfully with very little training data.

---

* This work was sponsored in part by Lockheed Martin Management and Data Systems.

As more training data is given to the system, the accuracy of the model should improve. To measure this quantitatively, the hybrid representation has been incorporated into a change detection system that attempts to identify significant changes in a scene over time. Images of the scene are taken periodically with arbitrary viewpoints and illumination angles. Currently, the change detection system attempts to detect the disappearance or removal of objects from the scene. Objects of interest are modeled with coarse CAD models, and these are used to detect the objects in later images. The hybrid model representation is well-suited to this domain, since images are input only periodically, and the wide range of imaging conditions precludes the use of pixel-level comparison.

Combining model-based and view-based representations has significant advantages over either method alone, especially when the scene is complex. Using only geometry to predict object appearance leads to performance degradation when scene or imaging conditions give rise to poor segmentations. However, segmentation behaviors caused by sensor imaging effects, surface albedoes, and unmodeled surface features are accounted for implicitly through the training imagery. Furthermore, these appearance characteristics are localized to specific object features. Other recent work [6, 2, 7] has concentrated on localization, since this allows geometric indexing, robustness under occlusion and figure-ground discrimination. However, Ikeuchi's system is specialized to range data, and the other systems do not use 3D models, leading to the need for prototypical views and relatively large training sets. By using the 3D model to provide geometric constraints, we avoid these difficulties under many circumstances.

The next section describes the change detection system, and how segmentation models are incorporated into it. Section 3 presents results and analysis comparing change detection with and without segmentation models created over a range of image training sets. The results indicate that segmentation models improve change detection performance, particularly on difficult problems.

2 Applying Segmentation Models

In previous work [4, 3], we describe an approach to modeling segmentations of geometric representations. Segmentation models are constructed by associating segmentation information from a training set of images with projected geometric features. This requires that geometric models of objects in the scene are available, and that the images used in construction are registered (calibrated) to the scene. The extracted image segmentation information is in the form of straight line segments, and the average gradient across each segment. Extracted segments are correlated to projected model edges, and used to update segmentation models attached to each object edge. Updating is performed to yield a Bayes estimate of the probability of detecting an edge segment. The aspect graph representation is used to partition the space of viewpoint and illumination parameters, assuming a single light source.

Segmentation models provide a formulation for estimating segmentation behavior based on previous imagery. To investigate how segmentation models could be used effectively to improve the performance of higher-level vision systems, we have developed a change detection system that identifies changes in time-sequenced images (not video) of man-made polyhedral structures such as buildings, roads, new construction, etc. in outdoor scenes. In this scenario, images of a fixed location are taken from aerial sensors over a period of time. Other efforts in this area rely on model geometry to predict appearance [1], giving rise to false alarms when geometric features are not detected in an image. The goal of our system is to identify true changes in structures while ruling out apparent differences due to non-geometric effects.

The change detection system architecture is shown in Figure 1. Initially, segmentation models are created using the model geometry of structures in a scene and registered (calibrated) training images of the scene. When a new image of the scene is presented to the system, it is registered to the geometric models or other scene features using manual or automatic techniques. Image registration itself is a well-studied area, and is beyond the scope of this work.

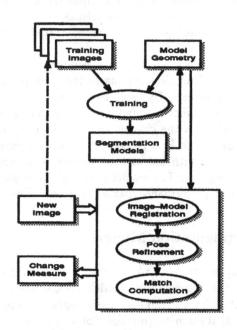

Fig. 1. Change detection process flow.

Many existing image registration algorithms, manual or otherwise, typically result in at least 2 pixels of error because of unmodeled sensor distortions and noise. To compensate for this, the system performs a local 2D translational pose refinement step that adjusts the position of a single object model with respect to the image [4].

The next stage of the algorithm, match computation, is a virtual operation; the pose refinement stage actually computes the match score that is the output change measure. The match metric is computed using the segmentation model.

Our previous work demonstrated the utility of using the segmentation model over pure model geometry for pose adjustment [4], and we have continued to develop this approach.

The model matching procedure is similar to that used in segmentation model construction. An image of the scene is segmented and the appropriate segmentation models of each object edge are computed based on prior knowledge of the viewpoint and illumination directions. Since the search does not compute object rotations, these values remain constant throughout the search. The segmented lines are correlated with the projected model edges for a given model position. Edge attributes are computed from the image for each visible model edge (the edge *profile*), and a match score between the single-image attributes and the prior edge segmentation models is calculated. The match score is then used to guide the hierarchical search by ranking position hypotheses. For details, see [4].

3 Learning Analysis

Every learning-based system is dependent on its training data. However, by combining training images with known geometry we hope to reduce the burden of each representation while retaining their advantages. We desire to produce a system that performs well on very small training sets that are highly unconstrained, i.e. no assumptions are made about the distribution of viewpoints or illumination angles in the training images.

To assess the effectiveness of our representation, we have compared the hybrid system to one using geometry alone on the same change detection problems. The images in Figure 3 show four aerial views of an example scene containing a difficult building obscured by trees. The building has a flat white roof, but it is occluded by tall trees and their shadows. We have a data set containing 24 images of this scene, which includes a number of other buildings of varying detectability. Four more of these images are shown in Figure 4; note that in three of these images the object model is close to the image, but further pose refinement is necessary to align the model edges with the image.

Using six images for training on the displayed building model, the results of the geometric and hybrid change detection systems are plotted in Figure 2. The training images were selected temporally; the training images are the first 6 of 24 images taken at random intervals over a period of four months. Two of the six training images are the upper right and lower left images in Figure 4.

In Figure 2, the dashed line connects the change levels computed on each image by using geometry alone, while the solid line plots the change levels computed using the segmentation model metric. For the segmentation model case, the scores on the training images (1 - 6) are plotted to show the best expected performance given the training data. All images except I_{12} show the building; for I_{12} the building was shifted to another part of the scene to simulate change (I_{12} is the bottom right image in Figure 4). Note that the peak in change level at I_{12} is clearly more distinctive in the segmentation model case.

The data shows that, in this case, the training data improves performance significantly on the images without change while producing a change level nearly equal to the purely geometric model on the image showing change. Note that the absolute values of the change levels are not important – it is the separation between change and no-change image results that matter.

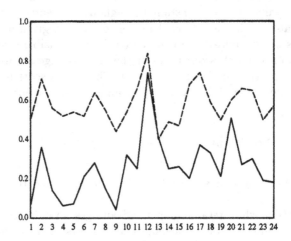

Fig. 2. Image-by-image comparison of the match level over a set of images.

To assess the system's learning capabilities on small training sets, we analyzed its performance as the number of training images varies from 0 to 10 (Figure 3 shows four of these training images). After training on the building model shown, the system was tested at each number of training images on the 14 images not included in any training set. Four of these test images are shown in Figure 4 (three of these show no change), and four more images showing change are in Figure 5. The 14 change images were "created" by shifting the building model in each of the no-change images. The building was placed by hand intentionally close to other buildings and distractors to increase the difficulty of detecting change. No system parameters (including segmentation parameters) were adjusted specifically for or during the experiments.

The results are shown in Figure 6. This graph shows the average level of change computed for three cases – on training images, on test images showing no change, and on test images showing change – as the number of training images is increased. For the cases showing no change (respectively change), the desired score is 0 (respectively 1). Ideally, the change test curve should converge to 1 and the no-change test curve should converge to 0 as more images are added to the training set. The critical metric on the graph is the level of discrimination between change and no change, which corresponds to the separation of their curves (i.e. their ratio).

This divergence is apparent in the graph, especially as the first few training

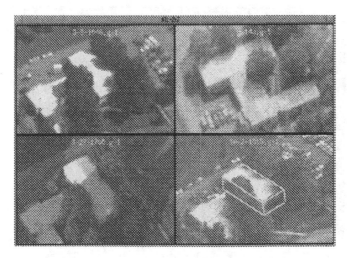

Fig. 3. Four of ten training images used in computing the learning curves. The building model is shown overlaid on the image in the bottom right pane.

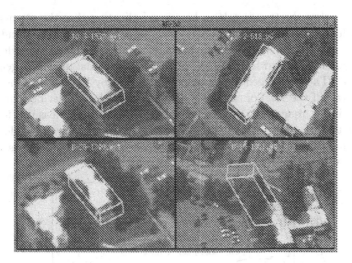

Fig. 4. Three images showing no change, and one showing "simulated" change by building displacement (lower right). The building model is overlaid in white.

images are added. The case of 0 training images corresponds to using only model geometry; in this case, the discrimination between the change and no-change images is negligible (.05). With one training image, the difference increases to .14, and it increases to .17 at 10 training images. However, the ratio of the change to no-change values increases steadily and reaches a maximum of 1.6 at 10 training images. Thus, on this data set, the system demonstrates rapid improvement when using small numbers of training images.

As expected, the training data produces the best performance. The test im-

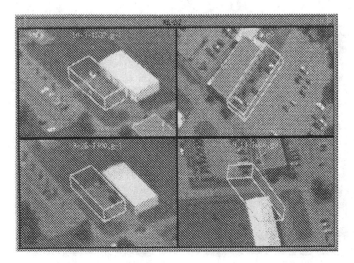

Fig. 5. Four images showing change (lower right). The building model was placed near similar buildings to act as distractors.

ages closely follow the training data, however, indicating that the segmentation model has learned characteristics of the training set that are also found in the test set. On the test images showing change, the level of change score decreases as more images are added to the training set because of the increased generality of the model. The model generality can be measured by the performance on its training data – if the level of change is nonzero, then there must be a difference between training images *in the same mode*. Thus this value measures the variance in the training data in a quantitative, task-based way that provides direct feedback on expected learning performance. This value inherently measures the difficulty of visually interpreting the scene in a model-based way.

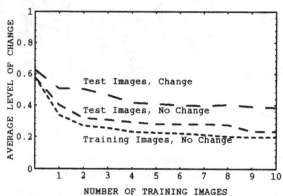

Fig. 6. Average change level plotted against the number of training images.

In many of the change images, the pose adjustment step aligned the building model with the distractor building. Despite this, the system was still able to produce a low match score because the distractor building does not have the same appearance characteristics as the source building.

It is also worth noting that the variance of the level of change scores impacts system performance. In the experiments, the standard deviations of these scores typically did not overlap, i.e. the average no-change value plus one s.d. was less than the change value minus one s.d.

4 Conclusion

This paper describes how our method for integrating view-based and model-based representations is used in a higher-level system performing change detection. By mapping segmentation features from small sets of training images onto 3D geometric models, the system provides data-driven prior information to improve appearance prediction. Model geometry is used to constrain the large number of parameters affecting segmentation behavior, so that performance improvement is apparent on small training sets. Segmentation models also yield a measure of the difficulty of scene interpretation, which is useful for predicting system performance.

By merging appearance characteristics and 3D geometry, the hybrid representation enhances model matching in domains where prior 3D models are available. The work presented here is necessarily limited in scope, but the techniques and principles described could be expanded and generalized to incorporate many problem domains, such as 3D object recognition and image registration.

There are many open issues in segmentation modeling that we are pursuing. The edge-based models used here can be generalized to two-dimensional, surface-based models for richer scene description. Additional edge attributes could provide useful information, and the evidential framework should be expanded to include edge, face and object information in a common representation. Such a representation could then be applied to study the effects of imaging parameters, interpolation across object aspects, and other interesting problems.

References

1. M. Bejanin, A. Huertas, G. Medioni, and R. Nevatia. Model validation for change detection. *Proceedings of the ARPA Image Understanding Workshop*, November 1994.
2. M. Costa and L. Shapiro. Scene analysis using appearance-based models and relational indexing. *Proceedings of the International Symposium on Computer Vision*, November 1995.
3. A. Hoogs and R. Bajcsy. Segmentation modeling. *Proceedings of IAPR Conference on Computer Analysis of Images and Patterns*, September 1995.
4. A. Hoogs and R. Bajcsy. Model-based learning of segmentations. *Proceedings of ICPR*, Vol. 4, August 1996.
5. H. Murase and S. Nayar. Learning object models from appearance. *Proceedings of AAAI: Recognition*, July 1993.
6. A. Pope and D. Lowe. Learning object recognition models from images. *Proceedings of the 4th ICCV*, May 1993.
7. M. Wheeler and K. Ikeuchi. Sensor modeling, probabilistic hypothesis generation, and robust localization for object recognition. *IEEE Transactions on Pattern Analysis and Machine Intelligence*, 17(3), March 1995.

Stereo Processing of Image Data from the Air-Borne CCD-Scanner WAAC

Ralf Reulke

German Aerospace Research Establishment (DLR)
Institute of Space Sensor Technology, Rudower Chaussee 5, D-12484 Berlin
e-mail: Ralf.Reulke@dlr.de

ABSTRACT

The use of the three-line CCD-scanner WAAC (Wide-Angle Airborne Camera) on board an aircraft introduces considerable problems to derive the digital elevation model (DEM), due to the disturbing movements of the aircraft (roll, yaw, pitch, ground speed and altitude variation).

To overcome this problem, it is necessary to measure and incorporate in the retrieving algorithm exact attitude data for each measured line. This concerns especially the angular attitude data.

The camera WAAC was rigidly mounted in the tail of a Do 228. A gyro block was mounted directly on the camera base plate. As the attitude angles given by the inertial navigation system (INS) do not have the necessary accuracy of one pixel IFOV (one arcminute), a separate fibre gyro system should be used instead of the installed INS.

Using this data set the images can be geometrically corrected assuming a flat ground surface. This procedure increases the image quality (e.g. streets become straight lines) and can enhance the number of matched points.

After finding corresponding points (stereo matching) in two image strips (e.g. nadir and forward line) stereo reconstruction and derivation of digital elevation model is possible.

1. INTRODUCTION

The Wide-Angle Optoelectronic Stereo Scanner (WAOSS) was one of the remote sensing instruments of the unsuccessful Mars-96 orbiter. The basic scientific objectives were the global imaging of Mars' surface and atmosphere for the investigation of meteorological, climatological and related surface phenomena and changes.

These special mission objectives, an anticipated highly elliptical orbit and the camera design required a very flexible concept of sensor-signal processing and evaluation algorithms. The development and tests of algorithms with current data are necessary, and for this purpose the airborne-sample WAAC (Wide-Angle Airborne Camera) of the satellite camerawas manufactured and flown (see [1], [2]).

The use of WAAC on board aircraft instead of satellites causes changes in measurement and evaluation procedures, particularly the attitude instabilities of the aircraft (roll, yaw, pitch, ground speed and altitude variation) which are much more pronounced than those of the satellite. A stabilizing aircraft platform was not used for reasons of simplicity, size, weight and funding.

The CCD-line scanner technology with in track stereo capability is a new technology. The following system were flown on aeroplane in the last few year: MEOSS [3], DPA [4], TLS [5] and HRSC [6].

Concerning the data evaluation of an image strip recorded from an airborne camera the principal disadvantage of a line sensor is that the attitude parameters are required for each row in the image.

Due to the inaccuracy of the INS, conventional photogrammetric methods use additional control points on the Earth's surface to improve the attitude parameters. The application of this approach to images from a line sensor needs a model for the flight path and an iterative improvement for the attitude parameters with additional ground control points (GCP) (see [7]).

Because of the extremly high accuracy requirements for the attitude measurements, the evaluation algorithms described in this paper require no or only a small number of GCPs.

An optical fibre gyro system was installed to overcome the problems of attitude inaccuracy. This measurement approach makes a platform unnecessary.

Using attitude and camera calibration data, the image strips of the nadir, the backward and the forward looking line can be geometrically corrected assuming a flat underlying surface. This image product is equivalent to an undisturbed flight over a flat surface. If these strips are processed on the same mean flight path, the correction procedure is equivalent to the conventional interior and exterior orientation and gives epipolar-like images.

The geometrically corrected images are a necessary preprocessed data product for further 3-D processing and allow also a simple coordinate measurement with a photogrammetric workstation, for example. This procedure also increases the image quality (e.g. streets become straight lines) and influences the number of matched points. The independence of additional GCPs and straightforward algorithms make an on-line approach for DEM generation or DEM improvement onboard the aircraft possible.

2. ATTITUDE DETERMINATION

Known methods are able to evaluate 3-D models from line scanner data only if the attitude disturbances of the aircraft are small compared with the instantaneous field of view (IFOV) of the camera (e.g.[3] and [7]).

To overcome this problem it is necessary to measure and incorporate exact attitude data in the retrieving algorithm. This concerns especially the angular attitude data.

The camera was rigidly mounted in the aircraft's tail and the aircraft axis coincided with the camera axis. The INS was placed in the nose of the Do 228. Deviations between the aircraft motion and the camera motion were expected, and the accuracy of measurements were in the order of a few arcminutes. To compensate for this, a gyro block was mounted directly on the camera base plate. As the attitude angles given by the INS do not have the necessary accuracy, an optical fibre gyro system has been used instead of the installed INS.

Supplementary measurements with GCPs, for example, are necessary for the precise determination of camera orientation in the aircraft and of angle offsets.

During the imaging the following attitude data was recorded: The altitude (GPS and Inertial Navigation System INS of the aircraft), the ground speed (GPS and INS), the flight path angle (GPS), yaw, pitch and roll angle (INS), and the angular velocity (3 fibre optical gyros).

The gyro sampling rate was 1 KHz (compared to the INS which had a rate of 100 Hz). The possible measurement range is 0 to 400 degrees/s and the resolution 50 degrees/h.

The angles are calculated from the velocities by integration of the angular velocity. The accuracy of the angles is more precise than the IFOV of a sensor pixel.

The evaluation procedure is explained in [8]. The integration of the angular velocity leads to an unavoidable drift rate caused by stochastical velocity errors, which were then corrected off-line by comparing gyro data and INS data. The usual drift rate for 1000 recorded lines is 15 pixel.

Fig. 1 Image of the WAAC camera

Fig. 1 shows a small part of an image of the WAAC camera, which was recorded during a flight over the city of Braunschweig. The original image size is 5184 • 25000 pixels and size of this cut is 700 • 1000. The ground pixel size is about 0.6 m • 0.6 m.

The influence of the aircraft motion is clearly visible in the left margin of the image. The two plots shows the roll and pitch angular changes. The interval for roll angle (white) is [-1.8°,2.5°], for the pitch (black) [4.7°,5.7°] and for the yaw [175°,178°].

The roll movement is clearly correlated to the image disturbances. The changes in pitch can be observed in the thick changes of the white colums.

3. CORRECTION OF AIRCRAFT ATTITUDE INSTABILITIES

With the knowledge of the attitude parameters a correction of the image strip is possible. The algorithm is processed in two steps:
1. Determination of the geometric relations in the object-space for each CCD-line pixel results from the disturbed aircraft movement

This task is equivalent to ray-tracing from real position and direction into the digital elevation model (DEM). As the original DEM is unknown, an ideal DEM is assumed as a reference plane at $z = z_i$. The following calculation is made for the object-space.

Geometric relations are as follows:

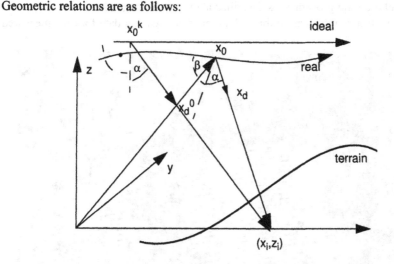

Fig. 2   Aircraft attitude instabilities

The intersection point with the surface is

$$x_i = x_0 + t \cdot x_d \tag{3-1}$$

with x_0 the current camera location and x_i the intersection point with reference plane at z_i

The actual disturbed direction vector x_d is related to the undisturbed by a rotation matrix **M**.

$$x_d = \mathbf{M} \cdot x_d^0 \tag{3-2}$$

M contains the disturbance of the flight path: roll, pitch and yaw, as described before. In the example (Fig. 2) the angle β corresponds to the pitch of the aircraft and the angle α is the stereo angle.
2. Back projection from the intersection point with the reference plane into the image plane of the camera moving on an ideal (undisturbed) linear flight path.

The simplest approach is the projection of the object-point into the image space or focal plane on an ideal flight path not affected by disturbances. By this procedure the data will be sorted and corrected.

The back projection procedure only works for some well known, but simple, flight trajectories. It is necessary to find a functional dependence for the projection point of the vector on an ideal flight trajectory x_0^k (see Fig. 2). So for the whole measured swath the number of parameters for describing the external orientation can be reduced, and they are determined by the flight parameters such as velocity v and height h_f.

Fig. 3 shows the corrected image (1000 • 1000 pixels) of the image in Figure 1. The figure shows that blurring effects in the uncorrected image can be fully compensated by the pixel ordering procedure as described above. This image also shows the necessity of exact attitude information for avaluation of line scanner data for each measured CCD-line.

Fig. 3   Aircraft attitude correction of the image in Fig. 1

Correction errors can occur, when the reference hight in the correction algorithm due to different object height, for example, is wrongly choosen (see some remaining small wavy structures along the borders of buildings in Fig. 3).

4. STEREO RECONSTRUCTION

The following procedure was used for the reconstruction of a digital elevation model (DEM). The first step is finding conjugated points in at least two different images using a matching algorithm. With the knowledge of their coordinates in the focal plane and of the attitude of the sensor, so-called pixel rays can be defined.

$$x_{in} = x_{0n} + t \cdot x_{dn} \tag{4-1}$$

x_{0n} current camera location,

x_{dn} camera's direction vector,

t unknown parameter,

where n is the number of the current line.

Under ideal circumstances the three coordinates of an object-point in the terrain are given by the intersection point of these rays. Because of the discretization errors caused by the finite resolution of the camera, there is no intersection point. So an error criterion must be defined to determine the vector with the smallest distance between the rays. This vector gives the 3D-coordinates of the reconstructed point.

$$\hat{x}_i = x_0^0 + t_0 \cdot x_d^0 + \xi_0 = x_0^1 + t_1 \cdot x_d^1 + \xi_1 \tag{4-2}$$

$\hat{x}_i$ estimated intersection point

ξ_n deviation between estimated and real intersection point.

The intersection points may be calculated by minimizing the error $\Im = \sum \xi_i^2$.

Equation (4-2) demonstrates the idea for two rays. The estimated intersection point between the rays and the terrain is defined as the vector, where the error is minimal. Fig. 4 shows this approach for a CCD-line scanner.

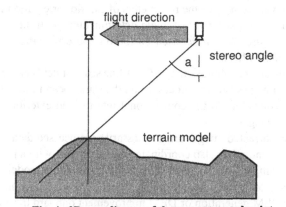

Fig. 4 3D-coordinates of the reconstructed point

After the matching procedure, pixels of different CCD-lines are related to the same object point. Pixel rays were defined, and the three coordinates of the point in the object-space were obtained. If a sufficient number of terrain points could be calculated, a DEM can be retrieved by a two-dimensional interpolation algorithm. This procedure was executed with simulated image data and yielded excellent results.

Fig. 5 shows the result of the described procedure. Two image strips of a flight over a mountainous area were evaluated and a DEM was generated. The corrected nadir image strip was laid over the DEM.

Fig. 5   DEM with the corrected nadir image

5. CONCLUSIONS

The Wide-Angle Optoelectronic Stereo Scanner WAOSS developed for the Mars-96 orbiter and the Wide-Angle Airborne Camera WAAC was used on board aircraft . Direct and accurate measurements of attitude parameters (roll, yaw, pitch, ground speed and altitude variation) give the parameters of exterior orientation for each measured CCD-line. The high accuracy of angle determination is strongly related to stochastic drift. The future work is focused on the on-line drift correction and precise determination of angle offset.

Using attitude and camera calibration data CCD-line scanner data can be geometrically corrected. If these image strips are processed on the same mean flight path, the correction procedure is equivalent to the conventional interior and exterior orientation and gives epipolar-like images.

The geometrically corrected images are a necessary preprocessed data product for further 3-D processing and allow also coordinates measurements with a photogrammetric workstation, for example, and the generation of digital elevation models (DEM).

The independence from additional GCPs and straightforward algorithms make an on-line approach for DEM generation or DEM improvement onboard the aircraft possible without human interaction.

6. REFERENCES

[1] D.Oertel, R.Reulke, R.Sandau, M.Scheele, T.Terzibaschian, A Flexible Digital Wide-Angle Optoelectronic Stereo Scanner, International Archives of Photogrammetry and Remote Sensing, Vol. XXIX, Part B1, Commission I, 1992, pp. 44

[2] R.Sandau, A.Eckardt, "The Stereo Camera Family WAOSS/WAAC for Spaceborne / Airborne Applications", International Archives of Photogrammetry and Remote Sensing, Vol. XXXI, part B1, pp. 170-175, 1996

[3] M.Lehner, R.Gill, Photogrammetrical Adjustment of Triple Stereoscopic Imagery of an Airborne CCD scanner, Optical 3-D Measurement Techniques, edited by A.Grün/H.Kahmen, pp.123, Karlsruhe 1989

[4] F.Müller, O.Hofmann, A.Kaltenecker, "Digital Photogrammetric Assembly (DPA) Point Determination Using Airborne Three-Line Camera Imagery: Practical Results", International Archives of Photogrammetry and Remote Sensing, Vol. 30, Part 3/2, Commission III, 1994, pp. 592-598

[5] S. Murai, "Stereoscopic Imagery with an Airborne 3-Line Scanner (TLS)", International Archives of Photogrammetry and Remote Sensing, Vol. 30, Part 5W1, pp. 20-25, 1994

[6] J.Albertz, F.Scholten, H.Ebener, C.Heipke, G.Neukum, "The Camera Experiments HRSC and WAOSS on the Mars 94 Mission", International Archives of Photogrammetry and Remote Sensing, Vol. XXIX, Part B1, Commission I, pp. 130-137, 1992

[7] C. Heipke, W. Kornus, A.Pfannenstein, "The Evaluation of MEOSS Airborne Three-Line Scanner Imagery: Processing Chain and Results", PE & RS, Vol.:LXII, pp.293-300, 1996

[8] M.Scheele, T.Terzibaschian, Attitude and positioning measurement system used in the airborne testing of the original Mars Mission Wide-Angle Optoelectronic Stereo Scanner WAOSS, International Archives of Photogrammetry and Remote Sensing, Vol. 30, Part 1, 1994, pp. 47

An Adaptive Method of Color Road Segmentation

ChengFa Fan, ZhuMing Li, XiuQing Ye, and WeiKang Gu

Department of Communication and Electronic Engineering

Zhejiang University, Hangzhou, 310027

P. R. China

Fax: (86-571)7951358   Tel: (86-571)7951529

Abstract - This paper presents a new adaptive and robust method for color road segmentation.   The method shows good performance in segmentation of road with shadows. The color features of road are obtained in a sample region and are expressed as two scatter diagrams of red-green and blue-green.   A fitting and predicting approach is used to extend the features to the whole image.   The pixels in the image are then classified as road pixels or non-road ones based on the extended characters.

1   Introduction

Road segmentation is a challenge to mobile robot system in complex real-road environments due to the diversity of natural scenes.   In previous mobile robot systems, a lot of attempts have been made on road segmentation.   Several systems use edge detection method [2, 3, 4], but the assumption of strong edges to distinguish the road from its surroundings is not always valid.   Methods of region extraction or fusing of region and edge are adopted in some other ALV (autonomous land vehicle) or UGV (unmanned ground vehicle) systems to achieve robust performance [2, 3, 10, 11, 12, 13].   We have developed a system to distinguish road from non-road backgrounds combining edge and region information and this paper shows the region branch.   The edge branch is described in another paper.

In order to distinguish road pixels from non-road ones, features of road should be collected firstly.   Features are usually obtained in two ways.   The first way is to store road and non-road features in a pre-defined database and then retrieves them from the database when applying classification [1, 2, 3, 5, 7, 10].   The database is refreshed by the information of the previous one or several frames.   In this way a lot of training has to be made and lots of classes have to be stored before classification. It is sensitive to changes of the circumstance and a well-defined model has to be made to cover most of the possibilities [2].   The second way is to get features from the current frame by defining a sample region that always lies in the road [11]. These features describe the road properly and will not be influenced by circumstance changing.   More over, it is an adaptive way because features used to segment road are found in current image frame.

Usually roads with shadows are difficult to be distinguished correctly from their surroundings.   Lots of efforts have been made to deal with shadows.   Some typical methods use several classes to describe road and several other classes to non-road objects [2, 3, 5, 7].   In later stage of these methods, shadowed and unshadowed road classes are merged to form a complete road description.   In most cases several discrete classes can indicate shadowed and un-shadowed road approximately. However, when there are heavy shadows on the road, more classes are needed to achieve a good approximation and these methods show their drawbacks due to their limit class number. To overcome this difficulty, we propose a continuous expression of the road features for each intensity level in our method.

Processing speed should be considered in road segmentation methods. Traditional intensity threshold method provides fast processing speed but its performance gets worse when facing shadowed environment. Some of the recent researches apply color segmentation based on color 3-D cluster analysis [8, 9]. When applying them on road segmentation good performance can also be got, but it is time and memory consuming and does not fit for real-time color road segmentation. Not surprisingly, the performance and speed criteria are likely to be satisfied simultaneously if 2-D cluster analysis is applied. A concise and efficient way of 2-D cluster is to use the scatter diagrams [1] derived from the red, green and blue components of the image.

We get features in a sample region and express it in two scatter diagrams. Then an approach of fitting and predicting is used to extend the features to the whole image region. At last the classification is applied based on the extended features. In section 2 of the paper, the general idea of the method is presented. Section 3 describes the implementation. Experiment results are discussed in section 4. Section 5 is the conclusion.

2 Element of the Method

Roads that are evenly illuminated can be easily recognized in color or monochrome video images. For example, a single threshold method can obtain good result. While a real ALV or UGV vision system has to face a variety of road conditions, it is not a little the case any more. As mentioned in section 1, when there are shadows on the road, some road pixels show more like surrounding pixels other than road ones in their luminance intensity. Color normalization method is only a linear transform such that it can only deal with slightly shadowed case [5]. John F. Gilmore et. al., in their UGV system, use intensity based image processing and wipe out the bad effect of shadows through texture [11], but we find that texture of shadows in road and background objects are sometimes similar to each other. In Carnegie Mellon's Navlab, the SCARF vision system uses a supervised classification method where several classes are used to describe road appearance and some other classes to describe non-road objects. In their method when there are shadows on the road, the road classes correspond to some discrete image intensity values. If we provide a class for each intensity value, the approximation for the shadowed road is likely to be more precisely.

Assuming that there is only a unique type of incident light with different intensity I, the spectrum of the incident light can be expressed as $IS(\lambda)$. We also assume the reflect coefficient of road be $\alpha(\lambda)$ and the response curve of the camera for red, green and blue component be $a(\lambda, I)$, $b(\lambda, I)$ and $c(\lambda, I)$, respectively. Let the red, green and blue output of the camera be $\mathbf{C}_0 = (R_0, G_0, B_0)$, then

$$R_0 = \int_\lambda IS(\lambda)\alpha(\lambda)a(\lambda, I)d\lambda \equiv f_R(I)$$

$$G_0 = \int IS(\lambda)\alpha(\lambda)b(\lambda,I)d\lambda \equiv f_G(I)$$

$$B_0 = \int IS(\lambda)\alpha(\lambda)c(\lambda,I)d\lambda \equiv f_B(I) \tag{1}$$

In the same image, the form of f_R, f_G and f_B depend only on $\alpha(\lambda)$, or the material. The real output is polluted by noise because of the diversity of the road material, the slightly change of the incident light, the noise of the camera, etc. Let the real output be $C = (R, G, B)$, then

$$C = (R,G,B) = (R_0 + n_R, G_0 + n_G, B_0 + n_B) \tag{2}$$

From (1) we can find that

$$R_0 = f_R(f_G^{-1}(G_0)) \equiv \gamma(G_0),$$

$$B_0 = f_B(f_G^{-1}(G_0)) \equiv \beta(G_0). \tag{3}$$

From (2), using Taylor's Theorem, we have

$$\gamma(G) = \gamma(G_0 + n_G) = \gamma(G_0) + \gamma'(G_0)n_G + o(n_G)$$

$$\beta(G) = \beta(G_0 + n_G) = \beta(G_0) + \beta'(G_0)n_G + o(n_G).$$

Omitting the higher term $o(n_G)$ and combining with (2) and (3), we can get

$$R - \gamma(G) = n_R - \gamma'(G_0)n_G$$

$$B - \beta(G) = n_B - \beta'(G_0)n_G. \tag{4}$$

If n_R, n_G and n_B are Gaussian noise, then $R - \gamma(G)$, $B - \beta(G)$ also satisfy Gaussian Distribution. Now it can be seen that red and blue components are nonlinear functions of green component, taken green component as an independent variable. Given a group of sample points, which stand for a nonlinear function polluted by noise, we can choose a polynomial to fit them. A 3-level polynomial is selected to fit the slowly changed two nonlinear functions. As mentioned in part I, a sample region is firstly selected, then two scatter diagrams of red-green and blue-green are obtained from the sample region. Centers of red or blue values with respect to each green value are calculated and two groups of sample points are generated. Fitting is used to get the nonlinear functions and then predicting is applied to extend the features to a wider range. The noise ranges are also calculated. At last segmentation can be applied based on the characters, namely, the functions and the noise ranges.

3 Realization of the Method

3.1 Features Fitting

The first step of the proposed method is to collect road features in a sample region. J. F. Gilmore et. al. use a triangle (region A) shown in figure 1 as the sample region [11]. Similarly, we use a trapezium (region B) as the sample region based on the knowledge that the nearby scene before the camera is road. Two scatter diagrams (or 2-D color histograms), $\{H_{ij}^{r-g}\}$ and $\{H_{ij}^{b-g}\}$, are obtained from the sample region where $H_{ij}^{r,b-g}$ denote the number of points in the sample region with red-value/blue-value i and green-value j. Because of the similarity of H_{ij}^{r-g} and H_{ij}^{b-g} we prefer to use $\{H_{ij}\}$ in later analyses. The center of red or blue value corresponding to each green value is

$$c_j = (\sum_i i \cdot H_{ij}^{0.5}) / \sum_i H_{ij}^{0.5}, \qquad (5)$$

and the variance is

$$\sigma_j = ((\sum_i (i - c_j)^2 H_{ij}^{0.5}) / \sum_i H_{ij}^{0.5})^{0.5}. \qquad (6)$$

The sample set $\{(x, y_x)\}$ is selected as

$$(x, y_x) = (j, c_j) \quad \text{if} \quad \sum_i H_{ij}^{0.5} \neq 0. \qquad (7)$$

The polynomial for fitting is selected as $p(x) = a_0 x^3 + a_1 x^2 + a_2 x$ such that its curve runs through the original point of the color coordinate. Assuming the sample set contains n elements, the fitting error is

$$E = \sum_x (y_x - a_0 x^3 + a_1 x^2 + a_2 x)^2 \omega_x, \qquad (8)$$

where $\omega_x = (\sum_i H_{ix})^{0.5}$ is the weight parameter.

Let $\dfrac{\partial E}{\partial a_0} = \dfrac{\partial E}{\partial a_1} = \dfrac{\partial E}{\partial a_2} = 0,$ it is easy to get that

$$\mathbf{a} = (\mathbf{AB})^{-1} \mathbf{Ay},$$

where

$$\mathbf{a} = \begin{bmatrix} a_0 \\ a_1 \\ a_2 \end{bmatrix}, \ \mathbf{A} = \begin{bmatrix} \omega_1 x_1^3 & \omega_2 x_2^3 & \cdots & \omega_n x_n^3 \\ \omega_1 x_1^2 & \omega_2 x_2^2 & \cdots & \omega_n x_n^2 \\ \omega_1 x_1 & \omega_2 x_2 & \cdots & \omega_n x_n \end{bmatrix}, \ \mathbf{B} = \begin{bmatrix} x_1^3 & x_1^2 & x_1 \\ x_2^3 & x_2^2 & x_2 \\ \cdots & \cdots & \cdots \\ x_n^3 & x_n^2 & x_n \end{bmatrix}, \ \mathbf{y} = \begin{bmatrix} y_1 \\ y_2 \\ \cdots \\ y_n \end{bmatrix}. \qquad (9)$$

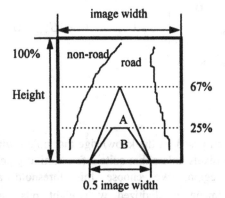

Figure 1   Road image and sample regions
A: triangular region; B: trapezoid region

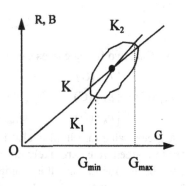

Figure 2   Scatter Diagrams,
fitting and predicting

The green value range in the sample region is also obtained for later predicting usage. The minimal value is set to the green value to where the lower pixels represent 3% of the pixels in the sample region. The maximal value is the green value to where the higher pixels represent other 3% of the pixels in the sample region. These two parameters are denoted as G_{min} and G_{max}.

3.2 Features Predicting

Usually features in the sample region represent most of the color distribution possibilities on the road, but occasionally there are regions on the road whose green value is beyond the limit of the sample region. A robust method needs a predicting mechanism to provide better classification result. In mono-spectral processing it is not feasible to predict for its 1-D (one dimension) property. In 2-D case, it is feasible but also should be careful. A direct thought is to take the fitting feature as the color features out of the green value limit of the sample region. Unfortunately, the fitting curve fits the points in the sample region well, but sometimes runs far away from the real center of color distribution of road points in other regions. Hence another predicting approach is developed.

The weight centroid of the color distribution cloud in the scatter diagram for the sample region is firstly calculated, then the slope K of the line from original point to the centroid (Figure 2). The slopes of the osculatories of the fitting curve on point $(G_{min}, p(G_{min}))$ and $(G_{max}, p(G_{max}))$ are denoted as K_1 and K_2, respectively. The predicted curves $l(G)$ and $h(G)$ for low and high terminals are

$$l(G) = (\omega_1(G)K + (1 - \omega_1(G))K_1)(G_{min} - G) + p(G_{min}),$$

$$h(G) = (\omega_2(G)K + (1 - \omega_2(G))K_2)(G - G_{max}) + p(G_{max}), \tag{10}$$

where

$$\omega_1 = \begin{cases} \dfrac{G_{min} - G}{G_{max} - G_{min}} & G \geq 2G_{min} - G_{max} \\ 1 & G < 2G_{min} - G_{max} \end{cases}, \quad \omega_2 = \begin{cases} \dfrac{G - G_{max}}{G_{max} - G_{min}} & G < 2G_{max} - G_{min} \\ 1 & G \geq 2G_{max} - G_{min} \end{cases}.$$

The final two fitting curves can be described as

$$\gamma, \beta(G) = \begin{cases} l_{R,B}(G) & G < G_{min} \\ p_{R,B}(G) & G_{min} \leq G < G_{max} \\ h_{R,B}(G) & G_{max} \leq G \end{cases}.$$

A green value threshold is also induced based on the knowledge that pixels with very low green value are likely not road pixels unless that quite a few similar pixels have been found in the sample region. We choose this threshold as $G_{th} = \min(G_{min}, 0.4G_{MAX})$ where the image is digitized with eight bits and $G_{MAX} = 255$.

3.3 Segmentation and post processing

The segmentation can be applied on the image based on the road features. Now two fitting curves $\gamma(G)$, $\beta(G)$ and average noise ranges σ_R, σ_B are found. Two binary images I^R and I^B are calculated separately as

$$I_{ij}^R = \begin{cases} 1 & \text{if } \gamma(G_{ij}) - 2\sigma_R \leq R_{ij} \leq \gamma(G_{ij}) + 2\sigma_R \\ 0 & \text{otherwise} \end{cases} \tag{11}$$

$$I_{ij}^B = \begin{cases} 1 & \text{if } \beta(G_{ij}) - 2\sigma_B \leq B_{ij} \leq \beta(G_{ij}) + 2\sigma_B \\ 0 & \text{otherwise} \end{cases} \tag{12}$$

A logical AND operation is applied to the two images to create a binary image with the road pixels set to 1. Then a dilation operation is applied on the binary image to fill up holes. A pruning algorithm [11] is used twice to wipe out small and isolate region. After that a supplemental dilation operation is applied and at last the road boundary is traced to get a linking road region.

4 Experiment Result

We have tried this method on hundreds of images, including campus avenues of Zhejiang University, paved roads of nearby parks, earth roads of the countryside and real street roads. Over 90 percent of the segmentation results can be accepted from human's view. The remnant less than 10% ill-segmented cases concludes 3 types of situations: it happens that there are non-road objects in the sample region; the image is two much saturated in luminance; it happens that the color features of some surrounding objects are similar to the road ones. This method is especially good when pixels in the sample region can almost represent all of the road pixels. In some ill-segmented cases, non-road objects are classified as road but they do not connect to the main road region and can be wiped out by post processing described in the end of previous section.

Figure 3 and figure 4 exhibit some of the experiment results. Figure 3 shows the segment process. Figure 3(a), (b) and (c) present the original color components of a road scene with heavy shadows. Figure 3(d) and (e) show the scatter diagrams and the fitting curve. Figure 3(f), (g) and (h) show the binary images after elementary classification, pruning and tracing, respectively.

Figure 4 tells the segment result of our method and intensity threshold method as a comparison for a scene with road, grass, sky and trees. Figure 4(f) shows the classify result using our method in which road is correctly found, not disturbed by the shadows of trees and similar luminance intensity of grass nearby. Figure 4(g) and (h) shows the results using traditional intensity based classification method (pixels with intensity between two thresholds are classified as road). In figure 4(g), grass is classified as road. In figure 4(h), road with shadows is stated as non-road. The comparison result tells that the proposed method can classify different objects by color when they have the same intensity and can find road in shadow by predicting while traditional intensity based method can not.

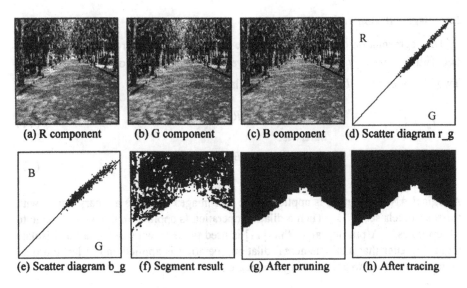

Figure 3   Segmentation process of the proposed method

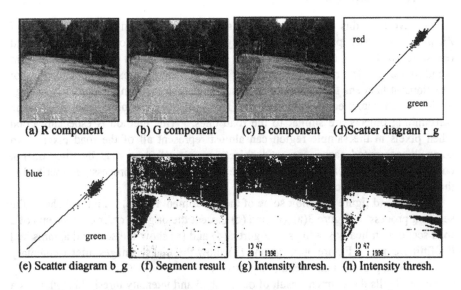

Figure 4   Road segmentation of a scene with grass, sky, trees and shadows.
Intensity threshold: (g) T_1=110, T_2=220,   (h) T_1=150, T_2=220.

5 Conclusion

Our intention is to solve the difficult problem of road segmentation in mobile robot system.   We studied several existing road classification approaches and propose a simple but robust and adaptive road segmentation method.   The method is based on the analysis of two scatter diagrams of red-green and blue-green obtained separately in a sample region.   The analysis gets road features in the sample region and extends the features to a wider range. The extended features are assumed to represent the whole road features and the classification is simply a searching in a lookup table derived from them.

The method performs well when dealing with shadowed roads as well as clean and evenly illuminated road in most situations. In some few cases, i.e., in the 3 types of situations discussed in experiment result, images are ill-segmented. We should also seek other constrains to solve this problem. For instance, we can avoid the influence of non-road object in sample region by selecting the sample region more carefully as those described in [1]. The other alternative is to combine this region method with edge approach to supplement the drawbacks of each other . Our other work has approved helpful of this simple idea.

References

[1] M. A. Turk, D. G. Morgenthaler, K. D. Gremban, and M. Marra, "VITS -- A Vision System for Autonomous Land Vehicle Navigation," *IEEE Trans. on PAMI*, vol. 10, no. 3, May 1988.

[2] C. Thorpe, M. H. Herbert, T. Kandade, and S. A. Shafer, "Vision and Navigation for the Carnegie-Mellon Navlab," *IEEE Trans. on PAMI*, vol.10, no. 3, May 1988.

[3] C. Thorpe, M. Herbert, T. Kanade, and S. Shafer, "Toward Autonomous Driving: The CMU Navlab, Part I -- Perception, " *IEEE Expert*, August 1991.

[4] A. Broggi, "Parallel and Local Feature Extraction: A Real-Time Approach to Road Boundary Detection," *IEEE Trans. on Image Processing*, vol. 4, no. 2, February 1995.

[5] K. Liu, Y. X. Liu, J. Y. Yang, Y. Q. Cheng, and N. C. Gu, "A new color road segmentation method," *SPIE vol. 1613 Mobile Robots VI (1991)*.

[6] Q. Zhang, R. Jiang, K. Liu, and J. Y. Yang, "Analysis of color and Range Image Using PDS," *SPIE vol. 1831 Mobile Robots VII (1993)*.

[7] L. J. Liu, Y. G. Wu, K. Liu, J. Y. Yang, "Color Road Segmentation for ALV Road Following," *SPIE vol. 2058 Mobile Robots VIII (1993)*.

[8] S. H. Park, I. D. Yun, and S. U. Lee, "Color Image Segmentation Based on the 3D Clustering and Morphological Operations," *ACCV'95 Second Asian Conference on Computer Vision*, December 5-8, Sigapore.

[9] M. S. Cho, B. H. Kang, S. J. Kang, J. W. Kim, "Color Image Segmentation Applying 3-D Cluster Analysis," *ACCV'95 Second Asian Conference on Computer Vision*, December 5-8, Sigapore.

[10] Spiegle, P. McIngvale, K. Olson, A. Scales, and K. Larsen, "Autonomous Road Navigation for Unmanned Groud Vehicles," *SPIE vol. 2463 Mobile Robots X (1995)*.

[11] John F. Gilmore, Harold Forbes, Kevin Payne, and Khalid Elibrary, "The Unmanned Guided Vehicle System," *SPIE vol. 2463 Mobile Robots X (1995)*.

[12] G.L. Foresti, V. Murino, C.S. Regazzoni and G. Vernazza, "Distributed spatial resoning for multisensory image interpretation," *Signal Processing 32 (1993)*, pp. 217-215.

[13] V. Murino, C. S. Regazzoni, G. L. Foresti, and G. Vernazza, "A multilevel Fusion Approach to Object Identification in Outdoor Road Scene," *International Journal of Pattern Recognition and Artificial Intelligence*, vol. 9, no. 1, (1995), pp. 23-65.

Optical Flow Detection Using a General Noise Model for Gradient Constraint

Naoya Ohta

Department of Computer Science, Gunma University,
1-5-1 Tenjin-cho, Kiryu 376 Gunma, Japan

Abstract. In the usual optical flow detection, the gradient constraint, which expresses the relationship between the gradient of the image intensity and its motion, is combined with the least-squares criterion. From a statistical point of view, this means assuming that only the time derivative of the image intensity contains Gaussian noise. However, it is more reasonable to assume that all the derivatives are observed with Gaussian noise because they are equally computed from pixels containing noise and approximated by finite difference. In this paper, we study a new optical flow detection method based on the latter assumption. Since this method requires the knowledge about the covariance matrix of the noise, we also discuss a method for its estimation. Our experiments show that the proposed method can compute optical flow more accurately than the conventional method.

1 Introduction

Detecting optical flow from motion images is one of the most fundamental processes for computer vision, and many techniques for it have been proposed in the past [2]. One of the most basic approaches is the use of the *gradient constraint*, which is expressed in the form

$$\bar{E}_x u + \bar{E}_y v + \bar{E}_t = 0, \tag{1}$$

where $\bar{E}_x$, $\bar{E}_y$ and $\bar{E}_t$ are partial derivatives of the image intensity with respect to the space coordinates x, y and time t, respectively; u and v are the x- and y-component of the flow, respectively. In order to utilize the constraint, however, the following two problems should be overcome:

1. The gradient constraint alone is insufficient to determine the flow uniquely at each pixel.
2. The gradient constraint imposes only an approximate relationship between the flow and the image derivatives.

The first problem is often referred to as the *aperture problem*. In order to obtain additional constraints, we need some assumption about the spatial distribution of the flow. The following are typical examples:

Constant flow assumption: The flow is constant over a small image patch [6].

Smoothness assumption: The flow changes smoothly over the entire image [3].

The second problem arises for several reasons. First of all, electronic noise of the imaging device invalidates Eq. (1). Furthermore, we must approximate the image derivatives by the value obtained by applying a digital kernel in actual computation. A traditional way to cope with this problem is to introduce an error term δ into Eq. (1) in the from

$$E_x u + E_y v + E_t = \delta, \tag{2}$$

and to apply the least-squares criterion. Here, symbols without bars represent actually computed values.

Combining the least-squares criterion with the constant flow assumption, we obtain the following minimization problem:

$$J_{LS} = \sum_{\text{patch}} \delta^2 \rightarrow \min. \tag{3}$$

The symbol $\sum_{\text{patch}}$ denotes summation over the pixels in the image patch over which the flow is assumed to be constant. We can view the above minimization as the following statistical estimation. Suppose the time derivative E_t is observed with additive Gaussian noise δE_t but E_x and E_y are free from noise:

$$E_x = \bar{E}_x, \quad E_y = \bar{E}_y, \quad E_t = \bar{E}_t + \delta E_t. \tag{4}$$

The minimization (3) is equivalent to *maximum likelihood estimation* for this noise model (see Sect. 14.1 of Ref. [8], for example). However, this noise model is inconsistent with the fact that the observed image intensity contains noise and all the derivatives are computed by digital kernels. Although some modified models have been proposed [5, 9], only noise in the time derivative was considered. Hence, it is more reasonable to assume that all the derivatives are observed with Gaussian noise in the form

$$E_x = \bar{E}_x + \delta E_x, \quad E_y = \bar{E}_y + \delta E_y, \quad E_t = \bar{E}_t + \delta E_t. \tag{5}$$

In the next section, we present an optical flow detection method based on this noise model coupled with the constant flow assumption.

2 Flow Detection

2.1 Definitions

We define the true intensity gradient vector $\nabla \bar{E}_\alpha$, the observed gradient vector ∇E_α, the three-dimensional representation of the flow u and the three-dimensional Gaussian noise vector e_α as follows:

$$\nabla \bar{E}_\alpha = (\bar{E}_{x\alpha} \ \bar{E}_{y\alpha} \ \bar{E}_{t\alpha})^\top, \tag{6}$$

$$\nabla E_\alpha = (E_{x\alpha} \ E_{y\alpha} \ E_{t\alpha})^\top, \tag{7}$$

$$u = (u \ v \ 1)^\top, \tag{8}$$

$$e_\alpha = (\delta E_{x\alpha} \ \delta E_{y\alpha} \ \delta E_{t\alpha})^\top. \tag{9}$$

In the above equations, the subscript α runs over the pixels in the patch over which the flow is assumed to be constant, and the superscript $\top$ denotes transpose. We assume that the noise e_α has zero mean and the same covariance matrix V_e at all pixels. In terms of the above vectors, the noise model is

$$\nabla E_\alpha = \nabla \bar{E}_\alpha + e_\alpha, \tag{10}$$

and the gradient constraint has the form

$$(\nabla \bar{E}_\alpha, u) = 0, \tag{11}$$

where $(\cdot, \cdot)$ denotes the inner product of vectors. Equation (2) can be rewritten as

$$(\nabla E_\alpha, u) = \delta_\alpha. \tag{12}$$

Under the noise model of Eq. (10) and the constraint (11), the maximum likelihood estimator $\hat{u}$ of the flow u is given as the solution of the following minimization (refer to Chap. 7 of Ref. [4]):

$$J = \sum_{\alpha=1}^{n} \frac{(\nabla E_\alpha, u)^2}{(u, V_e u)} \to \min. \tag{13}$$

Here, n is the number of the pixels in the patch.

2.2 Minimization Computation

Since the function J given in Eq. (13) is independent of the scale of u, the minimization can be conducted over a sphere in three dimensions. Using the three-dimensional vector

$$v = (p \quad q \quad r)^\top \tag{14}$$

and the matrix

$$M = \sum_{\alpha=1}^{n} \nabla E_\alpha \nabla E_\alpha^\top, \tag{15}$$

we can rewrite the problem (13) as follows ($\| \cdot \|$ denotes the norm of a vector):

$$J = \frac{(v, Mv)}{(v, V_e v)} \to \min, \quad \|v\| = 1. \tag{16}$$

The vector v can be converted into the vector u by normalizing the third component into 1: $u = v/r$. The solution of (16) is obtained by solving the following generalized eigenvalue problem of M with respect to V_e (refer to Sect. 2.2 of Ref. [4]):

$$Mv = \lambda V_e v. \tag{17}$$

The solution of this generalized eigenvalue problem is obtained by the following procedure:

Step 1: Compute the eigenvalues μ_i and the corresponding *unit* eigenvectors m_i of matrix V_e ($i = 1, 2, 3$).

Step 2: Compute

$$P = \sum_{i=1}^{3} \frac{m_i m_i^\top}{\sqrt{\mu_i}}. \tag{18}$$

Step 3: Compute the eigenvalues λ_i and the corresponding eigenvectors n_i ($i = 1, 2, 3$) of matrix PMP.

Step 4: The generalized eigenvalues and the corresponding unit generalized eigenvectors of Eq. (17) are given by λ_i and $Pn_i/\|Pn_i\|$, respectively.

The solution $\hat{v} = (\hat{p} \ \hat{q} \ \hat{r})^\top$ of the problem (16) is given by the generalized eigenvector for the smallest generalized eigenvalue of Eq. (17). The estimate $\hat{u}$ of the flow is given by $\hat{v}/\hat{r}$.

2.3 Reliability of the Detected Flow

It is very important to know the reliability of the detected flow [7, 2]. The reliability is measured by the covariance matrix V_u of the estimated flow $\hat{u}$, which can be evaluated as follows (refer to Sect. 7.1 of Ref. [4] for the details):

1. Approximate the function J given in Eq. (13) by a quadric function $\tilde{J}$ in u and ∇E_α around their true values $\bar{u}$ and $\nabla \bar{E}_\alpha$.
2. Let $u = \bar{u} + \Delta u$, and determine the value Δu that minimizes $\tilde{J}$ as a linear function in ∇E_α.
3. Evaluate the covariance matrix $V_u = E[\Delta u \Delta u^\top]$.

It can be shown that the resulting covariance matrix has the form

$$V_u = \left(\sum_{\alpha=1}^{n} \frac{(P_k \nabla \bar{E}_\alpha)(P_k \nabla \bar{E}_\alpha)^\top}{(\bar{u}, V_e \bar{u})} \right)^-, \tag{19}$$

where

$$P_k = \text{diag}(1, 1, 0). \tag{20}$$

Here, $(\cdot)^-$ denotes the (*Moor-Penrose*) *generalized inverse*, and $\text{diag}(a, b, c)$ denotes the diagonal matrix with diagonal elements a, b and c in that order. It can be proved that Eq. (19) gives a theoretical bound, called the *Cramer-Rao lower bound*, on attainable accuracy (refer to Sect. 14.2 of Ref. [4]).

In an actual computation, the true values $\bar{u}$ and $\nabla \bar{E}_\alpha$ in Eq. (19) are approximated by the estimated value $\hat{u}$ and the observed values ∇E_α, respectively. Since the third component of u is 1, the elements in the third column and the third row of V_u are all zero. Here, the covariance matrix of $(\hat{u} \ \hat{v})^\top$ is given in the form

$$V[u, v] = (\hat{u}, V_e \hat{u}) \begin{pmatrix} M_{11} & M_{12} \\ M_{21} & M_{22} \end{pmatrix}^{-1}, \tag{21}$$

where M_{ij} is the (i, j) element of the matrix M defined by Eq. (15).

3   Estimation of the Noise Covariance Matrix

In order to compute the estimator $\hat{u}$, we need to know the covariance matrix V_e of the noise e_α. Since the scale of V_e does not affect the solution of the minimization (13), the absolute magnitude of V_e is not very important. However, we need to know at least the ratios between the elements.

Although we do not know the true value $\nabla \bar{E}_\alpha$, the covariance matrix V_e can be estimated from the distribution of the residual δ_α if the true flow $\bar{u}$ is known. Let $\bar{u}_\alpha = (\bar{u}_\alpha \ \bar{v}_\alpha \ 1)^\top$ be the true flow at the αth pixel. From Eqs. (10), (11) and (12), the residual δ_α can be written as

$$\delta_\alpha = (\nabla \bar{E}_\alpha + e_\alpha, \bar{u}_\alpha) = (e_\alpha, \bar{u}_\alpha). \tag{22}$$

Since e_α is a Gaussian random variable of mean zero, so is δ_α. Its variance is given by

$$s_\alpha = (\bar{u}_\alpha, V_e \bar{u}_\alpha). \tag{23}$$

If we observe δ_α, $\alpha = 1, \cdots, m$, the log-likelihood l is given by

$$l = -\frac{m \log 2\pi}{2} - \frac{\tilde{l}}{2}, \tag{24}$$

where

$$\tilde{l} = \sum_{\alpha=1}^{m} \left(\log s_\alpha + \frac{\delta_\alpha^2}{s_\alpha} \right). \tag{25}$$

The maximum likelihood estimator of V_e is obtained by minimizing $\tilde{l}$ under the constraint that V_e is positive definite.

4   Experiments

We compared accuracy of the flow detected by the proposed method with that by the conventional method by experiments. We used a simulated image sequence because we need to know the ture flow for the accuracy evaluation. The image sequence was made by the following steps:

1. We took 16 shots of a static scene shown in Fig. 1(a) and averaged them to remove noise. The averaged image is regarded as a noise-free image.
2. We gave motions to the image by shifting it in a computer in horizontal direction and then reduced the size of the shifted images to 1/4. The resultant motions are multiples of 0.25 pixel.
3. Next, we extracted the camera noise. We took 20 images of a white panel and averaged them. The averaged image was subtracted from each original image. The resultant images contain the camera noise. One of them is shown in Fig. 1(b).
4. Each noise image was added to the shifted images. The resultant images constitute the simulated image sequence whose true motions are 0, ±0.25, ±0.5, ... pixels in horizontal direction.

Fig. 1. Images used in the experiments.

The image motions are only homogeneous and it might sound non realistic. Such motions are reasonable, however, because the fundamental aim of the experiments is to compare goodness of the noise models (4) and (5).

We need value of the covariance matrix V_e of the noise e_α to apply the proposed method. We estimated it from the simulated image sequence by the method described in Section 3. In the estimation, we assumed that the off-diagonal elements of V_e were zero, namely, the x-, y- and t- components of the noise were independent. We also assumed that $E_{x\alpha}$ and $E_{y\alpha}$ had the same variance. Minimization of Eq. (25) was carried out by Powell's method (refer to Sect. 10.5 of Ref. [8]). The estimated value is

$$V_{e1} = \begin{pmatrix} 2.075 & 0 & 0 \\ 0 & 2.075 & 0 \\ 0 & 0 & 0.344 \end{pmatrix}. \tag{26}$$

The optical flow was computed in three ways:

(1) the proposed method with the estimated covariance matrix V_{e1} of Eq. (26);
(2) the proposed method with covariance matrix $V_{e2} = I$ (the unit matrix);
(3) the least-squares method represented by Eq. (3).

When computing the flow, we used 5×5-pixel patches in which the flow was assumed constant. Here, we index the patches by β. The covariance matrix $V[u_\beta, v_\beta]$ of the flow $(\hat{u}_\beta \; \hat{v}_\beta)^\top$ detected in the βth patch was evaluated by computing Eq. (21), and its largest[1] eigenvalue η_β was used as a measure of uncertainty [7]. For the flow detected by the conventional least-squares, the covariance matrix was computed by the following formula.

$$V[u, v] = \left(\sum_{\alpha=1}^{n} \begin{pmatrix} E_{x\alpha}^2 & E_{x\alpha}E_{y\alpha} \\ E_{x\alpha}E_{y\alpha} & E_{y\alpha}^2 \end{pmatrix} \right)^{-1} \tag{27}$$

---

[1] In Ref. [7], the smallest eigenvalue of $V[u, v]^{-1}$ is used as a reliability measure with the same meaning.

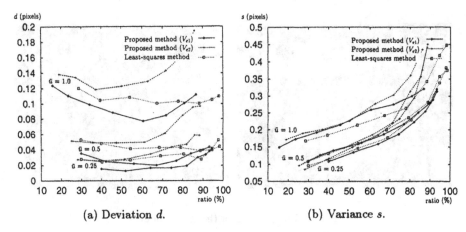

(a) Deviation d.   (b) Variance s.

Fig. 2. Accuracy versus number of the flows.

We should not evaluate a flow detection method solely by the *average* accuracy; a method which gives very accurate flow for a small number of pixel is also valuable even though the overall accuracy is low [2, 7]. For this reason, we evaluated the methods in relations between the accuracy and the percentage of the flows selected by their uncertainty measure. We set a threshold value θ and selected such flows that $\eta_\beta < \theta$ and rejected the rest. As θ becomes smaller, the selected flows can be regarded as more accurate, although the number of them decreases.

The accuracy of the selected flows was evaluated in two measures: the deviation from the true flow

$$d = \sqrt{(E^*[\hat{u}_\beta] - \bar{u})^2 + (E^*[\hat{v}_\beta] - \bar{v})^2} \tag{28}$$

and the uncertainty

$$s = \sqrt{\operatorname{tr}(V^*[\hat{u}_\beta, \hat{v}_\beta])}, \tag{29}$$

where $(\bar{u}, \bar{v})$ is the true flow and $(E^*[\hat{u}_\beta]\ E^*[\hat{v}_\beta])^\top$ and $V^*[\hat{u}_\beta, \hat{v}_\beta]$ are the sample mean and the sample covariance matrix of the selected flows, respectively. Figure 2 shows the results (graph (a) and (b) are for measure d and s, respectively).

We can observe the followings:

1. The flow estimated by all the methods has positive bias, which is approximately proportional to the true flow (Fig. 2(a)).
2. The bias is the smallest for the proposed method with covariance matrix V_{e1}. This is especially so when the motion is small.
3. There are no significant differences among the variance for the three methods (Fig. 2(b)). However, the least-squares method yields a good estimate when the motion is large, whereas the proposed method with covariance matrix V_{e1} is advantageous when the motion is small.

5 Discussions

Our experiments have confirmed that the proposed method can produce accurate optical flow for a sub-pixel motion if the covariance matrix of the noise is appropriately estimated. This fact is very important because the gradient-based approach is used to detect a sub-pixel motion. In order to detect a large motion, the gradient-based approach alone is not sufficient; we need to combine various techniques such as utilization of the image pyramid [1] and iterative estimation [6].

Although the use of an appropriate covariance matrix of the noise is crucial, it could be different from image to image, and the values given by Eq. (26) are not universal. From our experience, we have found that it depends more upon the kernel for the derivative computation than the images themselves.

Acknowledgments

The author thanks Professor Kenichi Kanatani of Gunma University for helpful discussions and encouragements. This work was in part supported by the Ministry of Education, Science, Sports and Culture, Japan under a Grant in Aid for Scientific Research C (No. 09680352).

References

1. Anandan, P.: A computational framework and an algorithm for the measurement of visual motion. Int. J. Comput. Vis. 2 (1989) 283–310
2. Barron, J.L., Fleet. D.J., Beauchemin, S.S.: Performance of optical flow techniques. Int. J. Comput. Vis. 12(1) (1994) 43–77
3. Horn, B.K.P., Schunk, B.G.: Determining optical flow. Artificial Intelligence 17 (1981) 185–203
4. Kanatani, K.: Statistical Optimization for Geometric Computation: Theory and Practice. Elsevier Science, Amsterdam (1996)
5. Kearney, J.K., Thompson, W.B., Boley, D.L.: Optical flow estimation: an error analysis of gradient-based methods with local optimization. IEEE Trans. Patt. Anal. Mach. Intell. 9(2) (1987) 229–244
6. Lucas, B, Kanade, T.: An iterative image registration technique with an application to stereo vision. Proc. 7th Int. Joint Conf. Artif. Intell., Vancouver Canada (1981) 674–679
7. Ohta, N.: Image movement detection with reliability indices. IEICE Trans. E74(10) (1991) 3379–3388
8. Press, H.P., Flannery, B.P., Teukolsky, S.A., Vetterling, W.T.: Numerical Recipes in C. Cambridge University Press Cambridge (1988)
9. Simoncelli, E.P., Adelson, E.H., Heeger, D.J.: Probability distributions of optical flow. Proc. Conf. Comput. Vis. Patt. Recog., Maui Hawaii (1991) 310–315

Algorithmic Solution and Simulation Results for Vision-Based Autonomous Mode of a Planetary Rover *

Marina Kolesnik[+], Gerhard Paar[++]

[+]Max-Planck Institute für Aeronomie, Lindau, Germany
[++]JOANNEUM RESEARCH, Graz, Austria

Abstract. A vision based navigation (VBN) system is chosen as a basic tool to support autonomous operations of a planetary rover during space missions. The rover equipped with a stereo vision system and perhaps a laser ranging device shall be able to maintain a high level of autonomy under various illumination conditions and with little a priori information about the underlying scene. Within the LEDA Moon exploration project currently under focus by the European Space Agency, in autonomous mode the rover should perform on-board absolute localization, digital elevation model (DEM) generation, obstacle detection and relative localization, global path planning and execution.

Focus of this paper is to simulate some of the path planning and path execution steps. Using a laboratory terrain mockup and an accurate camera mounting device, stereo image sequences are used for 3D scene reconstruction, risk map generation, local path planning, and position update by landmarks tracking. It is shown that standalone landmark tracking is robust enough to give navigation data for further stereoscopic reconstruction of the surrounding terrain. Iterative tracking and reconstruction leads to a complete description of the rover path and its surrounding with an accuracy high enough to meet the specifications for unmanned space exploration.

1 Introduction

Significant progress in outdoor autonomous vehicle development has been made during the last decade. The implementation of the recent technological achievements to planetary rovers is a challenging task outlined for LEDA [Age95]. Three roving control modes must be supported by the Vision-Based Navigation (VBN) system according to LEDA objectives:
- teleoperating mode of ground based piloting
- medium autonomy mode: autonomous piloting and ground based navigation
- full autonomy mode or autonomous navigation

This paper outlines the concept of fully autonomous mode for the rover equipped with a passive imaging sensors setup. In Section 2 a closed-loop algorithmic solution suitable for on-board implementation is described. Section 3 presents simulation results of rover operations with the help of a Lunar terrain mockup.

2 Closed Loop Solution for Autonomous Navigation

2.1 Specific Conditions

On a planetary surface immediate human intervention is in fact impossible. the Rover's VBN system must overcome a set of specific conditions and requirements:

---

* This work was supported in part by the Austrian Science Foundation (FWF) under grants S7003-MAT and M00265-MAT, and JOANNEUM RESEARCH.

- Automatic initial calibration of the vision sensors
- Absence of accurate reference points or landmarks for precise self calibration
- Low angle viewing conditions
- No a priori information about underlying terrain

On the other hand slow rover motion allows to move in a stop/thinking mode which gives enough time to perform necessary on-board operations.

2.2 Stereo Vision System Arrangement

Experiences of several rover teams have proved that parallel geometry of stereo cameras is difficult to arrange and maintain. For example, an autonomous walker "Dante" based on stereo vision and designed for Antarctic applications is described in [Ros93]. A special platform has been developed to adjust and maintain parallel geometry for three stereo cameras. Evidently, the parallel stereo geometry can be easily distorted after landing and due to the day/night time temperature variations. Therefore a 3D stereo reconstruction approach based on arbitrary stereo geometry looks more preferable and reliable. Consequently, the calibration procedure for the stereo system is composed from two steps:

intrinsic parameters	extrinsic parameters
focal length, principal point, lens distortion	position and pointing
done on ground, refined during operation	updated between motion steps
slight changes: temperature, vibrations	change considerably during motion

2.3 Navigational Steps

We propose the following approach to accomplish the rover's autonomous mode:

1. Initialisation phase to initialise the rover's operational units:
   - Initial calibration of the sensor units
   - Self-localisation of the rover position
2. Operational phase, implemented in the cycle:

   Stop-thinking mode   Path execution mode

   Stereo image acquisition   Consecutive image acquisition
   DEM reconstruction   Landmarks tracking
   Risk map generation   Update rover position
   Local path generation   Reflex obstacle detection

Initial Localisation of the Rover Position After rover deployment on the planetary surface a local map with respect to the lander can be used on-board for further rover operations. Its initial localisation accuracy depends upon the available surface map taken from orbit. The rover position and orientation shall be constantly controlled on-board in the local coordinate system utilizing the vision system.

Initial Calibration of the Sensors The initial calibration of all measuring units must be done after rover deployment on the surface. The initial stereo cameras calibration accuracy depends on the reference points' 2D coordinates accuracy (in the images) and their 3D coordinates accuracy (on the scene). According to [UP95], a very accurate and stable relative orientation between the cameras can be accomplished by utilising results from stereo matching.

3D DEM Reconstruction A stereo matching algorithm (Hierarchical Feature Vector Matching, (HFVM [PP92]) based on the arbitrary geometry of the vision system has been developed for terrain reconstruction. It is based upon the idea of creating a feature vector for each pixel (surrounding) and comparing these features in image pyramids from low to fine resolution levels. Each pixel of the resulting disparity map defines a parallax between corresponding points on the left and right images of the stereo pair. The major HFVM steps are:

1. Build the pyramid
2. Compute the feature images for each pyramid level
3. Match the top level of the pyramid (lowest resolution)
4. Filter the resulting disparity map
5. Check matching consistency by backmatching (left-to right, right-to-left)
6. Interpolate undefined and inconsistent disparities
7. Use the result as prediction to match the next higher resolution

The *Locus reconstruction* approach is used for 3D coordinates reconstruction from the disparity map [BP93, KK92]. The elevation at an arbitrary DEM position is found by intersecting a hypothetical line at this location with the object's surface. The line is projected into the stereo images (left and right Locus). The image information at the Locus location characterises a profile curve on the object's surface. The corresponding location on the other image is found by mapping the Locus using given disparities. The elevation is determined by the most consistent intersection between the Locus and the profile curve projection taking into account the known height on the curves. Output of the Locus reconstruction approach are height values on arbitrary DEM locations (e.g. in a regular grid) projected on the horizontal plane in the given world coordinate system.

Risk Map Generation A risk map is generated on the basis of the Locus DEM. Steep slopes and elevated areas on the DEM are marked as hazardous and unsuitable for the rover motion by comparing to some predefined safe threshold.

Local Path Planning There are several algorithms to build up a safe local path [PLSN93] on the basis of a DEM. Our idea is to consider pixels located outside of hazardous areas on the Risk map as the nodes of a directed graph. The path is generated from a predefined start point and should reach a specified target point on the local map in such a way to minimise the length of a sub-graph (Dijkstra algorithm [Dij59], see [Kol95] for details).

Natural Feature (Landmark) Tracking Displacements between landmarks (Interest Points) in subsequent images are used for calibration update [PSP95]. The Interest Points are extracted using a derivative of the Moravec operator. We propose to use a hierarchical approach to identify correspondent Interest Points on the subsequent frames since a prediction of the motion between consecutive frames is hard to obtain. First, the disparities of all image pixels are calculated in coarse resolution using HFVM. The coarse disparities are then used as prediction to calculate the exact correspondences only for the Interest Points.

Calibration Update Having two consecutive images (Frames N and $N + 1$), the camera position and pointing parameters for Frame $N + 1$ can be obtained in the following way:

The 3D Interest Points coordinates on Frame N are calculated from the DEM. Having their 2D coordinates from tracking between Frame N and Frame $N + 1$, a calibration can be performed for Frame $N + 1$. The calibration method which keeps intrinsic camera parameters fixed and gains only an update for extrinsic camera parameters has been described in [PU93, GT90]. It takes the particular conditions (3D noise, flat terrain, many points) into consideration.

The major parameters during Interest Points tracking are the number of landmarks and their back projection error as a measure of the calibration consistency. The most inconsistent Interest Points are rejected and are not used for calibration. Experiments showed that the optimum number of landmarks necessary for the reliable calibration is at about 200 points, however, a smaller set of landmarks ($<$200) still leads to robust tracking. In the common case, about 20% of the Interest Points are rejected (backprojection error more than 1 pixel) which verifies the usability of the Moravec operator and HFVM.

3 Simulation Results and Illustrations

The hardware used for simulating rover operations on the Moon surface (CamRobII) includes an accurate robot holding two cameras that can be moved each with 5 degrees of freedom (x, y, z; rotation around x and z) within a 2m x 1m x 1.5m wide volume. CamRobII is controlled by software running on a SPARC workstation. It enables camera motion on interactive-command basis to capture images and to store video data together with the camera position and orientation data (=ground truth).

A Lunar terrain model was placed in a 1.6 m x 2 m bed mockup. The correspondence between the CamRobII- and the world coordinate system is given by a transformation matrix. Initial coordinates for the very first position of the left stereo camera and the stereo basis for the stereo pairs in the sequence are taken as known. The relative orientation of the stereo configuration based on stereo correspondences was performed using a fixed baseline between the cameras to obtain a certain scale factor. The following sequence of the operations has been performed in a loop:

Stereo sequence image acquisition $\longrightarrow$ Stereo matching and DEM generation $\longrightarrow$ Landmarks tracking and calibration update.

Risk map generation and local path planning was done independently from the tracking simulation to demonstrate the robustness of the proposed approach for the Moon like terrain.

3.1 Input Image Acquisition

A sequence of 40 stereo pairs was taken with a scale factor of about 1:10 to reality. Both stereo cameras are set close to the mockup surface and directed slightly downward (15-20 degrees) to obtain a convergent low viewing angle perspective stereo pair of the mockup terrain. After the first pair is acquired both cameras are iteratively moved forward (15 mm in x, 30 mm in y for each step) to catch stereo pairs (Figure 1) on a straight path.

681

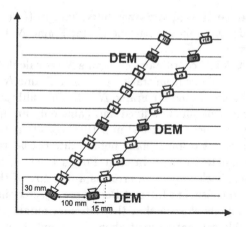

- Camera position and frame number

Fig. 1. Every fourth stereo pair is used for the reconstruction of a new DEM

Fig. 2. Ortho image merged from ten stereo configurations. Occluded and undefined areas are marked as white. Pixel size $1mm \times 1mm$, the scene covers 900×1400 mm.

3.2 DEM Generation

A general elevation model of the mockup terrain (ortho DEM) is generated using each fourth stereo pair (the *basic stereo pairs*). They are matched automatically, 3D Locus reconstruction is applied. DEM resolution is 1 mm in x and y. The frequency of stereo reconstruction is defined by the necessary overlap (about 70%) between subsequent reconstructed ortho DEMs. Intermediate images taken by the left camera are used for tracking and calibration update. The ortho DEMs calculated from the basic stereo pairs are merged to generate the entire ortho DEM for the underlying terrain. Figure 2 depicts the merging result of the ten ortho images calculated from ten subsequent basic stereo pairs.

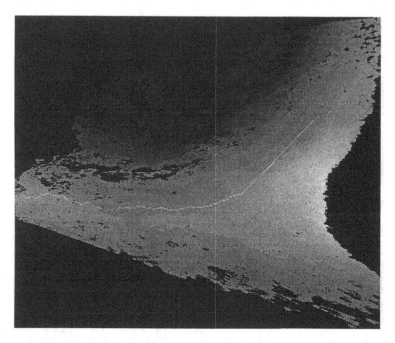

Fig. 3. Local path put on the DEM (elevations are grey coded, bright areas are high). Unknown and hazardous areas are marked black.

3.3 Path Planning

A safe local path which has been generated using the reconstructed DEM on the basis of the DEM slopes is shown on Figure 3. Hazardous areas unsuitable for the rover motion are marked on the DEM as black. The start and destination points for the rover path have been specified by an operator.

3.4 Tracking and Calibration Update

The calibration update results have been compared with the actual CamRobII coordinates. The image sequence is composed from 10 basic stereo pairs and 3 intermediate left image frames between each of them (40 frames). Relative calibration based on stereo matching [PU93] is used to calculate the position and pointing of the right camera for each basic stereo pair. Landmark tracking [PSP95] is used to maintain the coordinates of the 4 intermediate frames. Each fourth intermediate frame composes the left image for the next basic stereo pair starting the next calibration loop.

The actual CamRobII trajectory used for the stereo sequence acquisition is a straight line (Figure 1). The stereo basis (SB) for the basic stereo pairs is equal to 97 mm. An example of the tracking paths between the corresponding Interest Points on 4 subsequent even frames is shown on Figure 4. Figure 5 displays the trajectory calculated on the basis of tracking and calibration update for the cameras. The coordinates of the positions which composed the trajectory are presented in Table 1. They show that discrepancy between CamRobII coordinates and tracking positions have not been accumulated along the path. The

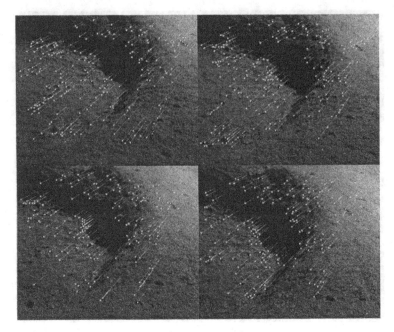

Fig. 4. Four consecutive image frames and landmark tracking paths

fact that Y offset values are always above 30 mm is explained by the uncertainty in the scale factor chosen with the estimated stereo baseline.

4 Conclusions

A closed-loop vision-based algorithmic solution for autonomous rover navigation mode is presented here. The solution integrates the following algorithmic components into the calculating chain: Consecutive image stereo data acquisition $\longrightarrow$ 3D Digital Elevation Model reconstruction of a terrain $\longrightarrow$ Risk map generation $\longrightarrow$ Local path planning $\longrightarrow$ Landmarks tacking $\longrightarrow$ Rover position update(calibration update). The algorithmic steps have been simulated with the help of an accurate robot placing a camera above a Lunar terrain mockup. Emphasis was put on the critical onboard calculations in providing fully autonomous and robust rover operations.

The following statements can be drawn on the basis of simulation sessions, using an accurate camera motion device as ground truth: 3D autonomous rover navigation on a Moon-like terrain is feasible. The accuracy and robustness of each algorithmic component of the VBN system was shown on a long (40 frames) consecutive image sequence. Stable calibration update results without outside intervention demonstrated high performance of the algorithms involved. The results are very important for the accurate local path execution by the rover. Assessments for the necessary computation efforts showed that the existing algorithms can be used for on-board implementation. On-ground software can duplicate on-board calculations at a higher level of accuracy to create a virtual reality environment of the planetary surface. Reflex obstacle detection is the remaining and not integrated aspect in the presented simulation which needs further justi-

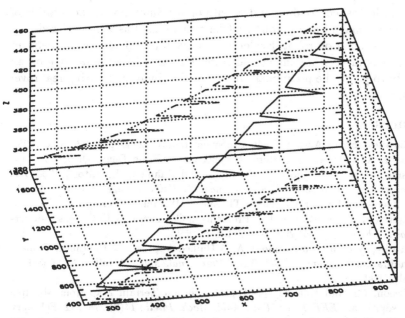

Fig. 5. Camera trajectory (~40 frames) as calculated on the basis of landmarks tracking. The positions of the second stereo camera are included (every fourth frame).

Frame	X	Y	Z	X offset	Y offset	SB
I1	256.165	417.157	331.636			96.996
I2	271.647	449.668	332.519	15.482	31.957	
I3	286.810	482.161	333.775	15.163	32.493	
I4	302.299	514.559	335.101	15.488	32.398	
I5	317.487	547.343	337.435	15.187	32.784	96.998
I6	333.371	579.563	339.031	15.884	32.220	
I7	349.014	612.294	339.976	15.643	32.730	
I8	364.387	644.058	341.940	15.373	31.764	
I9	380.025	676.150	342.851	15.638	32.091	96.996
I10	396.148	708.317	344.676	16.123	32.167	
I11	411.807	740.070	346.625	15.658	31.753	
I12	427.618	772.065	348.582	15.811	31.995	
I13	443.073	803.346	350.698	15.454	31.281	96.989
I14	458.708	834.988	353.727	15.635	31.642	
I15	474.679	866.187	356.595	15.970	31.199	
I16	490.766	897.497	359.578	16.087	31.309	
I17	506.651	928.671	362.590	15.884	31.174	96.976
I18	523.134	959.598	365.552	16.483	30.927	
I19	539.555	990.485	368.521	16.420	30.887	
I20	551.078	1022.09	371.993	11.523	31.608	
I21	567.620	1053.04	374.986	16.542	30.944	96.966
I22	584.219	1084.25	378.625	16.599	31.210	
I23	600.645	1115.20	381.423	16.426	30.948	
I24	616.792	1146.29	384.380	16.146	31.092	
I25	633.345	1177.35	388.005	16.553	31.058	96.951
I26	649.787	1208.06	391.497	16.442	30.710	
I27	666.145	1238.78	395.024	16.357	30.724	
I28	682.960	1269.23	398.441	16.814	30.444	
I29	698.848	1300.17	401.788	15.888	30.939	96.929
I30	716.479	1330.16	404.807	17.631	29.990	
I31	733.025	1361.22	409.718	16.545	31.056	
I32	749.612	1391.04	411.801	16.587	29.816	
I33	766.314	1420.93	414.673	16.701	29.892	96.904
I34	783.056	1451.25	419.980	16.742	30.320	
I35	799.582	1481.52	425.081	16.525	30.267	
I36	816.465	1511.60	429.738	16.882	30.083	
I37	833.774	1541.02	433.723	17.309	29.420	96.8825
I38	850.856	1570.43	438.901	17.082	29.410	
I39	867.135	1589.88	441.959	16.278	19.455	
I40	884.723	1619.29	446.964	17.588	29.406	

Table 1. Calibration update results on the basis of landmarks tracking. SB: stereo basis calculated.

685

fication. Another aspect we did not deal with in our simulation is the rotation of the consecutive image frames around y (the camera axis) as well as closed-loop control of the simulation robot. Evidently, relative rotation between subsequent image frames is always present in real rover motion. The authors hope on an opportunity to prove the preliminary results of the simulation during real rover experiments.

References

[Age95] European Space Agency. Leda assessment report: Leda-rp-95-02, June 1995.

[BP93] Bauer, A. and Paar, G. Stereo Reconstruction From Dense Disparity Maps Using the Locus Method. In Gruen, A. and Kahmen, H., editors, *Proc. 2nd Conference on Optical 3-D Measurement Techniques*, pages 460–466, Zürich, Switzerland, October 4-7 1993. ETH Zürich, Wichmann Verlag.

[Dij59] Dijkstra, E.W. A Note on Two Problems in Connection with Graphs. *Numerical Mathematics*, 1(5):269–271, October 1959.

[GT90] Grosky, W.I. and Tamburino, L.A. A Unified Approach to the Linear Camera Calibration Problem. *IEEE Trans. Patt. Anal. Mach. Intell.*, 12(7):663–671, July 1990.

[KK92] Kweon, I.S. and Kanade, T. High-Resolution Terrain Map from Multiple Sensor Data. *IEEE Trans. Patt. Anal. Mach. Intell.*, 14(2):278–292, February 1992.

[Kol95] Kolesnik, M. Vision and Navigation of Marsokhod Rover. In *Proc. ACCV'95*, pages III–772 – III–777, Dec. 5-8 1995.

[PLSN93] Proy, C., Lamboley, M., Sitenko, I., and Nguen, T.N. Improving Autonomy of Marsokhod 96. In *Proc. 44th Congress of the IAF*, Graz, Austria, Oct 16-22 1993. IAF. IAF-93-U.6.584.

[PP92] Paar,G. and Pölzleitner,W. Robust Disparity Estimation in Terrain Modeling for Spacecraft Navigation. In *Proc. 11th ICPR*. International Association for Pattern Recognition, 1992.

[PSP95] Paar. G., Sidla, O., and Pölzleitner, W. Natural Feature Tracking for Autonomous Navigation. In *Proc. 28th International Dedicated Conference on Robotics, Motion and Machine Vision*, Stuttgart, Germany, October 1995. ISA-TA.

[PU93] Pölzleitner, W. and Ulm, M. Robust dynamic 3d motion estimation using landmarks. In *Optical Tools for Manufacturing and Advanced Automation, Videometrics II*, 1993.

[Ros93] Ross,B. A Practical Stereo Vision System. In IEEE Computer Society, editor, *1993 IEEE Computer Society Conference on Computer Vision and Pattern Recognition*, pages 148–153, New York, June 15-18 1993. IEEE Computer Society Press.

[UP95] Ulm, M. and Paar. G. Relative Camera Calibration from Stereo Disparities. In *Proc. 3rd Conference on Optical 3-D Measurement Techniques*, Vienna, Austria, October 2-4 1995. ISPRS.

A Framework for Feature-Based Motion Recovery in Ground Plane Vehicle Navigation

J.M. Sanchiz[†], F. Pla[*], J.A. Marchant[‡]

[†]*University Jaume I. Department of Computer Science. Castelló. Spain*
[‡]*Silsoe Research Institute. Silsoe. Beds. UK*

Abstract. This paper describes a feature point matching strategy and motion recovery applied to vehicle navigation. A transformation of the image plane is used that keeps the motion of the vehicle parallel to the transformed plane. These allows us to define linear tracking filters to estimate the real-world positions of the features. The correspondences between features are first selected by similarity, taking into account the smoothness of motion and rigidity of the scene. Further processing brings out the rest of correspondences. The method is applied to a real application consisting of an autonomous vehicle navigating in a crop field.

1 Introduction

The general problem of motion analysis from image sequences can be stated as to finding the motion parameters that the camera co-ordinate system has to undergo in two consecutive images in order to match the projection of the scene viewed in both images [1,2,3]. Recovering a general 3D motion is an ill-defined problem and, due to the speed-scale ambiguity, only the direction of the translation can be recovered if no a-priori knowledge is applied. The computation of the motion parameters is based on a former computation of the projected motion, that is, the motion of the projected scene on the image plane, which is approached by either the computation of the optical flow, basically using differences in the intensity of two consecutive images, or by the computation of feature correspondences [1,4]. This latter method seems more reliable for real-time applications since, once the features have been selected, the amount of data to process is significantly reduced.

In real-world applications some constrains are usually applied to the general problem: for example, if the motion is known, then a 3D map of the scene can be recovered; or there exist some landmarks on the scene whose real-world positions are known; or some of the motion parameters are fixed, a rotation angle or a component of the translation vector. The latest situation is usually the case in autonomous vehicle navigation, Fig. 1. The camera height, v, and tilt angle, φ, are fixed. Also the roll angle is fixed and set to zero, since the vehicle is assumed not to roll. It is often assumed that the features lie on the ground plane. This configuration can be found in applications like road-following and indoor or outdoor navigation. Feature tracking is a usual approach to motion estimation although the nature of the extracted features depends on the type of scenes we deal with. Feature point tracking is a common approach, where the points are usually extracted from grey-level images by some corner detector. This is the configuration that will be followed in the present work.

2 Problem Statement

The relation between the camera and the world co-ordinate systems can be seen in Fig. 2. At time instant 0 (first frame) the x_w axis is defined to be aligned with the x_c axis, in further frames this alignment will no

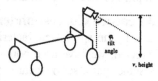

Fig. 1. Camera on a vehicle

longer exist (in general), given that the vehicle rotates and translates over the ground. With this camera configuration the vehicle 2D motion is expressed as a 3D motion in the camera co-ordinate system. To avoid this, and constrain the camera motion to be two-dimensional, we introduce the *virtual image plane*, Fig. 3, parallel to the plane of motion. The co-ordinate system of this virtual camera is defined by applying a rotation of angle φ around the x_c axis. The advantage of using the virtual image plane is that there will be no translation along the z_{vc} direction, this will be convenient when finding the feature correspondences and tracking the features through a sequence of images, since it allows us to define *linear* Kalman filters to estimate the real-world positions of the feature points. Features can be detected on the original image, then they are transferred to the virtual image plane, so one does not have to transfer the whole image, but just the selected features, saving processing time. Once the camera is calibrated [5] the transformation is fixed and it can be performed by a look-up table. The co-ordinates of the two planes are related by:

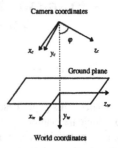

Camera coordinates

Ground plane

World coordinates

Fig. 2. Camera and world co-ordinates

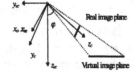

Real image plane

Virtual image plane

Fig. 3. Virtual camera

$$\left(x_{vc} \quad y_{vc} \quad z_{vc}\right)^{\mathrm{T}} = \left(f\frac{x_c}{y_c\sin\varphi + f\cos\varphi} \quad f\frac{y_c\cos\varphi - f\sin\varphi}{y_c\sin\varphi + f\cos\varphi} \quad f\right)^{\mathrm{T}} \tag{1}$$

Let $r_{k,k-1}$ and $t_{k,k-1}$ express the real-world frame-to-frame motion, $r_{k,k-1}$ is a 2D rotation matrix and $t_{k,k-1}$ a 2D translation vector. These motion parameters relate the projections, in two consecutive frames, of a pair of matched features on the virtual image plane:

$$\begin{pmatrix} x_{vc}^{proj} \\ y_{vc}^{proj} \end{pmatrix}_k = r_{k,k-1}\begin{pmatrix} x_{vc}^{proj} \\ y_{vc}^{proj} \end{pmatrix}_{k-1} + \frac{f}{z_{vc}}t_{k,k-1} \tag{2} \quad \begin{pmatrix} x_{vc}^{proj} \\ y_{vc}^{proj} \end{pmatrix}_{k-1} = r_{k,k-1}^{-1}\left[\begin{pmatrix} x_{vc}^{proj} \\ y_{vc}^{proj} \end{pmatrix}_k - \frac{f}{z_{vc}}t_{k,k-1}\right] \tag{3}$$

where the superscript *proj* indicates co-ordinates of the projected features, and z_{vc} is the real depth of the feature, whose value is known from the calibration and equal to the camera height, since the features lie on the ground. The absolute motion of the vehicle can be computed recursively from the frame-to-frame motion. Let R_k (2D rotation matrix) and T_k (2D translation vector) be the absolute motion of the vehicle in world co-ordinates, then:

$$R_k = R_{k-1}r_{k,k-1}^{-1} ; \quad R_0 = I \tag{4}$$
$$T_k = T_{k-1} - R_k t_{k,k-1} ; \quad T_0 = 0$$

The problem consists of finding $r_{k,k-1}$ and $t_{k,k-1}$ through a sequence of images. This implies to select the features and to track them estimating their real-world positions.

3 Feature Correspondence

Due to the special characteristics of the application to which this work is mainly directed, we have used points as features to be tracked. For the rest of the work we assume that a set of feature points is available for every image. The method to find the correspondence and motion recovery is independent of the way the feature points were extracted, the only requirement is that they are stable and that some characteristics of the points can be provided. In our approach we assume a measurement of similarity on pairs of features is available, we combine it with the assumptions of smoothness of motion and rigidity of the scene, and we exploit the special configuration of the camera used in vehicle navigation to define the virtual image plane. The correspondence problem has to be solved as a first step to feature tracking, by which we mean to assign a new feature observation to a tracker (tracking filter) that estimates the real position of the feature, based on previous observations; and to the problem of estimating the camera motion from the set of correspondences (point correspondences in this case). Each feature has a tracking filter (we will use Kalman filters) assigned to it, that estimates its most likely position from all previous observations. The data to be estimated are the projection on the virtual image plane, $(x_{vc}^{proj}, y_{vc}^{proj})$. Features observed in image k-1 already have an associated tracker. Features appearing in image k have not been associated to a tracker yet, but they will be after the correspondence is found, then the new observation will be used to update the corresponding tracker. Features in image k for which no correspondence is found are assumed to be new, and a new tracker is initiated for them. The matching is found from features in frame k to all the estimated positions of the features that have been observed before, we will call these present tracks. The method can cope with poor feature extraction: a feature that is not detected during some frames will be assigned to its corresponding track when detected again. When a feature is not detected a change of co-ordinates is made to its estimated position to update it on the virtual plane. The tracks are filtered before computing the matching to reject those falling outside the present field of view.

The correspondence problem can be stated as follows:

Let n_k be the number of detected features in image k, and let n_{k-1} be the estimated positions from the trackers at time k-1 that survive the filtering (tracks in image k-1). Solving the correspondence consists of finding a backward mapping, $\Psi: i \in [1.. n_k] \rightarrow j \in [0,1.. n_{k-1}]$ and a forward mapping $\Theta: j \in [1.. n_{k-1}] \rightarrow i \in [0,1.. n_k]$ following some criteria, and so that:

$\Psi(i)$ is the corresponding track in image k-1 to feature i in image k. $\Psi(i)=0$ means feature i has no corresponding in image k-1 (new appearing feature). $\Theta(j)$ is the corresponding feature in image k to track j in image k-1. $\Theta(j)=0$ means feature j has no corresponding in image k (it has disappeared from the field of view or has not been detected).

The criteria to find the mapping have to satisfy the following constraints:

3.1 Similarity between features

A general procedure to give a measure of similarity between features consists of computing a vector of characteristics for every feature, $(c_{1i}, c_{2i}, ..., c_{Ni})^T$. The meaning of these characteristics is highly dependent on the method used to detect them (details on the characteristics that have been used in our application will be given in the results section). Then a distance between features can be defined as:

$$d_{ij}^2 = \sum_{l=1}^{N} w_l^2 (c_{li} - c_{lj})^2 \qquad (5)$$

where d_{ij} is the distance between feature i in image k and feature j in image k-1, and w_l is the weight associated to characteristic l (l=1...N).

3.2 Smoothness of motion

The previous estimated frame-to-frame motion, $r_{k-1,k-2}$ and $t_{k-1,k-2}$, is used to search for correspondences. Given a feature i in image k, the search area to find its corresponding in image k-1 is located by back-projecting (3) its co-ordinates on image k-1, but an estimation of the depth, z_{vc}, is needed in equation (3), so we make the assumption that the depth is equal to the camera height, i.e. the features lie on the ground. This is a quite reasonable approach for most autonomous navigation applications where the camera is pointed to the ground, the same idea was used by Liu et al. in [2] to recover motion from line and point correspondences assumed to be on the ground plane. Finally the search area is set as an ellipse of centres a and b, whose main axis is orientated along the motion epipolar line. Co-ordinates of point a are set by back-projecting the co-ordinates of the feature using a decreased value for z_{vc}, and point b is set by back-projecting it with an increased value of z_{vc} (30% of increasing/decreasing has been used in our application).

The distance used can be considered a modified Euclidean distance since points inside a fixed distance threshold do not fall inside a circle, but inside an ellipse orientated in the direction of the epipolar line, direction which has been found using the previous value of the motion parameters. By this approach we favour the searching for correspondences in the direction of the motion, which is mainly forward although can have some rotational or transversal translation component.

3.3 Rigidity of the scene

The rigidity of the scene constrains the correspondences which arise from the same motion of the camera, this means that, ideally, the values of $r_{k,k-1}$ and $t_{k,k-1}$ in (2) and (3) have to be the same for all features. Once some candidate correspondences have been found by selecting similar features in the search areas, a Hough Transform-like technique is applied to further select those features having very close values of $r_{k,k-1}$ and $t_{k,k-1}$.

4 Procedure to find the correspondence

Following the criteria explained above leads to obtaining an initial set of correspondences of present features to existing tracks, and a first guess for the frame-to-frame motion. After selecting the coherent correspondences a better value for the motion parameters can be obtained by minimisation, the rest of correspondences can then be computed by back-projecting the still non-matched features using the recovered motion, and by finding the most similar track in image k-1. The complete method to find all the correspondences can be expressed as follows:

4.1 Compute candidate matches

Build a distance matrix, dm. Each entry, dm[i,j], represents the distance (or dissimilarity) between feature i in image k and track j in image k-1. Then find candidate correspondences as those pairs (i,j) in which position dm[i,j] is at the same time minimum in its row and its in column. This means that track j is the most similar to feature i, and feature i is the most similar to track j.

4.2 Select the coherent matches by a Hough Transform-like technique

For each candidate correspondence, give values to ψ and compute the set of points in the 3D parameter space, (ψ, t_x, t_y), using equation (2). Apply a clustering [6] to the set of points (ψ, t_x, t_y) and find the biggest cluster, $(\psi_0, t_{x,0}, t_{y,0})$, which gives a first guess for the motion parameters, $r_{0;k,k-1}$ and $t_{0;k,k-1}$. Mark the correspondences that originated the points that support the biggest cluster as a coherent match and discard the others. A similar technique was used by Sanchiz et al. in [7] to find correspondences of blobs.

4.3 Compute the motion by minimisation

Find a best value of the frame-to-frame motion by minimisation [3], $r_{1;k,k-1}$ and $t_{1;k,k-1}$. The coherent correspondences are used to prepare two sets of 3D points in world co-ordinates. The projected co-ordinates of all the features are known, and the depths, z_{vc}, are set to the camera height. The real-world positions in virtual camera co-ordinates of a feature are then:

$$(x_{vc}, y_{vc}, z_{vc})^\mathbf{T} = (x_{vc}^{proj}\frac{z_{vc}}{f}, y_{vc}^{proj}\frac{z_{vc}}{f}, z_{vc})^\mathbf{T} \tag{6}$$

4.4 Apply a further filtering to the present matches

The matches that still represent a big variation from the motion parameters that were found by minimisation are rejected. Once the motion parameters are known, the new observed depth of a feature can be found by triangulation. From equation (2), and assuming that the module of the translation due to this correspondence is the same as the module of the translation found by minimisation, we can solve for z_{vc} ($z_{vc} = z_{vc}$ (obs)) and $t_{k,k-1}$ ($t_{k,k-1} = t_{k,k-1}$ (obs)):

$$z_{vc(obs)} = f \frac{|t_{1;k,k-1}|}{\left|\begin{pmatrix} x_{vc}^{proj} \\ y_{vc}^{proj} \end{pmatrix}_k - r_{1;k,k-1}\begin{pmatrix} x_{vc}^{proj} \\ y_{vc}^{proj} \end{pmatrix}_{k-1}\right|} \tag{7}$$

$$t_{k,k-1(obs)} = \frac{z_{vc(obs)}}{f}\left[\begin{pmatrix} x_{vc}^{proj} \\ y_{vc}^{proj} \end{pmatrix}_k - r_{1;k,k-1}\begin{pmatrix} x_{vc}^{proj} \\ y_{vc}^{proj} \end{pmatrix}_{k-1}\right] \tag{8}$$

The filtering rejects the correspondences that produce a big variation in the depth or a big variation in the direction of the translation. In our application we have rejected variations in depth bigger than 30%, and variations of more than 5 degrees in the direction of the translation.

4.5 Find the final matching

The rest of the matches are found by computing $z_{vc\,(obs)}$ and $t_{k,k-1\,(obs)}$ for all possible correspondences of still non-matched features. Computing the distance for those whose $z_{vc\,(obs)}$ and $t_{k,k-1\,(obs)}$ values are inside the limits, and successively picking up the most likely correspondence.

4.6 Find the final value of the motion parameters

A new (and definitive) value of the frame-to-frame motion, $r_{k,k-1}$ and $t_{k,k-1}$, is found by minimisation [3] using all the correspondences.

5 Tracking Features

A tracking filter is initiated for every new feature appearing in the scene. Its function is to estimate the position of a feature from its set of observations, and from the estimated motion parameters. The data to estimate are the co-ordinates of the projection of a feature on the virtual image plane. Fixing the depth, z_{vc}, to the camera height, the real-world position can be computed from equation (2). The Kalman filter [8] is used as a tracker, it estimates the best value, in a least-squares sense, of a state vector from a set of Gaussian noisy measurements in dynamic linear systems. Precisely the frame-to-frame motion can be expressed as a linear system if we use the co-ordinate axes of the virtual vertical camera. The Kalman filter equations are:

System: $\qquad$ $\mathbf{x}_k = \mathbf{\Phi}_{k-1}\,\mathbf{x}_{k-1} + \mathbf{\Gamma}_{k-1}\,\mathbf{u}_{k-1} + \mathbf{v}_k;\ \mathbf{v}_k \in N(0,\mathbf{R}_1)$ $\qquad$ (9)

Measurement: $\qquad$ $\mathbf{y}_k = \mathbf{C}_k\,\mathbf{x}_k + \mathbf{e}_k;\ \mathbf{e}_k \in N(0,\mathbf{R}_2)$ $\qquad$ (10)

Initial state: $\qquad$ $E[\mathbf{x}_0] = \mathbf{x}_{0|0};\ \mathrm{cov}[\mathbf{x}_0] = \mathbf{P}_0;\ E[\mathbf{v}_k\,\mathbf{e}_k^{\mathrm{T}}] = \mathbf{0}$

Prediction at k-1: $\qquad$ $\mathbf{x}_{k|k-1} = \mathbf{\Phi}_{k-1}\,\mathbf{x}_{k-1|k-1} + \mathbf{\Gamma}_k\,\mathbf{u}_k$ $\qquad$ (11)

$\qquad\qquad\qquad\qquad$ $\mathbf{P}_{k|k-1} = \mathbf{\Phi}_{k-1}\,\mathbf{P}_{k-1}\,\mathbf{\Phi}_{k-1}^{\mathrm{T}} + \mathbf{R}_1$ $\qquad$ (12)

Prediction at k: $\qquad$ $\mathbf{K}_k = \mathbf{P}_{k|k-1}\,\mathbf{C}_k^{\mathrm{T}}\,[\mathbf{C}_k\mathbf{P}_{k|k-1}\mathbf{C}_k^{\mathrm{T}} + \mathbf{R}_2]^{-1}$ $\qquad$ (13)

$\qquad\qquad\qquad\qquad$ $\mathbf{x}_{k|k} = \mathbf{x}_{k|k-1} + \mathbf{K}_k\,[\,\mathbf{y}_k - \mathbf{C}_k\,\mathbf{x}_{k|k-1}\,]$ $\qquad$ (14)

$\qquad\qquad\qquad\qquad$ $\mathbf{P}_{k|k} = [\,\mathbf{I} - \mathbf{K}_k\,\mathbf{C}_k\,]\,\mathbf{P}_{k|k-1}$ $\qquad$ (15)

The state vector is defined as: $\qquad$ $x_k = \left(x_{vc}^{proj}\quad y_{vc}^{proj}\right)_k^{\mathrm{T}}$ $\qquad$ (16)

The transition matrix, $\mathbf{\Phi}_k$, is used to express the rotation, and the input part in equation (9) is used to model the translation:

$$\mathbf{\Phi}_k = \begin{pmatrix} \cos\psi & -\sin\psi \\ \sin\psi & \cos\psi \end{pmatrix}_k \ (\text{or } \mathbf{\Phi}_k = \mathbf{r}_{k,k-1}) \quad (17) \qquad \mathbf{\Gamma}_k = \mathbf{I};\ \mathbf{u}_k = \begin{pmatrix} t_x/f \\ t_y/f \end{pmatrix}_k \qquad (18)$$

The transition from state k-1 to state k is:

$$\begin{pmatrix} x_{vc}^{proj} \\ y_{vc}^{proj} \end{pmatrix}_k = \begin{pmatrix} \cos\psi & -\sin\psi \\ \sin\psi & \cos\psi \end{pmatrix}_{k-1} \begin{pmatrix} x_{vc}^{proj} \\ y_{vc}^{proj} \end{pmatrix}_{k-1} + \begin{pmatrix} t_x/f \\ t_x/f \end{pmatrix}_{k-1} \qquad (19)$$

The measurements from the visual information are the co-ordinates of the projected features, $(x_{vc}^{proj}, y_{vc}^{proj})$, so the measurement matrix is the 2x2 identity matrix, $\mathbf{C}_k=\mathbf{I}$. The covariance matrices are initiated as: $\mathbf{P}_0 = \mathbf{R}_1 = \mathbf{R}_2 = \sigma^2\,\mathbf{I}$, where σ is set to a fraction of the field size.

6 Experimental results

In our application the context is an autonomous vehicle that navigates in an outdoor crop field. The scenes we deal with consist of a perspective view of a piece of crop field where only natural objects (plants) appear. The purpose of the application is to spray on the plants or weeds automatically, thus, the vehicle is equipped with a bar of nozzles to perform the spraying. Images are segmented [9] to divide the scene into three classes, regions of class "soil", "plant" and "weed". The same images are used to identify the plants and to compute the motion parameters, which are used to place the images on a map of the field, built up while the vehicle moves [10]. Exploring the map along the nozzle bar allows us to open those nozzles that are over a plant or weed. The motion estimation is intended to be passed to the vehicle control system, thus closing the loop and trying to perform an autonomous row following.

Features are detected as dominant points in the contours of the regions of class "plant". A contour following algorithm was applied to code the boundaries, and the dominant points were found by a neural network-based algorithm for dominant point detection [11]. The tracking method explained in this paper has been tested with several image sequences obtained from a camera mounted on a manually driven vehicle, Fig. 1, undergoing a zigzag motion. The camera height was $v=1200$ mm., the tilt angle was $\varphi=66$ degrees and from a previous calibration [5] of the camera the lens focal length was $f=40$ mm.

Since we use dominant points in contours as features, two characteristics that give satisfactory results for similarity measurements are the convexity and the orientation of the contour in a small area around the point, so we fix $N=2$ in equation (5). Assuming that both characteristics have the same importance we fix $w_1=1$ and $w_2=1$ (5).

Fig. 4. Convexity, α, and orientation, β, at a dominant point.

From a dominant point, $\mathbf{p}$, two points are found at either side of the contour, $\mathbf{p}_a$ and $\mathbf{p}_b$, so that the distance between $\mathbf{p}$ and $\mathbf{p}_a$, and between $\mathbf{p}$ and $\mathbf{p}_b$ is as close as possible to a given value (10% the contour length has been used). The angles of convexity and orientation are computed as shown in Fig. 4.

An example of the matching process can be seen in Fig. 5. The rate of successful correspondences was over 95% through a whole sequence of thirty images, this rate was determined by manually identifying the correct correspondences, the incorrect, and the missed ones, in every image of the sequence. From the frame-to-frame motion, $\mathbf{r}_{k,k-1}$ and $\mathbf{t}_{k,k-1}$, the absolute position and orientation of the vehicle, $\mathbf{R}_k$ and $\mathbf{T}_k$, are found from (4). In order to measure the accuracy of the motion estimation, $\mathbf{R}_k$ and $\mathbf{T}_k$ were used to place every image over the ground plane, thus building a map of the crop at a desired scale. Since the images overlap in a certain amount, a majority voting scheme was used to determine the classification of the pixels, counting the times that a pixel is assigned a certain class. As the plants are aligned in rows, every plant was manually assigned to a certain row, and straight lines were fitted to every row, using the centres of the blobs of class "plant" as the data for the fit. The root-mean-square (r.m.s.) error of the fit, the parallelism and the distance between neighbouring lines (and its comparison with the real-world distance) are measurements that indicate the accuracy of the map, and so of the estimated motion. Fig. 6 shows a map drawn in a 256x512 image at a scale of 15 mm. per pixel, the last field of view and position of the nozzle bar are outlined. The r.m.s. error of the fit was below 30 mm. for the three lines respectively, the angle between neighbouring lines was below 0.5 degrees and the distance (measured at the centre of the map) was 413.2 and 408.3 mm. (400 mm. is the approximate real-world distance between rows in the crop). It has also to be noted that the lines were fitted to the centres of the blobs of class "plant", which are not exactly over the lines that pass through the centre of the crop rows, nevertheless the results are quite satisfactory.

7 Conclusions

A strategy to solve the correspondence problem and the tracking of features has been presented. The method is intended for autonomous navigation applications, where a general constraint is that the motion is undergone on the ground plane. The similarity between features and the smoothness of motion are taken into account to provide an initial matching. The matches that are coherent with the rigidity of the scene are selected by a Hough Transform. The motion is computed by minimisation, and used, together with the similarity between features, to obtain the final correspondence. A Kalman filter is defined for each feature to estimate its real

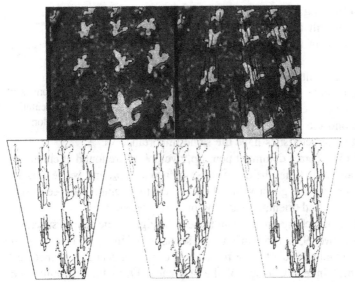
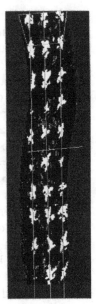

Fig. 5. From left to right and top to bottom: Two consecutive images (contours and dominant points outlined, correspondence superimposed on the second one). Both images transferred and overlapped on the virtual image plane, initial matching. Selected matches after applying the Hough Transform-like technique. Final correspondence.

Fig. 6. Map built from the recovered motion, with lines fitted the rows.

position. The method has been applied to a real-world application.
(Work supported by the Spanish Ministry of Science, CICYT TIC95-0676-C02-01)

References

[1] Y.F. Wang, N. Karandikar, K. Aggarwal, "Analysis of video image sequences using point and line correspondences", *Pattern Recognition.*, vol. 24, pp. 1065-1084. 1993.

[2] Y. Liu, T.S. Huang, O.D. Faugeras, "Determination of camera location from 2D to 3D line and point correspondences". IEEE Transactions on PAMI, vol. 12, no. 1, pp. 28-37. 1990.

[3] S. Umeyama, "Least-squares estimation of transformation parameters between two point patterns", IEEE Transactions on PAMI, vol. 13, pp. 376-380, 1991.

[4] G.L. Scott, H.C. Longuet-Higgins, "An algorithm for associating the features of two images". Proceedings of the Royal Society London, B 224, pp. 21-26. 1991.

[5] R.Y. Tsai, "An efficient and accurate camera calibration technique for 3D machine vision", Proceedings of the IEEE Conference CVPR'86, pp. 364-374. 1986.

[6] P. Trahanias, E. Skordalaskis, "An efficient sequential clustering method", Pattern Recognition, vol. 22, pp. 449-453, 1989.

[7] J.M. Sanchiz, F. Pla, J.A. Marchant, R. Brivot, "Structure from motion techniques applied to crop field mapping", Image & Vision Computing, vol. 14, no. 5, pp. 353-363. 1996.

[8] Y. Bar-Shalom, T.E. Fortmann, "Tracking and data association", edited by W.F. Armes. Academic Press, Inc. Math. & Science in Engineering, vol. 179. ISBN 0-12-079760-7. 1988.

[9] R. Brivot, J.A. Marchant, "Segmentation of plants and weeds using infrared images", IEE Proceedings, Vision, Image and Signal Processing, vol. 143, no. 2, pp. 118-124, 1996.

[10] J.M. Sanchiz, J.A. Marchant, F. Pla, A. Hague, "Real-Time visual sensing for task planning in a field navigation vehicle", Real-Time Imaging (in press). 1996.

[11] J.M. Sanchiz, J.M. Iñesta, F. Pla, "A neural network-based algorithm to detect dominant points from the chain-code of a contour", Proc. of the 13th ICPR, vol. IV, pp. 330-334. 1996.

Terrain Reconstruction from Multiple Views

Georgy L. Gimel'farb[1] and Robert M. Haralick[2]

[1] International Research and Training Center for Information
Technologies and Systems (NAS and ME of Ukraine)
Kiev-22, 252022 Ukraine
[2] Intelligent Systems Laboratory, University of Washington
Seattle, WA 98195, U.S.A

Abstract. A two-stage approach is discussed for reconstructing a dense
digital elevation model (DEM) of the terrain from multiple pre-calibrated
images taken by distinct cameras at different time under various illumination. First, the terrain DEM and orthoimage are obtained by independent voxel-based reconstruction of the terrain points using simple
relations between the corresponding image gray values. As distinct from
other approaches, possible occlusions and changing shadow layouts are
taken into account implicitly by evaluating a confidence of every reconstructed terrain point. Then, the reconstructed DEM is refined by
excluding occlusions of more confident points by less confident ones and
smoothed with due account of the confidence values. Experiments with
RADIUS model-board images show that the final refined and smoothed
DEM gives a feasible approximation to the desired terrain.

1 Introduction

Photogrammetric image processing has a significant place in today's robotics,
cartography, and remote sensing [2,4]. It includes, in particular, the calibration
of imaging cameras and the DEM reconstruction from the calibrated images.
The calibration estimates, by using visually or automatically detected ground
control points (GCP) with known world 3D coordinates, cameras model parameters that relate to where any 3D point will project on each imaging plane
(see, for instance, [4,5,11]). Here, we address the problem of multi-view DEM
reconstruction using a set of the pre-calibrated images.

The DEM reconstruction is most extensively studied in binocular stereo. As
does the majority of other inverse photometric problems, stereo belongs to the
class of ill-posed mathematical problems [7] because, even without a noise, there
always exist several 3D surfaces that produce the same stereo pair. Adequate
regularizing heuristics sometimes permit making the DEMs reconstructed by
stereo close enough to the desired surface [2,3,6]. One way to help decrease the
ill-conditionedness is to use multiple views [1].

In a few known works on the multiple-view reconstruction of dense terrain DEMs a prior knowledge about or restrictions on illumination, reflectance
(albedo), and smoothness of the surface are involved to simplify the problem

[9,10]. But, generally, terrains have arbitrary shapes with discontinuities and varying albedo. The images are sensed by several cameras with various resolutions, positions, and orientations, at different times when positions of some mobile objects may change, and under distinct illuminations giving changing shadow layouts. This results in a wide scatter of gray values representing the same surface point in the images. Our goal is to judge how to compute, under these conditions, a rough but plausible approximation to the dense DEM of arbitrary terrain if we presume no prior knowledge about the terrain features but can use simultaneously all the sensed image signals.

2 Methodology

We exploit a voxel representation of a 3D surface $\mathbf{Z} = \{Z(X,Y) : (X,Y) \in \mathbf{Q}\}$ over a supporting domain $\mathbf{Q}$ in the plane OXY in the world coordinate system $OXYZ$. Let the voxels $\langle (X_i, Y_j, Z_{ij}) : (X_i, Y_j) \in \mathbf{Q}_{IJ}, Z_{ij} \in \mathbf{H} \rangle$ represent the digital surface $\mathbf{Z}$ over an equi-spaced lattice $\mathbf{Q}_{IJ} = \{(X_i, Y_j) : i = 0, \ldots, I-1; j = 0, \ldots, J-1\}$. The set $\mathbf{H}$ of heights is a set of K equi-spaced values, $\mathbf{H} = \{Z_k : k = 0, \ldots, K-1\}$. For simplicity, we restrict the consideration to cubic voxels whose faces are aligned normal to the axes of the world coordinate system. Figure 1 shows a X- or Y-section of the 3D space where each voxel is represented by three sides of a square depicted by boldface lines with "bullet" ends. Either side is the cut of the voxel face which can form part of a visible surface. Black arrows show viewing directions, and "H", "VR", and "VL" denote, respectively, the horizontal upper face, visible to cameras with higher Z-positions, and vertical faces, visible to cameras with greater or smaller X- or Y-positions (that is, placed to the right or to the left of the voxel). Generally, the actual visibilty of these faces as well as admissible X- or Y-transitions, depicted by thin lines in Figure 1, between the visible neighboring faces have to be taken into account.

The calibration yields a projective correspondence between the 3D point coordinates (X, Y, Z) and the 2D image point coordinates $(x_{[t]}, y_{[t]})$ for every camera $t \in \mathbf{T} = \{1, \ldots, T\}$. If $G_{ij} \equiv G(X_i, Y_j, Z_{ij})$ and $g_{[t]} \equiv g_{[t]}(x_{[t]}, y_{[t]})$ are, respectively, the gray values in the 3D terrain point (X_i, Y_j, Z_{ij}) and in the corresponding 2D point $(x_{[t],ij}, y_{[t],ij})$ of the image $\mathbf{g}_{[t]}$ received by the camera t then $\mathbf{G}_{IJ} = \{G_{ij} : i = 0, \ldots, I-1; j = 0, \ldots, J-1\}$ is a terrain orthoimage.

Our methodology produces a simple two-stage DEM reconstruction. In the first stage, every position $(X, Y) \in \mathbf{Q}_{IJ}$ is examined. For each height $Z \in \mathbf{H}$, there is a corresponding 2D perspective projection of the 3D point (X, Y, Z) on each of the T images. For each image for which the 2D perspective projection of (X, Y, Z) lies on the image, there is an observed gray value. This produces the gray values $g_1, \ldots, g_S$. Let g_{min}, g_{max}, and g_{med} be, respectively, the minimum, the maximum, and the median of these gray values. We define the dissimilarity of the S gray values by $\hat{d} = (\max\{0, \varepsilon_{min} \cdot g_{max} - \varepsilon_{max} \cdot g_{min}\})^2$, where ε_{min} and ε_{max} are given numbers which bound the admissible variations in the surface albedo and transfer factors for the cameras. Some other tested measures, say, $\hat{d} = \max\{0, \varepsilon_{min} \cdot g_{max} - g_{med}, g_{med} - \varepsilon_{max} \cdot g_{min}\}$, gave worse results in our

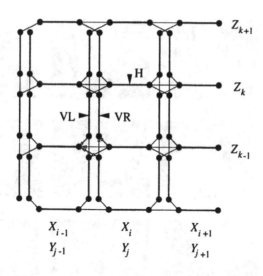

Fig. 1. Transitions between the voxel faces

experiments. We choose the height Z giving the smallest value of dissimilarity. And we assign the gray value g_{med} as the gray value at position (X, Y) of the ortho-image. For a confidence measure we use the range $\hat{R} = g_{max} - g_{min}$.

At the second stage, the reconstructed DEM is refined by checking possible occlusions of its voxels. If any less confident voxel occludes the more confident one from the viewing camera then the height of the occluding voxel is cut so as to exclude the occlusion. The confidence values are used, also, for the adaptive moving-window median smoothing of the refined DEM. The window contains only the points that have the same or higher confidence as the central window point and form a continuous region around it. In spite of simplicity, the proposed approach gives promising results for real pre-calibrated images.

3 Basic Features of Multiple Terrain Views

These features are evident from the RADIUS model-board image sets [8]. Figure 2 shows reduced examples from the set "M" containing 40 digital images, each of size 1350 columns × 1035 rows. These images were taken with different resolution (compare, for example, M16 and M20 or M35 and M36), at different times, and under the distinct illuminations.

The terrain smoothness varies arbitrarily and there are notable surface discontinuities, say, for the platform in the stadium or for the buildings. Only a central part of the model board is covered by all the views. Other parts are viewed only by different subsets of the cameras, down to two cameras per point. Due to occlusions, the image gray values collected for a 3D point which could be visible to several cameras, may correspond to different surface points. There are differently placed mobile objects such as cars in different parts of the images.

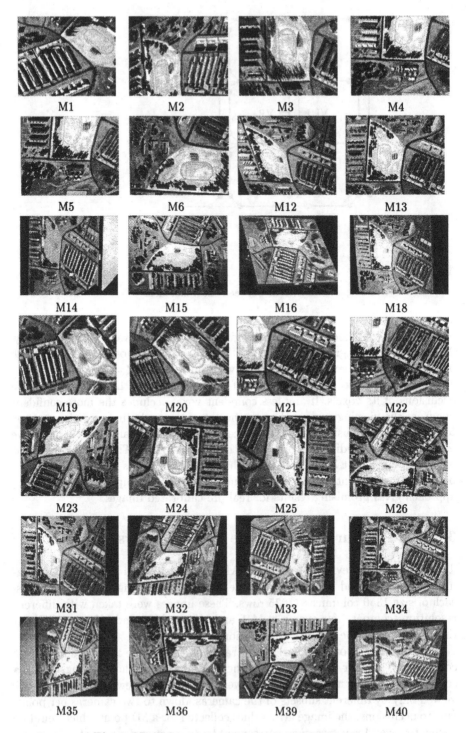

Fig. 2. Model board images from the RADIUS "M" set

Along with changes of the albedo of the surface points, the overall illumination itself varies from one to another subset of the images so that these subsets have different layouts of shadows and distinct contrasts for the same objects (say, for the walls of the buildings or the stadium's platform). Also, the calibration errors result in matching neighboring but different surface points.

If the point is not occluded and sensed under the same illumination, the signals form, mostly, a cluster which depends only on variations in the surface albedo and cameras transfer factors. There can be several such clusters that correspond to different illuminations and changes of the shadow layouts. At the same time, the signals for the points occluding the current one from some cameras are more or less uniformly distributed over the gray range.

It is obvious that the smaller the signal range, the more plausible that there are no outliers, namely, signals for the occluded or shadow points. Thus, the signal range evaluates the confidence of the heights $\hat{Z}$ found by minimizing the dissimilarities $\hat{d}$ for every model voxel over the supporting domain $\mathbf{Q}$.

4 Experimental Results and Conclusions

The experiments were carried out with the above-mentioned set "M" of the RADIUS images. The voxel lattice has the size $580(I) \times 580(J) \times 60(K)$ with the coordinate ranges $X_0 = -5$, $X_{I-1} = 53$, $Y_0 = -13$, $Y_{J-1} = 45$, $Z_0 = -1.5$, and $Z_{K-1} = 4.5$ units. Figure 3 shows the range image and the orthoimage of the reconstructed DEM. By comparing the orthoimage with Figure 2 one can

Fig. 3. Reconstructed DEM (a) and its orthoimage (b)

conclude that main features of this model-board scene are represented in the reconstructed DEM and its orthoimage. But, there are notable errors, mostly, in the less confident areas (most them have large Z-values being white in the DEM range image).

Figure 4,a displays the image of the confidence values: the darker the point, the higher the confidence, that is, the more narrow the signal scatter. As one might expect, less confident voxels are concentrated around buildings and vegetation, that is, in most occluded areas and areas where the shadow layouts are changing under different illumination. These errors are excluded by a subsequent

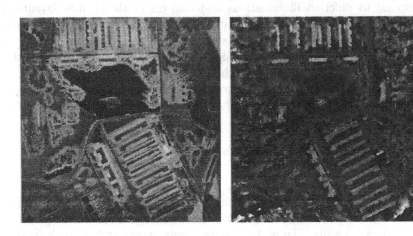

Fig. 4. Confidences for the reconstructed DEM (a) to get the refined DEM (b)

refinement and smoothing with the moving window 9×9, as shown in Figures 4,b and 5,a. Figure 5,b presents the smoothed refined DEM with overlaid outlines of the real roofs of the buildings. It is easily seen that the resulting DEM has good correspondence to the ground truth.

Reconstruction errors are estimated by comparing the DEM with the known 138 ground control points and 497 auxiliary passpoints used for the cameras calibration [11]. Figure 6,a gives positions of them in the reconstructed DEM. Here, cross sizes indicate relative error values. It should be noted that most ground control points are placed at the corners of the roofs and of the foundations of the buildings. These places are most difficult for our simplified reconstruction which searches for a single voxel per a planar position (X_i, Y_j) minimizing the dissimilarity between the corresponding signals so that chooses only one arbitrary voxel between the roof and the foundation along a visible wall of the building. The terrain discontinuities where the voxels to be found have the same planar position need some other processing techniques taking into account all the visible voxel faces and admissible transitions between them (see Figure 1).

Bounds $[\varepsilon_{min}, \varepsilon_{max}]$ in the range $[1.0, 1.0] \ldots [0.7, 1.3]$ change the final error rate within 10-15% para to the best results obtained with the bounds $[0.9, 1.1]$.

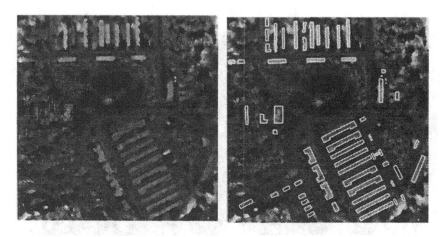

Fig. 5. Smoothed refined DEM (a) with overlaid roof outlines (b)

These latter results are summarized in Table 1 giving mean values ("mae") and standard deviations ("std") of the absolute DEM height errors relative to the control points and their cumulative histograms. In total, 69.6% of the GCPs and

Table 1. Precision of the DEM reconstruction

DEM	138 GCPs						497 passpoints					
	mae	std	≤ 0.1	≤ 0.2	≤ 0.3	≤ 0.6	mae	std	≤ 0.1	≤ 0.2	≤ 0.3	≤ 0.6
reconstructed	0.49	0.64	47	73	82	96	0.60	0.85	200	255	283	329
refined	0.39	0.42	50	65	78	108	0.40	0.45	176	245	292	371
smoothed	0.28	0.34	67	81	96	118	0.31	0.37	211	293	330	398

66.4% of the passpoints have the absolute error less than 0.3, that is, less than 5% of the height range in our model. Thus, the proposed approach, in spite of its simplicity, yields rather good close approximation to the desired dense DEM. The overall quality of the resulting DEM can be checked qualitatively also by estimating the visibility of the terrain points in terms of numbers of the cameras that view every point. Such a "visibility" pattern of the final DEM is shown in Figure 6, b. Here, the signals are proportional to the numbers of the viewing cameras: the more black the point, the less the number in the range $2 \ldots 40$. It is apparent that the reconstructed DEM, in spite of some local errors, reflects most characteristic features of the observed scene. In total, this visibility pattern is consistent with the one expected by visual perception of the initial images.

Our experiments show that a feasible approximation to the dense DEM of the terrain viewed by a set of the calibrated cameras can be obtained by independent reconstruction of each terrain point. The confidence values for the chosen voxels

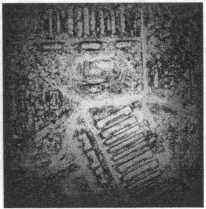

Fig. 6. Control points (*a*) and visibility pattern (*b*)

are crucial in excluding most part of the errors from the reconstructed DEM. Of course, the obtained rough representation of the viewed terrain needs to be further refined by more elaborate techniques. But, it possesses basic features of the observed terrain and therefore can be useful in practice.

References

1. Agouris, P., Schenk, T.: Automated aerotriangulation using multiple image multi-point matching. Photogramm. Eng. Remote Sens. 62:6 (1996) 703-710
2. Baker, H. H.: Surfaces from mono and stereo images. Photogrammetria 39:4-6 (1984) 217-237
3. Gimel'farb, G. L.: Symmetric bi- and trinocular stereo: tradeoffs between theoretical foundations and heuristics. Computing Suppl. 11 (1995) 1-19
4. Haralick, R. M., Shapiro, L. G.: Computer and Robot Vision. Addison-Wesley Publ. (1993) Vol. 2
5. Haralick, R. M., Thornton, K. B.: On robust exterior orientation. Robust Computer Vision: Quality of Vision Algorithms (W. Förstner, S. Ruwiedel, Eds.). Herbert Wichmann Verlag: Karlsruhe (1992) 41-49
6. Jenkin, M. R. M., Jepson, A. D., Tsotsos, J. K.: Techniques for disparity measurement. CVGIP: Image Understanding 53:1 (1991) 14-30
7. Kireytov, V. R.: Inverse Problems of Photometry. Computing Center, Acad. Sci. USSR, Siberian Branch: Novosibirsk (1983) [*In Russian*].
8. RADIUS Model Board Imagery and Groundtruth. CD-ROM, Vol. 1 and 2. ISL, Univ. of Washington: Seattle, USA (1996)
9. Schultz, H.: Shape reconstruction from multiple images of the ocean surface. Photogramm. Eng. Remote Sens. 62:1 (1996) 93-99
10. Shekarforoush, H., Berthod, M., Zerubia, J., Werman, M.: Sub-pixel Bayesian estimation of albedo and height. Int. J. Computer Vision 19:3 (1996) 289-300
11. Thornton, K. B.: Accurate Image-Based 3D Object Registration and Reconstruction. Dissertation (Ph.D.), Univ. of Washington (1996)

Detecting Motion Independent of the Camera Movement through a Log-Polar Differential Approach *

Jose A. Boluda[1], Juan Domingo[2], Fernando Pardo[1], and Joan Pelechano[2]

[1] Departament d'Informàtica i Electrònica, Universitat de València
[2] Institut de Robòtica, Universitat de València
C/ Doctor Moliner, 50. 46100 Burjassot. Spain.

Abstract. This paper is concerned with a differential motion detection technique in log-polar coordinates which allows object motion tracking independently of the camera ego-motion when camera focus is along the movement direction. The method does not use any explicit estimation of the motion field, which can be calculated afterwards at the moving points. The method, previously formulated in Cartesian coordinates, uses the log-polar coordinates, which allows the isolation of the object movement from the image displacement due to certain camera motions. Experimental results on a sequence of real images are included, in which a moving object is detected and optical flow is calculated in log-polar coordinates only for the points of the object.

1 Introduction

The problem of real-time motion detection and tracking from a moving camera is an important issue in artificial vision. Typical applications are obstacle avoidance and time-to-impact computation. The constrains for real-time implementation of these are the large amount of data to be processed and the high computational cost of the algorithms employed.

Motion estimation techniques fit into three categories: feature-based, optical flow and differential techniques [8]. The feature-based techniques are based on extraction and matching of interesting points [6]. A typical problem is instability of the extracted features due to noise or occlusions.

The optical flow computation is a powerful approach but suffers from a very high computational cost, since regressions on the neighbors of a pixel must be applied to solve the optical flow equation. The accuracy of these techniques increases as their computational cost does [2].

The differential techniques typically compute spatially and/or temporal derivatives for all the image. They suffer also from a very high computational cost but due to its parallel nature are very suitable for a parallel implementation [8].

* This work has been supported in part by the Spanish government (CICYT project TAP95-1086-C02-02) and has been partially developed at the Machine Vision Lab,Dept. of E.E., University of Virginia (USA).

On the other hand, the use of the Cartesian coordinates to describe images, due to the usual camera layout, seems not to be the natural way for the optical flow computation, especially when the prevalent movement is along the optical axes. In fact it has been shown that in this case the use of the log-polar coordinates simplifies the computation of the optical flow [10,5].

The differential approach used in this paper was previously formulated in Cartesian coordinates [4]. Our translation of the restrictions applied to the log-polar mapping will show the utility of this representation, which automatically discards the image displacement due to the camera ego-motion. Moreover, the information representation reduces the amount of data to be processed, thus making the algorithm suitable for real-time implementations.

2 The Log-Polar Representation

In humans, the retina exhibits a non-uniform photo-receptor distribution: more resolution at the center of the image and less at the periphery, which allows a selective reduction of the information. The advantages of the log-polar representation for this kind of active vision systems have been widely studied [9,10].

Special hardware that includes the log-polar transformation has been developed: a CMOS sensor [7] and a camera using this sensor [3] have been utilized for the experimental testing of the algorithm proposed. The sensor used has two different areas: the retina, which follows the log-polar law and the fovea which follows a linear-polar scale, in order to avoid the singularity at the origin and some scaling problems [7].

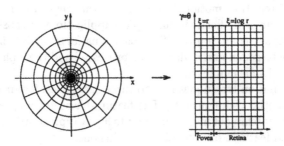

Fig. 1. The log-polar transformation

$$\begin{cases} \xi = \begin{cases} r & r \le r_0 \text{ (fovea)} \\ \log r & r > r_0 \text{ (retina)} \end{cases} \quad where \begin{cases} r = \sqrt{x^2 + y^2} \\ \theta = \arctan\left(\frac{y}{x}\right) \end{cases} \\ \theta = \theta \end{cases} \quad (1)$$

being r_0 is the radius of the fovea circle.

Figure 1 shows graphically the log-polar representation. This representation is a compromise between resolution and amplitude of the field of view, or in other words, between computational cost and width of view field.

3 The Original Algorithm and its Adaptation to Log-polar Mapping

The method for boundary motion detection employed here was developed in Cartesian coordinates by Chen and Nandhakumar [4]. Their formulation in the Cartesian plane (x, y) and our transformation to the log-polar plane (ξ, θ) is as follows:

Let $E(x, y, t)$ $(E(\xi, \theta, t))$ be the time-varying image sequence, $T(\Sigma)$ the projection of a surface Σ in the projection plane Π and $\partial T(\Sigma)$ the border points of Σ. The assumptions for the method are:

- E is piecewise linear with respect to x and y (to ξ and θ) on any point belonging to the projection of a surface in the image plane, which means:

$$\frac{\partial^2 E}{\partial x^2} = \frac{\partial^2 E}{\partial y^2} = 0 \qquad\qquad \frac{\partial^2 E}{\partial \xi^2} = \frac{\partial^2 E}{\partial \theta^2} = 0 \qquad (2)$$

almost everywhere on $T(\Sigma) - \partial T(\Sigma)$
- The motion of the scene is smooth with respect to time, which means $\forall x, y (\forall (\xi, \theta)) \in T(\Sigma)$

$$\frac{\partial^2 x}{\partial t^2} = \frac{\partial^2 y}{\partial t^2} = 0 \qquad\qquad \frac{\partial^2 \xi}{\partial t^2} = \frac{\partial^2 \theta}{\partial t^2} = 0 \qquad (3)$$

Now, the optical flow equation has to be used. It can be written as:

$$v_x \frac{\partial E}{\partial x} + v_y \frac{\partial E}{\partial y} = \frac{\partial E}{\partial t} \qquad\qquad v_\xi \frac{\partial E}{\partial \xi} + v_\theta \frac{\partial E}{\partial \theta} = \frac{\partial E}{\partial t} \qquad (4)$$

It is easy to prove that for the log-polar scaling (retina) and also for the linear-polar (fovea) the optical flow equation has exactly the same functional form, being the only difference the name of the used variables, as it has been written in (4). [1]

With these conditions, it can be proved [4] that $\forall x, y \notin \partial T(\Sigma)$

$$\frac{\partial^2 E}{\partial t^2} = 0 \qquad (5)$$

Thus, it is possible to detect motion boundaries through the computation of the second order temporal derivative of the image. When the absolute value of

---

[1] It remains to be proved that for any conformal mapping the optical flow equation would have also the same form.

this magnitude at a given pixel exceeds a threshold value (ideally 0) this point should be marked. The threshold depends on the smoothness quality and on the camera movement. The method includes the ego-motion of the camera in the x and y coordinates (under the log-polar mapping, in the ξ and θ coordinates). The only restriction for this ego-motion is condition (3).

The first assumption, optimal smoothing, can be achieved through a direct linearization. This method, though effective, requires solving many linear equations and it is computationally expensive, so this condition is approximately achieved through a smoothing in the cortical plane that has been made by using a convolution mask. This approximation could be compensated by increasing the threshold value to retain points whose second spatial derivatives are both about a threshold.

The second assumption, motion of the scene is smooth with respect to time, is a strong constrain, and it can be accepted only if a sufficiently high image rate is used, so in a sequence of 3 very close images, any movement can be approximated to a linear one.

Let us detail the meaning of 3 for the case of log-polar coordinates. For the angular coordinate θ the condition is clearly a rotation with a constant angular velocity. This assumption implies that the camera can have that class of rotational movement around its optical axis. Apart from that, the velocity condition for the radial coordinate ξ expressed in (3) will have to be translated to Cartesian coordinates to understand which will be the discarded movement. Using (1) and (3), the differential equations in the Cartesian plane for the fovea and retina are found to be:

$$\begin{cases} (x\dot{x} + y\dot{y})^2 = (x^2 + y^2)(\dot{x}^2 + \dot{y}^2 + x\ddot{x} + y\ddot{y}) & \text{for the fovea} \\ (x\dot{x} + y\dot{y})(2x\dot{x} + 2y\dot{y}) = (x^2 + y^2)(\dot{x}^2 + \dot{y}^2 + x\ddot{x} + y\ddot{y}) & \text{for the retina} \\ (x\dot{y} - y\dot{x})(2x\dot{x} + 2y\dot{y}) = (x^2 + y^2)(x\ddot{y} - y\ddot{x}) & \text{for both} \end{cases} \quad (6)$$

It can be checked that the general solution of (6) for retina and fovea are respectively:

$$\begin{cases} x = Ae^{b+vt}\cos(\theta_0 + \omega t) \\ y = Ae^{c+vt}\sin(\theta_0 + \omega t) \end{cases} \qquad \begin{cases} x = (r_0 + v_0 t)\cos(\theta_0 + \omega t) \\ y = (r_0 + v_0 t)\sin(\theta_0 + \omega t) \end{cases} \quad (7)$$

These equations represent exponential (linear) spirals where A, b, c, v, θ_0, ω, r_0 and v_0 are constants. ω is interpreted as an angular velocity, that should be constant, as it has been pointed out before. The other velocity, v, is interpreted as the translational velocity of the camera; r_0 and v_0 are constants related with the initial object position and translational velocity; θ_0 is a constant related to the initial angular position.

In the case of the angular movement it is clear that an ego-rotation of the camera around the optical axes with a constant angular velocity is included as an automatically not detectable image displacement, and now it is to be stated which kind of translational camera movements will not be detected automatically with the algorithm. Let us see how the size of an approaching object grows in the image plane (Fig. 2), assuming a pinhole model for the camera, $d' = f\frac{d}{z}$ where d is the object size, f is the equivalent focal of the camera, z is the distance from the object to the focal point and d' is the image in the focal plane.

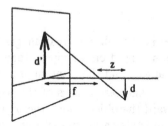

Fig. 2. Size of an approaching object

Assuming a lineal translation of the camera in the direction of the optical axis, the z parameter will decrease linearly as $z = z_0 - v_t t$. From the pinhole model is obtained:

$$d' = f\frac{d}{z_0 - v_t t} \qquad (8)$$

where v_t is the translational velocity along the optical axis direction. The radial dependence is exponential for the retina solution and linear for the fovea (7). Comparing this results with the linear approaching of an object (8) it is possible to see that the growing size in the image plane is very similar. In fact, normalizing and developing in Taylor series around the origin the solution for the retina, the polynomial expressions are very similar.

$$\frac{1}{1-x} = \sum_{n=0}^{\infty} n! x^n = 1 + x + 2x^2 + \ldots$$
$$e^x = \sum_{n=0}^{\infty} x^n = 1 + x + x^2 + \ldots \qquad (9)$$

For the fovea this approximation holds since the expression has the form $1+x$. The difference between them starts in the third order term. The assumption made by our method consists on considering that the object growing in the retinal plane (or image plane) corresponds to the solution of (6) for the retina and fovea. This approach will be shown to be accurate enough, as experimental results will prove.

According with the considerations stated up to now, the overall algorithm is as follows:

1. Acquisition of the images $E(\xi, \theta, t)$.
2. Low-pass filtering for image smoothing.
3. Second order temporal differentiation and thresholding.

With this simple algorithm it is possible to track motion boundaries with a log-polar camera with translation along the optical axis and rotation around it. The method does not use any previous motion field estimation. Image variation due to the camera ego-motion is taken away, as long as it accomplishes the former constraints.

4 Experimental Results

The algorithm was tested in the lab with real images taken with the log-polar camera. The camera was mounted on a support that allows the translation in the direction of the optical axis and the rotation around it.

Figure 3 shows the first and last images of the sequence. In this sequence, the camera advances toward the object along the optical axis, rotating around it at the same time. There is also a moving object (truck toy) at the center of the image, which is moving independently from left to right.

Fig. 3. First and last images of the sequence, and the detected motion

The rest of the objects are also approaching due to camera ego-motion, but their apparent motion is not detected in the final processed image (see Fig. 3). This is the expected result, since it has been justified that this motion in the image plane is similar enough to (7), which were the solutions to the smooth motion condition in log-polar coordinates for the fovea and retina.

After the acquisition, a low pass filter is applied in order to smooth the image. The mask applied has been a 5x5 pixel with unity value. Once smoothing has been performed, the second order temporal differentiation is made and a threshold is applied in order to binarise the image. The threshold is experiment dependent and it should be adjusted taking into account the strength of the edges of the image. Result for the last frame is shown at Fig. 3, right.

Most of the moving object has been tracked, and the motion due to camera rotation and translation has been completely discarded. All the experiments performed have shown that taking images very quickly the approximation of smooth motion is valid.

5 Optical Flow Computation

Even though optical flow determination was not the primary objective of our algorithm, an important reduction in the time needed to calculate it can be achieved because of two different reasons:

- The number of pixels in the log-polar representation is significantly smaller than in a Cartesian image with similar information content.
- It is only in those pixels which have been identified as belonging to the object where the calculation of the optical flow is interesting

This approach simplify the optical flow computation since most of the background shows either a radial or angular flow due to camera movement, which might confuse other algorithms for object motion determination, or no flow at all, due to the absence of borders, which generates null spatial derivatives in the Horn equation [1].

Equation (4) is a single equation with two unknowns for each point. The method we have used to solve it consists on applying linear regression using neighbor pixels, which are supposed to be points of the object. So, it is reasonable to suppose that all of them have approximately the same velocity vector, and it is meaningful to apply regression on them. This increments the computational cost, but the improvement in robustness is worth in this case. The computational cost is $\mathcal{O}(6pn^2)$ where p is the number of pixels and n the window size used for regression.

Fig. 4. Left: Optical flow in the cortical plane. Right: Same image with magnified center

Results in Fig. 4 show the optical flow field calculated for the last frame of the sequence where 456 pixels belonging to the object were detected. The two figures show the optical flow in the Cartesian plane (center magnified). From this figure it is appreciated that the car is moving from left to right but it is also affected by the camera ego-rotation. The optical axis crosses the image plane at a point slightly at the right of the center of the car.

709

6 Conclusions

A differential method for motion tracking has been applied in the log-polar plane. The method is an application of the algorithm developed in Cartesian coordinates by Chen and Nandhakumar [4]. The combination of the log-polar mapping with this method has proved to be useful in the case of a camera with translation in the optical axis and/or rotation around it.

The first assumption, image smoothness, is approximately achieved by using a low pass filter. The second assumption, (motion of the scene should be smooth) has been achieved through the use of the described camera ([3]) that delivers frames at a high rate (one hundred images per second, in log-polar native format). The log-polar camera will be placed at the top of a moving head of an autonomous navigation system, oriented in the direction of the translational movement.

References

1. Barron J. L., Fleet D. J. , Beauchemin S. S., and Burkitt T. A.: Performance of optical flow techniques. Proc. of the 1992 IEEE Conf. on Computer Vision and Pattern Recognition. Champaigne, Illinois, USA, (1992) 237–242
2. Beauchemin S.S. and Barron J.L.: The computation of the optical flow. ACM Computing Surveys, Vol. 27, N° 3. September (1995)
3. Boluda J.A., Pardo F., Kayser T., Pérez J.J. and Pelechano J.: A new foveated space-variant camera for robotic applications. 3rd. IEEE International Conference on Electronics, Circuits and Systems. Rhodes, Greece, October (1996) 680–683
4. Chen W. and Nandhakumar N.: A simple scheme for motion boundary detection. Proc. IEEE Int. Conf. on Systems, Man and Cybernetics. San Antonio, Texas. USA (1994)
5. Daniilidis K.: Computation of 3D-motion parameters using the log-polar transform. Proc. Inter. Conf. on Computer Analysis of Images and Patterns, Prague, Czech Republic, September (1995) 82–89
6. Domingo J., Ayala G. and Diaz M.E.: A method for multiple rigid-object motion segmentation based on detection and consistent matching of relevant points in image sequences. Proc. of the Int. Conf. on Acoustic, Speech & Signal Processing, ICASSP97 Munich, Germany. Vol IV, (1997) 3021–3024
7. Pardo F., Boluda J.A., Pérez J.J, Felici S., Dierickx B., and Scheffer D.: Response properties of a foveated space-variant CMOS image sensor. International IEEE Symposium on Circuits and Systems. Atlanta, USA; (1996) 373–376
8. Pitas I.: Parallel Algorithms for Digital Image Processing, Computer Vision and Neural Networks. Edited by I. Pitas. Printed by John Wiley & Sons Ltd. (1993)
9. Tistarelli M. and Sandini G.: Dynamics aspects in active vision. CVGIP: Image Understanding, 56 N° 1 (1992) 108-129
10. Tistarelli M. and Sandini G.: On the advantages of polar and log-polar mapping for direct estimation of time-to-impact from optical flow. IEEE Trans. Pattern Analysis and Machine Intelligence, 14, (1993) 401–410

Coordinate-Free Camera Calibration

Jens Arnspang*   Knud Henriksen   Niels Olsen

Department of Computer Science, University of Copenhagen
Universitetsparken 1, DK-2100 Copenhagen Ø, Denmark.

Abstract. A method for calibration of the optic center and the focal length of a pin-hole camera is suggested, where the camera parameters are constructed geometrically in the image plane. Box shaped objects of unknown dimensions, like furniture, buildings or a textured ground plane with trees, lamp poles or people are examples of sufficient calibration objects. Experiments are reported with relative errors in the range of one percent for suitable calibration views.

1 Introduction

Within computational vision, existing paradigms for determining entities in a spatial scene often require knowledge of camera parameters. A dominant camera model throughout the literature is the *pin-hole* model, which is defined by three parameters: the position of the optic center $o = (x_0, y_0)^T$ in the image plane and the focal length f of the camera. More elaborate camera models may be found in the literature. Methods for calibrating stereo cameras may be found in (Henriksen 1987), and models for certain photographic wide angle lenses, deliberately engineered to deviate from true perspective image formation may be found in (Fleck 1994). However, the mainstream research in computer and robot vision, as surveyed in (Ballard 1982), (Horn 1986), (Arnspang 1988), (Henriksen 1990), (Arnspang 1991), (Haralick 1992), and (Faugeras 1993), use the simpler pin hole camera model, to which the discussion in this paper is restricted.

A few properties of spatial lines $\mathcal{L}$ and planes Π and their projection ℓ and π onto the image plane, and among these the concept of horizons $\hbar$ and poles p, will be revisited in section 2. In section 3 it is shown, that in any view of a plane Π, the optic center o is restricted to lie on a line which is perpendicular to the horizon $\hbar$ of the plane and which passes through the pole p of the plane. Two such views of planes, where their horizons are not parallel are therefore sufficient for determining the optic center of the camera. In section 4 it is shown how the focal length of the camera may be constructed from the horizon, the pole and the optic center by means of simple geometric construction. In section 5 it is shown how single views of boxes, e.g. buildings, furniture or classic calibration objects may be used for determining both the optic center and the focal length of the camera. In section 6 experiments with boxes, buildings, and living rooms

---

* Jens Arnspang can be reached by Phone: +45 35 32 14 00, email: arnspang@diku.dk, or http://www.diku.dk/users/arnspang/

of unknown dimensions, pose and positions are reported. No measurements have been made in the spatial scene in order to prepare for the calibration. The relative errors of the results are in the range of one percent.

2 Properties of Lines, Planes, Horizons and Poles

A few properties of lines and planes and their projections onto the image plane of a pin-hole camera are shortly revisited. For proofs see (Arnspang 1989b), (Henriksen 1989), and for a deeper insight into classic projective geometry see (Coxeter1961).

In this paper the image plane is conceived as an infinite plane and not restricted to a certain frame size. Consider a 3D spatial plane Π which is not perpendicular to the optic axis of the camera. In this case the projection π of the spatial plane Π onto the image plane will have a horizon $\hbar$ in the image plane. If the optic axis of the camera is not parallel to the spatial plane Π, spatial lines normal, i.e. perpendicular, to the plane Π will project onto half-lines in the image plane with a common end point p, which is named the *pole* of the spatial plane Π. Formulated in terms of gradient space coordinates $(P, Q)^T$ as defined in Horn (1986), there are the following relations between the horizon, the pole and the camera focal length, see (Arnspang 1989b):

Let $(-P, -Q, 1)^T$ be a surface normal to a spatial plane Π and let f be the focal length of the camera.

Then the equation of horizon $\hbar$ in terms of image coordinates $(x, y)^T$ is given by

$$Px + Qy + f = 0 \tag{1}$$

and the image coordinates for the pole p are

$$p = f \begin{pmatrix} P \\ Q \end{pmatrix} \tag{2}$$

Note that a scene may contain several planes each with their set of surface normals. In that case an image of the scene may contain several horizons and poles.

3 Construction of the Optic Center

In this section it is shown that the optic center in the image plane may be constructed from two views of a 3D spatial plane. Consider figure 1, showing two sets of horizons and poles $\{\hbar_1, p_1\}$ and $\{\hbar_2, p_2\}$. The spatial planes in the view are assumed to have surface normals $(-P_1, -Q_1, 1)^T$ and $(-P_2, -Q_2, 1)^T$, and the camera is assumed to have focal length f. The horizon $\hbar_1$ has the equation $P_1x + Q_1y + f = 0$. A normal vector for the horizon $\hbar_1$ is therefore given by

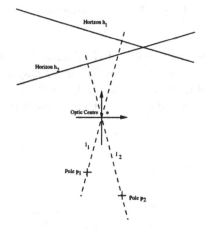

Fig. 1. Construction of the camera optic center.

$(P_1, Q_1)^T$. A line ℓ_1 perpendicular to the horizon $\hbar_1$ and which passes throgh the pole $\boldsymbol{p}_1 = f(P_1, Q_1)^T$ may therefore be parametrized as follows.

$$\ell_1(t) = \begin{pmatrix} x(t) \\ y(t) \end{pmatrix} = (f + t) \begin{pmatrix} P_1 \\ Q_1 \end{pmatrix} \tag{3}$$

Substituting $t = -f$ in equation (3) it is seen that $\ell_1(-f) = (0,0)^T$. This means that the line perpendicular to the horizon and through the pole will always contain the optic center $\boldsymbol{o}$. Analogously, a line ℓ_2, perpendicular to the horizon $\hbar_2$ and through pole $\boldsymbol{p}_2$ also contains the optic center. If the lines ℓ_1 and ℓ_2 are not parallel, their intersection $\boldsymbol{o}$ is then necessarily the optic center of the camera in the image plane. Algorithm 1 shows in detail how to determine the optic center.

Algorithm 1. *Determination of the optic center $\boldsymbol{o}$ of a pin-hole camera.*

1. *Identify the horizon $\hbar_i$ and pole $\boldsymbol{p}_i$ of a 3D spatial plane Π_i.*
2. *Draw a line ℓ_i perpendicular to the horizon $\hbar_i$ and through the pole $\boldsymbol{p}_i$.*
3. *Repeat steps 1 and 2 for one or more planes with different orientations.*
4. *The optic center $\boldsymbol{o}$ of the camera is given as the intersection of the lines ℓ_i; $i = 1, \ldots, n$.*

It is now obvious, that objects needed for the construction may be present in both indoor and outdoor scenes, e.g. a table and its legs, a room with a tiled floor and lamps hanging down from the ceiling, or a textured ground plane with trees or people. Horizons may be determined from texture and tiles, see (Arnspang 1989a), and may also be determined from motion of curves in a landscape, see (Arnspang 1989b). Poles may simply be determined as the intersection point of a set of projected surface normals, e.g. lamp poles on a road. The two sets of horizons and poles, necessary for the determination of the optic center, may be present in one scene or in two or more views of the same scene.

4 Construction of the Focal Length

Having determined the optic center of the camera from views of planes, using their horizons and poles, the camera focal length f may also be determined from simple geometric considerations. Consider figure 2, showing a view of a plane with horizon $\hbar$ and pole p and assume the optic center o of the camera has been determined.

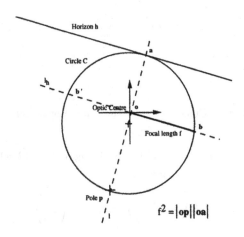

Fig. 2. Construction of the camera focal length.

The focal length f of the camera may then be determined as follows.
Algorithm 2. *Determination of the focal length f of a pin-hole camera.*

1. *Draw a line ℓ through the pole p and the optic center o – that line is also perpendicular to the horizon $\hbar$, see algorithm 1.*
2. *Let the point a be the intersection of the line ℓ and the horizon $\hbar$.*
3. *Draw a circle C through the pole p and the point a which has the horizon $\hbar$ as tangent.*
4. *Draw a line $\ell_\hbar$ through the optic center o parallel to the horizon $\hbar$.*
5. *The intersections of the circle C and the line $\ell_\hbar$ are denoted b and b'.*
6. *The focal length f equals the length of the line segment ob and also ob'.*

The correctness of algorithm 2 may be argued as follows. Since the line ℓ is perpendicular to the horizon $\hbar$, the intersection point a of ℓ and $\hbar$ is also the touching point of circle C and its tangent $\hbar$. Consequently, the line segment pa is a diameter in the circle C and therefore the triangle $\triangle abp$ has a right angle at point b. Using Pythagoras' theorem on the triangles $\triangle abo$, $\triangle bpo$, and $\triangle abp$

we have

$$| ab |^2 = | ao |^2 + | ob |^2 \qquad (4)$$

$$| bp |^2 = | bo |^2 + | op |^2 \qquad (5)$$

$$(| ao | + | op |)^2 = | ab |^2 + | bp |^2 \qquad (6)$$

implying

$$| ob |^2 = | ao | \, | op | \qquad (7)$$

Using a normalized version of equation (1) for the horizon $\hbar$, we note that the distance from the optic center to the horizon is

$$| oa | = \frac{f}{\sqrt{P^2 + Q^2}} \qquad (8)$$

and using equation (2) we note that the distance from the optic center to the pole is

$$| op | = f \sqrt{P^2 + Q^2} \qquad (9)$$

implying

$$f^2 = | ao | \, | op | \qquad (10)$$

The equations (7) and (10) finally yield

$$f = | ob | \qquad (11)$$

The correctness of algorithm 2 have thus been argued.

5 Box Calibration

In this section it is shown, how box objects may be used to compute both the optic center o and the focal length f of the camera. Consider figure 3, showing a box with the visible sides A, B, and C. The box produces three sets of horizons and poles: $W_A = \{\hbar_A, p_A\}$, $W_B = \{\hbar_B, p_B\}$ and $W_C = \{\hbar_C, p_C\}$. The three vanishing points p_A, p_B, and p_C each occur as the pole in one of the sets W and as part of a horizon in the two other sets. For each set the construction lines and circles for constructing the optic center according to section 3 and for constructing the focal length according to section 4 are shown in figure 3. Finally the 'Focal Length Circle' is added, with the estimated optic center o as its center and the estimated focal length f as its radius. Note, that views of two sides of the box may be sufficient to calibrate both the optic center o and the focal length f of the camera. Also note, that views of two such sides may be degenerate in the sense, that the horizons produced may be parallel or not well defined. In general, a third visible side allows a choice to use the two sides, whose horizons and poles are best defined in the image.

In classic calibration methods, which use boxes as calibration objects, the coordinates of a number of 'landmarks' on the box are often required a priori. The calibration method described in this paper requires no knowledge of any coordinates. Any box shaped object may be used.

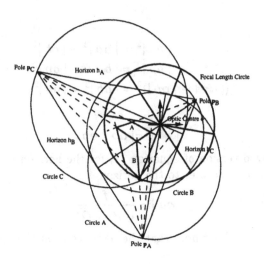

Fig. 3. The box calibration construction scheme.

6 Computational Experiments

An interactive program was implemented in which the lines used for the calibration were marked, and both the optic center and the focal length were computed, using the techniques developed in sections 3 and 4. The constructions of the optical center and the focal length rely on circles and perpendicular lines. Therefore, it is important that the image coordinate system is orthonormal. Thus, for the experiments to work the aspect ratio must be known, and it is assumed that pixel rows are perpendicular to pixel columns. Further, if the focal length is wanted in, say *mm*, the true pixel spacing must also be known. These quantities are characteristic of the digitizing circuitry, and are not liable to change when new optics are mounted. In this paper these quantities are not considered part of the calibration. In the experiments both high quality photographic images and low cost CCD camera images were used. For the laboratory scenes an 8 *mm* Ernitec lens was used, and for the natural scenes a 21 *mm* Leitz lens was used. The inter-pixel distances for the scanned cameras images are shown in table 1.

Inter-pixel Distances		
Camera	**rows (*mm*)**	**columns (*mm*)**
Burle TC355ACX	0.008368	0.012386
Leicaflex SL66	0.02771	0.02734

Table 1. Inter-pixel distances.

Figure 4 shows the actual images used for the calibration experiments. In the figure the computed horizons, poles, and the resulting 'focal length circles' are drawn as overlays. The results of the calibration of the optic center and the

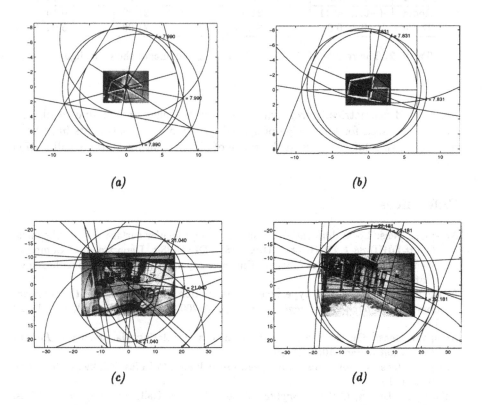

(a) (b)

(c) (d)

Fig. 4. (a) Laboratory scene: Box1. CCD camera image. (b) Laboratory scene: Box2. CCD camera image. (c) Natural scene: Room. Photographic image. (d) Natural scene: Terrace. Photographic image.

focal length are shown in table 2. Note, that when the lines used for computing a pole are almost parallel, the pole is ill-determined and the resulting estimates are not very accurate. This is clearly seen in the examples 'box2' and 'terrace' – figure 4 (b) and (d). In order to evade this situation an active calibration scheme should be used to get the best possible view of the scene. In the examples 'box1' and 'room' – figures 4 (a) and (c) – the poles are much better determined and the relative errors of the results are an order of magnitude smaller.

7 Conclusion

Revisiting the fundamentals of the pin-hole camera image formation it has been shown how both the optic center and the focal length of such a camera may be

Results of the Calibration				
Scene	Optic Center	Focal Length	True Focal Length	Rel.Err (%)
Box 1	(0.0, 0.0)	7.99	8.00	0.12
Box 2	(0.2, 0.0)	7.83	8.00	2.12
Room	(−0.1, −0.1)	21.04	21.00	0.19
Terrace	(−0.2, 0.6)	22.18	21.00	5.62

Table 2. The results of the calibration using the images shown in figure 4.

constructed geometrically from views of natural scenes or box shaped objects without the need for knowledge of spatial coordinates. Experiments have been reported with relative errors in the range of one percent for suitable calibration views.

References

Arnspang, Jens: Local differential kinematics of multiocular surface vision. Technical Report DIKU-88-1, DIKU-88-2, DIKU-88-3, DIKU-88-6, Department of Computer Science, University of Copenhagen, Universitetsparken 1, DK-2100 Copenhagen Ø, Denmark, 1988.

Arnspang, Jens: Shape from hypertexel virtual motion. In *Proceedings of the 6th Scandinavian Conference on Image Analysis*, pages 700–704, Oulu, Finland, June 19–22 1989.

Arnspang, Jens: On the use of the horizon of a translating planar curve. *Pattern Recognition Letters*, 10(1):61–69, July 1989.

Arnspang, Jens: *Motion Constraint Equations in Vision Calculus*. Elinkwijk Drykkerij, Utrecht, 1991.

Ballard, D., Brown, C.M.: *Computer Vision*. Prentice-Hall, Inc., Eaglewood Cliffs, New Jersey, 1982.

Coxeter, H.S.M.: *The Real Projective Plane*. Cambridge University Press, 2nd edition, 1961.

Faugeras, Olivier: *Three-Dimensional Computer Vision*. The MIT Press, Cambridge, Massachusetts, 1993.

Fleck, Margaret M.: Perspective projection: the wrong imaging model. Technical Report TR 95-01, Department of Computer Science, University of Iowa, 1994.

Haralick, Robert M., Shapiro, Linda G.: *Computer and Robot Vision*, volume I+II. Addison-Wesley Publishing Company, Reading, Massachusetts, 1992.

Henriksen, Knud, Kjærulff, Mikael C.: Camera calibration. Technical Report DIKU-87-17, Department of Computer Science, University of Copenhagen, Universitetsparken 1, DK-2100 Copenhagen Ø, Denmark, November 1987.

Henriksen, Knud, Arnspang, Jens: Direct determination of the orientation of a translating 3d straight line. *Pattern Recognition Letters*, 10(4):251–258, October 1989.

Henriksen, Knud: Projective geometry and straight lines in computational vision. Technical Report DIKU-90-5, Department of Computer Science, University of Copenhagen, Universitetsparken 1, DK-2100 Copenhagen Ø, Denmark, April 1990.

Horn, Berthold K.P.: *Robot Vision*. The MIT Press, Cambridge, Massachusetts, 1986.

A Passive Real-Time Gaze Estimation System for Human-Machine Interfaces

H. Klingspohr[1], T. Block, and R.-R. Grigat[1]

[1] Technische Universität Hamburg-Harburg, Technische Informatik I, Harburger Schloßstraße 20, 21079 Hamburg, telefax: ++49 40 7718-2911, telephone: ++49 40 7718-2816, email: klingspohr@tu-harburg.de

Abstract. A gaze detection system with real-time properties is described. In contrast to most systems available today it is fully passive and only relies on the images taken by a CCD camera without special illumination.

To avoid the computational intensive algorithms usually associated with Hough transforms for ellipses (e.g. [Bal81], [TM78]) a two-stage approach is presented, in which the Hough transform yields a segmented edge image used for a two-dimensional fitting algorithm. The algorithm has been implemented; experimental results are reported.

1 Introduction

The functionality of human–computer interfaces is steadily improving. Devices for intentional input by keyboard, graphic tablets, speech and gestures are available. It has been shown that the information of subconscious gaze direction of a computer user can be exploited for further improvements in user-friendliness [Jac90]. Gaze direction has been investigated by many groups with active infrared illumination of the eye and analysis of the reflections ([LKMEE93], [FWH90], [Hut93], [Car88]). In our approach the eyes are imaged by a CCD camera without special illumination.

Other potential applications of gaze analysis are simulators for cars, ships etc. to find out where the trainee is looking at to increase the resolution at the region of interest.

Handicapped persons can use intentional eye movements to control technical devices as writing machines, TV sets etc..

In some setups the camera is located on a head mounted facility close to the eyes ([Sch93], [YTS95]). In our investigations a camera is located below a computer monitor as shown in figure 1; the user's head must remain stationary. The viewing distance to the monitor is about 50 cm, the eye to camera distance about 35 cm. In the general case the position and orientation of the head in space has to be analysed before the direction of gaze can be calculated from the iris position relative to the head [Pet96].

The experiments reported in this paper concentrate on the second step, the estimation of the iris position. The setup (figure 1) is used with a stabilisation support for the head. The test person looks at the monitor and the algorithm analyses the direction of gaze and moves a cursor to the estimated position. In such a closed loop setup the user can immediately perceive accuracy and speed of the algorithm in real time [Blo96].

In the following section the algorithms used are explained and resulting images of the subsequent steps are given. In section 3 experimental results of the real time setup are presented.

2 Algorithmic Approach

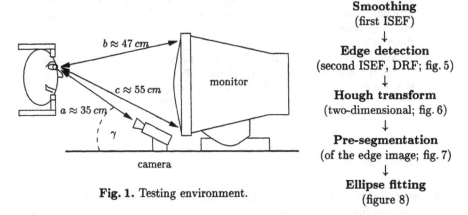

Fig. 1. Testing environment.

Smoothing
(first ISEF)
↓
Edge detection
(second ISEF, DRF; fig. 5)
↓
Hough transform
(two–dimensional; fig. 6)
↓
Pre-segmentation
(of the edge image; fig. 7)
↓
Ellipse fitting
(figure 8)

Fig. 2. Basic approach.

The outline of the iris is a circle on the surface of a sphere (the eyeball). Due to the high contrast between iris and sclera we use it as an easily detectable and significant feature. As a result of the projection onto the camera plane the iris contour degenerates into an ellipse. Therefore we need an algorithm optimised for the estimation of ellipse parameters.

The outline of the iris is often disrupted by reflections or partly hidden by eyelashes or shadows. Hence, the estimation of the ellipse parameters has to be very robust. We use the well-known Hough transform [Lea92] to calculate a region of interest for the iris edge before fitting an ellipse.

The main steps of the algorithm are shown in figure 2. Corresponding images for the three example images in figure 4 are shown in figures 5 to 8. The steps of the algorithm are discussed in more detail in the subsequent sections.

2.1 Smoothing Filter and Edge Detection

The edge detection process has to be fast and accurate. We use a cascade of first order recursive low-pass filters as proposed by [CZS90] which can be implemented with four multiplications per pixel. The first infinite size exponential filter (ISEF) removes noise and smoothes the image. A second filter calculates a band-limited second derivative of the image; the zero-crossings of this Laplacian image are candidates for edge points. This method is known as 'difference of recursive filters' (DRF, see [SC92]).

To provide additional information for the subsequent processing stages we calculate the local edge direction with a conventional 3×3 Sobel operator. As we are looking for the iris edge it is possible to remove certain edge points at this stage: We assume that the eye is never fully opened, i.e. that eyelids cover part of the iris (and its outline). As the upper eyelid usually covers a larger part than the lower lid we use the angles φ_t and φ_b to mark the edge pixel that cannot be part of the iris outline.

$$\frac{\varphi_t}{2} < |\varphi| < 180° - \frac{\varphi_b}{2} \quad (1)$$

$$-180° \leq \varphi < 180° \quad (2)$$

Fig. 3. Schematic model of a partially opened eye.

$\varphi = 0°$ denotes a light $\rightarrow$ dark transition from top to bottom. Typical values are $\varphi_t = 90°$ and $\varphi_b = 20°$; this means that roughly one third of all edge pixels is discarded during this stage.

Figure 5 shows typical edge images; the iris outlines are interrupted by spot lights, while eyebrows, skin structure etc. produce additional edge segments.

2.2 The Hough Transform

The Hough transform ([Hou62], [Bal81]) is a widely known and robust algorithm for detecting curves in an edge image (the 'primal sketch'). As ellipses have five free parameters we have to cope with a five-dimensional problem.

Drawbacks of a high-dimensional parametric estimation are mainly the slow execution speed and the excessive memory requirements. As the movement of the iris centre has only two degrees of freedom (pan and tilt), a reduction of the number of dimensions is possible.

Instead of using the full Hough transform for ellipses we use the more robust variant for circles which reduces the number of dimensions. The precision lost during this step is immaterial as the Hough parameters are not used for the final estimated coordinates, see section 2.3.

The standard implementation of the Hough transform for circles requires a three-dimensional accumulation space (centre coordinates (x, y) and radius r); since the exact radius of the circle is not needed for our application we can map this space to the x–y plane of possible centre coordinates, using a restricted range of radii $r_{min} \ldots r_{max}$. Thus it is sufficient to use a two-dimensional matrix to accumulate the x and y coordinates of the detected circles.

Figure 6 shows the results of the transform. Each maximum gives a first estimation for the centre coordinates of the iris. Even the highly distorted left image yields an unique maximum in parameter space.

2.3 Segmentation Stage

Based on information obtained in the Hough transform stage the edge image can be segmented. The global maximum at (x_c, y_c) in parameter space is generated by edge points inside the two circles defined by the centre coordinates and the radii r_{min} and r_{max}. Any edge points outside this area cannot belong to the iris outline and are therefore rejected.

The remaining edge points are checked for correct gradient direction. Based on the estimation of the centre coordinates, the geometric direction to the centre can be calculated for each edge point (x_i, y_i):

$$\varphi_{geo} = \arctan\left(\frac{y_i - y_c}{x_i - x_c}\right) \tag{3}$$

A comparison between geometric and gradient direction $|\varphi_{geo} - \varphi_{ed}| \overset{!}{\leq} \varphi_{max}$ removes points that do not belong to the iris outline. With the parameter φ_{max} it is possible to compensate for angular errors caused by the non-circular form of the iris outline.

As can be seen in figure 7, all edges that do not belong to an iris outline have been successfully filtered out. Whereas the iris outline of the centre image is nearly complete, large parts of the outline in the left image are missing because the iris' highly eccentric ellipse shape did not fit entirely in the ring.

2.4 Ellipse Fitting

The available resolution is limited, especially when using consumer type cameras. The resolution of a camera image can be estimated based on the geometry of the scene. In our test environment (see figure 1) the resolution is ≈ 0.8 mm per pixel.

To retrieve a maximum of information from the image, sub-pixel algorithms can be employed for edge detection, Hough transform or ellipse fitting. Since the first two stages are only used for pre-segmentation, sub-pixel analysis for the ellipse fitting algorithm has been investigated.

With an ideal segmentation of the object a direct fitting to the data is possible. The algorithm proposed by Fitzgibbon et al. [FPF96] minimises a squared arithmetic distance measure based on general conics. This leads to a generalised eigenvalue problem, incorporating the constraint $4ac - b^2 = 1$ into the normalisation factor. This guarantees that even distorted or disrupted edge segments always yield an ellipse of the form $ax^2 + bxy + cy^2 + dx + ey + f = 0$. To exclude edge points that are not part of the iris outline we use an iterative algorithm as proposed by [Str94].

The algorithm is well behaved. Figure 8 shows that the ellipse fits the real iris outline almost exactly.

Fig. 4. Typical images of eyes with different directions of gaze.

Fig. 5. Edge images generated with the operator described in section 2.1.

Fig. 6. Parameter space calculated with the Hough transform (see section 2.2).

Fig. 7. Segmented edges used for fitting (see section 2.3).

Fig. 8. Fitted ellipses superimposed on the original images (see section 2.4).

2.5 Calculation of Gaze Direction

The current calibration procedure is as follows: At first a region which includes the eye region has to be marked manually. By looking for two circles with identical radii – again using the Hough transform – both eyes are found and a more accurate region of interest is set. Then, the test person is instructed to look at the corners of the monitor and the corresponding iris positions are saved.

During the tracking loop, the calculated iris coordinates are mapped linearly onto monitor screen coordinates. A test picture provides numerous landmarks for the user to look at. As the human eye only focuses points with approximately 1° resolution [Car88], the overall precision cannot be better than approximately 9 mm on the screen with the given setup.

Due to the high level calibration method the algorithm is insensitive with respect to additional distortion caused by the camera viewing angle from below the monitor.

3 Experimental Results

The described algorithm has been implemented on a Sun SparcStation 10/40 (SPECfp95 1.0) in an experimental setup according to figure 1. The system displays a cursor on the monitor at the calculated gaze position, enabling the user to compare the cursor position with his intentional view. Using C/C++ with only simple optimisations a cycle time of one second is achieved; figure 9 shows the computation time distribution for a characteristic eye image.

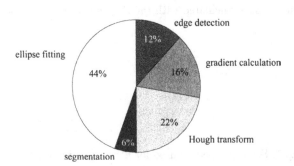

Fig. 9. Typical computation time distribution

The edge operator described in section 2.1 fulfills both speed and accuracy constraints. Edges are detected within a wide range of slopes. The operator is highly insensitive to noise while being considerably fast (approximately 100 ms for a 100 × 100 pixel image).

The execution times of the Hough transform algorithm, segmentation and gradient calculation scale linearly with the number of edge points.

The current ellipse fitting algorithm is implemented in Matlab; a re-implementation in a low-level language like C will reduce the computation time considerably.

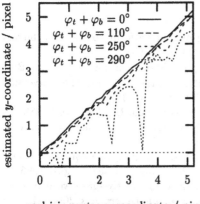

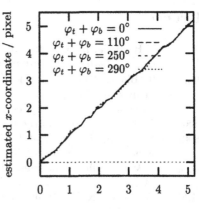

Fig. 10. Simulation of an eye moving in y-direction with different angles of opening

Fig. 11. Simulation of an eye moving in x-direction with different angles of opening

To test the accuracy of the algorithm we did a simulation of an idealised eye (black circle on white background) with a varying amount of the iris edge hidden behind eye lids. The results can be seen in figures 10 and 11.

When the simulated eye was fully opened ($\varphi_t + \varphi_b = 0°$), both x and y coordinates of the iris center could be estimated with an accuracy of much better than one pixel. When less of the iris edge is visible, the accuracy in the x direction is unaffected. However, the accuracy in the y-direction gets progressively worse when less and less edge pixels are available. The estimated y-coordinates are usually too small which is caused by the chosen allowed opening angles $\varphi_t > \varphi_b$.

The overall observed accuracy of the implementation is about 2 cm on the computer monitor. We believe that the accuracy can be further increased by algorithm improvement, for example with an enhanced edge detection algorithm using adaptive filters as proposed by Herpers et al. [HWM+96].

4 Summary and Further Work

A robust and real time capable system for gaze direction analysis by passive image analysis has been implemented on a Sun SparcStation. One estimation per second has been achieved without special optimisation or dedicated hardware.

A cursor on a monitor screen is moved to the point observed by the test person. With this setup an accuracy of 2 cm in a viewing distance of about 50 cm has been achieved.

Future work is directed to improvements in robustness, precision, the compensation for head movements and a more automatic alignment procedure. Specialised architectures for real time gaze estimation analysis with cycle times faster than 100 ms are under investigation.

References

[Bal81]   D.H. Ballard. Generalizing the Hough transform to detect arbitrary
            shapes. *Pattern Recognition*, 13(2):111–122, 1981.

[Blo96]   T. Block. Analyse der Blickrichtung des Menschen mittels passiver Bild-
            analyse zur Computersteuerung. Master's thesis, Technische Universität
            Hamburg-Harburg, Arbeitsbereich TI1, October 1996.

[Car88]   R.H.S. Carpenter. *Movements of the eyes*. Pion, London, April 1988.

[CZS90]   S. Castan, J. Zhao, and J. Shen. New edge detection methods based on
            exponential filter. 10th ICPR, 1990.

[FPF96]   A.W. Fitzgibbon, M. Pilu, and R.B. Fisher. Direct least squares fitting of
            ellipses. *Proc. Int. Conf. on Pattern Recognition, Wien*, August 1996.

[FWH90]   L.A. Frey, K.P.jr. White, and T.E. Hutchinson. Eye-gaze word process-
            ing. *IEEE Transactions on Systems, Man, and Cybernetics*, 20(4):944–950,
            August 1990.

[Hou62]   P.V.C. Hough. Method and means for recognising complex patterns. US
            Patent 3069654, 1962.

[Hut93]   T.E. Hutchinson. Eye-gaze computer interfaces. *Computer, Long Beach*,
            26(7):65–67, July 1993.

[HWM+96]   R. Herpers, L. Witta, M. Michaelis, J. Bruske, and G. Sommer. Detektion
            und Verifikation von charakteristischen Bildpunkten in Gesichtsbildern.
            DAGM, 1996.

[Jac90]   R.J.K. Jacob. What you look at is what you get: eye movement-based
            interaction techniques. *ACM Human Factors in Computing Systems, Em-
            powering People, CHI Conference Proceedings*, 9:11–18, April 1990.

[Lea92]   V.F. Leavers. *Shape Detection in Computer Vision Using the Hough
            Transform*. Springer, 1992.

[LKMEE93]   T. Lehmann, A. Kaupp, D. Meyer-Ebrecht, and R. Effert. Automatische
            Schielwinkelmessung durch Hough-Transformation und Kreuz-Kovaranz-
            Filterung. In S.J. Pöppel and H. Handels, editors, *Mustererkennung 1993*,
            pages 237–244. Springer, 1993.

[Pet96]   E. Petraki. Analyse der Blickrichtung des Menschen und der Kopf-
            orientierung im Raum mittels passiver Bildanalyse. Master's thesis, Tech-
            nische Universität Hamburg-Harburg, Arbeitsbereich TI1, 1996.

[SC92]   J. Shen and S. Castan. An optimal linear operator for step edge detection.
            CVGIP: Graphical Models and Image Processing, 54(2):112–133, March
            1992.

[Sch93]   W.E. Schroeder. Head-mounted computer interface based on eye tracking.
            *Proceedings of the SPIE - The International Society for Optical Engineer-
            ing*, 2094(3):1114ff, 1993.

[Str94]   M. Stricker. A new approach for robust ellipse fitting. In *Proc. of the
            third Intern. Conference on Automation, Robotics and Computer Vision
            (Singapore)*, volume 2, pages 940–945, November 1994.

[TM78]   S. Tsuji and F. Matsumoto. Detection of ellipses by a modified Hough-
            transform. *IEEE Trans. Comput.*, 27, 1978.

[YTS95]   D. Young, H. Tunley, and R. Samuels. Specialised Hough transform and
            active countour methods for real-time eye tracking. Technical Report
            CSRP no. 386, School of Cognitive and Computing Sciences, University
            of Sussex, July 1995.

An Active Vision System for Obtaining High Resolution Depth Information

W. Fellenz[*], K. Schlüns[**], A. Koschan[*], and M. Teschner[***]

[*]Technical University of Berlin, Dept. of Computer Science, FR 3-11 Franklinstr. 28-29 10587 Berlin, Germany fellenz@cs.tu-berlin.de koschan@cs.tu-berlin.de

[**]University of Auckland Computer Science Dept. Tamaki Campus Private Bag 92019 Auckland, New Zealand karsten@cs.auckland.ac.nz

[***]University of Erlangen-Nuremberg, Telecommunications Institute, Cauerstrasse 7 91058 Erlangen, Germany teschner@nt.e-technik.uni-erlangen.de

Abstract. A low-cost active vision head with ten degrees of freedom is presented that has been build from off-the-shelf parts. To obtain high resolution depth information of fixated objects in the scene a general purpose calibration procedure is proposed which estimates intrinsic and extrinsic camera parameters including the vergence axes of both cameras. To produce enhanced dense depth maps a hierarchical block matching procedure is presented that employs color information. To simplify the development of controlling strategies for the head a modular hierarchy is applied that distributes various tasks among different levels employing basic capabilities of the components of the head.

1 Introduction

Biological and engineering active vision systems share mechanisms that allow to accomplish visual tasks by varying view parameters like direction of gaze, vergence, focus, zoom and aperture. By adjusting these gaze parameters in a three dimensional world, the process of image acquisition can be voluntarily controlled, thereby constraining the acquired views. Fundamental advantages of actively controlling extrinsic camera parameters are the reduced computational burden for transmitting and processing visual information allowing the real-time computation of a depth map, the cooperation of visual behaviors like focus, stereo and vergence to overcome the limitations of a single fixed sensor [1, 7, 9], and the use of dynamic fixations for real-time tracking of a moving object employing visual feedback [2, 4, 8].

In this contribution a further advantage of active vision strategies will be outlined that allows us to obtain high resolution depth maps in the immediate surround of the stereo head using active fovealization of a visual target. The principal processing scheme is outlined in Fig. 1 showing a top view of the stereo head for different common viewing angles. For a fixed baseline and wide viewing angles the parallel orientation of both cameras suffices to compute a raw depth map of the scene (a). If the viewing angle of both cameras is narrowed without appropriate adjustment of the vergence or version angles (b) near objects are no longer visible in both images making them not fusable. The convergent fovealization of an object with both cameras (c) now allows the extraction of a high resolution depth map if the version angles are known. However, this enhancement in depth resolution is only limited by the zooming capabilities of the cameras and does not require a higher resolution of the sampling array.

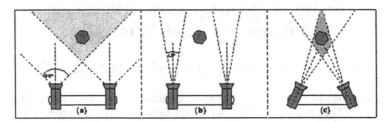

Figure 1: Convergent fovealization for high resolution sampling.

The assembled stereo head and its mechanical hardware will be presented first, followed by a description of the calibration procedure for the intrinsic and extrinsic camera parameters. A fast chromatic stereo procedure to obtain high resolution depth information will be presented next followed by some preliminary results

2 Mechanical Hardware

Apart from financial limitations, our main design objectives for the active stereo head were a hardware controlling scheme which uses a common interface protocol, a modular design of the head using off-the-shelf parts that can be assembled without much effort. The controlling hierarchy should be modular, exploiting the intrinsic competencies of the low level modules by higher level controlling strategies. The camera which is shown in Fig. 2 has been build up from the following off-the-shelf parts:

- a Pan-Tilt Unit from Directed Perception (PTU-46-17.5) with angular resolution of 3.086 arc minute and maximum speed of 150°/second,
- two small PAL color video cameras from Sony (EVI-311) with integrated 8x-zoom (f=5.9 (44.3°-H) to f=47.2 (5.8°-H), F1.4), Aperture, Focus, and 1/3" CCD-chip (752(H) x 582(V)) controllable by a RS-232 Interface Board (IF-51),
- two Vergence Stepper Motors/Indexer from IMS/SCT which have angular resolution of 50.800 counts per revolution at a torque of 0.7 Nm and optical encoders with a resolution of 500 counts per revolution.

Figure 2: The Active Vision System TUBI at TU Berlin.

The cameras are mounted directly to the motor shafts to minimize slippage using self-made connectors which allow to adjust the optical center to the axes. The parts were assembled using a light-weight vergence platform, build from an aluminum profile that is mounted directly on the pan-tilt unit. Additional modifications of the CCD-cameras were necessary to integrate connector sockets for RS-232, video and voltage supply.

The video data is collected by two Matrox Meteor frame-grabbers on a Pentium computer using routines under the Linux operating system. To communicate with the head via RS-232 protocol an internal interface-board is used to control all five serial lines of the head asynchronously from the host. The complete setup including the color cameras, two micro-steppers with indexers, the pan-tilt unit and two frame-grabbers costs about $10.000 with reasonable own work on the head.

3   Camera Calibration for an Active Stereo System

For an active stereo system it is not only important to estimate the intrinsic and extrinsic parameters of the cameras. Additionally, it is essential to know the vergence axes of the cameras to simplify the rectification process. Generally, for every single camera the vergence axis (axis of rotation) is not located in the optical center of the camera.

3.1   Estimating the Vergence Axis

The vergence axis of each camera is estimated by employing a general purpose 3-D calibration procedure. Among the many existing calibration methods, the Tsai method [10] has some advantages. It has in possession a known behavior and a well-known stability, since several research groups with independent realizations extensively analyzed this approach. For calculating a vergence axis, a sequence of at least two images of a calibration object is taken).

In the following description we use only two of the five coordinate systems introduced by Tsai, for simplification reasons, namely the camera coordinate system (X_k, Y_k, Z_k) and the world coordinate system (X_w, Y_w, Z_w). The equation

$$P_k = R \cdot P_w + T, \quad \text{with} \quad R \in \Re^{3\times3}, \quad T \in \Re^3$$

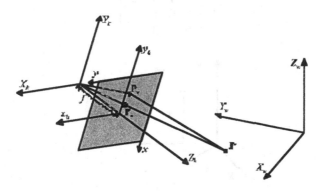

Figure 3: Camera model proposed by Tsai.

describes the relation between (X_k, Y_k, Z_k) and (X_w, Y_w, Z_w), see Fig. 3. Then, the transformation of the two 3D coordinate systems is simply

$$P_{k1} = \mathbf{R}_1 \cdot P_w + \mathbf{T}_1, \text{ with } \mathbf{R}_1 \in \mathfrak{R}^{3 \times 3}, \ \mathbf{T}_1 \in \mathfrak{R}^3 \text{ and}$$

$$P_{k2} = \mathbf{R}_2 \cdot P_w + \mathbf{T}_2, \text{ with } \mathbf{R}_2 \in \mathfrak{R}^{3 \times 3}, \ \mathbf{T}_2 \in \mathfrak{R}^3.$$

Using homogenous coordinates each of these transformations can be merged to one 4×4 matrix:

$$P_{k1} = \mathbf{M}_1 \cdot P_{k1} \text{ and } P_{k2} = \mathbf{M}_2 \cdot P_{k2} .$$

Now, the relation between both camera coordinate system points becomes

$$P_{k1} = \mathbf{M}_1 \cdot \mathbf{M}_2^{-1} \cdot P_{k2} .$$

The matrix product describes the general movement of the (rotating) camera. Furthermore, there exists a priori knowledge about the camera movement. Using a mechanical pre-calibration it can be assumed, that the camera rotates about an axis parallel to the image plane (projection plane, XY-plane) and aligned with the Y-axis of the (X_k, Y_k, Z_k) coordinate system, see Fig. 4.

The task is to estimate t_x and t_z which are unequal to zero, due to focus adjustment and camera shift. It is possible to describe the movement of the camera by using the model mentioned above. The movement consists of three steps:

1. Translating the vergence axis to bring it into conformity with the Y-axis:

$$\mathbf{A}_1 = \begin{pmatrix} 1 & 0 & 0 & t_x \\ 0 & 1 & 0 & 0 \\ 0 & 0 & 1 & t_z \\ 0 & 0 & 0 & 1 \end{pmatrix}.$$

2. The rotation itself:

$$\mathbf{A}_2 = \begin{pmatrix} \cos\alpha & 0 & \sin\alpha & 0 \\ 0 & 1 & 0 & 0 \\ -\sin\alpha & 0 & \cos\alpha & 0 \\ 0 & 0 & 0 & 1 \end{pmatrix}.$$

3. Inverting of step 1:

$$\mathbf{A}_3 = \begin{pmatrix} 1 & 0 & 0 & -t_x \\ 0 & 1 & 0 & 0 \\ 0 & 0 & 1 & -t_z \\ 0 & 0 & 0 & 1 \end{pmatrix}.$$

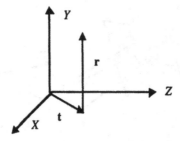

Figure 4: Model of the vergence axis.

The matrix $A = A_3 \cdot A_2 \cdot A_1$ is now an alternative representation of the movement of the camera. In addition, it is known that $M = A$.

$$A = \begin{pmatrix} \cos\alpha & 0 & \sin\alpha & -t_x + t_x\cos\alpha + t_z\sin\alpha \\ 0 & 1 & 0 & 0 \\ -\sin\alpha & 0 & \cos\alpha & -t_x\sin\alpha - t_z + t_z\cos\alpha \\ 0 & 0 & 0 & 1 \end{pmatrix}.$$

As shown before, M is completely known by applying the camera calibration method twice. For the estimation of t_x and t_z two components are chosen from the matrices to form an equation system (m_{ij} is an element of M).

$$-t_x + t_x\cos\alpha + t_z\sin\alpha = m_{03},$$
$$-t_x\sin\alpha - t_z + t_z\cos\alpha = m_{23}.$$

Comparing further matrix elements we get

$$\cos\alpha = m_{00} = m_{22} \quad \text{and} \quad \sin\alpha = m_{02} = -m_{20}.$$

Now

$$t_x = \frac{-m_{03} + m_{03}m_{00} - m_{23}m_{02}}{2 - 2m_{00}} \quad \text{and} \quad t_z = \frac{-m_{23} + m_{23}m_{00} + m_{03}m_{02}}{2 - 2m_{00}}$$

are solutions for t_x and t_z.

In that way the location of the vergence axes and the rotation angle are estimated. If the results of the camera calibration are not reliable it is possible to replace $\cos\alpha$ and $\sin\alpha$ using more than only one matrix element. Herewith errors are reduced that are related to numerical instabilities.

First results of the rectification procedure suggest that the estimation of the vergence axes is very reliable. Unfortunately, because of the differences between a real camera and the assumed camera model it is not possible to get the precision of the proposed method directly. Therefore, the following theoretical error propagation was analyzed. It is assumed that the camera calibration is influenced by noise. The error is modeled by adding Gaussian noise to the transformation matrix M. Fig. 5 shows that erroneous camera calibration has only little effect if t_x and t_z are small. For that reason it is useful to apply the approach more than once.

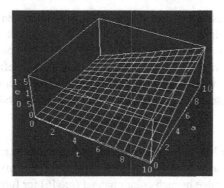

Figure 5: Error analysis, a: noise added to each matrix element of M (in percentage), t: t_x and t_z in mm, ($t_x = t_z$ to simplify the analysis), e: resulting error of estimating t.

4 Stereo Analysis Using Chromatic Block Matching

A fast stereo algorithm that produces dense depth maps is needed to obtain depth information with high resolution for an active vision system. Most of the existing stereo techniques are either fast and produce sparse depth maps or they compute dense depth maps and are very time consuming. In an earlier investigation [6], we found a hierarchical implementation of a Block Matching technique using color information to be very suitable for obtaining fast and precise dense depths maps. The use of color information enhances the quality of the results by 25 to 30 percent compared to the quality of the results obtained when using only intensity information. Thus, we employ color information to produce dense depth maps of higher quality.

A hierarchical implementation of the algorithm in an image pyramid enables an efficient realization of the matching process. However, our implementation on a single processor does not reach real-time demands at the moment. Currently, we are only approaching real-time requirements when using images of size 256×256 pixels and a parallel implementation on 10 processors [5]. Further investigations are necessary to speed up the algorithm. The principle of the chromatic Block Matching algorithm and its hierarchical implementation is outlined below.

Block Matching is based on a similarity check between two equal sized blocks ($n \times m$-matrices) in the left and the right image (area-based stereo). The mean square error MSE between the pixel values inside the respective blocks defines a measure for the similarity of two blocks. We propose to employ an approximation of the Euclidean distance to measure color differences. The left color image F_L and the right color image F_R may be represented in the RGB color space as $F_L(i,j) = (R_L(i,j), G_L(i,j), B_L(i,j))$ and $F_R(i,j) = (R_R(i,j), G_R(i,j), B_R(i,j))$. The MSE is defined with $n = m = 2k+1$ as

$$MSE_{color}(x, y, \Delta) = \frac{1}{n \cdot m} \sum_{i=-k}^{k} \sum_{j=-k}^{k} (|R_R(x+i, y+j) - R_L(x+i+\Delta, y+j)|^2$$
$$+ |G_R(x+i, y+j) - G_L(x+i+\Delta, y+j)|^2$$
$$+ |B_R(x+i, y+j) - B_L(x+i+\Delta, y+j)|^2),$$

where Δ is an offset describing the difference $(x_L - x_R)$ between the column positions in the left and in the right image. The block (of size $n \times m$) is shifted pixel by pixel inside the search area. Using standard stereo geometry the epipolar lines match the image lines. The disparity D of two blocks in both images is defined by the horizontal distance, showing the minimum mean square error. Furthermore, the search area in the right image is limited by a predefined maximum disparity value d_{max}:

$$D = \min_{|\Delta| \le d_{max}} \{MSE_{color}(x, y, \Delta)\}.$$

This chromatic Block Matching algorithm can be enhanced in robustness and in time efficiency by using a quad pyramid. The values of the pixels are obtained by calculating the mean value in each color channel (see [6] for further details).

5 Preliminary Results

Although not all visual capabilities have been integrated into one system, the performance of its parts can be studied. A requirement for tracking objects in real-time is the knowledge of their position in depth. The problem of finding the point in space which corresponds to the image points u_i and v_i in both images is commonly known as triangulation. If all calibration matrices are known, it is possible in principle to calculate the intersection of the two rays in space corresponding to the image points. In practice, these lines are not guaranteed to cross caused by discretization error and noise [3].

In the presented system the calibration matrices are calculated using the calibration method suggested by Tsai [10] and the triangulation of both image points is computed using the pseudoinverse of the resulting matrix equations. Fig. 6(a) and 6(b) shows the result of backprojecting the image points of 75 virtual points around the calibration object into the scene using the known depth of the points.

Fig. 6(c) shows the result of backprojecting the ideal image points of both images using the triangulation method. Further inspection of the marked points reveals small offsets of the outer marks. Fig. 7(a) and 7(b) illustrate the difference between moving the object and moving the camera to get temporal edges and points on a calibration object for on-line calibration. Although the background noise is lower when only the object is moved, the detected calibration pattern are similar in both cases.

Figure 6: Backprojection of 75 virtual points into the scene using their known depth values (a) in the left image and (b) in the right image. (c) Backprojection of the ideal image points.

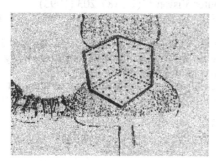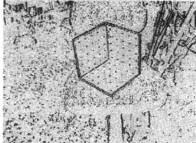

Figure 7: (a) Object movement and (b) camera movement for calibration.

6 Conclusion

A low-cost stereo active-vision head was presented that uses a modular cooperating control strategy. Furthermore, computationally efficient techniques for calibrating the vergence axes of an active vision system and for obtaining dense depth maps by employing color information were presented. Additional tests and research activities are necessary for a more detailed investigation of the techniques. Further results will be presented soon.

In summary, we should like to emphasize that an active vision system always allows to improve the resolution of depth information. Therefore, we believe that precise results can be efficiently obtained when combining hierarchical chromatic Block Matching with active fovealization.

Acknowledgment

This work was funded by the Deutsche Forschungsgemeinschaft (DFG).

References

[1] Ahuja, N., Abbot, A. L.: Active Stereo: integrating disparity, vergence, focus, aperture, and calibration for surface estimation. IEEE Trans. on PAMI 15(10), 1007-1029 (1993)

[2] Bradshaw, K.J., McLauchlan, P.F., Reid, I.D. Murray, D.W.: Saccade and pursuit on an active head-eye platform. Image and Vision Computing 12, 155-163 (1994)

[3] Hartley, R.I., Sturm, P.: Triangulation. Proc. Int. Conf. on Computer Analysis of Images and Patterns 1995, pp. 190-197

[4] Hayman, E., Reid, I.D., Murray, D.W.: Zooming while tracking using affine transfer. Proc. British Machine Vision Conf. 1996

[5] Koschan, A., Rodehorst, V.: Towards real-time stereo employing parallel algorithms for edge-based and dense stereo matching. Proc. Workshop on Computer Architectures for Machine Perception 1995, pp. 234-241

[6] Koschan, A., Rodehorst, V., Spiller, K.: Color stereo vision using hierarchical block matching and active color illumination. Proc. Int. Conf. on Pattern Recognition 1996, Vol. I, pp. 835-839

[7] Krotkov, E., Bajcsy, R.: Active vision for reliable ranging: cooperating focus, stereo, and vergence. Int. J. of Computer Vision 11(2), 187-203 (1993)

[8] Pahlavan, K., Uhlin, T, Eklundh J.-O.: Dynamic fixation. Proc. Int. Conf. on Computer Vision 1993, pp. 412-419

[9] v. Seelen, W., Bohrer, S., Engels, C., Gillner, W., Janssen, H., Neven, H., Schöner, G., Theimer, W.M., Völpel, B.: Visual Information processing in neural architecture. Proc. DAGM-Symposium Mustererkennung 1994

[10] Tsai, R. Y.: An efficient and accurate camera calibration technique for 3d machine vision, Proc. Int. Conf. on Computer Vision and Pattern Recognition 1986, pp. 364-374

Author Index

Springer
and the
environment

At Springer we firmly believe that an
international science publisher has a
special obligation to the environment,
and our corporate policies consistently
reflect this conviction.
We also expect our business partners –
paper mills, printers, packaging
manufacturers, etc. – to commit
themselves to using materials and
production processes that do not harm
the environment. The paper in this
book is made from low- or no-chlorine
pulp and is acid free, in conformance
with international standards for paper
permanency.

Springer

Lecture Notes in Computer Science

For information about Vols. 1–1229

please contact your bookseller or Springer-Verlag